ASYMPTOTIC GIANT BRANCH STARS
IAU SYMPOSIUM NO. 191

COVER ILLUSTRATION: “ AGB star”, a painting by Katrien Kolenberg, from a drawing by Erwin Sedlmayr, see page iv

Published on behalf of the
INTERNATIONAL ASTRONOMICAL UNION

by
Astronomical Society of the Pacific
First published 1999

Managing Editor, D. H. McNamara
Production Manager, Enid Livingston

EDITORIAL/PUBLISHING OFFICE:
Managing Editor
PO Box 24463
211 KMB Brigham Young University
Provo, UT 84602-4463
USA

(801) 378-2298 Phone
(801) 378-2265 Fax
pasp@astro.byu.edu E-mail

CATALOG/BOOK ORDERS:
IAU Publications
390 Ashton Avenue
San Francisco, CA 94112-1722
USA

(415) 337-1100 Phone
(415) 337-5205 Fax
catalog@aspsky.org E-mail
www.aspsky.org Web Site

Printed by BookCrafters, Inc.
Chelsea, Michigan

Library of Congress Catalog Card Number: 99-62044
ISBN: 1-886733-90-2

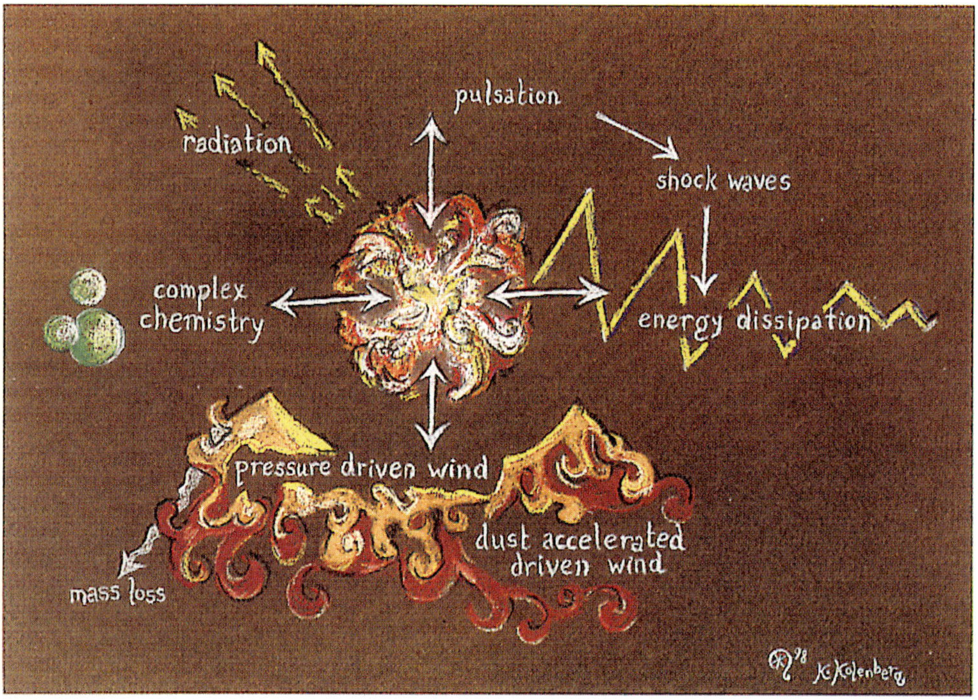

"**AGB star**", a painting by Katrien Kolenberg (Astronomical Institute, Leuven University), adapted from a drawing by Erwin Sedlmayr (1989, in "From Miras to Planetary Nebulae: Which Path for Stellar Evolution ?", eds. M.-O. Mennessier & A. Omont, Editions Frontières, p. 179)

INTERNATIONAL ASTRONOMICAL UNION
UNION ASTRONOMIQUE INTERNATIONALE

ASYMPTOTIC GIANT BRANCH STARS

Proceedings of the 191st Symposium
of the International Astronomical Union
held in Montpellier, France
27 August -1 September 1998

Edited by:

T. LE BERTRE
Paris Observatory, France

A. LÈBRE
Montpellier University, France

and

C. WAELKENS
Catholic University of Leuven, Belgium

Publisher

Information on other IAU Symposium proceedings is given at the back of this volume

Table of Contents

Part 3. Formation, Composition, and Processing of Dust

Part 4. Circumstellar Envelopes

Part 5. Non-Spherical Mass Loss, Binarity, Post-AGB Evolution

Part 6. AGB Stars as a Population of Various Galaxies

Part 7. Summary

Preface

The 191st I.A.U. Symposium was held in Montpellier (France) from August 27 to September 01, 1998. Although several conferences on subjects related to AGB stars had already occurred, it was the first time that a full I.A.U. Symposium was dedicated to these objects. Major observational breakthroughs, in particular in the fields of high-angular resolution, many new developments in theoretical models, and a lively meeting between experts on meteorites and on AGB stars marked this conference. It was attended by 245 participants from 27 different countries who presented 202 contributions. We were pleased to note the presence of many bright young scientists, an indication that the field of "old" stars has a promising future !

This book contains the 30 reviews and the 48 contributed talks organised by topics. The poster contributions appeared in the Abstract Booklet published by the *Groupe de Recherches en Astronomie et Astrophysique du Languedoc* (University of Montpellier) and can be consulted on the ADS Abstract Service. A scientific summary of the conference has been presented in the Publications of the Astronomical Society of the Pacific (1999, PASP 111, 125-126).

Several events have been connected to the Conference. In an effort to make our discipline better known to the general public, exhibits from ESA and from the Observatory of Paris were presented in the Congress Hall. Also, on Friday August 28, the Chairman of the SOC gave a well attended public lecture. A Concert of Baroque Music was given to the participants on August 29 within a nicely restored medieval building. On Sunday 30, a tour to the Grotte des Demoiselles, or the "Fairies Cave" (a network of natural caves, 30 km away from the city), let most participants speechless in the presence of such underground marvels. The Conference banquet was held outdoors, in the beautiful park of the XVIIth century Flaugergues castle. Finally, several tours of the Montpellier city, conducted by the local history expert, J.P. Lafon, were organised for the participants and their family.

The Scientific Organising Committee consisted of:

A. Alksnis (Latvia),	B. Gustafsson (Sweden),
H. Habing (The Netherlands),	S. Kleinmann (U.S.A.),
D. Lambert (U.S.A.),	T. Le Bertre (France),
J. Lépine (Brazil),	M.O. Mennessier (France),
A. Omont (France),	D. Schönberner (Germany),
A. Tielens (The Netherlands),	C. Waelkens (Belgium),
P. Whitelock (South Africa) and	P. Wood (Australia).

The I.A.U. sponsoring bodies were the Division IV (Stars) and the Commissions 29 (Stellar Spectra), 36 (Theory of Stellar Atmospheres), 27 (Variable Stars) and 34 (Interstellar Matter).

In Montpellier, the Symposium was organised by the *Groupe de Recherches en Astronomie et Astrophysique du Languedoc* (GRAAL) of the Montpellier University of Sciences and Technology (CNRS and UMII). The Local Organising Committee consisted of:

D. Barthès,	M. Belmas,	M. Chadid,	R. Coll,
P. Collard,	G. Jasniewicz,	J.P. Lafon,	P. de Laverny,
A. Lèbre,	A.M. Le Squeren,	C. Magnan,	M.O. Mennessier and
C. Mercier.			

The AGB Symposium has been possible thanks to the generous support of:
- the International Astronomical Union,
- the European Space Agency,
- the French Academy of Sciences,
- the French Ministery of Education (Ministère de l'Education Nationale et de l'Enseignement Supérieur),
- the French Foreign Office (Ministère des Affaires Etrangères),
- the Region Languedoc-Roussillon,
- the Montpellier District,
- the Centre National de la Recherche Scientifique (INSU and GdR 968),
- the University of Montpellier (UMII),
- the ISTEEM (Institut des Sciences de la Terre, de l'Eau et de l'Espace de Montpellier),
- the Pôle Universitaire Européen de Montpellier et du Languedoc-Roussillon,
- the IAU French National Committee for Astronomy (CNFA) and
- the Observatory of Paris.

We address our warmest thanks to all the individuals and to all the institutions who have contributed to the success of this Symposium.

T. Le Bertre, A. Lèbre and C. Waelkens

Leuven, February 1, 1999

Participant List[1]

Carlos ABIA, University of Granada, Spain ⟨cabia@goliat.ugr.es⟩

Michael D. ALBROW, University of Canterbury, Christchurch, New Zealand ⟨m.albrow@phys.canterbury.ac.nz⟩

J. ALCOLEA, National Astronomical Observatory, Alcalá de Henares, Spain ⟨j.alcolea@oan.es⟩

David R. ALEXANDER, Wichita State University, Wichita, KS, USA ⟨dra@twsuvm.uc.twsu.edu⟩

Yann ALIBERT, Ecole Normale Supérieure, Lyon, France ⟨yalibert@ens-lyon.fr⟩

Andrejs ALKSNIS, Radioastrophysical Observatory, Riga, Latvia ⟨aalksnis@acad.latnet.lv⟩

Rodrigo ALVAREZ, Free University of Brussels, Belgium ⟨ralvarez@astro.ulb.ac.be⟩

Sachiko AMARI, Washington University, St. Louis, MO, USA ⟨sa@howdy.wustl.edu⟩

Anja C. ANDERSEN, Astronomical Observatory, Copenhagen, Denmark ⟨anja@astro.ku.dk⟩

Ivan ANDRONOV, Odessa State University, Odessa, Ukraine ⟨astro@paco.odessa.ua⟩

W. AOKI, University of Tokyo, Japan ⟨waoki@mtk.ioa.s.u-tokyo.ac.jp⟩

Bernhard ARINGER, University of Vienna, Austria ⟨aringer@astro.ast.univie.ac.at⟩

Claudio ARLANDINI, Research Centre of Karlsruhe, Germany ⟨claudio@ik3.fzk.de⟩

Martin ASPLUND, Nordita, Copenhagen, Denmark ⟨martin@nordita.dk⟩

Mike J. BARLOW, University College, London, United Kingdom ⟨mjb@star.ucl.ac.uk⟩

Dominique BARTHES, University of Barcelona, Spain ⟨barthes@pchpc4.am.ub.es⟩

Paul BERLIOZ-ARTHAUD, Observatory of Lyon, France ⟨pberlioz@cumulus.univ-lyon1.fr⟩

John BIEGING, Steward Observatory, University of Arizona, Tucson, AZ, USA ⟨jbieging@as.arizona.edu⟩

Thomas BLÖCKER, Max-Planck Institute for Radioastronomy, Bonn, Germany ⟨bloecker@speckle.mpifr-bonn.mpg.de⟩

Joris BLOMMAERT, European Space Agency, Villafranca, Spain ⟨jblommae@iso.vilspa.esa.es⟩

---

[1] Affiliations are as of August 1998; e-mail addresses have been updated in January 1999 to the best of our knowledge.

Valentin BUJARRABAL, National Astronomical Observatory, Alcalá de Henares, Spain ⟨ bujarrabal@oan.es ⟩

Maurizio BUSSO, Observatory of Torino, Italy ⟨ busso@to.astro.it ⟩

Jan CAMI, SRON, Groningen, Netherlands ⟨ cami@astro.rug.nl ⟩

Kenneth G. CARPENTER, NASA Goddard Space Flight Center, Greenbelt, MD, USA ⟨ carpenter@stars.gsfc.nasa.gov ⟩

Simon CASASSUS, Nuclear & Astrophysical Laboratory, Oxford, United Kingdom ⟨ s.casassus1@physics.oxford.ac.uk ⟩

Piero CAU, UMIST, Manchester, United Kingdom ⟨ pac@saturn.phy.umist.ac.uk ⟩

Merieme CHADID, University of Montpellier II, France ⟨ chadid@graal.univ-montp2.fr ⟩

Karine CHAMAILLARD, Observatory of Paris, Meudon, France ⟨ Karine.Chamaillard@obspm.fr ⟩

Corinne CHARBONNEL, Midi-Pyrénées Observatory, Toulouse, France ⟨ corinne@ast.obs-mip.fr ⟩

Isabelle CHERCHNEFF, UMIST, Manchester, United Kingdom ⟨ imc@europa.phy.umist.ac.uk ⟩

Maria Rosa CIONI, Observatory of Leiden, Netherlands ⟨ mrcioni@strw.leidenuniv.nl ⟩

Martin COHEN, University of California, Berkeley, CA, USA ⟨ mcohen@astro.berkeley.edu ⟩

Vincent COUDE du FORESTO, Observatory of Paris , Meudon, France ⟨ foresto@obspm.fr ⟩

Michelle CREECH-EAKMAN, University of Denver, CO, USA ⟨ mce@phoenix.phys.du.edu ⟩

Merce CROSAS, Harvard-Smithsonian Center for Astrophysics, Cambridge, MA, USA ⟨ mcrosas@cfa.harvard.edu ⟩

Katia CUNHA, University of Texas, El Paso, USA ⟨ katia@baade.physics.utep.edu ⟩

Francesca d'ANTONA, Astronomical Observatory of Rome, Italy ⟨ dantona@coma.mporzio.astro.it ⟩

Aditya DAYAL, Caltech, Pasadena, CA, USA ⟨ adayal@ipac.caltech.edu ⟩

Leen DECIN, Catholic University of Leuven, Belgium ⟨ Leen.Decin@ster.kuleuven.ac.be ⟩

Teije de JONG, SRON, Utrecht, Netherlands ⟨ t.dejong@sron.ruu.nl ⟩

Patrick de LAVERNY, University of Montpellier II, France ⟨ laverny@obs-nice.fr ⟩

Herwig DEJONGHE, Astronomical Observatory, Gent, Belgium ⟨ Herwig.Dejonghe@rug.ac.be ⟩

Philip DIAMOND, National Radio Astronomical Observatory, Socorro, NM, USA ⟨ pdiamond@jb.man.ac.uk ⟩

Ulrich DIRKS, Technical University of Berlin, Germany
⟨ ulidnidg@mailszrz.zrz.tu-berlin.de ⟩

Inma DOMÍNGUEZ, University of Granada, Spain ⟨ inma@goliat.ugr.es ⟩

Thomas DRIEBE, Max-Planck Institute for Radioastronomy, Bonn, Germany
⟨ driebe@speckle.mpifr-bonn.mpg.de ⟩

Emmanuel DUFOUR, Observatory of Grenoble, France
⟨ Emmanuel.DUFOUR@obs.ujf-grenoble.fr ⟩

Dieter ENGELS, Observatory of Hamburg, Germany
⟨ dengels@hs.uni-hamburg.de ⟩

Nicolas EPCHTEIN, Observatory of Nice, France ⟨ epchtein@obs-nice.fr ⟩

Kjell ERIKSSON, Astronomical Observatory, Uppsala, Sweden
⟨ kjell.eriksson@astro.uu.se ⟩

Sandra ETOKA, N. Copernicus University, Torun, Poland
⟨ Sandra@astro.uni.torun.pl ⟩

Katrina EXTER, N. Copernicus Astronomical Center, Warsaw, Poland
⟨ kme@camk.edu.pl ⟩

Michael W. FEAST, University of Cape Town, South Africa
⟨ mwf@uctvax.uct.ac.za ⟩

Bruce FEGLEY, Washington University, St. Louis, MO, USA
⟨ bfegley@levee.wustl.edu ⟩

Jason FERGUSON, Wichita State University, Wichita, KS, USA
⟨ jfergus@wsuhub.uc.twsu.edu ⟩

Axel FLEISCHER, Technical University of Berlin, Germany
⟨ axel.fleischer-dr@defpa.mail.abb.com ⟩

David FONG, University of Illinois, Urbana, IL, USA ⟨ d-fong@astro.uiuc.edu ⟩

Manuel FORESTINI, Observatory of Grenoble, France
⟨ Manuel.Forestini@obs.ujf-grenoble.fr ⟩

Adam FRANKOWSKI, N. Copernicus Astronomical Center, Torun, Poland
⟨ torque@fox.ncac.torun.pl ⟩

Roberto GALLINO, University of Torino, Italy ⟨ gallino@ph.unito.it ⟩

Pedro GARCÍA-LARIO, European Space Agency, Villafranca, Spain
⟨ pgarcia@iso.vilspa.esa.es ⟩

Andreas GAUGER, Max-Planck Institute for Radioastronomy, Bonn, Germany ⟨ gauger@speckle.mpifr-bonn.mpg.de ⟩

Alfred GAUTSCHY, University of Basel, Switzerland
⟨ gautschy@astro.unibas.ch ⟩

Eric GÉRARD, Observatory of Paris, Meudon, France ⟨ gerard@obspm.fr ⟩

Leo GIRARDI, Max-Planck Institute for Astrophysics, Garching bei München, Germany ⟨ leo@mpa-garching.mpg.de ⟩

I.S. GLASS, South African Astronomical Observatory, South Africa
⟨ isg@saao.ac.za ⟩

Alfred GLASSGOLD, New York University, New York, NY, USA
⟨ al.glassgold@nyu.edu ⟩

Tim GLEDHILL, University of Hertfordshire, Hatfield, United Kingdom
⟨ tmg@star.herts.ac.uk ⟩

D. GONZÁLEZ DELGADO, Observatory of Stockholm, Sweden
⟨ delgado@astro.su.se ⟩

Malcolm GRAY, University of Bristol, United Kingdom
⟨ malcolm.gray@bristol.ac.uk ⟩

Martin GROENEWEGEN, Max-Planck Institute for Astrophysics, Garching bei München, Germany ⟨ groen@mpa-garching.mpg.de ⟩

Jean GUIBERT, Observatory of Paris, Meudon, France
⟨ Jean.Guibert@obspm.fr ⟩

Harm HABING, Observatory of Leiden, Netherlands
⟨ habing@strw.leidenuniv.nl ⟩

Christopher HANIFF, Mullard Radio Astronomical Observatory, Cambridge, United Kingdom ⟨ cah@mrao.cam.ac.uk ⟩

Patrick HARRINGTON, University of Maryland, College Park, MD, USA
⟨ jph@astro.umd.edu ⟩

Osamu HASHIMOTO, Gunma Astronomical Observatory, Maebashi, Japan
⟨ osamu@astron.pref.gunma.jp ⟩

Christiane HELLING, Technical University of Berlin, Germany
⟨ chris@physik.tu-berlin.de ⟩

Thomas HENNING, Astrophysical Institute, Jena, Germany
⟨ henning@astro.uni-jena.de ⟩

Fabrice HERPIN, Observatory of Bordeaux, France
⟨ herpin@observ.u-bordeaux.fr ⟩

Falk HERWIG, Astrophysical Institute, Potsdam, Germany
⟨ fherwig@astro.physik.uni-potsdam.de ⟩

Kenneth HINKLE, National Optical Astronomy Observatories, Tucson, AZ, USA ⟨ hinkle@noao.edu ⟩

Susanne HÖFNER, Astronomical Observatory, Copenhagen, Denmark
⟨ hoefner@astro.univie.ac.at ⟩

Sacha HONY, Astronomical Institute, Amsterdam, Netherlands
⟨ hony@astro.uva.nl ⟩

Bruce J. HRIVNAK, Valparaiso University, Valparaiso, IN, USA
⟨ bhrivnak@exodus.valpo.edu ⟩

Josef HRON, University of Vienna, Austria ⟨ hron@astro.ast.univie.ac.at ⟩

Trung HUA, Laboratory of Space Astronomy, Marseille, France
⟨ hua@astrsp-mrs.fr ⟩

Patrick J. HUGGINS, New York University, New York, NY, USA
⟨ patrick.huggins@nyu.edu ⟩

Jarrod HURLEY, Institute of Astronomy, Cambridge, USA
⟨ jhurley@ast.cam.ac.uk ⟩

Icko IBEN, Jr., University of Illinois, Urbana, IL, USA ⟨ icko@astro.uiuc.edu ⟩

Željko IVEZIĆ, Princeton University, Princeton, NJ, USA
⟨ ivezic@astro.princeton.edu ⟩

Hideyuki IZUMIURA, Okayama Astrophysical Observatory, Asakuchi, Japan
⟨ izumiura@oao.nao.ac.jp ⟩

Gérard JASNIEWICZ, University of Montpellier II, Montpellier, France
⟨ gjasniew@graal.univ-montp2.fr ⟩

Christophe JEAN, Astrophysical Institute of Liège, Belgium
⟨ jean@astro.ulg.ac.be ⟩

Emmanuel JEHIN, University of Liège, Belgium ⟨ E.Jehin@ulg.ac.be ⟩

Kyung Sook JEONG, Technical University of Berlin, Germany
⟨ jeong@physik.TU-Berlin.de ⟩

Francisco JIMÉNEZ ESTEBAN, European Space Agency, Madrid, Spain
⟨ fje@laeff.esa.es ⟩

Karin JONSELL, Uppsala Astronomical Observatory, Sweden
⟨ Karin.Jonsell@astro.uu.se ⟩

Uffe Gråe JORGENSEN, Astronomical Observatory, Copenhagen, Denmark
⟨ uffegj@nbi.dk ⟩

Alain JORISSEN, Astronomical & Astrophysical Institute, Brussels, Belgium
⟨ ajorisse@astro.ulb.ac.be ⟩

Eric JOSSELIN, Astrophysical Institute of Paris, France ⟨ josselin@oan.es ⟩

Michael JURA, UCLA, Los Angeles, CA, USA ⟨ jura@clotho.astro.ucla.edu ⟩

Kay JUSTTANONT, Observatory of Stockholm, Sweden ⟨ kay@astro.su.se ⟩

Claudine KAHANE, Observatory of Grenoble, France
⟨ kahane@gag.observ-gr.fr ⟩

Amanda KARAKAS, Monash University, Clayton, VIC, Australia
⟨ amanda@flash.maths.monash.edu.au ⟩

Margarita KAROVSKA, Harvard-Smithsonian Center for Astrophysics, Cambridge, MA, USA ⟨ karovska@cfa.harvard.edu ⟩

Joel H. KASTNER, Massachusetts Institute of Technology, Cambridge, MA, USA ⟨ jhk@juggler.mit.edu ⟩

Hans Ulrich KÄUFL, European Southern Observatory, Garching bei München, Germany ⟨ hukaufl@eso.org ⟩

Athol KEMBALL, National Radio Astronomical Observatory, Socorro, NM, USA ⟨ akemball@nrao.edu ⟩

Franz KERSCHBAUM, University of Vienna, Austria
⟨ kerschbaum@astro.univie.ac.at ⟩

Eric KETO, Harvard-Smithsonian Center for Astrophysics, Cambridge, MA, USA ⟨ keto@dogstar.harvard.edu ⟩

Alexander F. KHOLTIGIN, St. Petersburg University, Russia ⟨afk@astro.spbu.ru⟩

László KISS, University of Szeged, Hungary ⟨l.kiss@physx.u-szeged.hu⟩

Chiyoe KOIKE, Kyoto Pharmaceutical University, Japan ⟨koike@cr.scphys.kyoto-u.ac.jp⟩

Takashi KOZASA, Kobe University, Japan ⟨kozasa@kobe-u.ac.jp⟩

Sun KWOK, University of Calgary, Alberta, Canada ⟨kwok@iras.ucalgary.ca⟩

Jean-Pierre J. LAFON, University of Montpellier II, France ⟨lafon@graal.univ-montp2.fr⟩

Ariane LANÇON, Astronomical Observatory of Strasbourg, France ⟨lancon@astro.u-strasbg.fr⟩

Fredrik LARSEN, Stockholm Observatory, Sweden ⟨fredrik@astro.su.se⟩

John LATTANZIO, Monash University, Clayton, VIC, Australia ⟨johnl@flash.maths.monash.edu.au⟩

Thibaut LE BERTRE, Observatory of Paris, France ⟨lebertre@obspm.fr⟩

Miska LE LOUARN, European Southern Observatory, Garching bei München, Germany ⟨lelouarn@eso.org⟩

Anne-Marie LE SQUEREN, University of Montpellier II, France ⟨lesquere@graal.univ-montp2.fr⟩

Agnès LÈBRE, University of Montpellier II, France ⟨lebre@graal.univ-montp2.fr⟩

Thomas LEBZELTER, University of Vienna, Austria ⟨lebzelter@astro.univie.ac.at⟩

Brian LEWIS, Arecibo Observatory, Porto Rico, USA ⟨blewis@naic.edu⟩

Stéphane LIBERATORE, Observatory of Paris, France ⟨Stephane.Liberatore@obspm.fr⟩

Jeffrey LINSKY, JILA, University of Colorado, Boulder, CO, USA ⟨jlinsky@jila.colorado.edu⟩

T. LLOYD EVANS, South African Astronomical Observatory, South Africa ⟨tle@saao.ac.za⟩

Alex LOBEL, Armagh Observatory, Northern Ireland, United Kingdom ⟨ajrl@star.arm.ac.uk⟩

Katharina LODDERS, Washington University, St. Louis, MO, USA ⟨lodders@levee.wustl.edu⟩

Rita LOIDL, University of Vienna, Austria ⟨loidl@astro.univie.ac.at⟩

Bruno LOPEZ, Observatory of Nice, France ⟨lopez@obs-nice.fr⟩

Cécile LOUP, Astrophysical Institute of Paris, France ⟨loup@iap.fr⟩

Robert LUCAS, IRAM, Grenoble, France ⟨lucas@iram.fr⟩

Maria LUGARO, Monash University, Clayton, VIC, Australia ⟨marial@thala.maths.monash.edu.au⟩

Xavier LURI, University of Barcelona, Spain ⟨xluri@mizar.am.ub.es⟩

Julie LUTZ, University of Washington, Seattle, WA, USA
⟨ jlutz@orca.astro.washington.edu ⟩

Pierre MAGAIN, University of Liège, Belgium ⟨ magain@astro.ulg.ac.be ⟩

Christian MAGNAN, University of Montpellier II, France
⟨ magnan@graal.univ-montp2.fr ⟩

Arturo MANCHADO, Astrophysical Institute of Canarias, Tenerife, Spain
⟨ amt@iac.es ⟩

Massimo MARENGO, Harvard-Smithsonian Center for Astrophysics, Cambridge, MA, USA ⟨ mmarengo@cfa.harvard.edu ⟩

Paola MARIGO, Max-Planck Institute for Astrophysics, Garching bei München, Germany ⟨ marigo@lunatix.pd.astro.it ⟩

M. R. W. MASHEDER, Radioastronomical Observatory, Dwingeloo, Netherlands ⟨ mrwm@jive.nfra.nl ⟩

Peter MATA, FBIS, Battle, E. Sussex, United Kingdom ⟨ pdmata@aol.com ⟩

S. MATSUMOTO, University of Tokyo, Japan
⟨ matsumot@mtk.ioa.s.u-tokyo.ac.jp ⟩

Mikako MATSUURA, Inst. Space & Astronautical Science, Sagamihara, Japan
⟨ mikako@astro.isas.ac.jp ⟩

Janet MATTEI, American Assoc. of Variable Star Observers, Cambridge, MA, USA ⟨ jmattei@aavso.org ⟩

Nicolas MAURON, University of Montpellier II, France
⟨ mauron@graal.univ-montp2.fr ⟩

D.H. McNAMARA, Brigham Young University, Provo, UT, USA
⟨ mcnamara@physc2.byu.edu ⟩

Margaret MEIXNER, University of Illinois, Urbana, IL, USA
⟨ meixner@astro.uiuc.edu ⟩

Djamel MEKARNIA, Observatory of Nice, France ⟨ mekarnia@obs-nice.fr ⟩

Norayr MELIKYAN, Byurakan Observatory, Armenia ⟨ nmelikia@bao.sci.am ⟩

Garrelt MELLEMA, Stockholm Observatory, Sweden ⟨ garrelt@astro.su.se ⟩

Marie-Odile MENNESSIER, University of Montpellier II, France
⟨ menes@graal.univ-montp2.fr ⟩

Joanna MIKOLAJEWSKA, N. Copernicus Astronomical Center, Warsaw, Poland ⟨ mikolaj@camk.edu.pl ⟩

Frank MOLSTER, Anton Pannekoek Astronomical Institute, Amsterdam, Netherlands ⟨ frankm@astro.uva.nl ⟩

John MONNIER, University of California, Berkeley, CA, USA
⟨ monnier@physics.berkeley.edu ⟩

Mustapha MOUHCINE, Astronomical Observatory of Strasbourg, France
⟨ mouhcine@newb6.u-strasbg.fr ⟩

Nami MOWLAVI, Observatory of Geneva, Switzerland
⟨ Nami.Mowlavi@obs.unige.ch ⟩

Nathan NETZER, ORT Braude College, Karmiel, Israel ⟨ netzer@braude.ac.il ⟩

Yuen Keong NG, Padova Astronomical Observatory, Italy ⟨yuen@eos6.sron.nl⟩

Gilles NICCOLINI, Observatory of Nice, France ⟨niccolin@iram.es⟩

Arlette NOELS, University of Liège, Belgium ⟨noels@astro.ulg.ac.be⟩

Pierre NORTH, University of Lausanne, Switzerland
⟨Pierre.North@obs.unige.ch⟩

Lars-Åke NYMAN, European Southern Observatory, La Silla, Chile
⟨lnyman@eso.org⟩

Keiichi OHNAKA, Technical University of Berlin, Germany
⟨ohnaka@export.physik.TU-Berlin.de⟩

Devendra OJHA, Tata Institute, Mumbai, India ⟨ojha@tifr.res.in⟩

Hans OLOFSSON, Stockholm Observatory, Sweden ⟨hans@astro.su.se⟩

Alain OMONT, Astrophysical Institute of Paris, France ⟨omont@iap.fr⟩

Roger OSTERBART, Max-Planck Institute for Radioastronomy, Bonn, Germany ⟨osterbart@mpifr-bonn.mpg.de⟩

Geneviève PARMENTIER, University of Liège, Belgium
⟨G.Parmentier@ulg.ac.be⟩

M. PARTHASARATHY, Indian Institute of Astrophysics, Bangalore, India
⟨partha@iiap.ernet.in⟩

Mikhail PASHCHENKO, Sternberg Astronomical Institute, Moscow, Russia
⟨pachenko@sai.msu.su⟩

Beate PATZER, Technical University of Berlin, Germany
⟨patzer@physik.TU-Berlin.de⟩

Roberto PEREIRA ORTIZ, Leiden Astronomical Observatory, Netherlands
⟨ortiz@strw.leidenuniv.nl⟩

Mario PERINOTTO, Department of Astronomy, Florence, Italy
⟨mariop@arcetri.astro.it⟩

Guy PERRIN, Observatory of Paris, Meudon, France
⟨perrin@hplyot.obspm.fr⟩

Bertrand PLEZ, Physics Institute, Lund, Sweden ⟨plez@graal.univ-montp2.fr⟩

Karen POLLARD, University of Canterbury, Christchurch, New Zealand
⟨k.pollard@phys.canterbury.ac.nz⟩

Onno POLS, Astrophysical Institute of Canarias, Tenerife, Spain
⟨onno@ll.iac.es⟩

Luciana POMPEIA CAVALCANTI, University of São Paulo, Brazil
⟨pompeia@iagusp.usp.br⟩

Daniel POOLEY, University of Canterbury, Christchurch, New Zealand
⟨d.pooley@phys.canterbury.ac.nz⟩

Solange V. RAMIREZ, Ohio State University, Columbus, OH, USA
⟨solange@astronomy.ohio-state.edu⟩

Thomas RAUSCHER, University of Basel, Switzerland
⟨tommy@quasar.physik.unibas.ch⟩

Maarten REYNIERS, Catholic University of Leuven, Belgium ⟨ maartenr@mailhost.ster.kuleuven.ac.be ⟩

Anita RICHARDS, NRAL Jodrell Bank, Univ. of Manchester, Macclesfield, United Kingdom ⟨ amsr@jb.man.ac.uk ⟩

Heike RICHTER, Technical University of Berlin, Germany ⟨ richter@pils.physik.tu-berlin.de ⟩

Georgij RUDNITSKIJ, Sternberg Astronomical Institute, Moscow, Russia ⟨ gmr@sai.msu.su ⟩

Nils RYDE, Astronomical Observatory, Uppsala, Sweden ⟨ ryde@astro.uu.se ⟩

Raghvendra SAHAI, Jet Propulsion Laboratory, Pasadena, USA ⟨ sahai@grandpa.jpl.nasa.gov ⟩

Carmen SÁNCHEZ CONTRERAS, National Astronomical Obs., Alcalá de Henares, Spain ⟨ sanchez@oan.es ⟩

Detlef SCHÖNBERNER, Astrophysical Institute, Potsdam, Germany ⟨ DeSchoenberner@aip.de ⟩

Klaus-Peter SCHRÖDER, Technical University of Berlin, Germany ⟨ schroder@weizen.physik.tu-berlin.de ⟩

Mathias SCHULTHEIS, Astrophysical Institute of Paris, France ⟨ schulthe@iap.fr ⟩

Hugo SCHWARZ, Nordic Optical Telescope, Santa Cruz de La Palma, Spain ⟨ hschwarz@not.iac.es ⟩

Maartje SEVENSTER, Mt Stromlo Observatory, Australia ⟨ msevenst@mso.anu.edu.au ⟩

Yvonne SIMIS, Leiden Observatory, Netherlands ⟨ simis@strw.leidenuniv.nl ⟩

Guy SIMON, Observatory of Paris, France ⟨ simon@obspm.fr ⟩

Philippe SIVAGNANAM, Observatory of Paris, Meudon, France ⟨ siva@obspm.fr ⟩

Verne V. SMITH, University of Texas, El Paso, USA ⟨ verne@balmer.physics.utep.edu ⟩

Noam SOKER, University of Haifa at Oranim, Israel ⟨ soker@physics.technion.ac.il ⟩

Angela SPECK, University College London, United Kingdom ⟨ aks@star.ucl.ac.uk ⟩

Matthias STEFFEN, Astrophysical Institute, Potsdam, Germany ⟨ MSteffen@aip.de ⟩

Kate Y.L. SU, University of Calgary, Alberta, Canada ⟨ ysu@iras.ucalgary.ca ⟩

Hiroshi SUTO, Universities of Hawaii, USA, and Kyoto, Japan ⟨ suto@crl.go.jp ⟩

Allen V. SWEIGART, NASA Goddard Space Flight Center, Greenbelt, MD, USA ⟨ sweigart@bach.gsfc.nasa.gov ⟩

Karoly SZATMÁRY, JATE University, Szeged, Hungary ⟨ k.szatmary@physx.u-szeged.hu ⟩

Ryszard SZCZERBA, N. Copernicus University, Torun, Poland
⟨ szczerba@ncac.torun.pl ⟩

Marian SZYMCZAK, N. Copernicus University, Torun, Poland
⟨ msz@astro.uni.torun.pl ⟩

T. TANABE, University of Tokyo, Japan ⟨ ttanabe@mtk.ioa.s.u-tokyo.ac.jp ⟩

Anne THOUL, University of Liège, Belgium ⟨ Anne.Thoul@ulg.ac.be ⟩

Alexander TIELENS, Kapteyn Astronomical Institute, Groningen, Netherlands ⟨ tielens@astro.rug.nl ⟩

Christopher A. TOUT, Institute of Astronomy, Cambridge, United Kingdom
⟨ cat@ast.cam.ac.uk ⟩

Yitzchak TUCHMAN, Hebrew University of Jerusalem, Israel
⟨ tuchma@vms.huji.ac.il ⟩

Peter TUTHILL, University of California, Berkeley, CA, USA
⟨ gekko@ssl.berkeley.edu ⟩

Gerard T. VAN BELLE, Jet Propulsion Laboratory, Pasadena, CA, USA
⟨ gerard@huey.jpl.nasa.gov ⟩

Sophie VAN ECK, Institute of Astrononomy & Astrophysics, Brussels, Belgium ⟨ svaneck@astro.ulb.ac.be ⟩

Huib Jan VAN LANGEVELDE, JIVE, Dwingeloo, Netherlands
⟨ huib@jive.nfra.nl ⟩

Jacco Th. VAN LOON, Astronomical Institute, Amsterdam, Netherlands
⟨ jvanloon@astro.uva.nl ⟩

Hans VAN WINCKEL, Catholic University of Leuven, Belgium
⟨ Hans.VanWinckel@ster.kuleuven.ac.be ⟩

Jean-Louis VIDAL, University of Montpellier II, France
⟨ vidal@graal.univ-montp2.fr ⟩

Eva VILLAVER SOBRINO, Astrophysical Institute of Canarias, Tenerife, Spain ⟨ villaver@ll.iac.es ⟩

Kevin VOLK, University of Calgary, Alberta, Canada ⟨ volk@iras.ucalgary.ca ⟩

Nikolai VOSHCHINNIKOV, St. Petersburg University, Russia
⟨ avv@common.aispbu.spb.su ⟩

Astrid WACHTER, Technical University of Berlin, Germany
⟨ wachter@pils.physik.tu-berlin.de ⟩

Christoffel WAELKENS, Catholic University of Leuven, Belgium
⟨ christoffel@ster.kuleuven.ac.be ⟩

Roni WALDMAN, Hebrew University of Jerusalem, Israel
⟨ waldman@cc.huji.ac.il ⟩

George WALLERSTEIN, University of Washington, Seattle, WA, USA
⟨ wall@orca.astro.washington.edu ⟩

Rens WATERS, University of Amsterdam, Netherlands ⟨ rensw@astro.uva.nl ⟩

Gerd WEIGELT, Max-Planck Institute for Radioastronomy, Bonn, Germany
⟨ weigelt@mpifr-bonn.mpg.de ⟩

Klaus WERNER, University of Tübingen, Germany
⟨ werner@astro.uni-tuebingen.de ⟩

Bengt E. WESTERLUND, Astronomical Observatory, Uppsala, Sweden
⟨ bengt.westerlund@astro.uu.se ⟩

Patricia WHITELOCK, South African Astronomical Observatory, South Africa ⟨ paw@saao.ac.za ⟩

Walter WINDSTEIG, University of Vienna, Austria
⟨ windsteig@auro.astro.univie.ac.at ⟩

Robert F. WING, Ohio State University, Columbus, OH, USA
⟨ wing@astronomy.ohio-state.edu ⟩

Jan Martin WINTERS, Technical University of Berlin, Germany
⟨ winters@physik.TU-Berlin.de ⟩

Peter WOITKE, Technical University of Berlin, Germany
⟨ woitke@physik.tu-berlin.de ⟩

Peter WOOD, Mt Stromlo & Siding Spring Obs., Canberra, Australia
⟨ wood@mso.anu.edu.au ⟩

Atara YA'ARI, Hebrew University of Jerusalem, Israel
⟨ atara@saba.fiz.huji.ac.il ⟩

Issei YAMAMURA, University of Amsterdam, Netherlands
⟨ yamamura@astro.uva.nl ⟩

Jeremy YATES, University of Hertfordshire, Hatfield, United Kingdom
⟨ jyates@star.herts.ac.uk ⟩

John S. YOUNG, Mullard Radio Astronomical Obs., Cambridge, United Kingdom ⟨ J.S.Young@mrao.cam.ac.uk ⟩

Boris YUDIN, Sternberg Astronomical Institute, Moscow, Russia
⟨ yudin@sai.msu.su ⟩

Laimons ZACS, VIRAC, Riga, Latvia ⟨ zacs@latnet.lv ⟩

Albert ZIJLSTRA, UMIST, Manchester, United Kingdom
⟨ aaz@iapetus.phy.umist.ac.uk ⟩

Introduction

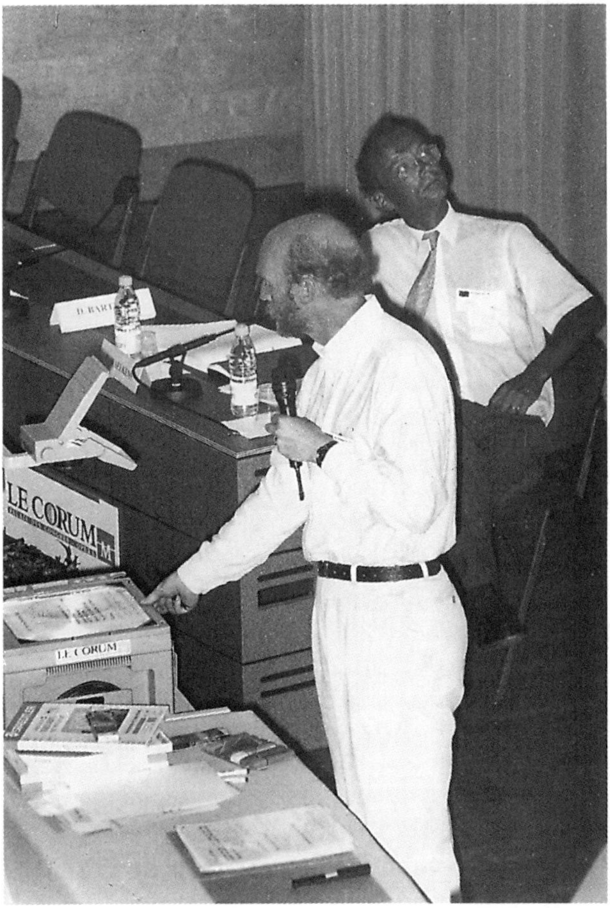

Hans Olofsson setting the stage[1].

[1]credit for all the photographs taken during the conference and reproduced in this volume: 30 × 40 IMAGE, Montpellier, France

Asymptotic Giant Branch Stars
T. Le Bertre, A. Lèbre and C. Waelkens, eds.
I.A.U. Symposium 191, 1999

The AGB-star Phenomenon: Setting the Stage

Hans Olofsson

Stockholm Observatory, SE–133 36 Saltsjöbaden, Sweden

Abstract. The asymptotic giant branch (AGB) evolution of stars is interesting from many points of view. It is the final evolutionary phase for the large majority of all stars in the Universe that have left the main sequence. During this stage they contribute to the chemical evolution of galaxies, contribute to the integrated starlight of many galaxies, may provide the only interstellar matter there is in some galaxies, and can, due to their high luminosities and ages, be used as important probes of galactic structure and dynamics. It is also the final evolutionary stage of our own star, the Sun, and hence touches on aspects beyond the immediate astronomical interest. Furthermore, these stars provide us with fascinating systems, where intricate interplays between physical and chemical processes take place, that we simply would like to understand. In this introductory outline our present knowledge of the AGB-star phenomenon is reviewed.

1. AGB-stars

Red giant stars have effectively divided themselves into two parts: a small, very hot ($\sim 10^8$ cm, $\sim 10^8$ K) core that is strongly gravitationally bound, and a huge, much cooler mantle ($\sim 10^{13}$ cm, $\sim 10^3$ K at the surface) where the external parts are only weakly gravitationally bound. We are here only interested in those red giants that have experienced central He-burning, and now populate the asymptotic giant branch (AGB). These are much rarer than those red giants that have only experienced central H-burning (the RGB-stars). The AGB-stars experience substantial mass loss from the surface, and huge circumstellar envelopes (CSEs) of gas and dust are formed with extents approaching 10^{19} cm and temperatures below 10 K in the external parts. This can be regarded as the third part of the star, and hence an AGB-star stretches more than 11 orders of magnitude in size and 7 orders of magnitude in temperature. The evolution of the star on the AGB depends on an intricate interplay between processes taking place in the three parts.

In this outline only the pure AGB-stars are considered, but it will set the stage also for the understanding of other fascinating aspects of the AGB-star phenomenon: binarity (Jorissen these proceedings, hereafter th.pr.), post-AGB objects (Van Winckel th.pr.), galactic structure (Dejonghe & Van Caelenberg th.pr.), population studies (Lançon th.pr.), chemical evolution (Gustafsson & Ryde 1998). A limited number of subjectively chosen references are given. More detailed reviews, including many of those in these proceedings, are mentioned in the text.

2. The core

The core consists of an electron-degenerate C/O-mixture [where the relative number of C and O depends on the still uncertain rate for the $^{12}C(n,\gamma)^{16}O$ reaction, C/O$\sim$0.1–1 depending on mass], where no nuclear-burning processes take place (at the border towards supergiant stars electron-degenerate O/Ne-cores exist). It will eventually appear on the scene as a small, dense, hot (but gradually cooling) stellar remnant, a white dwarf (WD). In the classical models only stars with an initial mass less than 8–9 $M_\odot$ will have an electron-degenerate C/O-core (convective over-shooting may decrease this by $\sim$30%, Bressan et al. 1993), and hence will not experience further nuclear burning processes, unless the core mass, M_c, exceeds 1.4 $M_\odot$ and gravitational collapse ensues. Due to the short lifetime of an AGB-star M_c does not increase significantly during evolution ($\sim 10^{-7}\,M_\odot\,yr^{-1}$), and the core mass of even a very massive AGB-star will not reach the 1.4 $M_\odot$-limit. [The fact that the descendants of AGB-stars, WDs, have sharply peaked mass distributions in the range 0.5–0.6 supports a short AGB-lifetime, and strongly indicates that substantial loss of mass must take place during the AGB (Weidemann 1990).] Only stars with a mass above $\sim$0.5 $M_\odot$ will enter the AGB following exhaustion of central He (Bressan et al. 1993), but only stars above $\sim$0.8 $M_\odot$ will reach the AGB within $\sim$14 Gyr. Thus, the AGB-stage is the final evolutionary stage for the large majority (>95%) of all stars in the Universe that have left the main sequence.

3. The H-mantle

Above the core lies a He-shell and above that a H-mantle, and these are very active regions. He-burning provides the energy during the early evolution on the AGB (E-AGB). Eventually, H ignites at the base of the H-mantle and from now on H-burning supplies the energy, except for very short periods ($<10^3$ yr) of He-burning in the form of a thermo-nuclear runaway [He-shell flashes, or thermal pulses (TPs)] that repeats semi-periodically on time scales of $\sim 10^{4-5}$ yr (longer the smaller the mass; Wagenhuber & Groenewegen 1998). This defines the thermally pulsing AGB (TP-AGB; $\sim$5–20% of the E-AGB time scale). The sudden increase in energy release during the TP extinguishes the H-burning and the outer convective parts of the H-mantle may extend into the inter-shell region (enriched in ^{12}C), leading to a dredge-up of material to the surface, the 3rd dredge-up [the process that produces C-stars; see Wallerstein & Knapp 1998, for a review on this stellar type]. In the inter-shell region ^{13}C (still of uncertain origin) appears to be the neutron source, and a production of s-process elements will take place here (Lambert et al. 1995; Straniero et al. 1997). Finally, in the massive AGB-stars ($\geq 4\,M_\odot$) the convective part of the H-mantle may penetrate into the H-burning shell, leading to hot bottom burning (HBB; Boothroyd et al. 1993). HBB will effectively convert ^{12}C into ^{13}C and finally ^{14}N (also ^7Li and ^{26}Al are produced, while e.g. ^{19}F and ^{18}O are destroyed).

The TP-induced dredge-up occurs easier in high-M stars than in low-M stars. Theoretically, for the moment, the lower mass limit to C-stars lies at $M_{\rm MS}{\sim}1.5\,{\rm M}_\odot$ for solar composition (Straniero et al. 1997). HBB prevents the more massive stars to become C-stars, and the upper mass limit to C-stars is ${\sim}4\,{\rm M}_\odot$. Since it must be easier to achieve C>O in low-Z stars we expect the relative number of M- and C-stars to be a strong function of metallicity.

Our physical understanding of this region is good, but not perfect (Iben & Renzini 1983; Frost & Lattanzio 1996a; Blöcker th.pr., Lattanzio & Forestini th.pr.). Two outstanding problems remain. First, the "crude" treatment of the convection using the mixing-length theory (the turbulence is local and instantaneous, characterized by one-size eddies, and a free parameter is introduced, the mixing length scale). Particular problems arise at the boundaries between convective and non-convective regions [the over-shoot problem (Frost & Lattanzio 1996b), and the mixing]. During the TPs we have regions where the dynamical, thermal, and nuclear burning time scales are equal, and hence many common approximations become doubtful. When coupled to pulsation another, much shorter, time scale is introduced. Here we mention some important uncertainties. The dredge-up efficiency dependences on mass, metallicity, and time are crucial but poorly understood, e.g., some additional mixing may be required to produce low-M C-stars (Herwig et al. 1997), or maybe degenerate TPs will do it (Frost et al. 1998b), and also the composition (the relative abundances of He, C, O) of the dredged-up matter is uncertain. If mass loss decreases the H-mantle mass, $M_{\rm e}$, sufficiently, HBB is quenched and high-M stars may become luminous C-stars with low ^{12}C/^{13}C-ratios (Frost et al. 1998a). There is also a possibility that HBB will produce luminous M-stars with low ^{12}C/^{13}C-ratios (e.g., evolved OH/IR-stars, Delfosse et al. 1997), and it may lead to stars that are highly enhanced in N. The second major problem is the mass loss. A credible stellar evolution model must take this into account. The main problem is that it cannot yet be calculated from first principles, and hence some recipe that relates $\dot{M}$ to other stellar parameters must be used. Usually this results in substantially increasing mass loss with time, that is strongly modulated during the TPs, and final core masses that agree well with the mass-distribution of WDs (Blöcker 1995; Schröder et al. 1998). The AGB evolution is terminated when $M_{\rm e}$ decreases below a certain value, ${\sim}0.001\,{\rm M}_\odot$. This is almost certainly determined by the mass loss for all AGB-stars. Thus, we can expect a complex interplay between dredge-up, HBB, and mass loss that we still do not understand, and in the end the evolution of these giants depends on the details. We will not have reliable synthetic AGB evolution models before the main problems are solved (Groenewegen & de Jong 1993; Marigo et al. 1996).

There is no full-proof direct observational evidence for the existence of TPs. Wood & Zarro (1981) interpret period change rates as due to the behaviour during a TP. Olofsson et al. (1996, 1998a) have suggested that the very geometrically thin gas-shells around a few C-stars are due to a recent TP. Sakurai's object could be a very recent born-again AGB-star, since drastic changes in elemental composition over short time scales (months) are apparent (Asplund et al. 1997; Asplund th.pr.).

The luminosity, L, of the star is determined by the rate of H-burning (except during a TP). There exists a linear relation (as long as the core radius depends linearly on M_c) between L and M_c (which depends very little on other parameters, at least, after a few TPs, Wagenhuber & Groenewegen 1998), $L \sim 57000\,(M_c - 0.5)\,L_\odot$, and hence $L_{max} \sim 56000\,L_\odot$ ($M_{bol} \sim -7.1$). However, it should always be remembered that the L-M_c relation applies only during a limited period between TPs (<50% of the time). Consequently, also an estimate of M_c using it may be wrong (by more than 10%). HBB in the more massive stars will destroy the linear relation and produce higher L than given by the L-M_c relation (Boothroyd & Sackman 1992).

At some point the H-mantle becomes unstable, and it will start to pulsate. There exists as yet no detailed model of the pulsation of AGB-stars, including e.g. a time-dependent treatment of convection (Tuchman th.pr.). The period calculations are sensitive to the outer boundary conditions (very difficult to define for a pulsating AGB-star). The growth rates are very sensitive to the details of the convection, as well as to the boundary conditions. It is not yet finally settled which stars pulsate in which mode. In principle, the problem requires only measurements of the stellar radius, R, and the period, P, and the use of the fundamental pulsation equation, $P = QR^{3/2}M^{-1/2}$, where Q is to be obtained from the model. However, the radius is not well defined in the case of an AGB-star, and it varies by as much as a factor of two with wavelength. The theoretical definition is usually in terms of the Rosseland mean optical depth. Taken at face values interferometric measurements of Miras at $\sim$0.8 μm point to radii too large for fundamental-mode pulsation, if $M < 1.5\,M_\odot$ (Haniff et al. 1995), while interferometer observations at 2.2 μm are not so conclusive (van Belle et al. 1997; Feast th.pr.). However, there are strong arguments in favour of fundamental mode pulsation for Miras in the form of parallel K-log P relations in the LMC (Wood & Sebo 1996; Wood et al. th.pr.). The direct measurements may be reconciled with this if the measured radii are too large, e.g., due to scattering in the upper layers of the atmosphere.

4. The atmosphere

The outer part of the H-mantle is cool enough ($\lesssim$3500 K) that formation of molecules will take place, and it is their presence that determines the spectral appearance of the stars, at least in the near-IR, and to some extent also the structure of their atmospheres (e.g., line blanketing). Our physical understanding of the atmosphere and the formation of the photospheric lines is good (Gustafsson 1989; Gustafsson & Jørgensen 1994; Plez th.pr.). The "classical" sources of uncertainties are effective temperatures, surface gravities, sphericity, opacities, LTE analysis, and the approximate treatment of convection. The realization that the atmosphere is much more extended than previously thought and non-hydrostatic (systematic flows and shocks) is gradually introduced into models (see Sect. 5). Therefore, non-LTE effects are very likely present. Thermal instabilities, due to molecule and grain formation, may lead to inhomogeneities. Dust is possibly an important opacity source, and this requires a detailed knowledge of grain formation.

The determination of elemental abundances are far from trivial and elaborate models have been developed over the last decades. The spectra are extremely crowded, and usually a continuum level is difficult to find. Impressive results have nevertheless been obtained on elemental abundances (Smith & Lambert 1990; Smith th.pr.), as well as on physical conditions [in the s-process region, Lambert et al. (1995), and in the HBB region, Smith et al. (1995)]. Atmosphere models also provide relations between important quantities like L and T_{eff} and broad-band colours (Bessel et al. 1998). Still, discrepancies between model and observations and, not so uncommonly, between different models applied to the same objects exist [e.g., Ohnaka & Tsuji (1998) and references therein].

There are lots of crucial observations to be obtained here, such as very high spectral resolution data in the near-IR region for larger samples. Interferometer data at different wavelengths provide information on the stratification of the atmosphere, as well as on inhomogeneities at the surface. Evidence is gathering that the AGB-stars are asymmetric, but it is not clear whether the asymmetries are due to spots in the photospheres, clouds in the upper atmospheres, or to non-radial pulsations (Haniff et al. 1992; Karovska et al. 1997; Weigelt et al. 1996). An interesting effect of non-uniform disks is that the positions, parallaxes, and proper motions of AGB-stars measured by HIPPARCOS could have considerable uncertainties (e.g., 1000 $R_{\odot}$ at 50 and 250 pc corresponds to 100 and 20 mas, respectively, considerably larger than the position resolution).

5. The extended atmosphere

The pulsation of the star will modulate the radiated power, but also deposit mechanical energy into the weakly gravitationally bound outer parts, the main effect being an atmosphere with a considerably increased scale height compared to a hydrostatic one. Shocks will develop, and at high enough altitudes grain condensation takes place in the post-shock gas. It appears that gas particles reach the escape velocity, under a broad range of conditions, due to the mechanical energy input augmented by radiation pressure on grains (and perhaps molecules).

Our physical understanding of this region is very limited (Höfner th.pr.). A complex interplay between different (non-linear) physical and chemical processes with different characteristic time scales takes place. In particular, grain formation is poorly understood in detail, but will have a significant effect on the thermal, hydrodynamical, and chemical structure of the upper atmosphere. The region is very likely inhomogeneous. Nevertheless, progress on dynamical stellar atmospheres is being made, in particular for C-stars where the grain formation is better understood (Fleischer et al. 1992; Höfner & Dorfi 1997). Synthetic IR-light curves show complex (e.g., super-P) behaviours (Winters et al. 1994b), and the brightness distributions have considerable radial structure (Winters et al. 1995). Synthetic near-IR colour calculations have given results in accordance with observations (Alvarez & Plez 1998). Also, synthetic spectra, in the 2-12 μm range suitable for comparison with ISO-data, are being produced (Hron et al. 1998). The discrepancies are substantial and much remains to be done, but the prospects are excellent. Models with improved treatments of opacity show significant changes in the upper density structure (Höfner et al. 1998). Proper treatments of cooling in the post-shock gas and non-equilibrium processes are still lacking, and so is a proper coupling to a 'real' pulsating star. It is not clear to what extent the magnetic field is of importance for the structure and the dynamics (Hartquist & Dyson 1997). Finally, we note the lack of similar models for M- and S-stars.

The importance of grain formation for the evolution of AGB-stars cannot be over-emphasized. It is therefore unfortunate that our knowledge of this complicated process is so rudimentary. The choice stands between a thermodynamic or a kinetic approach. In C-stars one possibility is that reactions between C_2H_2-molecules eventually lead to soot particles (Cherchneff & Cau th.pr.; Allain et al. 1997). However, many uncertainties remain, and other possible condensation seeds are SiC-, TiC-, and pure C-particles (Lodders & Fegley th.pr.). In an O-rich environment TiO_2 may be a possible nucleation species (Jeong et al. th.pr.). Also the radiative characteristics are uncertain (Henning th.pr.).

Crucial observations are provided by near- to mid-IR spectra, in particular of "simple" regions, and long time-series data. Detailed, multi-wavelength light-curves provide important constraints (de Laverny et al. 1997). High-resolution observations provide information on atmosphere extension (Tuthill et al. 1995), dust formation zones (Danchi et al. 1994; Ivezić & Elitzur 1996; Monnier th.pr.), inhomogeneities (Lopez et al. 1997; Weigelt et al. 1998), and departures from spherical symmetry (Lopez th.pr.). With new large telescopes we can look forward to detailed images of the extended atmospheres, including time variability. Multi-epoch images of SiO masers provide a very exciting view of the structure and dynamics of the upper atmospheric layers (Diamond & Kemball th.pr.), and similar multi-epoch imaging of stellar H_2O masers are also undertaken (Richards et al. th.pr.). There remains to determine whether we see matter or pattern motion. A recent very interesting development is the finding of a radio photosphere by Reid & Menten (1997). A model places it at $\sim 2R_*$, i.e., within the dust-condensation zone, where the density is $\sim 10^{12}\,\mathrm{cm}^{-3}$ and the temperature $\sim 1600\,\mathrm{K}$. They detected no variations with time ($<\pm 15\%$), and this puts limits on the shocks passing through this region. Likewise, there may exist a quasi-static layer of molecular gas in the upper atmosphere (Tsuji et al. 1997).

A very important aspect of the dynamical atmosphere models is the possibility of deriving mass loss rates from first principles (Sedlmayr & Dominik 1995; Fleischer et al. th.pr.). It appears that the levitation mechanism itself (shocks or possibly acoustic waves) can drive mass loss (and it may dominate for low-$\dot{M}$ objects), but it is strongly augmented by the radiation pressure on dust, and possibly molecules, when this becomes effective. Models that give mass loss rates as a function of the fundamental stellar parameters have been presented for stationary (Winters et al. 1994a) and dynamical winds (Bowen & Willson 1991; Arndt et al. 1997). However, Höfner et al (1998) have shown that more realistic opacities in the atmospheric models may result in (considerably) lower $\dot{M}$. We are still only in the beginning of obtaining reliable quantitative theoretical mass loss rates.

6. The circumstellar envelope

An AGB-star will eventually lose mass at a considerable rate in the form of a slow ($\sim 15\,\mathrm{km\,s^{-1}}$) wind, and this forms a CSE of escaping gas- and dust-particles. The gas is to a large extent molecular with a composition that reflects the chemistry of the star. The molecules are located in envelopes of different sizes (the by far largest ones being those of CO and H_2), or shells of different radii and widths, depending on their resistance to UV photons and the formation routes. The dust grains exist way beyond the molecular envelopes.

Our understanding of the CSE is relatively good (Habing 1996; Olofsson 1996; Bujarrabal th.pr.). However, the exact geometry of a CSE is uncertain, and the radial density distribution is determined by the (uncertain) mass loss history of the star, and it may be affected by the fact that one medium (dust) streams through another (gas), or a faster wind runs into a slower one. Furthermore, the details of the density and temperature structure of the gas is unknown. In the footsteps of the dynamical atmosphere models time-dependent hydrodynamical models that give the density, temperature, velocity, and chemical structures of CSEs are now being developed (Steffen et al. 1998, th.pr.).

Dust emission models have so far been based on stationary, smooth, spherical winds. The gross features of the spectral energy distributions can be well modelled using dirty amorphous silicates for O-rich CSEs (including the 10 and 18 μm features; Le Sidaner & Le Bertre 1996) and amorphous carbon for C-rich CSEs (featureless; Groenewegen et al. 1998). Very important results are the tight $\dot{M}$-colour (in particular K–L) relations, although they differ for the two chemistries (Le Bertre & Winters 1998). On the contrary the correlations between $\dot{M}$ and far-IR colours are poor. The more self-consistent models, i.e., those coupled to dynamical atmosphere models, have not yet been applied to large samples, but a detailed study of the C-star AFGL 3068 shows good agreement with a large data set (Winters et al. 1997).

The 11.3 μm emission feature observed in C-rich CSEs can be well modelled if a small amount of SiC dust is added (Lorenz-Martins & Lefèvre 1994). It seems that the discrepancy between AGB- and meteoritic-SiC grains has now been resolved. The presence of a 13 μm emission feature in O-rich CSEs (Sloan et al. 1996) points to the existence of crystalline grains, possibly α-Al_2O_3. Relatively narrow features beyond 20 μm in O-rich CSEs also suggest the presence of crystalline material (its fraction appears to increase with $\dot{M}$) (Waters et al. 1996, th.pr.). The carriers of the 21 μm feature in C-rich post-AGB CSEs (Kwok et al. 1995, th.pr.), and the 30 μm feature in C-rich (AGB and post-AGB) CSEs (Omont et al. 1995) have not yet been identified, although MgS may play a role for the latter (Szczerba et al. 1997). In very cold O-rich post-AGB objects features, attributed to ice, appear at $\sim$43 and 60 μm. There are post-AGB objects which show both O- and C-rich dust features (Waters th.pr.; Cohen et al. th.pr.). It is conceivable that in these objects there exists an O-rich (circumbinary) disk, and a C-rich AGB-CSE remnant.

Classification systems of dust features, that can be compared with different stellar populations, have been developed for the 8–14 μm range for O-rich (Hron et al. 1997) and C-rich CSEs (Sloan et al. 1998). These are interesting developments, but problems remain in the subtraction of the stellar- and dust-continuum.

Considerable progress has been done in our understanding of the origins of the molecules detected in CSEs (Glassgold 1996, th.pr.). The photospheric species are produced in a stellar atmosphere equilibrium chemistry, but the abundance varies with height in the atmosphere, and there may be effects due to shock waves, to chromospheric radiation, and to the formation of dust (Willacy

& Cherchneff 1998). Eventually, the photospheric species are photodissociated by the interstellar UV field [this requires a knowledge of the photodissociation properties of the molecule, the UV properties of the dust, and the properties (in the case of dissociation in lines, also the details) of the interstellar UV field]. The photodissociation will also initiate a circumstellar chemistry, where neutral-neutral reactions appear to be the most effective (Millar & Herbst 1994; Willacy & Millar 1997). In addition, formation on the surfaces of dust grains may be important for some species (e.g., CH_4, NH_3).

There are presently ~60 different molecular species detected in AGB-CSEs (Olofsson 1997). For the study of AGB stellar evolution the OH masers have been of utmost importance. There is a possibility that even more results will be derived from them, since several lines in the pump cycle of the 1612 MHz line have now been detected (Justtanont et al. 1996; Sylvester et al. 1997). For the study of the properties of the CSEs, and the determination of reliable mass loss rates, the CO line emission has proven to be most important. The detections of highly excited J-lines from a number of molecular species by ISO will lead to significant improvements in our understanding of the inner regions of the CSE (Barlow th.pr.). The detailed testing of the circumstellar chemistry can in principle only be done for one object for the moment, IRC +10216. To a first approximation the images clearly show the expected distribution of molecules with different origins (Guélin et al. 1996; Lucas & Guélin th.pr.), but they also show spatial overlaps in emissions from species which from a chemical point of view are expected to be well separated (Guélin et al. 1993). This shows how important maps are for the interpretation.

A major problem in the interpretation of the molecular line data is the possibility of an inhomogeneous medium. Evidence for this comes from e.g. detailed light curves (Whitelock et al. 1997), high-resolution optical/IR images (Sect. 4), SiO maser maps (Boboltz et al. 1997), images of various molecular line emissions towards IRC+10216 (Guélin et al. 1996), and geometrically thin CO shells of a number of C-stars (Olofsson et al. 1996, 1998a). CO maps of evolved PNe indicate that already the AGB-CSE must have been inhomogeneous (Huggins et al. th.pr.). It seems reasonable to assume that the wind is already initially inhomogeneous at some level, but the further evolution of the density structure is not known.

7. Properties of AGB-stars

7.1. Identification

It is not an altogether simple task to isolate the AGB-stars within the much larger pool of red giants and other chemically peculiar stars. We can expect a major contamination with RGB-stars among the hardly variable and the low-luminosity AGB-stars, and a not negligible contamination with supergiants among the most luminous AGB-stars. There are ways (though not full-proof) to distinguish between these groups, e.g., JHK-diagrams, and comparison of CO radio line and 60 μm fluxes (Josselin et al. 1998). There are also various ways of distinguishing between M- and C-stars if detailed spectra are not available, e.g., data from narrow filters centred on the TiO and CN absorption features, colour-

colour diagrams (e.g., J-K vs [12]-[25]; Le Bertre et al. 1994), and circumstellar molecular radio line intensity ratios (Olofsson et al. 1998b).

There is also the more intricate problem of distinguishing between 'intrinsic' AGB-stars and 'extrinsic'-stars, i.e., stars with some AGB-characteristics due to mass overflow (Jorissen th.pr.). Here we find the 'extrinsic' S-stars which lie on the RGB (Van Eck et al. 1998), and the giant CH- (pop II) and Ba-stars (pop I). The dwarf C-stars probably also belongs to this category (maybe the dominant form of C-stars in the Galaxy). One should note here that many criteria for identifying 'extrinsic' stars do not work for extragalactic samples.

The J-stars are C-rich, but it has been speculated that they became C-enriched during or after the He-core flash (constitute ~10% of the giant C-stars). In some cases (O-rich) matter, lost at the flash, may have been captured in a disk around a companion, hence forming a 'silicate C-star' (Lloyd Evans 1990). The R-type C-stars could be horizontal branch stars, for some reason C-enriched, and possibly the precursors of J-stars.

7.2. Masses

Distribution studies in the Galaxy suggest that the M-type semiregulars and intermediate-period ($300^d < P < 400^d$) Miras belong to the "thin disk" population and have progenitors in the 1.0–1.2 $M_\odot$ range (Jura & Kleinmann 1992a,b; Kerschbaum & Hron 1992). This is supported by kinematic studies, which also suggest that the mass increases with the period (Feast 1989). The short-period M-Miras ($P < 300^d$) belong to the "thick disk" population (Jura 1994). The obscured OH/IR-stars have on average higher-M progenitors, in the range 2–3 $M_\odot$ (Blommaert et al. 1994; Wood et al. 1998). If the upper mass limit to C-stars is ~4 $M_\odot$ we expect OH/IR-stars with $P > 1000^d$ to be more massive than 4 $M_\odot$ since no C-stars with $P > 800^d$ have been found, unless the long period is obtained through extensive mass loss (Habing 1996).

The C-star mass range appears narrower. Distribution and kinematic studies suggest that the progenitor masses may be somewhat higher, 1.2–1.6 $M_\odot$, than for M-stars of the same periodicity (Claussen et al. 1987; Guglielmo et al. 1998). An upper limit of ~4 $M_\odot$ is consistent with the results from studies of obscured C-stars with high expansion velocity CSEs (Barnbaum et al. 1991; Kastner et al. 1993), but the exact limit is uncertain.

7.3. Radii and effective temperatures

Interferometer data suggest $R \sim 400$–$600\,R_\odot$ for Miras, with the non-Miras being on average ~200 $R_\odot$ smaller (Haniff et al. 1995; van Belle et al. 1997). These results are still dependent on the uncertain distances, and the models, since the angular sizes have been converted to photospheric sizes. For the effective temperature van Belle et al. find on average $T_{\rm eff} \sim 2200$–2700 K for Miras, with the non-Miras being on average ~600 K warmer in all groups (the spread in each group is considerable; see also Perrin et al. 1998). van Belle et al. (1996) also find that $T_{\rm eff}$ change with phase in the expected manner, and that R decreases linearly with increasing $T_{\rm eff}$, and Burns et al. (1998) have detected periodic size variations for the M-Mira R Leo.

7.4. Luminosities

It has been long realized that in the optical surveys of the LMC there is a lack of C-stars brighter than $M_{\rm bol}$~-6, and in general a deficit of AGB-stars brighter than this (Frogel & Blanco 1990; Hughes & Wood 1990; Costa & Frogel 1996). HBB and mass loss can explain this, at least qualitatively. However, the possibility exists that high-L sources can be found among the obscured objects missed by optical surveys. For instance, in the Galaxy the most luminous objects are the very long-P OH/IR-stars, $M_{\rm bol,max}$~-7 to -7.5 and $1000^d<P<2800^d$ (van Langevelde et al. 1990). The preliminary analysis of a small sample of IRAS sources in the LMC indicate that the luminosity function for these objects peaks at $M_{\rm bol}$~-5.8, compared to ~-4.9 in the optical surveys, and there are at least three C-stars brighter than $M_{\rm bol}$=-6.0 (van Loon et al. 1998; Groenewegen & Blommaert 1998). On the other side of the coin, we have the faint C-stars ($M_{\rm bol}\leq$-1.8) detected in the SMC (Westerlund 1995; could they be 'extrinsic' C-stars?).

The fundamental pulsation equation provides a relation between L and P (Sect. 3), which is observationally confirmed in the form of relatively tight P-L relations up to ~500^d (Groenewegen & Whitelock 1996). For longer periods there are significant deviations (Reid et al. 1995; Zijlstra et al. 1996; Blommaert et al. 1998; Wood et al. 1998). The effects of metallicity are uncertain, but probably small, although Wood et al. (1998) argue that a metallicity effect may be present in a Galactic Centre sample of OH/IR-stars. In the same sample the long-P objects lie distinctly below the predicted L, which could be an $\dot{M}$-effect: substantial mass loss moves the star to longer periods with little change in luminosity. It seems reasonable that the P-L relation mainly reflects a mass sequence, i.e., evolutionary effects are small.

7.5. Elemental abundances

Smith & Lambert (1990) present results for the CNO-abundances of M-, MS-, S-, SC-, and C-stars, the average C/O-ratios are 0.41 (M), 0.49 (MS), 0.57 (S), 0.95 (SC), and 1.15 (C) with some considerable scatter within each group. The s-processes elements also show a clear increase as the C/O-ratio increases in the M to C sequence. Some peculiarities remain and may indicate systematic problems in the analysis, e.g., the markedly higher abundance of N in MS- and S-stars compared to the M- and C-stars, and the low C/O-ratio in C-stars (some PNe have C/O well above 1.5 and up to ~4). The next step here is definitely to get the same type of data, e.g., on CNO-abundances, for the stars in the Magellanic Clouds.

A most beautiful finding is that information about the elemental abundances of AGB-stars lies at our doorstep, enclosed in the meteorites (Zinner 1998). These contain Al_2O_3 and SiC grains, which carry information about O-isotopes, ^{26}Al, ^{12}C/^{13}C, ^{14}N/^{15}N, and a large number of s-process elements. In particular, isotope patterns can be determined to an impressive accuracy (Zinner & Amari th.pr.).

7.6. Space densities and galactic distribution

Groenewegen et al. (1995b) estimate that in the solar neighbourhood the relative numbers of M-, S-, and C-stars are C/M$\sim$0.2 and S/C$\sim$0.3 for stars with $P<400^d$, but for the rarer, more obscured objects the C/M-ratio is more like 0.5 (Jura & Kleinmann 1989). In the LMC the C/M-ratio is in the range 0.6–2.2 for optically visible AGB-stars (Hughes & Wood 1990), while for the obscured objects it lies in the range 0.2–0.8 and decreases with luminosity (van Loon et al. 1998).

The radial distributions of M- and C-stars are very different in the Galaxy. The surface density of 'thin disk' M-stars follow an exponential disk with a scale length of $\sim$4 kpc (Habing 1988; Blommaert et al. 1993). The C-star distribution is essentially constant within $\sim$3 kpc of the Sun (Jura & Kleinmann 1990; Guglielmo et al. 1998). Thus, there is a marked trend in the C/M-ratio with galactocentric distance, very likely due to the effects of metallicity. Lloyd Evans (1984) showed that in the LMC the luminosity where C-stars become abundant increases with decreasing cluster age (hence increasing metallicity). The same metallicity effect is present in M31 (Brewer 1996).

7.7. Mass loss

There remains no doubt that the AGB-evolution is accompanied by substantial mass loss (observationally detectable when it exceeds $\sim 10^{-8}\,M_\odot\,yr^{-1}$). However, when it comes to the details like $\dot{M}(t, M_{\rm MS}, Z, \theta, \phi)$ and dependences on pulsational behaviour things become more difficult. In the following discussion one should bear in mind that these dependences are not easily separated.

For the short-term behaviour we have information at different time scales: a density modulation in the CSE of the M-Mira IK Tau corresponds to a time scale of $\sim$10 yr (Hale et al. 1997), CO emission from extreme OH/IR-stars suggests considerably increased mass loss over the last $\sim 10^3$ yr (Delfosse et al. 1997), extended 1.3mm emission from IRC+10216 indicates several periods of enhanced mass loss during the last 10^3 yr (Groenewegen et al. 1997), detached CO and dust shells around C-stars suggest episodic mass loss in the range $(1\text{–}20)\times10^3$ yr (Olofsson et al. 1996; Izumiura et al. 1997), extended 60 μm emission also indicates time variations (for some reason the semiregulars dominate among the extended sources; Young et al. 1993), and the HST-image of the Egg nebula shows a large number of concentric rings, interpreted as a mass loss rate modulation occuring every 150–450 yr during the final AGB evolution (Sahai et al. 1998b).

For the long-term behaviour we are mainly restricted to statistical studies. Habing (1996) suggests a time dependence since there are optical Miras and obscured OH/IR-stars with the same L, i.e., for a given L (and hence M) stars may have different $\dot{M}$, and among AGB-stars with the same L, but with different (long) P, the longer-P stars tend to have the highest $\dot{M}$ (assuming of course that the longer P is attributed to $\dot{M}$). On the other hand, correlations between $\dot{M}$ and P indicate an M-dependence. The tight P-L relations for Miras suggest that very intense mass loss, that significantly alters M and hence P, must be of relatively short duration. Thus, it appears that on average $\dot{M}$ increases with time, but that the achievable maximum mass loss rate increases with $M_{\rm MS}$.

We do not know when the onset of mass loss occurs. It is likely that we will find the information among the low-$\dot{M}$ ($<10^{-7}\,M_{\odot}\,yr^{-1}$) objects (Kerschbaum & Olofsson 1998). At the other end, it is clear that the mass loss drops by orders of magnitude when the star leaves the AGB. However, the details are unknown, and there are considerable uncertainties in the interpretation (Meixner et al. 1997; Steffen et al. 1998; Schröder et al. 1998).

The mass loss characteristics dependence on metallicity mainly stems from studies of OH/IR-stars in the Galactic Centre, the Bulge, the anti-centre and in the LMC. It appears from these that both $\dot{M}$ and the gas expansion velocity increase (weakly) with metallicity (Zijlstra et al. 1996; Blommaert et al. 1998; Wood et al. 1998). Groenewegen et al. (1995a) studied three stars (in the Galaxy, LMC, and SMC) with similar P and L and derived CSE dust optical depths ratios of ~15:10:1, very likely a metallicity effect.

The geometry of the mass loss is another important property for understanding the mass loss mechanism. The conclusion drawn from observations of sufficiently high sensitivity and spatial resolution is that the AGB mass loss is essentially isotropic (Bowers & Johnston 1990; Neri et al. 1998; Olofsson et al. 1996, 1998a; Groenewegen et al. 1997), though there are exceptions among the low-$\dot{M}$ sources where examples of bipolar outflows have been found (Kahane & Jura 1996; Kahane et al. 1996). This picture of over-all spherical symmetry contrasts with the appearance of post-AGB objects where clear axi-symmetry is the rule (Alcolea et al. 1996; Bujarrabal et al. 1998; Sahai et al. 1998a). A current acceptable picture is one of isotropic mass loss on the AGB, except perhaps at the final stages. During post-AGB evolution a high-velocity wind, maybe inherently bipolar or channelled by a (weak) equatorial density enhancement in the AGB envelope, will effectively lead to an axi-symmetric CSE, where the brightness contrast may be much higher than the total density contrast. In some objects strong equatorial density enhancements in the form of disks (of unknown origin) seem to be present.

8. Prospects

The prospects for progress are very good. On the theoretical side the computers become increasingly faster, but this is partly counteracted by the realization that the models must include a multitude of complex physical and chemical phenomena, and the very likely importance of significant departures from spherical symmetry. On the observational side sensitive high spectral resolution observations, high spatial resolution observations (in all wavelength regimes), and the sensitive surveys performed (e.g., DENIS, 2MASS, and by-products of projects like MACHO; Epchtein th.pr.; Loup et al. th.pr.; Wood et al. th.pr.) will significantly increase our knowledge of the behaviour of AGB-stars. In particular, the effects of metallicity (Groenewegen th.pr.; Zijlstra th.pr.) will be clarified. It is important in this context to remember that AGB-stars in the LMC/SMC were close to the detection limit for IRAS, meaning that there is far from a complete inventory of the mass-losing AGB populations in these objects. It is also very important that the effects of binarity are sorted out (Iben 1991; Jorissen th.pr.), in particular, when interpreting results from external galaxies.

Acknowledgments. I am grateful to the Swedish Natural Science Research Council (NFR) and the French Foreign Office for travel support.

References

Alcolea J., Bujarrabal V., Sánchez Contreras C., 1996, A&A 312, 560

Allain T., Sedlmayr E., Leach S., 1997, A&A 323, 163

Alvarez R., Plez B., 1998, A&A 330, 1109

Arndt T.U., Fleischer A.J., Sedlmayr E., 1997, A&A 327, 614

Asplund M., Gustafsson B., Lambert D.L., Rao N.K., 1997, A&A 318, 521

Barnbaum C., Morris M., Zuckerman B., 1991, AJ 102, 289

Bessel M.S., Castelli F., Plez B., 1998, A&A 333, 231

Blommaert J.A.D.L., van der Veen W.E.C.J., Habing H.J., 1993, A&A 267, 39

Blommaert J.A.D.L., van der Veen W.E.C.J., Van Langevelde H.J., Habing H.J., Sjouwerman L.O., 1998, A&A 329, 991

Blommaert J.A.D.L., Van Langevelde H.J., Michiels W.F.P., 1994, A&A 287, 479

Blöcker T., 1995, A&A 297, 727

Boboltz D.A., Diamond P.J., Kemball A.J., 1997, ApJ 487 L147

Boothroyd A.I., Sackmann I.-J., 1992, ApJ 393, L21

Boothroyd A.I., Sackmann I.-J., Ahern S.C., 1993, ApJ 416, 762

Bowen G.H., Willson L.A., 1991, ApJ 375, L53

Bowers P.F., Johnston K.J., 1990, ApJ 354, 676

Bressan A., Fagotto F., Bertelli G., Chiosi C., 1993, A&AS 100, 647

Brewer J.P., 1996, PASP 108, 379

Bujarrabal V., Alcolea J., Neri R., 1998, ApJ 504, 915

Burns D., Baldwin J.E., Boysen R.C., et al., 1998, MNRAS 297, 462

Claussen M.J., Kleinmann S.G., Joyce R.R., Jura M., 1987, ApJS 65, 385

Costa E., Frogel J.A., 1996, AJ 112, 2607

Danchi W.C., Bester M., Degiacomi C.G., Greenhill L.J., Townes C.H., 1994, AJ 107, 1469

Delfosse X., Kahane C., Forveille T., 1997, A&A 320, 249

de Laverny P., Geoffray H., Jorda L., Kopp M., 1997, A&AS 122, 415

Feast M.W., 1989, in Evolution of Peculiar Red Giant Stars, H.R. Johnson & B. Zuckerman (eds.), IAU Coll. 106, Cambridge University Press, Cambridge, p. 26

Fleischer A.J., Gauger A., Sedlmayr E., 1992, A&A 266, 321

Frogel J.A., Mould J., Blanco V.M., 1990, ApJ 352, 96

Frost C.A., Cannon R.C., Lattanzio J.C., Wood P.R., Forestini M., 1998a, A&A 332, L17

Frost C.A., Lattanzio J.C., Wood P.R., 1998b, ApJ 500, L355

Frost C.A., Lattanzio J.C., 1996a, in Stellar Evolution: What Should Be Done?, A. Noels et al. (eds.), Proc. of the 32nd Liege Coll., p. 307

Frost C.A., Lattanzio J.C., 1996b, ApJ 473, 383
Glassgold A.E., 1996, ARA&A 34, 241
Groenewegen M.A.T., Blommaert J.A.D.L., 1998, A&A 332, 25
Groenewegen M.A.T., de Jong T., 1993, A&A 267, 410
Groenewegen M.A.T., van der Veen W.E.C.J., Lefloch B., Omont A., 1997, A&A 322, L21
Groenewegen M.A.T., Whitelock P.A., 1996, MNRAS 281, 1347
Groenewegen M.A.T., Whitelock P.A., Smith C.H., Kerschbaum F., 1998, MNRAS 293, 18
Groenewegen M.A.T., Smith C.H., Wood P.R., Omont A., Fujiyoshi T., 1995a, ApJ 449, L119
Groenewegen M.A.T., van den Hoeck L.B., de Jong T., 1995b, A&A 293, 381
Guélin M., Lucas R., Cernicharo J., 1993, A&A 280, L19
Guélin M., Lucas R., Neri R., 1996, in Science with Large Millimetre Arrays, P.A. Shaver (ed.), Springer, Berlin, p. 276
Guglielmo F., Le Bertre T., Epchtein N., 1998, A&A 334, 609
Gustafsson B., 1989, ARA&A 27, 701
Gustafsson B., Jørgensen U.G., 1994, A&AR 6, 19
Gustafsson B., Ryde N., 1998, in The Carbon Star Phenomenon, R. Wing (ed.), IAU Symp. 177, Kluwer, Dordrecht, in press
Habing H.J., 1988, A&A 200, 40
Habing H.J., 1996, A&AR 7, 97
Hale D.D.S., Bester M., Danchi W.C., et al., 1997, ApJ 490, 407
Haniff C.A., Scholz M., Tuthill P.G., 1995, MNRAS 276, 640
Haniff C.A., Ghez A.M., Gorham P.W., Kulkarni S.R., Matthews K., Neugebauer G., 1992, AJ 103, 1662
Hartquist T.W., Dyson J.E., 1997, A&A 319, 589
Herwig F., Blöcker T., Schönberner D., El Eid M., 1997, A&A 324, L81
Hron J., Aringer B., Kerschbaum F., 1997, A&A 322, 280
Hron J., Loidl R., Höfner S., Jørgensen U.G., Aringer B., Kerschbaum F., 1998, A&A 335, L69
Hughes S.M.G., Wood P.R., 1990, AJ 99, 784
Höfner S., Dorfi E.A., 1997, A&A 319, 648
Höfner S., Jørgensen U.G., Loidl R., Aringer B., 1998, A&A 340, 497
Iben I., Jr., 1991, ApJS 76, 55
Iben I., Jr., Renzini A., 1983, ARA&A 21, 271
Ivezić Ž., Elitzur M., 1996, MNRAS 279, 1019
Izumiura H., Waters L.B.F.M., de Jong T., Loup C., Bontekoe Tj.R., Kester D.J.M., 1997, A&A 323, 449
Josselin E., Loup C., Omont A., Barnbaum C., Nyman L.-Å., Sèvre F., 1998, A&AS 129, 45
Jura M., 1994, ApJ 422, 102

Jura M., Kleinmann S.G., 1989, ApJ 341, 359
Jura M., Kleinmann S.G., 1990, ApJ 364, 663
Jura M., Kleinmann S.G., 1992a, ApJS 79, 105
Jura M., Kleinmann S.G., 1992b, ApJS 83, 329
Justtanont K., de Jong T., Helmich F.P., et al., 1996, A&A 315, L217
Kahane C., Audinos P., Barnbaum C., Morris M., 1996, A&A 314, 871
Kahane C., Jura M., 1996, A&A 310, 952
Karovska M., Hack W., Raymond J., Guinan E., 1997, ApJ 482, L175
Kastner J., Forveille T., Zuckerman B., Omont A., 1993, A&A 275, 163
Kerschbaum F., Hron J., 1992, A&A 263, 97
Kerschbaum F., Olofsson H., 1998, A&A 336, 654
Kwok S., Hrivnak B.J., Geballe Th.R., 1995, ApJ 454, 394
Lambert D.L., Smith V.V., Busso M., Gallino R., Straniero O., 1995, ApJ 450, 302
Le Bertre T., Epchtein N., Guglielmo F., Le Sidaner P., 1994, Ap&SS 217, 105
Le Bertre T., Winters J.M., 1998, A&A 334, 173
Le Sidaner P., Le Bertre T., 1996, A&A 314, 896
Lloyd Evans T., 1984, MNRAS 208, 447
Lloyd Evans T., 1990, MNRAS 243, 336
Lopez B., Danchi W.C., Bester M., et al., 1997, ApJ 488, 807
Lorenz-Martins S., Lefèvre J., 1994, A&A 291, 831
Marigo P., Bressan A., Chiosi C., 1996, A&A 313, 545
Meixner M., Skinner C.J., Graham J.R., Keto E., Jernigan J.G., Arens J.F., 1997, ApJ 482, 897
Millar T.J., Herbst E., 1994, A&A 288, 561
Neri R., Kahane C., Lucas R., Bujarrabal V., Loup C., 1998, A&AS 130, 1
Ohnaka K., Tsuji T., 1998, A&A 335, 1018
Omont A., Moseley H.S., Cox P., et al., 1995, ApJ 454, 819
Olofsson H., 1996, Ap&SS 245, 169
Olofsson H., 1997, in Molecular Astrophysics: Probes & Processes, E. van Dishoeck (ed.), IAU Symp. No. 178, Kluwer, Dordrecht, p. 457
Olofsson H., Bergman P., Eriksson K., Gustafsson B., 1996, A&A 311, 587
Olofsson H., Bergman P., Lucas R., Eriksson K., Gustafsson B., Bieging J.H., 1998a, A&A 330, L1
Olofsson H., Lindqvist M., Nyman L.-Å., Winnberg A., 1998b, A&A 329, 1059
Perrin G., Coudé du Foresto V., Ridgway S.T., et al., 1998, A&A 331, 619
Reid M.J., Menten K.M., 1997, ApJ 476, 327
Reid I.N., Hughes S.M.G., Glass I.S., 1995, MNRAS 275, 331
Sahai R., Hines D.C., Kastner J.H., et al., 1998a, ApJ 492, L163
Sahai R., Trauger J.T., Watson A.M., et al., 1998b, ApJ 493, 301
Schröder K.-P., Winters J.M., Sedlmayr E., 1998, A&A in press

Sedlmayr E., Dominik C., 1995, Sp. Sc. Rev. 73, 211
Sloan G.C., Le Van P.D., Little-Marenin I.R., 1996, ApJ 463, 310
Sloan G.C., Little-Marenin I.R., Price S.D., 1998, AJ 115, 809
Smith V.V., Lambert D.L., 1990, ApJS 72, 387
Smith V.V., Plez B., Lambert D.L., Lubowich D., 1995, ApJ 441, 735
Steffen M., Szczerba R., Schönberner D., 1998, A&A 337, 149
Straniero O., Chieffi A., Limongi M., Busso M., Gallino R., Arlandini C., 1997, ApJ 478, 332
Sylvester R.J., Barlow M.J., Nguyen-Q-Rieu, et al., 1997, MNRAS 291, L42
Szczerba R., Omont A., Volk K., Cox P., Kwok S., 1997, A&A 317, 859
Tsuji T., Ohnaka K., Aoki W., Yamamura I., 1997, A&A 320, L1
Tuthill P.G., Haniff C.A., Baldwin J.E., 1995, MNRAS 277, 1541
van Belle G.T., Dyck H.M., Benson J.A., Lacasse M.G., 1996, AJ 112, 2147
van Belle G.T., Dyck H.M., Thompson R.R., Benson J.A., Kannappan S.J., 1997, AJ 114, 2150
Van Eck S., Jorissen A., Udry S., Mayor M., Pernier B., 1996, A&A 329, 971
van Langevelde H.J., van der Heiden R., van Schooneveld C., 1990, A&A 239, 193
van Loon J.Th., Zijlstra A.A., Whitelock P.A., et al., 1998, A&A 329, 169
Wagenhuber J., Groenewegen M.A.T., 1998, A&A 340, 183
Wallerstein G., Knapp G.R., 1998, ARA&A 36, 369
Waters L.B.F.M., Molster F.J., de Jong T., et al., 1996, A&A 315, L361
Weidemann V., 1990, ARA&A 28, 103
Weigelt G., Balega Y., Blöcker T., Fleischer A.J., Osterbart R., Winters J.M., 1998, A&A 333, L51
Weigelt G., Balega Y., Hofmann K.-H., Scholz M., 1996, A&A 316, L21
Westerlund B.E., Azzopardi M., Breysacher J., Rebeirot E., 1995, A&A 303, 107
Whitelock P, Feast M., Marang F., Overbeek M.D., 1997, MNRAS 288, 512
Willacy K., Cherchneff I., 1998, A&A 330, 676
Willacy K., Millar T.J., 1997, A&A 324, 237
Winters J.M., Dominik C., Sedlmayr E., 1994a, A&A 288, 255
Winters J.M., Fleischer A.J., Gauger A., Sedlmayr E., 1994b, A&A 290, 623
Winters J.M., Fleischer A.J., Gauger A., Sedlmayr E., 1995, A&A 302, 483
Winters J.M., Fleischer A.J., Le Bertre T., Sedlmayr E., 1997, A&A 326, 305
Wood P.R., Habing H.J., McGregor P.J., 1998, A&A 336, 925
Wood P.R., Sebo K.M., 1996, MNRAS 282, 958
Wood P.R., Zarro D.M., 1981, ApJ 247, 247
Young K., Phillips T.G., Knapp G.R., 1993, ApJ 409, 725
Zijlstra A.A., Loup C., Waters L.B.F.M, Whitelock P.A., van Loon J.Th., Guglielmo F., 1996, MNRAS 279, 32
Zinner E., 1998, ARE&PS 26, 147

Part 1. Basic Facts, Structure, Evolution, Nucleosynthesis

Basic facts, ... (Nami Mowlavi)

Asymptotic Giant Branch Stars
T. Le Bertre, A. Lèbre and C. Waelkens, eds.
I.A.U. Symposium 191, 1999

Structure and Evolution of AGB Stars

Thomas Blöcker

Max-Planck-Institut für Radioastronomie, Bonn, Germany

Abstract. An overview of the structure and evolution of Asymptotic Giant Branch stars is presented. We focus upon the occurrence of thermal pulses, hot bottom burning and mass loss and summarize briefly the corresponding nucleosynthesis and mixing events. The interplay between these important evolutionary features is discussed.

1. Introduction

Single stars of *low* and *intermediate-mass* ($\sim$ 0.8 to $6 - 8M_\odot$) proceed after completion of central hydrogen and helium burning through the Asymptotic Giant Branch (AGB) phase before they finally end their evolution as white dwarfs consisting of carbon and oxygen (Weidemann & Koester 1983). The upper mass limit depends mainly on the size of the convective cores during the main sequence phase and thus is subject to the uncertainties of the convection prescription. Less massive stars ($\lesssim 0.8M_\odot$) cannot complete central hydrogen burning within a Hubble time whereas the more massive ones ($\gtrsim 6 - 8M_\odot$) are able to reach advanced burning stages, leading either (within a small mass range) to the formation of ONe white dwarfs after central carbon burning (Ritossa et al. 1996) or to the further evolution towards a supernova explosion.

After completion of central hydrogen burning on the main sequence, hydrogen burns in a shell around the He core. Due to core contraction the envelope expands. The star evolves towards larger radii and lower effective temperatures and ascends the Red Giant Branch (RGB). For low-mass stars, $M \lesssim 2M_\odot$, the He cores become electron-degenerate. Accordingly, they ignite central helium burning at the tip of the RGB under degenerate conditions (helium core flash) whereas intermediate-mass stars do it non-degenerately. During the evolution along the RGB the envelope convection moves downwards reaching layers which have previously experienced H-burning (*1st dredge-up*) leading to the mixing of processed material to the surface (Fig. 1).

After the subsequent central He burning the core consists of carbon and oxygen and is surrounded by two shells, one burning He and the other H. The star becomes a red giant a second time and ascends the Asymptotic Giant Branch. At first, He shell-burning dominates (Early AGB). For stars with $M \gtrsim 4M_\odot$ the outer convection can penetrate the hydrogen-exhausted layers leading to a considerable reduction of the core mass and to a second enrichment of the surface with CNO cycled material (*2nd dredge-up*). Later, hydrogen burning takes over and the helium-burning becomes recurrently unstable (Thermally Pulsing AGB). Fig. 1 illustrates the extent of the convective regions and burning

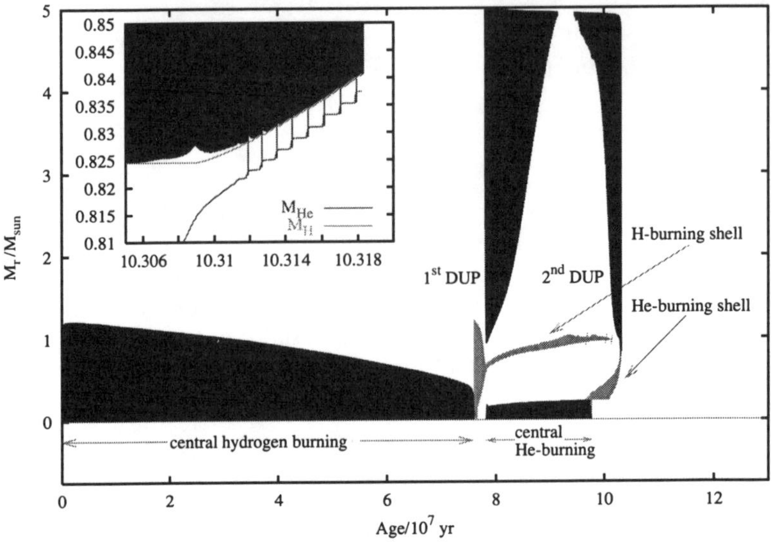

Figure 1. Mass-wise extension of convective regions and burning shells for a $5M_{\odot}$ model vs. time. Dark shaded regions refer to convectively unstable regions, light ones to the burning shells. The inlet shows the evolution on the upper AGB. The "spikes" correspond to the thermal pulses, $M_{\rm H}$ and $M_{\rm He}$ are the masses of the hydrogen and helium-exhausted core, respectively.

shells for different stages of evolution. The hydrogen-exhausted core of an AGB star is highly degenerate, it contains a substantial fraction of the stellar mass (~ 0.5 to $1M_{\odot}$) but the radial extent of its edge is only about 10^{-4} of that of the stellar surface. The C/O centre is surrounded by a thin layer of some $10^{-2}\,\rm M_{\odot}$ occupied by the He and H burning shells. Correspondingly, this core can be thought of as being a very hot white dwarf which grows in mass by accreting nuclear-processed matter from the envelope. Neutrino losses provide effective cooling of the core and postpone the ignition of carbon burning until the core has reached $1.4M_{\odot}$. However, mass loss prevents the cores from growing to such a critical size and terminates the AGB evolution well beforehand.

In the following sections we focus upon the evolution along the TP-AGB which is determined by the thermal instabilities of the helium burning shell (*thermal pulses*), the penetration of the convective envelope into the hydrogen burning shell for more massive objects (*hot bottom burning*), and strong and steeply increasing *mass losses*.

2. Mass loss

Stars evolving along the AGB suffer from continuously increasing mass loss. Observations indicate rates of $10^{-7}\,\rm M_{\odot}/yr$ for small-period Mira stars and up to $10^{-4}\,\rm M_{\odot}/yr$ for luminous long-period variables (Wood 1997). These winds

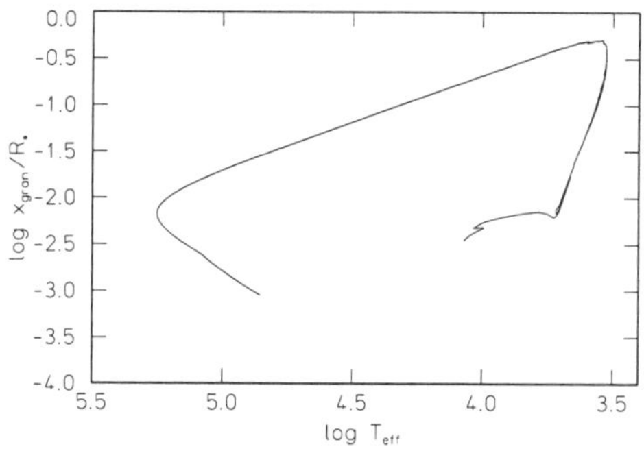

Figure 2. "Evolutionary track" for the diameter of surface granules given in stellar radii from the main sequence through the AGB towards the stage of white dwarfs for a $3M_\odot$ sequence with 20 pulses on the AGB

are most likely dust-driven supported by shock waves generated by the pulsating stellar envelope (Winters 1998) and lead at larger mass-loss rates to the complete obscuration of the star by a dusty circumstellar envelope. For reviews, see e.g. Habing (1996) and Winters (1998).

AGB stars exhibit a large-scale surface convection which may lead to inhomogeneities in the dust formation process and clumpy outflows (e.g. Weigelt et al. 1998). In contrast to their progenitors, the circumstellar shells of AGB successors often expose prominent features of asphericity. This change in the circumstellar shells geometry is supposed to take place either when the star moves off the AGB or already during its evolution along the upper AGB. It has been shown by Schwarzschild (1975) that for red giants the dominant convective elements may become so large that only a few of them can occupy the surface at any time leading to large temperature variations on the surface and concomitant brightness fluctuations. Hydrodynamical simulations for convection zones of main sequence stars and subgiants by Freytag et al. (1997) seem to support this scenario. They found a tight correlation between the photospheric pressure scale height, $H_{\rm p0}$, and the granule size, $x_{\rm gran}$, viz. $x_{\rm gran} \approx 10 H_{\rm p0}$ covering more than two orders of magnitudes in gravity. If one assumes that such a correlation also holds in the red giant regime, it indeed turns out that during the course of evolution, the granules become so large that the whole surface can be occupied by at most a few granules. Fig. 2 gives an "evolutionary track" for the granule sizes estimated from the correlation of Freytag et al. (1997) for a $3M_\odot$ sequence.

To consider mass loss in stellar evolution calculations one still has to rely on empirical relations or adaptions of (mass-loss) model grids. Recent approaches used in AGB model calculations are based on an empirical mass-loss – pulsation-period relation (Vassiliadis & Wood 1993), on a semi-empirical prescription derived from Bowen's (1988) hydrodynamical calculations of pulsating Mira atmospheres (Blöcker 1995a), and on a fit of wind models for carbon-rich stars (Arndt et al. 1997; Schröder et al. 1998). It is noteworthy that these prescriptions predict mass-loss rates that increase rapidly with luminosity and are considerably larger than those which would follow from the Reimers's formula (1975). Comparisons between the models of Vassiliadis & Wood (1993) and Blöcker (1995a) are given in Habing (1996), Wood (1997) and Blöcker (1998). Both sets are, within the observational errors, consistent with the empirical initial-final mass relation of Weidemann (1987). Note, however, that the uncertainties of such empirical relations are rather high, especially for large initial masses.

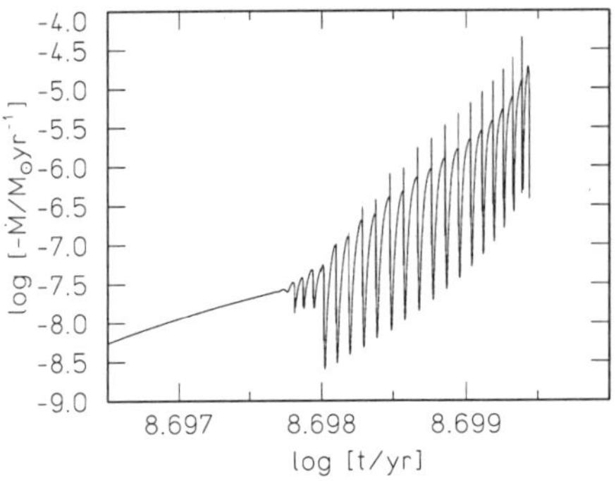

Figure 3. Evolution of the AGB mass-loss rate for a 3 $M_{\odot}$ model sequence (20 thermal pulses, Blöcker 1995a). The final mass is 0.625 $M_{\odot}$.

During the thermal pulses an AGB star does not only suffer from strong variations of its interior luminosity contributions but also from modulations of its surface luminosity and radius leading to corresponding mass-loss variations. Fig. 3 shows as an example the whole mass-loss evolution along the AGB for a 3 $M_{\odot}$ model sequence. One clearly sees how thermal pulses lead to order-of-magnitude modulations or short interruptions of the otherwise monotonically increasing mass-loss rate. The corresponding path in the two colour diagram is triggered by these flash-induced mass-loss modulations leading to recurrent loops. Detached dust shells can develop and during the pulses the stars can spend about 10 % of their life away from the main IRAS two-colour relation of dusty AGB stars (Steffen et al. 1998). Thus, mass-loss modulations by thermal pulses seem to account for the observed detached shells and loops in the IRAS two-colour diagram (van der Veen & Habing 1988; Olofsson et al. 1990; Zijlstra et al. 1992; Izumiura et al. 1996).

Finally, mass loss terminates the AGB evolution when the envelope mass is reduced to a few $\approx 10^{-2}\,M_{\odot}$. The star moves off the AGB (Schönberner 1979) evolving towards the regime of central stars of planetary nebulae and finally reaches the stage of white dwarfs. Even beyond the AGB the evolution depends on the preceding mass-loss history. Since mass loss determines the life-time on the AGB, it determines not only the final mass but also the internal temperature-density structure reached at the tip of the AGB and, thus, the timescales for the evolution into a white dwarf (Blöcker 1995a, b).

3. Hot bottom burning

As the stars evolve towards larger luminosities along the AGB, their envelope convection extends further downwards. For $M > 4...5\,M_{\odot}$ (depending on metallicity) the bottom of the convective envelope can even reach and dip into the H-burning shell (hot bottom burning, HBB; Iben 1975; Scalo et al. 1975). The consequences for stellar evolution are far reaching: the bottom of the envelope burns with temperatures of typically 30 to $80 \cdot 10^6$ K and the processed material is immediately convected to the surface. HBB models evolve rapidly to very high luminosities and no longer obey Paczyński's (1970) classical core-mass luminosity relation (Blöcker & Schönberner 1991).

The CNO-cycling of the envelope in HBB models destroys ^{12}C and builds up ^{14}N and ^{13}C. Accordingly, the typical signature for HBB is a low $^{12}C/^{13}C$

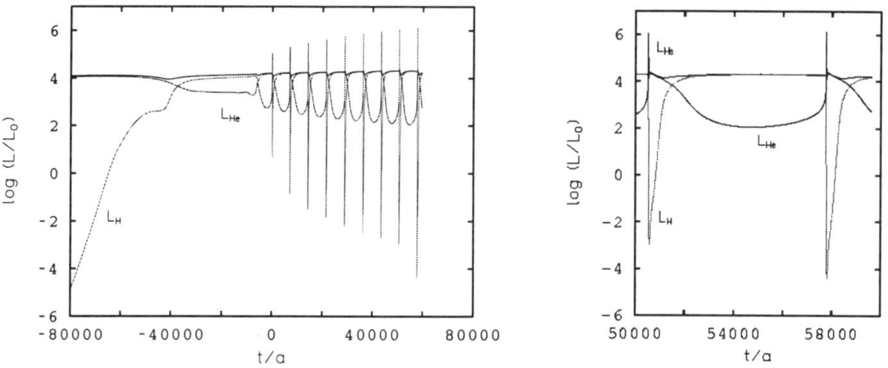

Figure 4. **Left:** Surface luminosity and shell-source contributions on the AGB for a $5M_\odot$ model (Blöcker 1995a). **Right:** The last two pulses on an expanded scale.

ratio and the delay or even prevention of the carbon-star phase (e.g. Iben 1975; Renzini & Voli 1981; Boothroyd et al. 1993). HBB leads to the formation of lithium-rich stars (Scalo et al. 1975) via the Cameron & Fowler (1971) mechanism which first synthesizes ^{7}Be and then convects it to cooler regions where ^{7}Li is finally formed. Observations in the Magellanic Clouds show that Li-rich stars are only found between $M_{\rm bol} \approx$-6 to -7 (Smith & Lambert 1989, 1990; Plez et al. 1993) and the results from evolutionary calculations (Sackmann & Boothroyd 1992; Forestini & Charbonnel 1997) are in very good agreement with these findings. However, Li-rich galactic carbon stars are observed at much lower luminosities, $M_{\rm bol} \approx$-3.5 to -6, indicating that possibly additional mixing processes operate in these stars (Abia & Isern 1997) such as, e.g., the so-called cool bottom processing by deep circulation (Wasserburg et al. 1995).

4. Thermal pulses

Early numerical calculations of AGB models showed that burning hydrogen and helium simultaneously in two shells is thermally unstable in the sense that the helium-burning shell repeatedly enters into run-away situations (Schwarzschild & Härm 1965; Weigert 1966). For more details, see e.g. Iben & Renzini (1983). Fig. 4 illustrates the typical luminosity evolution of both shells for a $5M_\odot$ model after completion of central He burning until the onset of the first pulses. With the onset of a thermal pulse the luminosity of the He-burning shell increases rapidly within a short time of about 100 yrs to $10^6...10^7$ L$_\odot$. During this phase the He shell is forced to become convectively unstable due to the huge amount of energy produced: a pulse-driven convective zone develops which comprises almost the whole intershell region and mixes products of He burning, in particular C (and O), outwards (see Figs. 1 and 5). Simultaneously, the H-shell is pushed outwards into cooler domains and extinguishes. After the flash, the envelope convection proceed downwards, may penetrate deeply into those intershell layers formerly enriched by the flash-driven intershell convection, and mixes this material to the surface. This is the *3rd dredge-up* which may change the envelope C/O ratio

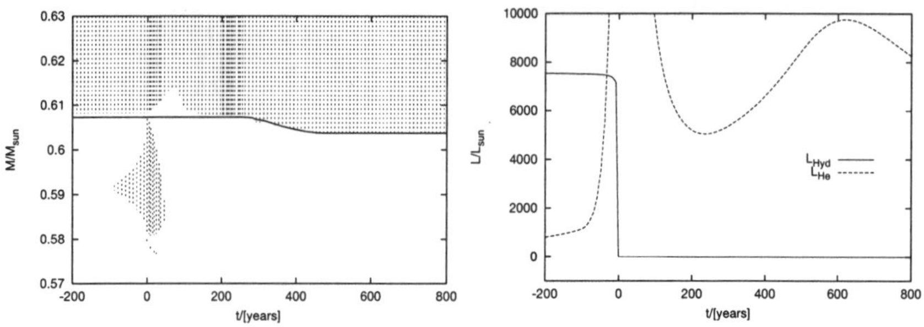

Figure 5. Extension in mass of convective regions (left) and luminosity contributions of the shell sources (right) vs. time for the 12th pulse of a 3M$_\odot$ AGB star (Herwig et al. 1997). Time is set to zero at maximum L_{He} ($\sim 10^6\,\mathrm{L}_\odot$). The solid line denotes the border of the hydrogen exhausted core. Shaded regions indicate convective mixing. He burning takes place between $M_{\mathrm{r}} = 0.585$ and $0.570\,\mathrm{M}_\odot$.

from smaller than unity to larger than unity, i.e. lead to the formation of carbon stars. Fig. 5 shows the dredge-up episode of a 3 M$_\odot$ AGB model. In this example the flash-driven convective zone exists for 150 yr. Dredge-up starts 250 yr after this convective shell has disappeared. Hydrogen re-ignites 5 000 yr later, and the next pulse commences after further 60 000 yr of evolution. Recently, Frost et al. (1998a) reported that very deep dredge-up may lead to the ignition of He under partially degenerate conditions.

H burning remains virtually extinguished during the first $\sim$15% of the cycle time, and He burning provides the surface luminosity. Then, H burning is restored and controls the evolution. Typically, a thermal-pulse cycle lasts several 10^3 up to 10^5 yrs depending on core mass. The large energy excess liberated during a pulse is temporarily stored as potential energy and re-radiated later on a much longer time scale. Therefore, the changes at the stellar surface, i.e. of its radius and luminosity, are much smaller than the many order-of-magnitude variations of the shell luminosities themselves but may lead to considerable modulations of other quantities such as, e.g., mass-loss rates (Fig. 3).

The intershell region of AGB stars is also the site of s-process nucleosynthesis (e.g. Gallino et al. 1998). Two possible reactions provide the necessary neutrons: ^{22}Ne(α,n)^{25}Mg and ^{13}C(α,n)^{16}O. The temperatures reached in the intershell zone during thermal pulses appear to be too low to activate the ^{22}Ne(α,n)^{25}Mg reaction. Instead, the s-process is most likely driven by the ^{13}C(α,n)^{16}O neutron source. This requires the mixing of protons into ^{12}C-rich layers in order to produce sufficient amounts of ^{13}C via ^{12}C(p,γ)^{13}N(β^+,ν)^{13}C. Iben & Renzini (1982) found that the protons can diffuse from the bottom of the convective envelope into the intershell zone due to semiconvection. However, this scenario is restricted to low metallicities (Iben 1983). To follow the s-process under typical conditions one has to artificially ingest a certain amount of ^{13}C at the proper time (e.g. Straniero et al. 1995).

The question of burning and mixing along the AGB is intimately linked to the so-called *carbon star mystery* (Iben 1981): observations of the LMC and

SMC prove that most of the carbon (and s-process element) enriched AGB stars are found at rather low luminosities ($M_{\rm bol} > -6$) indicating that they are *lower mass* stars (cf. Smith et al. 1987, Frogel et al. 1990). On the other hand, AGB models mostly predict the 3$^{\rm rd}$ dredge-up for much higher luminosities, i.e. larger (core) masses.

The lack of luminous carbon stars is likely to be explained in terms of hot bottom burning and mass loss. HBB can turn the carbon enriched envelopes of massive (and therefore luminous) models into nitrogen-rich ones. Furthermore, HBB models do not obey Paczyński's (1970) classical core-mass luminosity relation but evolve rapidly towards large luminosities. On the other hand, the steeply increasing AGB mass-loss rates do not allow a very long evolution on the upper AGB and may terminate the evolution before the carbon-star stage is reached (Iben 1981; D'Antona & Mazzitelli 1996). Finally, the continuous erosion of the stellar surface by stellar winds leads to obscuration by dust which may hide luminous carbon stars from optical surveys (Iben 1981). Indeed, van Loon et al. (1998) observed obscured carbon stars in the Magellanic Clouds which are brighter than $M_{\rm bol} = -6$. Whether luminous carbon stars can be formed depends crucially on the competition between dredge-up and HBB which both require minimum envelope masses to operate and thus depend on the assumed mass-loss evolution (and treatment of convection). Frost et al. (1998b) found that HBB ceases earlier than dredge-up during the superwind phase when the star is expected to be dust-enshrouded. Dredge-up continues and leads to the formation of a carbon star.

The second problem concerns the difficulties in obtaining dredge-up for models of the observed mass range. The amount of dredge-up found in evolutionary calculations depends sensitively on metallicity, core and envelope masses (Wood 1981), and is in any case too small to account for the observations. Additionally, numerical details may play an important role (Frost & Lattanzio 1996). The cause of the large uncertainties concerning the efficiency of dredge-up is in particular our inability to treat convection properly. Accordingly, this affects the question of how to form ^{13}C as well.

Synthetic AGB calculations, therefore, utilize stellar parameters known from evolutionary AGB model sequences but take the minimum core mass for dredge-up, $M_{\rm H}^{\rm min}$, and the dredge-up parameter λ (ratio of dredged-up mass to growth of $M_{\rm H}$ per flash cycle), as adjustable parameters (among others) in order to match the observations, i.e. the luminosity function of LMC carbon stars. For example, Marigo et al. (1996) and van den Hoek & Groenewegen (1997) find $M_{\rm H}^{\rm min} = 0.58\,{\rm M}_\odot$ and $\lambda = 0.65$ and 0.75, resp., as best fit contrasting the results of evolutionary calculations ($M_{\rm H}^{\rm min} \gtrsim 0.65\,{\rm M}_\odot$ and $\lambda \approx 0.25$, see Wood 1997).

Due to these uncertainties of stellar evolution calculations and their weakness to face the observations, it is often concluded that mixing may take place outside the formally convective boundaries (Iben 1976; D'Antona & Mazzitelli 1996; Wood 1997). Recently, Herwig et al. (1997) introduced overshoot based on the results of hydrodynamical simulations of convection by Freytag et al. (1996) in their treatment of convection. The parametrisation of exponentially decreasing velocities of convective elements beyond the classical convective border then leads to some extra partial mixing (Blöcker et al. 1998). This method provided for AGB stars with rather *small* core masses of $\sim 0.6\,{\rm M}_\odot$ *both* a sufficient amount of dredge-up (note, however, that Straniero et al. (1997) found dredge-up for low core masses without additional mixing) *as well as* the production of ^{13}C. It should be noted that a ^{13}C pocket formed during the dregde-up phase will not survive until the onset of the next pulse. Due to the high temperatures reached in the intershell region in the course of evolution it is already burnt during the interpulse phase and will not be engulfed by the next flash-driven convection zone (Straniero et al. 1995). Accordingly, the s-process takes place in a radiative environment (Gallino et al. 1998).

Further consequences of this additional mixing are the drastically changed intershell abundances of (^{4}He,^{12}C,^{16}O) = (40/40/16) by mass (see Herwig et al. 1998b) compared to standard calculations which give (70/26/1). It is noteworthy that these modified intershell abundances are close to the observed surface abundances of Wolf-Rayet central stars of planetary nebulae and PG 1159 stars. The evolutionary history of these H-deficient post-AGB stars is not well understood. One scenario discussed is that of a final thermal pulse after the cessation of H-burning on the white dwarf cooling track which predicts surface abundances close to those of the intershell region (Iben & MacDonald 1995). On the other hand, there is evidence that the observed abundance pattern has already developed immediately after the star has moved off the AGB (Werner et al. 1998). In any case, it appears promising that overshoot models may provide a better understanding of these objects.

Regardless of the mixing approach invoked to be in accordance with the observations, the occurrence of effective dredge-up has another important consequence. As reported by Herwig et al. (1998a) models with efficient dredge-up ($\lambda \rightarrow 1$) and lower core masses ($\lesssim 0.8 M_\odot$, i.e. lower than those associated with hot bottom burning) violate Paczyński's (1970) classical core-mass luminosity relation. Even when the core-mass growth is completely compensated by dredge-up the luminosity still increases. Homology relations show that, in principle, the luminosity depends on *both* core mass *and* core radius. For core masses $\lesssim 0.8 M_\odot$ one typically obtains $L \sim M_{\rm H}^2/R_{\rm H}$. For instance, models without dredge-up show (after some pulses) an almost linear core-radius decrease with increasing core mass leading to the classical linear core-mass luminosity relation. On the other hand, strong dredge-up of $\lambda = 1$ cancels the core-mass dependence while the core radius continues to decrease. Accordingly, the luminosities evolve to higher values than predicted by Paczyński's (1970) relation.

5. Conclusions

Along the AGB mixing episodes are able to dredge up products of the interior nucleosynthesis to the surface whereas simultaneously strong stellar winds erode effectively the stellar envelopes leading to an enrichment of the interstellar medium with processed material. The envelope's chemical evolution and thus the composition of the stellar ejecta depends crucially on the treatment of convection and mass loss. The course of mass-loss evolution is not well known, and due to the drawbacks of the local treatment of convection additional mixing processes seem to be required during various phases of evolution. The important interplay between mass loss, mixing and nucleosynthesis and its implications for the structure and evolution of AGB stars remains to be investigated in more detail.

References

Abia C., Isern J., 1997, MNRAS 289, L11

Arndt T.U., Fleischer A.J., Sedlmayr E., 1997, A&A 327, 614

Blöcker T., 1995a, A&A 297, 727

Blöcker T., 1995b, A&A 299, 755

Blöcker T., 1998, Astrophys. Sp. Sci 255, 267

Blöcker T., Schönberner, D., 1991, A&A 244, L43

Blöcker T., Herwig F., Schönberner D., El Eid M., 1998, The Carbon Star Phenomenon, IAU Symp. 177, ed. R. Wing, Kluwer, Dordrecht, in press

Boothroyd A.D., Sackmann I.J., Ahern S.C., 1993, ApJ 416, 762

Bowen G.H., 1988, ApJ 329, 299

Cameron A.G.W., Fowler W.A., 1971, ApJ 164, 111

D'Antona F., Mazzitelli I., 1996, ApJ 470, 1093

Forestini M., Charbonnel C., 1997, A&AS 123, 241

Freytag B., Ludwig H.-G., Steffen M., 1996, A&A 313, 497

Freytag B., Holweger H., Steffen M., Ludwig H.G., 1997, in Science with the VLT Interferometer, F. Paresce (ed.), Springer, Berlin, p. 316

Frogel J.A., Mould J.R., Blanco V.M., 1990, ApJ 352, 96

Frost C.A., Lattanzio J.C., 1996, ApJ 473, 383

Frost C.A., Lattanzio J.C., Wood P.R., 1998a, ApJ 500, 355

Frost C.A., Cannon R.C., Lattanzio J.C., Wood P.R., Forestini M., 1998b, A&A 332, L17

Gallino R., Arlandini C., Busso M., Lugaro M., Travaglio C., Straniero O., Chieffi A., Limongi M., 1998, ApJ 497, 388

Habing H.J., 1996, A&AR 7, 97

Herwig F., Blöcker T., Schönberner D., El Eid M., 1997, A&A 324, L81

Herwig F., Schönberner D., Blöcker T., 1998a, A&A 340, L43

Herwig F., Blöcker T., Schönberner D., 1998b, this volume, p. 41

Iben I., Jr., 1975, ApJ 196, 525

Iben I., Jr., 1976, ApJ 208, 165

Iben I., Jr., 1981, ApJ 246, 278

Iben I., Jr., 1983, ApJ 275, 65

Iben I., Jr., MacDonald J., 1995, in White dwarfs, eds. D. Koester & K. Werner, *Lecture Notes in Physics* 443, Springer, Berlin, p. 48

Iben I., Jr., Renzini A., 1982, ApJ 263, L23

Iben I., Jr., Renzini A., 1983, ARAA 21, 271

Izumiura H., Hashimoto O., Kawara K., Yamamura I., Waters L.B.F.M., 1996, A&A 315, L221

Marigo P., Bressan A., Chiosi C., 1996, A&A 313, 545
Olofsson H., Carlström U., Eriksson K., Gustafsson B., Willson L.A., 1990, A&A 230, L13
Paczyński B., 1970, Acta Astron. 20, 47
Plez B., Smith V.V., Lambert D.L, 1993, ApJ 418, 812
Reimers D., 1975, Mem. Soc. Sci. Liège 8, 369
Renzini A., Voli M., 1981, A&A 94, 175
Ritossa C., Garcia-Berro E., Iben I., Jr., 1996, ApJ 460, 489
Sackmann I.J., Boothroyd A.I., 1992, ApJ 392, L71
Scalo J.M., Despain K.H., Ulrich R.K., 1975, ApJ 196, 805
Schönberner D., 1979, A&A 79, 108
Schröder K.-P., Winters J.M., Arndt T.U., Sedlmayr E., 1998, A&A 335, L9
Schwarzschild M., 1975, ApJ 195, 137
Schwarzschild M., Härm R., 1965, ApJ 142, 855
Smith V.V., Lambert D.L., 1989, ApJ 345, L75
Smith V.V., Lambert D.L., 1990, ApJ 361, L69
Smith V.V., Lambert D.L., McWilliam A., 1987, ApJ 320, 862
Steffen M., Szczerba R., Schönberner D., 1998, A&A 337, 149
Straniero O., Gallino R., Busso M., Chieffi A., Raitieri C.M., Salaris M., Limongi M., 1995, ApJ 440, L85
Straniero O., Chieffi A., Limongi M., Busso M., Gallino R., Arlandini C., 1997, ApJ 478, 332
van den Hoek L.B., Groenewegen M.A.T., 1997, A&AS 123, 305
van der Veen W.E.C.J., Habing H.J., 1988, A&A 194, 125
van Loon J.T., Zijlstra A.A., Whitelock P., et al., 1998, A&A 329, 169
Vassiliadis E., Wood P.R., 1993, ApJ 413, 641
Wasserburg G.J., Boothroyd A.I., Sackmann I.J., 1995, ApJ 447, L37
Weidemann V., 1987, A&A 188, 74
Weidemann V., Koester D., 1983, A&A 121, 77
Weigelt G., Balega Y.Y., Blöcker T., Fleischer A.J., Osterbart R., Winters J.M., 1998, A&A 333, L51
Weigert A., 1966, Z. Astrophys. 64, 395
Werner K., Dreizler S., Rauch T., Koesterke L., Heber U., 1998, this volume
Winters J.M., 1998, Astrophys. Sp. Sci. 255, 257
Wood P.R., 1981, in Physical Processes in Red Giants, eds. I. Iben & A. Renzini, Reidel, Dordrecht, p. 135
Wood P.R., 1997, in Planetary Nebulae, IAU Symp. 180, H.J. Habing & H.J.G.L.M. Lamers (eds.), Kluwer, Dordrecht, p. 297
Zijlstra A.A., Loup C., Waters L.B.F.M., de Jong T., 1992, A&A 265, L5

Asymptotic Giant Branch Stars
T. Le Bertre, A. Lèbre and C. Waelkens, eds.
I.A.U. Symposium 191, 1999

Nucleosynthesis in AGB Stars

John Lattanzio

Dept of Mathematics and Statistics, Monash University, Australia

Institute of Astronomy, Madingley Rd, Cambridge, England

Manuel Forestini

Laboratoire d'Astrophysique, Université Joseph Fourier, Grenoble, France

Abstract. We give a qualitative review of the nucleosynthesis occurring in AGB stars. We summarise some new calculations of intermediate mass stars which include all thermal pulses until the star is about to leave the AGB, as well as a detailed nucleosynthesis network. We will show that hot bottom burning delays, rather than prevents, the formation of carbon stars; those that form are not optically visible.

1. Introduction

Although only a brief phase compared to the overall stellar lifetime, the Asymptotic Giant Branch is crucial because it is here that stars experience substantial nucleosynthesis, which is the topic of this review.

The AGB phase is the last phase of evolution for low and intermediate mass stars (about 1–8$M_\odot$). The evolution leading up to this phase has been discussed in detail in Lattanzio & Boothroyd (1997) and Lattanzio (1998). Briefly, stars begin by burning hydrogen in their cores. Following the exhaustion of their central hydrogen supply, they become red-giants. During this phase the stars develop a deep convective envelope which extends inwards and can reach into regions where partial hydrogen burning has taken place. This is called the "first dredge-up" event, and it results in polluting the stellar surface with the products of hydrogen burning.

The ascent of the giant branch is terminated by ignition of the central helium supply, either degenerately in a core flash (in the case of low mass stars, below about 2.5$M_\odot$) or non-degenerately (for higher masses). The subsequent evolution is characterized by helium burning in a convective core and a steadily advancing hydrogen burning shell. For low mass stars the evolution is complicated by semiconvection, but this is a detail we do not need for our present purposes. The fusion of helium produces ^{12}C, and this carbon is in turn subject to alpha capture to form ^{16}O. Eventually the helium supply is totally consumed, leaving a core of carbon and oxygen (the exact proportions of which depend on the infamously uncertain rate for the $^{12}C(\alpha,\gamma)^{16}O$ reaction). In any event,

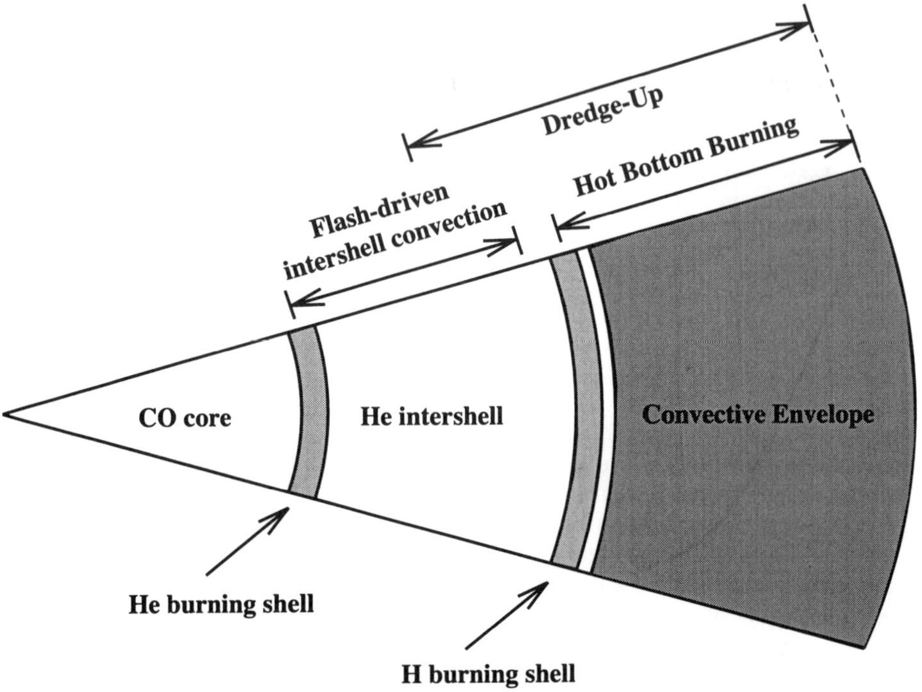

Figure 1. Schematic of the interior of an AGB star.

following the exhaustion of central helium the star begins to ascend the giant branch again. It is now called the "second" or "asymptotic" giant branch.

When a star reaches the AGB it has the structure shown in Fig. 1. If its mass is above about $4M_\odot$ the shifting of helium burning from the core to a shell causes the extinguishing of the hydrogen shell. In lower mass stars this stops the inward advance of the convective envelope, but in more massive stars the envelope penetrates the extinct hydrogen shell, and products of hydrogen burning are mixed to the stellar surface, in the "second dredge-up" episode. Note that low mass stars do not experience the second dredge-up.

Perhaps the most important aspect of AGB stars lies in the nucleosynthesis which occurs during their evolution. During this phase they show an electron degenerate carbon-oxygen core, and two nuclear shells: one burning helium and one burning hydrogen. The subsequent nucleosynthesis comes from two sources: the thermal instability of the helium shell (known as "thermal pulses" or "shell flashes") and, for masses above about $5M_\odot$, nuclear burning at the bottom of the convective envelope (known as "hot bottom burning", hereafter HBB).

2. Thermally pulsing AGB evolution

For most of the evolution on the AGB, the helium shell is largely inactive. But periodically the shell suffers a thermal runaway and generates enormous quantities of energy, at rates up to $10^8 L_\odot$ or so, for short periods of time (of order a year). This energy release results in a "flash-driven convection zone" which extends from the helium shell almost to the hydrogen shell, as shown in Fig. 1. This mixes the ^{12}C produced by the helium shell throughout this convective region. As the helium flash dies down, the energy deposited in the star causes a substantial expansion and cooling. The hydrogen shell is extinguished and the convective envelope reaches in beyond its old location, and (for later pulses) into the carbon-rich region left behind by the flash convection. This mixes carbon to the surface of the star, and is known as the "third dredge-up". It is responsible for the production of N-type carbon stars. As the star now contracts again, the hydrogen shell is re-ignited and the star begins an extended "interpulse" phase of order 10^4 years (but depending primarily on the mass of the hydrogen exhausted core, hereafter called the core-mass). During this period the hydrogen shell provides essentially all of the star's luminosity. Eventually there is another flash, and the cycle repeats. We will give a qualitative summary of AGB nucleosynthesis below, but details may be found in Forestini & Charbonnel (1997).

Another important consequence of thermal pulses is the production of elements beyond iron, produced by slow neutron capture, and known as *s*-process elements. For a detailed review see Gallino, Busso & Lugaro (1997). For our purposes here, we simply note that some partial mixing of hydrogen from the envelope into the carbon-rich intershell region is required. These protons are captured by ^{12}C to produce ^{13}C which later releases neutrons via the $^{13}C(\alpha,n)^{16}O$ reaction. Early calculations by Iben & Renzini (1982a,b) found that a small semiconvective zone formed at the bottom of the convective envelope, during the dredge-up phase. But other calculations have failed to reproduce the process. (Normal, complete mixing will not suffice, incidentally, because the number of protons must be small, or else the CN cycle will go to completion and produce ^{14}N rather than ^{13}C.) Nevertheless it, or something very much like it, almost certainly occurs. Observations of these stars clearly show the *s*-process enhancements that these ideas would predict. Quantitative calculations are performed by Gallino and co-workers (e.g. see Gallino, Busso & Lugaro 1997) and are very successful, although the amount of ^{13}C present is a parameter which must be specified in advance. Other possible mixing mechanisms are overshoot, and partial mixing (see Herwig et al. 1997), or possibly something like shear mixing.

A related issue is the production of ^{19}F within these stars. Jorissen, Smith & Lambert (1992) discovered that the $^{19}F/^{16}O$ ratio in AGB stars increases with the $^{12}C/^{16}O$ ratio, which implicates thermal pulses in the origin of this ^{19}F. They investigated many possible scenarios, and this was followed by Forestini et al. (1992) who investigated the most promising scenario in more detail, and to whom we defer for details (see also Mowlavi, Jorissen & Arnould 1996). The mechanism is shown in Fig. 2. Here, some ^{13}C produces neutrons via the $^{13}C(\alpha,n)^{16}O$ reaction, and some of these neutrons are captured by ^{14}N to produce ^{14}C and protons. These protons, plus possibly some from $^{26}Al(n,p)^{26}Mg$, are then captured by ^{18}O and the sequence $^{18}O(p,\alpha)^{15}N(\alpha,\gamma)^{19}F$ produces the observed ^{19}F, which is then dredged to the surface in the usual way following the pulse. For all except those stars with the highest abundances of ^{19}F it appears that the amount of ^{13}C left from the CN cycling H-shell is sufficient.

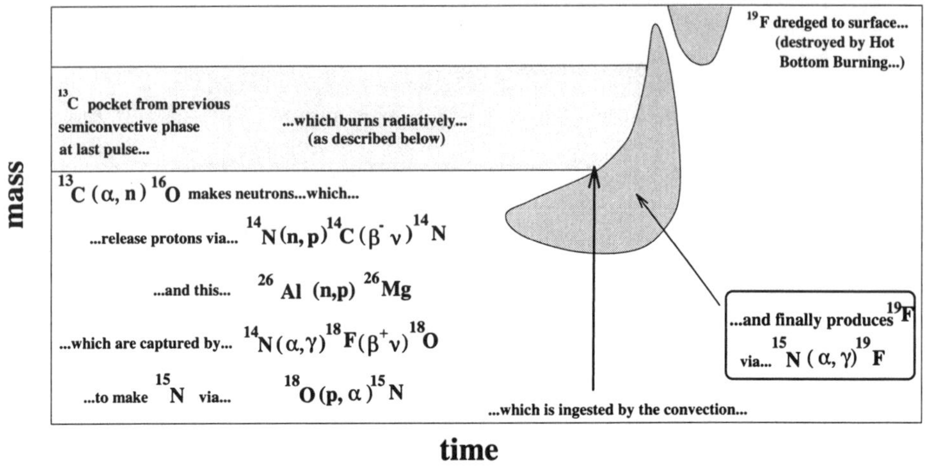

Figure 2. Production of ^{19}F in AGB stars.

Yet for those stars showing more enhanced ^{19}F we may need to invoke some extra-mixing, just as is required for the *s*-process production.

For stars above about $5M_{\odot}$ a second important phenomenon is the occurrence of hot bottom burning. Here the convective envelope is so deep that it penetrates into the top of the hydrogen burning shell, so that nucleosynthesis actually occurs in the envelope of the star. Temperatures can reach as high as almost 10^8K. This results in the activation of the CN cycle within the envelope, and the consequent processing of ^{12}C into ^{13}C and ^{14}N. This prevents the star from becoming a carbon star, because the primary carbon dredged to the envelope is then transmuted into nitrogen, keeping the C/O ratio less than unity (e.g. Boothroyd et al. 1993).

Perhaps the most striking consequence of hot bottom burning is the production of ^7Li by the Cameron-Fowler Mechanism. Sackmann & Boothroyd (1992) showed that ^7Li was produced as soon as the temperature T_{bce} at the bottom of the convective envelope exceeded 50×10^6K, and produced abundances which match the Magellanic Cloud observations very nicely (see below). Another consequence of HBB is the substantial production of ^{26}Al by the Mg-Al cycle. The amount produced can be substantial, and this is implicated in explaining the $^{26}Al/^{27}Al$ isotopic ratios seen in meteorite grains (for a review please see Zinner 1997).

3. Complete AGB evolution for intermediate mass stars

We have recently completed detailed evolutionary calculations for stars of intermediate masses from the ZAMS through to the end of the AGB evolution (Frost 1998). These calculations covered three compositions: Z= 0.004 (corresponding to the Small Magellanic Cloud), Z= 0.008 (Large Magellanic Cloud) and Z= 0.02 (solar neighbourhood). For each composition we have considered three masses, $M/M_{\odot} = 4, 5$ and 6. Subsequent nucleosynthesis has been calculated

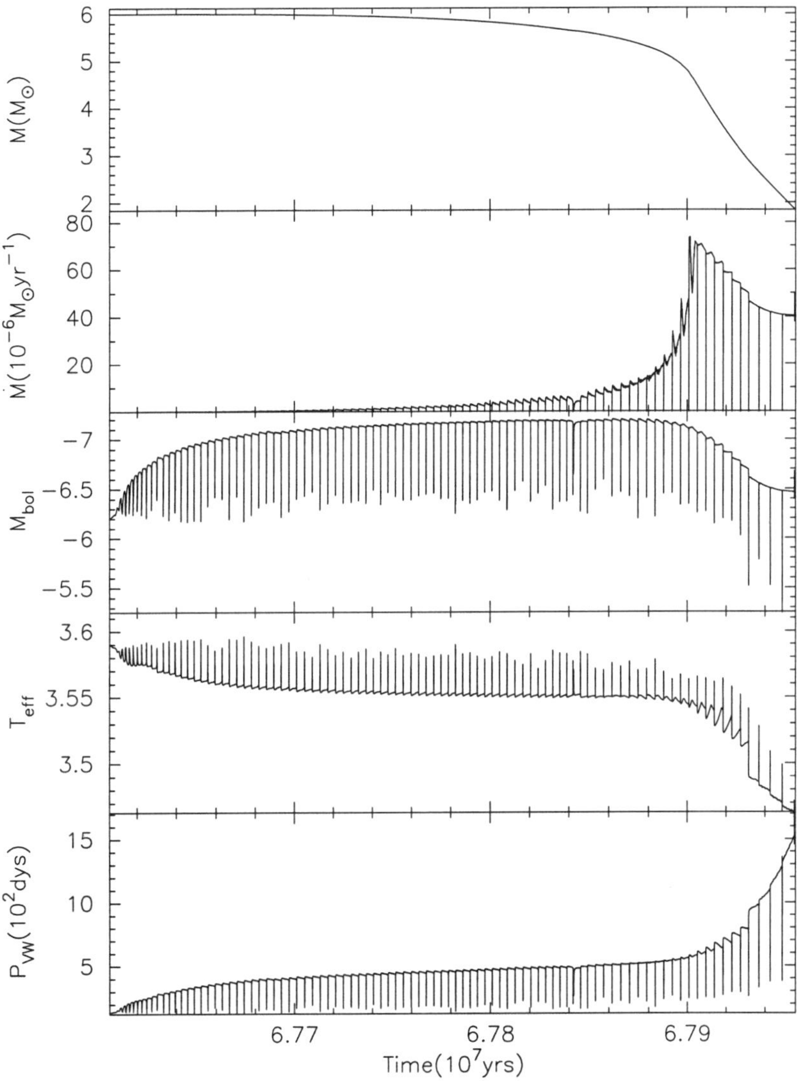

Figure 3. AGB evolution for a $6M_{\odot}$ model with $Z = 0.004$. The upper panel shows the variation in total mass; the second panel shows the variation in the mass-loss rate; next is the bolometric magnitude, then the effective temperature; finally we show the period as calculated by the formula of Vassiliadis & Wood (1993).

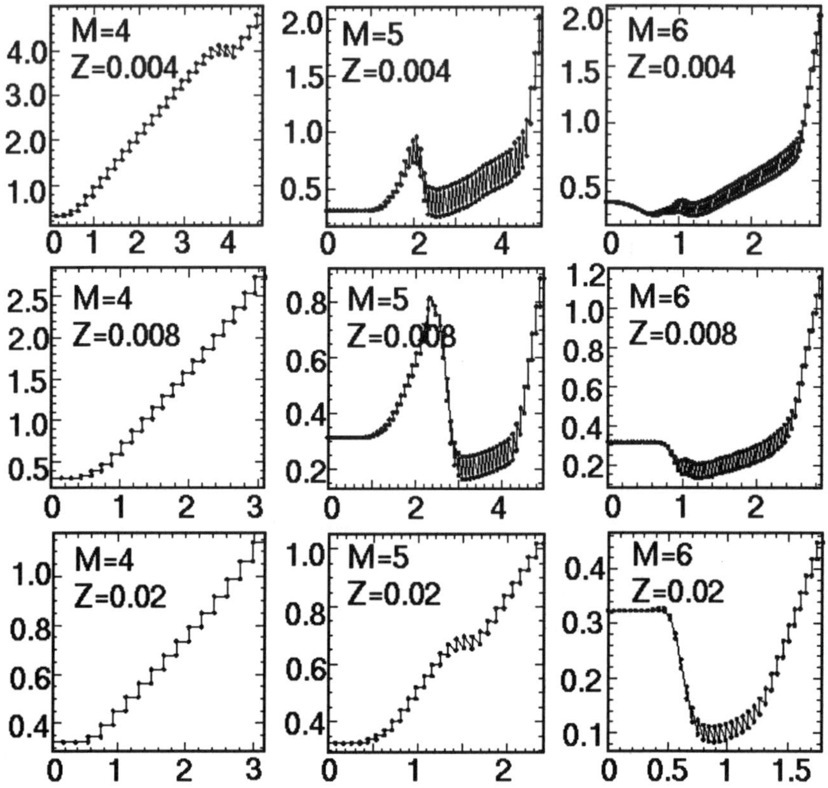

Figure 4. Surface value of C/O for the nine cases discussed in the text. The x-axis is the time elapsed (in 10^5 years) since the first thermal pulse.

using the Monash University Nucleosynthesis Code (based on that of Cannon 1993). Results from these calculations have not yet been published, with the exception of one preliminary report (Lattanzio et al. 1996) and two short papers (Frost et al. 1998; Frost, Lattanzio & Wood 1998). A brief summary will be given here, and details will be published elsewhere.

3.1. Model calculations

Models were calculated with the Mt Stromlo Stellar Structure Code, updated to include the OPAL opacities. Mass-loss was included using the prescription of Vassiliadis & Wood (1993), but without the modification for M above $2.5M_\odot$. As an example of the AGB evolution, Fig. 3 shows a summary of the case for the $6M_\odot$ model with $Z = 0.004$. We note a crucial aspect of this evolution: mass-loss is relatively small until near the end of the evolution, when the rate becomes very high. We will see the consequences of this below.

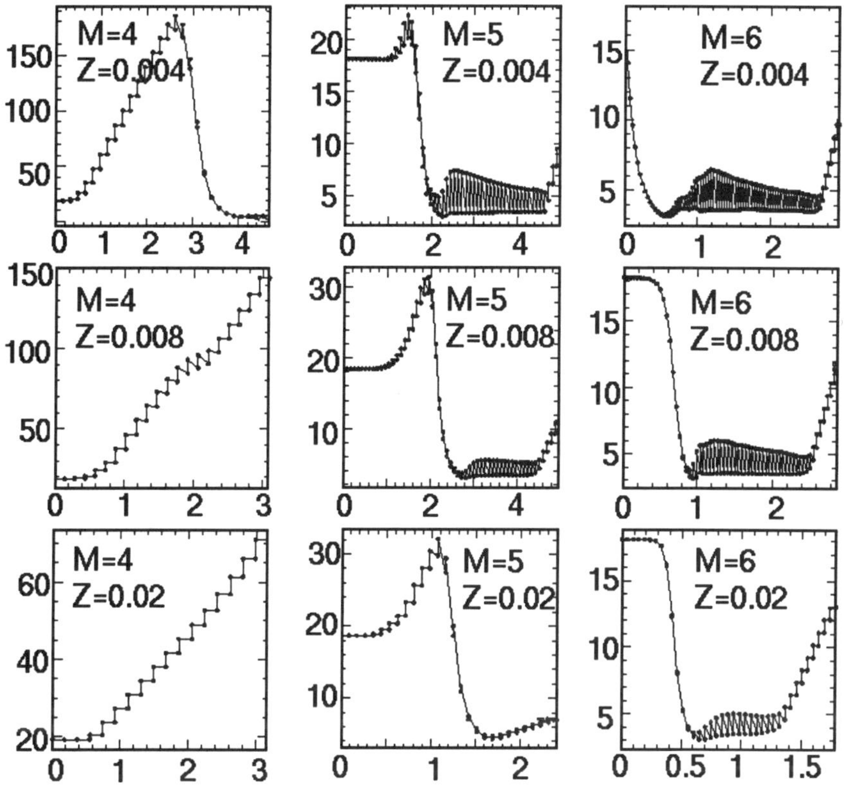

Figure 5. Same as Fig. 4 but for $^{12}C/^{13}C$.

The nucleosynthesis calculations include all elements up to sulphur and also a small *s*-process network based on the iron-peak. We have also developed the extrapolation code of Forestini & Charbonnel (1997) to include the results of these detailed calculations. For many variables we have found it best to interpolate within envelope mass, rather than the traditional core-mass. Some of the results presented below actually come from the synthetic code (and agree with the detailed calculations to within 10%).

3.2. The brightest carbon stars

Fig. 4 shows the surface C/O ratio for the nine models considered. The repeated effects of dredge-up are clearly visible in the discontinuous changes in the ratio at each dredge-up event. The effect of HBB can also be seen as a smooth decrease in the ratio between dredge-up events. Note that there is no HBB in the three $4M_\odot$ models, and the C/O ratio increases monotonically for these cases (in fact, there is a small amount of HBB in the $Z = 0.004$ case at a time of about 4×10^5 years).

The effects of HBB are clearly seen in the 5 and $6M_\odot$ models. Consider firstly the $5M_\odot$ cases, where the HBB is important at low metallicities (due to

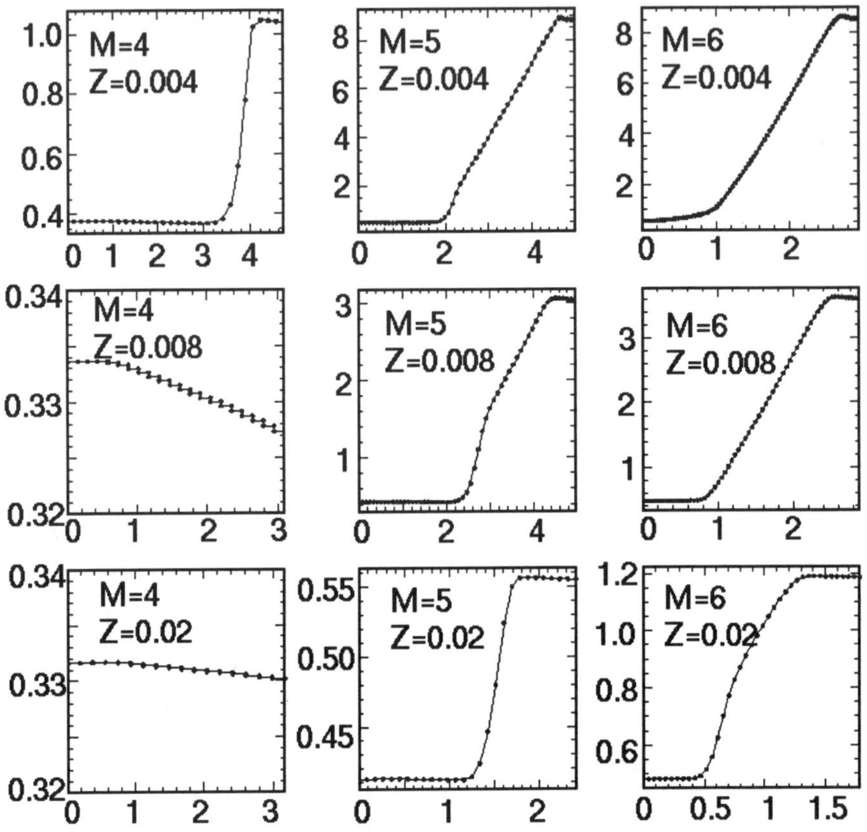

Figure 6. Same as Fig. 4 but for N/O.

higher temperatures in these cases). Although dredge-up begins in the M5Z004[1] case and produces a C/O ratio almost unity, the start of HBB then reduces this value dramatically and keeps it below unity until an AGB age of about 4×10^5 years. This is also seen in the M5Z008 case, where the C/O ratio reaches 0.8 before HBB reduces it to about 0.2. Even in the M5Z02 case we see the start of HBB at an AGB age of 1.7×10^5 years. Yet in each of these three cases the C/O ratio finally increases. This is due to mass-loss, which decreases the envelope mass and hence the dilution of any material added by dredge-up.

The continued decreasing of the envelope mass finally reduces the weight of the envelope to such an extent that HBB is terminated. Yet dredge-up continues. This is a new result from these calculations, and it is very important for understanding the final stages of AGB evolution. Hence the C/O ratio again begins to increase as ^{12}C is added to the envelope without being burned by HBB. This effect shows itself most dramatically when the mass decreases the

[1] MaZb refers to the case with $M = aM_\odot$ and $Z = b$.

most, which is near the end of the AGB lifetime (see Fig. 3). Note that HBB is quite extreme in the $6M_\odot$ cases, where the C/O ratio is kept very small (due to HBB) until the end of the star's lifetime, when it can indeed exceed unity. Thus we predict a population of bright ($M_{bol} \simeq -6.6$) carbon stars, but these would not be optically visible because of their high mass-loss. This result was reported in Frost et al. (1998) and seems to explain the infra-red population of carbon stars reported by van Loon et al. (1998) and earlier by Kastner et al. (1993).

HBB produces ^{13}C by the CN cycling at the bottom of the envelope. This is shown in Fig. 5. The timescales are such that the entire envelope is processed by HBB and this can result in the ^{12}C/^{13}C ratio reaching its equilibrium value of 3 – 3.5, as is seen in many cases in Fig. 5. In fact, one may use this ratio, with care, as an indicator of HBB. An observed low value is strong evidence for HBB, whereas a high value is evidence for dredge-up of ^{12}C without HBB. Intermediate values are harder to interpret.

A very important consequence of HBB in these stars is the production of primary ^{14}N. This is shown in Fig. 6. Substantial enhancement factors can be obtained, especially for the more massive low metallicity models. These stars are likely to be an important source of ^{14}N in the Galaxy, especially at earlier times when the metal content of intermediate mass stars is low.

HBB produces ^{7}Li, which is therefore a good indicator of the presence of HBB, although in the more massive and lower metallicity models we can find substantial Li production even before the first thermal pulse. Hence models such as M6Z004 begin their thermally-pulsing AGB with high Li abundances. Note also that when the initial supply of ^{3}He is used up, there is again a decrease of the Li content. The models agree very well with the observations of Smith & Lambert (1989, 1990) and Smith et al. (1995) which show Magellanic Cloud AGB stars of M_{bol} between -6 and -7 to have[2] $\log \varepsilon(^7\mathrm{Li})$ in the range 2.2 to 3.8. This was also shown (first!) by Sackmann & Boothroyd (1992).

During the evolution of the models discussed above, we found a different kind of thermal pulse would occasionally occur. This resulted from the prolonged occurrence of deep dredge-up on the AGB. This cools the intershell region and extinguishes the helium burning before the tail of the helium shell has burned fully, and results in a long tail extending deeply into the star, and experiencing partially degenerate conditions. Eventually this tail ignites in a "degenerate pulse" and produces a shell flash of unusually high temperatures, as reported in Frost, Lattanzio & Wood (1998). The nucleosynthesis occurring in these cases is currently under investigation.

4. Conclusion

We have made an extensive investigation of the evolution and nucleosynthesis of intermediate mass stars. This is not to say that there are no uncertainties: the details of how to calculate dredge-up remain a serious problem. But we have the tools for a consistent investigation of the behaviour of the stars, given this caveat. Intermediate mass stars were investigated first, but it is well known

---

[2] $\log \varepsilon(^7\mathrm{Li}) \equiv \log\{n(^7\mathrm{Li})/n(\mathrm{H})\} + 12$).

that the majority of the carbon stars and *s*-process enriched stars are of lower masses, about $1 - 3M_\odot$. These are the subject of current work.

Acknowledgments. We are particularly grateful to the organisers for the meeting, and JL also wishes to thank them for their financial assistance.

References

Boothroyd A.I., Sackmann I.-J., Ahern S.C., 1993, ApJ 416, 762

Cannon R.C., 1993, MNRAS 263, 817

Forestini M., Charbonnel C., 1997, A&AS 123, 241

Forestini M., et al., 1992, A&A 261, 157

Frost C.A., 1998, Ph. D thesis, Monash University

Frost C.A., Lattanzio J.C., Wood P.R., 1998, ApJ 500, 355

Frost C.A., et al., 1998, A&A 332, L17

Gallino R., Busso M., Lugaro M., 1997, in Astrophysical Implications of the Laboratory Study of Presolar Materials, T.J. Bernatowicz & E. Zinner (eds.), AIP, p. 115

Herwig F., Blöcker T., Schönberner D., El Eid M., 1997, A&A 324, L81

Iben I., Jr., Renzini A., 1982a, ApJ 259, L79

Iben I., Jr., Renzini A., 1982b, ApJ 263, L23

Jorissen A., Smith V.V., Lambert D.L., 1992, A&A 261, 164

Kastner J.H., et al., 1993, A&A 275, 163

Lattanzio J.C., 1998, in Stellar Evolution, Stellar Explosions and Galactic Chemical Evolution, A. Mezzacappa (ed.), IOP, p. 299

Lattanzio J.C., Boothroyd A.I., 1997, in Astrophysical Implications of the Laboratory Study of Presolar Materials, T. J. Bernatowicz & E. Zinner (eds.), AIP, p. 85

Lattanzio J.C., Frost C.A., Cannon R.C., Wood P.R., 1996, Mem. Astro. Soc. Italia 67, 729

Mowlavi N., Jorissen A., Arnould M., 1996, A&A 311, 803

Sackmann I.-J., Boothroyd A.I., 1992, ApJ 392, L71

Smith V.V., Lambert D.L., 1989, ApJ 345, 375

Smith V.V., Lambert D.L., 1990, ApJ 361, L69

Smith V.V., Lambert D.L., Plez B., Lubowich D.A., 1995, ApJ 441, 735

van Loon J.Th., et al., 1998, A&A 329, 169

Vassiliadis E., Wood P.R., 1993, ApJ 413, 641

Zinner E., 1997, in Astrophysical Implications of the Laboratory Study of Presolar Materials, T.J. Bernatowicz & E. Zinner (eds.), AIP, p. 3

Asymptotic Giant Branch Stars
T. Le Bertre, A. Lèbre and C. Waelkens, eds.
I.A.U. Symposium 191, 1999

The role of convective boundaries

F. Herwig

Institut für Physik, Universität Potsdam, Germany

T. Blöcker

Max-Planck-Institut für Radioastronomie, Bonn, Germany

D. Schönberner

Astrophysikalisches Institut Potsdam, Germany

Abstract. We investigate the influence of convective overshoot on stellar evolution models of the thermal pulse AGB phase with $M_{\rm ZAMS} = 3{\rm M}_{\odot}$. An exponential diffusive overshoot algorithm is applied to all convective boundaries during all evolutionary stages.

We demonstrate that overshooting at the bottom of the pulse-driven convective zone, which forms in the intershell during the He-shell flash, leads to more efficient third dredge-up. Some overshoot at the bottom of the convective envelope removes the He-H discontinuity, which would otherwise prohibit the occurrence of the third dredge-up for this stellar mass. However, no correlation between the amount of envelope overshoot and the efficiency of the third dredge-up has been found.

Increasingly efficient third dredge-up eventually leads to a carbon star model. Due to the partial mixing efficiency in the overshoot region a ^{13}C-pocket can form after the third dredge-up event which may be crucial for n-capture nucleosynthesis.

1. Introduction

Current stellar evolutionary models of the AGB stage fail to reproduce a number of observationally based constraints. The observed carbon star luminosity function requires that the third dredge-up in AGB stars occurs efficiently at core masses of about $M_{\rm H} = 0.6{\rm M}_{\odot}$ (Groenewegen et al. 1995; Marigo et al. 1996).

For the explanation of s-process elements on the surface of AGB stars, dredge-up is an important prerequisite as well. Obviously, dredge-up enables the transport of processed material to the surface. But dredge-up is also needed in order to establish a region in the star where the H-rich convective envelope has immediate contact with the ^{12}C-rich intershell region. Additionally to sufficient dredge-up a mechanism must be identified to ingest protons into the intershell region in order to produce the neutron source ^{13}C via the reaction chain $^{12}{\rm C}({\rm p},\gamma)^{13}{\rm N}(e^{+}\nu)^{13}{\rm C}$ (Gallino et al. 1998).

A certain class of post-AGB stars (about 20% of all central stars of planetary nebulae) show a hydrogen – deficient surface composition. The evolutionary origin of the typical abundance pattern of these stars [e.g. (He/C/O)=(33/50/17) in mass fractions for the PG1159 stars, Werner et al. (1998)] are not well understood. One possible evolutionary scenario establishes a link between these surface abundances with the intershell abundance of the AGB stars (Iben & McDonald 1995). Current models, however, typically predict an abundance pattern like ($^{4}He/^{12}C/^{16}O$)=(70/26/1) for the intershell region.

In Sect. 2. we will describe the motivation and method of overshooting which we have applied to stellar evolution models of AGB stars. In Sect. 3. we focus on the effect of overshooting of the pulse-driven convective zone, which evolves in the intershell during the He-flash, on the intershell abundance. We will also discuss the role of overshooting at the bottom of the envelope convection zone for the third dredge-up. This will then lead to the presentation of the ^{13}C-pocket which also results from the overshooting treatment (Sect. 4.).

2. Overshooting with decreasing mixing efficiency

Overshooting is the mixing of material beyond the boundary of convection. The impact of overshooting has been studied for different stellar conditions in order to address various problems in stellar evolution models. For example, Shaviv & Salpeter (1973) and Maeder (1975) have studied the general properties of overshooting at the boundary of a convective core. The effect of overshooting at the bottom of the convective envelope has been studied by Alongi et al. (1991) for red giant branch stars and by Iben (1976) for AGB stars.

Mixing due to overshoot has often been treated by the formal extension of the instantaneously mixed convective region (Schaller et al 1992). However, hydrodynamic simulations (Freytag et al 1996) of the surface convection zone of A-stars and white dwarfs show prominent convective rolls and downdrafts which extend beyond the boundary of convection and lead to a partial mixing of this region. The exponential decay of the turbulent velocity field and the diffusive character of the associated mixing are common features of this overshoot. These results have motivated the ansatz for an exponentially declining diffusion coefficient for the overshoot region in order to describe the mixing beyond the convective boundary (Herwig et al 1997). Further on, we assume that the exponential diffusive description of overshoot is applicable to the deep stellar interior. For this region we have scaled the exponential coefficient f by fitting the main sequence width. We then assumed that the resulting exponential coefficient of $f = 0.016$ can serve as an approximation for the convective boundaries encountered in AGB stellar models.

This method of *exponential diffusive overshoot* has been applied to all convective boundaries during the AGB evolution (and before). This includes the pulse driven convective zone in the intershell layer which results from the huge energy release during the He-shell flash.

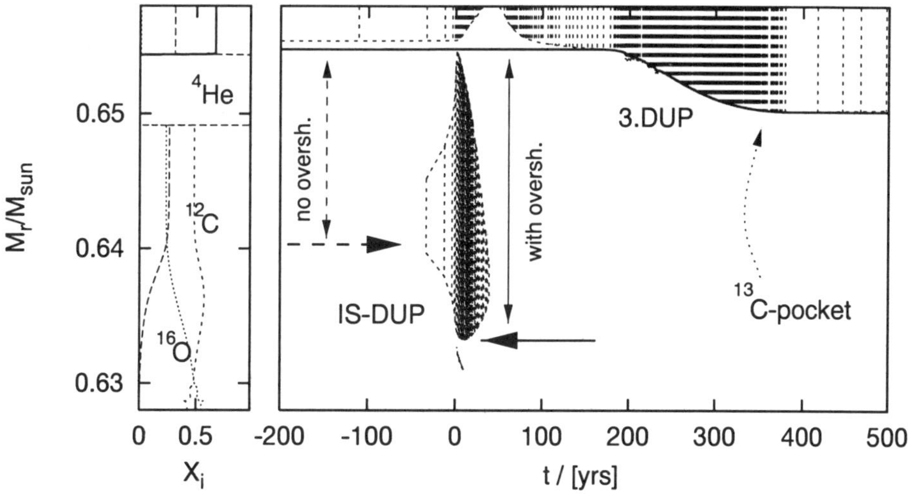

Figure 1. Isotopic abundance profiles of ^{4}He, ^{12}C and ^{16}O at the end of the interpulse phase (left) and the evolution of convective regions during a thermal pulse (at $t = 0$yr) including the third dredge-up (main panel) at the 9$^{\rm th}$ TP of a $M_{\rm ZAMS} = 3{\rm M}_\odot$ sequence. In the main panel the regions covered with vertical dashed lines are convectively unstable (convective envelope at the top and the pulse-driven convective zone in the center of the figure). The density of these lines indicate the variation of the time resolution. The dashed arrows (labeled no oversh.) indicate the extend that a pulse-driven convective zone during a TP with this core mass would have without application of overshoot to this convective region. Label IS-DUP:*intershell dredge-up*.

3. The dredge-up in the pulse-driven convective zone and the third dredge-up

While overshoot at the top boundary of the pulse-driven convective zone has no noticeable effect, mixing beyond the bottom boundary leads to a decrease of the ^{4}He abundance in the intershell. At the same time the intershell region is enriched with ^{12}C and ^{16}O. This change of composition leads to an increase of opacity and therefore the radiative gradient $\nabla_{\rm rad}$ (which is proportional to the opacity) is lifted with respect to the adiabatic gradient $\nabla_{\rm ad}$, and the further downwards extension of the pulse-driven convective zone is supported. This process drives an even deeper penetration of the pulse-driven convective zone into the core region. In Fig. 1 the left panel shows the abundance profiles in the mass range of the two burning shells shortly before the flash occurs, which is shown in the main panel. The region between the mass coordinates $M_{\rm r} \approx 0.64{\rm M}_\odot$ and $\approx 0.65{\rm M}_\odot$ shows the already altered intershell abundances (compared to standard calculations) formed during previous thermal pulses. The almost pure He layer above $M_{\rm r} \approx 0.65{\rm M}_\odot$ results from H-shell burning during the previous interpulse

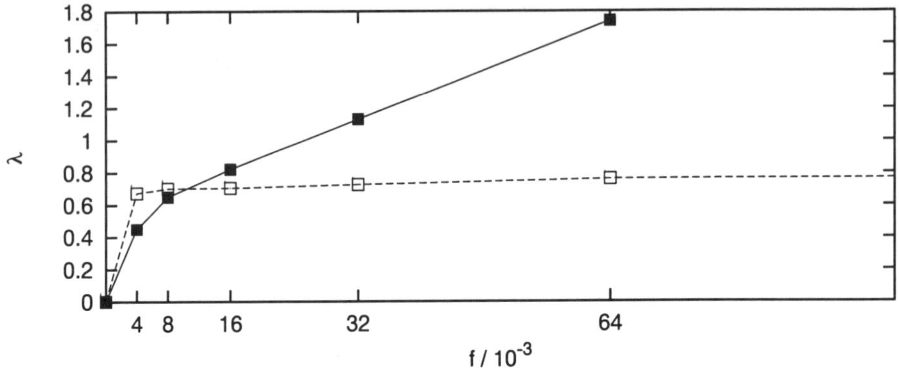

Figure 2. The open symbols (dashed line) show λ [λ = (core mass decrease by dredge-up) / (core mass growth by hydrogen burning)] for a series of test sequences started at $t \approx +100$yr after the 8th TP flash peak (when the pulse-driven convective zone has already disappeared), with different values of f (during all previous evolution $f = 0.016$ has been used). Each test sequence has been evolved through the next pulse with the f value indicated. The filled squares (solid line) show λ after this next pulse. The difference is that here also the pulse-driven convective zone has been computed with the changed f value.

phase. Below $\approx 0.64 M_{\odot}$ the ^{4}He abundance drops continuously according to the preceding He-shell burning. The pulse-driven convective zone of a TP computed without overshoot would not extend into this He-poor region as can be seen from the dashed arrow in the main panel of Fig. 1. In contrast to that the pulse-driven convective zone actually shown in the main panel is computed with overshoot and does extend much deeper into the core. It covers a mass fraction which is larger by about a factor of 1.5 compared to the pulse-driven convective zone without overshoot. One might call this process *intershell dredge-up* (IS-DUP).

If overshooting at the bottom of the pulse-driven convective zone is considered, the intershell abundance evolves with pulse number. While the ^{4}He abundance decreases steeply over the first pulses at the expense of ^{12}C, the ^{16}O abundance increases steadily. After about a dozen pulses the ^{4}He abundance has recovered from a relative minimum (corresponding to a relative maximum of ^{12}C) and all three abundances level at about (He/C/O)=(40/40/16). These changed intershell abundances (compared to standard models) may eventually contribute to a better understanding of the above mentioned abundance patterns of H-deficient post-AGB stars.

After the pulse-driven convective zone has disappeared IS-DUP has led to an intershell abundance which contains less ^{4}He compared to models without overshoot. Then, following the arguments given in the previous section, the efficiency of the third dredge-up is enhanced. This can be seen from Fig. 2. If a larger overshoot efficiency is applied to the pulse-driven convective zone (filled symbols) less ^{4}He is present in the intershell region after the TP and the third dredge-up is more efficient. Thus, we find that the application of overshoot to

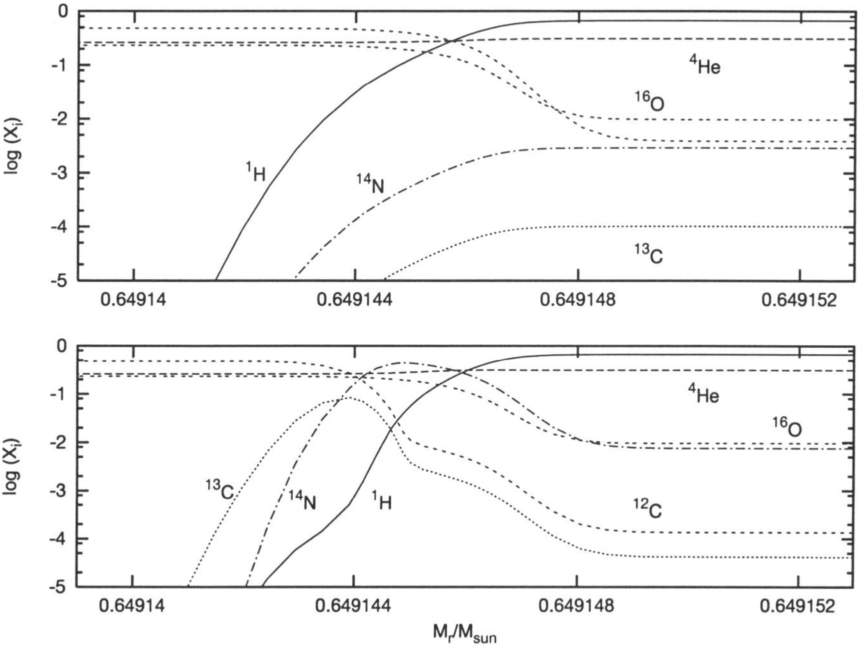

Figure 3. After the $8^{\rm th}$ TP of a $M_{\rm ZAMS} = 3{\rm M}_\odot$ sequence: isotopic abundance profiles at the interface of ^{12}C-rich intershell and H-rich envelope shortly after the end of the third dredge-up (top) and about 1800yr later when the ^{13}C-pocket has developed (bottom).

the pulse-driven convective zone during the He-shell flash increases the amount of third dredge-up occurring immediately after the TP.

But what is the role of overshoot at the bottom of the convective envelope? The open symbols in Fig. 2 each represent the dredge-up efficiency λ when f has been changed *after* the pulse-driven convective zone has disappeared. This means that all these dredge-up events occur with identical intershell abundance (the same overshoot has been applied to the previous pulse-driven convective zone). Only the efficiency of the overshoot at the bottom of the convective envelope is different. One can see that the amount of overshoot at this convective boundary is not related to the dredge-up efficiency. The only effect of this overshoot is the removal of the H-He discontinuity which would otherwise prohibit the third dredge-up in our models ($\lambda = 0$ for $f = 0$ in Fig. 2). We conclude that some non-zero overshoot at the bottom of the convective envelope is necessary in our models to make the third dredge-up at all possible (However, Straniero et al (1997) did find the third dredge-up for similar core masses without invocation of overshoot.) while overshoot at the bottom of the pulse-driven convective zone has an effect on the efficiency of the third dredge-up. Our model sequence shows the third dredge-up with increasing efficiency from the third thermal pulse on until ($\lambda > 1$). A carbon model star is found after the 12th TP.

Finally we note, that the efficient dredge-up leads to deviations from the linear core mass - luminosity relation originally described by Paczyński (1970). The deviations occur at lower core masses than those associated with hot bottom burning. They can be understood by the analysis of homology relations as described by Herwig et al (1998).

4. The ^{13}C-pocket

The overshoot effects described so far are possibly not strongly depending on the overshoot description. However, *partial* mixing beyond the bottom of the envelope convection becomes crucial when the third dredge-up comes to an end. Then the bottom of the H-rich envelope is immediately neighbored by the ^{12}C-rich intershell region and due to the exponential diffusive overshoot a thin layer forms where protons and ^{12}C coexist (Fig. 3, top panel). When the temperatures rise in this region due to resuming contraction after the TP a ^{13}C-pocket (^{13}C-rich layer) is formed (Fig. 3, bottom panel). It typically contains about $3 \cdot 10^{-7} M_{\odot}$ of ^{13}C. It may, in the further course of the pulse, serve as the neutron source for the formation of heavy elements. While the qualitative modeling of such a pocket is quite encouraging the ^{13}C formation must be investigated more quantitatively in the future.

Acknowledgments. F. H. acknowledges funding by the *Deutsche Forschungsgemeinschaft, DFG* (grants Scho 394/13 and La 587/16).

References

Alongi M., Bertelli G., Bressan A., Chiosi C., 1991, A&A 244, 95

Freytag B., Ludwig H.-G., Steffen M., 1996, A&A 313, 497

Gallino R., Arlandini C., Busso M., Lugaro M., Travaglio C., Straniero O.A., Limongi M., 1998, ApJ 497, 388

Groenewegen M., van den Hoek L., de Jong T., 1995, A&A 293, 381

Herwig F., Blöcker T., Schönberner D., El Eid M.F., 1997, A&A 324, L81

Herwig F., Schönberner D., Blöcker T., 1998, A&A 340, L43

Iben I., Jr., 1976, ApJ 208, 165

Iben I., Jr., McDonald J., 1995, in White Dwarfs, D. Koester and K. Werner (eds.), No. 443 in LNP, Springer, Heidelberg, p. 48

Maeder A., 1975, A&A 40, 303

Marigo P., Bressan A., Chiosi C., 1996, A&A 313, 545

Paczyński B., 1970, Acta Astr. 20, 47

Schaller G., Schaerer D., Meynet G., Maeder A., 1992, A&AS 96, 269

Shaviv G., Salpeter E., 1973, ApJ 184, 191

Straniero O., Chieffi A., Limongi M., Busso M., Gallino R., Arlandini C., 1997, ApJ 478, 332

Werner K., Dreizler S., Rauch T., Koesterke L., Heber U., 1998, this volume

Asymptotic Giant Branch Stars
T. Le Bertre, A. Lèbre and C. Waelkens, eds.
I.A.U. Symposium 191, 1999

Dredge-Up in Asymptotic Giant Branch Stars

N. Mowlavi

Observatoire de Genève, CH-1290 Versoix, Switzerland

Abstract. The third dredge-up phenomenon in asymptotic giant branch stars, responsible for the formation of C stars, is discussed based on detailed evolutionary model calculations. The structural readjustment of the star as dredge-up proceeds is analyzed, and its consequences on dredge-up predictions discussed. The question of how to obtain dredge-up in asymptotic giant branch models is also addressed. It is stressed, in particular, that the modeling of dredge-up requires some sort of extra-mixing to be applied below the envelope if the local Schwarzschild criterion is used to delimit convective zones.

1. Introduction

The third dredge-up (3DUP) phenomenon is the process by which the envelope of asymptotic giant branch (AGB) stars[1] penetrates into the C-rich layers, bringing to the surface the ashes of helium burning nucleosynthesis (see, e.g., Lattanzio & Forestini in this volume or Mowlavi 1998b for a general review on the structural and chemical evolution of those stars). It constitutes a key issue in our understanding of the chemical abundances observed at the surface of AGB stars. It is responsible, in particular, for the formation of carbon stars. However, despite the efforts of AGB modelers since more than two decades, reliable predictions of 3DUP are still lacking. Many AGB model calculations even fail to reproduce efficiently, if at all, that phenomenon (for example Vassiliadis & Wood 1993, Wagenhuber & Weiss 1994, Blöcker 1995, Forestini & Charbonnel 1997). The predictions of those which do succeed in obtaining it (for example Straniero et al. 1997, Herwig et al. 1997, Mowlavi 1998a), on the other hand, do not agree with each other. This situation is rather confusing even to experts in AGB modeling, and raises some fundamental questions. Why those contradictory predictions? Can we hope AGB modelers to agree one day on the dredge-up predictions? If not, what are the parameters, whether numerical or physical, responsible for the scatter in those predictions? The lively discussions which animated this 191st IAU Symposium around those questions, among others, revealed that they are not trivial to be answered.

[1]The term 'envelope' of AGB stars designates the convective region in their interior which extends from about the photosphere down to layers with temperatures higher than several million degrees. It must not be confused with the 'circumstellar envelope' used in the terminology of observers.

The thermal instabilities (called 'pulses') which periodically develop in the He-burning shell of AGB stars provide an adequate stage for the operation of the 3DUP. After such a thermal pulse, the structural readjustment of the star eventually leads to the deepening of the envelope into the H-depleted regions. While this scenario is well known since twenty years, the physics involved in the dredge-up process has not yet been fully discussed. Yet, a good understanding of that process obviously is a necessary prerequisite to the analysis of model predictions. An effort towards that direction has recently been done by Mowlavi (1998a). It is the purpose of this contribution to highlight some of the basic physical considerations which govern the 3DUP process, and to present their expected consequences on 3DUP predictions.

2. Modeling third dredge-up

The question of how to obtain dredge-up in AGB models is related to the delicate question of the determination of convection boundaries. From a numerical point of view, the Schwarzschild criterion is used, to my knowledge, in all AGB models published to date (including my own calculations). This is a local criterion, and consequently raises the question of whether to apply extra-mixing or not beyond the convection boundaries. It is a crucial point in the modeling of third dredge-up. According to my calculations, models strictly using the Schwarzschild criterion (i.e. without *any* extra-mixing beyond the convection boundaries so defined) do not experience dredge-up. Such models are even unphysical, as they lead to the development of a discontinuity in the abundance profiles at the edge of the convective envelope. In contrast, models with extra-mixing do lead to efficient dredge-up. Using an extra-mixing procedure of some sort seems thus to be a prerequisite to obtain dredge-up in AGB models.

How much extra-mixing should be used could be an important question too. Fortunately, however, the dredge-up characteristics are found to be rather insensitive to the amount of extra-mixing put in the models. That important conclusion obtained in Mowlavi (1998a) is confirmed by model calculations, and can be understood on grounds of simple physical arguments summarized in Sect. 4. This fact enables us to analyze the dredge-up predictions without worrying much about the details of the extra-mixing. It should be stressed, however, that a diffusion algorithm should be used for the mixing of the chemical elements. Not only does it lead to smooth abundance profiles at the H-He transition zone, but it also helps to the numerical calculation of the AGB star structure. Indeed, using instantaneous mixing is known to lead to convergence difficulties (Frost & Lattanzio 1997).

Extra-mixing has also important consequences on the production of ^{13}C in the intershell regions (Herwig et al. 1997), and on the consequent synthesis of ^{19}F and s-process elements. The yields of those nuclei, however, *are* expected to depend on the extra-mixing parameters, as stressed in Mowlavi (1999).

The extra-mixing procedure used in the models presented here considers an exponentially decreasing bubble velocity field below the convective envelope, and uses a diffusion algorithm to perform the mixing of chemical elements. More details on those models can be found in Mowlavi (1998a).

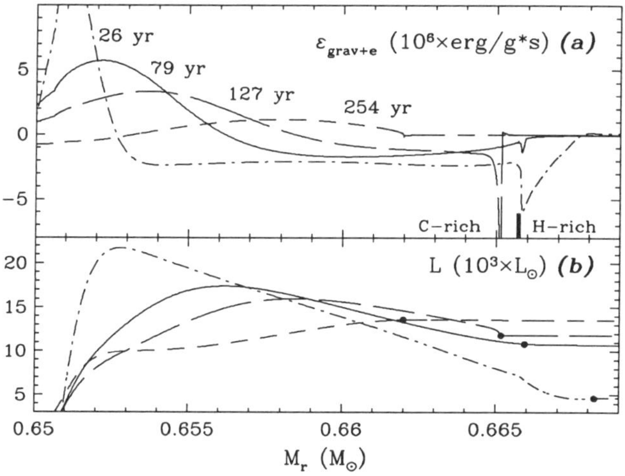

Figure 1. *(a)* Gravothermal energy profiles at four different times, as labeled on the curves, after the 19th pulse of a 3 $M_\odot$ star computed with extra-mixing below the convective envelope. The time origin is taken at maximum pulse extension. The extents of the H- and C-rich regions before dredge-up are also indicated, the He-rich region being identified by the thick vertical line separating the H- and C-rich layers; *(b)* Same as *(a)*, but for the luminosity profiles. The filled circles on the curves locate the bottom of the convective envelope.

3. The afterpulse phase

The afterpulse phase is defined as the phase which follows a thermal pulse and during which the star undergoes various structural and thermal readjustments. Those readjustments result from the outwards evacuation of a positive gravothermal energy which develops at the bottom of the former pulse (Paczyński 1977). In the 19th afterpulse phase of a 3 $M_\odot$ model star, this gravothermal energy wave reaches the H-rich layers in about 250 yr [see Fig. 1*(a)*]. The luminosity at the bottom of the envelope increases concomitantly with time, reaches a maximum value, and then decreases as the gravothermal energy evacuates outwards [Fig. 1*(b)*]. The increase in the luminosity around the bottom of the convective envelope renders those layers more unstable against convection, and is responsible for the penetration of the envelope in the deeper layers. Concomitant to the luminosity, the convective envelope reaches a maximum depth, and then retreats back to higher mass locations.

When the envelope reaches the H-depleted layers, a discontinuity in the hydrogen abundance profile develops in models using the Schwarzschild criterion (e.g. Iben 1976). It is this discontinuity that prevents a further deepening of the envelope *if no extra-mixing is used* (Mowlavi 1998a, 1999). The situation is however unphysical, and a (diffusive) extra-mixing *is* required as recalled in Sect. 2.

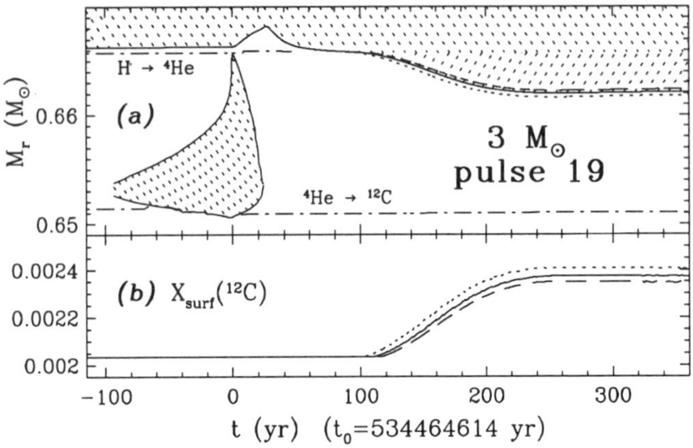

Figure 2. *(a)* Evolution of the convection zones (shown in hatched) during and after the 19th pulse in the models of the 3 $M_\odot$ star computed with extra-mixing. The solid, dashed and dotted lines differentiate models computed with different extra-mixing parameters (see Mowlavi 1998a). The extent of the overshooting regions is less than 10^{-4} $M_\odot$ and cannot be distinguished in the figure; the dashed-dotted lines locate the layer of maximum nuclear energy production in the H and He burning shells, as labeled in the graph. The time origin is identical to that in Fig. 1. *(b)* Same as *(a)*, but for the surface ^{12}C mass fraction.

4. The third dredge-up process

Efficient dredge-up is obtained in AGB models using extra-mixing. Model calculations performed with various extra-mixing extents and efficiencies reveal that the dredge-up characteristics are independent of those extra-mixing parameters. Three such sets of calculations are displayed in Fig. 2 for the 19th afterpulse of the 3 $M_\odot$ model star. The dredge-up rate is found to be equal to 4.2×10^{-5} $M_\odot$/yr in all three sets. That the dredge-up rate is insensitive to the extra-mixing parameters can be understood from the following reasoning.

An AGB star can basically be thought, for our purpose, as being composed of a H-depleted core and a H-rich envelope. The core is very dense and compact with a radius of about $0.10 - 0.15$ $R_\odot$ and a mean density of $10^5 - 10^7$ g cm^{-3}. It is for a great part degenerate in the electrons, property which determines the structural evolution of its interior. In contrast, the envelope is rather dilute with a mean density of about 10^{-6} g cm^{-3}, and extends up to the photosphere several hundreds of solar radii away from the center of the star (see, e.g., Mowlavi 1995). Convection is its dominant mode of energy transport, and constrains its structure. The envelope structure thus mainly depends on the conditions near the stellar surface rather than on the stellar interior properties.

As dredge-up proceeds, H-depleted matter is added from the core into the envelope. That matter is lifted to higher radius, and its potential energy increases. The characteristic time-scale of that process is determined by the thermal relaxation time-scale of those layers engulfed in the envelope and which must expand to fit the envelope structure. This time-scale imposes a maximum rate at which dredge-up can proceed. An estimation of that maximum dredge-up rate, r_{dup}, is given by (see Mowlavi 1998a):

$$r_{dup} \simeq 3.2 \times 10^{-8} \, \frac{R_c \, L_c}{M_c}, \tag{1}$$

where M_c and R_c are the mass and radius of the H-depleted core, and L_c the luminosity at the core edge. All quantities in Eq. 1 are given in solar units, and r_{dup} is expressed in $M_\odot$/yr. For the 19th pulse of the 3 $M_\odot$ star displayed in Fig. 2, $M_c \simeq 0.66\ M_\odot$, $R_c \simeq 0.12\ R_\odot$ and $L_c \simeq 10^4\ L_\odot$. Equation 1 then leads to $r_{dup} \simeq 5.8 \times 10^{-5}\ M_\odot$/yr. This is close enough to the value obtained from the evolutionary model calculations.

The mixing time scale of matter across the convection boundary is estimated to be much shorter than the dredge-up rate (Mowlavi 1999). This ensures that the dredge-up rate obtained in AGB models is essentially independent on the extra-mixing parameters. The 3DUP predictions are rather determined by the thermal relaxation time scale of the envelope.

5. Third dredge-up characteristics

Equation 1 predicts that the maximum dredge-up rate is proportional to the luminosity at the bottom of the envelope. This is valid to first approximation if we neglect the change in R_c/M_c. If we further assume that the luminosity is constant in the envelope, then we expect a linear relation between the dredge-up rate and the surface luminosity. Dredge-up calculations performed on *selected* afterpulses of the 3 $M_\odot$ star actually confirm that expectation[2] (Mowlavi 1998a).

The dredge-up efficiency is traditionally expressed by a number λ, which is the ratio of the amount of matter dredged-up relative to the increase in core mass due to hydrogen burning during two successive pulses. A value of $\lambda = 1$, for example, implies a constant core mass from one pulse to the next. Model calculations reveal that λ, too, increases linearly with the surface luminosity[2]. Moreover, the dredge-up efficiency is predicted to be rather high, reaching unity at $L \simeq 13000\ L_\odot$ in a 3 $M_\odot$ star. High values of λ are indeed expected from the analysis of the C-star populations in the Magellanic Clouds (Groenewegen & de Jong 1993). Our predictions are moreover compatible with those obtained by Herwig et al. (1997) who use an overshooting prescription in their models very similar to ours. The dredge-up efficiencies obtained by those authors are even higher, at equivalent luminosities, than ours. Certainly, the inclusion of extra-mixing in AGB model calculations opens new horizons in the field of AGB modeling.

---

[2]The feedback of the dredge-ups on the evolution of AGB stars, which is not considered in this paper, may alter some 'standard' results of AGB evolution. Such is the case, for example, for the core mass-luminosity relation (Herwig et al. in this volume). The possible consequences of such a feedback on the dredge-up characteristics themselves are discussed in Mowlavi (1998a).

6. Conclusion

Obtaining dredge-up, especially in models of low-mass stars, was still a challenge at the beginning of this decade. Sackmann & Boothroyd (1991) suggested at that time that the solution may reside in the use of new radiative opacities then available, and of higher mixing length parameters for convection. Straniero et al. (1997) even suggest that the difficulty is purely of numerical origin, and that the use of higher spatial and temporal resolution in the models could solve the difficulties of obtaining dredge-up.

The recent studies on the role of extra-mixing in obtaining dredge-up open new perspectives for the study of 3DUP. Herwig et al. (1997) obtain efficient dredge-up in models using extra-mixing. Mowlavi (1998a) further shows that the dredge-up would not occur in AGB models using *only* the Schwarzschild criterion *without* any extra-mixing, and that an extra-mixing procedure of some sort is necessary. He shows, moreover, that the dredge-up characteristics are essentially independent of the extra-mixing parameters, provided of course that such an extra-mixing is applied. This allows AGB modelers not to worry too much about the details of the extra-mixing, at least as far as the dredge-up characteristics are concerned. Hopefully, consistent *and reproducible* predictions of third dredge-up in AGB stars will soon become available.

References

Blöcker T., 1995, A&A 297, 727

Forestini M., Charbonnel C., 1997, A&AS 123, 241

Frost C.A., Lattanzio J.C., 1996, ApJ 344, L25

Groenewegen M.A.T., de Jong T., 1993, A&A 267, 410

Herwig F., Blöcker T., Schönberner D., El Eid M., 1997, A&A 324, L81

Iben I., Jr., 1976, ApJ 208, 165

Mowlavi N., 1995, Ph.D. Thesis, Université Libre de Bruxelles[3]

Mowlavi N., 1998a, A&A, submitted

Mowlavi N., 1998b, eds. M. Arnould et al., in *Tours Symposium on Nuclear Physics III*, AIP Conf. Proc., New-York, p. 507[3]

Mowlavi N., 1999, *Nuclei in the Cosmos V*, eds. N. Prantzos and S. Harissopulos, Editions Frontières, France, in press[3]

Paczyński B., 1977, ApJ 214, 812

Sackmann I.-J., Boothroyd A.I., 1991, in 'Evolution of Stars: The Photospheric Abundance Connection', IAU Symp. 145, eds. G. Michaud and A. Tutukov, Kluwer Acad. Publ., p. 275

Straniero O., Chieffi A., Limongi M., Busso M., et al., 1997, ApJ 478, 332

Vassiliadis E., Wood P.R., 1993, ApJ 413, 641

Wagenhuber J., Weiss A., 1994, A&A 286, 121

---

[3]Postscript file available by anonymous ftp on 'obsftp.unige.ch' in 'pub/mowlavi', or on the WWW at 'http://obswww.unige.ch/~mowlavi/publications'.

Asymptotic Giant Branch Stars
T. Le Bertre, A. Lèbre and C. Waelkens, eds.
I.A.U. Symposium 191, 1999

Improved synthetic TP-AGB models

P. Marigo

Max-Planck-Institut für Astrophysik, Garching bei München, Germany

Abstract. We present recent improvements and results of an extensive analysis of the TP-AGB phase performed by means of a synthetic model (Marigo 1998a, b; Marigo et al. 1998a, b). The improvements concern: i) the use of a homogeneous and accurate set of analytical relations (Wagenhuber & Groenewegen 1998); ii) a new treatment of envelope burning in the most massive TP-AGB stars ($M > 3.5M_{\odot}$) to account for the possible break-down of the core mass-luminosity relation; iii) a better treatment of the third dredge-up to infer if and when the process takes place.

Extensive calculations of synthetic TP-AGB models have been carried out over the mass range ($0.8M_{\odot} \div 5M_{\odot}$) and for three sets of initial metallicity ($Z = 0.019$, $Z = 0.008$, $Z = 0.004$). The formation of carbon stars is investigated addressing the following issues: a) the reproduction of the observed luminosity functions of carbons stars in both Magellanic Clouds, and b) the formation of very bright and optically obscured carbon stars.

1. Introduction

Synthetic TP-AGB models may be a powerful tool of investigation, thanks to quick computing and easy testing of different prescriptions. These characteristics make them complementary to complete but extremely time-consuming calculations of stellar models. To this respect, we report here on recent effort made to improve the degree of accuracy of the synthetic approach, and discuss some applications.

2. A consistent treatment of envelope burning

Synthetic TP-AGB calculations have so far adopted the erroneous assumption that the core mass-luminosity ($M_c - L$) relation determines the surface luminosity even of stars experiencing efficient envelope burning (or hot-bottom burning; e.g. Renzini & Voli 1981; Groenewegen & de Jong 1993; Marigo et al. 1996). Marigo et al. (1998a) and Marigo (1998ab) have made the first effort to overcome this inadequacy, by developing a solution scheme, based on envelope integrations, in which the equation of energy balance is included and integrated with the aid of proper boundary conditions. The method turns out to be reliable as it reproduces with great accuracy the results of full calculations (i.e. Blöcker & Schönberner 1991; Blöcker 1995; see Fig. 1).

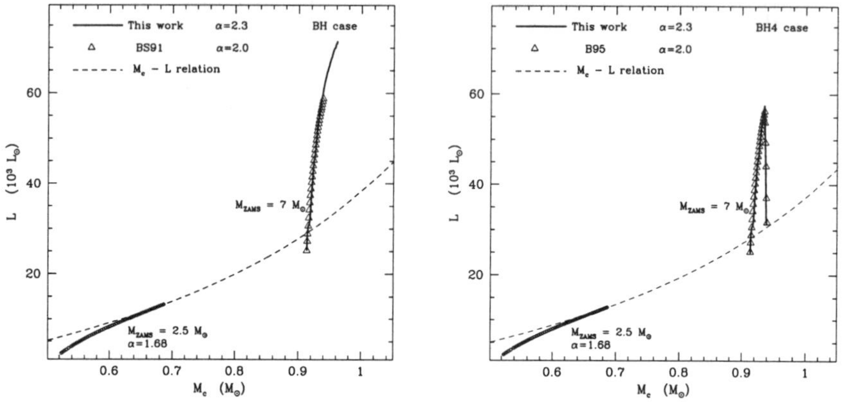

Figure 1. Quiescent luminosity evolution as a function of the core mass for a $7M_\odot$ TP-AGB model of solar composition, according to two prescriptions for mass loss: 1) Baud & Habing 1983, (BH case, left panel); 2) the same as the former case up to the activation of a constant mass-loss rate of $10^{-4}M_\odot$ yr^{-1}, (BH4 case, right panel). The predictions of full calculations are denoted by triangles [Blöcker & Schönberner 1991 (BS91); and Blöcker 1995 (B95)] at the stage of the pre-flash luminosity maximum. The results of the present synthetic calculations are shown (solid line), starting from the same initial conditions at the first thermal pulse. See Marigo (1998b) for more details.

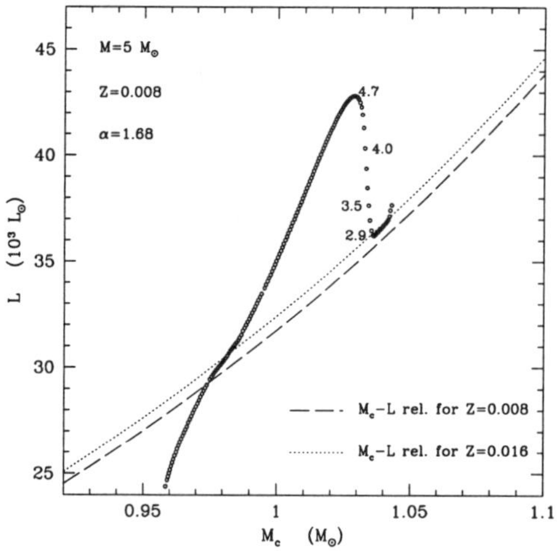

Figure 2. Quiescent luminosity evolution as a function of the core mass for a $5M_\odot$, $Z = 0.008$ model experiencing envelope burning. A few values of the current stellar mass are indicated in $M_\odot$. The mass-loss prescription is according to Vassiliadis & Wood (1993).

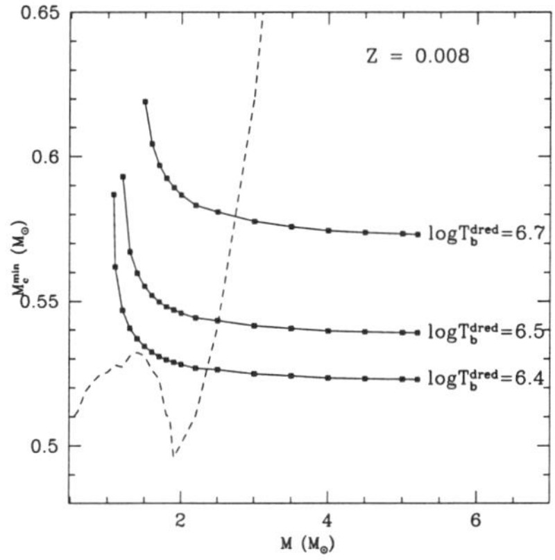

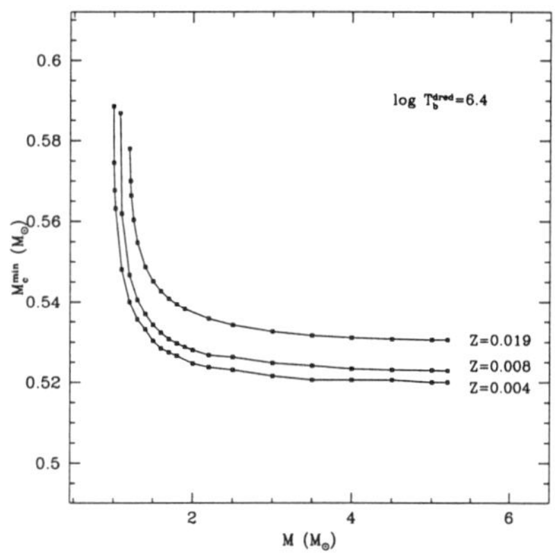

Figure 3. Minimum core mass $M_c^{\rm min}$ derived from envelope integrations as a function of stellar mass and metallicity, for different choices of the temperature parameter $T_b^{\rm dred}$. The core mass at the first thermal pulse, for models with initial metallicity $Z = 0.008$, is shown (left panel; dashed line).

The typical evolution of the quiescent luminosity, as a function of the core mass, described by a TP-AGB star with envelope burning is shown in Fig. 2, which refers to a ($5M_\odot$, $Z = 0.008$) model. The luminosity first increases above the $M_c - L$ relation, reaches a maximum at the onset of the superwind regime, then quickly drops as envelope burning weakens because of the significant reduction of the envelope mass, and eventually approaches the $M_c - L$ relation, which is followed till the end of the evolution. Note that the final recovering of the $M_c - L$ relation is consistent with a current envelope metallicity nearly twice larger ($Z = 0.016$) than the original one ($Z = 0.008$), due to the effect of both dredge-up and envelope burning.

3. The third dredge-up: if and when from envelope integrations

The usual treatment of the third dredge-up in synthetic TP-AGB models is based in the use of two free parameters, namely: the minimum core mass $M_c^{\rm min}$ for convective dredge-up to occur; and the efficiency $\lambda = \Delta M_{\rm dred}/\Delta M_c$, i.e. the fraction of the core mass increment over an inter-pulse period which is dredged-up to the surface at the subsequent thermal pulse.

For the sake of simplicity, both parameters have been often assumed constant in synthetic calculations, regardless of the stellar metallicity and mass (e.g. Groenewegen & de Jong 1993; Marigo et al. 1996). On the contrary, full modelling of the third dredge-up shows that their properties vary according to many factors, i.e. treatment of convection, core and envelope mass, metallicity, pulse strength, numerical details (see, for instance, Mowlavi, in this volume).

In this context, the synthetic treatment of the third dredge-up is improved as briefly described. According to the indications from complete calculations

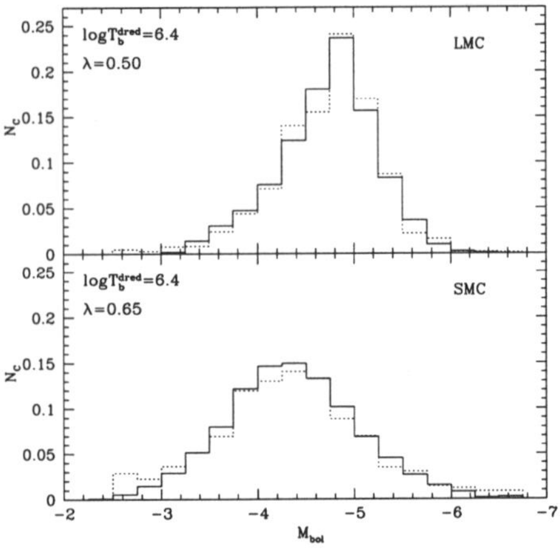

Figure 4. Carbon star luminosity functions in the Large and Small Magellanic Clouds (LMC and SMC). Observations (dotted line) are those by Costa & Frogel (1996) for the LMC, and Rebeirot (1993) for the SMC. Theoretical best fits (solid line) are obtained adopting the dredge-up parameters as indicated, and a constant star formation rate history. See Marigo et al. (1998b) for more details.

of thermal pulses (Wood 1981; Boothroyd & Sackmann 1988), the occurrence of the third dredge-up requires that the base of the convective envelope attains a minimum temperature, $T_{\rm b}^{\rm dred}$, at the stage of the post-flash luminosity maximum. Then, we relax the usual criterion based on a constant $M_{\rm c}^{\rm min}$, and replace it with one on the base temperature to infer if and when dredge-up takes place (see also Wood 1981). To this aim, envelope integrations are performed, in which the post-flash luminosity maximum is given by the core mass-peak luminosity relation by Wagenhuber & Groenewegen (1998). In this way, we are able to predict both the onset and the shut-down of the third dredge-up as a function of the stellar mass M, metallicity Z, and pulse strength.

The behaviour of $M_{\rm c}^{\rm min}(M, Z)$, as derived from envelope integrations, is displayed in Fig. 3 for different choices of the temperature parameter $T_{\rm b}^{\rm dred}$. All the curves share a common trend, i.e. $M_{\rm c}^{\rm min}$ steeply increases at lower masses, and flattens out to a nearly constant value towards higher masses. It is worth noticing that the onset of dredge-up is favoured at lower metallicities, in agreement with the results from complete calculations.

4. The carbon star luminosity functions in the Magellanic Clouds

The calibration of both dredge-up parameters is carried out in order to fulfil the constraint given by the observed luminosity functions of carbons stars in the Magellanic Clouds (see Marigo et al. 1998b). It turns out that $T_{\rm b}^{\rm dred}$ essentially controls the faint end of the distribution (together with the effect due to the long-lived luminosity dip induced by thermal pulses in lower mass TP-AGB stars). Once the proper value of $T_{\rm b}^{\rm dred}$ is singled out, the efficiency parameter is changed

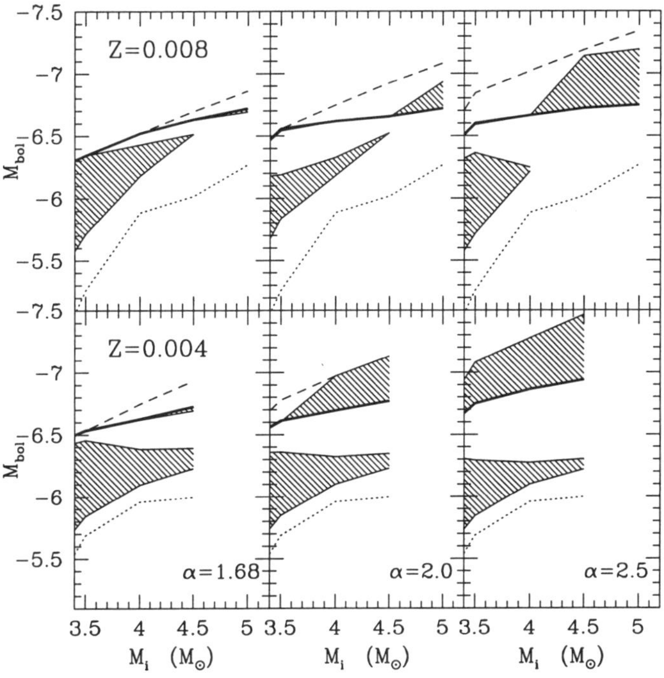

Figure 5. Transition luminosities as a function of the initial mass defining: the onset of the TP-AGB phase (dotted line), the carbon star domain (shaded areas), the AGB termination (thick solid line), the maximum luminosity attained during the TP-AGB evolution (dashed line). Each frame is characterised by the pair (Z, α). See the text for more details.

so as to reproduce the location of the peak of the distribution. Indeed, both features (i.e. the faint wing and the peak) are strong calibrators of the dredge-up parameters, whereas other properties of stellar populations (i.e. history of the star formation rate; initial mass function) are found to play a minor role.

It is interesting to notice that the best fits, displayed in Fig. 4, are obtained adopting the same value of the temperature parameter ($\log T_{\rm b}^{\rm dred} = 6.4$), and invoking a larger efficiency of dredge-up for the SMC ($\lambda = 0.65$) than for the LMC ($\lambda = 0.50$). Both findings are in a qualitative agreement with the expectations from full calculations of the third dredge-up, i.e. $T_{\rm b}^{\rm dred}$ is constant and dredge-up is favoured at lower metallicities.

5. The formation of massive and luminous carbon stars

The improved treatments of envelope burning and dredge-up allow to investigate the formation of high-luminosity carbon stars (Marigo, in preparation). More massive TP-AGB stars ($M > 3.5 M_\odot$) can make the transition to the carbon star domain as the third dredge-up dominates over envelope burning in the final part of their evolution. These stages are characterised by quite high values of the luminosity (due to the over-luminosity effect produced by envelope burning), and heavy mass-loss rates (after the onset of the super-wind) (cf. Frost et al. 1998; see also Lattanzio & Forestini, this volume).

Fig. 5 shows the predicted luminosity range spanned by carbon stars, for initial masses in the range $3.5 \div 5.0 M_\odot$, as a function of metallicity Z, and mixing-length parameter α. These models can experience one and/or two transition(s) to the the carbon star domain. The earliest one occurs as envelope burning is growing in efficiency, but still weaker than the third dredge-up. Indeed, this first transition can be even prevented by the occurrence of envelope burning. The latest transition usually takes place in the final stages when envelope burning is weakening and the third dredge-up becomes dominant again.

However, this standard picture may break down in TP-AGB models in which envelope burning is so efficient (e.g. for high α and low Z) and operating for a time long enough to activate the ON cycle. Under these circumstances, envelope burning can even favour the formation of carbon stars, increasing the surface C/O through significant destruction of oxygen (and not through dredge-up of carbon). We can notice that the predicted luminosity range of the brightest carbon stars is larger for higher M, α, and lower Z, and in most cases extends above the AGB termination luminosity as a consequence of the over-luminosity effect produced by envelope burning.

Acknowledgments. The author acknowledges the IAU for travel funding.

References

Baud B., Habing H.J., 1983, A&A 127, 73

Blöcker T., Schönberner D., 1991, A&A 244, L43 (BS91)

Blöcker T., 1995, A&A 297, 727 (B95)

Boothroyd A.I., Sackmann I.-J., 1988, ApJ 328, 671

Costa E., Frogel J.A., 1996, AJ 112, 2607

Frost C.A., Cannon R.C., Lattanzio J.C., et al., 1998, A&A 332, L17

Groenewegen M.A.T., de Jong T., 1993, A&A 267, 410

Marigo P., 1998a, PhD Thesis, University of Padova

Marigo P., 1998b, A&A, in press

Marigo P., Bressan A., Chiosi C., 1996, A&A 313, 545

Marigo P., Bressan A., Chiosi C., 1998a, A&A 331, 564

Marigo P., Girardi L., Bressan A., 1998b, A&A submitted

Rebeirot E., Azzopardi M., Westerlund B.E., 1993, A&AS 97, 603

Renzini A., Voli M., 1981, A&A 94, 175

Vassiliadis E., Wood P.R., 1993, ApJ 413, 641

Wagenhuber J., Groenewegen M.A.T., 1998, A&A 340, 183

Wood P.R., 1981, in Physical Processes in Red Giants, I. Iben Jr. and A. Renzini (eds.), D. Reidel Publishing Co., Dordrecht, p. 135

Asymptotic Giant Branch Stars
T. Le Bertre, A. Lèbre and C. Waelkens, eds.
I.A.U. Symposium 191, 1999

Presolar grains from meteorites: AGB star matter in the laboratory

Ernst Zinner and Sachiko Amari

McDonnell Center for the Space Sciences and Physics Department, Washington University, St. Louis, MO 63130, USA

Abstract. Primitive meteorites contain dust grains that predate the Solar System, formed in stellar atmospheres and thus represent samples of ancient stardust. Among the presolar grain types identified so far, corundum (Al_2O_3) and silicon carbide (SiC) are inferred to originate from AGB stars. Corundum grains carry the signatures of core H burning in their O isotopes and of shell H burning during the AGB phase in the form of extinct ^{26}Al. In presolar SiC, most of which originated from carbon stars, the C and N isotopes and ^{26}Al reflect core and shell H burning and shell He burning. In addition, many elements that carry the isotopic signature of neutron capture have also been measured. Most individual grains show excesses in ^{29}Si and ^{30}Si, but the contribution from neutron capture is only a minor effect and the major effect is due to galactic heterogeneity. Noble gases and the elements Ba, Nd, Sm, and Dy are measured in "bulk samples", collections of many grains. Their measured isotopic patterns are well reproduced by models of the s-process in AGB stars. Recently, the isotopic analysis of Sr, Zr and Mo in single SiC grains has been made possible by resonance ionization mass spectrometry. These measurements also point to low-mass AGB stars as the most likely sources. Specifically, large ^{96}Zr depletions in some grains indicate that the ^{22}Ne(α,n) source was not active in the grains' parent stars.

1. Introduction

Until recently it was believed that the only samples of extrasolar matter that could be studied in terrestrial laboratories were galactic cosmic rays, and even those are quite elusive. Information about distant stars was mostly obtained in the form of electromagnetic radiation. This situation changed fundamentally when it was discovered that primitive meteorites contain presolar dust grains (Lewis et al. 1987; Bernatowicz et al. 1987), specks of stardust that formed in expanding stellar atmospheres and supernova ejecta and survived the formation of the Solar System. These grains can be isolated from the meteorites and studied in detail in the laboratory, providing new and important information on nucleosynthesis and stellar evolution, mixing in AGB stars and supernova explosions, condensation in stellar envelopes, and the chemical evolution of the Galaxy (Anders & Zinner 1993; Ott 1993; Bernatowicz & Zinner 1997; Zinner 1998).

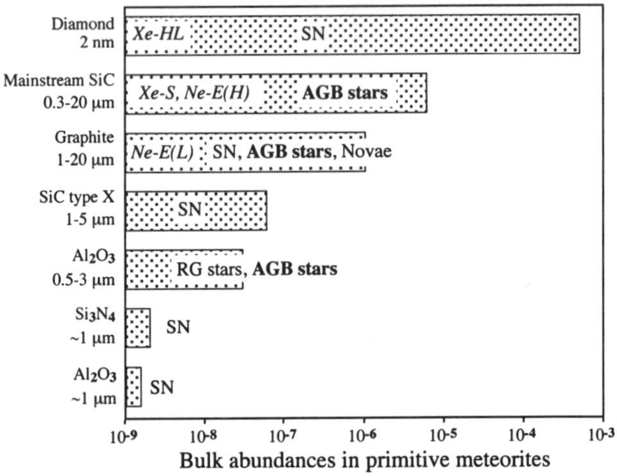

Figure 1. Abundances and sizes of presolar dust grains extracted from primitive meteorites. Inferred stellar sources and names of the anomalous noble gases carried by the grains are also indicated.

Fig. 1 lists the types of presolar grains discovered to date, together with their sizes, abundances, and likely stellar sources. Also given are the anomalous (or "exotic") noble gas components that are carried by some grain types. The presence of these noble gas components led to the discovery of presolar grains (see e.g., Anders & Zinner 1993, Tang et al. 1988). We note that all grain types are high temperature condensates. This and, even more so, the fact that the isotopic compositions of essentially all the elements that could be measured in the grains are completely different from those found in the Solar System leave little doubt that these grains represent true stellar material and preserved the isotopic compositions of their celestial sources. With the exception of diamond, presolar grains are large enough that isotopic ratios can be measured in individual grains, mostly by ion microprobe mass spectrometry (Hoppe & Ott 1997; Nittler 1997; Zinner 1998), recently also by micro-RIMS (resonance ionization mass spectrometry) (Nicolussi et al. 1997). Single grain measurements have revealed huge variations, which, for the elements C and O, range over almost four orders of magnitude. Presolar grains thus yield information on the isotopic compositions of individual stellar sources.

Although stellar sources have to be inferred from the isotopic compositions of the grains, this can usually be done fairly reliably, especially in cases where many different elements can be analyzed. Two types of grains are inferred to have originated from AGB stars: the majority of corundum (Al_2O_3) and silicon carbide (SiC) grains. A small fraction ($\sim$1%) of each of these presolar grain types apparently comes from supernovae (see Fig. 1), however we will not discuss those. An AGB origin of graphite grains rests mainly on noble gas analysis (Amari et al. 1995) and diagnosis is less certain for individual grains.

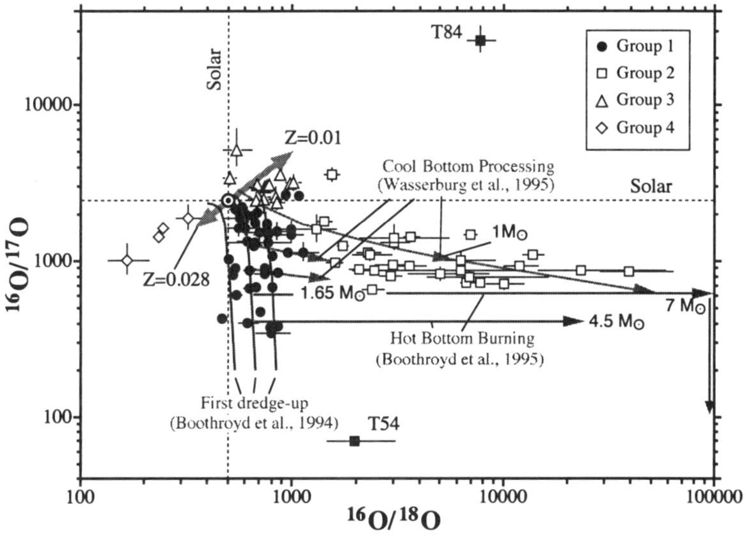

Figure 2. Oxygen isotopic ratios measured in single corundum grains. Four groups are distinguished by their isotopic ratios (Nittler et al. 1997), grains T54 and T84 are ungrouped. Also indicated are processes (CNO burning and first dredge-up, hot bottom burning, and cool bottom processing) that change the O isotopic ratios on the surface of RG and AGB stars. Arrow from Z=0.01 to Z=0.028 indicates the range of O ratios expected for this metallicity range. Broken lines in this and subsequent figures depict solar isotopic ratios.

2. Corundum grains

A little more than 100 presolar corundum grains have been identified to date and analyzed for their O, a fraction also for their Mg-Al and Ti isotopic compositions (Hutcheon et al. 1994; Nittler et al. 1997; Choi et al. 1998b, 1998a). Because presolar corundum in the size range indicated in Fig. 1 constitutes only ~1% of meteoritic corundum grains (most are of Solar System origin and are isotopically normal), isotopic imaging in the ion microprobe that allows the automatic analysis of many grains plays a crucial role in the identification of anomalous grains (Nittler et al. 1997).

Fig. 2 shows the O isotopic ratios measured in most presolar oxide grains. Nittler et al. (1997) divided the grains into four groups according to their O isotopic ratios. Grains of group 1 and 3 show the signature of H burning via the CNO cycle in low- and intermediate-mass stars and subsequent mixing of core material into the envelope by the first (in more massive stars also the second) dredge-up (Dearborn 1992; Boothroyd et al. 1994; El Eid 1994; Lattanzio & Boothroyd 1997). These processes result in ^{17}O enrichments and moderate ^{18}O depletions. Variations in the $^{16}O/^{17}O$ ratio mostly reflect differences in stellar mass whereas variations in the $^{16}O/^{18}O$ ratio can only be explained by differences in the original isotopic ratios of the stars, corresponding to different metallicities according to a galactic evolution model (Timmes et al. 1995). The three lines labeled "First dredge-up" in Fig. 2 depict the predicted ratios for stars of different masses and metallicities (Boothroyd et al. 1994, Boothroyd & Sackmann 1998). Giant stars with a range of masses and metallicities must have contributed oxide grains to the Solar System and the O isotopic ratios of group 1 and especially group 3 grains provide information on the galactic evolution of the O isotopes (Nittler 1997).

In addition to agreeing with theoretical expectations, the O isotopic compositions of group 1 and 3 grains are similar to those observed in red giant and AGB stars (Harris & Lambert 1984; Harris et al. 1987; Smith & Lambert 1990). AGB stars are the more likely sources since the large mass loss such stars undergo is expected to favor the formation of grains. Additional information that lets us distinguish between RG and AGB stars is provided by the presence of large ^{26}Mg excesses from the decay of ^{26}Al in many grains. More than half of group 1 and 3 grains have such excesses; inferred ^{26}Al/^{27}Al ratios range up to 0.01 (Nittler et al. 1997). Since ^{26}Al is produced at the higher temperatures characteristic of shell burning of H (Forestini et al. 1991), an AGB origin is indicated at least for these grains.

Group 2 grains have large ^{18}O depletions that cannot be explained as the result of first and second dredge-up because such an explanation would require unreasonably low metallicities. Hot bottom burning in AGB stars with $M>4M_\odot$ would effectively destroy ^{18}O, but such stars are predicted to have lower ^{16}O/^{17}O ratios than those observed in group 2 grains (Boothroyd et al. 1995). However, a model of slow circulation of envelope material through hot regions near the H-burning shell ("cool bottom processing") successfully reproduces the group 2 oxygen data (Wasserburg et al. 1995; Denissenkov & Weiss 1996).

Although an AGB origin either from stars with very high metallicities or from stars with ^{18}O dredge-up during early thermal pulses (Boothroyd & Sackmann 1988) has been considered for group 4 grains, a SN origin has also been proposed (Choi et al. 1998a). Grain T84 with its large ^{16}O excess is definitely a SN candidate (Nittler et al. 1998), and T54 possibly comes from a more massive AGB star with hot bottom burning (Nittler et al. 1997).

3. Light elements in presolar SiC

In contrast to corundum, all meteoritic SiC appears to be of presolar origin and most of it is believed to originate from carbon stars because: 1. The distribution of ^{12}C/^{13}C ratios of single SiC grains is similar to that observed in carbon stars (see Fig. 3 in Bernatowicz & Walker 1997); 2. AGB stars are the main contributors of carbonaceous dust to the interstellar medium (Whittet 1992); 3. Circumstellar dust shells of carbon stars show the 11.2 μm emission line of SiC (Little-Marenin 1986); and 4. SiC grains carry s-process isotopic signatures in the heavy elements (Hoppe & Ott 1997) and AGB stars are believed to be the major source of the s-process elements (Gallino et al. 1998).

Fig. 3 shows the C and N isotopic ratios measured in single SiC grains. Based on the C, N, and Si isotopes, several groups can be distinguished (Hoppe & Ott 1997, Zinner 1998). The mainstream grains account for ~95% of all grains. They and most Y and Z grains probably come from AGB stars (Zinner

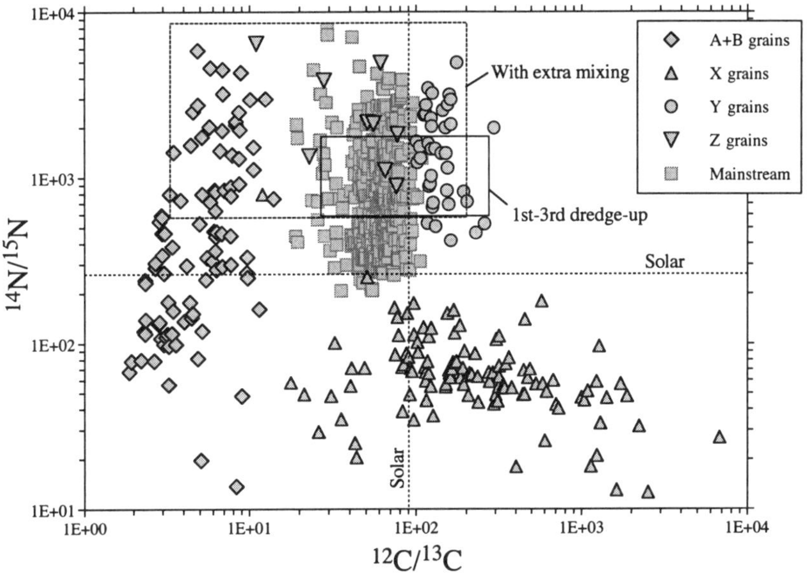

Figure 3. Carbon versus nitrogen isotopic ratios measured in single SiC grains from the Murchison carbonaceous meteorite. Five groups can be distinguished on the basis of the C, N, and Si isotopic ratios. Also indicated are the theoretically expected ranges resulting from H and He burning and subsequent dredge-up and from extra mixing (cool bottom processing) in low- to intermediate-mass carbon stars.

1998) and only SiC grains of type X (~1%), characterized by ^{15}N and ^{28}Si excesses, are inferred to have a SN origin (Nittler et al. 1995). Note that the distribution of the grains in Fig. 3 does not correspond to their natural abundances but that, because of special searches by ion imaging, rare grain types are overrepresented. The C and N isotopic ratios of most mainstream grains are generally in agreement with theoretical expectations for carbon stars (CNO cycle during core H burning followed by first and second dredge-up and shell He burning followed by the third dredge-up) (Iben & Renzini 1983; Bazan 1991; El Eid 1994; Frost & Lattanzio 1996). This is especially the case if extra mixing (cool bottom processing) is taken into account (Charbonnel 1995, Wasserburg et al. 1995). However, low $^{14}N/^{15}N$ ratios, especially in type A+B grains, which have low $^{12}C/^{13}C$ ratios, still do not have a satisfactory explanation (Hoppe et al. 1996; Amari et al. 1997; Huss et al. 1997). We will not discuss these any further.

Most SiC grains of inferred AGB origin have large ^{26}Mg excesses from the decay of ^{26}Al (Hoppe et al. 1994; Huss et al. 1997) and bulk samples have large ^{22}Ne excesses (Lewis et al. 1990, 1994). ^{26}Al is produced by shell H burning and is mixed into the envelope by the third dredge-up (Forestini et al. 1991) while ^{22}Ne is produced from ^{14}N by shell He burning. Ne isotopic ratios in SiC are very similar to those predicted for He-shell material (Gallino et al. 1990).

In contrast to C and N, the Si isotopes are not much affected by H burning but mostly by neutron capture in the He-burning shell (Gallino et al. 1990,

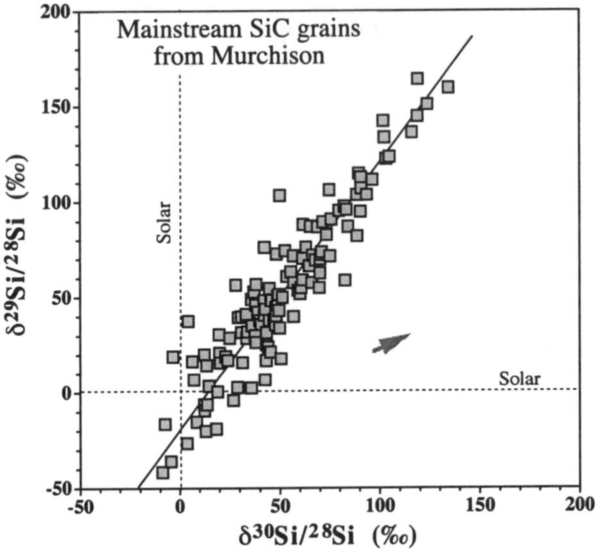

Figure 4. Silicon isotopic ratios measured in individual mainstream SiC grain (Hoppe et al. 1994). Ratios are plotted as δ-values, deviations in permil from the solar isotopic ratios. The solid line is the slope 1.34 correlation line for mainstream SiC grains. The short arrow is the theoretically predicted shift in isotopic composition of the envelope of carbon stars that results from dredge-up of material that experienced neutron capture in the He shell.

1994; Brown & Clayton 1992). Fig. 4 shows Si isotopic compositions measured in single SiC grains from the Murchison meteorite (Hoppe et al. 1994). The range of isotopic ratios is much larger than the shift expected from neutron capture, which is represented by the arrow, and thus cannot be explained by nucleosynthesis in a single AGB star. The most likely explanation is that many different AGB stars with a range in initial Si isotopic ratios contributed SiC to the Solar System (Clayton et al. 1991; Alexander 1993). This range in turn is attributed to galactic evolution and spatial heterogeneity of the Si isotopes (Timmes & Clayton 1996; Clayton 1997; Clayton & Timmes 1997; Alexander & Nittler 1998; Lugaro et al. 1999). Similarly to the Si isotopes, the Ti isotopes in SiC seem to be affected by both neutron capture in AGB stars and galactic chemical evolution, however only a limited amount of data exists (Hoppe et al. 1994; Alexander & Nittler 1998).

4. Heavy elements in presolar SiC

All the heavy elements analyzed so far show the isotopic signature of the s-process (Figs. 5 and 6). Elements measured in bulk samples include Kr and Xe (Lewis et al. 1994), as well as Sr (Podosek et al. 1998), Ba (Ott & Begemann 1990; Zinner et al. 1991, Prombo et al. 1993), Nd and Sm (Zinner et al. 1991; Richter et al. 1993), and Dy (Richter et al. 1994). Recently, the development of micro-RIMS (resonance ionization mass spectrometry) has made it possible to

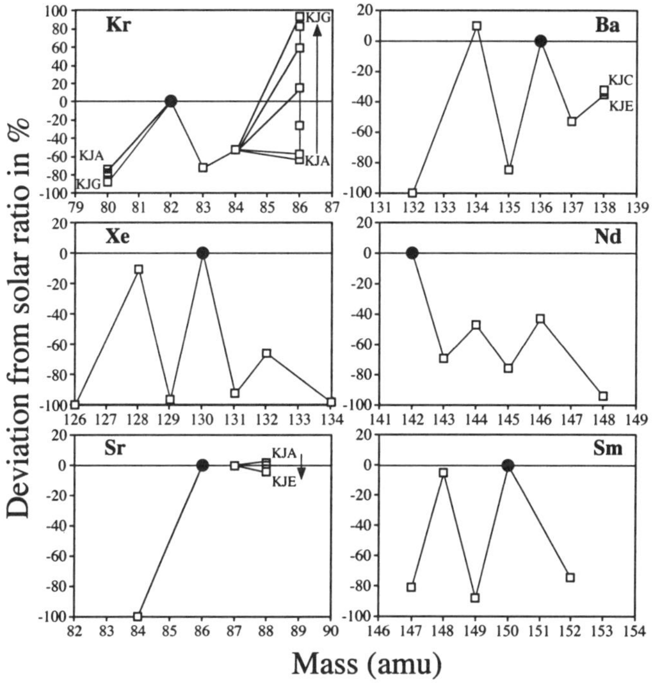

Figure 5. Isotopic patterns of heavy elements measured in bulk samples of SiC from the Murchison carbonaceous chondrite. Plotted are the s-process components extracted from the measurements (see Hoppe & Ott 1997). Isotopic ratios are relative to the reference pure s-process isotopes plotted as solid circles and are normalized to the solar ratios. Different data points for the heaviest isotopes of Kr, Sr, and Ba are for different grains size fractions of SiC; the sizes of their grains increase from KJA to KJG.

measure Sr (Nicolussi et al. 1998a), Zr (Nicolussi et al. 1997) and Mo (Nicolussi et al. 1998b) in single SiC and graphite grains.

The measured isotopic compositions of heavy elements can be interpreted as a mixture of an s-process component (the G component), reflecting the composition of the He-shell of AGB stars, and the N component, reflecting the initial composition of the envelope and possible terrestrial contamination (see, e.g., Hoppe & Ott 1997). The G components of Kr, Xe, Sr, Ba, Nd, and Sm are plotted in Fig. 5. This component can be determined if a pure p- or r-process isotope can be measured. In each case the isotopic ratios are normalized to a s-only isotope and the solar ratio. With the exception of Dy (there might be still experimental problems) the s-process compositions of the heavy elements are well reproduced by the "classical model" of the main component (Käppeler et al. 1989) and by the refinements of AGB stellar models (Gallino et al. 1997). Discrepancies between grain data and earlier theoretical predictions for Ba and Nd (Gallino et al. 1993) could largely be resolved by improved measurements of

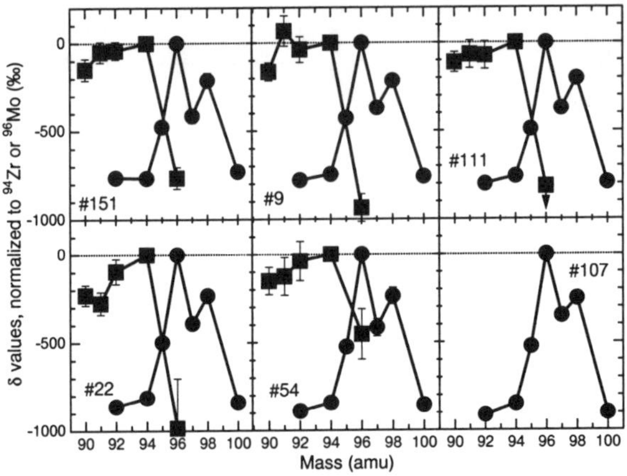

Figure 6. Zirconium (squares) and Mo (circles) isotopes measured by resonance ionization mass spectrometry in individual SiC grains from Murchison (Nicolussi et al. 1997, 1998b). Ratios are plotted as δ-values, deviations in permil from the solar isotopic ratios. All grains shown have s-process isotopic patterns consistent with an origin in AGB stars. Figure courtesy of Andy Davis.

the relevant neutron capture cross sections (e.g., Voss et al. 1994, 1995, Wisshak et al. 1997).

The SiC data allow the determination of different s-process parameters. ^{88}Sr/^{86}Sr and ^{138}Ba/^{136}Ba ratios vary with grain size (Fig. 5), indicating differences in mean neutron exposure τ_0. Inferred values for τ_0 range from 0.14 to 0.17 mb^{-1}, only about half the value derived for the Solar System (Gallino et al. 1993). The large variation in the ^{86}Kr/^{82}Kr ratio among different grain size separates is attributed to differences in neutron density, which affects the branching point ^{85}Kr (Lewis et al. 1994).

The s-process signatures of Zr and Mo measured in single SiC grains (Fig. 6) are dominated by large ^{96}Zr depletions and by large depletions of the p-only isotopes ^{92}Mo and ^{94}Mo and of the r-only isotope ^{100}Mo (Nicolussi et al. 1998b). Intermediate depletions are found for the s+r isotopes ^{95}Mo, ^{97}Mo, and ^{98}Mo. Since branching at ^{95}Zr ($t_{1/2} \approx 64$ days) affects the production and destruction of ^{96}Zr, its abundance sensitively depends on neutron density. Whereas at the low neutron densities expected from the ^{13}C(α,n) source (Gallino et al. 1998) ^{96}Zr is efficiently destroyed, activation of the ^{22}Ne(α,n) source during late thermal pulses is predicted to restore some of this isotope. The virtual absence of ^{96}Zr in several grains means that the ^{22}Ne(α,n) source was not active in their parent stars.

Acknowledgments. We thank Roberto Gallino for his continuing advice and Andy Davis for providing figure 6. This work was supported by NASA and NSF.

References

Alexander C.M.O'D., 1993, Geochim. Cosmochim. Acta 57, 2869

Alexander C.M.O'D., Nittler L.R., 1998, ApJ submitted

Amari S., Nittler L.R., Zinner E., Lewis R.S., 1997, Met. & Planet. Sci. 32, A6

Amari S., Lewis R.S., Anders E., 1995, Geochim. Cosmochim. Acta 53, 1411

Anders E., Zinner E., 1993, Meteoritics 28, 490

Bazan G., 1991, Ph.D. thesis, University of Illinois, Urbana-Champaign

Bernatowicz T., Fraundorf G., Ming T., Anders E., Wopenka B., Zinner E., Fraundorf P., 1987, Nature 330, 728

Bernatowicz T., Walker R.M., 1997, Physics Today 50 (12), 26

Bernatowicz T.J., Zinner E., 1997, Astrophysical Implications of the Laboratory Study of Presolar Materials, AIP, New York, p. 750

Boothroyd A.I., Sackmann I.-J., 1988, ApJ 328, 653

Boothroyd A.I., Sackmann I.-J., 1998, ApJ in press

Boothroyd A.I., Sackmann I.-J., Wasserburg G.J., 1994, ApJ 430, L77

Boothroyd A.I., Sackmann I.-J., Wasserburg G.J., 1995, ApJ 442, L21

Brown L.E., Clayton D.D., 1992, ApJ 392, L79

Charbonnel C., 1995, ApJ 453, L41

Choi B.-G., Huss G.R., Wasserburg G.J., 1998a, Met. & Planet. Sci. 33, A32

Choi B.-G., Huss G.R., Wasserburg G.J., 1998b, Lunar Planet. Sci. XXIX, Abstract #1898

Clayton D.D., 1997, ApJ 484, L67

Clayton D.D., Obradovic M., Guha S., Brown L.E., 1991, Lunar Planet. Sci. XXII, 221

Clayton D.D., Timmes F.X., 1997, ApJ 483, 220

Dearborn D.S.P., 1992, Phys. Rep. 210, 367

Denissenkov P.A., Weiss A., 1996, A&A 308, 773

El Eid M., 1994, A&A 285, 915

Forestini M., Paulus G., Arnould M., 1991, A&A 252, 597

Frost C.A., Lattanzio J.C., 1996, in Stellar Evolution: What Should Be Done?, A. Noel, D. Fraipont-Caro, M. Gabriel, N. Grevesse & P. Demarque (eds.), Université de Liège, Liège, p. 307

Gallino R., Arlandini C., Busso M., Lugaro M., Travaglio C., Straniero O., Chieffi A., Limongi M., 1998, ApJ 497, 388

Gallino R., Busso M., Lugaro M., 1997, in Astrophysical Implications of the Laboratory Study of Presolar Materials, T.J. Bernatowicz & E. Zinner (eds.), AIP, New York, p. 115

Gallino R., Busso M., Picchio G., Raiteri C.M., 1990, Nature 348, 298

Gallino R., Raiteri C.M., Busso M., 1993, ApJ 410, 400

Gallino R., Raiteri C.M., Busso M., Matteucci F., 1994, ApJ 430, 858

Harris M.J., Lambert D.L., 1984, ApJ 285, 674

Harris M.J., Lambert D.L., Hinkle K.H., Gustafsson B., Eriksson K., 1987, ApJ 316, 294

Hoppe P., Amari S., Zinner E., Ireland T., Lewis R.S., 1994, ApJ 430, 870

Hoppe P., Kocher T.A., Strebel R., Eberhardt P., Amari S., Lewis R.S., 1996, Lunar Planet. Sci. XXVII, 561

Hoppe P., Ott U., 1997, in Astrophysical Implications of the Laboratory Study of Presolar Materials, T.J. Bernatowicz & E. Zinner (eds.), AIP, New York, p. 27

Huss G.R., Hutcheon I.D., Wasserburg G.J., 1997, Geochim. Cosmochim. Acta 61, 5117

Hutcheon I.D., Huss G.R., Fahey A.J., Wasserburg G.J., 1994, ApJ 425, L97

Iben I., Jr., Renzini A., 1983, ARA&A 21, 271

Käppeler F., Beer H., Wisshak K., 1989, Rep. Prog. Phys. 52, 945

Lattanzio J.C., Boothroyd A.I., 1997, in Astrophysical Implications of the Laboratory Study of Presolar Materials, T.J. Bernatowicz & E. Zinner (eds.), AIP, New York, p. 85

Lewis R.S., Amari S., Anders E., 1990, Nature 348, 293

Lewis R.S., Amari S., Anders E., 1994, Geochim. Cosmochim. Acta 58, 471

Lewis R.S., Tang M., Wacker J.F., Anders E., Steel E., 1987, Nature 326, 160

Little-Marenin I.R., 1986, ApJ 307, L15

Lugaro M., Gallino R., Busso M., Amari S., Zinner E., 1999, in Nuclei in the Cosmos V, N. Prantzos (ed.), in press

Nicolussi G.K., Davis A.M., Pellin M.J., Lewis R.S., Clayton R.N., Amari S., 1997, Science 277, 1281

Nicolussi G.K., Pellin M.J., Lewis R.S., Davis A.M., Clayton R.N., Amari S., 1998a, Phys.Rev.Lett in press

Nicolussi G.K., Pellin M.J., Lewis R.S., Davis A.M., Clayton R.N., Amari S., 1998b, ApJ 504, 492

Nittler L.R., 1997, in Astrophysical Implications of the Laboratory Study of Presolar Materials, T.J. Bernatowicz & E. Zinner (eds.), AIP, New York, p. 59

Nittler L.R., Alexander C.M.O'D., Gao X., Walker R.M., Zinner E., 1997, ApJ 483, 475

Nittler L.R., Alexander C.M.O'D., Wang J., Gao X., 1998, Nature 393, 222

Nittler L.R., Hoppe P., Alexander C.M.O'D., Amari S., Eberhardt P., Gao X., Lewis R.S., Strebel R., Walker R.M., Zinner E., 1995, ApJ 453, L25

Ott U., 1993, Nature 364, 25

Ott U., Begemann F., 1990, ApJ 353, L57

Podosek F.A., Prombo C.A., Amari S., Lewis R.S., 1998, ApJ in press

Prombo C.A., Podosek F.A., Amari S., Lewis R.S., 1993, ApJ 410, 393

Richter S., Ott U., Begemann F., 1993, in Nuclei in the Cosmos, F. Käppeler & K. Wisshak (eds.), Institute of Physics, Bristol and Philadelphia, p. 127

Richter S., Ott U., Begemann F., 1994, in Proc. Europ. Workshop on Heavy Element Nucleosynthesis, E. Somorjai & Z. Fülöp (eds.), Hung. Acad. Sci., Debrecen, p. 44

Smith V.V., Lambert D.L., 1990, ApJS 72, 387

Tang M., Lewis R.S., Anders E., Grady M.M., Wright I.P., Pillinger C.T., 1988, Geochim. Cosmochim. Acta 52, 1221

Timmes F.X., Clayton D.D., 1996, ApJ 472, 723

Timmes F.X., Woosley S.E., Weaver T.A., 1995, ApJS 98, 617

Voss F., Wisshak K., Guber K., Käppeler F., Reffo G., 1994, Phys.Rev.C 50, 2582

Voss F., Wisshak K., Käppeler F., 1995, Phys.Rev.C 52, 1102

Wasserburg G.J., Boothroyd A.I., Sackmann I.-J., 1995, ApJ 447, L37

Wisshak K., Voss F., Käppeler F., Kazakov L., 1997, Nucl. Phys. A621, 270c

Zinner E., 1998, Ann. Rev. Earth & Planet. Sci. 26, 147

Zinner E., Amari S., Lewis R.S., 1991, ApJ 382, L47

Asymptotic Giant Branch Stars
T. Le Bertre, A. Lèbre and C. Waelkens, eds.
I.A.U. Symposium 191, 1999

The Analysis of AGB Star Photospheres

Verne V. Smith[1]
Dept. of Physics, University of Texas at El Paso, El Paso, TX 79968, USA

Abstract.
The presence of excess ^{12}C, ^{13}C, ^{14}N, Li, and the heavy s-process elements in the photospheres of AGB stars betray the operation of H- and ^{4}He-burning, hot-bottom burning, and slow neutron-capture nucleosynthesis. Careful abundance analyses of these various elements provides insight into both nucleosynthesis events and mixing processes which occur deep within the AGB star. Recent results for the s-process are discussed.

1. Introduction

Analyses of the cooler red giants which populate the asymptotic giant branch (AGB) are of interest for a couple of different reasons. From the standpoint of the evolution of stars and populations of stars, the AGB stars (and their close relatives the post-AGB stars reviewed in this volume by Van Winckel) are the link from the red-giant branch to the white dwarfs (the final resting place of all stars less massive than about $8M_{\odot}$). The AGB stars are also important within the context of chemical evolution, as nucleosynthesis during this phase of stellar evolution is responsible for a substantial fraction of ^{12}C in the Galaxy, as well as most of the heavy, s-process elements, and probably a sizable percentage of the light element ^{7}Li. Here we provide a brief review of recent results for the the s-process elements derived from the photospheric spectra of AGB stars and stars which have been polluted by mass transfer from AGB star photospheres .

2. The operation of the s-process on the AGB

As quasi-periodic shell ^{4}He-burning occurs during thermal pulses on the AGB, mixing episodes increase the atmospheric C/O ratio in these red giants, giving rise to the spectral sequence of M to MS to S to C. Carbon stars have been known for well over a century, while the intermediate C/O ratio S-stars were first isolated as a specific stellar type by Merrill (1922) and their spectral cousins the MS stars were discussed by Keenan (1954). This increase in C/O is accompanied, in general, by an increasing s-process overabundance, thus the MS, S, and C stars are heavy-element rich AGB stars.

---

[1]McDonald Observatory, University of Texas, Austin, TX 79912, USA

It has been known since the work of Schwarzschild & Harm (1967) and Sanders (1967) that the preferred astrophysical site for the production of many of the s-process nuclei was in a region associated with the H- and ^{4}He-burning shells found in asymptotic giant branch (AGB) stars. Much of the early work which connected the AGB stellar models to the details of the s-process nucleosynthesis was carried out by Iben and collaborators and is summarized nicely in the review by Iben & Renzini (1983).

Other warmer, heavy-element rich stellar classes are known; Bidelman & Keenan (1951) identified the barium (Ba) giants as being heavy-element (Ba and Sr) rich, and soon after it was realized that the CH stars discovered and classified by Keenan (1942) were metal-poor halo relatives of Ba stars. In more recent times, Bond (1974) found subgiant and near main-sequence relatives of the giant CH and Ba stars and dubbed them CH subgiants. Since Bond's discovery, many more true main-sequence s-process rich stars of spectral types F and G have been identified: Tomkin et al. (1989) or North, Bertet & Lanz (1994). Observationally defined s-process rich stars thus span the range in luminosity from main-sequence to extremely luminous AGB giants.

Since the s-process was identified with AGB evolution in the 1960's, the lower luminosity s-process rich Ba and CH giants and subgiants defied explanation for many years. McClure, Fletcher & Nemec (1980) broke the logjam of understanding these diverse groups when they discovered that all Ba stars were binaries with white-dwarf companions. The solution to the s-process nature of these stars thus centers on mass transfer; when the white dwarf now inhabiting these binary systems was a luminous AGB star and had mixed s-process rich material to its surface, it transferred some of this processed matter to its companion, which is now observed as an s-process rich star. Since the pioneering paper of McClure et al. (1980), follow-up studies (McClure 1984; McClure & Woodsworth 1990) have confirmed the mass-transfer scenario. In addition, it has been shown that the 40% of S-stars which do not show the lines of Tc I in their spectra are the cooler relatives of the Ba stars and result from mass transfer (Iben & Renzini 1983; Smith & Lambert 1988; Brown et al. 1990; Jorissen et al. 1993). Most recently, a number of the active symbiotic binaries have been identified as C-rich (Schmid & Nussbaumer 1993; Schmid 1994), while the yellow symbiotic stars AG Dra and BD-21° 3873 have both been shown to be s-process rich and, thus, technically Ba stars as well as symbiotics (Smith et al. 1996, 1997). There is now a coherent picture of the s-process rich classes of stars: the N-type carbon stars, along with the MS and S stars with Tc, are bonafide AGB stars which have recently dredged fresh s-process material to their surfaces. The other variety of s-process stars (the Ba and CH giants, subgiants, and main-sequence stars, the carbon star dwarfs, and now the C- and s-process rich symbiotic stars) result from mass-transfer in certain binary-star systems.

With so many varieties of stars which exhibit clear signatures of enhancements of the s-process in their abundance patterns, it is not surprising that a great deal of effort has been expended by observational spectroscopists in analyzing the details of s-process nucleosynthesis in the heavy-element rich stars. A number of detailed abundance analyses have been conducted on the various heavy-element rich stars over the last 15 years and can be summarized as follows: for the barium giants we have Tomkin & Lambert (1983, 1986), Smith (1984),

Kovacs (1985), Smith & Lambert (1984), Malaney & Lambert (1988), and Busso et al. (1995), while the main-sequence Ba stars have been studied by Tomkin et al. (1989) and North et al. (1994). The CH giants have been studied by Vanture (1992a, 1992b). The MS and S stars have been studied by Smith & Lambert (1990), Plez, Smith & Lambert (1993), Busso et al. (1995), and Lambert et al. (1995), while the most detailed analysis of the s-process in the carbon stars is still Utsumi (1985). Recent abundance studies have uncovered s-process rich post-AGB stars in the globular cluster ω Cen (Gonzalez & Wallerstein 1994), as well as s-process enhancements in the yellow symbiotics (Smith et al. 1996, 1997). Taken together, these abundance studies provide the data with which to probe the details of s-process nucleosynthesis in AGB stars.

One question that is basic to an understanding of the s-process in AGB stars is the nature of the neutron source. Since the realization that the slow capture of neutrons is a distinct stellar nuclear process, two candidate neutron producing reactions have been suggested: $^{13}C(\alpha,n)^{16}O$ and $^{22}Ne(\alpha,n)^{25}Mg$ (Cameron 1955; Greenstein 1954). These two separate sources have very different characters; the s-process in AGB stars is associated with ^{4}He-burning temperatures of T$\geq 10^8$ K, with the ^{13}C source being active near 1 x 10^8 K, while the ^{22}Ne source requires higher temperatures nearer to 2-3 x 10^8 K. The different temperatures required to drive the s-process via the different neutron sources will occur in different mass AGB stars, or at different points along the AGB. Historically, the ^{22}Ne source was favored as the most likely s-process candidate because the production of neon-22 can occur quite naturally as the result of successive α-captures onto ^{14}N in the He-burning shell of an AGB star ($^{14}N(\alpha,\gamma)^{18}F(e^+,\nu)^{18}O(\alpha,\gamma)^{22}Ne$), with the initial ^{14}N itself coming from CNO-cycle H-burning. The carbon-13 neutron source required an extra mixing mechanism to move protons into the He-burning shell to produce ^{13}C via $^{12}C(p,\gamma)^{13}N(\beta^+,\nu)^{13}C$, and it was not clear as to the physical nature of this extra mixing. However, the expected behavior of these neutron sources with metallicity is different and can be tested by the observations.

A convenient parameter to characterize the relative abundance distribution of the s-process is the neutron exposure, τ, defined as

$$\tau = \int_0^{t'} N_n(t)\ V(t)\ dt$$

where N_n is the neutron density and V(t) the relative velocity of neutrons and nuclei, with the integral taken over the interval of the s-process episode. A large neutron exposure leads to a larger amount of heavier nuclei produced (e.g. Ba, La, or Ce) relative to the lighter nuclei, such as Y or Zr.

As pointed out by Clayton (1988), the nature of the dominant s-process neutron source can be tested by using the heavy- to light-element abundance distribution of the s-process (a measure of τ) as a function of metallicity. For example if, during the s-process episode, the neutrons are in local equilibrium between their production and destruction, then the neutron density will be

$$N_n = (\Sigma\ N_i N_j \langle\sigma\rangle_{i,j})\ /\ (\Sigma\ N_k \langle\sigma\rangle_{k,n})$$

where the numerator is summed over all neutron-producing reactions between species i, j, with Maxwellian-averaged cross sections of $\langle\sigma\rangle_{i,j}$, and the denominator has the neutron destruction (absorbing) reactions, summed over all neutron absorbers, k, and again a Maxwellian-averaged neutron-absorption cross section of $\langle\sigma\rangle_{k,n}$. Making the simplifying assumption that the neutron-producing reac-

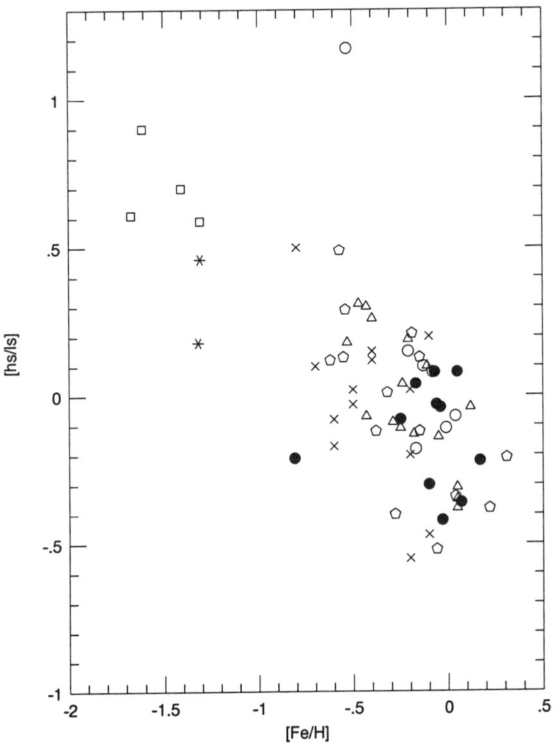

Figure 1. Published abundance results for the heavy s-process elements (Ba, La and Ce) relative to the light s-process species (Y and Zr) for the MS/S stars (filled circles, with Tc, and open circles, without Tc), Ba giants (open triangles), Ba dwarfs (open pentagons), CH giants (open squares), CH subgiants (crosses), and yellow symbiotics (six-pointed stars). The increase of heavy to light s-process (and hence the neutron exposure) with decreasing metallicity is clear.

tions come from α-captures, and that the dominant neutron absorbers are the seed nuclei of Fe, the above expression can be approximated as

$$N_n \propto (N_{He}N_j) \;/\; (N_{Fe}),$$

where $N_j = N(^{13}C)$ for the carbon-13 source and $N(^{22}Ne)$ for the neon-22 source. Since the ^{22}Ne in the He-burning layers comes from ^{14}N which, itself, comes from the initial ΣCNO which is $\propto N_{Fe}$, the expected neutron density (and neutron exposure) would be expected to be roughly independent of metallicity. Carbon-13, on the other hand, must result from the mixing of protons into a ^{12}C-rich region, where the carbon-12 comes from ^{4}He, which is almost independent of metallicity. Thus, if the structures of He-shells in AGB stars are not strong functions of metallicity, the numerator of the above expression is constant and N_n (and τ) is $\propto {N_{Fe}}^{-1}$.

The neutron exposure can be estimated using published abundance results and we show the results of such an exercise in Fig. 1, where we plot the quantity [hs/ls] versus [Fe/H] from a number of studies. The value of [hs/ls] stands for heavy-s to light-s and is an average overabundance of representative heavy s-process nuclei (Ba, La and Ce) relative to an average overabundance of the light s-process species Y and Zr, measured in standard spectroscopic bracket notation. There is a rather large scatter, although this is not surprising as we are comparing results from a large number of investigators using somewhat different techniques and different quality data. Nonetheless, there is a clear and significant trend of increasing [hs/ls] with decreasing metallicity.

The increasing value of [hs/ls] indicates a significant increase of neutron exposure with decreasing metallicity as predicted from Clayton's (1988) simple argument for the $^{13}C(\alpha,n)$ neutron source. Based on the observations, it would appear that the current best-candidate for the s-process neutron source in the AGB stars is carbon-13.

Acknowledgments. This research is sponsored in part by the National Science Foundation (AST96-18459).

References

Bidelman W.P., Keenan P.C., 1951, ApJ 114, 473

Bond H.E., 1974, ApJ 194, 95

Brown J.A., Smith V.V., Lambert D.L., Dutchover E., Jr., Hinkle K.H., Johnson H.R., 1990, AJ 99, 1930

Busso M., Lambert D.L., Beglio L., Gallino R., Raiteri C.M., Smith V.V., 1995, ApJ 446, 775

Cameron A.G.W., 1955, ApJ 121, 144

Clayton D.D., 1988, MNRAS 234, 1

Gonzalez G., Wallerstein G., 1994, AJ 108, 1325

Greenstein J.L., 1954, in *Modern Physics for Engineers*, edited by J.L. Ridenour, McGraw-Hill, New York

Iben I., Jr., Renzini A., 1983, ARA&A21, 271

Jorissen A., Frayer D.T., Johnson H.R., Mayor M., Smith V.V., 1993, A&A271, 463

Keenan P.C., 1942, ApJ 96, 101

Keenan P.C., 1954, ApJ 120, 484

Kovacs N., 1985, A&A 150, 232

Lambert D.L., Smith V.V., Busso M., Gallino R., Straniero O., 1995, ApJ 450, 302

Malaney R., Lambert D.L., 1988, MNRAS 235, 695

McClure R.D., 1984, ApJ 280, L31

McClure R.D., Fletcher J.M., Nemec J.M., 1980, ApJ 238, L35

McClure R.D., Woodsworth A.W., 1990, ApJ 352, 709

Merrill P.W., 1922, ApJ 56, 457

North P., Bertet S., Lanz T., 1994, A&A 281, 775

Plez B., Smith V.V., Lambert D.L., 1993, ApJ 418, 812

Sanders R.H., 1967, ApJ 150, 971

Schwarzschild M., Harm R., 1967, ApJ 150, 961

Schmid H.M., 1994, A&A 284, 156

Schmid H.M., Nussbaumer H., 1993, A&A 268, 159

Smith V.V., 1984, A&A 132, 326

Smith V.V., Lambert D.L., 1984, PASP 96, 226

Smith V.V., Lambert D.L., 1988, ApJ 333, 219

Smith V.V., Lambert D.L., 1990, ApJS 72, 387

Smith V.V., Cunha K., Jorissen A., Boffin H., 1996, A&A 315, 179

Smith V.V., Cunha K., Jorissen A., Boffin H., 1997, A&A 324 97

Tomkin J., Lambert D.L., 1983, ApJ 273, 722

Tomkin J., Lambert D.L., 1986, ApJ 311, 819

Tomkin J., Lambert D.L., Edvardsson B., Gustafsson B., Nissen P.E., 1989, A&A 219, 15

Utsumi K., 1985, in *Cool Stars with Excesses of Heavy Elements*, edited by M. Jashek and P.C. Keenan, Reidel, Dordrecht, p. 243

Vanture A.D., 1992a, AJ 103, 2035

Vanture A.D., 1992b, AJ 104, 1997

Asymptotic Giant Branch Stars
T. Le Bertre, A. Lèbre and C. Waelkens, eds.
I.A.U. Symposium 191, 1999

The modelling of M-giant spectra

Bertrand Plez[1]

Astronomiska observatoriet, box 515, S-75120 Uppsala and
Atomspektroskopi, Fysiska inst., box 118, S-221 00 Lund, Sweden

Abstract. I review the current status of cool O-rich giant star spectral modelling, stressing some specific problems like line profiles and molecular data completeness. I discuss recent progress in molecular line lists (TiO, VO) and describe recent attempts in modelling normal M-giants as well as AGB star spectra, using static and hydrodynamic models. I anticipate further advances in the near future, with better radiative transfer in hydrodynamic models, that will allow detailed comparisons of models and observations, opening the possibility of deriving stellar parameters from spectroscopic and interferometric observations.

1. Introduction

The modelling of cool giant spectra has greatly developed and improved in recent years, as this volume witnesses, as a consequence, in particular, of the new window opened in the IR by ISO, and of the recent great increase in molecular line databases content and accuracy.

Some specific modelling attempts for late-type giant spectra, esp. Miras, are discussed in this volume, or in the posters published electronically at the ADS, by Aringer et al. (1999, ISO-SWS of LPVs), Decin et al. (1999, ISO-SWS of cool giants), Ryde et al. (1999, ISO-SWS of AGB stars), Schultheis et al. (1999, TiO bands of Miras and semi-regular variables), for O-rich objects, and by Aoki et al. (1999, ISO-SWS of carbon stars), Hron et al. (1999, ISO-SWS of carbon variables), Winters et al. (1999, IR light curves, center-to-limb variations and CO line profiles in carbon stars), for C-rich objects.

In this review I would like to focus on other specific recent results and emphasize some particular problems and difficulties encountered when modelling M-giant spectra. I will also show that models are now capable of reproducing observations to a level of accuracy that could not be achieved 1 or 2 years ago. Progress has been rapid indeed.

Synthetic spectra are absolutely necessary for the interpretation of observations and the extraction of fundamental data such as chemical abundances, effective temperatures, radii, etc. They also permit the conversion of theoretical models to observables, and thereby open the possibility of first testing the models, before using them to derive fundamental stellar quantities. One very fine example is provided by Winters et al. (1999), with their modelling of IR light

[1]present address: GRAAL, Université Montpellier II, 34095 Montpellier cedex 5, France

curves and center-to-limb variations for the stars AFGL 3068 and IRC +10216. I do not think that the modelling has reached such a quality and self-consistency that we can unambiguously extract the luminosity, mass, $\log g$, etc. of a particular star from comparisons of theoretical and observed light curves (the coupling between the pulsation driving zone and the atmosphere must first be achieved), but we can foresee a not so remote future where this might be possible.

The computation of spectra demands a knowledge of the thermal structure (T and P vs. τ or R) usually provided by a model, but which in extremely simplified modelling could be that of a single layer. Then, assuming or not chemical equilibrium and LTE, partial pressures of the various species can be computed, which combined with continuous and line opacities allow the calculation of a synthetic spectrum. The quality of the result largely reflects the adequacy to the real situation of the model, of the opacities and of the hypotheses made (LTE, geometry, staticity/homogeneity, ...). Currently, most modelling assumes LTE, chemical equilibrium, plane-parallel or spherical symmetry, and in a few cases velocity fields (see Windsteig et al. 1999; Winters et al. 1999).

2. Intensity spectra

It is convenient to separate the discussion into intensity and flux spectra. Intensity spectra provide us with the variation of the spectrum with the position on the stellar surface. They are the tool for understanding high-angular resolution observations, and to relate, for example, intensity distributions at various wavelengths to a "radius" at some reference Rosseland depth or pressure. It must be emphasized that interferometric observations do not provide "the" radius of the star, esp. not in the case of LPVs. Baschek et al. (1991) and Scholz (1997) have discussed the radius concept in detail, and I recommend reading these papers. The problem may be illustrated as follows: assume we wish to know the radius where $\tau_{Ross} = 1$, as it is the reference radius we use in our atmospheric or pulsation model (through, for example: $L = 4\pi R^2 \sigma T_{\text{eff}}^4$ and/or $g = GM/R^2$). Interferometric measurements will provide us with the center-to-limb variation of the intensity in one or more passbands centered on various wavelengths. These usually differ by large amounts, both in angular extension and in shape, in cool giants, and esp. in LPVs. It is thus impossible to derive $R(\tau_{Ross} = 1)$ purely empirically. A model must be used, that must reproduce the intensity profiles observed in all passbands, with a single set of $(L, T_{\text{eff}}, M, ...)$ parameters. This is all too rarely done.

One of the reasons is probably that few models exist and have been discussed in detail, although in principle the calculation of intensity profiles can be done for any published model atmosphere grid, if appropriate opacities are available. The only systematic computation has been published by Hofmann & Scholz (1998) and by Hofmann et al. (1998), in their follow-up studies of the work of Bessell et al. (1996), for grids of static and hydrodynamic models of M giants and Miras. Their main conclusions are that: (i) it is necessary to use clean, well defined narrow passbands, that must avoid as much as possible the mixing of continuum and strong absorption bands. (ii) in Miras, the center-to-limb variations are very sensitive to details of the model; this makes the derivation of $R(\tau_{Ross} = 1)$ very difficult, but on the other hand offers the possibility to efficiently test the models and put strong constraints on them. This, however, requires high-accuracy interferometric measurements (typically at the 1% level or better on visibilities, and past the first zero). I wish to warn here again that uniform-disk or even fully-darkened radii derived from visibilities are often a poor approximation of the real intensity profile of Mira stars, and that it is not trivial to derive, e.g., a "pulsation" radius from these, that may be used in P-R-L relations. There is most of the time little similarity between the "photometric" radius derived from observations and the "pulsation" radius used in theoretical modelling. The atmospheric model is crucial in establishing the link betwen them, provided it reflects reality. For normal M-giant center-to-limb variations, Hofmann et al. (1998a, b) conclude that in the IR fully-darkened diameters are a reasonably good approximation, and might be used, with caution, in the absence of better information on the brightness profile.

Important progress should arise in this field from detailed spectrum calculations with better up-to-date opacities. Observers and modellers must work together, carefully defining bandpasses and using good models and opacities. This will allow meaningful information to be extracted from high-angular resolution observations. See Lopez (1999), Monnier (1999), Irrgang et al. (1999), Karovska (1999), Käufl et al. (1999), Perrin et al. (1999), Tuthill et al. (1999), van Belle et al. (1999), and Weigelt et al. (1999) in these proceedings for further description of interferometric techniques and recent results.

3. Flux spectra

Flux spectra may be more or less arbitrarily divided into (very) low-spectral resolution ($R <\sim 1\,000$) and (very) high-resolution ($R >\sim 20\,000$). The latter is used for abundance analyses and for the study of circumstellar shells (velocity structure). Low resolution spectra allow the exploration of the spectrum for, e.g., the identification of molecular bands (see the various ISO SWS and LWS studies in this volume: e.g. Aoki et al. 1999, Hron et al. 1999, Cohen et al. 1999). Synthetic low-resolution spectra can be used for the derivation of broad-band photometry and for calibrating $T_{\rm eff}$-colour relations. Population synthesis, narrow-band colours and spectrophotometric work require intermediate resolutions. Some abundance analyses (like $^{12}C/^{13}C$ using the CO ro-vibration bands in the IR, see Plez et al. 1993) may be carried out at relatively low spectral resolution (a few 1 000:s).

Estimates of the spectral energy distribution may be quickly calculated using mean or sampled opacities. This may lead to erroneous results, and whenever possible should be avoided. At least one control calculation should be done with a detailed line list at very high resolution (with a sampling much denser than the typical line width, i.e. $R > 100\,000$). Large errors are especially likely if the line density is low. A coarse sampling may well systematically miss a large number of lines, leading to an underestimate of the opacity and thus an overestimate of the flux. The converse is also possible if mostly line centers are sampled. The problem is not very severe at high line densities, like for TiO in the optical, where all kinds of lines, –strong and weak–, will be sampled by even a relatively small number of wavelengths.

4. Potential problems

When modelling the spectra of cool stars one is confronted with a number of questions and potential problems. The first is of course the adequacy of the model atmosphere being used: static or hydrodynamic, plane-parallel or spherical, with a variety of approximations (LTE, mixing-length theory for convection, etc.). I will not discuss this aspect which I developed elsewhere (Plez 1999). Other important issues are:

- the (in)completeness of line lists, missing oscillator strengths, uncertain line positions.
- the importance of pulsation, shocks and atmospheric extension.
- velocity fields, and line profiles.
- dust: present? under which form? opacities...
- circumstellar contributions.
- non homogeneity (spots, non-sphericity).
- non-LTE.
- non-equilibrium chemistry.
- ...?

Problems get more severe for Miras and tend to increase with decreasing T_{eff}. Let us very quickly mention a few of these problems before addressing the others in more detail.

Dust formation in O-rich stars is not well understood and most work on dust has been done in C-rich environments (cf. the reviews by Fleischer et al. 1999, Cherchneff & Cau 1999, and Lodders & Fegley 1999, and the works on O-rich dust formation of Jeong et al. 1999, and of Kosaza & Sogawa 1999, in this volume). On the other hand, molecular opacities appear better known in O-rich stars. Alexander et al. (1999) attempt to model the impact on the emergent spectrum of the depletion of some atomic and molecular species due to dust formation.

Non-LTE has been very little explored, and only for specific lines (a few emission lines by Luttermoser & Bowen 1992, or Li I by Kiselman & Plez 1995). The impact of non-equilibrium chemistry on spectra remains unstudied.

I will not extensively discuss or provide detailed answers to all the items listed above, but rather draw attention to a few of them, and to recent progress and to newly obtained results. Discussion of some of the others will be found elsewhere in these proceedings (for example, Winters et al. 1999 show the effect of pulsation on the IR spectrum).

4.1. Line profiles

Let me first illustrate the problems with line profiles and with line list completeness through a few examples. In our spectroscopic study of galactic S-stars

(Smith et al. in prep.) we found that the Li I 6707 Å line profile is of complex structure and cannot be modelled with a simple hydrostatic model. This is obvious in R=50 000 spectra of S Cas and other Miras. Moreover, the K I 7699 Å resonance line displays a similar doubly-peaked profile. Other lines are likely to be alike. There could be a circumstellar contribution (both in absorption and emission, as observed around some stars by e.g. Mauron & Caux 1992, Mauron 1997 and Plez & Lambert 1994). Alternatively the profiles could be affected by velocity gradients in the extended (shocked) atmosphere. These hypotheses can be checked through time-monitoring of the profiles and through a comparison with higher excitation lines. Bessell et al. (1996) computed profiles for a few test lines in their hydrodynamic models for Mira. Their figures 13 through 15 nicely illustrate the possible importance of velocity fields in Mira atmospheres on their emergent spectra. Inspection of these figures readily shows that, as expected, even the equivalent width of the lines is affected compared to the hydrostatic conditions. Abundance analyses will be in error if not caring for this effect. More higher resolution spectra must be accumulated for a sample of LPV's and non-variable cool giants, and compared to static and hydrodynamic model spectra.

4.2. Line lists completeness

Another potential source of problem is the continuum placement. Early spectral abundance analyses of M giants (RGB or AGB) in the optical used the local continuum and synthetic spectra computed with line lists for TiO compiled from laboratory observations. Still worse, some analyses were based on equivalent widths. This resulted in inconsistent and systematically low abundances. Luck & Lambert (1982) pointed out that this was probably due to unaccounted "pseudo-continuous" TiO absorption, which they modelled as a continuous opacity proportional to the TiO partial pressure. This was not very satisfactory but lead to much more coherent results. The use of a TiO line list computed up to much higher quantum numbers and with many more faint transitions included allowed us to account for this extra opacity and derive consistent abundances (Plez et al. 1993). In the case of a model typical for the MS stars at the tip of the Small Magellanic Cloud AGB ($T_{eff} = 3350$ K) the local "continuum" around the Li I 6707 Å doublet is at 30% of the true continuum which is not reached anywhere near in the spectrum. Two thirds of that depression are due to TiO lines which are unobserved in the laboratory... The key here is to use as complete a molecular line list as possible, and use lines of the same element from very different spectral regions thus differently affected by molecular blanketing. It is thus crucial to use the subordinate 8126 Å line (and 6103 Å when possible) in conjunction with the 6707 Å doublet when deriving Li abundances in cool giants.

To conclude, although abundances have been, and are derived, that seem consistent and mostly follow our expectations in terms of nucleosynthesis and late stellar evolution, there is a need to check line profiles at higher spectral resolution (2 to 3 km/s) in Miras or irregular variables against non-variable stars for various types of spectral lines. Some test analyses could be conducted with hydrodynamic model atmospheres, including velocity fields, and compared to similar analyses done with hydrostatic models.

5. Recent results and progress

Despite, or maybe because of, these problems and others not discussed here, tremendous progress was achieved in the past few years. The development of atmospheric modelling is reviewed by Plez (1999) and by Höfner (1999) in this volume.

5.1. Molecular line lists

Some persistent questions concerning opacity sources have been solved very recently. To cite a few examples:

- VO: Karlsson et al. (1997) measured the lifetime of the three excited A, B and C states, providing the opportunity to calculate for the first time laboratory gf-values to replace the (not so good) astrophysical calibrations.

- TiO: The ab initio calculations of Langhoff (1997) are in good agreement with the laboratory determinations of Hedgecock et al. (1995). New line lists were recently published (Plez 1998; Schwenke 1998). Plez carried out detailed comparisons of his line list with laboratory data and M-giant spectra. The match with observations is good, with the exception, however, of the δ band for which the strength from the Langhoff calculations is too large in M-giants. I recommend laboratory determination of the lifetimes and the branching ratios of the singlet states, which unfortunately seem not easy to obtain accurately.

- H_2O: a new ab initio line list was recently proposed by Partridge & Schwenke (1997), which seems of good quality, but which has not yet been extensively tested in M-giant (and dwarf) model atmospheres.

However, for the modelling of very cool objects, and for S-stars, some metal-oxide opacities are still missing. E.g. for LaO, there exists no lifetime measurement of the excited A levels, and gf-values for the A-X transition (bands around 7400 and 7900 Å) can only be guessed.

5.2. Some illustrations

The leap forward in computing power and storage capabilities of the last 3-4 years means that very extensive line lists can be used in the modelling of stellar atmospheres and spectra. Large portions of spectra, like the ISO SWS range, or the whole optical spectrum, may now routinely be computed for a grid of models, at high-resolution, and including tens of millions of spectral lines, without the recourse to a CRAY. Beautiful results have emerged recently as a consequence of this (r)evolution.

Bessell et al. (1998) have shown that good UBVRIJHK Cousins-Johnson photometry can now be derived for K-M stars from modern model grids (the Uppsala recent grids of MARCS models) incorporating up-to-date opacities. Plez (1998) demonstrated that the optical spectrum of M0 to M5 III stars is well modelled with his TiO line list (see his figure 2). Another nice piece of work is that of Alvarez & Plez (1998). Their pioneering attempt to model Lockwood (1972) narrow-band colour observations of O-rich Miras opens the prospect of detailed testing of hydrodynamic models against observations. Lockwood (1972) observed a total of 256 Miras at various phases of their cycle through 5 photometric bands of about 50Å width between 0.75 and 1.1 μm, sampling some TiO and VO absorption bands. He noted that Miras describe loops in colour-colour diagrams during their cycle. This can be qualitatively understood by the running of a shock (or some perturbation) through the extended atmosphere. As different TiO and VO absorption bands form at different depths they are affected at different times, and phase lags appear between the "TiO" and "VO" colours (the average phase lag is about 0.1 here). Alvarez & Plez demonstrated that a sequence of current static MARCS model atmospheres and the new updated TiO and VO line lists reproduce the colours of ordinary M-giants observed by Lockwood. Thereafter they computed LTE radiative transfer in a time-series of hydrodynamical models for a typical LPV (from Höfner, priv. comm.) using the same line lists. Despite the limitations (LTE, one single model compared to an observed star defined by the average of 256 observed stars, no velocity fields taken into account in the radiative transfer ,...), the model colour variations are in surprisingly good agreement with the observations. Both the amplitude of variation and the phase-lag are reproduced. The absolute values of some of the model colours are however too large. We are still far from deriving stellar parameters from such modelling, but we are certainly getting closer to it. Such detailed modelling and comparison to high-quality observations (low- and higher-resolution spectra, narrow band colours) will allow extensive testing of the models. Also the modelling should include velocity fields, which will affect the spectrum, as shown by Bessell et al. (1996) and Windsteig et al. (1999).

These progresses open new perspectives in the modelling of M-giant spectra, and of cool stars in general, including LPVs. Note, however, that due to larger uncertainties in the existing opacities of carbon-bearing polyatomic molecules, as well as the yet unknown absorption sources of potential importance (e.g. C_2H), the modelling of carbon star photospheric spectra lags slightly behind. Although the work of Hron and his collaborators (1998, 1999) may well demonstrate that this is quickly changing.

6. Conclusions

We are now able to reproduce the spectra of normal M-giants (although whether the center-to-limb variations can be properly modelled remains to be seen through detailed comparisons with interferometric data). Attempts to model LPV spectra with currently affordable methods (i.e. a posteriori LTE radiative transfer in a temperature-density atmospheric structure computed with a hydro-code) have been successful beyond expectations. Some discrepancies remain, and it is certainly fundamental for our understanding to find out where they stem from, but the road is open to a quantitative use of the models to accurately model stellar populations or convert theoretical evolutionary sequences (T_{eff}-L) into observed H-R diagrams (colour-magnitude). And, maybe of more interest, we are at the beginning of this exciting new era where we finally are able to test theoretical hydrodynamic models of LPVs against observations of light curves, colours, spectra and center-to-limb variations. This will bring us to an in-depth understanding of (pulsating) AGB stars through, of course, the accurate determination of T_{eff}, R, abundances, M, etc., but also, of the physical mechanisms at work in these fascinating objects, their interactions (e.g. between the pulsation driving zone and the atmospheric filter) and their consequences (e.g.: why is the phase shift between TiO and VO colours about 0.1? This must reflect some general scaling property between the atmosphere's size, the velocity of the shock waves, the pulsation period, etc.). If we add to this the fact that hydrodynamic models will soon be computed with a consistent treatment of the radiation field, and that model spectra will include the effect of velocity fields and, at least partly, of non-LTE, we can bet on a rich harvest of results for the next AGB symposium!

Acknowledgments. Bengt Gustafsson is thanked for financial support during the course of this work. I thank S. Höfner, K.H. Hofmann, M. Scholz and T. Tsuji for sending (p)reprints. Financial support was provided by the Swedish NFR.

References

Alexander D.R., Ferguson J.W., Wing R.F., Johnson H.R., Haushchildt P.H., Allard F., 1999, this volume

Alvarez R., Plez B., 1998, A&A 330, 1109

Aoki W., Tsuji T., Ohnaka K., 1999, this volume

Aringer B., Kerschbaum F., Hron J., Posch T., Windsteig W., Jørgensen U.J., Höfner S., 1999, this volume

van Belle G., The PTI collaboration, Thomson R.R., 1999, poster contribution, I.A.U. Symposium No. 191

Baschek B., Scholz M., Wehrse R., 1991, A&A 246, 374

Bessell M.S., Castelli F., Plez B., 1998, A&A 333, 231 (erratum A&A 337, 321)

Bessell M.S., Scholz M., Wood P.R., 1996, A&A 307, 481

Cherchneff I., Cau P., 1999, this volume

Cohen M., Barlow M.J., Sylvester R.J., Liu X.-W., Cox P., Lim T., Schmitt B., Speck A.K., 1999, this volume

Decin L., Cohen M., Eriksson K., Gustafsson B., Huygen E., Morris P., Plez B., Sauval J., Vandenbussche B., Waelkens C., 1999, poster contribution, I.A.U. Symposium No. 191

Fleischer A.J., Winters J.M., Sedlmayr E., 1999, this volume

Hedgecock I.M., Naulin C., Costes M., 1995, A&A 304, 667

Hofmann K.H., Scholz M. 1998, A&A 335, 637

Hofmann K.H., Scholz M., Wood P.R., 1998, A&A 339, 846

Hron J., Loidl R., Höfner S., Jørgensen U.J., Aringer B., Kerschbaum F., Windsteig W. 1998, A&A 335, L69

Hron J., Loidl R., Höfner S., Jørgensen U.G., Aringer B., Kerschbaum F., Windsteig W., 1999, this volume

Höfner S., 1999, this volume

Irrgang P., Balega Y.Y., Gauger A., Osterbart R., Schniggenberg G., Weigelt G., 1999, poster contribution, I.A.U. Symposium No. 191

Jeong K.S., Winters J.M., Sedlmayr E. 1999, this volume

Karlsson L., Lindgren B., Lundevall C., Sassenberg U., 1997, J. Mol. Spectroscopy 181, 274

Karovska M., 1999, this volume

Kiselman D., Plez B., 1995, Mem. Soc. Ast. It. 66, 429

Kozasa T., Sogawa H., 1999, this volume

Käufl H.U., Stecklum B., Richichi A., Righter S., 1999, poster contribution, I.A.U. Symposium No. 191

Langhoff S., 1997, ApJ 481, 1007

Lockwood G.W., 1972, ApJS 24, 375

Lodders K., Fegley B., Jr., 1999, this volume

Lopez B., 1999, this volume

Luck R.E., Lambert D.L., 1982, ApJ 256, 189

Luttermoser D.G., Bowen G.H. 1992, in CSSS, 7th Cambridge workshop, ASP Conf. Series 26, 558

Mauron N., 1997, A&A 326, 300

Mauron N., Caux E., 1992, A&A 265, 711

Monnier J.D., 1999, this volume

Partridge H., Schwenke D.W., 1997 J. Chem. Phys. 106, 4618

Perrin G., Coudé du Foresto V., Ridgway S.T., Mariotti J.-M., Ruilier C., Mennesson B., Traub W.A., 1999, poster contribution, I.A.U. Symposium No. 191

Plez B., 1998, A&A 337, 495

Plez B., 1999, in "The C-star Phenomenon", ed. R.E. Wing, Kluwer

Plez B., Lambert D.L., 1994, ApJ 425, L101

Plez B., Smith V.V., Lambert D.L., 1993, ApJ 418, 812

Ryde N., Eriksson K., Gustafsson B., Olofsson H., Plez B., 1999, poster contribution, I.A.U. Symposium No. 191

Scholz M., 1997, in Fundamental Stellar Properties, eds. T.R. Bedding et al.

Schultheis M., Aringer B., Höfner S., Jørgensen U.J., 1999, poster contribution, I.A.U. Symposium No. 191

Schwenke D.W., 1998, Faraday discussion 109: Chemistry of molecules and grains in space, in press

Smith V.V., Plez B., Lambert D.L., et al., in preparation

Tuthill P.G., Monnier J.D., Danchi W.C., 1999, this volume

Weigelt G., Blöcker T., Hofmann K.H., Osterbart R., Balega Y.Y., Fleischer A.J., Winters J.M., 1999, this volume

Windsteig W., Aringer B., Lebzelter T., Höfner S., 1999, poster contribution, I.A.U. Symposium No. 191

Winters J.M., Le Bertre T., Keady J.J., 1999, this volume

Asymptotic Giant Branch Stars
T. Le Bertre, A. Lèbre and C. Waelkens, eds.
I.A.U. Symposium 191, 1999

Effective Temperature Scales of Red Giant Stars

David R. Alexander, Jason W. Ferguson

Dept. of Physics, Wichita State University, Wichita, KS 67260-0032, USA

Robert F. Wing

Astronomy Dept., Ohio State University, Columbus, OH 43210, USA

Hollis R. Johnson

Dept. of Astronomy, Indiana University, Bloomington, IN 47405, USA

Peter H. Hauschildt

Dept. of Physics & Astronomy, University of Georgia, Athens, GA 30602-2451, USA

France Allard[1]

Dept. of Physics, Wichita State University, Wichita, KS 67260-0032, USA

Abstract.

We have completed a grid of spherically symmetric AGB star atmospheres using the state of the art spectral synthesis code PHOENIX. Models are constructed for stars with masses of 1 $M_\odot$ and 1.5 $M_\odot$, spanning the range 10 to 3300 $L_\odot$ in luminosity and 2500 to 5200 K in effective temperature. We find that grains of Al_2O_3 and $CaTiO_3$ among other species form in atmospheres cooler than $T_{eff} = 3000$ K. In the coolest models the grains cause a weakening of the TiO absorption features in the red and near infrared of up to 30% through both a depression of the continuum and a depletion of the TiO number abundance. We use spectrophotometric observations from a number of catalogs to determine effective temperature – spectral class and effective temperature – color relationships. We also compare synthetic colors calculated from our models with observations of M giants on Wing's 8-color narrow-band system of classification photometry.

[1]Now at CRAL, Ecole Normale Superieure, Lyon, 69364 France Cedex 07

1. Introduction

A great deal can be learned about AGB stars from accurate models of their photospheres. A comparison of the emergent spectra predicted by the models with flux-calibrated spectral observations makes possible the determination of the effective temperature, surface gravity, radius, luminosity, and surface chemical composition of individual stars. Because the photosphere must be the point of origin of the winds which emanate from these stars, an understanding of the thermal structure of the photosphere is fundamental to an understanding of the winds. The input physics and radiative transfer included in detailed calculations of model photospheres are required for an improved understanding of the hydrodynamic effects caused by stellar pulsations. Significant progress has been made in this field in recent years (see Gustafsson & Jørgensen 1994, for a recent review of the field) as a result of the inclusion of more realistic physical assumptions in the models, made possible by the twin developments of improved input physics and the availability of ever faster computers and algorithms.

2. Model photospheres: computational methods and input physics

The models presented here have been computed with the multi-purpose stellar atmosphere code PHOENIX (Hauschildt, Allard & Baron 1999; Hauschildt & Baron 1998). This program can compute models in either plane-parallel or spherical geometry, with a fully self-consistent treatment of steady expansion, in either LTE or NLTE, and with incident illumination included. While many of these program characteristics will be important for a full understanding of the AGB phenomenon, for this report we consider only models which are in hydrostatic equilibrium, in LTE, computed in spherical geometry, and fully converged for energy conservation. It is interesting to note that a comparison of plane-parallel and spherical models shows that the flux of a 4 200 K spherical model is depressed by about 15% in the violet compared to a plane-parallel model. At 3 200 K, the differences between spherical and plane-parallel models over the entire visual spectrum reach 40% (this effect is also seen in dwarf OB stars; Aufdenberg, Hauschildt & Shore 1998). Because the plane-parallel models have more flux at short wavelengths than the spherical models, comparisons of plane-parallel models to observed spectra in the visual will result in an underestimation of the effective temperature, with a rapidly increasing effect for later spectral types.

The molecular opacities in all of our models are computed with the direct opacity sampling technique, with monochromatic opacities computed at 20 000 wavelengths. Our spectral line database now includes more than 400 million molecular and atomic lines, including H_2O, TiO, CO, CN, HCN, and C_2H_2. The abundance of each chemical species is computed with an equation of state which assumes thermodynamic equilibrium exists. More than 850 molecules and 600 solids and liquids can be included in a fully self-consistent calculation. The equation of state program is based upon the work of Allard (Allard & Hauschildt 1995; Alexander et al. 1997) and solves the equations of mass action for each species with full coupling between the gas phase and condensed species.

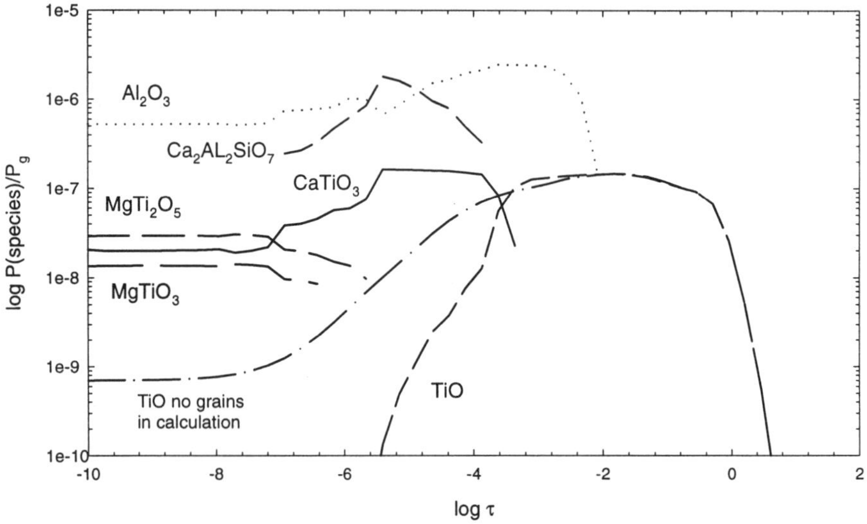

Figure 1. The fictitious partial pressures of condensed species in the photosphere of a 1.0 $M_\odot$ model with $T_{\rm eff} = 2\,500$ K and $\log g = -0.5$ are plotted versus the optical depth at 1.2 μm.

The opacity of solid grains is included with a Mie theory calculation for six different grain species, including two magnesium silicates, iron, corundum, and two carbon species. All grains are assumed to have the Mathis et al. (1977) power law size distribution determined from observations of the interstellar medium, which reasonably approximates the size distribution found by Dominik et al. (1993) in their calculation of dust formation in an M star wind.

3. Model photospheres: comparison of the grid with observations

For this report we have constructed a grid of 196 model photospheres which span the K and M spectral classes of red giant stars. All of these models have a solar composition, except that nitrogen has been enhanced by a factor of 2.5 and the carbon abundance has been reduced by a factor of 1.5 to match the abundances found by Smith & Lambert (1990) for 18 M giants. Future calculations will explore the effect of different C/O ratios and metallicities and will expand the grid to higher luminosities and masses.

Our results lead to the discovery that grains condense in the photospheres of many of the models we compute. All models cooler than $T_{\rm eff} = 3\,000$ K with luminosities greater than 500 $L_\odot$ show grain condensation. While Tsuji (1998) has reported the existence of carbon grains in the atmospheres of some carbon stars, this is the first calculation which shows condensation in the photospheres of oxygen-rich giant stars. Fig. 1 shows the fictitious partial pressure of condensed species in one of our coolest models, which corresponds roughly to the M10 composite spectrum presented by Fluks et al. (1994). The first species to condense, at a temperature around ~1 530 K, is Al_2O_3 at an optical depth of about 0.01. When $CaTiO_3$ condenses at an optical depth of 10^{-3}, the important gas phase absorber TiO virtually disappears from the model. The net effect of grains on the spectrum of this model is to slightly depress the continuum, due to the enhanced continuous opacity of the grains, and to significantly weaken the TiO features in the visual. The TiO bands at 0.59 μm and 0.62 μm are reduced by approximately 50% in strength compared to a model computed without grain condensation. Since TiO band strengths are the standard indicator of spectral type, this result may have profound effects on the comparison of spectral type, color, and effective temperature. Among the species which condense in this model, optical constants are available only for Al_2O_3. A proper interpretation of late type spectra will require optical constants for the two calcium species which condense at relatively high temperatures.

The models in the grid have been fit to observations of stars taken from eleven published catalogs, which contain more than 300 energy-calibrated spectra of K, M, S, and C stars. An automated procedure computes the root-mean-square (rms) deviation between a given observation and synthetic spectra for all the models in the grid smoothed to the same resolution as the observation. A plot of these rms deviations versus effective temperature usually shows a well defined minimum which accurately determines the effective temperature of the star. However, surface gravity is poorly determined by this technique because of its limited influence on the emergent spectrum. An example of the fit for the composite M3 spectrum of Fluks et al. (1994) is shown in Fig. 2. Stars with significantly different compositions or other peculiarities are easily identifiable from the large value of the minimum rms deviation. A direct comparison of the observed and computed spectra is used to confirm the appropriateness of the fit.

The models have also been compared to observations of 50 M giants obtained on the 8-color narrow band filter system of MacConnell, Wing & Costa (1992). Fig. 3 (a) shows that the observations of 5 M2 giants are well matched between models with effective temperatures of 3 700 K and 3 800 K, suggesting a range in temperatures for these stars of about 100 K. The TiO indices and color temperatures computed for the models agree well with the observations, as seen in Fig. 3 (b), for a wide range of stars from K0 to M3.

Fig. 4 shows the effective temperature scale determined from a comparison of these models with three sets of observations. Variations in the effective temperature determined for stars of a given spectral class from the Burnashev (1982) catalog may represent differences in the properties of stars classified with the same spectral class, caused for example by differences in chemical composition. These effects are significantly reduced in the $T_{eff} - (V - K)$ diagram. Differences in the effective temperature scale determined for the composite spectra of Fluks et al. (1994) and Silva & Cornell (1992) are more likely the result of the finite model grid used.

4. Conclusions

We present here the results of the computation of a new grid of model photospheres for oxygen-rich red giant stars. All models have been computed in spherical geometry, with hydrostatic equilibrium, LTE, energy conservation, and

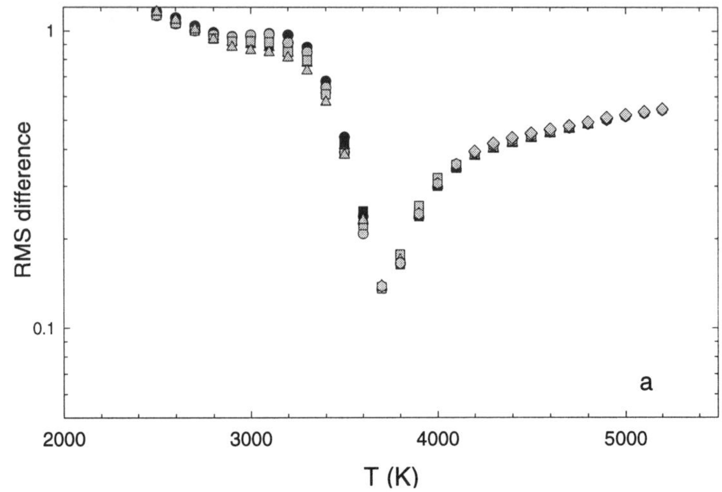

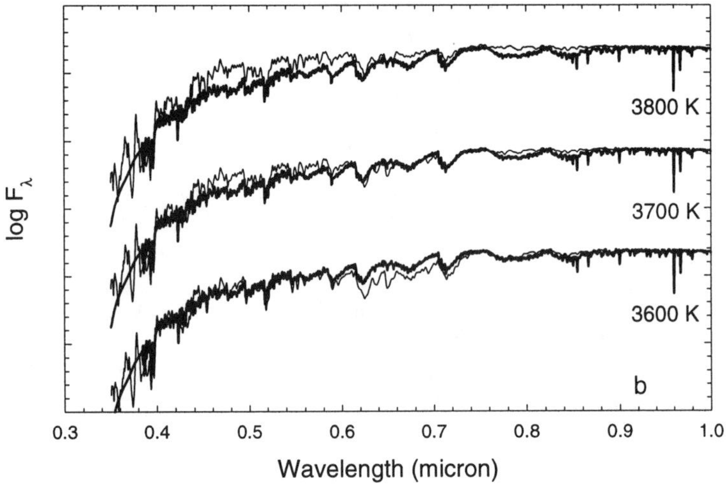

Figure 2. (a) The rms difference (χ^2) between the synthetic spectrum of an M3 giant observed by Fluks et al. (1994) and the models in the grid is plotted. Different symbols refer to models with differing L_* and M_*. The minimum at 3 700 K represents the best fitting models. Differences in $\log g$ are not well determined by this method. (b) The observed composite spectrum for M3 giants from Fluks et al. (1994) is plotted as the heavy line, along with models of three different effective temperatures. Differences caused by the temperature differences are obvious, although a human observer might have difficulty determining which model best fits the spectrum.

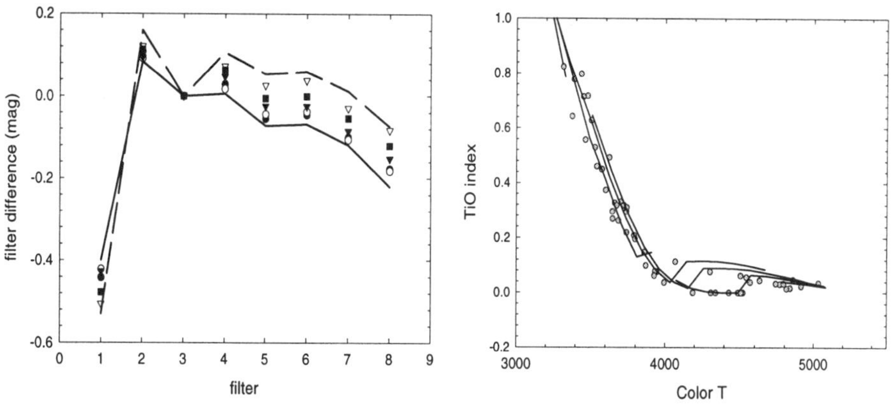

Figure 3. (a) Wing 8-color narrow band filter observations of 5 M2 standard giants are compared to synthetic colors computed from the models. The solid line are colors from a 3 800 K model, and the dashed from a 3 700 K model. All colors have been normalized to Wing filter 3. (b) The TiO index and color temperature computed from the 8-color Wing observations of 50 K & M giants are compared to the values computed from synthetic colors of the models. From left to right, successive curves represent models of progressively higher luminosity from the grid.

chemical equilibrium including both gas phases and condensation of solids and liquids. All models cooler than $T_{\rm eff} = 3\,000$ K show grain condensation (including Al_2O_3 and $CaTiO_3$) in the photosphere. This phenomenon profoundly affects the TiO band strength in the coolest models. The emergent spectra computed for these models have been compared to observations from a wide range of sources. The effective temperature scale determined from these comparisons agrees well with that determined from angular diameter observations (Ridgway et al. 1980). Future work will explore the effect of grain condensation, NLTE effects for atoms and the more abundant molecules, and the effects of changing the chemical composition.

Acknowledgments. Low Temperature Astrophysics research at Wichita State University is supported by NASA grants NCC5-168 and NAG5-3435. Some of the models reported here were computed from a generous allocation of computer time at the Center for Scientific Supercomputing at Kansas State University, supported by NSF grant CDA-9724289. PHH acknowledges support by NSF grant AST-9720704, NASA ATP grant NAG 5-3018 and LTSA grant NAG 5-3619 to the University of Georgia.

References

Allard F., Hauschildt P.H., 1995, ApJ 445, 433

Alexander D.R., Allard F., Tamani A., Hauschildt P.H., 1997, Ap&SS 251, 171

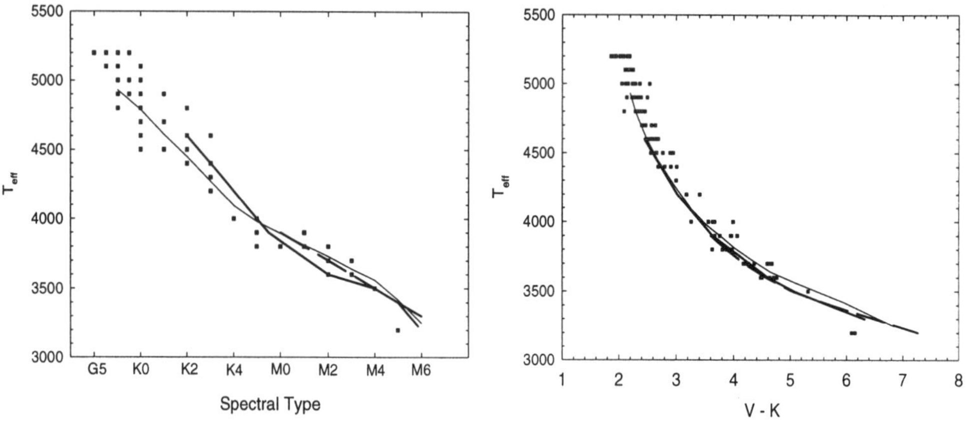

Figure 4. The effective temperature scale determined by fitting model spectra to observations from three catalogs as a function of (a) spectral class and (b) $V - K$ color. The square symbols are the Burnashev (1982) catalog, the thick solid line from Silva & Cornell (1992), and the dashed line from Fluks et al. (1994). The effective temperature scale determined by Ridgway et al. (1980) from angular diameter measurements is also given (thin solid line).

Aufdenberg J.P., Hauschildt P.H., Shore S.N., 1998, ApJ 498, 837

Burnashev V.I., 1982, Izv. Krymskoi Astro. Obs. 65, 94

Dominik C., Sedlmayr E., Gail H.-P., 1993, A&A 277, 578

Fluks M.A., Plez B., The P.S., De Winter D., Westerlund B.E., Steenman H.C., 1994, A&AS 105, 311

Hauschildt P.H., Allard F., Baron E., 1999, ApJ 512, 377

Hauschildt P.H., Baron E., 1998, J. Comp. App. Math., in press

Gustafsson B., Jørgensen U.G., 1994, A&A Rev. 6, 19

MacConnell D.J., Wing R.F., Costa E., 1992, AJ 104, 821

Mathis J.S., Rumpl W., Nordsieck K.H., 1977, ApJ 217,425

Ridgway S.T., Joyce R.R., White N.M., Wing R.F., 1980, ApJ 235, 126

Silva D.R., Cornell M.E., 1992, ApJ 81, 865

Smith V.S., Lambert D.L., 1990, ApJS 72, 387

Tsuji T., 1998, in *The Carbon Star Phenomenon*, ed. R.F. Wing, IAU Symp. 177, Kluwer Academic Publishers, in press

Asymptotic Giant Branch Stars
T. Le Bertre, A. Lèbre and C. Waelkens, eds.
I.A.U. Symposium 191, 1999

Lithium and s-process enrichment in massive galactic AGB stars

P. García-Lario

ISO Data Centre, Apartado de Correos 50727, Villafranca del Castillo, E-28080 Madrid, Spain

F. D'Antona

Osservatorio Astronomico di Roma, I-00040 Monteporzio, Italy

J. Lub

Sterrewacht Leiden, Postbus 9513, NL-2300 RA Leiden, The Netherlands

B. Plez

Astronomiska Observatoriet, Box 515, S-75120 Uppsala, Sweden

Atomspektroskopi, Fysiska Institution, Box 118, S-22100 Lund, Sweden

H.J. Habing

Sterrewacht Leiden, Postbus 9513, NL-2300 RA Leiden, The Netherlands

Abstract. We present the first results of a wide observational program intended to study the effects of hot-bottom burning and of the third dredge-up in the chemical evolution of the most luminous AGB stars in our Galaxy. The main goal is to provide new observational constraints to the models through the analysis of lithium and s-process element abundances in a carefully selected sample of galactic AGB stars. Our results indicate that, unlike observed in the Magellanic Clouds, the most massive AGB candidates in our Galaxy do not show any significant enrichment in s-process elements. In addition, only some of them are found to be Li-rich.

1. Introduction

Studies of AGB stars in the Magellanic Clouds (Plez et al. 1993; Smith et al. 1995) show a lack of high-luminosity C stars beyond the limit of $M_{bol} \sim -6$. This has been interpreted as the consequence of the transformation of ^{12}C into ^{14}N via the CN cycle in massive AGB stars, where the temperature of the convective envelope is hot enough ($T \geq 2 \times 10^7$ K) (Sackmann & Boothroyd 1992; Wood et al. 1983). These hot-bottom burning (HBB) models also predict the production of ^7Li, through the *^7Be transport mechanism* (Cameron & Fowler 1971), which

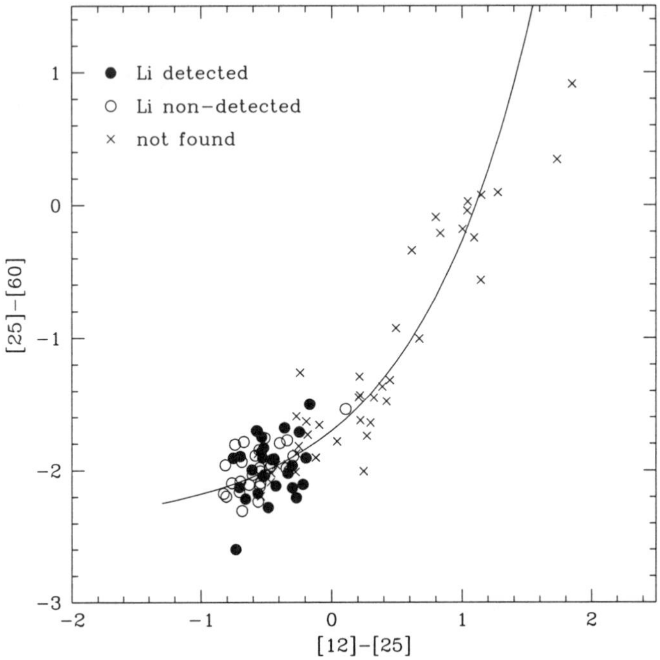

Figure 1. Distribution of the sources included in our sample in the IRAS two-colour diagram [12]−[25] vs. [25]−[60].

should be detectable for a short time in the atmosphere of the star, together with the s-process heavy elements formed by the capture of the neutrons produced during the thermal pulses, via the so called *third dredge-up*.

Extensive surveys have been carried out searching for lithium in AGB stars in the Magellanic Clouds (Smith & Lambert 1990; Smith et al. 1995) resulting in the detection of strong Li lines only in luminous O-rich AGB stars in the range $-7 \leq M_{bol} \leq -6$. These stars are also enriched in s-process elements, as expected.

Inside our Galaxy, only a handful of Li-rich stars have been found (Abia et al. 1993) and, unlike those detected in the Magellanic Clouds, they are not so luminous. Some of them are intermediate mass C-stars and S-stars with a modest luminosity. Recent model calculations show that a large Li abundance is expected only for the most massive AGB stars (4 to 7 $M_{\odot}$) (D'Antona & Mazzitelli 1996) which, in addition, should be O-rich. However, an observational problem exists with these stars. Most of them experience a strong mass loss and they are soon heavily obscured by thick circumstellar envelopes, making optical observations very difficult.

Here we present the first qualitative results of a wide observational programme based on high-resolution optical spectroscopy centered on a carefully selected sample of strongly reddened galactic O-rich AGB stars. The main goal is to search for the presence of lithium and s-process elements in the spectra of these stars, as the signature of hot-bottom burning and of the third dredge-up, respectively. The detailed abundance analysis will be presented elsewhere.

2. Selection of the sample

Sources were selected from the IRAS Point Source Catalogue on the basis of their far-infrared colours and their photometric behaviour. The best candidate stars are expected to be very bright infrared sources. They should be Mira-like variables with the longest periods (sometimes more than 500 days) and very large amplitude variations (10 magnitudes in the V band) with a strong OH maser emission. Fig. 1 shows the distribution of the sources selected in the far-infrared IRAS two-colour diagram. They fall along the AGB sequence previously defined by van der Veen et al. (1989), at places which correspond to different thickness of the circumstellar envelope. They span a wide range of periods (from only 300 days up to more than 1000 days) and OH expansion velocities (from just a few km s^{-1} to more than 20 km s^{-1}).

Both the period and the OH expansion velocities have been taken as distance independent mass indicators since recent studies indicate that both the expansion velocity of circumstellar envelopes of AGB stars showing OH maser emission and the period of variability are correlated with the mass of the progenitor star, the group of stars with the larger expansion velocities and the longer periods showing a galactic distribution which corresponds to a more massive population (Habing et al. 1994; Sevenster et al. 1995).

3. Observations

High-resolution (R$\sim$50 000) echelle spectra in the range 5500–9000 Å were obtained at the William Herschel Telescope (La Palma, Spain) with UES and at the 3.60 m ESO telescope (La Silla, Chile) with CASPEC during several runs in August 1996, February 1997 and July 1997. 96 sources were tried, and useful spectra were obtained for 55. The remaining 41 sources were either too red to obtain any significant information from the optical spectra or the counterpart was simply not found.

4. Results

Fig. 2 shows some of the spectra obtained in the region around the Li I line at 6707 Å. As we can see in this Figure, the Li I line is completely absent in some of the stars in our sample while it is found to be very strong in some others.

Table 1 displays the distribution of detections and non-detections as a function of the known periods and OH expansion velocities. Most of the AGB stars with the shorter periods are not detected in the Li I line while those stars with periods beyond 500 days (but still observable in the optical) are detected as Li-strong sources. Note, however, that for these objects we are very close to the limit of detection in the optical range, since the number of them too red or not found increase dramatically for periods longer than 500 days.

Similar statistics are obtained if we consider the expansion velocities derived from the OH maser emission. For low expansion velocities, below 6 km s^{-1}, there are no detections, while stars showing higher expansion velocities are predominantly Li-rich. Again, the same observational problems arise when we try to

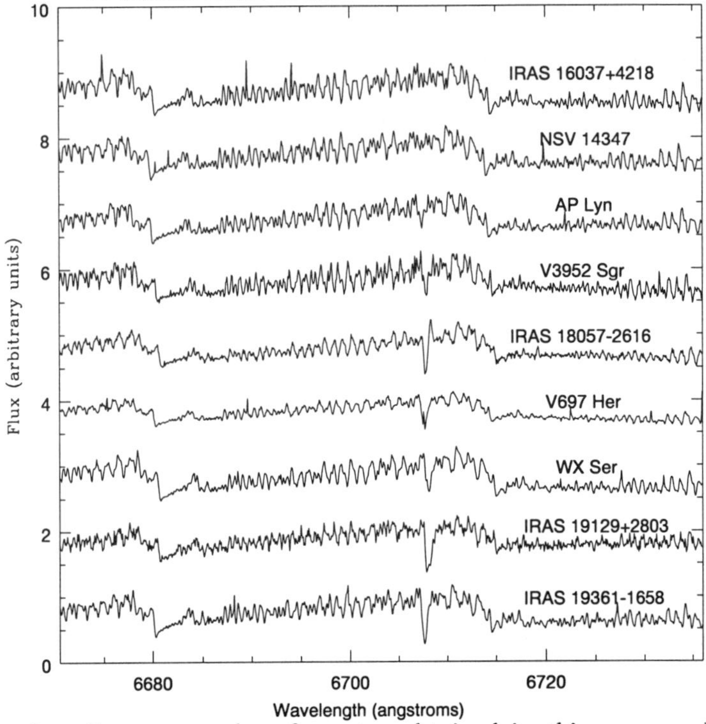

Figure 2. Some examples of spectra obtained in this survey where we show the different strengths found in the Li I λ6707 Å line.

observe the stars with the highest expansion velocities, completely obscured in the optical.

If the detection of lithium is taken as a signature of HBB and the period and the expansion velocities are accepted as valid mass indicators, our results confirm that a minimum mass is needed for the activation of HBB in AGB stars in the Galaxy. As it was previously observed in the Magellanic Clouds, this explains the absence of high luminosity C stars also in our Galaxy. However, unlike observed in the Magellanic Clouds, there are a few AGB stars in our sample with very long periods and very high expansion velocities for which lithium is not detected.

Concerning s-process elements, we used the ZrO band at 6474 Å as s-process enhancement indicator. Surprisingly, this band was not detected in any of the sources observed, nor the elemental lines corresponding to s-process elements in the range 7400-7600 Å seem to be enhanced with respect to what it is observed in normal supergiants. Thus, we conclude that the most luminous O-rich AGB stars in our Galaxy are not S-stars, again unlike observed in the Magellanic Clouds.

5. Discussion

Lithium is expected to be produced in massive AGB stars as soon as the star begins the thermally pulsing phase and until ^{3}He is completely exhausted.

Table 1. Distribution of detection/non-detections as a function of the distance-independent mass indicators used

Period (days)	Observed	Li detected	Li not detected	Not found
300–400	11	1	9	1
400–500	9	6	2	1
500–700	15	5	2	8
>700	14	1	0	13

V_{exp} (km s^{-1})	Observed	Li detected	Li not detected	Not found
<6	11	0	10	1
6–9	11	8	2	1
9–12	13	4	3	6
>12	39	4	4	31

Based on preliminary results derived from models recently developed by our group and which will be presented elsewhere, we obtain that for a solar metallicity and a 5–6 $M_{\odot}$ model, ^{3}He exhaustion takes place very rapidly as a consequence of the large temperature at the bottom of the convective envelope ($\sim 60\,10^6$ K). During the lithium production phase there is a strong enhancement of the Li abundance – $\log \epsilon(\mathrm{Li}) = +4.0$ – but it drops very soon after a few pulses, something which could explain the non-detection of lithium in some of the most massive objects observed.

For the same metallicity and 4 $M_{\odot}$, the stars undergo a few pulses before the production of lithium starts and this is not so huge – $\log \epsilon(\mathrm{Li}) = +2.5{-}3.5$ at $M_{bol} = -6$ – but it lasts for longer, decreasing again when ^{3}He is exhausted, so that the stars spend ~50% of their lifetimes as Li-rich.

A stronger mass loss in our Galaxy than in the Magellanic Clouds could also explain the different results obtained. Large temperature fluctuations are expected in the bottom of the envelope if mass loss is very strong, leading to strong fluctuations of the Li abundance within each thermal pulse, which might be burned out during interpulses. In spite of this, although the Li abundance goes up and down during the TP phase, the Li-rich phase would last longer and we would expect many more Li-rich stars than Li-poor (the ratio expected would be 8 to 2 which is exactly what we find in our survey).

The assumption of a stronger mass loss in our Galaxy looks reasonable considering the number of heavily obscured AGB stars known in our Galaxy compared to those in the Magellanic Clouds (whether this is a consequence of the different metallicity deserves further investigation). This would also be consistent with the even larger number of obscured AGB stars known in the galactic bulge.

The correct identification of the stars in our sample as AGB stars (against a possible classification of them as population I supergiants) is supported by the characteristic infrared photometric properties observed. The large amplitude variations are proto-typical of stars in the AGB phase, and would be completely unusual for a supergiant star, expected to show only irregular luminosity fluctuations.

In order to explain the absence of s-process elements enhancements in these stars one can argue that we might be observing them still in a very early stage of development as AGB stars, especially since the sources observed are not yet strongly obscured by their circumstellar envelopes. However, this would be inconsistent with the same objects being at the end of the Li production phase, as suggested by the absence of lithium in some of them, as a consequence of ^{3}He exhaustion. Alternatively, it could be that the Li production phase has not yet even been reached by these stars.

On the other hand, massive AGB stars in our Galaxy might evolve much faster than those in the Magellanic Clouds. Thus, there would only be time for a few thermal pulses to occur during the AGB lifetime, not enough to produce a significant change in the elemental abundances observed at the surface of the star. It should be noted that, contrary to the effects of hot-bottom burning, which are immediately detectable at the surface of the star, the s-process enhancement requires some kind of dredge-up to be active and it might be observable only after a long series of thermal pulses.

If our interpretion is correct, we would expect that the less luminous AGB stars show the stronger s-process enhancements. In fact, this is what was already found for AGB stars in the Magellanic Clouds in previous works (see Figure 4 in Smith et al. 1995).

Acknowledgments. This work was financed by the Spanish Ministerio de Educación y Cultura through the project PB94-1274.

References

Abia C., Isern J., Canal R., 1993, A&A 275, 96
Cameron A.G.W., Fowler W.A., 1971, ApJ 164, 111
D'Antona F., Mazzitelli I., 1996, ApJ 470, 1093
Habing H.J., Tignon J., Tielens A.G.G.M., 1994, A&A 286, 523
Plez B., Smith V.V., Lambert D.L., 1993, ApJ 418, 812
Sackmann I.-J., Boothroyd A.I., 1992, ApJ 392, L71
Sevenster M.N., Dejonghe H., Habing H.J., 1995, A&A 299, 689
Smith V.V., Lambert D.L., 1990, ApJ 361, L69
Smith V.V., Plez B., Lambert D.L. Lubowich D.A., 1995 ApJ 441, 735
van der Veen W.E.C.J., Habing H.J., Geballe T.R., 1989, A&A 226, 108
Wood P.R., Bessell M.S., Fox M.W., 1983, ApJ 272, 99

Asymptotic Giant Branch Stars
T. Le Bertre, A. Lèbre and C. Waelkens, eds.
I.A.U. Symposium 191, 1999

Results on AGB Stars from Infrared Surveys

N. Epchtein

Observatoire de la Côte d'Azur, Département Fresnel, BP 4229, F-06304 Nice Cedex 4, France

Abstract. In the last 3 decades, infrared surveys have discovered new classes of AGB stars, and significantly improved our knowledge of their statistical and physical properties, often in conjunction with surveys in other spectral ranges (mainly OH maser lines). After a short description of the major past and present infrared surveys and their most commonly used analysis tools, their impact is illustrated by a few examples such as the characterisation of the populations of the Bulge and Magellanic Clouds and the distribution of carbon-rich stars in the galactic disc. In conclusion, further all-sky surveys in the still unexplored spectral regions between 2 and 10 μm are strongly advocated as well as the development of ground-based telescopes dedicated to deeper surveys and monitoring of the variability.

1. Introduction

Large scale sky surveys are often at the basis of a major breakthrough in our knowledge of the structure and evolution of stars, galaxies and the Universe in general. They increase the sample of objects that belong to previously known classes, and consequently improve our knowledge of the statistical properties of these classes (e.g., their spatial number density distribution, luminosity/mass function, evolutionary status, variability properties, etc.). They discover, often unexpectedly, sets of objects with unknown properties that turn out to form new classes (e.g., pulsars, protostars, brown dwarfs, OH/IR sources, etc.). Finally they provide calibrated maps, either directly (isoflux maps), or through star counts (isoextinction maps). Surveys generally result in an assemblage of relatively scarce, but homogeneous, information for a great number of objects, preferably spread over large areas of the sky. Resulting data are gathered into photometric *catalogues* of up to billions sources measured through a limited number of wide-band filters. These catalogues must be exhaustive down to a given magnitude which defines the *completion* limit, and the existence of a source must have a high level of *reliability* down to this magnitude limit. In addition, they must be homogeneously calibrated, and the completeness limit must be as uniform as possible across the observed target. Most useful large scale surveys involve observations in at least two colours, or better in three or more colours. Two–colour surveys (e.g., the TMSS) allow the discrimination of unusually red objects and drawings of colour–magnitude diagrams, while three- (or more) colour surveys allow the design of colour–colour diagrams, the basic tools for the analysis of a photometric survey. I will essentially focus this review on near–infrared (hereinafter NIR) and mid-infrared photometric surveys of the full sky (or, at least, very large fractions of the sky) with a special attention to the currently on–going 2 micron surveys, DENIS (Deep Near Infrared Survey of the Southern Sky) and 2MASS (2 Micron All–Sky Survey).

2. Major infrared sky surveys

The major IR surveys covering at least half of the sky, and performed in at least 2 bands are listed in Table 1. They are briefly described in this section. More details on the history of IR surveys can be found in the excellent reviews by Price (1988, 1997).

Table 1. Past and on–going major astronomical infrared surveys

Year[a]	Name	λ (μm)	Sensitivity (magnitude)	Sky coverage	Number of sources	Major discoveries (AGB stars)
1965 1969	TMSS	0.85; 2.2	∼+3.5 (@ K)	+90° to -35°	∼ 5500	dusty envelopes of LTS, IRC +10216
1970 1976	AFGL[b]	4.2; 11.0 19.8; 27.4	∼+1.3 (@ 4.2 μm)	∼ 70%	∼ 2000	extreme AGB stars AFGL 3068; 2688...
1983 1986	IRAS	12,0; 25,0 60.0, 100.0	∼ +5.0 (@ 12 μm)	∼ 95%	∼ 250,000[c]	first all sky IR survey IR spec. class. (LRS)
1989 1990	DIRBE (COBE)	1.25 to 240	–	all-sky	–	Bulge shape bar(?)
1996 2001	DENIS	0.80;1.25 2.15	see Tab. 2	Southern sky	∼ 10^9(I)	complete census of galactic and Magel-
1997 2001	2MASS	1.25;1.65 2.15	see Tab. 2	all sky	a few 10^8	lanic AGB stars
1996 1999	MSX	8.3 to 21.4	∼ +7 (@ 10μm)	gal.plane Mag. Clouds	∼ 10^6	improve IRAS data in crowded areas

[a]in col. 1, first year refers to the beginning of the mission, and second, to the publication of the main catalogues

[b]space missions are underlined

[c]in the Point Source Catalog

The Two Micron Sky Survey (TMSS) and its associated catalogue the IRC mark the first important step in our modern knowledge of the infrared sky (Neugebauer & Leighton 1969). It discovered a few hundreds of new very red objects (Kleinmann & Payne–Gaposchkin 1979) and most importantly demonstrated that late-type giant stars are often surrounded by a thick envelope of dust which had prevented their detection so far. The most archetypical object discovered by this survey is the famous *'unusual'* carbon star IRC +10216 (Becklin et al. 1969) in the envelope of which most of the hydrocarbonaceous molecules were discovered later on by radioastronomers. The TMSS gave rise to a huge number of follow–up observations and studies during the past 30 years. It remained, unfortunately, uncompleted in the southern sky during 3 decades, except for a few modest attempts such as the *Valinhos* survey (Epchtein et al. 1987).

The Air Force Geophysical Laboratory Survey (AFGL) was a series of rocket flights (Price & Walker 1976) that surveyed the sky for the first time at mid–infrared wavelengths (4-20 μm). In the AGB star domain, it discovered several extreme objects such as CRL (or GL) 3068 (Lebofsky & Rieke 1977), which is, as yet, the reddest carbon star ever found.

The Infrared Astronomical Satellite (IRAS) was flown in 1983 and provided the first all–sky survey at IR wavelengths (10-100 μm range) (Neugebauer et al. 1984). It gave rise to the first comprehensive catalogue of IR sources (IRAS Science Team 1988) and led to a huge number of extremely interesting new results relevant to the AGB star domain which cannot be summarized in a few lines (see e.g., the review by Beichman 1987). Of particular interest was the collection of 8-22 μm low resolution spectra (LRS) (IRAS Science Team 1986; Volk & Cohen 1989; Volk et al. 1991) that provided a new stellar classification based on the strength of the dust (and gas) emission/absorption features in this spectral range (Walker & Cohen 1988; Walker et al. 1989; Sloan et al. 1998). Thanks to its sensitivity and completeness, IRAS provided an excellent data base to search for extreme late-type stars, and to investigate galactic structure parameters. To this aim it was followed up by many complementary observations at near–IR wavelengths (e.g., Whitelock & Feast 1984; Epchtein et al. 1990; Kwok & Hrivnak 1987) and in molecular maser lines (1612 MHz satellite line of OH, in particular, e.g., te Lintel Hekkert et al. 1991). It displayed the Bulge as mainly made of *'infrared stars'* (Habing 1988) and its shape. It showed also that very red mass–losing AGB stars are common and that *'a sequence'* can be delineated in colour–colour diagrams which might reflect their evolutionary status.

COBE/DIRBE is included in this list although, due to its large field of view (0.7°), it was not able (and was not designed) to pick up individual stars. The integrated flux measurements are useful to study the galactic structure and the variations of colours within the disc which can be interpreted in terms of stellar populations in the Galaxy. For instance, the Bulge emission essentially due to the red giant and AGB stars is extremely well delineated in the short wavelength bands (Arendt et al. 1994).

The Infrared Space Observatory (ISO) was not primarily intended to perform blind surveys, but substantial parts of the observing time was dedicated to cover targets of relatively large area. The ISO survey that is probably most relevant to AGB stars (excluding spectroscopic surveys that are reviewed by Barlow and Waters & Molster in this volume) is the ISOGAL program. ISOGAL is aimed at characterising the colours of point sources (at 7 and 15 μm) in several small areas that sample the Bulge and the inner disk of our Galaxy. The cross identification with the new 2 μm survey data DENIS (at IJK_s) allows to design a set of colour–colour diagrams which lead to characterise the various species of

AGB stars and to separate them from foreground red giants and Young Stellar Objects. First results of ISOGAL have been described by Pérault et al. (1996) and at this Symposium by Blommaert et al. (this volume) and Ojha et al. (poster contribution).

2MASS/DENIS. In the late eighties it became obvious that time was ripe to survey again the sky at 2 μm in order to improve the TMSS by a large factor (at least 10^4). This was made possible thanks to the release of large NIR panoramic arrays. Two projects, 2MASS (Skrutskie 1998) and DENIS (Epchtein et al. 1997; Epchtein 1998) are currently in progress and will end up in the early 2000s with the largest photometric databases ever produced in this spectral range. Their main characteristics can be compared in Table 2. At the sensitivity limit of these surveys, one may expect to detect all the AGB stars of the Galaxy and the Magellanic Clouds, except the most extreme ones (GL 3068 type). It should be noticed that the DENIS survey is the only attempt to survey the sky simultaneously in the NIR and optical (I band) ranges.

Table 2. *DENIS* and *2MASS* main characteristics

	DENIS	**2MASS**
Photometric bands	IJK_s	JHK_s
Sensitivity to point sources (3 σ)		
I,J,K_s	18.5, 16.3,14.0	-
J,H,K_s	-	17.1,16.4,15.6
Observing site (North)	-	Mount Hopkins, USA
(South)	La Silla, Chile	CTIO, Chile
Sky coverage	$\delta = +2$ °to -88 °	all sky
Telescope aperture	1 m	1.3 m
Pixel size (I)	1 ″	-
(J, H, K_s)	3 ″	2 ″
Image size	12′ × 12′	8.5′ × 8.5′
Position accuracy	0.6 ″	0.5 ″
Photometric accuracy	10 % or better	5 % (bright sources)
Completeness (at 10 σ)	0.99(goal)	0.99
Reliability (at 10 σ)	0.9995(goal)	0.9995
Survey strategy	strips 30 °	tiles 6 °
Status (as of Aug.98)	40 % achieved	40 % achieved (North) 15 % (South)

The Midcourse Space Experiment (MSX) is partly aimed at surveying the galactic plane (±5°) in 6 bands covering the 8 to 20 μm range, with deeper raster scans in selected regions (Price et al. 1998). This mission contributes to complete the IRAS mission by covering areas that were missed by IRAS, by extending the wavelength coverage and, above all, by improving the spatial resolution in crowded areas. This is of particular interest for the census of AGB stars in dense regions such as the galactic Bulge where IRAS was essentially confusion limited.

3. Some applications of IR survey data to AGB stars

The AGB phase is an essential step of stellar evolution, but it is still poorly understood because it involves intricate problems of physico–chemistry (molecule and dust grain formation), short lived phenomena such as mass loss, stellar winds, and far from equilibrium (thermal, mechanical, etc.), that were presented throughout this Symposium. Their high luminosity makes AGB stars excellent tracers of the stellar mass in the Galaxy. From the observer point of view, this high luminosity is a major advantage. The light can be analysed accurately even at large distances. This allows to obtain very-high-precision photometry (a few percents, see e.g. the HIPPARCOS data analysed by de Laverny et al. 1998), high angular resolution imagery (see the review by Lopez in this volume), high resolution spectroscopy, proper motions and derived distance measurements, and spatial distribution determinations in the solar neighbourhood (e.g., Knapik et al. 1998). One essential feature of AGB stars is that they are often surrounded by a dust (and gas) envelope (DGE) (see the complete monograph by Habing 1996) that dissipates into the ISM and enriches it with processed elements. Grains heated by the central star thermally radiate in the mid- and far-infrared. The IR range is therefore particularly well adapted to the study of AGB stars. The combination of HIPPARCOS and NIR photometric data leads to very accurate calibration of their luminosity (Alvarez et al. 1997). The NIR range is also more appropriate for high spatial resolution techniques allowing the mapping of envelopes of nearby stars (e.g., IRC +10216, Haniff & Buscher 1998). Finally, IR colour–colour diagrams and spectrophotometry (IRAS Science Team 1986) allow first-order classification of thick dust shell stars and evolutionary stages that were not previously detectable. The following Sect. is dedicated to the use and interpretations of these diagrams.

3.1. Infrared colour–colour diagrams

Near–Infrared colour-colour diagrams. A large number of JHK (LMN) photometric data have been obtained on many known AGB stars of the Galaxy, mainly Miras (e.g., Whitelock & Feast 1984) semi–regular variables (e.g., Kerschbaum & Hron 1994), carbon stars (e.g., Noguchi et al. 1981) and in the Magellanic Clouds (Sect. 3.4.). Efforts have been done also, to detect the counterparts of OH (mainly type II) maser sources which led to the new class of OH/IR objects (e.g., Hyland et al. 1972; Baud et al. 1975; Schultz et al. 1976; Glass 1978; Jones et al. 1982; Epchtein & Nguyen-Q-Rieu 1982), which are essentially an extension of the O–rich Miras towards longer periods (Le Bertre 1993), larger mass loss rates and consequently thicker dust envelopes. There are good correlations between the IR colours, the period, the 1612 MHz OH emission, and probably the mass of the star when it left the main sequence. Near IR colour–colour diagrams can be used to discriminate several types of stars such as O–rich against C–rich (Groenewegen 1998), but most accurately for objects with thick envelopes. Some confusion still remains for thin-envelope objects.

The IRAS [12–25] vs. [25–60] diagram was investigated by van der Veen & Habing (1988), and Volk & Cohen (1989) to discriminate between DGE objects with various shell opacities and chemical compositions, but some confusion remains in crowded areas of this diagram. Miras, OH/IR sources and protoplanetary nebulae are found along a sequence of increasing [12–25] index which has been interpreted as an evolutionary sequence (Olnon et al. 1984) or a sequence of increasing mass of the progenitor main sequence star (Lépine et al. 1995)

Combined near IR-IRAS diagrams (e.g., [12–25] vs. K–L). Using the results of a low sensitivity K-band survey of the southern galactic plane cross–identified with the IRAS data, Epchtein et al. (1987) have developed an efficient method of separation of C and M stars with thick DGE and showed the great reliability of the method with respect to the separation made only on the basis of IRAS colours. This method is based on the interpretation of the [12–25] vs. $[K-L]$ diagram and has been applied to pick up and classify a large number of extreme carbon stars that were not previously recognised from their optical or LRS spectra. Comparable results were recently obtained on the Magellanic AGB stars by Zijlstra et al. (1996; see Sect. 3.4.).

3.2. Mass–loss estimation from IR photometric data

Mass–loss is an essential feature of AGB stars which determines their evolution. Various semi–empirical methods have been developed based on near-infrared fluxes (Gehrz & Woolf 1971), optical CS absorption (e.g., Reimers 1977), millimetric CO emission (e.g., Knapp & Morris 1985), and far-infrared fluxes (Jura 1986). Recently, Le Bertre & Winters (1998) have shown that there is an excellent correlation of the mass–loss rate and the $[K-L]$ index. This relation holds independently for C–rich and O–rich stars. Comparable results are obtained with other NIR indices such as $[J-K]$. All-sky NIR surveys will therefore, in principle, provide an accurate estimation of the global return of matter to the interstellar medium by AGB stars.

3.3. The spatial distribution of carbon stars in the galactic disc

The exploitation of the IRC and IRAS data to derive the spatial number density distributions of AGB stars and their integrated mass–loss rate were first proposed by Claussen et al. (1987) and Thronson et al. (1987). Based on IRC data, they found a rather uniform distribution of the carbon stars up to a distance of about 1.5 kpc and a scale height of 200 pc ±50. Combining IRC and IRAS data, Jura & Kleinmann (1990) showed similar results up to a distance of about 2.5 kpc. The NIR observation of many poorly documented IRAS sources allowed Guglielmo et al. (1993, 1998) to identify a large number of new IR carbon stars (IRCS), pushing the limit of detection up to about 6 kpc and the completion distance to $\approx$ 4 kpc. They found a scale height of z $\approx$ 190 pc and a density of 22 IRCS per kpc^3 to be compared to the density of 40 kpc^{-3} found using HIPPARCOS data (Knapik et al. 1998), which includes only optical stars. The completeness limit of the IR sample is essentially given by the IRAS 25 μm sensitivity limit. The spatial density distribution of C-rich stars (optical or IR) is essentially constant in the galactic plane, but there is a deficiency of luminous carbon stars in the inner Galaxy. At large galactocentric distances, an AGB star more probably evolves towards an IRCS phase rather than through the OH/IR phase, a result that might be linked to the lower metallicity of the local interstellar medium. It is now important to keep on searching for new IRCS at larger distances to avoid biases due to the incompleteness of the samples. The combination of the 2MASS/DENIS surveys, the poorly documented IRAS sources (those many with just a 12 μm flux which cannot be classified at all), ISOGAL, and the MSX data will definitely improve our description of the spatial distribution of AGB stars in the Galaxy, and the variation of the O–rich/C–rich number density. Further important results would undoubtedly result from the projected all–sky mid–IR survey called *Next Generation Sky Survey (NGSS*; E. Wright, private communication, see Sect. 4.)

3.4. The AGB star populations of the Magellanic Clouds

The Magellanic Clouds have known distances and they are at sufficiently high galactic latitude to avoid confusion with galactic stars. So the study of the distribution and luminosity function of the AGB stars in the Magellanic Clouds is highly valuable especially to investigate the conditions of formation of the carbon–rich stars. Long-period variables in the MCs are excellent targets for the calibration of the Period–Luminosity relation of Mira and AGB stars (Hughes & Wood 1990). Optical (objective prism), and more recently photometric infrared surveys collected a large sample of Miras and AGB stars and carbon stars in the Clouds (e.g., Blanco & McCarthy 1983; Frogel & Blanco 1990; Westerlund et al. 1991; Costa & Frogel 1996). The first OH/IR star was reported by Wood et al. (1986). IRAS data have been recently exploited for a search of extreme AGB stars leading to the discovery of thick dust shell candidates (Zijlstra et al. 1996; Loup et al. 1997; van Loon et al. 1998). Preliminary K vs. $[J-K]$ colour–magnitude diagrams based on DENIS data (see Cioni et al., poster contribution, this Symposium) show that the different classes of late-type giant stars (red giants, O–rich E–AGB and TP–AGB, carbon rich AGB stars) can be easily separated. The cross–identification of these NIR data with longer-wavelength observations (IRAS, ISO, MSX) will provide an even more accurate classification and luminosity function determination (see Loup et al., this volume).

3.5. The AGB star populations of the Galactic Bulge and the inner Galaxy

The study of the late-type stellar content of the Galactic Bulge and the Galactic Center regions is essential for the understanding of the history of star formation and of the dynamical properties of the Galaxy and its central "object". Infrared observations are nevertheless strongly hampered in the most central regions by confusion due to crowding. In particular IRAS has been of little help to pick up individual stars in the inner kiloparsec. AGB star investigations in the inner Galaxy may contribute to solve some issues that deal with the star formation history near the center of the Galaxy and the metal enrichment. The Bulge population is essentially old, but a young population of long-period OH/IR stars has been singled out (Sevenster et al. 1995; Blommaert et al. 1998; Wood et al. 1998) which would argue in favour of a continuous star formation near the Galactic Center. The ISOGAL program which combines ISO data at 7 and 15 μm with DENIS data in selected areas of the Bulge will contribute to identify different populations of stars. Finally it will be of high interest to confirm (or infirm) the lack of luminous C-rich stars in the Bulge by combining near- and mid-infrared colours from DENIS/2MASS and MSX/ISO/NGSS.

4. Needs for future IR surveys

Table 3. Main future infrared space missions

Mission (countries)	Launch yr. (duration)	Covered area	Main instruments	wavelengths (μm)	Major objectives
in progress					
WIRE (USA)	1998 (4 months)	100 sq °	2 channel photometer	12 and 25	extragalactic and cosmology
SIRTF (USA)	2001 (3-4 yrs.)	observatory mode	IR camera IR spectrograph imaging photometer	3.6-10 6-40 20-180	multi objectives *Legacy program*
SOFIA (USA,D)	2001 (years)	observatory mode	Echelle spectro. FIR bolometer Mi-dIR camera	17-210 40-300 5-40	multi-objective + submillimeter heterodyne
IRIS (J)	2003 (1 yr.)	Observatory survey facility	NIR camera FIR surveyor	2-25 50-200	limited area surveys
in project					
NGSS [a] (USA)	?	all-sky survey	4 channel photometer	3-4.5;4.5-7 9-15;18-30	4-colour photometric survey
RESPIRE [b] (F)	?	all-sky survey	2 channel spectro-imager	1.8-3.6 3.6-6.5	spectrophotom. NIR survey (AGBs, ISM)

[a]Next Generation Sky Survey
[b]RElevé SPectroscopique InfraRouge Exhaustif

Large-scale NIR surveys are most valuable for the study of the physical and statistical properties of stars in general and AGB stars in particular. Even the low-sensitivity *TMSS* has been powerful enough to discover a new class of stars and detect for the first time thick dust envelopes around AGB stars. This was the starting point, together with OH surveys, of a considerable breakthrough in the understanding of mass–loss effects and gave rise to a complete renewal of the modelling of the late stages of stellar evolution towards the formation of planetary nebulae. With IRAS, DENIS/2MASS, MSX, the full sky is now covered in the short-wavelength (1-2.5 μm) and long-wavelength (8-100 μm) range with a sensitivity that allows the detection of basically all AGB stars in the Galaxy and Magellanic Clouds. Although there is no firm project aimed at performing new all–sky surveys after 2MASS/DENIS, most of the future (space) infrared missions (see Table 3) will offer some facilities to survey more deeply and recurrently selected sky areas. A large spectral window (2-12 μm) of extreme importance for AGB stars (and many other types of objects) still remains uncovered, however. There are, to my knowledge, only 2 small-size space projects that are aimed to fill this gap, the NGSS project for a 4–colour photometric survey, and a French proposal called RESPIRE which involves a 2-channel low resolution spectro–imager with special attention to survey the dust and PAHs spectral features in stellar envelopes. Although ISO has obtained a wealth of very exciting spectroscopic results in this range (e.g., reviews by Barlow and Waters & Molster in this volume), an all-sky unbiased survey such as RESPIRE would also be quite advisable for a better infrared spectral classification of the most extreme types of stars. Searches for variable objects and long–term monitoring at NIR wavelengths (e.g., Le Bertre 1992, 1993) should be done at larger scale especially in the Magellanic Clouds and in the inner Galaxy (Wood et al. 1998). With the development of large NIR arrays (up to 2k) the construction of the equivalent of Schmidt telescopes for the NIR range should be most advisable to probe large sky areas more deeply (up to $K \approx 20$).

References

Alvarez R., Mennessier M.-O., Barthès D., Luri X., Mattei J.A., 1997, A&A 327, 656

Arendt R.G., Berriman G.B., Boggess N., et al., 1994, ApJ 425, L85

Baud B., Habing H.J., Matthews H.E., et al., 1975, Nature 258, 406

Becklin E.E., Frogel J.A., Hyland A.R., et al., 1969, ApJ 158, L133

Beichman C., 1987, ARAA 25, 521

Blanco V.M., McCarthy M.F., 1983, AJ 88, 1442

Blommaert J.A.D.L., van der Veen W.E.C.J., Van Langevelde H.J., et al., 1998, A&A 329, 991

Claussen M.J., Kleinmann S.G., Joyce R.R., Jura M., 1987, ApJS 65, 385

Costa E., Frogel J.A., 1996, AJ 112, 2607

de Laverny P., Mennessier M.-O., Mignard F., Mattei J.A., 1998, A&A 330, 169

Epchtein N., Nguyen-Q-Rieu, 1982, A&A 107, 229

Epchtein N., Le Bertre T., Lépine J.R.D., et al., 1987, A&AS 71, 39

Epchtein N., Lépine J., Le Bertre T., 1990, A&A 227, 82

Epchtein N., de Batz B., Capoani L., et al., 1997, The ESO Messenger 87, 27

Epchtein N., 1998, 3rd Euroconference on Near-Infrared surveys, ed. N. Epchtein, Kluwer, Astr. Sp. Sciences Lib. 230, p. 3

Frogel J.A., Blanco V.M., 1990, ApJ 365, 168

Gehrz R.D., Woolf N.J., 1971, ApJ 165, 285

Glass I.S., 1978, MNRAS 182, 93

Groenewegen M.A.T., de Jong T., 1998, A&A 337, 797

Groenewegen M.A.T., 1998, 3rd Euroconference on Near-Infrared surveys, ed. N. Epchtein, Kluwer, Astr. Sp. Sciences Lib. 230, p. 73

Guglielmo F., Epchtein N., Le Bertre T., et al., 1993, A&A 99, 31

Guglielmo F., Le Bertre T., Epchtein N., 1998, A&A 334, 609

Habing H.J., 1988, A&A 200, 40

Habing H.J., 1996, A&A Rev. 7, 97

Haniff C.A., Buscher D.F., 1998, A&A 334, L5

Hughes S.M.G., Wood P.R., 1990, AJ 99, 784

Hyland A.R., Becklin E.E., Frogel J.A., Neugebauer G., 1972, A&A 16, 204

IRAS Science Team, 1986, A&AS 65, 607 (Low Resolution Spectra, LRS)
IRAS Science Team, 1988, Point Source Catalog, eds. C.A. Beichman, G. Neugebauer, H.J. Habing, P.E. Clegg and T.J. Chester, NASA RP-1190
Jones T.J., Hyland A.R., Caswell J.L., Gatley I., 1982, ApJ 253, 208
Jura M., 1986, ApJ 303, 327
Jura M., Kleinmann S.G., 1990, ApJ 364, 663
Kerschbaum F., Hron J., 1994, A&AS 106, 397
Kleinmann S.G., Payne–Gaposchkin C., 1979, Earth and Extrater. Sc. 3, 161
Knapik A., Bergeat J., Rutily B., 1998, A&A 334, 545
Knapp G.R., Morris M., 1985, ApJ 292, 640
Kwok S., Hrivnak B.J., 1987, ApJ 312, 303
Le Bertre T., 1992, A&AS 94, 377
Le Bertre T., 1993, A&AS 97, 729
Le Bertre T., Winters J.M., 1998, A&A 334, 173
Lebofsky M.J., Rieke G.H., 1977, AJ 82, 646
Lépine J., Ortiz R., Epchtein N., 1995, A&A 299, 453
Loup C., Zijlstra A.A., Waters L.B.F.M., Groenewegen M.A.T., 1997, A&AS 125, 419
Neugebauer G., Leighton R.B., 1969, NASA SP-3047
Neugebauer G., Habing H.J., van Duinen R., et al., 1984, ApJ 278, L1
Noguchi K., et al., 1981, PASJ 33, 373
Olnon F.M., Baud B., Habing H.J., et al., 1984, ApJ 278, L41
Pérault M., et al., 1996, A&A 316, L165
Price S.D., Walker R.G., 1976, AFGL-TR-0208
Price S.D., 1988, PASP 100, 171
Price S.D., 1997, 2nd Euroconference on Near-Infrared surveys, eds. F. Garzon et al., Astr. Sp. Sciences Lib. 210, p. 3
Price S.D., Egan M.P., Shipman R.P., et al., 1998, 3rd Euroconference on Near-Infrared surveys, ed. N. Epchtein, Kluwer, Astr. Sp. Sciences Lib. 230, p.17
Reimers D., 1977, A&A 54, 485
Schultz G.V., Kreysa E., Sherwood W.A., 1976, A&A 50, 171
Sevenster M.N., Dejonghe H., Habing H.J., 1995, A&A 299, 689
Skrutskie M., 1998, 3rd Euroconference on Near-Infrared surveys, ed. N. Epchtein, Kluwer, Astr. Sp. Sciences Lib. 230, p. 11
Sloan G.C., Little–Marenin I.R., Price S.D., 1998, ApJ 115, 809
te Lintel Hekkert P., Caswell J.L., Habing H.J., et al., 1991, A&AS 90, 327
Thronson H.A., Latter W.B., Black J.H., 1987, ApJ 332, 770
van der Veen W.E.C.J., Habing H.J., 1988, A&A 194, 125
van Loon J.Th., Zijlstra A.A., Whitelock P.A., et al., 1998, A&A 329, 169
Volk K., Cohen M., 1989, AJ 98, 931
Volk K., Kwok S., Stencel R.E., Brugel E., 1991, ApJS 77, 607
Walker H.J., Cohen M., 1988, AJ 95, 1801
Walker H.J., Cohen M., Volk K., et al., 1989, AJ 98, 2163
Westerlund B.E., Azzopardi M., Breysacher J., Rebeirot E., 1991, A&AS 91, 425
Whitelock P., Feast M., 1984, MNRAS 211, 25p
Wood P.R., Bessell M.S., Whiteoak J.B., 1986, ApJ 306, L81
Wood P.R., Habing H.J., McGregor P.J., 1998, A&A 336, 925
Zijlstra A.A., Loup C., Waters L.B.F.M., et al., 1996, MNRAS 279, 32

Part 2. Pulsation, Mass Loss, Cool Envelopes

Michael Feast during a fundamental (or was it an overtone ?)
talk on the pulsation mode of Miras

Asymptotic Giant Branch Stars
T. Le Bertre, A. Lèbre and C. Waelkens, eds.
I.A.U. Symposium 191, 1999

Pulsation Modes in Mira and Semiregular Variables

Michael Feast

Department of Astronomy, University of Cape Town, Rondebosch, 7700, South Africa

Abstract. The most direct evidence for the pulsation mode of Mira and SR variables comes from the linear diameters derived from individual parallaxes and angular diameters, together with the observed periods. The data now available strongly suggest that most, if not all, Miras and SRs pulsate in the same mode. If comparison is made with conventional models, this mode is the first overtone. Uncertainties in deriving the pulsational radius of a Mira from the observations do not appear significant enough to affect this conclusion. Evidence from SRs and Miras in globular clusters supports this conclusion and shows that the pulsation mode does not depend on metallicity. Analysis of light curves leads to broadly the same conclusions, if less directly. The light curves indicate that periodicities other than the main one may also be present and that mode switching may occur in some stars.

1. Introduction

There has been a long debate regarding the mode of pulsation of Mira variables. Discussion has mainly focussed on whether they pulsate in the fundamental radial mode or in the first overtone. This question is closely related to the problem of the absolute magnitudes and effective temperatures of the Miras. There has been much less discussion regarding the mode of pulsation of the Semiregular (SR) variables. However in quite recent times there has been a good deal of work bearing on the question of pulsation modes for both the Miras and the SRs. The main pieces of new evidence come from the measurement of angular diameters of Miras and SRs by interferometry or occultations; parallaxes for individual members of these classes from HIPPARCOS; and, the detailed analysis of light curves.

2. Modes from linear diameters

The most direct way to determine the pulsational mode of a Mira or SR variable is to derive its linear diameter from its measured angular diameter and its parallax and then to use the $P\sqrt{\rho} = Q$ relation between the period (P) and the mean density (ρ) (or a more sophisticated form of this relation) to deduce the the pulsational constant (Q) and to compare this with theoretical prediction. This procedure is possible because, as a class, the Miras and SRs are old, low mass stars. Thus a mass near one solar mass can be adopted. Supergiant SRs (SRc variables) will not be discussed in this paper.

The first angular diameters of these stars were obtained from lunar occultation observations, but great progress has been made recently using interferometry. A problem arises at once with the Miras since they have very deep atmospheres and the measured diameter varies with wavelength. Models suggest that the measured angular diameter in the infrared (at about 2 μm), where most of the energy is being radiated, should approximate closely to that required in the pulsation analysis. Observations are being increasingly made in the infrared but the currently available data contains crucial observations which have been made at shorter wavelengths. These have been corrected for atmospheric extension using models. That these models (e.g. Bessell et al. 1996) are successful in predicting these corrections is shown by the general agreement of the radii of Miras determined in the infrared with those measured at shorter wavelengths and corrected (see Feast 1996).

There are certainly many problems remaining to be sorted out regarding the measured angular diameters of Miras. Different workers do not always agree too well with one another. Part of this might be observational. But there appears to be considerable variations in diameter with time for some of the variables, at least at the shorter (visual) wavelengths. Some of these changes appear to be unrelated to pulsation (or at least to the main pulsational period) (e.g. Tuthill et al. 1995). There is also considerable evidence for non-circularly symmetrical discs, at least at optical wavelengths. This may well be associated with large star spots or, perhaps, either non-radial oscillations or stellar rotation (see e.g. Haniff 1995 and references there, Lattanzi et al. 1997, and papers in the present volume). There is also the question of the variation of the radius with phase in the pulsation cycle itself. This is only just beginning to be studied though it is not expected to be large in the 2 μm region. All these factors certainly imply that the estimate of the true mean angular diameter of a Mira will have a greater uncertainty than the individual measurements from which it is derived. Nevertheless the available data should give a set of angular diameter measurements suitable for the analysis of pulsational modes, especially if the stars are considered as a group rather than individually.

Robertson & Feast (1981) used two infrared occultation diameters of Miras together with distances derived from a period-luminosity (PL) relation, to deduce that Miras were overtone pulsators. First overtone pulsation was assumed although it is in fact difficult to distinguish between this and higher overtones. Haniff et al. (1995) used interferometric diameters of 10 Miras together with a PL relation to confirm this conclusion. Tuthill et al. (1994) used an accurate ground based parallax of R Leo (Gatewood 1992) to reach to same result.

It is now possible to make progress both for the Miras and the SRs using individual parallaxes from the HIPPARCOS catalogue. Perhaps the main concern here is that one might get spurious parallaxes due to star spots on the Miras. The angular diameter of a Mira is typically about five times that of its parallax. If star spots cause the apparent photocentre of the star to move around then this might affect the derived parallax. Unless the motion of the photocentre mimicked the parallactic motion, this would simply be reflected in a larger than normal parallax error. Even if the parallax itself were affected this could either result in an overestimate or an underestimate of the true value and this would average out in a set of Miras. This does however suggest that care is necessary in dealing with individual stars.

Early release of HIPPARCOS data allowed van Leeuwen et al. (1997) to derive the linear diameters of eight Miras using the angular diameters of Haniff et al. (1995). Their results indicated that Miras with periods less than 400 days are overtone pulsators but that at least some of the Miras with longer periods pulsate in the fundamental. This result (i.e. both overtone and fundamental pulsators) was rather unexpected. Whitelock et al. (1998) have now discussed all the Miras and SRs which have measured angular diameters and also HIPPARCOS parallaxes. A little consideration shows that it is best to work with the reciprocal of the linear diameter (D) rather than with D directly. In the sample considered the percentage error in the parallax (π) can, in the best cases, be about the same as that in the angular diameter (ϕ), but it is often much greater. In these circumstances the uncertainty in $1/D(= \pi/\phi)$ remains approximately gaussian whilst that of D itself becomes quite skew and diagrams of D against log P, for instance, can be somewhat misleading. Working with 1/D means that one can use all the parallaxes (including negative ones, if they occur). This is important since provided the stars are not chosen by measured parallax, and since the uncertainty in 1/D depends on the error in the parallax (σ_π) and not on σ_π/π, one can combine the results for a sample of stars without introducing significant statistical bias (Lutz-Kelker type bias etc.).

Whitelock et al. (1998) use angular diameters from a variety of sources. Early data were summarized by Feast (1996) and to this has been added data from van Belle et al. (1996, 1997) and Bedding et al. (1998). Where necessary the angular diameters have been reduced to effective infrared diameters and corrected at least approximately for limb darkening. A comparison of theoretical predictions with a plot of 1/R (where R is the linear radius) against log P for Miras and SRs, and including both oxygen-rich and carbon-rich stars, indicates that at least the bulk of the stars are pulsating in the first overtone. With a first overtone relation from Fox & Wood (1982) viz.:

$$log P = 1.5\ log\, R - 0.5\ log\, M + log\, Q, \tag{1}$$

and adopting Q (the pulsation constant) = 0.04, one finds a mean mass (M) of 1.3 ± 0.2 solar masses. The actual value of the mean mass is of course sensitive to the adopted pulsation constant but the small uncertainty is an indication that the bulk of the stars are pulsating in the same mode whether they are oxygen or carbon rich, or Miras or SRs. The residuals in 1/R from the theoretical overtone relation are quite gaussian. The 40 stars [1] involved in the discussion include the Mira, R Cas, which was the clearest example of a long-period fundamental pulsator in the analysis of van Leeuwen et al. (1997). R Cas has indeed the largest fractional residual from the solution ($\Delta/\sigma = 2.8$ where Δ is the residual and σ the standard error of 1/R for R Cas). However in a sample of 40 stars, such a residual is not particularly abnormal. Whitelock et al. conclude that their analysis does not compel one to believe that either R Cas or χ Cyg ($\Delta/\sigma = 2.0$)

[1] R Aql and U Ori were in each case included twice, once with the Haniff et al. (1995) diameters and once with those of van Belle et al. (1996) which are rather different.

are anything other than overtone pulsators. Individual fundamental pulsators might exist in the present sample but only improved data can show this. The evidence now seems rather strong that at least the vast majority of the Miras and SRs pulsate in the first overtone.

3. Miras and SRs in globular clusters and metallicity effects

Feast (1996) argued that the SRs and Miras in the globular cluster 47 Tuc must all lie on a single evolutionary track. This can then be combined with a pulsation equation to derive the change of bolometric magnitude with period. Consistency of observations and theory can only be obtained if the SRs and Miras pulsate in the same mode.

Whitelock (1986) assembled the data (including infrared photometry) of SR and Mira variables in globular clusters. Her results show that the SR variables from metal-rich clusters have a distinctly different period-temperature relation (or equivalently, a different period-$(J-K)_o$ relation) from that in metal-poor clusters. This seemed a remarkable result at the time and possibly suggested that the metal-poor and metal-rich SRs pulsated in different modes. However one can now understand the result in terms of Wood's (1990) parameterization of the AGB. Combining his relation with the first overtone pulsation equation one obtains (e.g. Feast 1996):

$$log\, T_{eff} = -0.13\; log\, P - 0.073\; log\, Z + 0.038\; log\, M + const \qquad (2)$$

where T_{eff} is the effective temperature. This is a useful equation since the dependence on mass (M) is very small. The equation shows that a dependence on metallicity (Z) is predicted and that the metal independent quantity should be, $log\, T_{eff} + 0.073\; log\, Z$. Whitelock et al. (1998) show that the observations bear out this prediction indicating that both metal poor and metal rich SRs pulsate in the same mode.

4. The relative positions of SRs and Miras in the PL plane

Wood & Sebo (1996) discovered a number of SR variables in the LMC. These lie above the Mira PL relation in a luminosity - log P diagram (that is, at a given luminosity, the SRs have shorter periods than the Miras). Recently, Bedding et al. (1998) have used HIPPARCOS parallaxes of galactic SR variables to show that they too lie above the Miras in the PL plane. Wood & Sebo interpret their results as showing that the SRs and Miras pulsate in different modes, which they suggest are the first overtone and the fundamental. However one cannot necessarily argue that, because two stars have the same luminosity but different periods, they pulsate in different modes. Even if the masses of the stars are nearly the same, the two stars may have different surface temperatures and hence different radii. That such is indeed the case for at least the SRs and Miras in metal-rich globular clusters is shown by the work of Whitelock (1986). She found that in these clusters the SR variables defined a line in the PL plane which had a shallower slope than the Mira PL relation and lay above it. The two relations intersected at the position of the Miras in the clusters. It is quite clear in this case that the Whitelock track for the SRs is an evolutionary track ending at the position of the cluster Miras. The Mira relation on the other hand is a sequence of variables of different mass and/or metallicity. Thus the relative positions of SR and Mira variables in the PL plane do not require that these two classes of variables are pulsating in different modes. Furthermore, at least in the case of our Galaxy, SRs of period near 100 days cannot be considered as overtones of Miras with periods near 200 days since the galactic kinematics of these two groups are quite different (see Feast et al. 1972).

5. Evidence from light curves

Finally, on the observational side there is the direct analysis of the light curves. There are various problems here. Long series of observations are required and the analysis (especially in the case of SR variables) sometimes reveals many frequencies, the long term stability of which is uncertain. In addition the interpretation of the data can be uncertain since it depends on a comparison of observed period ratios with theoretical predictions and the latter are sensitive to the models (see for instance Barthès & Mattei 1997).

An extensive analysis of *o* Ceti itself has been published by Barthès & Mattei (1997). The main period is 332.9 days and the next most important 1503.8 days with evidence for many combinations of these two periods. They conclude that 332.9 days is probably the first overtone and 1503.8 days the fundamental although they do not entirely rule out other alternatives. Barthès & Tuchman (1994) found that the dominant mode in the Miras S CMi and χ Cyg was also the first overtone.

As regards the SR variables, particularly, it has to be borne in mind that brightness variations on characteristic time scales may well occur due to the effects of star spots or large convection cells on the stellar surface. In the case of the SRa variable V Boo (spectral type Me) Szatmáry et al. (1996) find a main periodicity of 258 days which they take as the fundamental, with a secondary period of 137 days which is then interpreted as an overtone. On the other hand for BS Lyr, Mantegazza (1996) finds a dominant period of 304.7 days which is taken as the first overtone with a secondary period of 272.3 days which is interpreted as the second overtone. Percy & Desjardins (1996) find three periods for W Boo, 25, 35 and 50 days. They suggest these are the 4th, 3rd and 2nd overtones with an interpretation as the 3rd, 2nd and 1st overtones less likely. There is also evidence for power switching between the modes. A very extensive analysis of the light curve of R Dor has recently been published by Bedding et al. (1998). The star shows two periods, 332 days, which they interpret as the 1st overtone, and 175 days, which they believe to be the 3rd overtone. They find that power shifts back and forth between these two modes. They also find evidence for weak chaos in the light curve.

6. Summary

The direct empirical evidence suggests that the main periodicity of most, perhaps all, Miras corresponds to the first overtone at least if classical pulsation models are considered. Other modes may also be present. The same is true for the SR variables with, in this case, evidence for exchange of power between modes (e.g R Dor).

A problem raised by these conclusions is that theorists have so far had considerable difficulty in producing strong shock waves in model atmospheres with radii as large as those implied by the angular diameter measurements. Such shocks are general considered to be necessary to explain the observed splitting of the CO absorption lines at certain phases in the light curve and also the excitation of the characteristic Mira emission-line spectrum. This remains an unsolved theoretical problem whether or not we interpret the data (as in this paper) using classical pulsation models or whether one adopts the model of Tuchman (this volume).

7. Acknowledgements

I am grateful to Dr Patricia Whitelock for many helpful discussions and for unpublished data. I am also grateful to several astronomers for preprints.

References

Barthès D., Mattei J.A., 1997, AJ 113, 373

Barthès D., Tuchman Y., 1994, A&A 289, 429

Bedding T.R., Zijlstra A.A., von der Lühe O., Robertson J.R., Marson R.G., Barton J.R., Carter B.S., 1997, MNRAS 286, 957

Bedding T.R., Zijlstra A.A., 1998, ApJ 506, L47

Bedding T.R., Zijlstra A.A., Jones A., Foster G., 1998, MNRAS 301, 1073

Bessell, M.S., Scholz M., Wood P.R, 1996, A&A 307, 481

Feast M.W., 1996, MNRAS 278, 11

Feast M.W., Woolley R., Yilmaz N., 1972, MNRAS 158, 23

Fox M.W., Wood P.R., 1982, ApJ 259, 198

Gatewood G., 1992, PASP 104, 23

Haniff C., 1995, in Astrophysical Applications of Stellar Pulsation, R.S. Stobie & P.A. Whitelock (eds.), ASP Conf. Ser. 83, p. 270.

Haniff C.A., Scholz M., Tuthill P.G., 1995, MNRAS 276, 640

Lattanzi M.G., Munari U., Whitelock P.A., Feast M.W., 1997,ApJ 485, 328

Mantegazza L., 1996, A&A 315, 481

Percy J.R., Desjardins A., 1996, PASP 108, 847

Robertson B.S.C., Feast M.W., 1981, MNRAS 196, 111

Szatmáry K., Gál J., Kiss L.L., 1996, A&A 308, 791

Tuthill P.G., Haniff C.A., Baldwin J.E., Feast M.W., 1994, MNRAS 266, 745

Tuthill P.G., Haniff C.A., Baldwin J.E., 1995, MNRAS 277, 1541

van Belle G.T., Dyck H.M, Benson J.A., Lacasse M.G., 1996, AJ 112, 2147

van Belle G.T., Dyck H.M., Thompson R.R., Benson J.A., Kannappan S.J., 1997, AJ 114, 2150 van Leeuwen F., Feast M.W., Whitelock P.A., Yudin B., 1997, MNRAS 287, 955

Whitelock P.A., 1986, MNRAS 219, 525

Whitelock P.A., et al., 1998, in preparation

Wood P.R., 1990, in From Miras to Planetary Nebulae: Which Path for Stellar Evolution ?, M.O. Mennessier & A. Omont (eds.), Editions Frontières, Gif-sur-Yvette, p. 67

Wood P.R., Sebo K.M., 1996, MNRAS 282, 958

Marie-Odile Mennessier, from HIPPARCOS to GAIA:
which path for stellar evolution ?

Asymptotic Giant Branch Stars
T. Le Bertre, A. Lèbre and C. Waelkens, eds.
I.A.U. Symposium 191, 1999

Physics and Evolution of LPVs from HIPPARCOS Kinematics

M.O. Mennessier and R. Alvarez[1]
GRAAL, CNRS and Université MontpellierII, F-34095 Montpellier Cedex 05, France

X. Luri
Departament d'Astronomia i Meteorologia, Universitat de Barcelona, E-08028 Barcelona, Spain

M. Noirhomme-Fraiture and E. Rouard
Institut d'Informatique, Université de Namur, B-5000 Namur, Belgium

Abstract. The HIPPARCOS satellite provides astronomical data for about one thousand Long Period Variable stars (LPVs) from which kinematics properties and luminosity calibrations in several bandpasses are deduced using an appropriate method. Several results are deduced: a classification of the LPVs and its relation with the classification from the light curves, a calibration of luminosities inducing properties (age, mass, etc.) along the AGB and a comparison of oxygen to carbon-rich stars.

1. Introduction

It is known that the evolution of LPVs strongly depends on their initial mass and metallicity. However a lot of questions remain open as, for instance, the evolution from O to C-rich stars, the faintest and brightest luminosities on the AGB (see the reviews by Lattanzio & Forestini, Blöcker, Mowlavi in this volume). Kinematical and photometric data allow to calibrate luminosities together with galactic properties. Using HIPPARCOS data we set out such results and their contribution to evolution problems.

2. Luminosity calibrations and mean kinematical parameters

2.1. Samples and analyzing method

The sample is composed of all LPVs observed by HIPPARCOS, i.e. the LPVs which are brighter than V=12.5 mag during more than 80% of the time (around 900 stars). They are Miras, Semi-Regular or irregular L variables, O-rich, C-rich

[1]Now at Institut d'Astronomie et d'Astrophysique, Université Libre de Bruxelles, B-1050 Bruxelles, Belgium

and S stars.

This sample is analyzed using the LM method of calibration (Luri, Mennessier et al. 1996) that needs apparent magnitudes and proper motions, and also uses available parallaxes (even not accurate ones) and radial velocities, and was implemented to:

- allow the treatment of non-homogeneous samples and determine the number of significant groups,
- take into account the effects due to sampling, interstellar absorption, observational errors,
- produce unbiased estimates of mean velocies (U, V, W), of axes of the velocity ellipsoid of each group ($\sigma_U, \sigma_V, \sigma_W$) and of their mean scale height (z), and calibrate the mean luminosity and its dispersion (M, σ_M) of each group,
- calculate the a posteriori probability of a star to belong to a group and classify the stars into the groups, and estimate individual distances and luminosities.

Several calibrations were done, using successively:
- V magnitudes (882 stars) (results in HIPPARCOS Venice'97 Symposium)
- K magnitudes (652 stars)
- IRAS 12 and 25 magnitudes (793 stars) defined as:

$12 = 3.63 - 2.5\log(F_{12})$ and $25 = 2.07 - 2.5\log(F_{25})$

Let us note that 608 stars belong to both K and IRAS samples.

2.2. Results of the calibrations

A statistical Wilk's test allows to determine the significant number of groups. We find: 6 groups for the V analysis, 3 for the K analysis and 4 for the two IRAS analyses. Crossing K and IRAS gives 7 non-empty groups. The groups differ from one another by kinematics (i.e. ages and main sequence masses) and/or luminosities.

Fig. 1 shows the estimated distributions of parameters for each group from the K analysis. Each branch of the diagram corresponds to one parameter: (U, σ_U), (V, σ_V), (W, σ_W), z, (M, σ_M). The axes of the velocity ellipsoid and the scale height are increasing from the diagrams D to TD and from the latter to ED. This can be interpreted as reflecting the disc (D), thick disc (TD) and extended disc (ED) galactic populations.

Fig. 2 shows a similar representation from the IRAS-12 analysis. The distributions of velocities and z are similar as from the IRAS-25 analysis. An increasing of the velocity ellipsoid and of the scale height is seen from the diagrams D to TDb and to ED with a similar interpretation as for the K analysis. However there is a fourth significant group (TDf) with kinematical and spatial distributions very similar to these of TDb but with different luminosity calibration: TDf is, in average, fainter (M12=−6.4 mag, M25=−6.5 mag) and TDb is brighter (M12=−7.9 mag, M25=−8.5 mag). So the analysis separates them because of envelope properties. Moreover D, TDb and ED have, in average, similar IRAS color indices: 12−25=0.6 mag while TDf has a mean index of 0.1 mag that points to a probable majority of stars with thin envelopes.

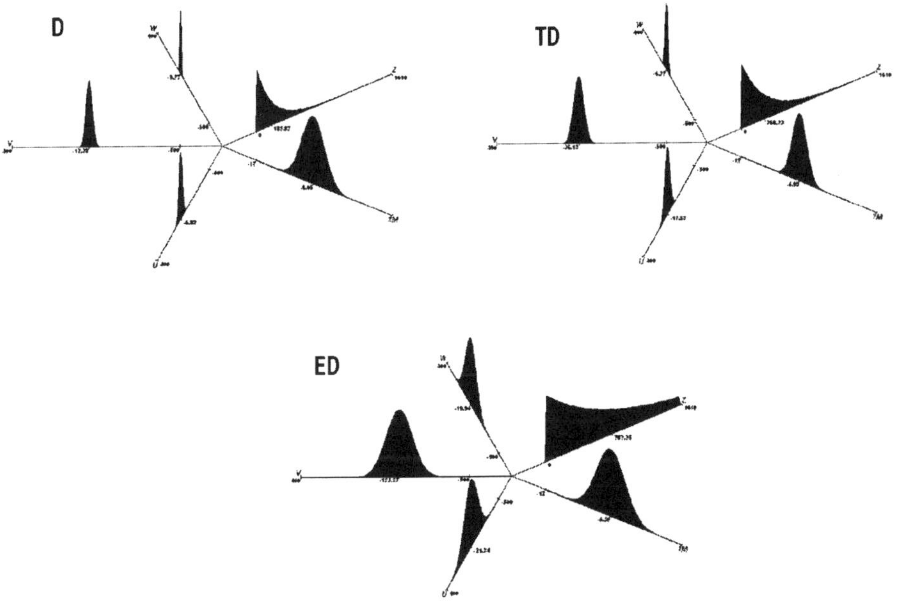

Figure 1. Distributions of estimated parameters from the K calibration

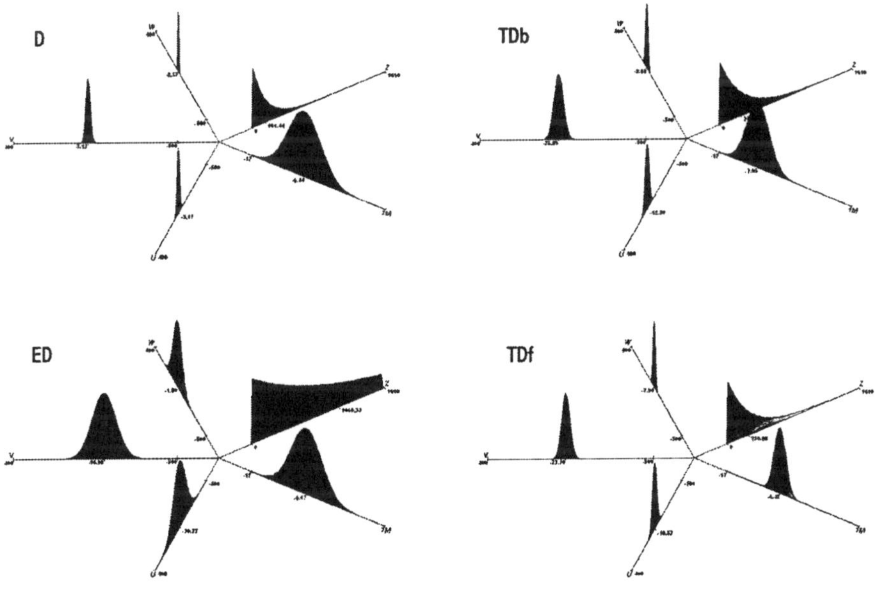

Figure 2. Distributions of estimated parameters from the IRAS calibration

3. Populations and variability types

3.1. Assignation of stars to groups

For each analysis we a posteriori assign stars to one of the groups. In the following we focus on the classification when K and IRAS analyses are crossed: this classification results from both stellar parameters and envelope properties.

Fig. 3 shows the correspondence analysis of variability types and of crossed K-IRAS groups characterized not by the classical contingency table but by the attraction-repulsion indices. Each graph corresponds to one group denoted G (G') where G is the group from K-analysis and G' is the group from IRAS-analysis. On each graph, each radius corresponds to one variability type (in clockwise): S-type SRb's and SRa's, C-rich, O-rich and S-type L variables, C-rich, O-rich and S-type Miras, C-rich SRb's and SRa's, O-rich SRb's and SRa's. On each radius the attraction-repulsion index is indicated. The index is larger or smaller than 1 for attractive and repulsive groups respectively. A faint polygon at the value 1 and strong dots for the significant correlations are drawn.

By increasing ages and decreasing main sequence masses, the galactic groups are mainly correlated with:
- C-rich and S-type SR's and L variables [D(D)]
- C-rich variables and —but in a less proportion— O-rich Miras [TD(D)]
- O-rich SR's and Miras and C-rich Miras [D(TDb)]
- O-rich SRa's and Miras [TD(TDb) and ED(ED)]

Finally the groups D(TDf) and TD(TDf) are related to O-rich SRb's and irregular L variables.

This confirms that:
- nearly no C-rich stars are found in the groups corresponding to the lowest main sequence masses: the dredge-up may be more efficient for more massive stars (Busso et al. 1995)
- SRa's seem close to the Miras while SRb's have a behaviour similar to L variables.

3.2. Estimates of individual luminosities

Individual allocation of one star to one group allows to estimate its most probable distance and so its most probable luminosity. Fig. 4 shows a luminosity–color diagram in function of the galactic groups:
- stars belonging to the disk population (i.e. with the highest main sequence mass) seem to go up higher on the AGB than stars belonging to the thick disk population,
- stars identified as belonging to the extended disk population go up along the AGB as high as disk population LPVs: the metallicity effect makes them brighter.

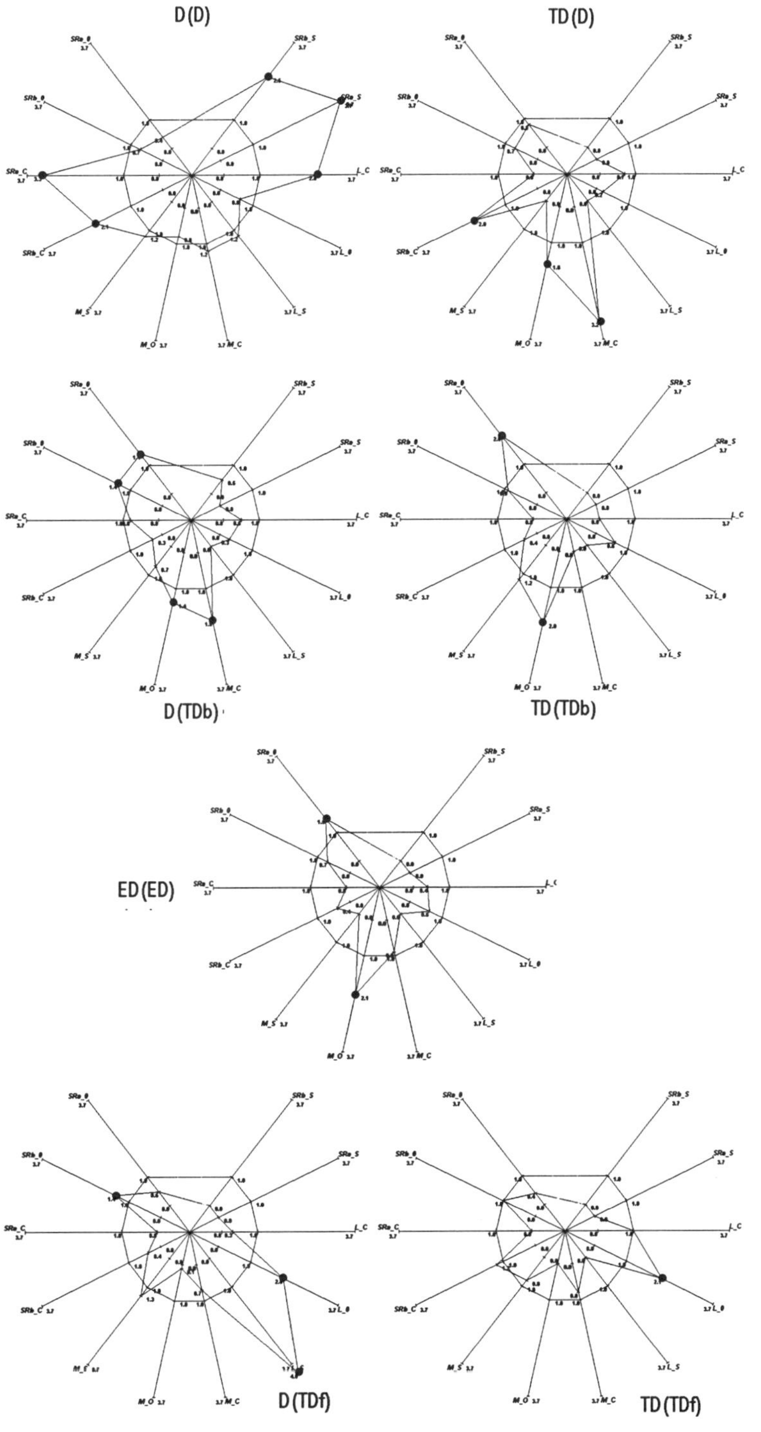

Figure 3. Correlations of variability types and K(IRAS) groups

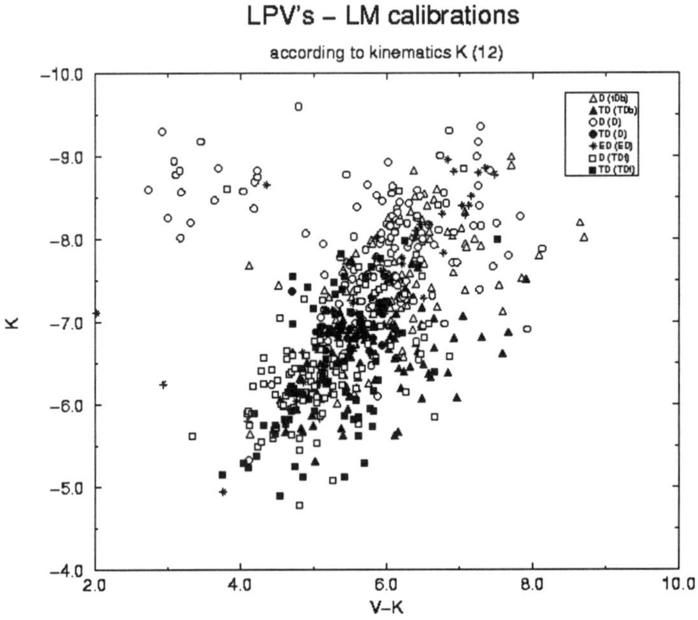

Figure 4. Color–luminosity diagram according to K(IRAS) groups

4. Conclusion

These first rough results that need to be refined show how much such considerations about luminosities joined to galactic properties could be a tool to constrain the models of evolution along the AGB.

But we must be very cautious with the conclusions because of possible biases due to a sampling effect: HIPPARCOS LPVs are the ones with the brightest optical counterparts. It will be necessary to include fainter LPVs which are bright infrared stars and LPVs without optical counterparts such as OH/IR stars. So it seems important to have an astrometrical and kinematical picture of the infrared sky in complement to photometric picture given by the IR surveys in course. The GAIA project is promising in this respect.

References

Busso M., Lambert D.L., Beglio L., et al., 1995, ApJ 446, 775

Luri X., Mennessier M.O., Torra J., Figueras F., 1996, A&A 117, 405

Asymptotic Giant Branch Stars
T. Le Bertre, A. Lèbre and C. Waelkens, eds.
I.A.U. Symposium 191, 1999

Theoretical pulsation models for Long Period Variables

Y. Tuchman
Racah Institute of Physics, The Hebrew University of Jerusalem, Jerusalem 91904, Israel

Abstract. Three issues, which are to my opinion the main obstacles in modeling the variability of Miras and Long Period Variables (LPVs), are presented and discussed:

a. The initial static models - Possible influences of the evolutionary details, the helium shell flashes and the use of the core-mass luminosity relation.

b. Modeling the convective energy transport - The mixing-length theory, its limitations and validity. First attempts for two-dimensional calculations of the convective flow in AGB stars.

c. The surface conditions - Introduction to the main difficulties and uncertainties. What should be done in the future.

Finally, the ancient problem, concerning the mode of pulsation, in Miras, is discussed using the recent radii observations confronted with the results of our full non-linear calculations.

1. Major problems in modeling Long Period Variables

1.1. The initial static configurations (Wagenhuber & Tuchman 1996)

This problem which, in practice, is considered of a secondary importance, might be crucial in the understanding of the general features of Miras or Long Period Variables (LPVs) as a group. All previous initial models for LPVs have been created using the, well known, core-mass luminosity (CML) relation (Paczyński 1971; Becker & Iben 1980). The real situation, however, is quite different. One should always remember that this core-mass luminosity (CML) relation was built for the evolutionary phase during which the star passes through the successive helium shell flashes. This relation was created to fit the luminosity around the peaks of the quiescent hydrogen-burning phase, between flashes.

The photospheric luminosity of a $1.5M_{\odot}$ as a function of time is presented in Fig. 1 (Wagenhuber & Tuchman 1996) for most of the relevant AGB phase. The core-mass (labeled M_C) as well as the luminosity, according to the CML relation ($L(M_c)$, dotted line), are presented as well.

At this point we are not interested in the very rapid flash phase (of about 1 000 years), even though, it might be very important as a trigger for mass loss (e.g. the creation of multiple shell planetary nebulae, Tuchman & Barkat 1980). Concentrating upon the slow inter-flash phases (each of about 10^5 years), the following conclusions should be drawn from Fig. 1:

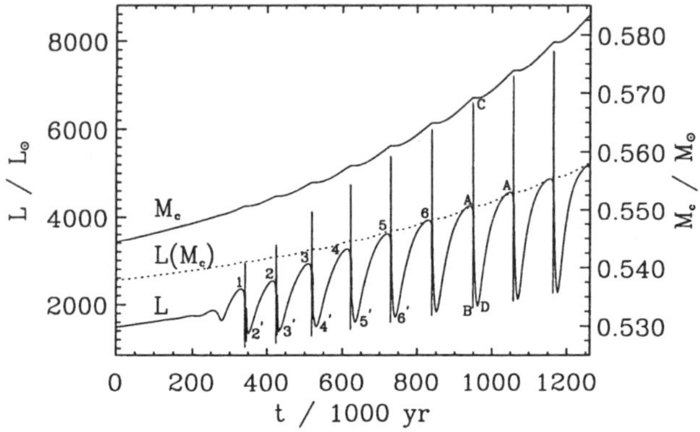

Figure 1. The photospheric luminosity evolution (L) of a $1.5M_{\odot}$ star along the AGB. The dotted line ($L(M_c)$) presents the commonly used evolution assuming the CML relation. Also shown is the evolution of the core mass M_c

a. The CML relation holds only for about 20% of the time. Elsewhere, in general, the real luminosity is *lower* than the one dictated by the CML relation.

b. The same luminosity can be obtained for a relatively large range of core masses (e.g. a luminosity of $2\,000L_{\odot}$ is obtained in *all* inter-flash periods).

c. In the first few flashes the luminosity is much below the values given by the CML relation.

d. The stellar luminosity might change by a factor of 2 or even more during one inter-flash period.

All these facts should be considered in any attempt to confront theoretical models of these stars with observations.

The stellar periods, like its luminosity, are not monotonically increasing as might be predicted from the CML relation (Fig. 2). As it can be seen from Fig. 2, the linear periods might change by almost a factor of *three* during one inter-flash time interval.

The first apparent conclusion caused by the differences between the real evolution and the one dictated by the CML relation (Fig. 2) is the relatively fast period changes. In our specific example, the period of the fundamental mode in a $1.5M_{\odot}$ star is increasing about 0.3 days per 100 years, to be compared with 0.02 days according to the CML relation. Much faster increase is expected for more massive cores ($> 0.8M_{\odot}$), with progenitors of about $4M_{\odot}$. In these cases a change of about 3 days per 100 years is expected (can it be confirmed with observations?).

A second conclusion is that stars might switch to a lower mode of pulsation while passing through the low luminosity range of the inter-flash periods. Stars located at these phases will probably appear as semi-regulars, in addition to the pre-flashes period.

A last important conclusion, derived from the real period evolution, is the possibility for a star with highly evolved core-mass (above $0.8M_{\odot}$) to have rel-

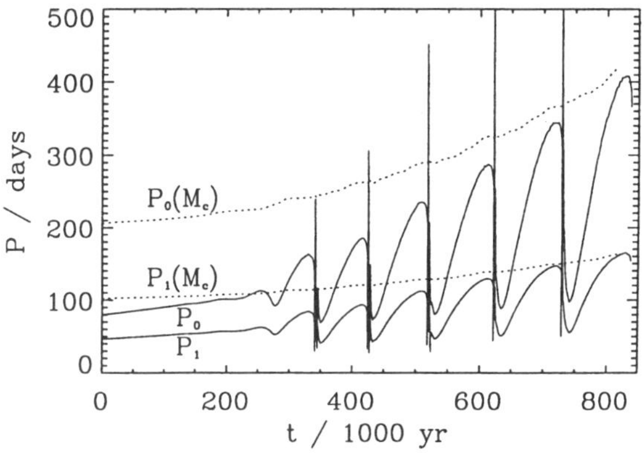

Figure 2. The period evolution, fundamental mode (P_0) and first overtone (P_1), calculated for the real luminosity sequence (full line) and for the one which assumes the CML relation (dotted line) as was presented in Fig. 1.

atively low luminosity, just starting an inter-flash period. This situation might explain the possibility to observe stars, with features characteristic for evolved stars, also in a region commonly believed to be occupied with low mass cores (low luminosity), as for example carbon stars (Lattanzio 1989).

1.2. Modeling the convection in LPVs (Asida & Tuchman 1997)

An accurate calculation of the convective energy transport in envelopes of AGB stars is one of the crucial demands in any attempt for modeling these stars. Its importance is caused by the fact that the convective region in these stars is very efficient and large (in mass and radius).

The method commonly used to calculate the convective flux in AGB stars is the well-known mixing length theory (MLT, Bohm-Vitense 1958) and its variants (Baker & Gough 1979). Four major defects are embedded in the MLT. The assumptions that the convective flux is: local, instantaneous, characterized by one-size eddies, and uses an unknown length-scale (the mixing-length parameter). Many attempts have been carried out in order to release one or more of these assumptions (e.g. Ulrich 1976; Stellingwerf 1982; Xiong, Cheng & Deng 1997; Canuto & Mazzitelli 1991). The success was always found to be only partial and in most cases the corrections were very complicated and difficult to be used. At this point a multidimensional (3D or at least 2D) nonlinear simulation was naturally required.

Previous two-dimensional (2D) calculations have already been performed for RR-Lyrae stars (Deupree 1977, 1985). However, the convection zone in these stars is narrow and its efficiency is relatively low, so that the convection is a small perturbation to the radiative structure. In these cases, the needed relaxation time, to follow the star thermal changes, from an initial radiative structure to the convective one, is therefore short enough to be traced by a two-dimensional calculation. This privilege does not exist in envelopes of AGB stars.

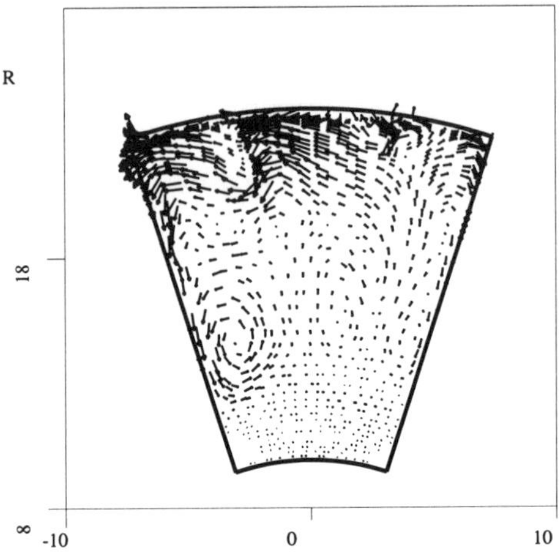

Figure 3. A typical velocity field of the convective flow in a 2D calculation

This problem, as will be explained below, practically eliminates the possibility for a simple direct calculation of the final 2D convective profile starting from an arbitrary initial configuration.

Two main time-scales are ruling the thermal relaxation of the stellar envelope at the AGB phase: the convective time-scale, i.e. the time required to reach a steady turbulence flow, which equals to several convective turnovers; and the thermal time-scale, i.e. the time needed for achieving thermal equlibirium.

The convective time-scale, for a typical AGB envelope, (the height of a convective cell divided by the mean convective velocity) is a few days. The thermal time-scale, on the other hand, is more than 50 years.

A typical time-step in a 2D code is about 100 seconds (mainly imposed by limitations of the relative energy changes). Each time-step needs about 1 second CPU time. Thus, for a typical AGB envelope, the convective flow evolves in a few CPU hours while the thermal relaxation takes about one year of CPU time. A full 2D simulation for an AGB envelope is therefore unrealistic with the present computational capabilities.

This difficulty was partially solved using the following procedure (Asida & Tuchman 1997): two dimensional calculations were followed only for about one year - long enough for the establishment of the convective flow, but still far below the thermal time-scale needed for achieving full energy balance. However, this period of time was long enough to estimate, for each calculated model, its deviation from an accurate energy balance. Thus, using different initial envelope models (created with different values for the mixing-length parameter), we could approach the desired steady configuration. There is no way to ensure the convergence of this procedure, simply because it is based on the assumption that one of the initial models (integrated using the MLT) is indeed very close to the final 2D structure. However, if it *does converge*, it will obviously lead to the correct profile.

The typical structure of the turbulence flow is presented in Fig. 3.

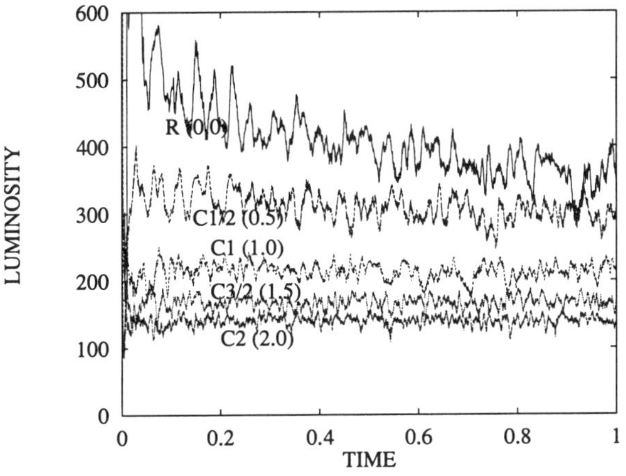

Figure 4. Outer luminosities in the 2D calculations for the different values of λ (0.,0.5,1.0,1.5 and 2.0) used in calculating the initial profile (upper curve for minimal λ, etc.)

Fig. 4 shows the surface luminosity evolution in a star of $1.2\,M_\odot$ star, with a core of $0.96\,M_\odot$, where a luminosity of $200L_\odot$ was assumed to enter into the envelope at its bottom. The surface luminosity evolution is shown for five different initial models which were integrated using the mixing-length theory with five different scale-lengths: 0, 0.5, 1., 1.5 and 2 times the local pressure scale height. Since the luminosity that enters the envelope was constantly equal to $200L_\odot$, it is clear from Fig. 4, that in those cases where the mixing-length parameter (λ) differs from ~ 1 the envelope is far from thermal equilibrium.

In those cases, where λ is smaller than 1, the outer luminosity is found to be much above the luminosity that goes into the envelope ($200L_\odot$) and the envelope is loosing energy. In these cases one can recognize a relatively fast decrease in the outer luminosity (Fig. 4). For the models created initially with λ larger than 1, the surface luminosity is below $200L_\odot$ and the envelope is accumulating energy. In these cases the change in the surface luminosity is too moderate to be recognized during a period of one year.

The similarity between the integrated 1D model, with $\lambda = 1$, and the explicit 2D simulation, is not restricted only to the outer luminosity but also to its detailed structure. Fig. 5 presents, as an example, the local energy flux in both models.

More details about the 2D calculations, methods and assumptions, can be found in Asida & Tuchman (1997).

The main conclusions that should be drawn from these preliminary 2D calculations are the following:

a. A static 1D model, integrated using the mixing-length theory for calculating the convective energy-flux, was found to be *very similar* with the exact 2D simulation. There is no proof that this result can be generalized to any convective envelope of an AGB star. Much more work is needed to clarify this point.

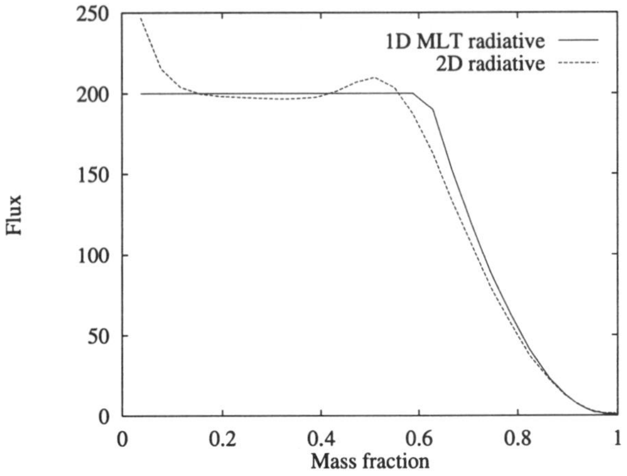

Figure 5. The detailed radiative luminosity, as a function of mass, for an AGB envelope, presented for the 2D converged model, compared with the best 1D integrated profile ($\lambda = 1$)(the convective luminosity= $200L_\odot$−radiative flux)

b. The value of the mixing-length parameter (λ) which was found to provide the best fit with the 2D calculations is one pressure scale-height. However, it should be emphasized that according to more recent 2D calculations, with much better space resolution, the best value for λ is found to be close to *1.5 rather than 1.*

It should be pointed that the results presented above have been obtained just for one model, at the beginning of the AGB phase. It is not clear, yet, whether the compatibility with the mixing length theory, on the whole, and the specific value fitted for the mixing-length parameter, are even nearly the same for other envelopes as well. This question should be clarified in the future.

The major drawback of the procedure we used for obtaining the final 2D configuration is that it *can not* be extended to dynamical calculations. Therefore, non-linear 2D calculations of the *variability* in Miras are still beyond our practical horizon.

1.3. The surface conditions

Unfortunately the modeling of pulsating LPVs is divided, for a long time, into the dynamical simulations of the envelope, on one side, and detailed atmospheric calculations on the other. This, unnatural, separation leads to the absence of adequate outer surface conditions for dynamical envelope simulations, and at the same time to a lack of reliable inner boundary conditions for the atmosphere calculations. In both cases synthetic prescriptions are used. For the envelope calculations one assumes a schematic dependence of the external pressure upon the surface radius, while, for the atmosphere calculations, a rigid piston moving with a dictated, time-dependent, velocity is assumed.

To my opinion, it is time for a combined (envelope + atmosphere) calculation. For the beginning, it might be sufficient to use very basic and simple methods, just in order to provide a reasonable evaluation for the mutual influence of these two adjacent media.

It is hard to guess the results of these combined calculations, but it is not unconceivable that it will lead to some considerable changes in the present theoretical estimates for the pulsating amplitude, spectra, mass loss and even for the non-linear periods of Miras and Long Period Variables.

2. On the mode of pulsation in Miras - Non-linear calculations confronted with radii observations (Ya'ari & Tuchman 1998)

In spite of the huge difference (a factor of two and more), between the linear values of the fundamental mode and the first overtone in Miras, a clear identification of the dominant pulsation mode, in these stars, is still uncertain. The problem is partially caused by observational uncertainties, as well as by our relative ignorance in estimating the total mass of the observed variables. However, the main contribution is in modeling the convective energy transport. The major ambiguity, in our models, is the exact value used as a scale-length in the mixing-length method (Bohm-Vitense 1958) for calculating the convective flux. For example, it turns out that the linear period of the *fundamental* mode, calculated using the mixing length theory with *one* pressure scale-height as the mixing-length parameter (λ), coincides with the *first overtone* period obtained for the same model but with $\lambda = 1.6$ pressure scale-height. This theoretical ambiguity can be resolved if a reasonable estimate of the surface radii of these Mira stars could be observationally determined. The reason is that unlike the *period-luminosity* relation, which has been shown to be very sensitive to the details in modeling the convective flux, the *period-radius* relation is found to be almost *independent* of the specific way convection is being treated. The new direct measurements of Miras radii (Haniff, Scholz & Tuthill 1995; van Belle et al. 1996; van Leeuwen et al. 1997) were very promising as a final decisive way to determine the actual mode of pulsation in these stars. Indeed, according to most of the investigations published since then, the *first overtone* is strongly favored as the dominant mode of pulsation (i.e. Haniff, Scholz & Tuthill 1995; Feast 1996; van Leeuwen et al. 1997).

To our opinion, *all* previous attempts confronting radii observations of Miras with the theoretical predictions, in order to determine their pulsation mode, have neglected two crucial facts that might *invert* their conclusions:

a. The observed data has always been compared to the theoretical *linear* values. Whereas the *nonlinear* periods, as well as the *nonlinear* radii, are found to be very different (Ya'ari & Tuchman 1996). Changes up to 30% should be expected.

b. All previous studies have ignored the variability in the surface radius caused by the stellar pulsation. Actually, it was assumed to be very small (less than 10%, Haniff, Scholz & Tuthill 1995). The exact phase at which the observation is performed was therefore ignored.

The incorrectness of this assumption is easily shown using the observed velocities in Mira variables; if ΔV is the observed velocity amplitude, the corresponding change in the surface radius, during one period (P), can roughly (assuming a linear approximation) be estimated to $(1/2)\Delta V P$. For a typi-

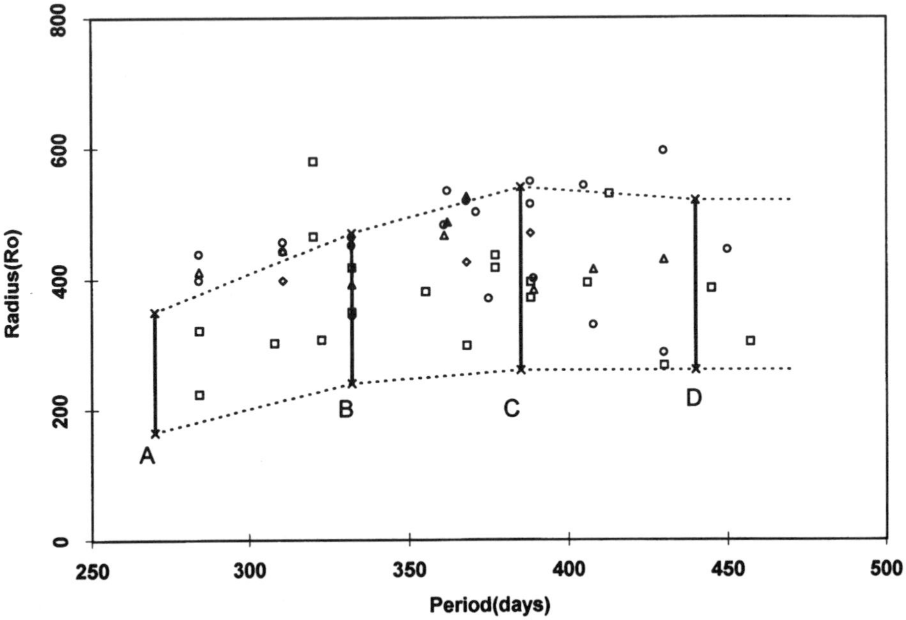

Figure 6. Effective observed radii, as a function of period, accumulated from the following sources: van Leeuwen, Feast, Whitelock & Yudin 1997 (circles), Feast 1996 (diamonds), Haniff, Scholz & Tuthill 1995 (triangles) and van Belle, Dyck, Benson & Lacasse 1996 (squares). The pulsational variability in the surface effective radii for the four chosen calculated models (labeled A-D) is shown as well

cal Mira $\Delta V = 20$ km/sec (Hinkle 1978; Hinkle, Scharlach & Hall 1984) and $P = 300$ days. Thus, the whole change in the surface radius, during one period, is about 400 solar radii. This is obviously a crude estimate but it is good enough to demonstrate the fact that the radius variability *can not* be ignored when comparing the observations with theory.

In the following, a detailed comparison between the observed radii of Mira variables and the theoretical *non-linear* calculated values, taking into account the two remarks above, is described.

Observed radii of Mira variables have been compiled from four different sources (van Leeuwen et al. 1997; Feast 1996; Haniff, Scholz & Tuthill 1995; van Belle et al. 1996). This data is presented in Fig. 6. Four theoretical models (with masses between 1.2 and 1.5 $M_{\odot}$) covering the relevant period range, have been selected, to be compared with this observed data. Their pulsational variability in the surface effective radii, minimum to maximum, are also marked in Fig. 6.

As can be seen from Fig. 6 *most* of the observed points are located between the boundaries connecting the minimum and maximum theoretical nonlinear radii of the Mira models pulsating in their *fundamental* mode. Only few observed radii are located above, but within the observational error of the theoretical range.

We may therefore conclude, in contrast to previous works, that the radii observations of Miras are *compatible* with the theoretical *non-linear* models pulsating in the *fundamental* mode.

References

Asida S.M., Tuchman Y., 1997, ApJ 491, L47

Baker N.H., Gough D.O., 1979, ApJ 234, 232

Becker S.A., Iben I., Jr., 1980, ApJ 237, 111

Bohm-Vitense E., 1958, Zs. f. Astrophysik 46, 108

Canuto V.M., Mazzitelli I., 1991, ApJ 370, 295

Deupree R.G., 1977, ApJ 211, 509

Deupree R.G., 1985, ApJ 296, 160

Feast M.W., 1996, MNRAS 278, 11

Haniff C.A., Scholz M., Tuthill P.G., 1995, MNRAS 212, 783

Hinkle K.H., 1978, ApJ 220, 210

Hinkle K.H., Scharlach W.W.G., Hall D.N.B., 1984, ApJS 56, 1

Lattanzio J.C., 1989, in Evolution of Peculiar Red Giant Stars, IAU Coll. 106, Cambridge University Press, p. 161

Paczyński B., 1971, Acta Astr. 21, 417

Stellingwerf R.F., 1982, ApJ 262, 330

Tuchman Y., Barkat Z., 1980, ApJ 242, 199

Ulrich R.K., 1976, ApJ 207, 564

van Belle G.T., Dyck H.M., Benson J.A., Lacasse M.G., 1996, AJ 114, 2150

van Leeuwen F., Feast M.W., Whitelock P.A., Yudin B., 1997, MNRAS 287, 1027

Wagenhuber J., Tuchman Y., 1996, A&A 309, 511

Xiong D.R., Cheng Q.L., Deng L., 1997, ApJS 108, 529

Ya'ari A., Tuchman Y., 1996, ApJ 456, 350

Ya'ari A., Tuchman Y., 1998, ApJ, submitted

The audience during one of the lively discussions
(here with Thomas Blöcker)

Asymptotic Giant Branch Stars
T. Le Bertre, A. Lèbre and C. Waelkens, eds.
I.A.U. Symposium 191, 1999

Multiperiodicity in semiregular variables

L.L. Kiss & K. Szatmáry

Department of Experimental Physics and Astronomical Observatory, JATE University, Szeged, Dóm tér 9, H-6720 Hungary

Abstract. We present a detailed period analysis for 98 red semiregular variables by means of Fourier and wavelet analysis of long-term visual observations carried out by amateur astronomers. The overwhelming majority of the studied stars show multiperiodic behaviour. We found two significant periods in 62 variables, while there are definite signs of three periods in 13 stars. 20 stars turned out to be monoperiodic with small instabilities in the period. Since this study deals with the general trends, we want to find only the most dominant periods.

The distribution of periods and period ratios is examined in the ($\log P_1, \log P_0/P_1$) plots. Three significant and two less obvious sequences are present which can be explained as the straight consequence of different pulsational modes. This hypothesis is supported by the multiperiodic variables with three periods. A clear distinction between C-rich and O-rich stars has been found in these diagrams suggesting a connection between the chemistry and pulsational characteristics.

1. Introduction

Mira and semiregular variables (SRVs) are pulsating low and intermediate mass red giants located on the asymptotic giant branch (AGB). The importance of these variables is highlighted by the fact that they are primary sources of the diffuse interstellar matter. The observed pulsational behaviours may lead to a better understanding of inner physical processes having crucial effects on the stellar evolution.

SRVs have amplitudes smaller than 2.5 mag. in V, while typical periods range from 25 to hundreds of days. Their basic properties (classification, temperature, luminosity, space distribution, important spectral features) were studied in a series of papers by Kerschbaum and his collaborators (e.g. Kerschbaum & Hron 1992, Kerschbaum & Hron 1994, Kerschbaum & Hron 1996). There has been an increasing evidence for a close relationship between the SRVs and Miras. Kerschbaum & Hron (1992, 1994) claimed that some stars are more closely related to Miras than the pure classification suggests. Szatmáry et al. (1996) found V Boo to have a dramatically decreasing amplitude over decades of time mimicking evolution from Mira to semiregular state. A similar phenomenon was found by Bedding et al. (1998) for R Dor, which implies that long period semiregulars may be a subset of Mira variables.

A detailed review on the mode of pulsation is given by Percy & Polano (1998), who showed that higher overtone pulsation is suggested by the observations (up to the third and fourth overtone). Wood (1998) presented 5 different period-luminosity sequences for the LMC red variables based on the MACHO photometric database, concluding similarly to Percy & Polano (1998) that even third and fourth overtones could be the dominant excited modes. Bedding et al. (1998) claim that the observed mode switching in R Dor occurs between the first and the third overtone. All these studies strengthen the idea that fundamental and first overtone pulsation in SRVs is an oversimplified assumption and the complex light variations may be due to many simultaneously excited modes.

The main aim of this study is to present a detailed light curve analysis for 98 SRVs based on long-term visual observations. The basic intention is to call for attention by demonstrating the general trends and the most interesting phenomena. A similar analysis was done by Andronov (1998).

2. Observations

The bulk of the analysed data were taken from three international databases of visual observations. These belong to the Association Française des Observateurs d'Etoiles Variables (AFOEV[1]), the Variable Star Observers' League in Japan (VSOLJ[2]) and the Hungarian Astronomical Association – Variable Star Section (HAA/VSS[3]).

The main selection criterion in choosing the sample was the length and the continuity of the light curves. In order to reach high resolution in the frequency domain, we usually kept only those stars with at least 10 years long data series. This corresponds to a frequency resolution (~length^{-1}) of $2.7 \cdot 10^{-4}$ cycles/day. But the usual length of the analysed data is about 50 years, occasionally 70-80 years (see the list of stars in Kiss et al. 1998).

In order to decrease the effects of scatter due to visual uncertainties, we averaged the light curves. The averaging procedure consists in taking 10-day bins and calculating the mean value from the individual points. Since the typical time scale in the sample of semiregular variables is about one hundred days this binning procedure does not smooth too much the light variations. A rough estimate of the resulting improvement in precision is as follows. The error of an individual observation is about ± 0.3 mag. For a given 10-day bin with 10 points within it, the standard error of the mean value will be $0.3/\sqrt{10} \approx 0.1$ mag. The amount of data and their distribution in our sample in most cases enable such precision. We have to mention that extremely deviant points in the original data were rejected during a close visual inspection at all light curves.

Two tests were made to check the reliability of the resulting mean light curves. The first is a comparison with available simultaneous photoelectric V-measurements. Although the spectral response function of the human eye differs

[1] `ftp://cdsarc.u-strasbg.fr/pub/afoev`

[2] `http://www.kusastro.kyoto-u.ac.jp/vsnet/gcvs`

[3] `http://www.mcse.hu/vcssz/data`

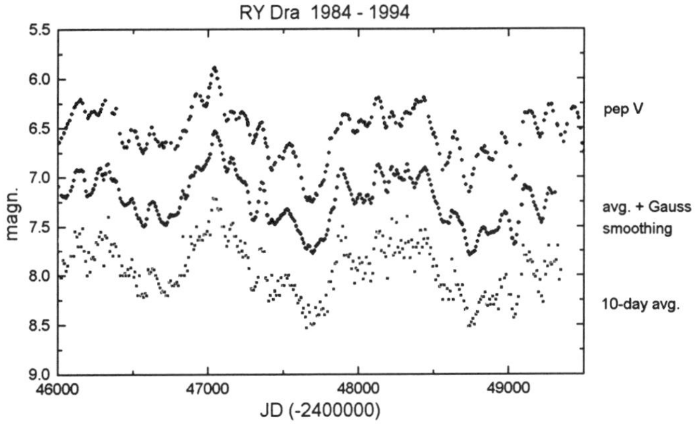

Figure 1. Comparison between the photoelectric and visual observations for RY Dra. See text for further explanation.

from the Johnson V filter, there were several attempts to calibrate the transformations. Stanton's (1981) formula is: $m_{vis.} = V + 0.182\ (B - V) - 0.15$. According to the available photoelectric observations of semiregular variables, their $B - V$ colours change with much smaller amplitudes than the V brightnesses do, thus it is a straightforward simplifying assumption that there is a constant shift between the photoelectric V and visual light curves. This is, of course, true only at a level of about 0.1 mag., which is in the range of the scatter of visual data.

A direct comparison is shown in Fig. 1 where we plotted visual data of RY Dra with simultaneous photoelectric V-measurements carried out by R.R. Cadmus (personal communication). The top curve is the photoelectric one, while the bottom curve is the corresponding 10-day mean of visual data. The middle curve is a noise filtered version of the lower curve, where noise filtering was done by a simple Gaussian smoothing with 8 days FWHM. Note, that while the visual curves were shifted in vertical direction for clarity, the distance between the smoothed visual and the photoelectric curves is the real difference caused by the colour effects. The observed average shift of 0.60 mag. is in good accordance with the predicted 0.44 mag. by Stanton's formula ($\langle B - V \rangle \approx$ 3.3 for RY Dra). The agreement between the visual and photoelectric curves are very good. A similar conclusion can be drawn using HIPPARCOS Tycho-V data (ESA 1997[4]): visual observations define light curves highly parallel to the photoelectric ones.

The second test was similar to that of Szatmáry & Vinkó (1992). We analysed artificial triply periodic data which simulated the typical observations: a 16 000 days long time-series with Gaussian noise added in order to adjust the signal-to-noise ratio to be about 100 and 1. Original periods and amplitudes could be determined with high accuracy even from the low quality data suggesting that the length of data is much more important than the actual S/N ratio.

---

[4] `http://astro.estec.esa.nl/Hipparcos`

3. Results

We calculated Discrete Fourier Transforms (DFT) of the finally adopted time-series. We did not try to extract as many periods as possible from the power spectra because the excited frequencies in semiregular variables are not so stable over time (see e.g. Mattei et al. 1998). The DFT may contain many peaks due to the cycle-to-cycle variations, therefore accepting those false peaks would be misleading. Our approach was to accept only the highest and most significant periodicities which were tested by whitening, cleaning and alias filtering. One-year alias peaks occur in many stars, while in some cases cross-production terms are present too (e.g. $f_0, f_1, f_0 \pm f_1$). We did an iterative period determination, where we checked the consistency of the fitted harmonics with the light curve itself after every step.

Among the studied 98 semiregulars we have found 20 purely monoperiodic stars and 62 stars with unambiguous multiperiodic behaviour (49 bi- and 13 triperiodic). We did not find any peak higher than the calculated noise level in 16 variables. We present the triply periodic SRVs in Table 1 with the determined periods and their amplitudes. The period uncertainty was estimated from the width of the peaks in the spectra at 90% of maximum. It has to be noted, that amplitude values have smaller significance because of the instability of the periods. This aspect was studied by wavelet analysis, which is a useful tool for studying temporal variations in the frequency content (see e.g. Bedding et al. 1998, Barthès & Mattei 1997, Szatmáry et al. 1996, Gál & Szatmáry 1995). Therefore, the actual amplitude values listed in Table 1 have only informative meaning about the relative strength of the periodicities found.

Table 1. Triply periodic variables. ΔT is the length of the analysed time series in days.

Star	$\langle m \rangle$	ΔT	P_0	A_0	P_1	A_1	P_2	A_2
U Cam	8.2	26800	2800(100)	0.13	400(30)	0.09	220(5)	0.09
RS Cam	8.7	25000	966(10)	0.17	160(1)	0.15	90(1)	0.12
ST Cam	7.3	28000	1580(10)	0.10	372(3)	0.12	202(2)	0.08
X Cnc	6.7	25800	1870(10)	0.08	350(3)	0.08	193(1)	0.09
Y CVn	5.7	28500	3000(100)	0.08	273(3)	0.06	160(2)	0.05
AF Cyg	7.2	26600	921(10)	0.08	163(1)	0.11	93(1)	0.11
TX Dra	7.6	26800	706(2)	0.10	137(1)	0.06	77(3)	0.07
UW Her	8.1	8600	1000(10)	0.09	172(1)	0.08	107(1)	0.09
Y Lyn	7.5	8500	1300(50)	0.41	611(10)	0.13	134(2)	0.12
Y UMa	8.6	32000	324(1)	0.16	315(1)	0.09	164(2)	0.06
RX UMa	10.6	33200	201(1)	0.37	189(1)	0.26	98(0.5)	0.16
V UMi	8.1	29000	737(10)	0.06	126(2)	0.04	73(0.5)	0.06
SW Vir	7.6	8500	1700(50)	0.15	164(1)	0.13	154(1)	0.20

In order to examine the general distribution of the obtained periods, we made pairs of periods in 62 multiperiodic variables and plotted the period ratios against the shorter periods in Fig. 2. Three very significant sequences can be

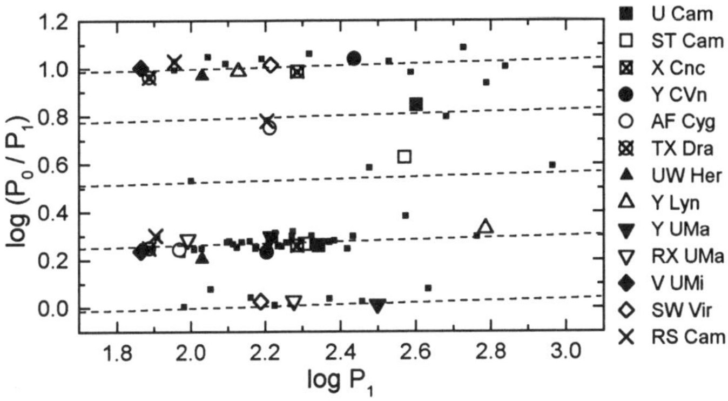

Figure 2. The Petersen diagram for all multiperiodic semiregulars. Triply periodic variables are plotted separately.

seen while two others may also occur. All of them are marked by dashed lines which were drawn as follows. We fitted a least-square linear trend to the most populated sequence and shifted that line to match the other sequences. It is very interesting how parallel these ridges are. One would expect such a separation between stars pulsating in different modes, assuming that the periodicities are due to pulsation. The most populated region contains stars with period ratios between 1.80 and 2.00. Mattei et al. (1998) pointed out that in their sample 63% of the multiperiodic stars (19 out of 30) fall in that range. Our larger sample supports this result very well, as 40 of 62 stars have period ratios around 1.90 ± 0.15 (65%).

The upper sequence in Fig. 2 is populated by stars with period ratios around 10. This is a well-known period ratio for semiregulars (e.g. Houk 1963, Percy & Polano 1998). The intermediate ratios were not discussed in the earlier papers, although they are present in the observational data to a smaller extent (see Figure 6 in Mattei et al. 1998).

The assumption of different modes can be tested with the triply periodic variables. The fact that their period ratios fit very well the five sequences is a supporting evidence. Recently, Bedding et al. (1998) have studied the mode switching in R Dor ($P_{long}/P_{short} = 1.81$), where they conclude that it probably pulsates in the first and third overtones. Furthermore, they suggest that all stars with similar period ratios do pulsate in the same modes. Wood et al. (1998) presented multiple structures in a diagram similar to Fig. 2 for the LMC red variables and again higher modes than fundamental and first overtone were suggested. We have to note that other possible explanations could not be excluded, as quasi-periodic cycles might be caused by other physical mechanisms than pulsation (e.g. duplicity, rotation, dust-shell dynamics).

We have tried to find connections between the periods, period ratios and several main physical properties, such as the infrared JHKL'M colours (Kerschbaum & Hron 1994), galactic latitude, mass-loss rates. Eventually no correlation has been found with these parameters. What we found is that there is a clear distinction between the O-rich and C-rich variables. Two regions can

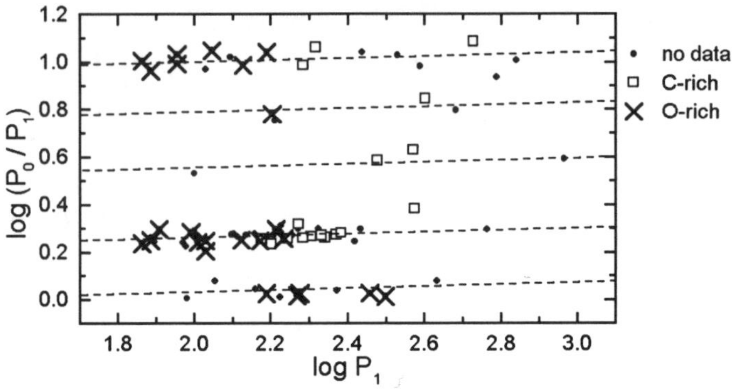

Figure 3. C-rich and O-rich occupy different regions in the Petersen diagram with quite narrow border around $P_1 \approx 160$ days.

be defined according to occurrence of stars with different abundances. There is no C-rich star in the lowest sequence while the border between the two types is quite narrow, centered around $P_1 \approx 160$ days. Therefore, there is likely some connection between the chemistry and pulsation which needs further theoretical investigation.

Acknowledgments. This research was supported by Hungarian OTKA Grants #F022249, #T022259 and Szeged Observatory Foundation. LLK gratefully acknowledges the kind help and financial support of the SOC. The NASA ADS Abstract Service was used to access data and references.

References

Andronov I.L., 1998, poster contribution, I.A.U. Symposium 191

Barthès D., Mattei J.A., 1997, AJ 113, 373

Bedding T.R., Zijlstra A.A., Jones A., Foster G., 1998, MNRAS 301, 1073

Gál J., Szatmáry K., 1995, A&A 297, 461

Houk N., 1963, AJ 68, 253

Kerschbaum F., Hron J., 1992, A&A 263, 97

Kerschbaum F., Hron J., 1994, A&AS 106, 397

Kerschbaum F., Hron J., 1996, A&A 308, 489

Kiss L.L., Szatmáry K., Cadmus R., Mattei J.A., 1998, A&A, submitted

Mattei J.A., Foster G., Hurwitz L.A., Malatesta K.H., Willson L.A., Mennessier M.O., 1998, in "HIPPARCOS - Venice'97", ESA SP-402, p. 269

Percy J.R., Polano S., 1998, ASP Conf. Series 135, 249

Stanton R.H., 1981, J. Amer. Assoc. Var. Star Obs. 10, 1

Szatmáry K., Vinkó J., 1992, MNRAS 256, 321

Szatmáry K., Gál J., Kiss L.L., 1996, A&A 308, 791

Wood P.R., et al., 1998, this volume

Asymptotic Giant Branch Stars
T. Le Bertre, A. Lèbre and C. Waelkens, eds.
I.A.U. Symposium 191, 1999

Fifteen Years of High Angular Resolution Studies of Mira's Atmosphere

Margarita Karovska

Harvard-Smithsonian Center for Astrophysics, 60 Garden Street, Cambridge, MA 02138, USA

Abstract.
I describe here results from high-angular resolution imaging studies of *o* Ceti (Mira). In 1983, we discovered that the atmosphere of the prototype of Mira-type variables is not symmetric. Since then, a number of multiwavelength high-angular resolution observations have confirmed the presence of asymmetries in Mira's atmosphere, and detected asymmetries in the atmospheres of other Mira-type variables. The high-angular resolution images of Mira obtained over the past fifteen years, including recent HST observations, show that the strength and shape of the asymmetries change as a function of wavelength and time. Plausible mechanisms for these asymmetries include hot spots, nonspherical pulsations, interaction with the companion and bipolar outflow. The presence of asymmetries in Miras could have serious impact on evolutionary models, and on the development of model atmospheres.

1. Introduction

o Ceti (Mira) is a pulsating cool giant (M2-M7 III) with a diameter of several hundred solar radii and a mass comparable to that of our Sun. The brightness of this prototype of long period large amplitude variable stars changes over seven magnitudes between its minimum and maximum, with a period of 332 days. Mira is losing mass at a rate of about 1×10^{-7} $M_{\odot}$/yr (e.g. Bowers & Knapp 1988) via a slow wind as it approaches the final stages of its evolution toward a planetary nebula. Mira has a companion at an angular distance of approximately 0.5 arcseconds. The companion is a white dwarf embedded in an accretion disk of variable luminosity.

Mira's system provides a unique laboratory for detailed study of the characteristics of pulsating giant atmospheres and wind accretion processes in interacting binaries, because it can be spatially resolved using high-angular resolution ground- and space-based observations, and the components can be studied individually.

2. High Angular Resolution Observations

Because of its large angular size, Mira can be easily resolved using high-angular resolution interferometric techniques. Pease (1931) first resolved this star using

Michelson interferometry and measured a diameter of 47 mas at 575 nm. Half a century later, Labeyrie et al. (1977) obtained the first high precision measurements of the diameter of Mira using the 5 m Palomar telescope and a new technique - speckle interferometry. Diameter measurements were made assuming spherical symmetry for the object. The speckle observations revealed a huge extension of Mira's atmosphere at optical wavelengths, from 50 mas to more than 100 mas. Measured diameters vary by factors as large as 2.5 over a few neighboring pseudo-continuum regions (Bonneau et al. 1982). These variations have been interpreted, quite roughly, in terms of optical depth variations due to the effects of pulsation on the structure of the extended atmosphere.

In November 1983 we carried out the first high-angular resolution imaging observations of this source using the Harvard-Smithsonian Astrophysical Observatory (CfA) speckle interferometry system. The observations were obtained at Mira's light curve phase 0.43 (Mattei, private communication) at the Steward Observatory 2.25 m telescope. We discovered a substantial asymmetry in Mira's images processed with Knox-Thompson image reconstruction techniques. The direction of the major axis of this asymmetry was nearly perpendicular to the direction toward the companion.

Follow up observations, carried out using larger aperture telescopes and therefore higher diffraction limited resolution, confirmed the presence of asymmetries in Mira's atmosphere. We carried out high-angular resolution speckle observations of Mira at several epochs using the 4-m telescopes at Kitt Peak and Cerro Tololo, and observed asymmetries at different wavelengths. For example, the observations carried out in 1988 at Cerro Tololo show an elongation in the 533 nm image with a position angle of the major axis of 135 ± 10 *degrees*, which is similar to the position angle of the asymmetry in the 775 nm image obtained at Kitt Peak one year earlier. However, the 1988 image shows some change in the geometry, and the degree of asymmetry does appear somewhat smaller. Since then, a number of observations using high-angular resolution techniques detected asymmetries in Mira and in other Mira-type variables using ground- and space-based observations (e.g. Karovska et al. 1991; Haniff et al. 1992).

Recently, we imaged the individual components of Mira's binary system using the HST Faint Object Camera (FOC) (Fig. 1). The multiwavelength images combined with the objective prism spectra provide a unique perspective of this accreting system and its components at wavelengths ranging from 150 nm to 550 nm (Karovska et al. 1997). We determined the spectral energy distribution of each component unambiguously at UV and optical wavelengths and obtained the first high spatial resolution images at UV wavelengths.

The high angular resolution of the COSTAR-corrected FOC f/96 camera, and the small pixel size provided an opportunity for exploring Mira's atmosphere at minimum light and at spatial scales as small as 30 milliarcseconds. The images resolved Mira at several optical wavelengths, and for the first time in the UV (see Fig. 2 and Fig. 3). The size of Mira's atmosphere appears smaller in the quasi-continuum spectral regions than in the regions containing TiO bands. If this is interpreted in terms of a stratification in the atmosphere of Mira, then depending on the spectral region in which the observations were made, we effectively probed different layers in the atmosphere.

Figure 1. Direct HST/FOC images of Mira A (right) and Mira B (left) in the F501N filter. The separation between the components is 0.58 arcseconds.

The most surprising result came from the first UV image of Mira recorded using a filter centered at 346 nm (Fig. 2). The deconvolved image shows a hook-like appendage extending from Mira in the direction of the companion (Karovska et al 1997; NASA and STScI Press release STScI-PRC97-26, 1997). This feature could be material from the giant being gravitationally drawn toward the accreting companion, a physical phenomenon never directly observed at this small spatial scale.

We detected asymmetries in the Mira multiwavelength images deconvolved using the images of the companion as point sources. For example, the 501 nm deconvolved image of Mira shows an asymmetry in the extended atmosphere in the south-east direction (Fig. 3). The apparent size of the major axis of the dominant asymmetry as measured using elliptical gaussian fitting to the deconvolved images is 56+/-1 mas, with a position angle of 175 +/- 2 degrees. The estimated ratio between the minor and the major axis of the asymmetry is 0.2.

This asymmetry is similar to the asymmetries previously detected using ground-based interferometry (e. g. Karovska et al. 1991; Haniff et al. 1992). HST multiwavelength images also detected low brightness arc-like features in Mira's circumstellar environment. These structures are reminiscent of the morphology of pre-planetary nebulae and appear to carry signatures of a bipolar outflow.

3. Discussion

High-angular resolution observations of Mira show direct evidence that the atmosphere of this star departs substantially from spherical symmetry. The appearance of the asymmetry changes as a function of time and wavelength.

The cause of the asymmetries in the Mira atmosphere has not yet been determined. Thermal, density, or dynamic instabilities could create asymmetries in brightness distribution on the surface of Mira, or geometric asymmetries with

Figure 2. The first UV image of Mira showing a hook-like appendage extending from Mira in the direction of the companion (NASA and STScI Press release STScI-PRC97-26, 1997). This deconvolved image has been magnified for visualization purpose.

orientation that changes with time. For example, dynamic instabilities apparent in some atmospheric models can produce "shock pairing" which in turn could produce departures from spherical symmetry (Willson & Bowen 1988). Nonradial pulsations could also create departures from spherical symmetry similar to the asymmetries observed in Mira. Convective instabilities (Schwarzschild 1975) could result in a formation of large supergranules or spots on the photosphere which if partially resolved could create asymmetries in the diffraction limited images.

Rapid rotation, or the companion and its orbital motion, could affect the upper atmosphere of Mira and create favorable conditions for a bipolar outflow. In fact, the HST images show structures that could be associated with a bipolar outflow. This is in agreement with radio observations which detected an extended bipolar outflow in the CO $J=2-1$ and $J=1-0$ emission of the circumstellar envelope of Mira (Planesas et al. 1990). The observed structure has been interpreted as partial collimation of the stellar wind due to an enhanced density of gas in the equatorial plane of the inner envelope of Mira.

The presence of substantial asymmetries in the atmospheres of Mira-type variables has an important effect on the measured quantities and the model atmospheres for these stars. For example, even without the asymmetries, it is difficult to determine directly the "true" photospheric diameter, the crucial parameter for establishing the effective temperature and the mode of pulsation because of the effects of the extended atmosphere (Willson & Bowen 1988). Currently, observers use model brightness distribution derived from various model atmosphere codes to correct the measured diameters and estimate the "photospheric" diameter and the mode of pulsation of Miras. These models are often very simplistic, and assume that the extended Mira atmosphere is spherically

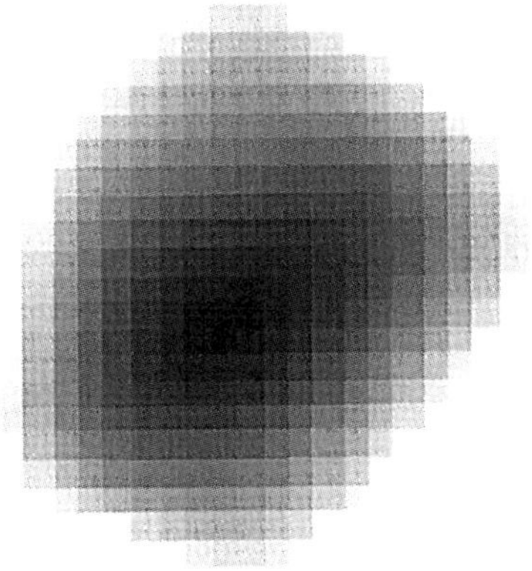

Figure 3. Deconvolved HST image of Mira at 501 nm showing substantial asymmetry in the south-east direction. The image has been magnified five times using bicubic spline interpolation for visualization purpose.

symmetric. Therefore, the results obtained using these models should be handled with care.

Continuing the high-angular resolution studies of the detailed structure of Mira's atmosphere, and of the origin and the nature of the asymmetries is crucial for further development and testing of model atmospheres and evolutionary models.

Acknowledgments. This work was supported in part by STScI grant GO-5822.01-94A to the Smithsonian Astrophysical Observatory. MK is a member of the AXAF Science Center and is partially supported by NASA, Contract #NAS8-39073.

References

Bonneau D., Foy R., Blazit A., Labeyrie A., 1982, A&A 106, 235

Haniff C.A, Ghez A.M., Gorham P.W., et al., 1992 AJ 103, 1667

Karovska M., Nisenson P., Papaliolios C., Boyle R.P., 1991, ApJ 374, L51

Karovska M., Hack W., Raymond J., Guinan E., 1997, ApJ 482, L175

Labeyrie A., Koechlin L., Bonneau D., Blazit A., Foy R., 1977, ApJ 218, L75

Pease F.G., 1931, Ergebn. Exacten. Naturwis. 10, 84

Planesas P., Kenney J.D.P., Bachiller R., 1990, ApJ 364, L9

Schwarzschild M., 1975, ApJ 195, 137

Wilson R.W., Baldwin J.E., Busher D.F., Warner P.J., 1992, MNRAS 257, 369

Peter Wood and Paola Marigo during the pannel discussion:
"Stellar evolution on the AGB: a critical view on the models"

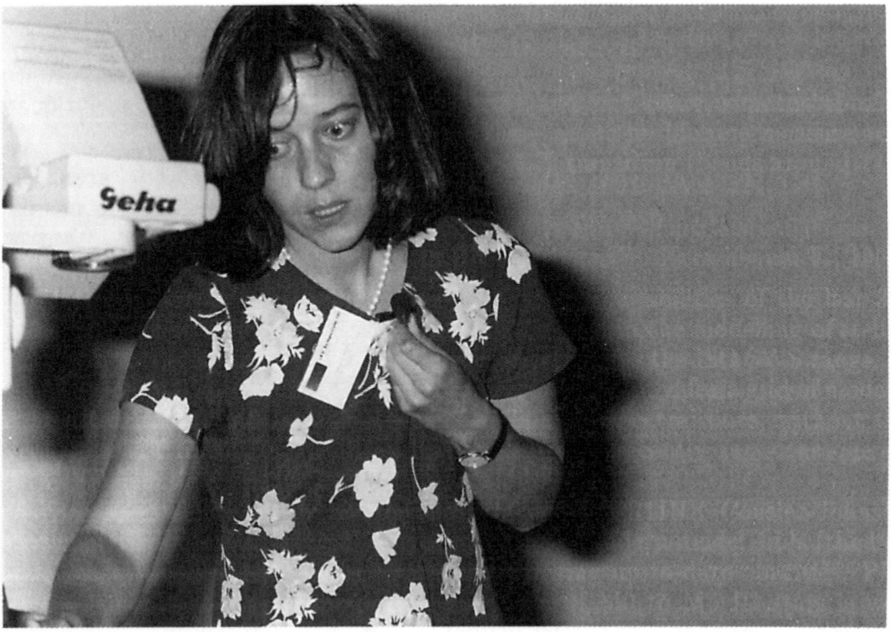

Asymptotic Giant Branch Stars
T. Le Bertre, A. Lèbre and C. Waelkens, eds.
I.A.U. Symposium 191, 1999

Measurements of the changes in angular diameter of Mira variables with pulsation phase

J. S. Young, J. E. Baldwin, R. C. Boysen, C. A. Haniff, D. Pearson, J. Rogers, D. St-Jacques, P. J. Warner and D. M. A. Wilson

Mullard Radio Astronomy Observatory, Cavendish Laboratory, Madingley Road, Cambridge CB3 0HE, UK

Abstract.
We report the direct detection of cyclic diameter variations in the Mira variable χ Cygni. Interferometric observations made between 1997 July and 1998 September, using the Cambridge Optical Aperture Synthesis Telescope (COAST) indicate periodic changes in the apparent angular diameter with amplitude 45 per-cent of the smallest value.

The measurements were made in a 50 nm bandpass centred on 905 nm, which is only moderately contaminated by molecular absorption features. To assess the effects of atmospheric stratification on the apparent diameter measured in this band, we have also measured near-infrared diameters for a sample of five Miras, in both the J-band (1.3 μm) and Wing's (1971) 1.04 μm band, which is expected to isolate essentially pure continuum emission. We present J-band visibility curves which indicate that the intensity profiles of the stars in the sample differ greatly from each other.

1. Introduction

Diameter changes of Mira-variable stars have been measured previously by van Belle et al. (1996), who found variations with phase within a sample of one or two-epoch diameter measurements of 18 stars, and by Tuthill, Haniff & Baldwin (1995), who made seven observations of *o* Ceti spread over a three year period.

Cyclic variations in the apparent diameter of an individual Mira were first detected by Burns et al. (1998, henceforth B98), who monitored R Leonis with COAST (Baldwin et al. 1998) and the William Herschel Telescope for almost two pulsation cycles.

In this paper we present COAST diameter measurements of χ Cygni, which exhibit phase-coherent changes similar to the variations previously seen in R Leonis.

2. Phase-coherent diameter changes

2.1. Observations

χ Cygni was monitored with COAST at two wavelengths: 905 nm and 1290 nm. Observations were made on 22 nights (between the dates of 1997 July 20 and 1998 September 19), covering a full pulsation cycle. The observation and data reduction procedures are outlined by Burns et al. (1997).

2.2. Results and discussion

In order to characterise the stellar size, we have fitted both uniform disk and Gaussian functions to the visibility amplitude data. At a wavelength of 905 nm, a Gaussian was a better fit to the long baseline data ($V < 30\%$), even near maximum light, and so only the results of the Gaussian fits are presented here. However, the phase-coherent modulation seen in the Gaussian diameters at 905 nm is also seen in the 905 nm uniform disk diameters, which suggests that most of the variation is due to a real change in the *scale* of the emitting region.

The apparent angular diameter of χ Cygni at 905 nm (Fig. 1) clearly varies in a phase coherent manner: the star appears smallest at visual phase 1.0 and largest near phase 0.5, the same as B98 found for R Leonis (see Fig. 2). The 1992 measurement of Haniff, Scholz & Tuthill (1995) is consistent with the diameter modulation seen during 1997 and 1998 and therefore suggests that the variation in χ Cygni may be coherent over five cycles. The peak-to-peak amplitude of the variation is approximately 45% of the smallest diameter. The amplitude of any variation at 1.3 μm is much smaller. B98 found that the diameter of R Leonis varied by ~35% at both 830 nm and 940 nm.

The flux transmitted by the 830, 905 and 940 nm filters is dominated by continuum emission, with only a small amount of atomic and molecular blanketing. There is however weak contamination by TiO absorption bands, whose strength varies with pulsation phase (Spinrad & Wing 1969). Hence the detected photons originate from several different layers within the stratified photosphere of the star, whose relative contributions vary with phase. It is clear that the diameter we measure in these wavebands does not simply tell us the position of the continuum-forming layers.

Hofmann, Scholz & Wood (1998) have calculated the apparent diameters of six model Mira variables covering a wide range of effective temperatures, at various wavelengths and pulsation phases. The physical size of their model stars is largest at phase ~0.2, therefore the apparent diameter in the continuum is larger at maximum than at minimum light for all six of the model series. At near-continuum wavelengths increased contamination by molecular bands tends to enlarge the star near minimum. Despite this effect, five of the six model series still appear larger at maximum, the exception being the P series (low-$T_{\rm eff}$, fundamental-mode pulsation), whose apparent diameter at 820 nm and 920 nm is 15–30% larger at minimum than at maximum. The change in our Gaussian fit between minimum and maximum should be slightly smaller than this, if the Hofmann et al. predictions for phase-dependent limb-darkening are correct. These differences are much less than the 45% we have observed for χ Cygni at 905 nm, which may suggest that the treatment of molecular opacity in the models is inadequate.

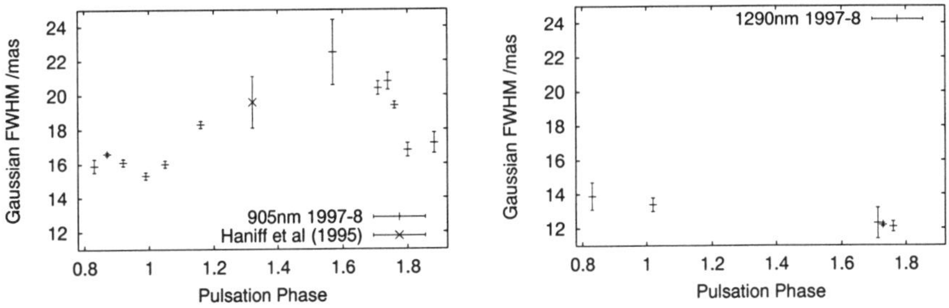

Figure 1. Apparent angular diameters (Gaussian full widths at half-maximum intensity) and 1σ errors for χ Cygni at 905 nm (left) and 1290 nm (right), plotted against the phase of the visual light curve (zero is maximum). The diagonal cross at phase 1.32 is the 1992 measurement at 902 nm of Haniff, Scholz & Tuthill (1995). The diameter increases slowly from phase 1.0 to phase ~1.5, then decreases rapidly between phases 1.6 and 1.8. A similar slow diameter increase during the first half of the pulsation cycle is seen in the B98 measurements of R Leonis (Fig. 2).

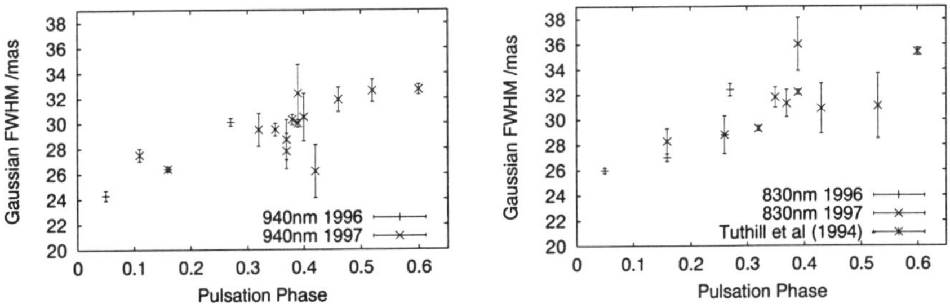

Figure 2. Apparent angular diameters (Gaussian full widths at half-maximum intensity) and 1σ errors for R Leonis at 940 nm (left) and 830 nm (right), plotted against the phase of the visual light curve (zero is maximum). Horizontal marks are 1996 data, and diagonal crosses are 1997 data. The 1996 and 1997 points are the result of Gaussian fits to the visibility amplitude measurements of B98 (who plot the results of uniform disk fits to the same data). The asterisk at phase 0.25 is the 1992 measurement of Tuthill et al. (1994). Note the different horizontal scale from Fig. 1.

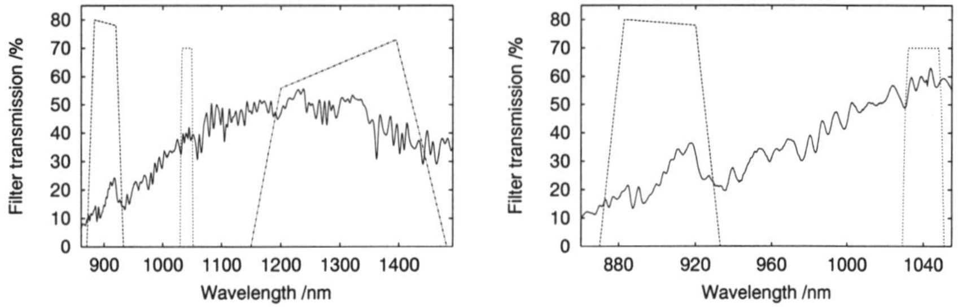

Figure 3. Wavebands used for observations, superimposed on the spectrum of an M8 giant star, from Pickles (1998). From left to right, the centre wavelengths and half-power bandwidths of the filters are 905(50) nm, 1040(19) nm and 1290(260) nm. The right plot shows the shorter-wavelength region in more detail.

3. Variation of apparent diameter with waveband

In order to determine the effect of atmospheric stratification on apparent diameters measured in different wavebands, we have begun a programme of diameter measurements in two bands where we expect most of the radiation to originate from a single layer: Wing's 1.04 μm continuum band (Wing 1971), and the J-band at 1.3 μm. Contemporaneous measurements were made in the 905 nm band. The filter bandpasses are shown schematically in Fig. 3. We have observed a sample of five Mira variables in these bands.

For those targets where we have data on a range of baselines, we find:

- Visibility amplitudes at 905 nm are all well fit by a Gaussian.
- At 1.3 μm a Gaussian model is generally a poor fit to the long-baseline data, but there are some striking differences between different stars in the sample, as illustrated by the visibility curves shown in Fig. 4.

In order to compare measurements at different wavelengths, we fitted a Gaussian model even when this model was a poor fit at long baselines. In such cases the long baseline visibilities were removed. We found that the apparent Gaussian FWHM varies substantially with wavelength. The scale of the variations is clearer if we normalise the 905 nm and 1040 nm diameters to the 1.3 μm diameter (which varies least with phase); these ratios are given in Table 1.

The apparent diameter at 905 nm is always larger than that at 1.04 μm or 1.3 μm, probably due to absorption by TiO in the outer atmosphere of the star. The ratio of the diameter at 905 nm to that at 1.3 μm appears to be phase-dependent.

The 1.04 μm continuum band diameters are also all *larger* than the J-band diameters. This may be due to a combination of the following:

- Molecular absorption in the outer atmosphere (which leads to a larger diameter being measured).

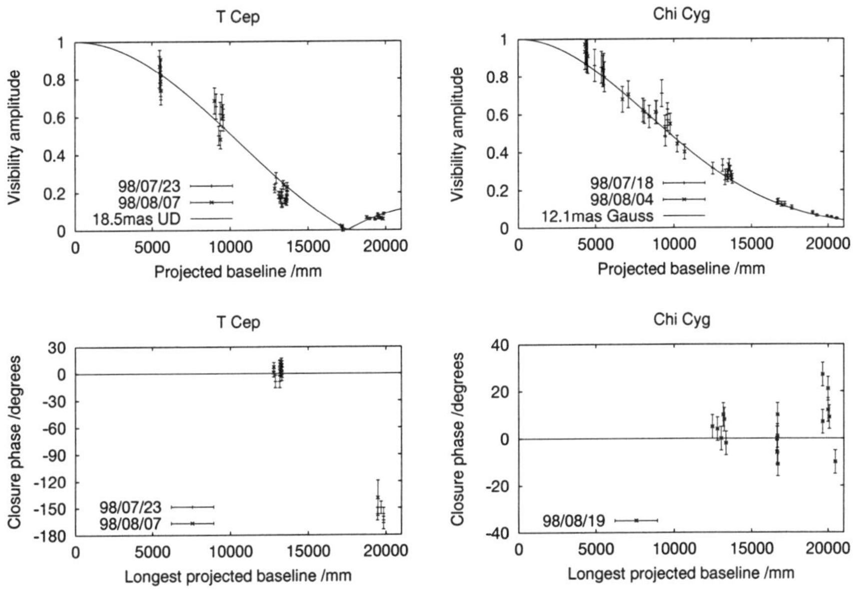

Figure 4. J-band visibility curves (top) and corresponding closure phases (bottom) for T Cephei (left) and χ Cygni (right). The visibility amplitude for T Cephei goes through a zero, and at the same baseline the closure phase changes from 0 to ~180 degrees, as expected for a centro-symmetric disk. There is evidence for some limb-darkening. In contrast, there is no null in the visibility function for χ Cygni, which is very well fit by a Gaussian model for the intensity profile.

Table 1. Ratio of apparent angular diameter D_λ (Gaussian FWHM) at 905 nm and 1040 nm to that at 1290 nm. The apparent diameter at 1290 nm is given in the 6th column.

Star	Date(s)	Phase	D_{905}/D_{1290}	D_{1040}/D_{1290}	D_{1290}
χ Cyg	97/07/19, 20	0.83	1.14 ± 0.07		13.9 ± 0.8
	97/09/22, 97/10/05	1.00	1.14 ± 0.04		13.4 ± 0.4
	98/07/18, 24	1.73	1.72 ± 0.04	1.57 ± 0.04	12.1 ± 0.1
	98/08/04	1.76	1.60 ± 0.04		12.1 ± 0.3
	98/08/04, 18	1.78	1.39 ± 0.05	1.14 ± 0.06	12.1 ± 0.3
R Cas	97/10/18, 22	0.52	1.32 ± 0.14		16.1 ± 0.6
	98/08/02, 05	1.19	1.15 ± 0.02		16.6 ± 0.2
T Cep	98/08/02, 07	0.23	1.12 ± 0.03		13.7 ± 0.2
o Cet	98/08/10, 18, 19	0.67	1.34 ± 0.03	1.16 ± 0.03	19.4 ± 0.3
T Cas	98/08/18, 19	0.60	1.21 ± 0.04	1.06 ± 0.05	7.8 ± 0.2

- Wavelength-dependent continuous opacity (an increase in opacity leads to a larger diameter being measured).
- Wavelength-dependent limb-darkening (increased limb-darkening makes the star appear smaller).

We will consider molecular absorption first. Our 1.04 μm bandpass may be contaminated by VO, but the VO band only covers at most 10% of the filter bandpass, so any increase in the observed apparent diameter due to VO will be small. A similar effect may enlarge the J-band diameter: the spectra of cool stars have a wide absorption feature at 1.4 μm due to H_2O . However, the increase in apparent diameter will be small because most of the photons in the wavelength range of the feature will be absorbed by telluric water and thus not detected.

It therefore seems unlikely that the observed apparent diameter ratios can be explained by spectral features. The dominant source of near-infrared continuous opacity in cool stars is H^-, and the total H^- opacity is greater at 1.04 μm than at 1.3 μm (Wing & Rinsland 1981). This, together with the effect of small changes in limb-darkening, may explain the observed diameter ratios.

4. Conclusions

We have measured phase-coherent variations in the apparent diameter of χ Cygni. The amplitude at a wavelength of 905 nm is 45% of the smallest diameter.

We have observed a sample of five Miras in two other, less-blanketed near-infrared bands. Brightness profiles inferred from J-band data show striking differences between the sample members, and the J-band diameters are systematically smaller than those measured in the 1.04 μm continuum window.

Our results provide important new constraints for theoretical models of Mira variables.

References

Baldwin J.E., et al., 1998, Proc. SPIE 3350, 736

Burns D., et al., 1997, MNRAS 290, L11

Burns D., et al., 1998, MNRAS 297, 462 (B98)

Haniff C.A., Scholz M., Tuthill P.G., 1995, MNRAS 276, 640

Hofmann K.-H., Scholz M., Wood P.R., 1998, A&A 339, 846

Pickles A.J., 1998, PASP 110, 863

Spinrad H., Wing R.F., 1969, ARA&A 7, 249

Tuthill P.G., Haniff C.A., Baldwin J.E., Feast M.W., 1994, MNRAS 266, 745

Tuthill P.G., Haniff C.A., Baldwin J.E., 1995, MNRAS 277, 1541

van Belle G.T., Dyck H.M., Benson J.A., Lacasse M.G., 1996, AJ 112, 2147

Wing R.F., Rinsland C.P., 1981, Rev. Mexicana Astron. Astrof. 6, 145

Wing R.F., 1971, in Proc. Conf. on Late-Type Stars (Kitt Peak Natl. Obser. Contr. No. 554), G.W. Lockwood and H.M. Dyck (eds.), p. 145

Asymptotic Giant Branch Stars
T. Le Bertre, A. Lèbre and C. Waelkens, eds.
I.A.U. Symposium 191, 1999

MACHO observations of LMC red giants: Mira and semi-regular pulsators, and contact and semi-detached binaries

P.R. Wood[1], C. Alcock[2,5], R.A. Allsman[3], D. Alves[4,5], T.S. Axelrod[1], A.C. Becker[2,6], D.P. Bennett[2,5,7], K.H. Cook[5], A.J. Drake[1], K.C. Freeman[1], K. Griest[2,8], L.J. King[2,7], M.J. Lehner[9], S.L. Marshall[5], D. Minniti[5,10], B.A. Peterson[1], M.R. Pratt[11], P.J. Quinn[12], C.W. Stubbs[2,6], W. Sutherland[13], A. Tomaney[6], T. Vandehei[2,8] and D.L. Welch[14]

Abstract. The MACHO data base has been used to examine light curves of all red giant stars brighter than $M_{\rm bol} \sim -2$ in a $0.5^\circ \times 0.5^\circ$ area of the LMC bar. Periods, often multiple, have been searched for in all stars found to be variable. Five distinct period-luminosity sequences have been found on the low mass ($M \lesssim 2.25 M_\odot$) giant branch. Comparison of observed periods, luminosities and period ratios with theoretical models identifies Miras unambiguously as radial fundamental mode pulsators, while semi-regular variables can be pulsating in the 1st, 2nd or 3rd overtone, or even the fundamental. All these variables lie on just 3 of the 5 distinct sequences, and they all appear to be on the AGB.

The fourth sequence contains red giants on the first giant branch (FGB) or at the red end of the core-helium burning loops of intermediate mass stars ($M \gtrsim 2.25 M_\odot$). The light curves of these stars strongly suggest that they are contact binaries, and they make up ~0.5% of stars within 1 mag. of the FGB tip. Stars on the fifth sequence show semi-regular, eclipse-like light curves. The light curves and periods of these stars suggest that they are in semi-detached binaries, transfering mass to an invisible companion via a stellar wind or Roche lobe overflow. They make up ~25% of AGB stars. If the existence of these red giant contact and semi-detached binaries is confirmed, then extant theories of binary star evolution will require substantial modification.

[1]Mount Stromlo and Siding Spring Observatories, Australian National University. [2]Center for Particle Astrophysics, University of California, Berkeley. [3]Supercomputing Facility, Australian National University. [4]Department of Physics, University of California, Davis. [5]Lawrence Livermore National Laboratory. [6]Departments of Astronomy and Physics, University of Washington. [7]Department of Physics, University of Notre Dame. [8]Department of Physics, University of California, San Diego. [9]Department of Physics, University of Sheffield. [10]Departamento de Astronomia, P. Universidad Católica, Santiago. [11]Center for Space Research, MIT. [12]European Southern Observatory. [13]Department of Physics, University of Oxford. [14]Department of Physics and Astronomy, McMaster University.

1. Introduction

The MACHO experiment (Alcock et al. 1992) has now accumulated light curve data for LMC stars over an interval of ~2 200 days without any substantial breaks. This is sufficiently long that periods can be derived for essentially all the red, long-period variables (LPVs). Furthermore, the long time span is ideal for picking out secondary periods. In this paper, we examine the light curves of all red variables in a small 0.5°×0.5° area of the LMC bar centred on $5^h28^m14.8^s$, -69°45′42″, J2000.

2. Data calibration and period searches

The MACHO data consists of light curves in non-standard red (R_M) and blue (B_M) bandpasses. Variables were searched for in the area of the (R_M,B_M-R_M) plane redder than the main-sequence and brighter than ~ 1 mag. from the tip of the FGB ($M_{\rm bol} \lesssim -2$). For each variable, up to three significant periods in the range $20 < P(d) < 1\,200$ were determined. A total of 1 431 red stars were identified as variable: 398 had no significant period, 608 had 1, 379 had 2 and 46 had 3. In addition, 107 Cepheids and 22 other blue variables were found in the field. For comparison with theory, B_M and R_M were transformed to V and I (Cousins) using photometry obtained with the 1 m telescope at Siding Spring Observatory (SSO). J and K images of the field were obtained with the 2.3 m telescope at SSO.

3. Results

Fig. 1 shows the period-luminosity (P-L) relation for the red variables, where there is a point in the figure for each period found (i.e. a star with more than 1 significant period will contribute more than 1 point to the figure). The quantity $I_W = I_0 - 1.38*(V-I)_0$ was chosen as the luminosity index as it is almost reddening free and the theoretical period-luminosity-colour relation for LPVs suggests that stars pulsating in a given mode should lie on the same (I_W, log$P$) relation regardless of mass. Stars with $(J-K)_0 > 1.4$ are assumed to be carbon stars and are shown as filled circles. The positions of the FGB tip and the minimum luminosity for thermally-pulsing AGB (TPAGB) stars are indicated.

It is clear that there are 4 parallel sequences (labelled A, B, C and D) above the TPAGB minimum in Fig. 1, and another sequence (E) extending down to the minimum luminosity $I_W \sim 14.25$ of the variable search (most of these sequences are visible in Cook et al. 1997). An examination of the luminosity functions of variable and non-variable stars on the giant branch shows that >80% of definite AGB stars (stars above the FGB tip) are variable. The luminosity function for variable stars below the FGB tip (and above the TPAGB minimum) looks like an extension of the luminosity function for definite AGB stars, suggesting that all the variable stars are on the AGB. A similar conclusion was reached by Alves et al. (1998). Note that the carbon stars are mostly confined to the brighter end of the observed luminosity range, and they lie mainly on sequences B and C.

Past studies (Feast et al. 1989; Hughes & Wood 1990) have shown that Mira variables lie on a distinct sequence in the (K, log$P$) plane. The present

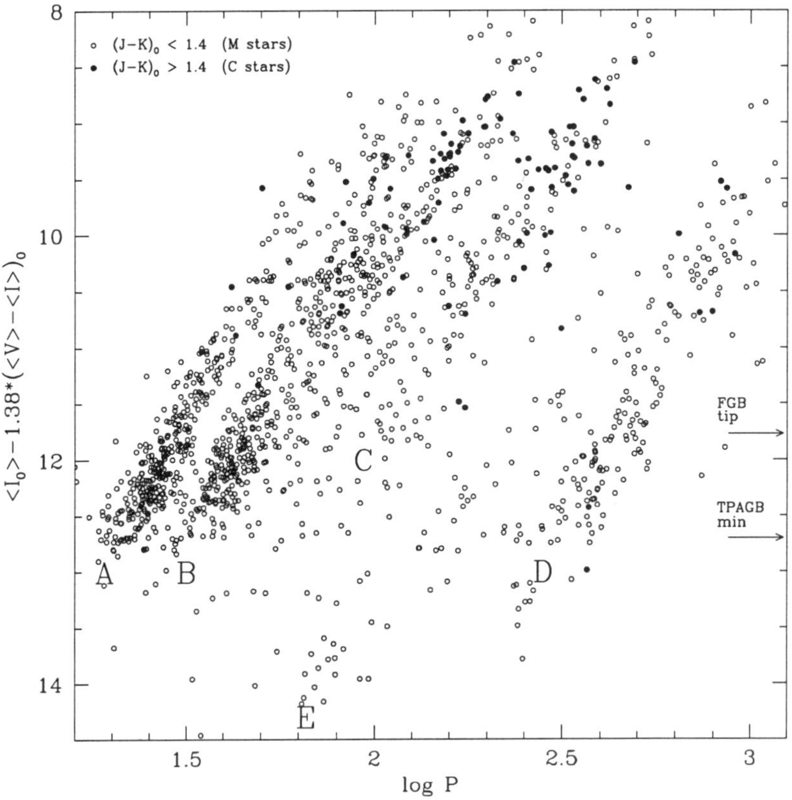

Figure 1. The period-luminosity relation for red variables. <*I*> and <*V*> are mean magnitudes, and *P* is in days. See text for details.

sample of stars has been plotted in the (K, log$P$) plane using single-phase K observations and from this plot we have identified sequence C as the sequence on which the Miras lie (this is also confirmed by the fact that the largest amplitude variables lie on this sequence, although not all stars on the sequence are of large amplitude). Wood & Sebo (1996) have shown that there is a second sequence of small-amplitude LPVs running parallel to the Mira sequence, and from the K observations we can identify sequence B with the second sequence of Wood & Sebo. Using HIPPARCOS data, Bedding & Zijlstra (1998) show that local semi-regular (SR) variables lie on both the Mira sequence and the second sequence of Wood & Sebo.

Examples of light curves of stars on each of the five sequences are shown in Fig. 2. Many of the light curves are semi-regular and multiperiodic. For those stars with two or more periods, the period ratio is plotted against P in Fig. 3.

There is a distinct group of variables in Fig. 3 with a period ratio of ~5-13. In these stars the longer period belongs to sequence D while the shorter period usually belongs to sequence B. A group of local SR variables showing similar period ratios is listed by Houck (1963). A second distinct group of stars in Fig. 3

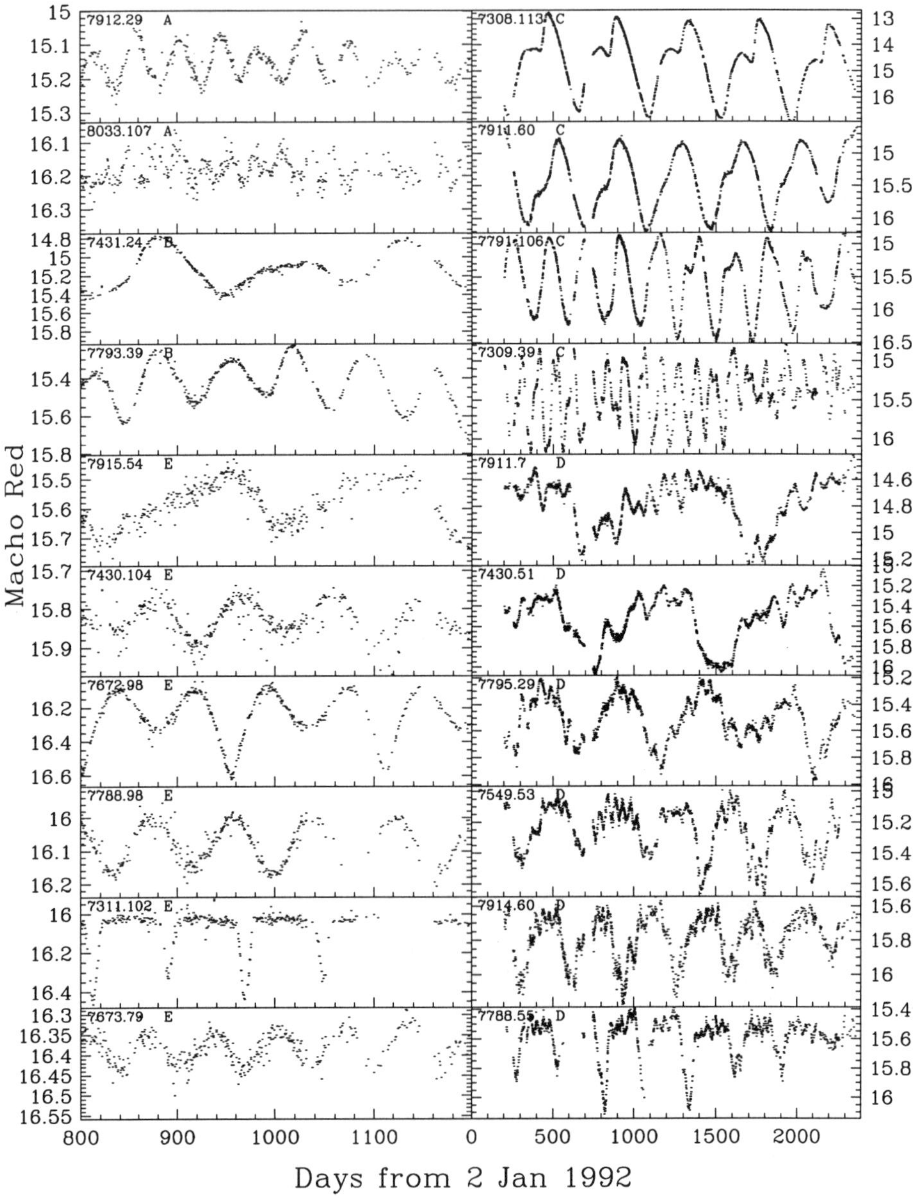

Figure 2. Examples of light curves. The sequence from which each variable comes is noted by one of the letters A...E. The luminosity and period in each sequence increase toward the top of the figure. Note the shorter time interval of the left column.

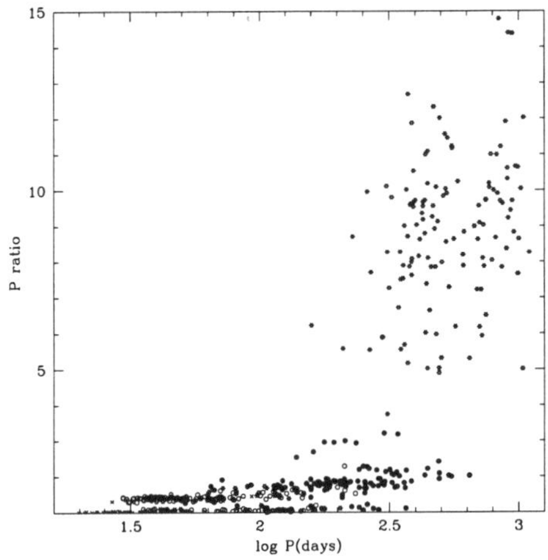

Figure 3. The ratio of the longer to the shorter period in multiperiodic stars plotted against the longer of the two periods.

has a period ratio of 1 to 4, with all periods belonging to sequences A, B or C. These stars are discussed in the next section.

4. The pulsating variables

Successful theoretical pulsation models for LPVs should be able to reproduce the observed P-L relations and the period ratios when the models lie on the observed giant branch. In this section, we try to reproduce the Mira P-L relation as a sequence of stars of different luminosity pulsating in the radial fundamental or first overtone mode. We note that Wood & Sebo (1996) found that they could reproduce the LMC Mira P-L relation assuming fundamental mode pulsation, but here we have the extra constraint of period ratios.

The results of the modelling, using the programs described in Wood & Sebo, are shown in Fig. 4. It is clear that fundamental mode pulsation is able to accurately reproduce the P-L relation of the Miras while simultaneously reproducing the observed period ratios reasonably well (considering the uncertainties in structure resulting from use of mixing-length theory). It is important to note that the solid circles in the period-ratio plot arise from stars in which the longer period lies on sequence C (the Mira sequence). Since the Miras are fundamental mode pulsators in these models, period ratios P_0/P_n, with $n > 0$, should match these points. Indeed, stars with period ratios >2.5 are explained by the ratios P_0/P_2 or P_0/P_3, and the sequence of stars with period ratios of 1.7-2.5 is explained by the ratio P_0/P_1. The SR variables become 1st, 2nd or 3rd overtone pulsators in these models, although some could be low-amplitude fundamental mode pulsators. Finally, we note that short P Miras require a smaller mass than long P Miras if the models are to simultaneously fit the P-L relation and the giant branch. This is consistent with masses estimated from the kinematics of Galactic Miras (Feast 1963).

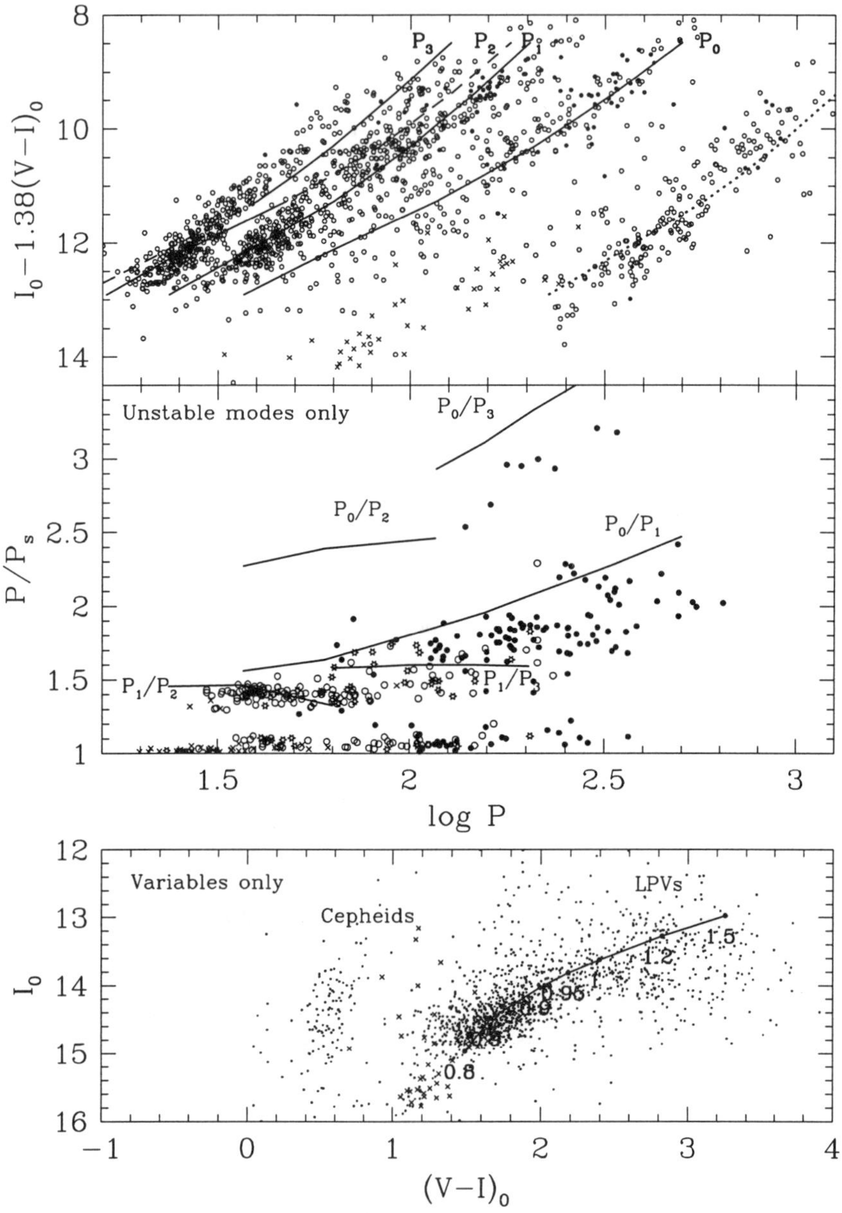

Figure 4. *Top panels*: the luminosity index I_W and the period ratio plotted against P. The solid (dashed) lines correspond to the unstable (stable) radial pulsation modes P_0, P_1, P_2 and P_3, as indicated. In the P-ratio plot, a line is shown only if both modes are unstable. The dotted line is the orbital period expected for an AGB star in a semi-detached binary (see text). *Bottom panel*: the theoretical AGB in the I_0, $(V\text{–}I)_0$ diagram. The mass ($M_\odot$) used at each luminosity is displayed. Crosses are stars with light curves like contact binaries.

Attempts were made to model the Miras as first overtone pulsators since this mode is suggested by direct radius measures of Miras (Haniff et al. 1995). However, first overtone models that fit the P-L relation are far too cool to fit the observed giant branch and, most importantly, it is not possible to produce Mira-to-secondary period ratios $\gtrsim$1.7 (as observed) from period ratios P_1/P_n, where n = 2 or 3 (such ratios are always $\lesssim$1.5, as in Fig. 4). The above results provide very strong evidence that Miras are fundamental mode pulsators.

5. Contact and semi-detached binaries

Six light curves from sequence E are shown in Fig. 2. Two of these light curves are immediately identifiable as binaries: 7311.102 is a detached, but relatively close, binary while 7672.98 is a contact binary. It is suggested here that all the other stars in this sequence are contact binaries. Most have quite symmetric light curves, like those shown by 7430.104, 7788.98 and 7673.79. These light curves are readily recognised among the other LPVs by their regularity and they are indicated by crosses in Fig. 4. These contact binaries lie on the FGB, where they make up ~0.5% of all stars in the search region (the brightest 1 mag. of the FGB); they also lie in the region occupied by intermediate mass stars ($M \gtrsim 2.25\ M_\odot$) at the red end of their core He burning loops.

Six light curves for stars on sequence D are shown in Fig. 2. These stars all have two distinct periods with a period ratio of 9±4 (Fig. 3). The higher frequency variation usually belongs to sequence B (1st overtone pulsation). The long secondary period (hereinafter LSP) belongs to sequence D. Some possible explanations for the LSPs are:

- Nonradial modes. The p-mode periods are shorter than those of the corresponding radial mode and they can not explain the LSPs. The g^- modes are dynamically unstable and lead to convection (rotation may stabilize some g^- modes, but it seems unlikely that such modes would fall on a well-defined P-L sequence). The oscillatory g^+ modes are evanescent in convective regions so it is unlikely they would be seen.
- Strange modes. Strange modes with roughly the correct period exist in these stars but all modes found were extremely damped and they should not be seen.
- The dust κ-mechanism (Winters et al. 1994; Höfner et al. 1995), a modulation caused by dust formation in the stellar wind. However, this mechanism seems to work only in the dense winds of (carbon) stars with $L \gtrsim 10^4 L_\odot$ whereas the LSPs are seen at all luminosities down to ~2 000 $L_\odot$.

The mechanisms above seem unable to explain the observed LSPs. The model proposed here is that the LSPs are due to an eclipse of the AGB star by a cloud of dusty accreted matter surrounding an orbiting, invisible companion star. The pulsating AGB star almost fills its Roche lobe and loses matter to the companion star via an enhanced stellar wind or Roche lobe overflow. The properties of the sequence D light curves, and their interpretation in terms of this model, are now given:

- The amplitudes range from the limit of detectability up to 0.8 mag.: the amplitude variations result from orbital inclination variation, as well as variations in the spatial extent and optical thickness of the cloud.
- The decline in the light curve is generally faster than the rise: this is consis-

tent with the comet-like appearance of the cloud of accreted wind material in the SPH simulations of Theuns & Jorissen (1993).

- The 1st overtone pulsation is visible at all phases of the LSP: this is consistent with the semi-detached model where only the AGB star is seen.
- The 1st overtone period does not appear to vary with phase: this indicates that the LSP does not cause significant changes in the radius of the AGB star, as would happen with an intrinsic pulsation (e.g. a strange mode).
- The LSPs are seen only after the onset of intrinsic pulsation: this indicates that the enhanced mass loss induced by pulsation is needed to build up a significant cloud of accreted material.
- Most importantly, the LSPs lie on a distinct P-L sequence (Fig. 1).

The last point allows a quantitative test of the semi-detached binary hypothesis. Assuming the binary system is made up of two stars each of mass equal to the Mira mass derived above from pulsation theory, and assuming that the AGB star fills its Roche lobe, it is possible to calculate the orbital period of the system. The orbital period derived in this way is shown in Fig. 4 by the dotted line. There is remarkably good agreement between the predicted and observed periods, providing strong support for the hypothesis.

The ultimate test of the reality of the existence of semi-detached and contact binary systems on the AGB and FGB will come from observations of radial velocity variations. If the semi-detached binary nature of ~25% of AGB stars is confirmed, then these binaries provide a natural explanation for the formation of bipolar planetary nebulae, but their existence in such large numbers will require a substantial modification to binary star evolution theory.

References

Alcock C., et al., 1992, in *Robotic Telescopes in the 1990s*, ed. A.V. Fillippenko, ASP Conf. Ser. 34, p. 193

Alves D., et al., 1998, in *Pulsating Stars - Recent Developments in Theory and Observation*, eds. M. Takeuti & D. Sasselov, Universal Academy Press, p. 17

Bedding T.R., Zijlstra A.A., 1998, ApJ 506, L47

Cook K.H., et al., 1997, in *Variable Stars and the Astrophysical Returns of Microlensing Surveys*, eds. R. Ferlet, J.P. Maillard & B. Raban, Editions Frontières, p. 17

Feast M.W., 1963, MNRAS 125, 367

Feast M.W., Glass I.S., Whitelock P.A., Catchpole R.M., 1989, MNRAS 241, 375

Haniff C.A., Scholz M., Tuthill P.G., 1995, MNRAS 276, 640

Höfner S., Feuchtinger M.U., Dorfi E.A., 1995, A&A 297, 815

Houck N., 1963, AJ 68, 253

Hughes S.M.G., Wood P.R., 1990, AJ 99, 784

Theuns T., Jorissen A., 1993, MNRAS 265, 946

Winters J.M., Fleischer A.J., Gauger A., Sedlmayr E., 1994, A&A 290, 623

Wood P.R., Sebo K.M., 1996, MNRAS 282, 958

Asymptotic Giant Branch Stars
T. Le Bertre, A. Lèbre and C. Waelkens, eds.
I.A.U. Symposium 191, 1999

Dynamical Modelling of AGB Star Atmospheres

Susanne Höfner

Niels Bohr Institute, Astronomical Observatory
Juliane Maries Vej 30, DK-2100 Copenhagen, Denmark

Abstract. Time-dependent dynamics is an important ingredient for understanding the atmospheres of pulsating AGB stars. The shock waves created by stellar pulsation modify the atmospheric structure and, consequently, influence the mass loss and the observable properties. So far, hydrostatic model atmospheres have been used in most cases to analyse photospheric spectra, neglecting the effects of dynamics. The recent progress in observational techniques, however, has demonstrated the importance of consistent time-dependent models. The main topics of this contribution are to discuss fundamental physical processes and technical problems in constructing dynamical model atmospheres, to review the present status of modelling and to indicate possible future developments.

1. Introduction

The atmospheres of AGB stars are strongly affected by time-dependent dynamical phenomena. Stellar pulsation creates shock waves which propagate through the atmosphere, changing its structure both on a local and a global scale (see Fig. 1). In the inner parts of the atmosphere the passing shocks cause a more or less periodic modulation of the structure. In the upper layers the dissipation of mechanical energy leads to a levitation, i.e. a density enhancement of up to several orders of magnitude compared to a hydrostatic atmosphere. Both of these effects influence the formation of molecules and dust grains and, consequently, the optical properties as well as the mass loss of these stars.

Up to now hydrostatic model atmospheres have been used in most cases to interpret observed photospheric spectra of AGB stars, neglecting the effects of dynamics on the atmospheric structure. While impressive results have been obtained with this approach for non- or weakly pulsating late-type stars, systematic discrepancies or misinterpretation may result in situations where dynamics plays a decisive role (see reviews by Gustafsson & Jørgensen 1994, Gustafsson 1998 and Plez, this volume).

On the other hand, for the last two decades dynamical models have been used to study mass loss by stellar winds and (more recently) optical properties of circumstellar dust shells, but mainly concentrating on objects with high mass loss rates and more or less optically thick dust envelopes (see Fleischer et al. and Winters et al., this volume). Due to computational reasons time-dependent dynamical models are generally based on a simple (grey) treatment of radiative transfer, in contrast to their hydrostatic counterparts. This approach works

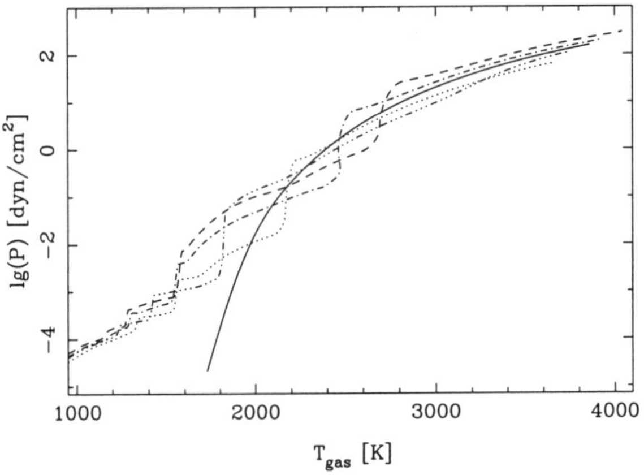

Figure 1. Influence of dynamics on the atmospheric structure (gas pressure vs. temperature): a dynamic model at different phases (0.50: dashed; 0.75: dash-dotted; 1.00: dotted; 1.25: dash-triple-dot) and the corresponding hydrostatic initial model (full line); phase 0.5 corresponds to minimum light, 1.0 to maximum light. The 'steps' in the pressure indicate shock waves. Their propagation through the atmosphere can be seen by tracking the progress of individual features from higher pressures and temperatures to lower ones (i.e. from right to left) with increasing time (phase 0.5 to 1.25). The periodic modulation of the inner parts of the atmosphere and the levitation of the outer layers are visible in this plot (model P7C14U4 of Höfner et al. 1998).

reasonably well if dust is the main opacity source but it is problematic when molecular features dominate (line blanketing).

The construction of realistic time-dependent dynamic models of the inner dust-free atmospheric layers which allow a quantitative prediction of observable properties for stars with optically thin circumstellar envelopes has to be considered as an open problem. However, in the last few years some progress has been achieved in this field and at least at a qualitative level the theoretical results are comparable with observations. Currently, several projects are devoted to improve the treatment of relevant physical processes.

2. Modelling methods: general concepts

In addition to the theoretical aim of understanding the atmospheres of AGB stars in terms of fundamental physics, the impressive progress of observational techniques achieved during the last few years requires a similar improvement of models. The simple picture of a (quasi-) static atmosphere surrounded by a stationary outflow is not adequate for a reliable interpretation of observational data with high spectral resolution (individual molecular bands originating in different atmospheric layers; line profile variations due to velocity fields), a wide wavelength coverage (simultaneous monitoring of molecular and dust features), high angular resolution (spatial structure of the atmosphere) and detailed monitoring of temporal variations of individual stars. This purpose requires consistent, time-dependent models which cover both the inner atmosphere and the stellar wind as well as the highly complicated transition region, taking the relevant physical processes into account.

Ideally, such models should be based on a self-consistent simultaneous solution of time-dependent dynamics (including a realistic treatment of stellar pulsation, shock waves and mass loss by a stellar wind), a detailed frequency-dependent radiative transfer (molecules, dust) and a full non-equilibrium treatment of micro-physical processes (chemistry, dust formation). However, present dynamic models are far from this ideal due to computational limitations and a lack of fundamental physical data. Even when using the assumptions of micro-physical equilibrium and LTE (with the possible exception of dust formation or thermal relaxation behind shocks) the remaining system of time-dependent dynamics and detailed frequency-dependent radiative transfer is far beyond the capacities of present-day computers. Therefore, in general a two-step approach is used to model atmospheres of long-period variables.

First, the variable structure of the atmosphere and circumstellar envelope is obtained by solving the equations of hydrodynamics together with a simple grey treatment of radiative transfer and a (consistent, time-dependent or parameterized) description of dust formation. At this stage, the pulsation of the star is simulated by a more or less sophisticated variable inner boundary which is located below the photosphere and accounts for the motions and luminosity variations induced by the pulsation. The immediate results of these dynamical calculations are mass loss rates, outflow velocities and dust-to-gas ratios. A comparison of the input physics and the results of present dynamical models produced by different groups can be found in the review by Woitke (1998). A short summary of results is given in Sect. 3.

In a second step, a detailed radiative transfer calculation based on the structures obtained in the dynamical computation yields the required observable properties: synthetic spectra at various resolutions and their variation with phase, photometric colours and light curves, spatial intensity distributions or line profile variations. An overview of these results is presented in Sect. 4. (see also Table 1).

3. Atmospheric structure and stellar wind

The pioneering dynamical calculations of Wood (1979) and Bowen (1988) demonstrated that the typical mass loss rates and wind velocities of LPVs are most probably due to the combined effects of stellar pulsation (shock waves) and radiative driving (dust). Furthermore, in both papers the considerable influence of thermal relaxation behind shocks (finite rate or immediate cooling) on the atmospheric structure and on the mass loss rate were discussed. The question of cooling rates (and the relevance of non-LTE processes in this context) is still a matter of debate (e.g. Woitke et al. 1996, 1998; Willson & Bowen 1998).

Fleischer et al. (1992) presented models for C-rich LPVs which include a consistent, time-dependent description of the dust formation process, demonstrating that the interaction of dynamics and dust formation can result in a shell-like structure of the circumstellar envelope. The strong coupling between

Table 1. Papers on synthetic observable properties and references for the corresponding dynamic models (modelling methods).

dynamic models	observable properties	
Wood (1979)	Bessell et al. (1989)	spectra, colours, radii (M)
	Scholz & Takeda (1987)	limb darkening, radii (M)
Wood (1990)	Bessell et al. (1996)	spectra, radii, line profiles (M)
	Scholz (1992)	line profiles (M)
Bowen (1988)	Beach et al. (1988)	limb darkening, radii
	Luttermoser & Bowen (1990)	atomic lines (non-LTE)
Fleischer et al. (1992)	Winters et al. (1994)	light curves (dust)
	Winters et al. (1995)	brightness profiles (dust)
	Winters et al. (1997)	model for AFGL 3068
	Gauger et al. (1998)	CO line profile variations
Höfner & Dorfi (1997)	Windsteig et al. (1997)	mid-/far-IR colours (dust)
	Windsteig et al. (1998)	CO line profile variations
Höfner et al. (1998)	Loidl et al. (1997, 1998, 1999)	molecular spectra (C)
	Aringer et al. (1997, 1999)	molecular spectra (M)
	Hron et al. (1997, 1998)	ISO-SWS spectra (C)
	Alvarez & Plez (1998)	colour variations (M)

thermodynamics, dust formation and dynamics may lead to instabilities (dust-induced κ-mechanism; Fleischer et al. 1995, Höfner et al. 1995) and to multi- or non-periodicity in the behaviour of the circumstellar envelope (e.g. Winters et al. 1994, Höfner & Dorfi 1997) since the dust component is governed by its own time-scales. First results of similar dynamic models for O-rich stars have been presented by Jeong et al. (this volume). For a review of dynamical wind models see Fleischer et al. (this volume).

Recently we have calculated a new set of dynamic models (Höfner et al. 1998) for the purpose of studying the time-dependent behaviour of IR molecular features. Instead of using a constant value to represent the gas opacity as in many earlier models (e.g. Bowen 1988, Fleischer et al. 1992, Höfner & Dorfi 1997) these new model atmospheres are based on Planck mean opacities computed from detailed frequency-dependent molecular absorption coefficients (SCAN data base, Jørgensen 1997). This has a considerable influence on the atmospheric background structure and consequently on the mass loss and observable properties.

Fig. 2 shows the pressure-temperature structures of several hydrostatic models with identical stellar parameters but using different treatments of the gas opacity and radiative transfer: the grey hydrostatic initial model[1] based on

[1] Our dynamical calculations start from models representing the hydrostatic limit case of radiation hydrodynamics. These initial models are defined by the following parameters: luminosity $L_\star$, stellar mass $M_\star$, effective temperature $T_\star$ and the carbon-to-oxygen ratio $\varepsilon_C/\varepsilon_O$

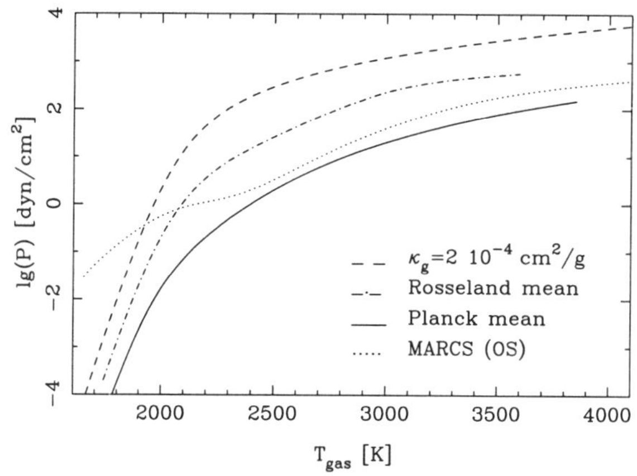

Figure 2. Influence of the radiative transfer and gas opacities on the atmospheric structure (gas pressure vs. temperature) of hydrostatic models: grey initial models based on Planck mean (full line) and Rosseland mean (dash-dotted) molecular absorption coefficients and on a constant value of the gas opacity (dashed) as well as a MARCS model atmosphere (dotted) based on full monochromatic opacity sampling of the same molecular data as in the mean opacity models. All models have the same stellar parameters ($L_\star = 7\,000\,L_\odot$, $M_\star = 1\,M_\odot$, $T_\star = 2\,880\,\mathrm{K}$, $\varepsilon_\mathrm{C}/\varepsilon_\mathrm{O} = 1.4$).

Planck means (full line) is in reasonable agreement with a MARCS model (dotted) which includes a frequency-dependent solution of radiative transfer, based on the same data for molecular absorption coefficients. In contrast, the model calculated with a constant gas absorption coefficient introduced by Bowen (1988) and used in many dynamic models (dashed line) shows a much higher density at a given temperature, leading to unrealistically strong molecular features. For comparison, a model based on Rosseland means of the same molecular data as used in the Planck mean and MARCS models is shown in Fig. 2 (dash-dotted). While the Rosseland mean is the correct mean to apply in the stellar interior (diffusion limit) it tends to underestimate the opacity in the atmosphere and leads to much higher densities than the Planck mean model.

The density differences in the hydrostatic initial models caused by different gas opacities are also reflected in the resulting wind properties of the dynamical models. The Planck mean models presented in Höfner et al. (1998) show reduced dust condensation and significantly smaller mass loss rates than those of Höfner & Dorfi (1997; constant gas opacity). Furthermore, the molecular opacities may contribute to driving the stellar wind. Fig. 3 shows an O-rich model calculated without dust formation which exhibits an outflow driven by radiation pressure on molecules with a typical wind velocity (12 km/s) and a mass loss rate of $\approx 10^{-6} M_\odot/\mathrm{yr}$. Note in this context that the driving by molecules only becomes efficient in the dynamical model where the radiation pressure acts on material which has already been lifted to cooler layers by the pulsation-induced

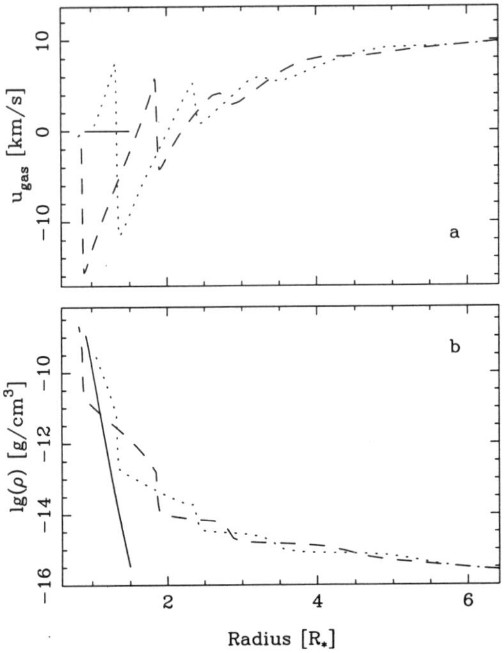

Figure 3. Radial structure of a dynamic model at different phases (0.50: dashed, minimum light; 1.00: dotted, maximum light) and of the corresponding hydrostatic initial model (full line); flow velocity (a) and gas density (b) as a function of the radius (given in units of the radius of the hydrostatic initial model). The wind of this O-rich model is driven by radiation pressure on molecules (no dust formation included). Model P7OU4 of Höfner et al. (1998).

shock waves. The corresponding non-pulsating initial model (full line) shows no tendency to form a radiation-driven wind.

Helling et al. (poster contribution, this conference) have investigated the influence of molecular opacities on stellar winds using Planck and Rosseland means to estimate upper and lower limits of the effects. The results for Planck mean models are in qualitative agreement with Höfner et al. (1998) while models based on Rosseland mean opacities largely resemble earlier models based on a small constant value of the gas absorption coefficient (e.g. Fleischer et al. 1992, Höfner & Dorfi 1997).

4. Observable properties of dynamic models

Regarding predictions of observable features, the existing dynamic models can be divided roughly into two groups: models for dust-enshrouded stars hidden in their circumstellar envelopes (IR carbon stars, OH/IR stars) on the one hand and models for stars with optically thin (or non-existent) circumstellar envelopes ('optical' AGB stars) on the other hand.

For the first group the observable properties are determined mostly by outflow characteristics (mass loss rate, velocity field, spatial structure of the circumstellar envelope) and by the properties of the dust grains (chemical composition, grain sizes). Typical examples of predicted quantities are spectral energy distributions and spatial brightness profiles of the circumstellar envelopes, dust features, IR photometric colours and light curves, as well as line profile variations due to velocity fields (e.g. Winters et al. 1994, 1995, 1997; Windsteig et al. 1997, 1998; Gauger et al. 1998). For a more detailed discussion see the contribution of Winters et al. (this volume).

In contrast, stars with circumstellar envelopes that are more or less optically thin at NIR wavelengths can give valuable insights into the complex interactions of the time-dependent atmospheric dynamics with the formation of molecules and dust grains. Molecular features allow us to probe the critical transition region between the photospheric layers and the circumstellar dust shell where the mass loss is initiated.

Predictions of observable properties for stars with optically thin envelopes (e.g. molecular spectra) require a realistic treatment of the physics that governs the atmospheric background structure and the shock waves (e.g. gas opacities and radiative transfer, see Sect. 3.). A detailed *quantitative* interpretation of observational data is beyond the scope of present time-dependent models because they are based on some more or less crude physical assumptions (e.g. grey radiative transfer, piston boundary) due to computational restrictions. However, they give us some *qualitative* understanding of intrinsically time-dependent phenomena which are not accessible with hydrostatic models.

Early exploratory studies of the effects of atmospheric extension on observable properties of M-type Mira stars were performed by Bessell et al. (1989) and Scholz & Takeda (1987). Using parameterized density structures similar to the models of Wood (1979) they obtained IR spectra and a phase and wavelength dependence of monochromatic radii in reasonable agreement with observations.

Limb darkening for extended atmospheres with shocks and the effects on angular diameter determinations by lunar occultation observations were studied by Beach et al. (1988) using the dynamic models of Bowen (1988) and simple radiative transfer. Luttermoser & Bowen (1990) discussed first results of non-LTE radiative transfer calculations for atomic lines.

Based on the pulsation models of Wood (1990), Bessell et al. (1996) investigated the photospheric structure and observable properties of Mira variables. The general appearance and the variability of the resulting synthetic spectra which were used to derive monochromatic radii are in good agreement with observations. The velocity fields lead to a complex time-dependent behaviour of absorption line profiles and influence the integrated line strengths (Scholz 1992). Certain features do not only show a variation with phase but also differ at a given phase from cycle to cycle, reflecting a non-periodic behaviour of the atmospheric layers where they originate.

Recently, several studies of observable properties based on the dynamic model atmospheres of Höfner et al. (1998) have been presented. Alvarez & Plez (1998) used a dynamic model to explain phase shifts between different observed narrow band colours of M-type Mira stars (modulation of the structure by propagating shocks) which could not be reproduced with hydrostatic model

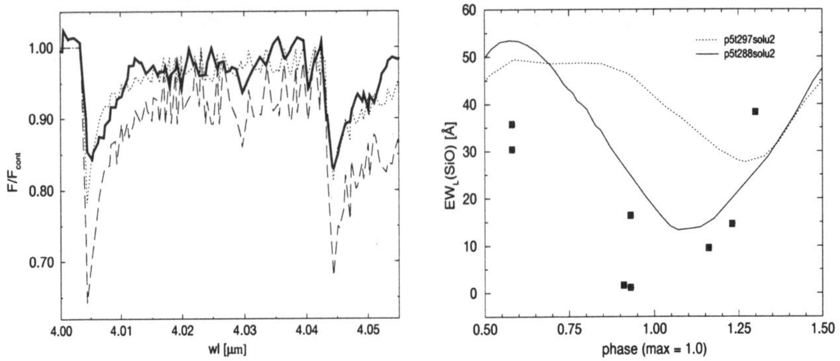

Figure 4. Theoretical and observed properties of SiO rotation-vibration bands: Left panel: a comparison of normalized spectra in the 4 μm region; observed spectrum of S Gru (full line) and synthetic spectra based on a dynamical model (dotted) and a hydrostatic atmosphere (dashed). Right panel: the corresponding equivalent width of the SiO bands as a function of the bolometric phase for two dynamical atmospheres (full and dotted lines) and as a function of visual phase for observed Mira stars (squares). From Aringer et al. (1999).

atmospheres. IR molecular spectra of carbon stars have been calculated by Loidl et al. (1997, 1998, 1999) and were compared to ISO-SWS observations by Hron et al. (1997, 1998 and this volume), demonstrating the potential of analyzing the stellar spectra and their variability with dynamic models. A confrontation of synthetic spectra of O-rich models (Aringer et al. 1997) with ISO-SWS data is presented in Hron et al. (1997) and Aringer et al. (this volume). At a higher resolution, the behaviour of SiO rotation-vibration bands around 4 μm has been studied by Aringer et al. (1999). While hydrostatic model atmospheres predict too strong features for very cool and extended objects, the experimental dynamic models are in principle able to explain the whole observed range of equivalent widths and their variations (see Fig. 4). Further constraints on atmospheric structure and velocity fields can be expected from a comparison of line profile variations of molecular lines with observations.

5. Recent developments and unsolved problems

Time-dependent dynamics is essential for understanding the atmospheres of pulsating AGB stars: it influences the structure both on global (levitation, wind) and local (shock waves) scales. Therefore, dynamic models are the only consistent way of studying temporal variations of observable properties. Existing dynamic model atmospheres give valuable insights into time-dependent phenomena and are in reasonable qualitative agreement with observations (Sect. 4.). However, due to a lack of computing capacities and of fundamental physical data they suffer from restrictions and approximations. Step-by-step improvements are necessary for a detailed quantitative understanding of observations.

A more realistic (i.e. non-grey) description of radiative transfer seems to be one of the most pressing problems due to the considerable effect on the atmospheric background structure (see Sect. 3.). While a detailed treatment with thousands of frequency points as in hydrostatic models is far beyond the capacities of present computers, it should be possible to include frequency-dependent radiative transfer with a very limited number of frequencies in the near future. By systematically reducing the number of opacity sampling points in hydrostatic models Helling & Jørgensen (1998) demonstrated that as few as 50 frequencies may be sufficient to obtain better results than with mean opacities.

Another important point is the treatment of stellar pulsation. In most state-of-the-art dynamic models the pulsation is simulated by a variable boundary (piston) which is located beneath the stellar photosphere and moving sinusoidally with given velocity amplitude and period. In contrast, Bessell et al. (1996) and Scholz (1992) based their investigations on photospheric models calculated with a variable inner boundary derived from the dynamic models of Wood (1990). These models include the driving zone of the pulsation but introduce a depth-dependent free parameter (artificial damping/growth factor) into the energy conservation equation which is adjusted to produce the desired non-linear pulsation amplitude. Despite these problems of modelling the pulsation itself this is an interesting approach to define a more consistent boundary condition for the atmospheric calculation.

Regarding the impressive progress achieved in non-linear pulsation models for other types of variable stars during the last few years (e.g. RR Lyrae double mode pulsations, Feuchtinger 1998) fundamental improvements for LPVs can be expected in the near future. This would be an important step towards the long-standing goal of a simultaneous, self-consistent modelling of pulsation, atmosphere and stellar wind of AGB stars.

Acknowledgments. I thank U.G. Jørgensen, R. Loidl, B. Aringer (who also provided the plots shown in Fig. 4), J. Hron, E.A. Dorfi and M. Feuchtinger for inspiring discussions and helpful comments. This work was supported by the Austrian Science Fund (FWF; grants J01283-AST, J1487-PHY and S7305-AST).

References

Alvarez R., Plez B., 1998, A&A 330, 1109

Aringer B., Kerschbaum F., Jørgensen U.G., et al., 1997, in First ISO Workshop on Analytical Spectroscopy, ESA SP-419, p. 249

Aringer B., Höfner S., Wiedemann G., et al., 1999, A&A, in press

Beach T.E., Willson L.A., Bowen G.H., 1988, ApJ 329, 241

Bessell M.S., Brett J.M., Scholz M., Wood P.R., 1989, A&A 213, 209

Bessell M.S., Scholz M., Wood P.R., 1996, A&A 307, 481

Bowen G.H., 1988, ApJ 329, 299

Feuchtinger M.U., 1998, A&A 337, L29

Fleischer A.J., Gauger A., Sedlmayr E., 1992, A&A 266, 321

Fleischer A.J., Gauger A., Sedlmayr E., 1995, A&A 297, 543

Gauger A., Winters J.M., Fleischer A., Keady J.J., 1998, in Cyclical Variability in Stellar Winds, L. Kaper and A.W. Fullerton (eds.), ESO Astrophysics Symposia, Springer, p. 309

Gustafsson B., 1998, Ap&SS 255, 241

Gustafsson B., Jørgensen U.G., 1994, A&AR 6, 19

Helling C., Jørgensen U.G., 1998, A&A 337, 477

Höfner S., Dorfi E.A., 1997, A&A 319, 648

Höfner S., Feuchtinger M., Dorfi E.A., 1995, A&A 297, 815

Höfner S., Jørgensen U.G., Loidl R., Aringer B., 1998, A&A 340, 497

Hron J., Aringer B., Loidl R., et al., 1997, in First ISO Workshop on Analytical Spectroscopy, ESA SP-419, p. 213

Hron J., Loidl R., Höfner S., Jørgensen U.G., Aringer B., Kerschbaum F., 1998, A&A 335, L69

Jørgensen U.G., 1997, in Molecules in Astrophysics: Probes and Processes, E.F. van Dishoeck (ed.), IAU Symp. 178, Kluwer, p. 441.

Loidl R., Hron J., Höfner S., et al., 1997, Ap&SS 251, 243

Loidl R., Hron J., Aringer B., Höfner S., Jørgensen U.G., 1998, Ap&SS 255, 289

Loidl R., Höfner S., Jørgensen U.G., Aringer B., 1999, A&A, in press

Luttermoser D.G., Bowen G.H., 1990, in Cool Stars, Stellar Systems, and the Sun, Sixth Cambridge Workshop, G. Wallerstein (ed.), A.S.P. Conf. Ser. 9, p. 491

Scholz M., 1992, A&A 253, 203

Scholz M., Takeda Y., 1987, A&A 186, 200

Willson L.A., Bowen G.H., 1998, in Cyclical Variability in Stellar Winds, L. Kaper and A.W. Fullerton (eds.), ESO Astrophysics Symposia, Springer, p. 294

Windsteig W., Dorfi E.A., Höfner S., Hron J., Kerschbaum F., 1997, A&A 324, 617

Windsteig W., Höfner S., Aringer B., Dorfi E.A., 1998, in Cyclical Variability in Stellar Winds, L. Kaper and A.W. Fullerton (eds.), ESO Astrophysics Symposia, Springer, p. 308

Winters J.M., Fleischer A.J., Gauger A., Sedlmayr E., 1994, A&A 290, 623

Winters J.M., Fleischer A.J., Gauger A., Sedlmayr E., 1995, A&A 302, 483

Winters J.M., Fleischer A.J., Le Bertre T., Sedlmayr E., 1997, A&A 326, 305

Woitke P., Krüger D., Sedlmayr E., 1996, A&A 311, 927

Woitke P., 1998, in Cyclical Variability in Stellar Winds, L. Kaper and A.W. Fullerton (eds.), ESO Astrophysics Symposia, Springer, p. 278

Wood P.R., 1979, ApJ 227, 220

Wood P.R., 1990, in From Miras to Planetary Nebulae: Which Path for Stellar Evolution?, M.O. Mennessier and A. Omont (eds.), Editions Frontières, Gif-sur-Yvette, p. 67

Asymptotic Giant Branch Stars
T. Le Bertre, A. Lèbre and C. Waelkens, eds.
I.A.U. Symposium 191, 1999

Atmospheric Structure and Mass Loss of O-rich Long Period Variables. A Confrontation of Models with ISO-SWS Observations

B. Aringer, F. Kerschbaum, J. Hron, T. Posch and W. Windsteig

Institut für Astronomie, Univ. Wien, Vienna, Austria

U.G. Jørgensen and S. Höfner

Niels Bohr Institute, Univ. Obs. Copenhagen, Copenhagen, Denmark

Abstract. We present ISO-SWS observations of cool oxygen-rich AGB variables and compare them to synthetic molecular and dust spectra based on hydrostatic and dynamic model atmospheres.

1. Modelling

1.1. Hydrostatic and dynamic model atmospheres

Synthetic spectra of cool AGB stars have usually been calculated using classical model atmosphere programs like the MARCS code (Gustafsson et al. 1975). These models are very sophisticated concerning wavelength dependent radiative transfer with a detailed treatment of molecular opacities, continuum absorption and convection. Nevertheless, it is obvious that the atmospheres of very cool AGB stars are dominated by such phenomena as pulsation, shock waves and heavy mass loss, which can definitely not be described by a hydrostatic approximation. As a consequence classical hydrostatic models fail to reproduce the strength and variations of certain molecular features in such objects (e.g. Aringer et al. 1997 for SiO).

We have calculated spectra based on hydrostatic MARCS models as well as on dynamic models for dust formation and mass loss developed by Höfner et al. (Höfner & Dorfi 1997; Höfner et al. 1998) where pulsation is simulated by a moving piston at the inner boundary. Using a given temperature-pressure structure we have computed chemical abundances of molecular species (equilibrium), continuum absorption (routines from MARCS code) and molecular opacities (directly from linelists or from opacity sampling). This is done by a program called COMA (**C**openhagen **O**pacities for **M**odel **A**tmospheres) producing opacities as a function of wavelength and radius, which are then used as an input for a spherical radiative transfer code. The following species have been included into the molecular opacities: CO, SiO, OH, H_2O, CO_2 from HITRAN and TiO. As an example of the results we show the normalized spectra of hydrostatic models with T = 2 600 K and T = 3 400 K in Fig. 1. It is obvious that the molecular absorption increases at lower temperatures.

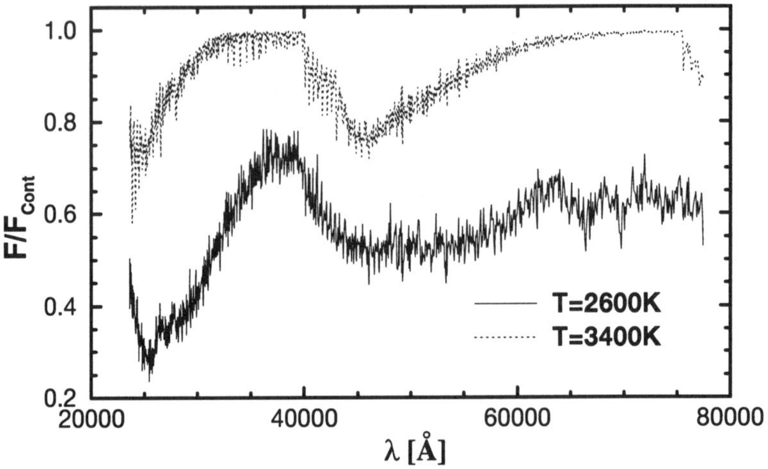

Figure 1. Normalized synthetic infrared spectra for two hydrostatic atmospheres with different temperatures (2600 and 3400 K), $\log(g\,[\mathrm{cm/s^2}]) = 0.0$, solar mass and chemical abundances. The resolution is about 900.

There are still some problems concerning the dynamic model atmospheres. First, they are calculated with a grey radiative transfer using mean opacities, which probably produces significant deviations from the correct structure, especially in the outer layers of the atmosphere where most of the water absorption comes from. Also, the piston as an inner boundary for the models does not represent a very satisfactory solution, because it introduces additional degrees of freedom. Finally, for the oxygen-rich case we did not include any dust in the computations.

1.2. Radiative transfer with DUSTY

In order to investigate the contribution of the circumstellar dust we used the publicly available radiative transfer code DUSTY (Ivezić et al. 1997). In contrast to the approach chosen by most authors we did not take blackbodies as the central light sources, but the synthetic spectra discussed above.

The upper panel of Fig. 3 illustrates the contributions of the individual constituents. The final model spectrum is divided by the theoretical photospheric continuum in order to visualize the corresponding deviations. Note the 30 % difference between the stellar spectrum and the theoretical continuum around 8 μm, a region that is used for the definition of the stellar continuum by many authors.

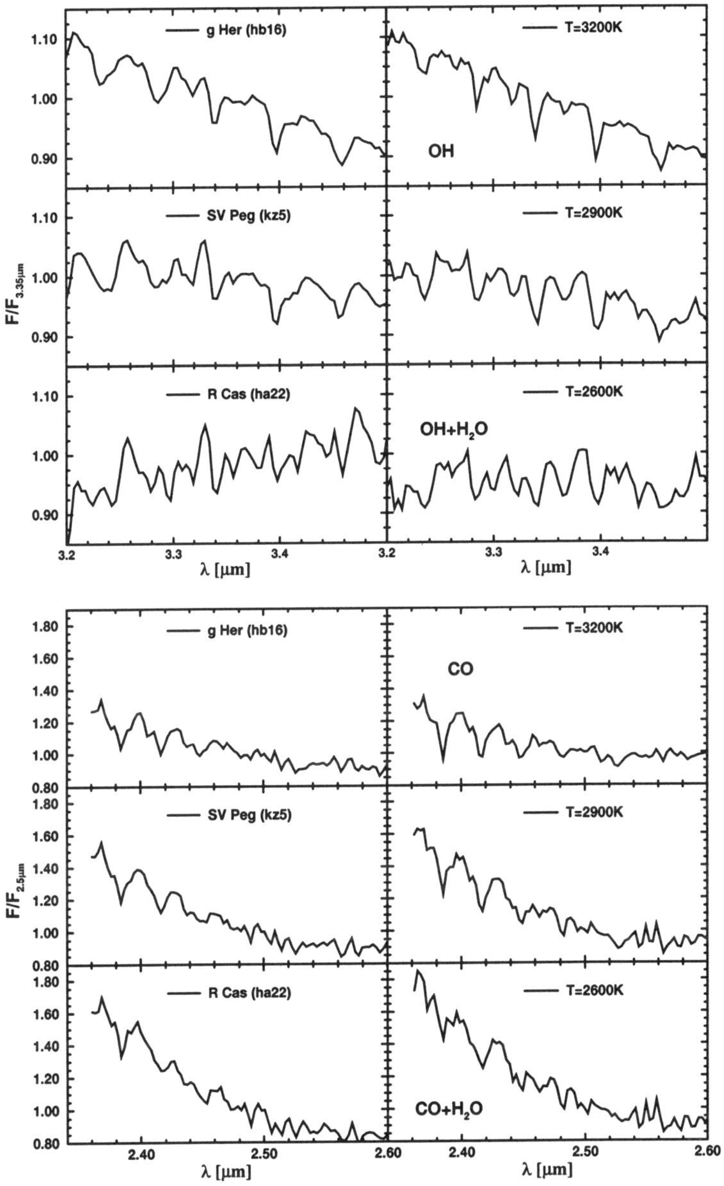

Figure 2. ISO-SWS observations of the AGB variables g Her, SV Peg and R Cas compared with synthetic spectra for hydrostatic atmospheres with different temperatures, $\log(g\,[\mathrm{cm/s^2}]) = 0.0$, solar mass and chemical abundances. The resolution is between 300 and 400.

2. Observations

Observational material of 5 open time ISO projects (*fkerschb.orichsrv/orichsrl/zzagb2pn* and *jhron.varlpv/varlpv2*) was used for our work. We tried to cover on one hand a large number of representative O-rich AGB variables of different pulsational properties (Mira, SRa, SRb, Lb, and OH/IR) and on the other hand to monitor a smaller subsample in order to study the spectral changes. In total we were able to observe 22 objects at least once. For 10 of them we could take between 2 and 7 individual spectra at different pulsational phases.

We used ISO-SWS (de Graauw et al. 1996) with full grating scans (2.4 μm to 44 μm, AOT01 and speed 2, typically) and a typical resolution between 400 and 600. The pipeline processed data products (OLP 6) were further reduced with ISAP and for the circumstellar modelling rebinned to a resolution of 100. For all spectra a significant offset between the two scan directions was found in band 2C (7 μm to 12 μm) which was corrected by rejecting the scan direction with the higher flux. The connection of the individual bands which introduces an additional uncertainty was done by multiplying the longer wavelength band in order to fit overlapping regions. For flux levels below 100 Jy the correction was applied additively.

3. Results

3.1. Stellar atmospheres

As it turned out, large fractions of our ISO-SWS spectra of O-rich AGB stars can be reproduced by a temperature sequence of hydrostatic models. This is demonstrated in Fig. 2 where we compare the observations of three AGB variables with synthetic spectra based on MARCS atmospheres. The temperatures have been selected to fit the overall energy distribution of the ISO-SWS spectra. It is obvious that the intensity of the CO, OH and water bands is described very well in all cases. Nevertheless, for some objects one has to adopt unrealistically low temperatures ($T < 2600$ K) in order to reproduce the observed data. In addition, the low resolution of our ISO spectra ($R < 1000$) does not always allow a distinction of the bandheads from the local continuum also dominated by molecular absorption. Thus, effects like the observed weakness of the SiO bands compared to theoretical predictions from hydrostatic models will not appear in our data.

We want to emphasize that due to their limited spectral resolution ISO-SWS data can only be reproduced using molecular linelists which are complete concerning the overall opacity. This is especially true for water where we have taken approximately 20 000 000 lines into account. At the moment such complete lists can only be obtained from theoretical calculations, and not from any experimental database.

Using dynamic models we have been successful in reproducing the bandheads of OH, CO and SiO, but we got much too weak water absorption. This can be explained by the large deviations introduced by the grey approximation for the radiative transfer in the outer layers of the atmosphere.

Finally, it should be mentioned that the CO_2 feature around 4.3 μm could not be fit at all. The observed bands are always much stronger than the synthetic

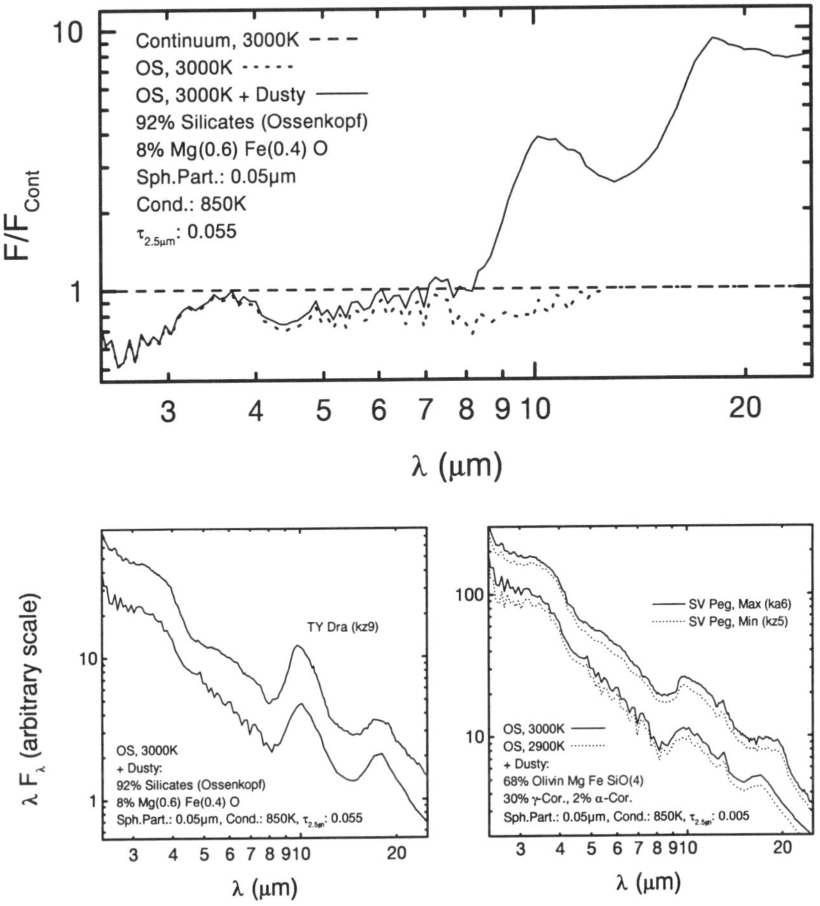

Figure 3. Combined models using opacity sampling spectra based on hydrostatic model atmospheres as the input for the DUSTY radiative transfer code.

ones. This might be due to an incomplete linelist or incorrect chemical data (molecular equilibrium abundances).

3.2. Circumstellar material

The lower panels of Fig. 3 show two examples of our combined results using opacity sampling spectra based on hydrostatic model atmospheres as the input for the DUSTY radiative transfer code. The Lb variable TY Dra (left) can be reproduced with a dust shell consisting of a mixture of "astronomical silicates" and Mg(0.6) Fe(0.4) O on top of a hydrostatic 3 000 K photosphere with $\log(g\,[cm/s^2]) = 0.0$. More information on the modelling is given in the figure. Also pure amorphous Olivine ($MgFeSiO_4$) could explain the observed spectral energy distribution.

On the right hand side ISO observations and calculated spectra for two pulsational phases of the Semiregular variable SV Peg are shown. The small variation can be explained by a change of the effective temperature (3 000 and 2 900 K, resp.) of the model atmosphere. The contribution of the dust component is assumed to be constant between the two phases. The model parameters are again given in the figure.

A few conclusions can be drawn from the objects analyzed so far: in order to reproduce the spectral energy distribution of circumstellar dust one requires atmospheric models reproducing the molecular absorption. The "canonical" 2 500 K blackbody gives wrong results for the dust component. As an example, in all objects with optically thin circumstellar shells one needs a significantly weaker dust emission to explain the observations since part of the 8–9 μm feature contrast originates from the underlying atmospheric molecular absorption by SiO and H_2O.

The spectra of Semiregular and Irregular variables hardly can be reproduced using only "astronomical silicates". Most of these objects show indications of corundum and/or olivines.

For the variable objects studied so far as a function of phase, e.g. the Mira star R Aql, we found no indication of changing dust features but a strong influence of the time-dependent H_2O and SiO absorption. It should be noted that due to the small size of our sample and the uncertainties of the models we cannot exclude variations of the dust emission. However, we want to emphasize again that such studies require modelling of the molecular "background".

Finally, we want to mention that the intensity of the rotation-vibration bands of CO_2 increases with a stronger 13 μm feature attributed to corundum (compare also with Ryde et al. 1999).

Acknowledgments. This project was financed by the Fonds zur Förderung der wissenschaftlichen Forschung under project number S7308-AST. The work of FK was supported by APART (Austrian Programme for Advanced Research and Technology) from the Austrian Academy of Sciences and the Österreichische Nationalbank under the Jubiläumsfonds-project number 6876. We are grateful to Dr. H. Mutschke, Jena, for supplying us with published and unpublished optical constants of various types of dust grains.

More information on this contribution can be found at:
http://rigel.astro.univie.ac.at/~aringer/agb98/agb98.html

References

Aringer B., Jørgensen U.G., Langhoff S.R., 1997, A&A 323, 202

de Graauw T., Haser L.N., Beintema D.A., et al., 1996, A&A 315, L49

Gustafsson B., Bell R.A., Eriksson K., Nordlund Å., 1975, A&A 42, 407

Höfner S., Dorfi E.A., 1997, A&A 319, 648

Höfner S., Jørgensen U.G., Loidl R., Aringer B., 1998, A&A 340, 497

Ivezić Ž., Nenkova M., Elitzur M., 1997, User Manual for DUSTY, Internal Report, University of Kentucky, accessible at http://www.pa.uky.edu/~moshe/dusty

Ryde N., Eriksson K., Gustafsson B., 1999, A&A 341, 579

Asymptotic Giant Branch Stars
T. Le Bertre, A. Lèbre and C. Waelkens, eds.
I.A.U. Symposium 191, 1999

Infrared molecular spectra of carbon stars observed by the ISO SWS

W. Aoki, T. Tsuji, K. Ohnaka

Institute of Astronomy, The University of Tokyo, Mitaka, Tokyo, Japan

Abstract.
We carried out observations of bright optical carbon stars with the ISO SWS, and detected the absorption features of CO, CS, CH, SiS and HCN in N-type and SC-type stars. The CH fundamental bands in 3-4 μm in N-type stars are stronger than those in SC-type stars. On the other hand, the first overtone bands of SiS at 6.6 μm were detected in WZ Cas (SC-type), while none of them were detected in N-type stars. These results can be explained well by the lower C/O ratio in SC-type stars than in N-type stars. The absorption features of HCN (ν_1, $\nu_2 + \nu_3$ and $\nu_1 - \nu_2$) are stronger in SC-type stars than in N-type stars. This fact can be explained by the cooler atmospheres of our SC-type stars due to the lower C/O ratio as well as due to the lower effective temperature, compared with the N-type stars in our sample. The CS bands and the band heads of the CO fundamental bands are weaker in the observed spectra than in those predicted by our model atmospheres. This result cannot be explained by the present model atmosphere and one possible explanation is the contribution by the emission of CO and CS in the outer atmosphere.

1. Introduction

Extensive efforts to obtain infrared spectra have been made for carbon stars, because of their importance in the study of the late stages of stellar evolution. However, the wavelength region and the number of objects studied so far are quite limited. The entire spectra from 2.5 to 45 μm have become available for the first time by the Short Wavelength Spectrometer (SWS, de Graauw et al. 1996) on board the Infrared Space Observatory (ISO)[1].

We carried out observations of various types of carbon stars with the ISO SWS, and investigated the vibration-rotation spectra of various molecules in the infrared region which are expected to be quite useful for the understanding of the photosphere and the outer atmosphere of cool evolved stars. In this paper, we report the results of the analysis of molecular absorption features in three N-type and three SC-type stars. SC-type stars are distinguished from N-type stars by the weakness of C_2 absorption in the optical region. This fact is generally

[1]Based on observations with ISO, an ESA project with instruments funded by ESA Member States (especially the PI countries: France, Germany, the Netherlands and the United Kingdom) with the participation of ISAS and NASA.

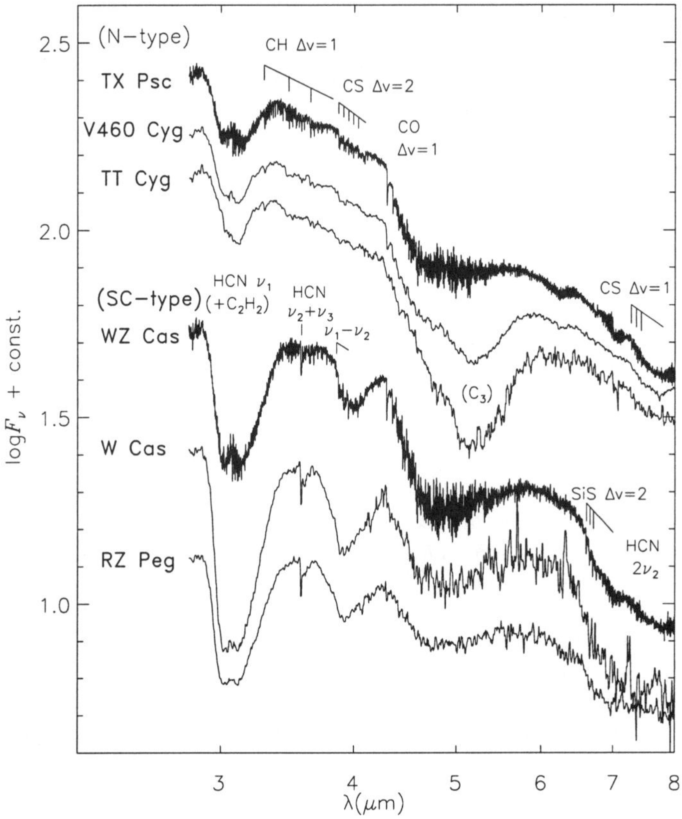

Figure 1. The observed spectra. The upper three are the spectra of N-type stars and the lower three are those of SC-type stars

interpreted that the carbon to oxygen abundance ratios (C/O ratios) in SC-type stars are smaller than those in N-type stars. We examine this effect in the molecular features of the infrared region.

2. Observation, data reduction and analysis

We observed TX Psc (N-type) and WZ Cas (SC-type) with the SWS06 at a resolution of 1000-2000. We observed other stars with the SWS01 (speed 1) at a resolution of about 200. The details of our observations are given in Table 1. The data reductions were done by the SWS Interactive Analysis software. The discrepancies in flux between different AOT bands were corrected by scaling with respect to the band 1B (2.6-3.0 μm). The flux discrepancies are at most 10%, except for that at 7 μm which is 40% in the worst case. This correction is not important for the present work on the individual molecular features, though the better flux calibration is necessary for the study of the spectral energy distribution. The spectra from 3 to 8 μm of these six carbon stars are shown in Fig. 1.

In the present work, we compare the observed spectra with synthetic spectra based on our model atmospheres. In the calculation of model atmospheres, the plane parallel geometry, hydrostatic equilibrium and LTE are assumed. The stellar parameters used to specify the model atmospheres are also given in Table 1. The details of the model calculation and the stellar parameters are discussed in Aoki et al. (1998) in which the molecular data used in our analysis are also described. An important point is that the carbon to oxygen abundance ratios (C/O ratios) of SC-type stars are close to unity (e.g. WZ Cas), and hence the backwarming by CN and C_2 is weak in SC-type stars. As a result the atmosphere of SC-type stars is cooler than that of N-type stars of the same effective temperature.

Table 1. Summary of observations and the stellar parameters

star	spectral type	AOT	obs. date	$T_{\rm eff}$[1]	C/O[2]	C/O[3]
TX Psc	N-type (C7,2)	SWS06	26 Nov 1996	3080	1.027	1.16
V460 Cyg	N-type (C6,3)	SWS01-1	11 Jan 1997	3230	1.062	
TT Cyg	N-type (C5,4)	SWS01-1	30 May 1997	3050	-	
WZ Cas	SC-type (C9,2JLi)	SWS06	22 Jul 1996	2980	1.010	
W Cas	SC-type (C7,1e)	SWS01-1	12 Jan 1997	3000	-	
RZ Peg	SC-type (C9,1e)	SWS01-1	10 Jun 1996	2650	-	

[1]The effective temperature were determined by Ohnaka & Tsuji (1996). The $T_{\rm eff}$ of SC-type stars are corrected by Aoki et al. (1998).

[2]The C/O ratio were determined by Lambert et al. (1986).

[3]The C/O ratio was determined by Ohnaka (1997).

3. Molecular features identified in the observed spectra

We identified the features of several molecules in the observed spectra, especially in the spectra obtained with the SWS06 at the higher resolution. In this section, we discuss the characteristics of molecular features corresponding to the spectral types, and the results of the comparison of the observed features with those predicted by our model atmospheres.

3.1. CH and SiS

As shown in Fig. 1, we identified the CH fundamental bands in the spectra of three N-type stars. The CH absorption features in the three SC-type stars are quite weak if any. The weakness of CH features in SC-type stars can be explained by the small abundance of free carbon atoms which are not locked up into CO molecules. This is the same characteristic as the weak C_2 and CN absorption in the optical spectra of SC-type stars, which is used as the classification criterion of carbon stars.

We also identified the SiS first overtone bands at 6.6 μm in WZ Cas, while none of them is seen in other stars (Fig. 1). In Fig. 2, we show the spectra of TX Psc and WZ Cas with the synthetic spectra of the SiS first overtone bands. The SiS band heads appear in the synthetic spectrum with C/O = 1.01 (i.e. the case of SC-type stars), while no feature appears in the synthetic spectrum with

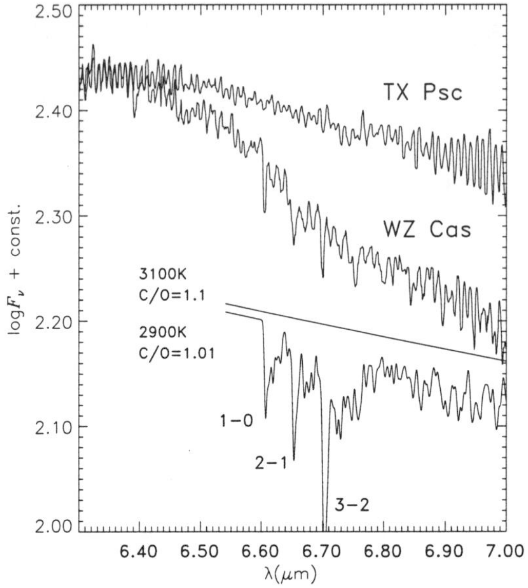

Figure 2. The 6-7μm region of the observed spectra of TX Psc and WZ Cas, and the synthetic spectra of the SiS first overtone bands. The broad absorption in the wavelength region longer than 6.5μm of the spectrum of WZ Cas is due to the $2\nu_2$ bands of HCN

C/O = 1.1 (i.e. the cases of N-type stars). The strong dependence of the SiS absorption on the C/O ratio is qualitatively explained as follows; in carbon rich case as N-type stars (e.g. C/O $\geq$ 1.02), carbon is abundant enough to consume almost all sulfur atoms by forming CS molecules. On the other hand, in SC-type stars (C/O $\sim$ 1.01), the abundance of carbon which is not locked up into CO molecules is too low to consume sulfur atoms by forming CS molecules, and the remaining sulfur can form SiS. The coolness of the atmospheres of SC-type stars also promotes the formation of SiS.

3.2. HCN vibration-rotation spectra

As shown in Fig. 1, we detected the well known 3μm band which is attributed to HCN and/or C_2H_2. We also detected the HCN $\nu_2 + \nu_3$ bands at 3.56μm in the three SC-type stars. The $\nu_1 - \nu_2$ bands at 3.9μm would appear in the spectrum of WZ Cas, though these features blend with the CS first overtone bands.

In Fig. 3, we show the synthetic spectra of the above three bands of HCN together with the observed ones for TX Psc and WZ Cas. The vibration-rotation spectra of $H^{13}CN$ and of C_2H_2 are not included in the calculation of the 3μm band, although the absorption of $H^{13}CN$ may be important in WZ Cas because of the lower $^{12}C/^{13}C$ ratio (see Table 1).

One important result is that the absorption of HCN in our SC-type stars is stronger than that in the N-type stars (see Fig. 1), in contrast with the weakness of absorption features of other carbon-bearing molecules (e.g. C_2, CN and CH) in SC-type stars because of the lower C/O ratio. This is because the absorption

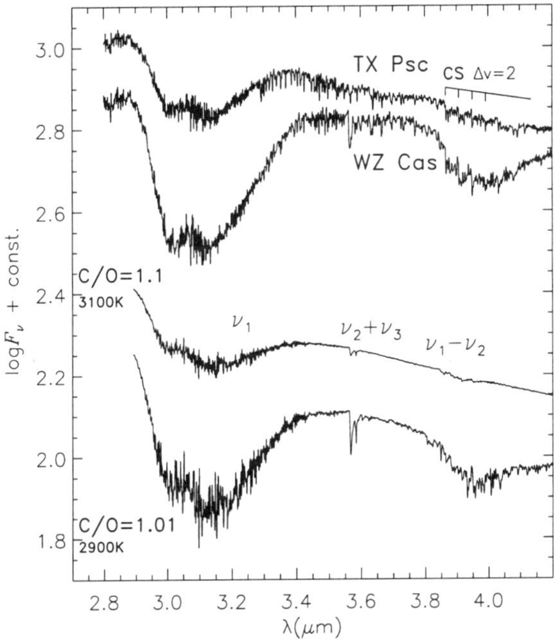

Figure 3. The observed spectra of TX Psc and WZ Cas, and the synthetic spectra of the vibration-rotation bands of HCN

of HCN is very sensitive to the temperature structure of the atmosphere. Since the atmospheres of SC-type stars are cooler than those of N-type stars with the same effective temperature, the HCN absorption in SC-type stars could be stronger than that of N-type stars, in spite of the lower C/O ratio. (However, we cannot conclude that the HCN absorption features in SC-type stars are generally stronger than those in N-type stars, because the HCN absorption is also sensitive to the effective temperature.)

3.3. CO and CS

The CS first overtone and CO fundamental bands are clearly identified (Fig. 1). However, the absorption in the observed spectra is weaker than in the spectra predicted by our model atmospheres as shown in Fig. 4. The fundamental bands of CS are seen at 7 μm in the spectra of TX Psc (Fig. 1), but they are very weak. Since these molecules are quite stable in the atmospheres of cool carbon stars, it is difficult to explain these facts by the small changes of the temperature structure of the model atmosphere or of the stellar parameters.

One possible reason for the weakness of these absorption features is the contribution of the emission of CO and CS in the outer atmosphere. Our observations of oxygen-rich giants with the ISO SWS (Tsuji et al. 1997) revealed that the absorption features of CO and SiO in M giants are weaker, while the absorption of H_2O and CO_2 are stronger, than those predicted by our model atmosphere, and we proposed the existence of warm molecular envelopes around the photospheres. We suggest now that warm molecular envelopes exist not only in oxygen-rich giants but also in carbon stars.

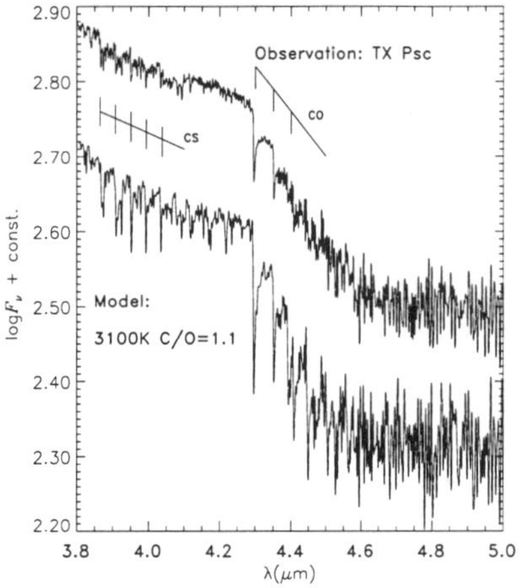

Figure 4. The observed spectrum of TX Psc and the synthetic spectrum of the CS first overtone and the CO fundamental bands

4. Discussion and concluding remarks

In addition to the features identified in the previous section, we identified HCN $2\nu_2$ bands at around $7\,\mu$m in SC-type stars, and C_3 at $5\,\mu$m in TT Cyg and V460 Cyg (see Fig. 1). C_2H_2 may contribute to the absorption at around 3 and $7\,\mu$m, but the contribution is not large in our sample. At the present stage, most of the absorption features in this wavelength region are identified, and the dependence on the spectral type is explained well for N-type and SC-type stars. The importance of the C/O ratio for the structures of atmospheres as well as for the molecular formation is shown by the analyses of the molecular features.

However, the strength of the absorption of CO and CS can not be explained well by our model atmospheres. For detailed and quantitative discussion on the molecular spectra of carbon stars, we need more realistic model atmospheres and more precise parameters of atmospheres, and may need to take the time variation of the spectra into account. The detailed models of the outer atmosphere are also required to study the implications of these features.

Acknowledgments. We are grateful to Prof. H. Okuda for taking the initiative of our ISO project. We would like to thank Dr. I. Yamamura for helpful support in the data reduction, and also thank the SWS Instrument Dedicated Team for making available the SWS Interactive Analysis software. W. A. and K.O. are supported by JSPS Fellowship for young scientists.

References

Aoki W., Tsuji T., Ohnaka K., 1998, A&A 340, 222

de Graauw Th., et al., 1996, A&A 315, L49

Lambert D.L., Gustafsson B., Eriksson K., Hinkle K.H., 1986, ApJS 62, 373

Ohnaka K., 1997, Ph. D. Thesis, The University of Tokyo

Ohnaka K., Tsuji T., 1996, A&A 310, 933

Tsuji T., Ohnaka K., Aoki W., Yamamura I., 1997, A&A 320, L1

Asymptotic Giant Branch Stars
T. Le Bertre, A. Lèbre and C. Waelkens, eds.
I.A.U. Symposium 191, 1999

Infrared spectra of C-type variables with ISO

J. Hron, R. Loidl

Institut für Astronomie, Univ. Vienna, Türkenschanzstraße 7, A-1180 Wien, Austria

S. Höfner[1], U.G. Jørgensen

Niels Bohr Institute, Astronomical Observatory, Juliane Maries Vey 30, DK-2100 Copenhagen, Denmark

B. Aringer, F. Kerschbaum, W. Windsteig

Institut für Astronomie, Univ. Vienna, Türkenschanzstraße 7, A-1180 Wien, Austria

Abstract. We present ISO-SWS spectra for several representative AGB carbon stars obtained at several phases of the pulsational cycle. For stars with significant mass loss we correct the spectra for the dust emission by using DUSTY models. We then compare these corrected data to synthetic spectra based on hydrostatic and dynamical model atmospheres. We also discuss the influence of the fundamental model parameters on the resulting synthetic spectra.

1. Introduction

ISO-SWS spectra of AGB stars allow to study both the molecular features and the dust emission of these stars. Since also most of the AGB stars are (strongly) variable, ISO-SWS data provide important constraints for hydrostatic and dynamic model atmospheres. Carbon stars are of special importance in this respect because the molecular features probably affected by dynamical phenomena fall in the ISO-SWS range and also the theory of dust formation is much better understood than in the O-rich case.

2. Model spectra

Synthetic spectra for carbon-rich AGB stars in the wavelength region between 0.5 and 12 μm were computed using the improved dynamical model atmospheres of Höfner et al. (1998; also this volume) and using the opacity sampling technique. The molecules CO, CH, C_2, CN, C_3, HCN and C_2H_2 were included. The models are characterized by the following set of parameters: stellar mass,

[1]also: Institut für Astronomie, Univ. Vienna

luminosity, effective temperature of the hydrostatic initial model and the carbon-to-oxygen abundance (C/O) ratio as well as the piston period P and velocity amplitude Δu_P. All abundances except that of carbon are assumed to be solar. Our 'standard' model has the following parameters: $M = 1\ M_\odot$, $L = 7000\ L_\odot$, $T_{eff} = 2880$ K, C/O = 1.4, P = 390 d and $\Delta u_P = 4$ km s^{-1}. Table 1 summarizes how the different model parameters change the spectra. A detailed discussion can be found in Loidl et al. (1998).

Table 1. Influence of different model parameters on spectra

parameter	features	variation
T higher	weaker	stronger
L higher	stronger	smaller
M smaller	stronger	smaller
C/O higher	stronger	stronger
Δu_P higher	stronger	stronger
no dust	stronger	

In Fig. 1 gas pressure versus temperature is plotted for the maximum (full) and the minimum (dashed) phase of our 'standard' dynamical model. The horizontal bars indicate the temperature ranges for the molecules CO, C_2, CN, HCN, C_3 and C_2H_2 where their features between 0.5 and 12 μm are approximately formed. The grey shaded area denotes the region of dust formation.

3. Observational data

Within our open time observing programs the spectral variations of a few representative AGB stars were monitored with the ISO-SWS (de Graauw et al. 1996). In the following table we list the major characteristics of our targets:

Table 2. Summary of targets

Name	Type	Spectrum	P[d]	dM/dt [10^{-6}]	Comment
T Dra	M	C6,2-C8,3	422	2	typical C-Mira
V Cyg	M	C5,3-C7,4	422	2	typical C-Mira
V CrB	M	C6,2	358	0.2	metal poor? C-Mira
R Scl	SRb	C6,5	370	<1	large amplit., detached shell
TT Cyg	SRb	C5,4e	118	0.01	small amplit., detached shell
V460 Cyg	SRb	C6,4	180	0.1	small amplit.

For the first five targets we have typically five observations distributed over the pulsational cycle. The spectra were full grating scans (2.4 μm to 44 μm) with a typical resolution of 400. The pipeline processed data (mostly OLP versions

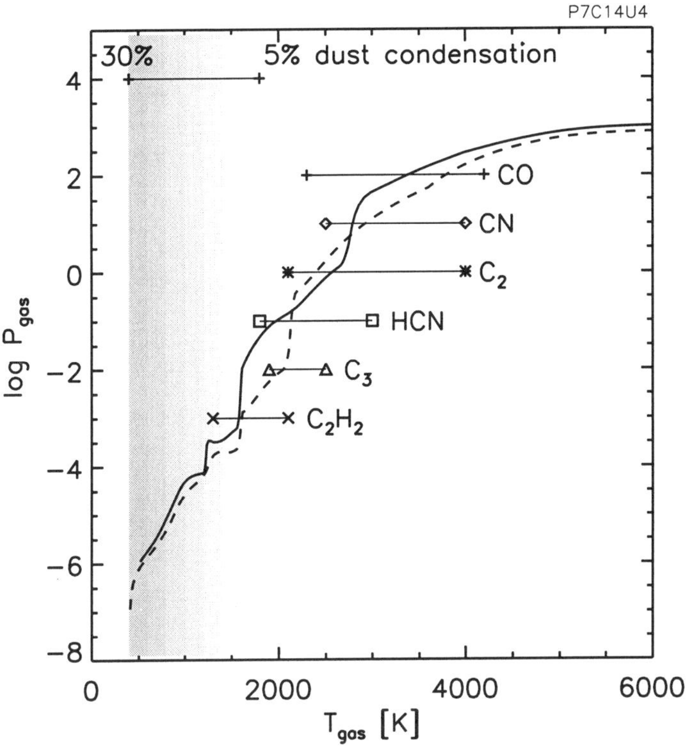

Figure 1. Gas pressure versus temperature for maximum (full) and minimum (dashed) phase of the 'standard' dynamical model. The horizontal bars indicate the temperature ranges for the molecules CO, C_2, CN, HCN, C_3 and C_2H_2 where their features between 0.5 and 12 μm are approximately formed. The grey shaded area denotes the region of dust formation.

6.xx) were further reduced with ISAP and then rebinned to a resolution of 100. For all spectra a significant offset between the two scan directions was found in band 2C (7 μm to 12 μm) which was corrected by rejecting the scan direction with the higher flux. The agreement between spectra of the same star taken within a few days was generally better than 10%. There was also good agreement with (quasi-simultaneous) ground based and IRAS photometry.

First results on R Scl have been presented by Hron et al. (1998) and will serve as a reference for the other targets.

4. (Almost) Dust-free carbon stars

The spectra of TT Cyg and V460 Cyg have been compared with R Scl and hydrostatic model atmospheres. For the comparison with the models, the ISO data were normalized by linear interpolation between the fluxes at several selected 'pseudo-continuum' points and then scaled to a continuum level of 0.8 to 0.9. The continuum points were selected on the basis of the synthetic spectra. This normalization is somewhat problematic shortward of 2.8 μm due to the lack of data points at shorter wavelengths. The results of the comparison can be summarized as follows:

- The depths of the 3 and 3.8 μm (C_2H_2) features relative to the 5.1 μm (C_3) feature suggest a sequence of decreasing temperature: V460 Cyg $\rightarrow$ TT Cyg $\rightarrow$ R Scl (maximum) $\rightarrow$ R Scl (minimum).
- The ratio of the strength of the 4.6 μm (CO) to the 5.1 μm (C_3) features indicates a sequence of increasing C/O from V460 Cyg ($1.1 < C/O \leq 1.4$) to TT Cyg and R Scl (both with C/O ≈ 1.4).
- The best fits are obtained for $T_{eff} = 3\,200$ K. This is in agreement with results from the Infrared Flux Method (Ohnaka & Tsuji 1996) but significantly hotter than what is expected from the spectral type/temperature relation derived from angular diameters (Dyck et al. 1996).
- Even for $T_{eff} = 3\,200$ K the hydrostatic models give too strong features at 3 and 3.8 μm and a too weak 4.6 μm feature.
- The region from 6 to 10 μm is generally not well reproduced.

The discrepant feature strengths shortward of 6 μm could be due to opacity problems or a non-hydrostatic T/P-structure (as e.g. caused by shock waves). At longer wavelengths, instrumental problems, the contribution from dust and opacity problems are possible causes. We note that contrary to Aoki et al. (1998), our models always predict a significant (50%) C_2H_2 contribution to the 3 μm feature.

5. Dusty carbon stars

For the comparison with the dynamic models, a correction of the observed spectra for the dust emission is of course necessary. For this purpose we have used the publicly available DUSTY code (Ivezić et al. 1997). Using selected synthetic spectra for the photospheric input radiation, a ratio of amorphous carbon to SiC of 4:1 and a $1/r^2$ dust density law, we fitted the full SWS spectra at selected phases. The best fits were obtained with a condensation temperature of 1 000 K and optical depths at 2.2 μm between 0.1 (maximum phases) and 0.3 to 0.5 (minimum phases). Since the amount of dust emission for a given ISO spectrum did not really depend on the input spectrum (as long as a realistic synthetic spectrum was used), we constructed smooth dust continua for several optical depths. The typical contribution of these continua to the total flux ranged from 40% at 4 μm to 90% at 10 μm. The dust continua were then subtracted from the observed spectra and these spectra were then normalized in the same fashion as described above. We note that due to this procedure, the SiC feature is not subtracted, only the SiC continuum. Furthermore, the use of a synthetic spectrum assumes that all the molecular features form inside the dust condensation radius, an assumption which seems to be justified on the basis of Fig. 1.

In Fig. 2 we show corrected and normalized spectra of T Dra for two phases together (thick lines) with synthetic spectra based on dynamic model atmospheres. The results from comparing our spectra with several models can be summarized as follows:

- There seems to be a sequence of decreasing T_{Max}: TT Cyg → V CrB → T Dra & V Cyg → R Scl.
- The weaker 5.1 μm feature in the Miras indicates a sequence of increasing C/O: V CrB & V460 Cyg → V Cyg & T Dra → TT Cyg & R Scl.
- The three Miras show similar variability patterns: the deepest features appear near visual phase 0.6, the amplitudes increase from the 5.1 μm to the 3 μm and the 3.8 μm features.
- The agreement with the dynamical models is comparable to what was found for R Scl. A model with 7 000 $L_\odot$ fits better than one with 10 000 $L_\odot$.
- The 14 μm feature is comparable to the 3 μm feature in depth and variation, which is in agreement with preliminary models for this feature.
- The SiC feature might have a slightly lower contrast near minimum light. Its shape differs systematically from the DUSTY models. This effect is most likely due to the influence of molecular features which means that the interpretation of the 10 μm feature requires a proper treatment of the molecules and the dust!

Acknowledgments. This work is supported by the Austrian Science Fund (FWF projects S7308 and J1487), the Danish Natural Science Research Council and the Austrian Academy of Sciences (APART programme). It is a pleasure to thank the ISO operations and instrument teams for their support.

References

Aoki W., Tsuji T., Ohnaka K., 1998, A&A 340, 222

de Graauw T., Haser L.N., Beintema D.A., et al., 1996, A&A 315, L49

Dyck H.M., van Belle G.T., Benson J.A., 1996, AJ 112, 294

Ivezić Ž., Nenkova M., Elitzur M., 1997, User Manual for DUSTY, Internal Report, University of Kentucky, accessible at http://www.pa.uky.edu/~moshe/dusty

Hron J., Loidl R., Höfner S., et al., 1998, A&A 335, L69

Loidl R., Höfner S., Jørgensen U.G., Aringer B., 1998, A&A, submitted

Ohnaka K., Tsuji T., 1996, A&A 310, 933

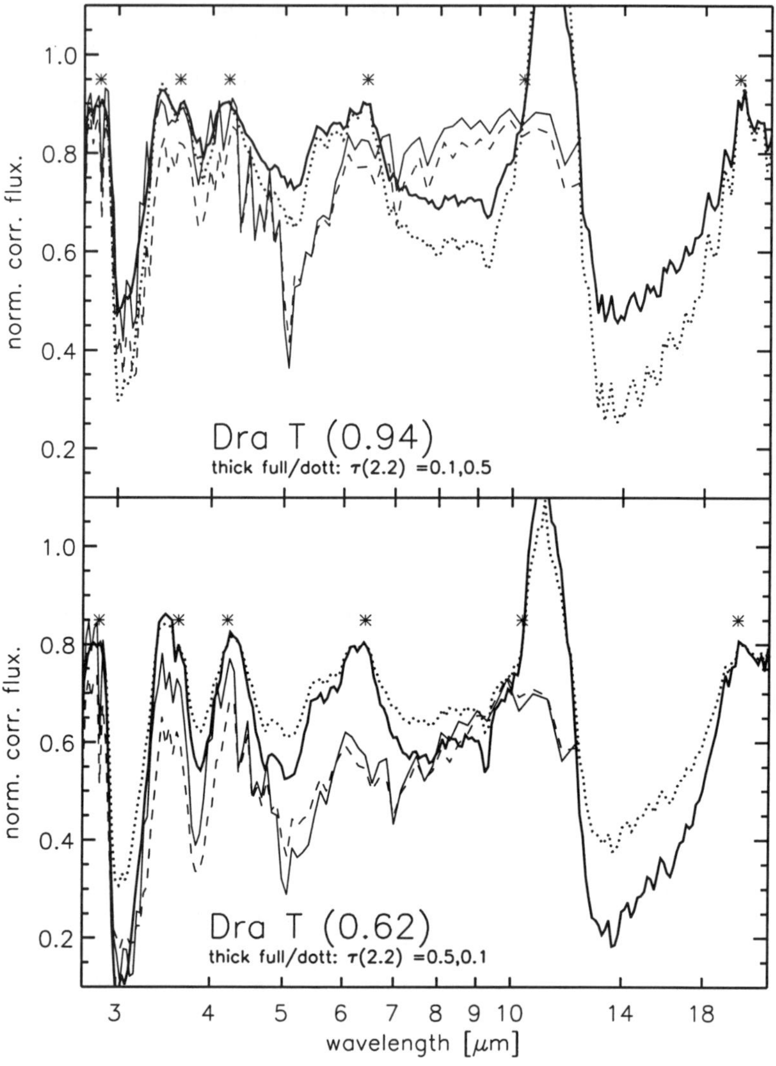

Figure 2. ISO-SWS spectra of T Dra taken near visual phases 0.94 (top) and 0.62 (bottom) compared to synthetic spectra. The observed spectra were normalized and corrected for continuous dust emission (Sect. 5) and are shown with thick full and dotted lines. The dust optical depths at 2.2 μm are indicated in the plot, the thick full lines result from the best DUSTY fit, the thick dotted lines give an indication for the influence of this parameter on the depth of the molecular features. The thin full lines are the synthetic spectra at similar phases for the 'standard' dynamical model described in the text, the thin dashed lines are for a model with 10 000 $L_\odot$. The asterisks mark the wavelengths used for the normalization.

Asymptotic Giant Branch Stars
T. Le Bertre, A. Lèbre and C. Waelkens, eds.
I.A.U. Symposium 191, 1999

Dynamical Modeling of Circumstellar Outflows

Axel J. Fleischer, Jan Martin Winters and Erwin Sedlmayr

Institut für Astronomie und Astrophysik, TU Berlin, Sekr. PN 8-1, Hardenbergstr. 36, 10623 Berlin, Germany

Abstract. We review dynamical models of circumstellar dust shells around long-period variables which include time-dependent hydrodynamics and a detailed treatment of dust formation, growth and evaporation. Important effects caused by the complex interaction between the dynamics of the pulsating atmosphere and the dust complex which only can be revealed in the dynamical approach are summarized. Special emphasis is given to the treatment of the dust and gas opacity.

1. Introduction

In order to arrive at a realistic picture of an AGB outflow it has been shown that an appropriate model has to meet at least the following requirements (Fleischer et al. 1992; Höfner et al. 1996):

1. The hydrodynamical description has to be time–dependent, to describe the variations of the velocity field and the occurrence of shock waves caused by the pulsation of the central star.
2. Since the formation of dust is a crucial phenomenon in these objects, it is necessary to use a theory which describes the formation of solid particles depending on the physical conditions, and which does not prescribe essential quantities such as the site where the dust forms.
3. Furthermore, the chemistry of the gas phase has to be modeled in order to know at each instant of time and at each radial position how much condensible material is available.
4. Finally, it is mandatory to treat the radiative transfer problem in order to describe the influence of the radiation of the central object.

In contrast to the classical atmosphere problem, the presence of dust introduces a number of non-linear couplings and interactions which are of essential importance for the whole problem of an AGB outflow or, in general, for every outflow where phase transitions can take place. A reliable theoretical modeling on the one hand requires a *physical* description of the various ingredients, and on the other hand, at least equally important, it requires taking into account all interactions among the different physical components.

It is evident from the above list that the problem is non-linear and strongly coupled. If one considers for instance the coupling between hydrodynamics and

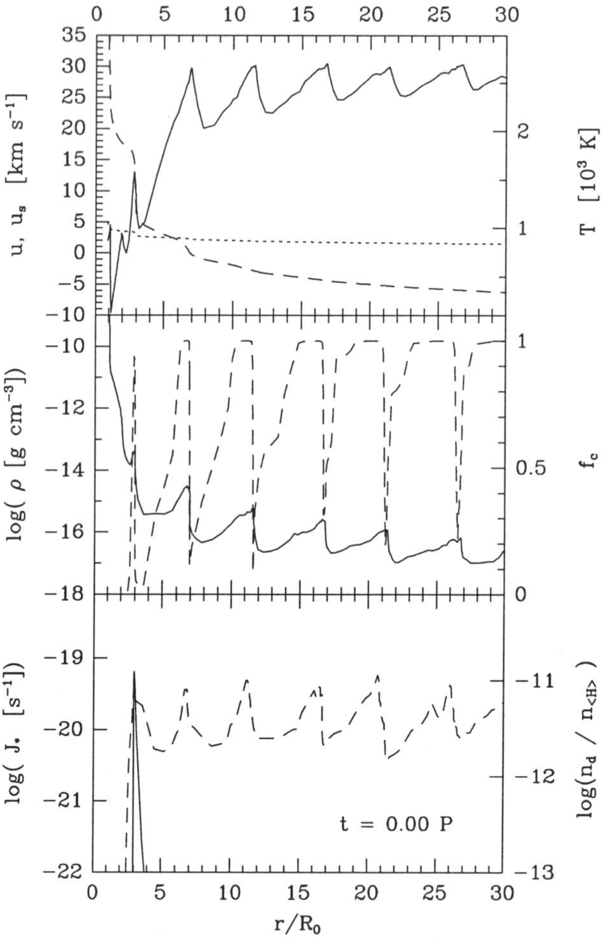

Figure 1. A typical radial structure of a dynamical model calculation

the dust complex: effective dust formation is possible only if the density is high enough; but at the same time, since radiation pressure on dust enters into the equation of motion, dust formation immediately influences the velocity structure, which in turn alters the density stratification.

The basic equations describing the dust forming circumstellar shell and the numerical methods applied for their solution are described in detail in Fleischer et al. (1992) and Fleischer (1994). These works deal exclusively with the case of carbon-rich circumstellar dust shells (CDS). First results concerning dynamical models of oxygen-rich CDS have been presented at this conference (Jeong et al., this volume). The pioneering works by e.g. Wood (1979) and Bowen (1988) give details about earlier research on this field.

For an introduction to the theory of dust formation we refer the reader to the series of papers by Gail and Sedlmayr. A good paper to start with is Gail & Sedlmayr (1988).

2. Phenomena revealed by the dynamical approach

The radial structure of a typical dynamical model is shown in Fig. 1. at the beginning of the pulsational period.

The parameters of this model are $L_\star = 1.0\ 10^4\ L_\odot$, $M_\star = 1.0\ M_\odot$, $T_\star = 2600\,K$, $\epsilon_C/\epsilon_O = 1.8$, P = 650 days, $\Delta u = 2$ km s^{-1}. In the following we refer to this set of parameters as our *standard* model.

It is evident from Fig. 1. that the dust is not distributed homogeneously across the shell but is concentrated in distinct layers such that the circumstellar dust shell exhibits an onion-like structure. Furthermore it can be seen that the dust quantities, e.g. the degree of condensation f_c (2nd panel, dashed line), are intimately correlated with the hydrodynamical quantities, e.g. the velocity u (upper panel, solid line). This suggests that there is a common mechanism which produces this structure.

2.1. Re-amplification of the shock waves

A more detailed analysis shows that the shocks propagating through the atmosphere are not produced by the interior pulsation alone but are a product of the re-amplification of the pulsational shocks which is caused by the radiative acceleration α on dust grains in the discrete dust layers. The re-amplification is the result of a 2-step process: first, pulsational waves run into the region where initial dust nuclei are formed. They compress the material and initiate a phase of effective dust growth. Second, once a relevant amount of material has condensed into dust grains these grains become dynamically important. Radiation pressure on these newly formed grains amplifies the pulsational wave and dominant shock waves are formed which clearly can be seen in Fig. 1.

Since the radiative acceleration α exceeds unity close to the star, the material is accelerated to velocities above the escape velocity v_{esc} already at radii around $4\,R_*$ and a massive outflow is initiated. Details can be found in Fleischer et al. (1992).

2.2. Backwarming effect

Due to its opacity, the dust also leads to a pronounced heating of the material inside the dust layers (backwarming), as is evident from the steps present in the temperature stratification around $3\,R_\star$ and $6\,R_\star$ (upper panel, dashed line). The backwarming effect also introduces a new time scale into this dust-forming atmosphere (see Fleischer et al. 1995, Höfner et al. 1995, and Sect. 2.3). This can be inferred from the nucleation rate $J_\star$, the number of dust nuclei formed per second (lower panel, solid line). $J_\star$ has a sharp edge towards the star, caused be the high gas temperature, which prevents further nucleation. This high temperature is caused by the backwarming of the new dust layer which is also formed in this region. If this dust layer moves outwards the temperature behind the layer decreases and a new dust nucleation cycle may start.

2.3. Exterior κ-mechanism

Considering very luminous models of carbon-rich CDS a new nonlinear effect shows up which is caused by the dust opacity. Small amplitude waves are initiated in the innermost shell region by the backwarming. These waves provide the density enhancement necessary for a new dust formation and growth cycle. It can be shown that this mechanism already starts in an initially hydrostatic atmosphere and is self-maintaining in the dynamical model.

For these models, the CDS mimics a pulsating atmosphere, although this effect is solely caused by an instability due to the dust formation process.

Details about this phenomenon can be found in Fleischer et al. (1995).

2.4. Multiperiodicity

An important property of the model shown in the preceding section is that its radial structure repeats after *one* pulsational cycle.

For models calculated with the same parameter set as our standard model except with a reduced carbon overabundance, it turns out that the radial structure of the model repeats only after *two* pulsational cycles. Even tiny details of the radial structure, e.g. the multiple shocks present in the inner region below $4\,R_*$ in this model, are reproduced after this period of time. Around $2.5\,R_*$ at $t = 1.0\ P$ and $t = 3.0\ P$ in this model, a detailed analysis shows that the radiative acceleration α causes a perturbation in the velocity structure which later on turns into a dominant shock wave that sweeps up the preceding pulsational shock. Due to the reduced amount of condensible material it takes two pulsational cycles to form a new dust layer. The effect of *multiperiodicity* strongly depends on the ϵ_C/ϵ_O ratio. Lowering this number causes longer intervals between the formation of two dust layers; an increase causes the opposite effect. The time interval between the formation of two dust layers is of course not necessarily an integer multiple of the pulsation period, e.g. models with ϵ_C/ϵ_O values between 1.8 and 1.5 form a new dust layer on time scales larger than 1 and smaller than 2 pulsation periods. Depending on the remaining model parameters it is also possible that a dust layer forms on time scales of 3, 4, ... times the pulsation period, or even twice per pulsation cycle.

2.5. Deriving a mass loss formula

Based on a larger grid of dynamical models Arndt et al. (1997) were able to derive a fitting equation which relates a given set of the six model parameters to the resulting mass-loss rate. This formula reflects the strong dependence of the mass loss rate on the stellar mass and the stellar temperature.

Schröder et al. (1998) used this mass loss formula in stellar evolution calculations. They revealed an attractive mechanism which may lead to the formation of detached shells in the late stages of the AGB evolution. This finding was only possible by using a mass loss formula which is based on physical models in the relevant range of stellar parameters.

2.6. Infrared appearance of the dynamical models

The consistent treatment of the relevant physical processes also has profound influence on the IR-appearance of the models described above. We will not discuss here the occurring effects but refer the reader to the contribution of Winters et al. (1998, this volume) and e.g. to the recent paper of Winters et al. (1997).

3. Influence of the gas opacity

In the dynamical model calculations published so far, molecules have been treated only as a chemical pool for dust formation and growth. Radiation pressure on molecules has not yet been used as a supporting mechanism to drive the wind mostly because the investigations were aimed at dust-enshrouded objects. If dust is present it dominates the dynamics because the gas opacity, which is orders of magnitudes smaller, can be neglected. However, the question arises about the influence of radiation pressure on gas in less enshrouded objects and in the phase were the dust is initially forming. Höfner et al. (1998a, b) have pointed out that the gas opacity plays an important role in determining the near IR properties of these objects.

Very recently model calculations have been carried out by Helling (1998) in order to investigate the role of the gas opacity in more detail. From the opacity sampled line lists from the group of U. Jørgensen she computed mean opacities $\bar{\kappa}^{gas}$ for the two limiting cases of Planck and Rosseland opacities. These mean opacities were included into the hydro-code, replacing the previously used constant opacity.

Already in the initial hydrostatic model the influence of the gas opacity can be seen. The grey hydrostatic model atmospheres with $\kappa_H = \bar{\kappa}^{gas}_{Planck}$ are less dense than the hydrostatic models with $\kappa_H = \bar{\kappa}^{gas}_{Ross}$ which has profound consequences for the dynamical development of the model. Furthermore, models calculated with $\kappa_H = \bar{\kappa}^{gas}_{Planck}$ show a gas pressure inversion for certain combinations of stellar parameters which prevents effective dust formation in the corresponding dynamical model because the shocks caused by the interior pulsation cannot run into the region containing the dust nuclei but are damped already in the region of the pressure inversion. This effect does not occur in models with $\kappa_H = \bar{\kappa}^{gas}_{Ross}$.

For the parameters of the standard model which has been discussed in Fleischer et al. (1992) and in Sect. 2, the following quantities result for the two limiting cases:

Table 1. Resulting quantities of the evolved models for different mean gas opacities

	$<\dot{M}>$ $[M_\odot \mathrm{yr}^{-1}]$	$<v_\infty>$ $[\mathrm{kms}^{-1}]$	ρ_d/ρ_g
Planck mean	$1.5\,10^{-7}$	6.9	$4.8\,10^{-4}$
Rosseland mean	$4.3\,10^{-6}$	21.5	$2.4\,10^{-3}$

As a result of the pressure inversion in the "Planck models" the degree of condensation reaches only 15 % so that the mass loss rate and the final outflow velocity remain small. For the Rosseland mean the radial structure and the dynamics of the models are more or less similar to the ones described in Sect. 2.

To summarize, these calculations show that radiation pressure on molecules may have a non-negligible influence on the dynamics and the driving of the wind even in the case of carbon-rich CDS. Therefore, already these investigations show that radiation pressure on molecules is an excellent candidate to initiate and to support the winds of oxygen-rich objects where dust formation alone might be too inefficient to maintain the outflow.

Although mean opacities are only a rough approximation for frequency-dependent opacities they are a very convenient tool to study the influence of radiation pressure on molecules on the dynamics of circumstellar outflows.

4. Influence of the grain opacity

The extinction properties of the dust component play a decisive role for the hydrodynamic and thermodynamic structure of a CDS and also determine its spectral appearance if relevant amounts of dust are formed.

In order to calculate these transport coefficients in the framework of Mie theory, the complex refractive index $m(\lambda)$ of the grain material has to be known.

However, the optical constants of the grains do not only depend on the dust material but also on the microscopic structure of the particles. In case of an ordered (crystalline) structure one finds a $1/\lambda^2$ - dependence of the FIR extinction, while a disordered (amorphous) structure gives a $1/\lambda$ - dependence. In the case of carbon dust, both structures, amorphous carbon and crystalline graphite, are possible.

The actual structure of the grains formed in the stellar wind is, of course, not known a priori but is a result of a complex interaction. Theoretical considerations predict the formation of inhomogeneous grains with crystalline cores surrounded by amorphous mantles (Gail & Sedlmayr 1984). Only the mantles interact with the radiation field, and hence, are accessible by observations.

Recent spectral modelling of observational results implies that the presence of amorphous grains rather than graphite is more likely in the circumstellar shells around C-stars (e.g. Griffin 1990; Orofino et al. 1990; Bagnulo et al. 1995).

Unfortunately, the optical properties of the grains, which decisively influence their physical environment, depend sensitively on their exact internal structure, e.g. the degree of crystallization and the hydrogen content, which in turn is sensitive to the physical conditions of their production.

Tables for the Planck–mean value of the dust opacity as a function of temperature were calculated from the data of the complex refractive index obtained in the laboratory. For details of the data we refer to the references given in Table 2. These tables were used in the dynamical model calculations and the resulting quantities (mass loss rates, terminal expansion velocities, dust-to-gas mass ratios) are given in Table 2.

It turns out that the mass loss rate and the dust-to-gas-ratio are unaffected by the table used to calculate the dust opacity. The opacity, however, directly determines the acceleration of the material, and hence, significantly influences the final outflow velocity. The overall structure of the outflow differs only slightly from the structure presented in Fig. 1.

Table 2. Resulting quantities of the evolved models for different data sets describing the complex refractive index of carbon grains. The parameters of the standard model are used.

table	$<\dot{M}>$ $[M_{\odot}\mathrm{yr}^{-1}]$	$<v_{\infty}>$ $[\mathrm{kms}^{-1}]$	ρ_d/ρ_g
AC1[1]	$1.1\,10^{-5}$	24.3	$4.2\,10^{-3}$
BE1[1]	$9.1\,10^{-6}$	32.5	$3.5\,10^{-3}$
HAPS[1]	$1.0\,10^{-5}$	33.1	$3.3\,10^{-3}$
FC21PS[1]	$1.1\,10^{-5}$	32.6	$3.5\,10^{-3}$
MARONAC[2]	$1.3\,10^{-5}$	27.8	$4.1\,10^{-3}$
POYHAC[3]	$1.0\,10^{-5}$	33.7	$4.0\,10^{-3}$
MWAC[4]	$9.4\,10^{-6}$	31.4	$3.6\,10^{-3}$
Graphite[5]	$1.2\,10^{-5}$	24.7	$4.1\,10^{-3}$

[1]: Rouleau & Martin (1991), [2]: Maron (1990), [3]: Preibisch et al. (1993), [4]: Mathis & Whiffen (1989), [5]: Draine (1985)

Although the dust opacity influences only marginally the dynamical behavior it has substantial influence on the optical appearance of the model, e.g. its position in a two-color diagram. This will be discussed in detail in a forthcoming paper (Winters & Fleischer 1998).

5. Concluding remarks

Dynamical models of carbon-rich outflows have reached a maturity in the last years which makes them ideal tools to interpret the observations. These models can still be improved; however, it is unlikely, that this will change the general picture we have now of these objects. Unfortunately, comparable studies of oxygen-rich outflows are not yet available, mainly because of the much more complex dust formation process in this case (see Jeong et al., in this volume)

Both, carbon-rich as well as oxygen-rich models are based on a simple approximation to simulate the interior pulsation. To combine dynamical atmospheric models with stellar pulsational models is one of the great challenges for stellar astrophysics in the next years.

References

Arndt T., Fleischer A.J., Sedlmayr E., 1997, A&A 327, 614

Bagnulo S., Doyle J.G., Griffin I.P., 1995, A&A 301, 501

Bowen G.H., 1988, ApJ 329, 299

Draine B.T., 1985, ApJS 57, 587

Fleischer A.J., 1994, PhD Thesis, TU Berlin (http://export.physik.tu-berlin.de)
Fleischer A.J., Gauger A., Sedlmayr E., 1992, A&A 266, 321
Fleischer A.J., Gauger A., Sedlmayr E., 1995, A&A 297, 543
Gail H.-P., Sedlmayr E., 1984, A&A 132, 163
Gail H.-P., Sedlmayr E., 1988, A&A 206, 153
Griffin I.P., 1990, MNRAS 247, 591
Helling C., 1998, PhD Thesis, TU Berlin, in preparation
Höfner S., Feuchtinger M.U., Dorfi E.A., 1995, A&A 297, 815
Höfner S., Fleischer A.J., Gauger A., Feuchtinger M.U., Dorfi E.A., Winters J.M., Sedlmayr E., 1996, A&A 314, 204
Höfner S., Jørgensen U.G., Loidl R., 1998a, in "ISO's View on Stellar Evolution", eds. L.B.F.M. Waters, C. Waelkens, K.A. van der Hucht and P.A. Zaal, Kluwer, Dordrecht, p. 281
Höfner S., Jørgensen U.G., Loidl R., Aringer B., 1998b, A&A 340, 497
Maron N., 1990, Ap&SS 172, 21
Mathis J.S., Whiffen G., 1989, ApJ 341, 808
Orofino V., Colangeli L., Bussoletti E., Blanco A., Fonti S., 1990, A&A 231, 105
Preibisch T., Ossenkopf V., Yorke H.W., Henning T., 1993, A&A 279, 577
Rouleau F., Martin P.G., 1991, ApJ 377, 526
Schröder K.P., Winters J.M., Arndt T.U., Sedlmayr E., 1998, A&A 335, L9
Wood P., 1979, ApJ 227, 220
Winters J.M., Fleischer A.J., 1998, in preparation
Winters J.M., Fleischer A.J., Le Bertre T., Sedlmayr E., 1997 A&A 326, 305

Asymptotic Giant Branch Stars
T. Le Bertre, A. Lèbre and C. Waelkens, eds.
I.A.U. Symposium 191, 1999

Making movies of stars: VLBA monitoring of the SiO masers around the Mira variable TX Cam

P. J. Diamond and A. J. Kemball

NRAO, PO Box 0, Socorro, NM, 87801, USA

Abstract. We have produced a movie of the time evolution of the v=1, J=1–0 43 GHz SiO masers in the inner circumstellar envelope of the isolated Mira variable TX Cam. Our observations, taken every two weeks with the VLBA, currently cover about 60% of the stellar cycle from $\phi \sim 0.6 \rightarrow \phi \sim 1.2$. They show that, in contrast to previous observations and theoretical models, the gas in which the masers are embedded is undergoing expansion during this period with a typical velocity of 3.65 ± 0.15 km/s. Some isolated features are seen to start falling back towards the star at around $\phi \sim 1.0$. We suggest that the $5 \rightarrow 10$ G magnetic fields we have previously detected at the position of the SiO masers may explain the departure from the expected behaviour.

1. Introduction

Circumstellar envelopes can be studied in many ways as the numerous papers presented at this meeting have shown. However most observational studies deal with data taken at a single epoch or coarsely sampled data studying the large-scale properties of the envelopes or particular chemical species. Theoretical studies based on the observations attempt to reproduce the gross properties of the envelopes and the underlying stellar structure and have only recently attempted to describe the dynamics involved. More detailed theoretical work has been hampered by the lack of high-resolution, high time-sampled data that reveal the true structure and evolution of the circumstellar envelopes.

With the advent of NRAO's Very Long Baseline Array (VLBA) we now have a sensitive, 10-station interferometer, observing year round with a spatial resolution $\sim 50 - 100$ times that of the Hubble Space Telescope. One of the stated goals of the VLBA is to enable frequent time monitoring of radio sources. Observations of molecular masers in the circumstellar envelopes of late-type stars are well suited to such an instrument.

Habing (1996) has produced an excellent review on the subject of circumstellar envelopes and describes in some detail the location and structure of OH, H_2O and SiO masers. Masers are detectable by VLBI techniques since they are compact high brightness temperature objects. They are thus very useful probes of the structure and dynamics of the clouds in which they reside. Habing (1996) sums up the available literature on SiO masers which show that they lie close to the stellar photosphere inside the dust condensation radius and, in most stars, form a beautiful ring-like structure (Diamond et al. 1994). One obvious conclusion, based on the observed rings, is that the SiO masers are tangentially amplified unlike the radial amplification seen in OH masers in the same stars. The SiO masers also exhibit strong linear polarization with a highly ordered position angle structure primarily tangential to the envelope (Kemball & Diamond 1997).

Little is known about the variability of SiO maser structure. The current state of knowledge has been gleaned principally from single-dish observations (e.g. Cho et al. 1996; Pijpers et al. 1994). There appears to be little connectivity of spectra and spectral features from cycle to cycle of the underlying stellar pulsation. From day to day changes appear minor and are mainly confined to simple flux density variability. However, within a stellar cycle there do appear to be characteristics of the variability that are common to most stars:

- the SiO maser flux density varies in the same manner as the optical flux but with a phase lag ($\Delta\phi \sim 0.2$);
- the brightness contrast between minimum and maximum SiO maser flux density varies from $5 \rightarrow 100$ depending on the star;
- 86 GHz SiO masers (v=1, J=2–1) are brighter for longer than 43 GHz masers (v=1, J=1–0);
- from $\phi \sim 0.2 \rightarrow 0.8$ the average velocity of the 43 GHz SiO spectrum may be red-shifted with respect to the stellar velocity implying that the gas is infalling (Cho et al. 1996).

Boboltz et al. (1997) showed, through VLBA observations at three epochs, that the SiO masers around the symbiotic Mira R Aquarii were falling back towards the star with an average velocity of 4.2 ± 0.9 km/s over a phase range of $0.78 \rightarrow 1.04$. This is in agreement with the statistical single-dish results of Cho et al. (1996) mentioned above and the theoretical models of Bowen (1988). One result from the Boboltz et al. (1997) observations was that in order to adequately follow individual maser components with time the optimum VLBI observing interval was 2 weeks. In this contribution we present preliminary results from such a monitoring program of the 43 GHz (v=1, J=1–0) SiO masers towards the isolated Mira variable TX Cam which has a period of 557 days and lies at a distance of 317 pc. Previous observations of the masers around TX Cam (Diamond et al. 1994; Kemball & Diamond 1997) have shown them to lie in a well-defined ring structure with a typical diameter of 30 milliarcsec and with a polarization structure which shows electric vectors lying predominantly tangential to the ring.

2. Observations

Multiple epoch VLBA observations at two week intervals of the SiO maser in TX Cam began in May 1997 and are still running. We hope to obtain 50 epochs to cover 1.25 cycles of the Mira's pulsation. In this paper we report on the preliminary results from the first 21 epochs. The AAVSO light curve of TX Cam is shown in Fig. 1 with the time of each VLBA observation marked with

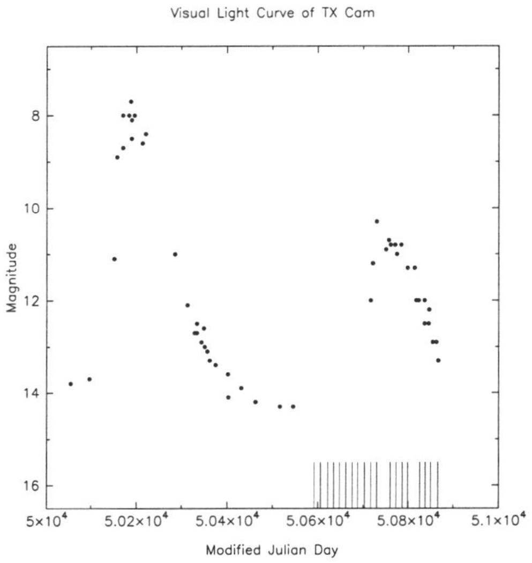

Figure 1. The light curve of TX Cam (courtesy of AAVSO). The vertical lines at the bottom of the plot represent the times of VLBA observations used to produce the movie.

a vertical line. The phase range covered by the current movie is from $\phi \sim 0.6$ to $\phi \sim 1.2$, that is similar to, but larger than, the phase range covered by the Boboltz et al. (1997) observations of R Aquarii mentioned above.

At each epoch we obtain 6 hours of data on TX Cam and correlate the data at the VLBA correlator in Socorro, NM producing 128 channel, 4 Stokes parameter data for all available baselines. Most observations include a single antenna of the VLA. In order to handle the enormous amount of data generated we have developed an automatic data calibration and imaging procedure within the AIPS package. This has proven highly effective and has enabled us to keep up with the influx of data.

Work is continuing on semi-automatic methods of image analysis which will result in our final data product: a large list of component positions, fluxes and polarization properties. However, it is possible to visualize the structural changes in the SiO maser emission by making a movie. This is constructed by first taking the Stokes I image cube of each epoch and running through the cube in velocity space, extracting the maximum value of each pixel and forming a composite image that represents the total structure well. The resultant images at each epoch have to be registered in some manner since we do not have absolute position measurements of the required accuracy (μarcsec level); this is done by assuming that there are no major changes between epochs and then cross-correlating the images and performing pixel level shifts to maximize the correlation. Such a method works well and justifies our desire to observe every two weeks. The final step is to interpolate across the occasional gap that occurs in our sequence of observations (due to operational limitations, bad weather etc.) and then form a new cube which now has time as the third axis.

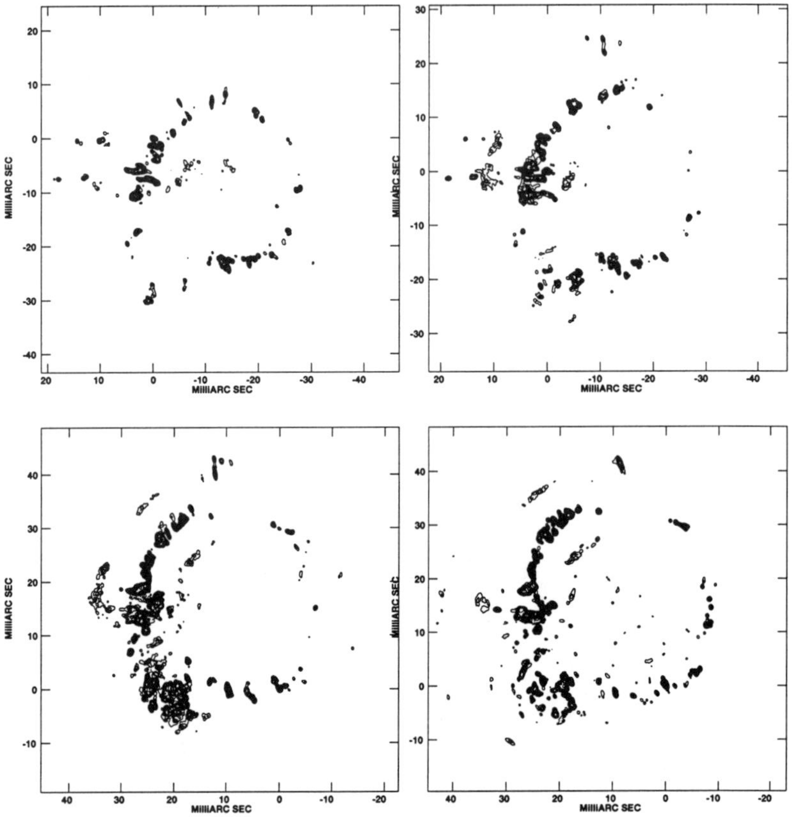

Figure 2. Four frames of the VLBA movie of TX Cam. The dates of observation are: top left - 24 May 1997; top right - 16 Aug 1997; bottom left - 5 Dec 1997; bottom right - 22 Feb 1998

3. The movie

Fig. 2 shows four frames from the movie shown during the conference. The principal result is that the dominant dynamical mode is one of expansion. The first epoch (24/5/97) shows a largely complete, almost circular ring, with an average diameter of 31 milliarcsec (9.8 a.u.). The final epoch (22/2/98) shows a structure that has transformed into an ellipse with an EW diameter of 34 milliarsec (10.8 a.u.) and a NS diameter of 36 milliarcsec (11.4 a.u.). The changes are highly significant being of order ten times larger than the VLBA's beamsize at 7mm and correspond to a 10% increase in size in the EW direction and 20% in the NS.

We can follow the motion of individual features. Gaussian fitting of maser components at each epoch will produce proper motion diagrams. Fig. 3 shows the data for one such feature in the NE quadrant of the ring. A fit to the data points shows that, in the 6 month period before the feature faded, it maintained a constant outflow velocity of 3.65 ± 0.15 km/s. This velocity is consistent with the width of the SiO maser spectrum.

The large complex of features that lies in the SE portion of the ring increased in flux during the observation period and shows evidence of peculiar motions with features sliding in non-radial directions. It is possible that this complex is indicative of non-symmetric mass-loss.

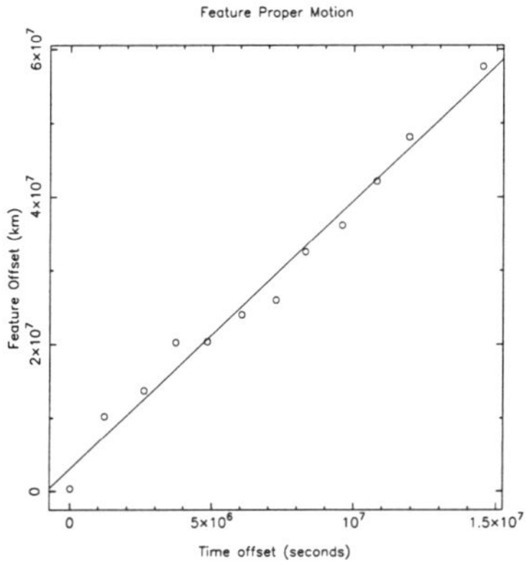

Figure 3. The proper motion of a feature in the NE quadrant of the TX Cam maser ring. The errors bars for each feature are smaller than the circles on the plot. The feature fades from view after 6 months. The straight line is a fit to the data which results in a feature proper motion of 3.65 ± 0.15 km/s.

4. Discussion

The continual expansion of the maser ring from $\phi \sim 0.6 \rightarrow \phi \sim 1.2$ is in direct contrast to the contraction observed in R Aquarii (Boboltz et al. 1997) and implied by the statistical single-dish observations of Cho et al. (1996). It also contradicts the theoretical model proposed by Bowen (1988) and recently expanded upon by Gray et al. (1998). The current theoretical understanding of the variations of the SiO masers is summed up by Gray et al. (1998) who state that flux and line profiles changes are expected as the masing objects move within a region with changing density, temperature and radiation fields. An abrupt change is expected once per stellar cycle due to the passage of the outward-moving shock front. Based on various observations of the SiO masers in R Aquarii Gray et al. (1998) suggest that the shock wave moves through the photosphere at a visual phase of $\phi \sim 0.7$ and that one would then expect to see contraction due to the star's gravitational field and flux increase of the maser emission from $\phi \sim 0.8 \rightarrow \sim 0.4$.

In the case of TX Cam we do observe a flux increase throughout the period of observation with a levelling off towards the end. However, as stated above the masers appear to be generally expanding at a more or less constant velocity. A few isolated maser features appear at $\phi = 1.0$ and begin to fall back towards the star but not along radial paths.

A major question is: what mechanism supports the observed continual expansion? The models of Bowen (1988) suggest quite strongly that after the impulse imparted by the passage of the shock wave gravity should eventually slow and then pull back the expanding gas. R Aquarii seems to follow this model yet TX Cam does not. We suggest that the stellar magnetic fields may play a crucial role in the dynamics of the inner envelopes.

Using circular polarization measurements Kemball & Diamond (1997) estimated the magnetic field strength in the vicinity of the SiO masers around TX Cam to be $Bsec\Theta \sim 5 - 10$ G, where Θ is the angle between the line of sight and the magnetic field. As was pointed out by Barvainis et al. (1987) the ratio of magnetic to thermal energies in the gas of the circumstellar envelope is > 1 for magnetic fields of this strength. Even for field strengths of ~ 0.1 G the overall dynamics of the gas will be strongly influenced by the magnetic fields. Future analysis of the polarization structure visible in our data should provide us with crucial information on the role of magnetic fields in the circumstellar environment.

Our VLBI monitoring data of SiO masers around TX Cam provides us with a wealth of data on the structure, motions, time evolution, mass-loss process, magnetic field structure, conditions and overall dynamics of the inner regions of the circumstellar envelope. We have only scratched the surface of this immense dataset in producing the movie.

Acknowledgments. The National Radio Astronomy Observatory is a facility of the National Science Foundation operated under cooperative agreement by Associated Universities, Inc. We acknowledge with thanks, data from the AAVSO International Database, based on observations submitted to the AAVSO by variable star observers worldwide.

References

Barvainis R., McIntosh G., Predmore C.R., 1987, Nature 329, 613

Boboltz D.A., Diamond P.J., Kemball A.J., 1997, ApJ 487, 147

Bowen G.H., 1988, ApJ 329, 299

Cho S., Kaifu N., Ukita N., 1996, AJ 111, 1987

Diamond P.J., Kemball A.J., Junor W., Zensus J.A., Benson J., Dhawan V., 1994, ApJ 430, 61

Gray M.D., Ivison R.J., Humphreys E.M.L., Yates J.A., 1998, MNRAS 295, 970

Habing H.J., 1996, Astron. Astroph. Rev. 7, 97

Kemball A.J., Diamond P.J., 1997, ApJ 481, 111

Pijpers F.P., Pardo J.R., Bujarrabal V., 1994, A&A 286, 501

Asymptotic Giant Branch Stars
T. Le Bertre, A. Lèbre and C. Waelkens, eds.
I.A.U. Symposium 191, 1999

Multiwavelength Studies of Mira Ceti-type Variable Stars

V.F. Esipov, M.I. Pashchenko, G.M. Rudnitskij, M.V. Kozin, E.E. Lekht, A.E. Nadjip, and S.V. Fomin

Sternberg Astronomical Institute, 13 Universitetskij prospekt, Moscow, 119899 Russia

Abstract. Since 1994, observations of a sample of about 20 Mira Ceti-type and semiregular variables have been carried out in three spectral ranges: radio (H_2O maser line $\lambda = 1.35$ cm), optical (spectroscopy and *UBV* photometry) and infrared (*JHKLM* photometry). Time series of the Hα emission intensity and H_2O line flux, covering several periods of the stars, have been obtained. Correlation of the intensity variations of the H_2O maser with optical variability in the maser stars RR Aql, U Ori, VX Sgr and others was confirmed. One of the most interesting results is the flare of the H_2O maser emission in R Leo, which happened in autumn 1997, 14 months after a flare of the Hα emission.

1. Introduction

The project of multiwavelength investigation of late-type variable stars grew up from monitoring observations of maser radio sources emitting in the rotational line 6_{16}–5_{23} of the H_2O molecule at $\lambda = 1.35$ cm. We started this program at the end of 1979 and, since then, we have been permanently observing variability of a large sample of H_2O masers, including about 30 late-type giants (Mira stars and semiregular variables).

To get a more complete picture of the stars variability, in 1994 we began a multiwavelength monitoring of a sample of late-type variables with (and without) H_2O maser emission. We were primarily interested in effects of shock waves in the atmospheres and envelopes of the stars. The diagnostics of the shocks is presumably the emission-line spectrum of the stars. Shock waves may influence the intensity of maser emission of OH, H_2O, and SiO in circumstellar envelopes — both directly (by impact of the shock on the maser generation region, Rudnitskij & Chuprikov 1990) and indirectly (through the radio continuum emission of the shock, amplified by masering molecules, Rudnitskij 1990).

There have been a number of studies of the H_2O-line variability in late-type stars (e.g., Schwartz et al. 1974, Cox & Parker 1979, Gómez Balboa & Lépine 1986, Woods et al. 1990); in particular, recently some interferometric monitoring of stellar H_2O masers has been done (Engels et al. 1998; Richards et al. 1998). In our single-dish H_2O observations, we do not attempt to compete with interferometry mapping of the masers. The main advantage of our project is long-term, multiwavelength study of a sample of stars, covering several cycles of their variability. A comparison of these data could yield a new insight on the problems of late-type stars variability, maser emission and shock waves.

In the framework of the project the following programs are carried out.

(1) Regular optical (*UBV*) monitoring of several Mira-type and semiregular variables (R Aql, R Leo, U Ori, U Her, RT Vir, R Cas, IK Tau, R Crt, RX Boo, W Hya).

(2) Optical spectroscopy of the stars.

(3) Radio spectral observations of maser emission of the same stars in 1.35-cm line of H_2O.

(4) IR observations of the stars.

(5) Observations of the stars at centimeter wavelengths for a systematic search for the post-shock radio continuum emission.

2. Observations

Radio observations in the H_2O line at $\lambda = 1.35$ cm are performed on the 22-meter radio telescope of the Lebedev Institute of Physics (Russian Academy of Sciences) in Pushchino, Moscow Region, by means of the 1.35-cm wavelength receiver with a helium-cooled FET amplifier and a 128-channel filter-bank spectrometer with a resolution of 0.1 km/s. Intervals between consecutive observing sessions normally do not exceed 1–2 months.

For optical spectroscopy, we use a diffraction spectrograph with a resolution of 0.25 Å/pixel. An echelle spectrograph with a resolving power of 10 000 has been tested. The spectra are recorded with ST-6 and ST-8 SBIG CCD cameras. This instrumentation is in operation on the 100- and 60-cm telescopes in Crimea.

IR observations are carried out on the Crimean telescopes, equipped with the IR photometer of the Sternberg Astronomical Institute.

In the radio continuum, we have observed a sample of 34 Miras and semiregular variables at $\lambda = 6$ and 3 cm on the Australia Telescope Compact Array (ATCA) in Narrabri (Chapman & Rudnitskij 1999).

3. Results

We have traced the evolution of the Hα emission in several stars (R Aql, R Leo, R Cas, U Her, χ Cyg and others). The appearance of the emission is not regular, not all light cycles are accompanied by it, contrary to the data of previous spectroscopic monitoring with a limited time coverage (e.g., Gillet et al. 1983, Fox, Wood & Dopita 1984, Woodsworth 1995, Castelaz et al. 1997). The emission flared at different light phases φ of the stars, from minimum to maximum, although phases $\varphi \sim 0.4 - 0.5$ seem to be preferred.

Comparing time series of optical and IR photometry, optical and H_2O line spectra, we confirm the correlation of the H_2O maser flux with the optical light curve for many stars studied, in particular, VX Sgr, U Ori and RR Aql (Berulis, Lekht & Pashchenko 1994; Berulis et al. 1998). Among our recent results, the most interesting are: detection of a superperiod in H_2O maser variations of U Ori; Hα-related flare of the H_2O maser R Leo; flare of the H_2O maser RT Vir; redistribution of the H_2O line flux between the groups of blue- and red-shifted

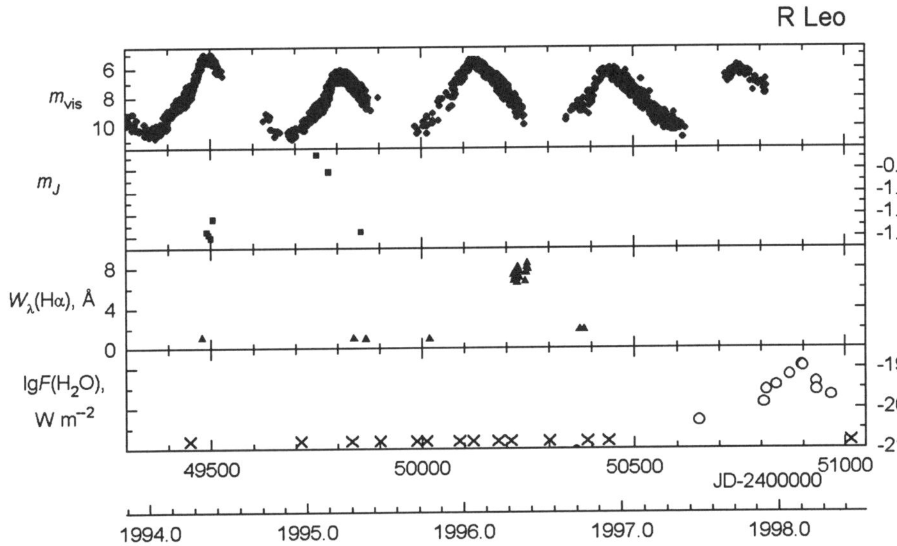

Figure 1. Data on R Leo. From top to bottom: visual light curve (AAVSO and AFOEV data); J-band stellar magnitudes; equivalent width of the Hα emission line; integrated flux of the H_2O maser emission (crosses are upper limits).

spectral features in bipolar outflows from the semiregular variables RT Vir and VX Sgr.

3.1. U Ori

This star was monitored in the H_2O line since March 1980, when it was below our detection limit (< 10 Jy). However, in October 1980 its H_2O peak flux density rose to about 1 000 Jy (Berulis et al. 1983). Since then, the H_2O maser in U Ori has shown flux variations with decreasing amplitude, closely following the light curve with a certain phase delay (Berulis, Lekht & Pashchenko 1994). The delay was not constant, but varied between 0.1 and $0.3P$ in an almost sinusoidal law with a period of about 9 years. We call it 'superperiod' of stellar activity.

3.2. R Leo

R Leo displayed pronounced H_2O flux variations, correlated with its light curve, in the early years of our H_2O observations (1980–1983) (Rudnitskij 1987), but in 1984–1997 we could not detect it at a 3σ level of 10 Jy. In autumn 1997 R Leo flared in the H_2O line, reached a peak flux density of about 150 Jy at the end of March 1998, but then faded again to less than 10 Jy in June 1998. This H_2O flare happened 14 months after the Hα flare observed by us in R Leo during summer 1996 (Fig. 1).

Two other examples of strong Hα flares that occurred when the stars were 'H_2O-quiet' (i.e. below our detection limit) are U Aur (in autumn 1997) and

R Cas (in September 1998). At present these two stars remain unobservable for us in the H_2O line, but we expect that they may soon flare up, too.

3.3. RT Vir

Our H_2O observations of RT Vir, obtained in 1986–1998 (for earlier spectra see Berulis et al. 1983, 1987, Mendoza-Torres et al. 1997), have shown that the H_2O maser emission is concentrated in three velocity intervals: $V_{LSR} <$ 15.5 km/s, 15.5 km/s$< V_{LSR} <$ 21 km/s, and $V_{LSR} >$ 21 km/s. Stellar velocity, determined from thermal molecular lines, is 18.2 km/s, i.e. close to the middle of the H_2O V_{LSR} interval. An analysis of the character of H_2O integral-flux variations in the above three groups of maser features yields a characteristic variability timescale of 5–6 years, i.e. much longer than typical timescale of optical brightness variations. Thus, a certain 'superperiod' in the maser activity of RT Vir can be distinguished, too. Anticorrelation of the integrated fluxes of the three spectral intervals is observed; it may be due to alternating increase of the intensities in the two jets of the bipolar outflow. In 1998 RT Vir has strongly flared in the H_2O line: the peak flux density in the blueshifted group of maser features reached 1000 Jy, the highest value ever observed in this star.

3.4. VX Sgr

VX Sgr is a semiregular M-type supergiant with a mean period of light variations of 732^d. Its H_2O spectrum is rich with features, numbering up to 15. The spectrum can be divided into three velocity segments, like in the case of RT Vir: $V_{LSR} = -5 \rightarrow 0$, $0 \rightarrow 10$ (this group of features is approximately centered on the stellar velocity $V_* = 5.5$ km/s) and $10 \rightarrow 20$ km/s. The central group remained more or less stable, while the side groups were varying, repeating, with some phase delay, the light variations of VX Sgr. Before 1987, the blueshifted group of features was stronger, but between 1987 and 1993, the red-shifted group increased its integrated flux and almost equaled the blueshifted one. Fig. 2 shows the evolution of the integrated flux of the halves of the VX Sgr H_2O spectrum; H_2O variations generally follow (with some, variable, delay) the light curve. The blue- and red-shifted groups are most probably connected with the two jets of the bipolar outflow (Berulis, Pashchenko & Rudnitskij 1998); the change of their relative intensities may be due to a reversal of the general magnetic field of the star (see discussion in Pashchenko & Rudnitskij 1999).

3.5. The radio continuum

Positive detections of Mira stars in the radio continuum are scarce (Knapp et al. 1995; Reid & Menten 1997). The stars detected were R Aql, W Hya, o Cet, χ Cyg, R Cas and R Leo. Typically the fluxes observed were a few tenths of a milliJansky. Jointly with Dr. Jessica M. Chapman we undertook on the ATCA a radio continuum ($\lambda = 6$ and 3 cm) aimed at detecting an eventual correlation between the radio continuum and the H_2O maser intensity. The expected mechanism of the correlation between optical brightness and H_2O maser emission, proposed by Rudnitskij (1990), involves amplification of the post-shock radio continuum by masering H_2O molecules. However, the search we have done yielded negative results for 33 Miras and semiregulars, except for the known source R Aqr (for details see Chapman & Rudnitskij 1999).

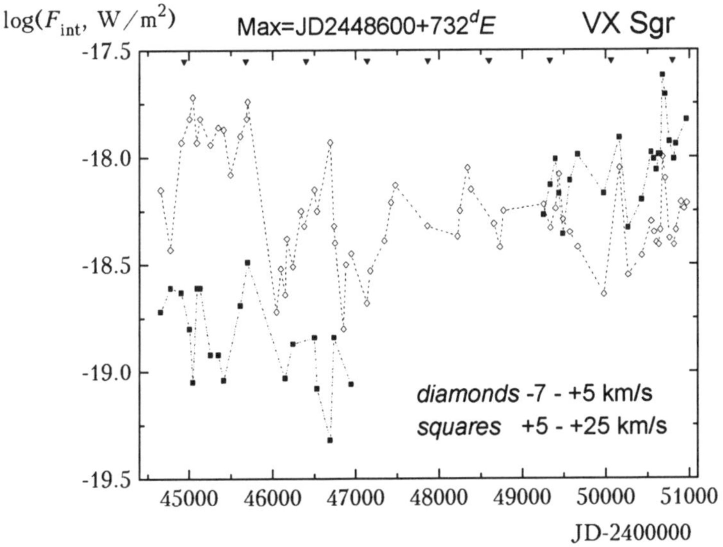

Figure 2. Flux variability of the H_2O maser in VX Sgr. Triangles mark light maxima of the star, computed with the light elements (Berulis, Pashchenko & Rudnitskij 1998), given in the upper part of the graph.

4. Conclusions

The preliminary conclusions we draw from our synchronous monitoring in the H_2O and Hα lines and from our (negative) results on the radio continuum are:

(1) The observed weak radio continuum from Miras must be strongly variable: we could not detect even the stars previously known to be radio continuum sources (o Cet, R Aql, W Hya: Knapp et al. 1995, Reid & Menten 1997), though the sensitivity was sufficient (typically 0.1–0.3 mJy at the 3σ level).

(2) The upper levels obtained imply that strong shocks with front velocities $v_s \geq 15 - 20$ km/s do not exist in Miras or they may be invisible at microwaves, because they do not propagate above the radius of the 'radio photosphere', $r_{\rm ph} \sim 2R_*$ (Reid & Menten 1997), strongly absorbing the radio emission, but transmitting the optical emission.

(3) The model for H_2O maser variability in late-type giants that involves amplification of the variable post-shock radio continuum by the masering molecules (Rudnitskij 1990) now seems less plausible than that assuming a direct impact of a weaker ($v_s \sim 5 - 10$ km/s) shock, reaching the masering region in the circumstellar envelope (Rudnitskij & Chuprikov 1990). The delay between the optical and H_2O maxima depends on the shock travel time and may be as long as a few stellar periods.

Acknowledgments. The authors are grateful to the staff of the Pushchino Radio Astronomy Observatory for their help with the H_2O observations. The AAVSO and AFOEV (SIMBAD) databases were used. This work is supported by the Russian Foundation for Basic Research (project code 96-02-18867), National Program in Astronomy and Fundamental Space Research (grant 1.4.3.1), and Program of Integration of the High School and Science (project 315).

References

Berulis I.I., Lekht E.E., Munitsyn V.A., Rudnitskij G.M., 1998, Astron. Reports 42, 346

Berulis I.I., Lekht E.E., Pashchenko M.I., 1987, Pis'ma AZh 13, 305

Berulis I.I., Lekht E.E., Pashchenko M.I., 1994, Astron. Lett. 20, 115

Berulis I.I., Lekht E.E., Pashchenko M.I., Rudnitskij G.M., 1983, Soviet Ast. 27, 179

Berulis I.I., Pashchenko M.I., Rudnitskij G.M., 1998, Astron. Astrophys. Trans. 15, in press

Castelaz, M.W., Luttermoser, D.G., 1997, AJ 114, 1584

Chapman J.M., Rudnitskij G.M., 1999, MNRAS, submitted

Cox G.G., Parker E.A., 1979, MNRAS 186, 197

Engels D., Winnberg A., Brand J., Walmsley C.M., 1998, this volume

Fox M.W., Wood P.R., Dopita M.A., 1984, ApJ 286, 337

Gillet D., Maurice E., Baade D., 1983, A&A 128, 384

Gómez Balboa A.M., Lépine J.R.D., 1986, A&A 159, 166

Knapp G.R., Bowers P.F., Young K., Phillips T.G., 1995, ApJ 455, 293

Mendoza-Torres J.E., Lekht E.E., Berulis I.I., Pashchenko M.I., 1997, A&AS 126, 257

Pashchenko M.I., Rudnitskij G.M., 1999, Astron. Reports, submitted

Reid M.J., Menten K.M., 1997, ApJ 476, 327

Richards A.M.S., Yates J., Cohen R.J., Bains I., 1998, this volume

Rudnitskij G.M., 1987, in Circumstellar Matter, IAU Symp. 122, I. Appenzeller & C. Jordan (eds.), Reidel, Dordrecht, p. 267

Rudnitskij G.M., 1990, in From Miras to Planetary Nebulae: Which Path for Stellar Evolution?, M.O. Mennessier & A. Omont (eds.), Editions Frontières, Gif-sur-Yvette, p. 268

Rudnitskij G.M., Chuprikov A.A., 1990, Soviet Ast. 34, 147

Schwartz P.R., Harvey P.M., Barrett A.H., 1974, ApJ 187, 491

Woods T.C., Little-Marenin I.R., Benson P.J., 1988, J. Amer. Assoc. Var. Star Observ. 17, 120

Woodsworth A.W., 1995, ApJ 444, 396

Part 3. Formation, Composition, and Processing of Dust

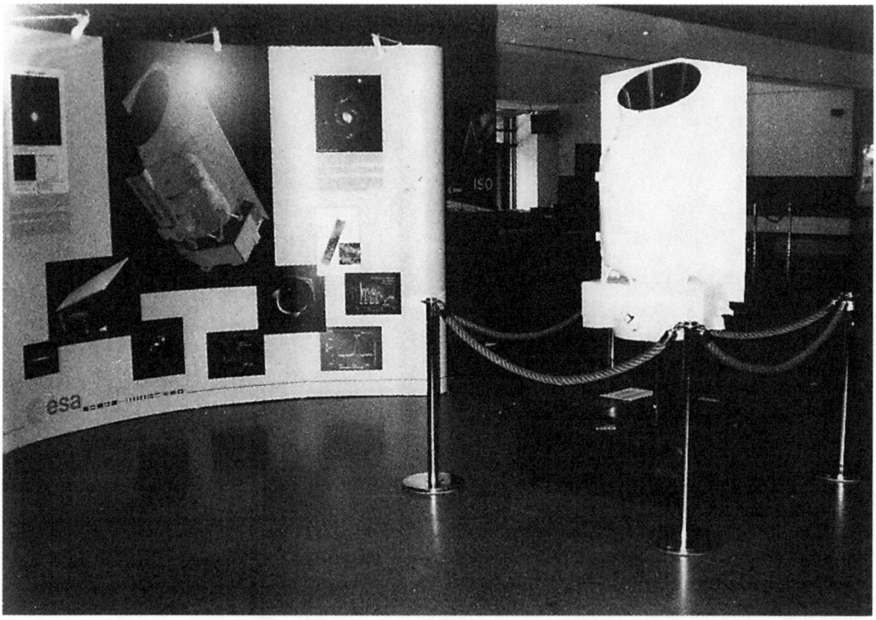

ISO in Montpellier

Asymptotic Giant Branch Stars
T. Le Bertre, A. Lèbre and C. Waelkens, eds.
I.A.U. Symposium 191, 1999

Crystalline silicates in AGB and post-AGB stars

L.B.F.M. Waters[1] and F.J. Molster

Astronomical Institute, University of Amsterdam, Kruislaan 403, NL-1098 SJ Amsterdam, The Netherlands

Abstract. We discuss ISO spectroscopy of oxygen-rich dust shells surrounding evolved stars. The dust that condenses in the outflows of stars on the Asymptotic Giant Branch consists mainly of amorphous silicates and simple oxides. For high mass loss rates, crystalline silicates begin to appear at modest abundance. These crystalline silicates are cold and Fe-poor. ISO spectroscopy for the first time allows quantitative mineralogy of oxygen-rich circumstellar dust. Crystalline silicates are found at high abundance in sources with peculiar (disk) geometry, such as long-lived circum-binary disks. Some C-rich post-AGB stars, notably the Red Rectangle and several nebulae surrounding [WC] central stars of Planetary Nebulae, also show crystalline silicates. We speculate on the origin and evolution of the crystalline dust component in evolved stars.

1. Introduction

The evolution of stars on the Asymptotic Giant Branch (AGB) is strongly influenced by mass loss via a slowly expanding, dusty stellar wind. Mass loss rates can be as high as 10^{-4} $M_{\odot}$/yr; at these rates, the stellar wind dominates the speed of evolution of the star (dictated by the rate at which the envelope is lost). Therefore, it is important to understand the physics of AGB mass loss, and its dependence on mass, chemical composition, rotation, and time. Our current understanding of the mass loss mechanism is still rather primitive, although recently impressive progress has been made (see e.g. the contributions by Fleischer et al., and by Höfner, in this volume). Most workers agree that dust formation in a pulsating, extended atmosphere drives AGB winds. While the dust which condenses in the outflow contains only about 1 per cent of the mass, it dominates the shape of the energy distribution due to its efficient absorption of stellar radiation: as mass loss increases, the dust in the outflow obscures the central star and most of the energy is radiated at (far-) IR wavelengths.

A critical question to address is the kind of dust which condenses, and the location where the dust first condenses from the gas phase. The detailed condensation sequence of dust in AGB stars is difficult to calculate from first principles, and observations are badly needed to constrain models. The *com-*

---

[1]SRON Laboratory for Space Research Groningen, P.O. Box 800, 9700 AV Groningen, The Netherlands

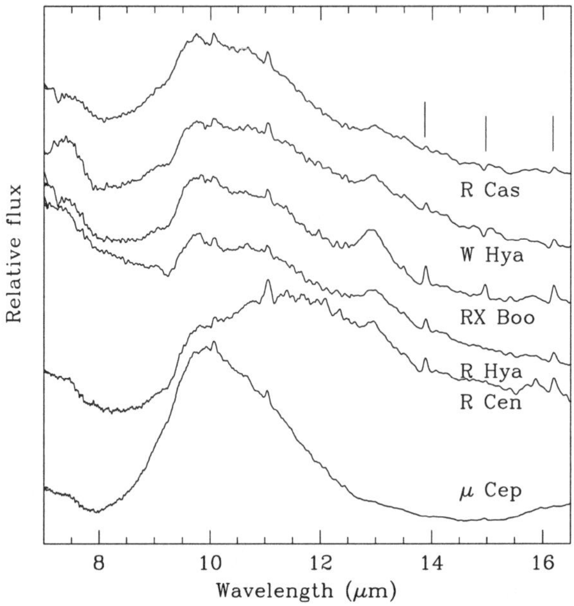

Figure 1. ISO-SWS spectra of oxygen-rich AGB stars with modest silicate emission (Justtanont et al. 1998). Note the large variety in appearance of the silicate band, the presence of the 13 μm emission feature, and the gas-phase CO_2 emission bands, indicated by tick marks

position of dust around AGB stars can be studied spectroscopically at infrared wavelengths, while its *spatial distribution* requires sub-arcsec IR imaging observations. In this paper, we discuss recent spectroscopic observations of oxygen-rich dust around AGB and post-AGB stars using the spectrometers on board of the Infrared Space Observatory (ISO; Kessler et al. 1996), that reveal an incredible richness in previously unknown solid state bands. These observations give new insight into the condensation of dust in the winds of AGB stars, and into their chemical evolution. We will concentrate on results obtained with the Short Wavelength Spectrometer (SWS; de Graauw et al. 1996). For a description of results obtained with the Long Wavelength Spectrometer (LWS; Clegg et al. 1996), we refer to the papers by Barlow (1998, and in this volume).

2. Oxygen-rich dust in AGB and post-AGB stars

A significant fraction of the Guaranteed Time (GT) observing programme of the SWS consortium was aimed at the study of the composition of the photospheres and winds of late type stars, both low and high mass. Here we focus on the results for oxygen-rich AGB stars, post-AGB stars and Planetary Nebulae (PNe). The SWS spectra cover the wavelength range between 2.4 and 45 μm, with a spectral resolution between several hundred and 2 500 (depending on the observing mode chosen). This spectral resolution is ideal to study the solid state emission and

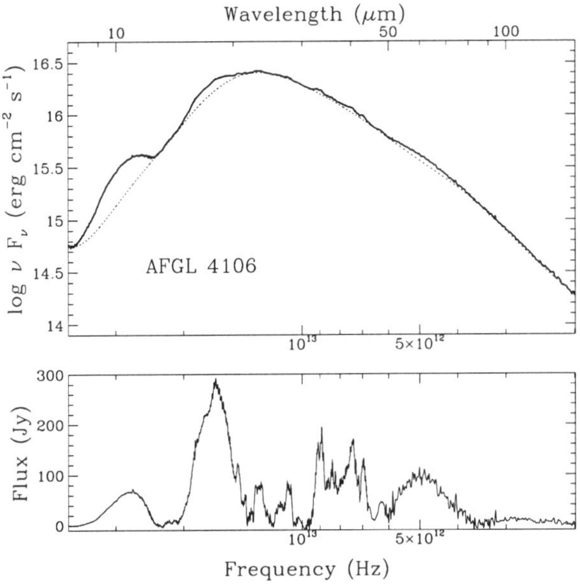

Figure 2. Top panel: joint SWS and LWS spectrum of AFGL 4106 (Molster et al. 1999). The dotted line represents the eye estimate of the continuum. Lower panel: continuum subtracted spectrum of AFGL 4106. Note the strong amorphous silicate bands near 10 and 18 μm, and the numerous narrow, weak emission bands at longer wavelengths due to crystalline silicates.

absorption, while the broad, uninterrupted spectral coverage allows a reliable determination of the continuum.

The SWS spectra of oxygen-rich AGB stars are dominated by solid state emission from amorphous silicates, showing emission at about 10 μm and 18 μm (Fig. 1). These bands were already well known from earlier observations, and are ascribed to the stretching and bending mode of Si-O and O-Si-O respectively. We note a rich variety in the appearance of the 10 μm and 18 μm emission, which sometimes is very broad. This probably reflects a rich variety in chemical composition of the amorphous silicates (see also Tielens 1990). Also prominent is, in cases with modest 10 μm emission, a bump at 13 μm, often ascribed to corundum (Al_2O_3) (e.g. Kozasa & Sogawa 1998). The appearance of this solid state feature seems to be correlated with *gas-phase* CO_2 emission at 15 μm and other wavelengths (see Ryde et al. 1998; Justtanont et al. 1998). Both features disappear for high mass loss rates, i.e. when the 10 μm silicate feature becomes very strong or shows a self-reversal. Kozasa & Sogawa (1998) propose that the Al_2O_3 grains form a silicate mantle which for high mass loss rates suppresses the 13 μm feature.

At longer wavelengths, the spectra of AGB stars with low mass loss rates are smooth and do not reveal prominent emission (sometimes a weak bump near 30 μm is observed). For stars with high mass loss rates, and for post-AGB stars and PNe with cool detached dust shells, the SWS (and LWS) spectra reveal a very rich structure of new, narrow solid state emission bands (Waters et al.

1996; Molster et al. 1998), which can be identified with crystalline silicates, both olivines ($Mg_{2-2x}Fe_{2x}SiO_4$) and pyroxenes ($Mg_{1-x}Fe_xSiO_3$). The Mg-rich end member of the olivines and pyroxenes (x = 0) are forsterite and enstatite respectively. So far, only very few stars have been investigated in detail. In Fig. 2, we show the continuum subtracted joint SWS and LWS spectrum of AFGL 4106, which is likely a very luminous post-red-supergiant (Molster et al. 1998, 1999). Table 1 contains a list of the detected features and identifications based on laboratory data published by Jäger et al. (1998), Koike et al. (1993), and Koike & Shibai (1998). The narrow emission features are weak, only a few per cent over continuum, and the amorphous dust component dominates: in AFGL 4106, only a few per cent of the dust is crystalline. Note also prominent bumps near 43 μm and 60 μm, that are attributed to crystalline H_2O ice (the 43 μm band is blended with crystalline pyroxene).

λ μm	F_{band} 10^{-14} Wm^{-2}	FWHM μm	FWHM/λ	I_{peak}/I_{cont}	Identification
10.8	$4 \cdot 10^2$	2.5	0.23	1.8	amorph. silicate
13.6	10.	.5	0.037	1.04	cryst. forsterite
14.2	6.5	.4	0.028	1.05	cryst. enstatite ? instrumental ?
16.1	40.	.7	0.043	1.06	cryst. forsterite
16.8	15.	.6	0.036	1.03	unidentified
17.6	$5 \cdot 10^2$	3.	0.17	1.2	amorph. silicate
18.1	75.	1.0	0.055	1.07	cryst. enstatite
19.2	50.	1.0	0.052	1.045	cryst. enstatite and cryst. forsterite
20.6	20.	0.48	0.023	1.035	cryst. enstatite ?
21.5	7.3	0.25	0.012	1.024	cryst. enstatite and cryst. forsterite?
22.9	25.	0.5	0.021	1.04	unidentified
23.6	30.	0.7	0.030	1.04	cryst. forsterite
24.1	.46	0.09	0.0037	1.004	unidentified
26.2	16.	1.05	0.040	1.02	cryst. forsterite
27.8	44.	1.15	0.041	1.05	cryst. forsterite and cryst. enstatite
32.8	20.	.6	0.018	1.07	cryst. enstatite
33.6	40.	1.0	0.030	1.08	cryst. forsterite
34.0	.96	.12	0.0035	1.016	cryst. enstatite ?
35.8	5.5	.54	0.015	1.03	cryst. enstatite
36.5	2.4	.28	0.0077	1.02	cryst. forsterite
40.4	13.	1.1	0.027	1.05	cryst. enstatite
41.1	1.1	.17	0.0041	1.02	unidentified
43.1	17.	1.2	0.028	1.07	cryst. enstatite and cryst. H_2O
47.8	9.	1.9	0.040	1.036	unidentified
61	68.	20.	0.33	1.05	cryst. H_2O

Table 1. The characteristics of the dust features in the spectrum of AFGL 4106

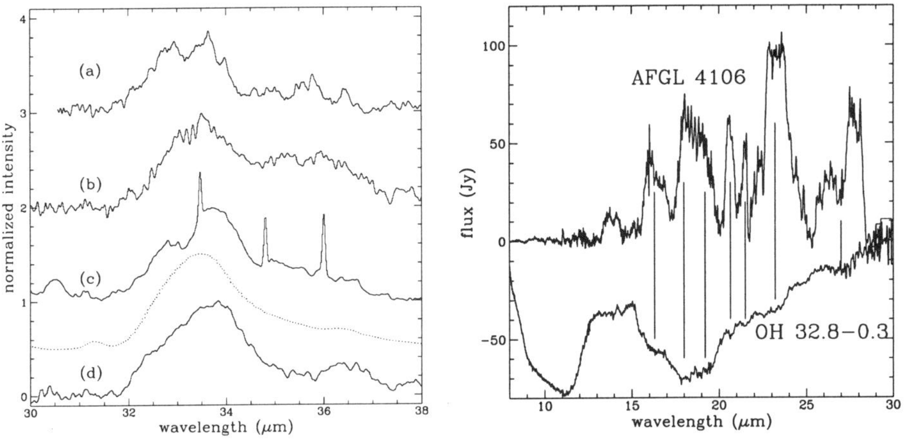

Figure 3. Left: continuum subtracted SWS spectra of some representative oxygen-rich dust envelopes with crystalline silicates. The spectra have been normalized for ease of comparison. Shown are AFGL 4106 (a), OH 26.5+0.5 (b), NGC 6302 (c), and the young star HD 100546 (d, Malfait et al. 1998). The dotted line is a laboratory transmission spectrum of forsterite (Jäger et al. 1998). Right: Continuum subtracted SWS spectra of AFGL 4106 and OH 32.8-0.3. Note that the emission bands seen in AFGL 4106 line up well with those seen in absorption in OH 32.8-0.3.

3. Composition and abundance of crystalline silicates

The wavelength of the emission bands of crystalline silicates is sensitive to the Mg/Fe ratio (Koike et al. 1993; Jäger et al. 1998). Comparison of the ISO data with laboratory spectra indicates that the crystalline silicates are Fe-poor and Mg-rich, with Fe/Mg ratios less than 0.1. Especially the 69 μm band, which falls in the LWS wavelength range, is sensitive to the Fe/Mg ratio, and shifts from 69 μm (for Fe/Mg = 0) to 73 μm for Fe/Mg = 0.05. The observed wavelength of this band is always close to 69 μm (Barlow 1998), which suggests that the crystalline olivine is in fact pure forsterite (no Fe). A similar conclusion can be drawn for the pyroxenes. The low Fe/Mg ratio is in contrast with the composition of the amorphous silicates, that are believed to be much more Fe-rich.

The relative abundance of olivines to pyroxenes can be determined by measuring the strength of the 33 μm pyroxene band relative to that of the 33.8 μm olivine band. We show several examples in Fig. 3. For AFGL 4106 Molster et al. (1999) find a ratio of pyroxenes to olivines between 1 and 3. The stars in the SWS GT programme cover a range in luminosities; the data suggest that more luminous sources (red supergiants) have a higher pyroxene to olivine ratio than lower luminosity AGB stars (Molster et al., in preparation).

The richness of the emission bands of individual dust components offers the unique opportunity to determine the temperature of these components, by comparing observed peak strengths to laboratory data. Care should be taken to properly define the continuum, which is in many cases not trivial due to strong blending of narrow bands. Comparison of the colour temperature of the olivine emission bands in AFGL 4106 with that of the underlying continuum shows that the olivine bands are significantly cooler ($\sim$ 80 K) than the continuum and the amorphous emission bands ($\sim$ 120 K). A similar trend is found for other sources. We conclude that crystalline silicates tend to be cool. The temperature difference suggests that we are dealing with separate grain components that are not in thermal contact with each other. This different temperature of amorphous and crystalline silicates may be the result of their different Fe/Mg ratio: Fe causes a significantly higher absorptivity at near-IR wavelengths, and hence higher thermal equilibrium temperatures than Fe-poor grains of the same size and shape.

From our spectral survey of late type stars, we find that crystalline silicates tend to appear only when the mass loss rate is high (either currently, or recently), i.e. when the colour temperature of the dust is below 200-300 K (there are some exceptions however). A beautiful example is the case of OH 32.8-0.3, which has a very high mass loss rate and shows strong absorption from amorphous silicates, but also in the crystalline silicate bands. In Fig. 3 we compare the continuum subtracted spectra of OH 32.8-0.3 and AFGL 4106, and find a good match between the emission bands in AFGL 4106 and those in absorption in OH 32.8-0.3. This demonstrates that crystalline silicates form in the outflows of late type stars.

The strength and appearance of crystalline silicate bands show a large degree of diversity from source to source, but certain trends can be distinguished. The strength of the bands (expressed in band over continuum ratio) varies from a few per cent (for sources as AFGL 4106) to more than 50 per cent (for objects as MWC 922, Barlow 1998; Voors, private communication). There is a clear trend that 'disk sources' (AC Her, van Winckel et al. 1998; 89 Her, Waters et al. 1993) show strong emission bands, while dust shells with less extreme geometry (HD 161796; AFGL 4106) have modest band strengths (less than 15 per cent over continuum). The interpretation of this trend is not clear; we will return to this point in Sect. 5.

4. Crystalline silicates in carbon-rich objects

The presence of crystalline silicates in circumstellar dust shells and their rich emission band spectrum at long wavelengths opens the possibility to determine the chemical composition of cool dust around evolved stars, both O-rich and C-rich. This has led to a number of remarkable detections of cool crystalline silicates surrounding carbon-rich AGB and post-AGB stars. Before the ISO data became available, warm O-rich dust was known to exist in a number of C-rich AGB stars (Willems & de Jong 1986; Little-Marenin 1986). Also, some O-rich PNe were known to show C-rich PAH bands (e.g. NGC 6302, Roche & Aitken 1986; see also Beintema 1998). Here, we discuss the case of the Red Rectangle and of the [WC] central stars of PNe.

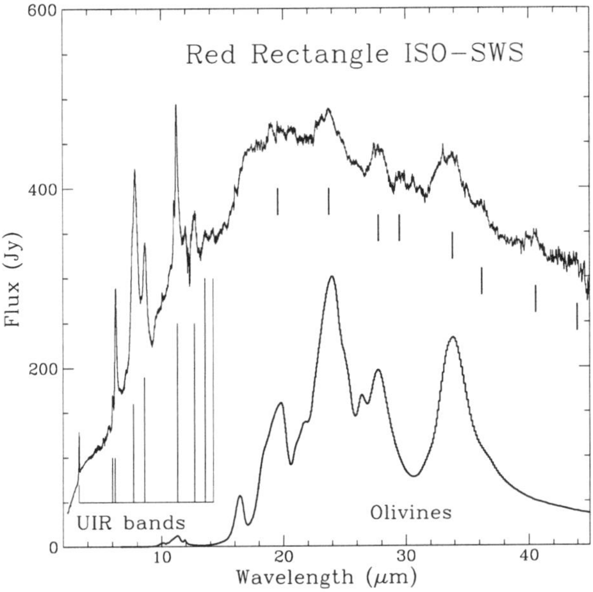

Figure 4. Full SWS grating scan (AOT01 speed 4) of the Red Rectangle nebula and its central star. The thick tickmarks indicate the positions of the strongest oxygen-rich bands. The lower solid line gives the emissivity of olivines with T = 120 K. Note also the presence of strong emission from the unidentified infrared bands (UIRs) at $\lambda < 15$ μm (indicated by the thin tickmarks), usually attributed to Polycyclic Aromatic Hydrocarbons (PAHs)

4.1. The Red Rectangle

The Red Rectangle (RR) is the 'standard' for the famous family of infrared emission (UIR) bands at 3.3, 6.2, 7.7, 8.6 and 11.3 μm (Russel et al. 1978), commonly attributed to Polycyclic Aromatic Hydrocarbons (PAHs; Léger & Puget 1984). It also shows prominent extended red emission (ERE) at optical wavelengths (Cohen et al. 1975), that are attributed to fluorescence from C-rich molecules (e.g. Witt & Schild 1988). However, recently, Witt et al. (1998) and Ledoux et al. (1998) showed that the ERE may be caused by very small silicate grains. Previous to ISO observations, the infrared spectral characteristics, and some of the optical properties of the nebula pointed to a rich C-based chemistry, and the RR was long considered as prototype for C-rich post-AGB stars. The X-shaped nebula (Cohen et al. 1975) hosts a central binary with a luminous post-AGB star with $T_{\rm eff} \sim 7500$ K and an unseen companion in a 318 day period orbit with eccentricity 0.38 (Waelkens et al. 1996). High-resolution optical and near-IR imaging revealed the presence of an optically thick circum-binary disk with reflection lobes above and below the disk (Roddier et al. 1995).

The ISO-SWS spectrum of the RR is shown in Fig. 4 and displays the familiar PAH bands. In addition, at longer wavelengths emission from crystalline olivines and pyroxenes can clearly be distinguished (Waters et al. 1998a). Also, absorption from gas-phase CO_2 is observed at 4.3 and 15 μm. The ISO observations indicate that about 4 10^{-5} $M_\odot$ of oxygen-rich dust is present, and from the absorption of the CO_2 and the distribution of PAH emission in the nebula, Waters et al. (1998a) conclude that the crystalline silicates must be in the circum-binary disk. The abundance of the oxygen-rich dust is difficult to estimate due to the simultaneous presence of emission from C-rich dust. Waters et al. (1998a) derive an abundance of 13 per cent, but this is likely a lower limit.

The origin of the oxygen-rich dust in the RR is probably related to the chemical evolution of the central star on the AGB. During the oxygen-rich phase of the AGB, mass somehow accumulated in the circum-binary disk and remained there while the central star evolved into a C-rich AGB star. Note that the orbital separation of the RR binary does not accomodate a luminous AGB star, and that some interaction must have taken place, however without destroying the binary. The old disk in the RR has properties that are reminiscent of the disks surrounding young pre-main-sequence stars: a highly gas-depleted disk with large grains (Jura et al. 1997) and a high abundance of crystalline silicates; it is tempting to conclude that the processes that modify the disk composition in young stars (which eventually lead to planet formation) also work in the RR disk (Waters et al. 1998a). Recently, the notion of planet formation has gained support from an observation by Jura & Turner (1998) of a cool dust clump in the outer regions of the RR disk with a mass comparable to that of Jupiter.

Circum-binary disks in evolved post-AGB stars are also observed in e.g. AC Her and 89 Her; in these cases also a high abundance of crystalline silicates is found, but no evidence for C-rich dust.

4.2. [WC] central stars of planetary nebulae

[WC] central stars of PNe (CSPNe) are a sub-class of C-rich PNe with a central star that shows a He-rich and C-rich atmosphere and no strong evidence for H. The central stars also exhibit a dense, fast outflow very similar to the population I counterparts. The infrared spectra of [WC] nebulae show strong emission from the familiar UIR bands, indicating that the dust in the nebula is also C-rich. The origin of the [WC] CSPNe is still not well understood, but is probably related to the occurrence of a thermal pulse at the end of the AGB or in the very early phase of the post-AGB evolution (Iben 1984; Zijlstra et al. 1991; see also Werner et al., in this volume).

The ISO-SWS spectra of two [WC] nebulae were analysed by Waters et al. (1998b), and show convincing evidence for the presence of crystalline silicates at wavelengths longward of 20 μm. Cohen et al. (1998; see also in this volume) and Barlow (1998) show that the C-rich [WC] nebula CPD-56°8032 exhibits strong emission from crystalline olivines and pyroxenes, as well as from crystalline H_2O ice. SWS spectra of several [WC] nebulae published by Szczerba et al. (1998) also reveal emission from crystalline silicates, although the authors do not comment on this in their paper. It appears that a large fraction of PNe with [WC] central stars show cool crystalline silicates. The question arises whether this is a property of PNe in general (i.e. independent from the chemical composition of the central star) or restricted to [WC] stars.

If the presence of cool oxygen-rich dust is correlated with the central star spectral type, the explanation must be found in the formation of [WC] CSPNe.

Waters et al. (1998b) have suggested that the late thermal pulse which removes the remaining H-rich layers from the atmosphere is also the thermal pulse which changes the chemistry from O-rich to C-rich. This transition must then have taken place very recently, typically less than 1 000 years ago, an perhaps even less than a few 100 years (see also Cohen et al. 1998). The [WC] CSPNe are only a minor fraction of all known CSPNe (about 50 are known, Gorny & Stasinska 1995). If we assume that the transition is related to a thermal pulse, and that at the end of the AGB thermal pulses occur at intervals of about 10^4 years (see e.g. Blöcker, in this volume), then 10 per cent of the thermal pulses may occur close enough to the termination of the AGB to be consistent with the presence of cool, oxygen-rich material in a C-rich inner nebula. However, this final thermal pulse also must be the one changing the chemistry, and this may be statistically implausible. Perhaps the efficiency of mixing is much higher in 'fatal thermal pulses' than in earlier thermal pulses. It is clear that a better overview of the occurrence of C-rich nebulae with O-rich cool dust is highly desirable.

Alternatively, the presence of oxygen-rich dust may be unrelated to the [WC] nature of the central stars (although so far the correlation seems to hold). In that case, the origin of the material may be related to the destruction of a pre-existing cloud of dust, similar in nature to the Kuiper belt or to the Oort Cloud in our Solar System (Cohen et al. 1998; see also Cohen et al., in this volume).

5. Discussion

The new solid state bands discovered in the ISO spectra of oxygen-rich AGB stars and related objects demonstrate the power of high quality infrared spectroscopy. We can now determine the mineral composition of specific dust components in the outflows and circumstellar disks of stars: mineralogy of oxygen-rich dust shells has now become possible. Several questions come to mind when confronted with the ubiquitous presence of crystalline silicates, and we address some of these below.

The first point to consider is the occurrence and chemical composition of crystalline silicates. The lack of crystalline silicates in dust shells of low mass loss AGB stars and RSG stars suggests that the density in the dust forming layers must be high in order for the crystalline silicates to be detected. Tielens et al. (1998) speculate that the crystalline Mg-rich silicates are the grains that form first in dusty outflows of *any* density, since their condensation temperature is high. At lower temperature (below or close to 1000 K), amorphous, Fe-rich silicates are formed by adsorption of Fe in the lattice of the Mg-rich grains. This reaction can only take place at temperatures below the glass temperature of silicates, and hence the grain becomes amorphous. For low mass loss rates, all Mg-rich crystalline silicates are transformed to amorphous grains by adsorption of Fe. In a high density environment, Fe may be incorporated in other dust species (for instance metallic Fe, Gail, private communication) at high temperature before the reaction with the silicates can occur, depleting the gas-phase Fe and leaving some crystalline grains without Fe. This could explain why crystalline silicates only appear in stars with very high mass loss rates.

A second point to address is the (very) high abundance of crystalline silicates in sources with stationary, circum-binary disks. As mentioned earlier, sources as the RR, 89 Her and AC Her, all binaries with wide orbits, show a high abundance of crystalline silicates when compared to sources with simpler geometries, i.e. detached, slowly expanding AGB remnants. A notable case is NGC 6302, an outflow source which shows prominent crystalline silicates (see Waters et al. 1996; Beintema 1998; Barlow 1998) and whose geometry is strongly bi-polar. The question arises whether the high abundance is the result of the longevity of the grain population, or due to an intrinsically high abundance of the crystalline grains formed in the AGB outflow.

In stationary disks, the grain population may become more crystalline with time, perhaps due to exposure to high energy UV photons, or by the interaction between a slow and a fast wind. We note here that it is not sufficient to change the lattice structure of the Fe-rich amorphous grains, but it also requires Fe removal from the lattice. This process may not occur at the low temperatures observed in circum-binary disks. Alternatively, the amorphous grains may be selectively removed because of the difference in radiation pressure between Fe-poor, cool (transparent) crystalline grains and Fe-rich, warm (opaque) grains. This mechanism can only be effective in an environment where gas and dust are not dynamically coupled. This issue is not yet settled.

In non-spherical outflow sources (such as NGC 6302), it is possible that the abundance of crystalline grains is different in the polar (low density) and equatorial (high density) regions. The subsequent dynamical evolution of the nebula, with the fast wind clearing out the polar regions more quickly than the equatorial regions, may increase the relative abundance of crystalline silicates.

Acknowledgments. The authors thank T. de Jong, D. Beintema, I. Yamamura, J. Bouwman, R. Voors, C. Kemper, A. de Koter, X. Tielens, A. Zijlstra, K. Justtanont, J. Cami, H. van Winckel, C. Waelkens, K. Malfait, Th. Henning, C. Jäger, M. Barlow, R. Sylvester, T. Lim and P. Cox for valuable discussions on crystalline circumstellar silicates. LBFMW acknowledges financial support from an NWO *Pionier* grant.

References

Barlow M.J., 1998, Astrophys. Sp. Sc. 255, 315

Beintema D.A., 1998, Astrophys. Sp. Sc. 255, 507

Clegg P., Ade P.A.R., Armand C., et al., 1996, A&A 315, L38

Cohen M., et al., 1975, ApJ 196, 179

Cohen M., Barlow M.J., Sylvester R.J., Liu X.-W., Cox P., Lim T., Schmitt B., Speck A.K., 1998, ApJ, in press

De Graauw Th., et al., 1996, A&A 315, L49

Gorny S.K., Stasinska G., 1995, A&A 303, 893

Iben I., Jr, 1984, ApJ 277, 333

Jäger C., Molster C., Dorschner J., Henning Th., Mutschke H., Waters L.B.F.M., 1998, A&A 339, 904

Justtanont K., Feuchtgruber H., de Jong T., et al., 1998, A&A 330, L17

Jura M., Turner J., Balm S.P., 1997, ApJ 474, 741
Jura M., Turner J., 1998, Nature 395, 144
Kessler M.F., et al., 1996 A&A 315, L27
Koike C., Shibai H., Tuchiyama A., 1993, MNRAS 264, 654
Koike C., Shibai H., 1998, ISAS report no. 671
Ledoux G., Ehbrecht M., Guillois O., et al., 1998, A&A 333, L39
Léger A., Puget J.L., 1984, A&A 137, L5
Little-Marenin I.R., 1986, ApJ 307, L15
Malfait K., Waelkens C., Waters L.B.F.M., et al., 1998, A&A 332, L25
Molster F.J., Waters L.B.F.M., van Loon J.Th., et al., 1998, Astrophys. Sp. Sc. 255, 469
Molster F.J., Waters L.B.F.M., Trams N.R., et al., 1999, A&A, submitted
Roche P.F., Aitken D.K., 1986, MNRAS 221, 63
Roddier F., Roddier C., Graves J.E., Northcott M.J., 1995, ApJ 443, 249
Russel R.W., Soifer B.T., Willner S.P., 1978, ApJ 220, 568
Ryde N., Eriksson K., Gustafsson B., Lindqvist M., Olofsson H., 1998, Astrophys. Sp. Sc. 255, 301
Szczerba R., Górny S.K., Tylenda R., 1998, Astrophys. Sp. Sc. 255, 515
Tielens A.G.G.M., 1990, in 'From Miras to Planetary Nebulae: Which path for stellar evolution?', eds. M.O. Mennessier & A. Omont, Editions Frontières, p. 186
Tielens A.G.G.M., Waters L.B.F.M., Molster F.J., Justtanont K., 1998, Astrophys. Sp. Sc. 255, 415
Van Winckel H., Waelkens C., Waters L.B.F.M., et al., 1998, A&A 336, L17
Waelkens C., Van Winckel H., Waters L.B.F.M., Bakker E.J., 1996, A&A 314, L17
Waters L.B.F.M., Waelkens C., Mayor M., Trams N.R., 1993, A&A 269, 242
Waters L.B.F.M., et al., 1996, A&A 315, L361
Waters L.B.F.M., et al., 1998a, Nature 391, 868
Waters L.B.F.M., et al., 1998b, A&A 331, L61
Willems F.J., de Jong T., 1986, ApJ 309,L39
Witt A.N., Schild R.E., 1988, ApJ 325, 837
Witt A.N., Gordon K.D., Furton D.G., 1998, ApJ 501, L111
Zijlstra A.A., Gayland M.J., te Lintel Hekkert P., et al., 1991, A&A 243, L9

Nicolas Epchtein surveying surveys

Asymptotic Giant Branch Stars
T. Le Bertre, A. Lèbre and C. Waelkens, eds.
I.A.U. Symposium 191, 1999

Laboratory Astrophysics of Circumstellar Dust

Th. Henning

Astrophysical Institute and University Observatory (AIU), Friedrich Schiller University, Schillergäßchen 3, D-07745 Jena, Germany

Abstract. This paper summarizes laboratory investigations of stardust analogue materials with special emphasis on the spectroscopic behaviour of silicates and carbonaceous materials. The review will also deal with the identification of the observed features in the dust spectra of AGB stars.

1. Introduction

Small solid particles form an important component of the circumstellar (CS) environment of AGB stars. Although they are only minor contributors to the total mass of the envelopes, they are of decisive importance for their thermal, dynamical, and chemical state. IR and submm/mm continuum radiation is produced by the thermalization of stellar light. The gas is heated through collisions between dust particles and the gas atoms and molecules. The momentum in the outflows is supplied by radiation pressure acting on dust grains and the collisional transfer of momentum to the gas. The surfaces of dust grains can act as catalytic sites for the formation of molecules. In the outer regions of the envelopes, dust grains play an important role for shielding the molecules from dissociative interstellar UV radiation and cosmic rays. AGB stars are the main sources of freshly nucleated interstellar grains. Therefore, these objects are important for the chemical evolution of grain populations in galaxies, whose properties influence the birth of the next generation of stars.

A deeper understanding of the complex interplay between hydrodynamics, radiation transfer, and chemistry has to be based on a better knowledge of the microphysics of CS dust. Guided by IR spectroscopy, submm/mm observations, and the analysis of stardust in primitive meteorites, laboratory astrophysics must contribute to the understanding of dust formation from the gas phase, the formation of molecules by surface reactions, and the optical behaviour of small solid particles. On the other hand, AGB star envelopes are interesting regions for the study of grain formation and molecular chemistry under rather extreme conditions, usually difficult to reproduce in terrestrial laboratories.

The review will mainly deal with the optical properties of stardust analogue materials. Without better optical data, the radiative transfer modelling of spectral energy distributions and the interpretation of absorption and emission features observed in the dust spectra of AGB stars will remain ambiguous.

For a more detailed discussion of dust properties and evolution in our galaxy, the interested reader is referred to recent reviews of this topic (Dorschner & Henning 1996; Henning 1997, 1998a, 1998b; Sandford 1996).

2. Stardust materials - An overview

A key parameter for dust formation in the outflows of AGB stars is the elemental C/O ratio because the atoms of the less abundant element are mostly tied up in the stable CO molecules and do not participate in dust formation (Dorschner & Henning 1996). Therefore, stars with oxygen-rich outflows ($C/O < 1$) produce refractory dust components which are different from the dust condensates of carbon stars ($C/O > 1$). The small transition group of S-type stars is characterized by a C/O ratio very close to 1. As a consequence of nucleosynthesis and dredge-up processes during AGB star evolution, "oxygen-rich" stars with masses smaller than about 4 $M_\odot$ will evolve into carbon stars and the composition of their molecular outflows will change. Other factors influencing the chemical state of the outflows apart from the C/O ratio can be the formation of molecules by grain surface reactions, the dissociation of some of the CO and N_2 molecules by chromospheric UV radiation, and the influence of shock chemistry. These factors have been suggested to be responsible for the presence of carbon-bearing molecules in the envelopes of individual oxygen-rich late-type stars (see, e.g., Willacy & Millar 1997 and references therein).

In oxygen-rich envelopes, the most abundant elements available for grain formation are O, Fe, Si, Mg, Al, and Ca. Without considering the kinetics of grain formation, we therefore expect the presence of silicates and other oxides, iron and silicon as well as FeSi grains. Spectroscopic evidence for the presence of amorphous silicates comes from the observations of two broad features at 10 and 18 μm in a large number of objects (see the reviews by Dorschner & Henning 1995; Habing 1996). If the feature appears in emission, self-absorption or absorption mainly depends on the optical depth of the envelope (Henning 1983). The Infrared Space Observatory (ISO) led to the clear spectroscopic detection of the presence of crystalline Mg-rich silicates with prominent features between 20 and 45 μm, which vary in strength from object to object (Waters et al. 1996; Cami et al. 1998; Jäger et al. 1998a). Evidence for the presence of Al_2O_3 comes from the observation of a relatively broad feature between 11 and 12 μm and a narrow feature at 13 μm (Onaka et al. 1989; Stencel et al. 1990; Sloan et al. 1996; Begemann et al. 1997 - see also for laboratory data, Justtanont et al. 1998). We should note that the identification of the 13 μm feature remains controversial and that only crystalline α-Al_2O_3 (corundum) would be a good candidate for the carrier (Begemann et al. 1997). The identification with crystalline α-Al_2O_3 is supported by the observed correlation of the 13 μm feature with emission around 20 μm (Aringer et al. 1998). Such a correlation is expected because α-Al_2O_3 has a second band around 20 μm. Based on KAO data, an additional feature at 7.15 μm has been detected, whose carrier remains to be identified (Goebel et al. 1994). In addition, there exists some spectroscopic evidence for the presence of other metal oxides (see Henning et al. 1995a), although the formation of iron oxides seems to be kinetically inhibited in AGB star envelopes (Gail & Sedlmayr 1998a). The formation of oxides other than aluminium oxides or silicates may, however, occur due to solid-state reactions (Jäger et al. 1998a).

Spectra of carbon-rich stars show two clear dust features peaking at about 11.2 and 30 μm whose strength increases as a function of optical depth (see, e.g., Omont et al. 1995, Speck et al. 1997, Yamamura et al. 1998). The 11.3 μm feature has been attributed to SiC grains (see Mutschke et al. 1998b), whereas the 30 μm feature has been identified with MgS (see Begemann et al. 1994). Features in IRAS LRS spectra at about 8.5 and 11.9 μm have been attributed to hydrogenated amorphous carbon grains (Baron et al. 1987; Goebel et al. 1995). However, such grains produced in an oxygen-free atmosphere do not show a strong band at 8.5 μm. Therefore, the identification of this band remains an open problem. It is striking that the PAHs as the suggested carriers of the ubiquitous "unidentified" IR bands (UIBs) are not observed in carbon-rich AGB stars, although they should be part of the formation route of carbon grains from the acetylene molecule. The lack of evidence for the presence of PAHs in the envelopes of AGB stars may be related to the absence of exciting UV and optical radiation and/or the fact that the PAHs are rapidly consumed during the dust formation process. In the latter case, PAHs would only be produced during later stages of stellar evolution or during processing of grains in the diffuse interstellar medium (see, e.g., Speck & Barlow 1997, Tielens 1997). It is also important to note that absorption features due to aromatic (3.3 μm) or aliphatic (3.4 μm) hydrocarbon grains are not observed. This may be due to the lack of such a component or a band which is weak compared with the continuum (see Schnaiter et al. 1998). The characteristic 3.4 μm absorption feature observed in some post-AGB stars and the diffuse interstellar medium could be the result of a further evolution of aliphatic hydrocarbon grains in a hydrogen-rich environment (Chiar et al. 1998; Schnaiter et al. 1998).

IRAS LRS data already showed that some carbon stars still have silicate features in their spectra, possibly as a fingerprint of their preceding evolutionary phase (Little-Marenin 1986; Willems & de Jong 1988; Chan & Kwok 1991; Kwok & Chan 1993). This detection was later confirmed by ground-based observations (LeVan et al. 1992). Various models have suggested that oxygen-rich material may be present in a long-lived disk which was left over from the "oxygen-rich" phase of the giant (see Jura 1997). Waters et al. (1998) indeed demonstrated the presence of an oxygen-rich circumbinary disk and carbon-rich material in the extended nebula of the Red Rectangle system.

The identification of spectral features with well-defined materials produced in the laboratory requires the derivation of observationally-based mass absorption coefficients or the complex dielectric function of the cosmic materials (see, e.g., Ossenkopf et al. 1992 for silicate data and Begemann et al. 1997 for aluminium oxide data). However, one has to realize that even in the case of optically thin envelopes radiation transfer effects complicate the derivation of such data. A "good" example is the ratio of the strengths of the 10 to 18 μm silicate features, which can only be reliably determined if the dust temperature distribution is given or self-consistently calculated (Henning et al. 1983). In addition, the presence of molecular absorption bands has to be taken into account when deriving the optical properties of CS dust from IR spectroscopy. Additional complicating factors come from the influence of shape/size distributions and coatings on the spectral appearance of solid-state features, which are especially important for very strong resonances. A textbook example for such a strong resonance is the IR band of crystalline SiC (Bohren & Huffman 1983; Papoular et al. 1998; Mutschke et al. 1998b). Finally, more than one grain component usually contributes to the continuum flux making the derivation of the optical properties for *one* component extremely difficult.

3. Grain formation in AGB star envelopes

The first steps of the kinetic route towards the formation of carbon grains from acetylene have recently been reviewed by Cherchneff (1998). However, specific experiments which relate IR spectra of carbonaceous grains to their formation conditions and internal structures have just started (see, e.g., Schnaiter et al. 1998). Our knowledge of how and which grains form in oxygen-rich environments is mostly guided by thermodynamical arguments or classical nucleation theory (see, e.g., Gail & Sedlmayr 1998a, b). However, non-equlibrium effects (Nuth 1996), kinetic barriers in low-density regions of the outflows (Gail & Sedlmayr 1998a for a discussion of "rusting" of iron in AGB star envelopes), and the details of solid-state reactions (Stencel et al. 1990; Jäger et al. 1998a) must be considered before we can reach a better understanding of the formation of grains in oxygen-rich envelopes. The formation of crystalline silicates by annealing of amorphous structures seems to be possible if the grains are hot enough in the dust formation zone ($T \geq 900$ K) (Gail & Sedlmayr 1998a, b). A discussion of the different possible explanations why the crystalline silicate grains in AGB star envelopes are Mg-rich can be found in Tielens et al. (1998). The formation of hydrated silicates by the reaction with water vapour was generally assumed to be inefficient at the low densities of the outer envelopes. However, the ISO data have to be carefully checked for the signatures of such materials. Grain formation may well start with the nucleation of the most stable clusters which could be TiC clusters (carbon-rich envelopes; Cherchneff et al. 1998) and TiO clusters (oxygen-rich envelopes; Gail & Sedlmayr 1998c) with the subsequent formation and growth of other species on the surfaces of these clusters. In addition, strong evidence exists for the presence of high-density clumps in the AGB star envelopes. In such regions the growth of much larger particles may be possible. Direct evidence for the formation of such large grains comes from the presence of micron-sized SiC stardust grains in primitive meteorites.

4. Laboratory data

The laboratory measurement of optical data for astronomically relevant materials requires three steps: production of the material, detailed characterization of its structure, and the measurement of transmission or reflection data.

A wide variety of techniques has been applied for the preparation of cosmic dust analogues including laser evaporation and pyrolysis (silicon, silicates, carbides, carbonaceous materials), joint evaporation of Mg or FeO and SiO (or O_2 and SiH_4) (silicates), evaporation from an electric arc or resistively heated rods with quenching afterwards (silicates, carbonaceous materials), quenching melts (silicates), sol-gel reactions (silicates), plasma deposition techniques (carbon films), and combustion and pyrolysis of organic material (carbonaceous grains). The structure of the material has to be carefully characterized in order to relate optical properties to the structural composition (e.g. silicate band positions as a function of Mg/Fe content, slope of the FIR absorption of carbonaceous grains as function of the sp^2 to sp^3 hybridization ratio). Relevant techniques are SEM/TEM (scanning/transmission electron microscope) analysis and EDX (energy dispersive X-ray diffraction) analysis for silicates (Jäger et al. 1998a) and EELS (electron energy loss spectroscopy) and HRTEM (high-resolution electron microscopy) for carbonaceous materials (Henning & Schnaiter 1999). The last step is the derivation of optical constants or absorption and scattering efficiencies. In the case of measurements on small-particle systems, size, shape, and agglomeration effects have to be carefully taken into account.

The following laboratory sections will concentrate on silicates and carbonaceous materials. Data for other oxides can be found in Begemann et al. (1997) and Henning et al. (1995a). Recent results for SiC grains have been compiled by Papoular et al. (1998) and Mutschke et al. (1998b). These two studies clearly demonstrate that shape effects as well as free charge carriers are much more important than crystal type in determining SiC spectra. A recent experimental study of MgS and FeS can be found in Begemann et al. (1994).

4.1. Silicates

Much of the earlier work on cosmic silicates has been summarized by Dorschner & Henning (1986) and Jäger et al. (1994) and many useful references can be found in these papers.

One of the questions of silicate mineralogy is the problem of non-crystalline silicates having regular silicate-oxygen tetrahedra and consisting of sufficiently homogeneous phases or showing deviations from stoichometric proportions and considerable compositional variations from point to point. Basic data for the first class of amorphous silicates have been provided by Jäger et al. (1994), Dorschner et al. (1995), and Mutschke et al. (1998a). Nuth & Hecht (1990) introduced the term "chaotic" silicates for the second class of materials. Most experimental data for such solids, also designated as Mg-SiO smokes, have been gathered by B. Donn and J. Nuth (see, e.g., Nuth 1996). The structure of these silicates could be revealed by a combination of SEM/TEM and EDX techniques. However, such data are not always available for the chaotic silicates. Phase separation has been observed in silicate melts during cooling (Jäger et al. 1998a) leading to systems with different well-defined phases, which may finally look like "chaotic" materials on coarser scales.

A special problem of CS silicate dust has always been its "dirtiness". CS silicate grains have to be much more absorptive at near-infrared (NIR) wavelengths than most of the terrestrial minerals or glasses. In order to explain the observed spectral energy distributions, Jones & Merrill (1977) coined the term "dirty silicates" for materials with high absorption at NIR wavelengths (high value of the imaginary part k of the refractive index; k=0.1). Two possibilities exist to produce such dirty silicates: (a) an admixture of finely dispersed impurities of iron or iron oxides to the silicates (Ossenkopf et al. 1992) or (b) the presence of Fe^{2+} cations in the silicates leading to transitions of the 3d electrons in the iron cations (Dorschner et al. 1995). In the latter case a systematic

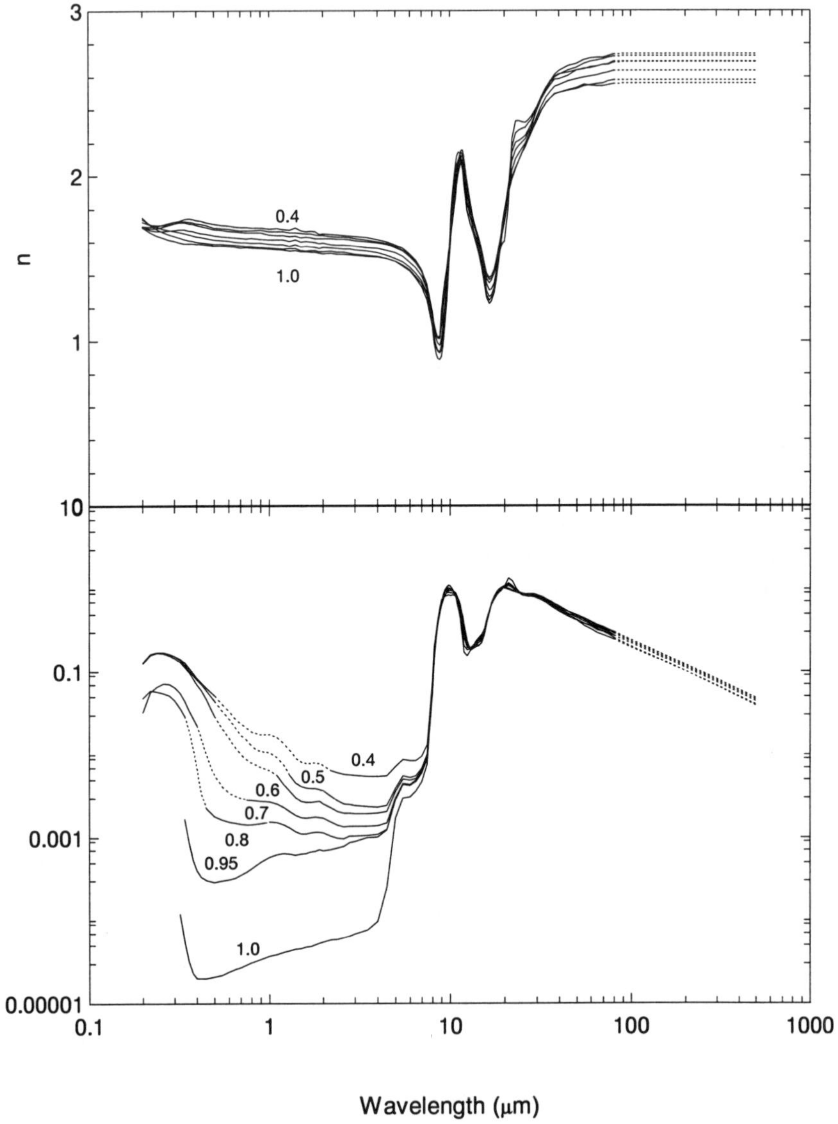

Figure 1. Optical constants n (top) and k bottom of pyroxene glasses of the compositions $Mg_xFe_{1-x}SiO_3$ for x=0.4, 0.5, 0.6, 0.7, 0.8, 0.95, and 1.00. In both graphs, dotted lines indicate regions of interpolations or extrapolations. After Dorschner et al. 1995.

increase of the NIR absorptivity with iron content has been found in pyroxene glasses (see Fig. 1).

Spectroscopy with ISO clearly revealed the presence of crystalline silicates. Two main groups of astronomically relevant minerals exist: olivines and pyroxenes (see Fig. 2). Olivines can be considered as solid solutions of Mg_2SiO_4 and Fe_2SiO_4; the mixture has the sum formula $Mg_{2x}Fe_{2-2x}SiO_4$ with x between 1 and 0. The olivines are neso-silicates (island silicates) because they consist of isolated SiO_4 tetrahedra which are linked by divalent cations. The pyroxenes form a large group of crystalline silicates belonging to the inosilicates (chain silicates). The two SiO_4 tetrahedra form chains, i.e. each tetrahedron shares two of its oxygen atoms with its neighbours. Like in the case of olivines, the symbolic formula $Mg_xFe_{1-x}SiO_3$ with x between 1 and 0 can be used.

Optical data for these two classes of crystalline silicates with varying Fe/Mg ratios have been provided by Koike et al. (1993) and Jäger et al. (1994, 1998a). An assignment of the bands to the different modes (as far as this is known) is given by Jäger et al. (1998a). IR spectra of pyroxenes are usually more complicated because the polymerization leads to additional bands. The strong olivine bands at 23.5 and 33.5 μm as well as a variety of pyroxene bands could be detected in ISO spectra (Waters et al. 1996). The laboratory data (Jäger et al. 1998) show a clear correlation between Fe/Mg ratio and band position. The bands are shifted towards longer wavelengths with increasing Fe content. The features observed with ISO always point to the presence of the magnesium-rich end members of the silicates. A useful indicator for the presence of Mg-rich olivine is a feature at 69.7 μm which is not present in iron-rich silicates. It is important to note that the observed features are even narrower than the features of the crystalline silicates measured in the laboratory. This may point to the formation of extremely pure crystals in the AGB star envelopes.

4.2. Carbonaceous solids

Carbon forms the basis of a rich molecular chemistry in CS envelopes of carbon-rich AGB stars (see, e.g., Oloffson 1997) and leads to the possibility of an enormous structural variety of dust grains (see Henning & Schnaiter 1999, Henning & Salama 1998) because of the unique property of this element to occur in different hybridization states. Carbon atoms in the sp^3 hybridization state are the basis of solids with diamond-like structure. The sp^2 hybridized carbon atoms form planar graphitic structures whose size and distribution depend on the condensation conditions. Apart from the distinct hybridization states, carbon atoms in mixed hybridization states exist which lead to curved structures. In the perfect case these curved structures are closed shells (fullerites). In non-crystalline carbonaceous grains, all kinds of mixtures of, for instance, sp^2 and sp^3 hybridization as well as mixed hybridization states are possible. This leads to carbonaceous materials of different short-, medium-, and long-range order (Papoular et al. 1996; Michel et al. 1998; Jäger et al. 1999; see Henning & Schnaiter 1999 for a review).

Carbonaceous grains dominate the FIR and submm/mm emission of the envelopes of carbon-rich AGB stars. The submm/mm dust continuum radiation is used for the determination of the total mass of the envelopes. This estimate strongly depends on the value of the dust opacity at the wavelength of

Silicate structure

Pyroxenes: $MgFeSi_2O_6$ or $Mg_xFe_{1-x}SiO_3$

Solid solution series of $MgSiO_3/FeSiO_3$:

$MgSiO_3$ Enstatite 0-5 Ma% FeO
Bronzite 5-15 Ma% FeO
Hypersthene >15 Ma% FeO
Fe-Hypersthene
$FeSiO_3$ Ferrosilite <10 Ma% MgO

Chain silicates: $[SiO_3]^{2-}_{\infty}$

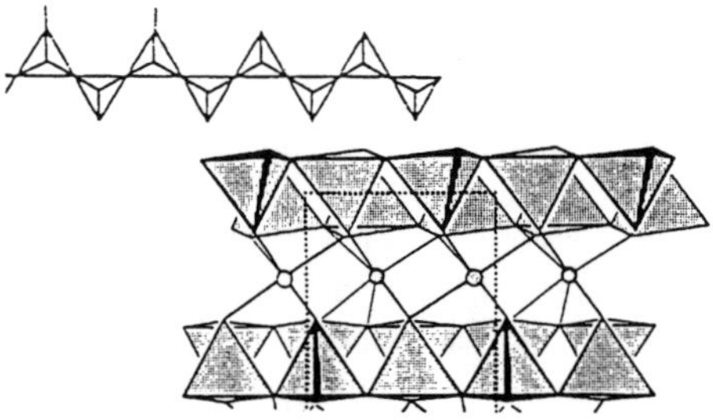

Olivines: $Mg_{2x}Fe_{2-2x}SiO_4$

Solid solution series of Mg_2SiO_4/Fe_2SiO_4:

Mg_2SiO_4 Forsterite <10 Ma% FeO
Olivine 10-30 Ma% FeO
Hortolonithe
Fe-Hortolonithe
Fe_2SiO_4 Fayalite <10 Ma% MgO

Neso silicates: $[SiO_4]^{4-}$

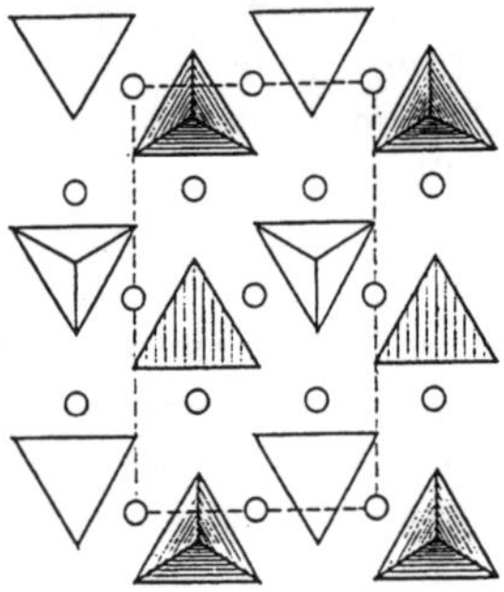

Figure 2. Structure of olivines and pyroxenes

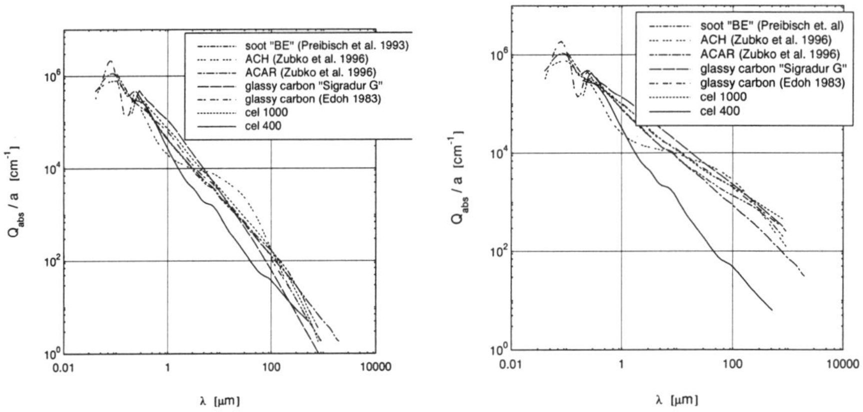

Figure 3. Absorption efficiencies divided by particle radius calculated for spheres in vacuum from the optical data obtained for the materials cel400 (sp^2 content= 67%) and cel1000 (sp^2 content= 80%) compared with the results for other soot materials (Preibisch et al. 1993; Zubko et al. 1996) and glassy carbon (Edoh 1983) (left figure). The same calculations for a CDE distribution of particle shapes (right figure). After Jäger et al. (1998b).

the observed flux. Based on the "observed" dust absorption emissivity law for these particles ($Q(\lambda) \propto \lambda^{-\beta}$ with $\beta \simeq 1$; see, e.g., Sopka et al. 1985, Martin & Rogers 1987, Bagnulo et al. 1998 - see also for a discussion of the combined effects of optical properties and geometrical structure of the envelope), it was generally concluded that the carbon grains in these envelopes have a structure characterized by heavily disturbed graphitic zones and larger contents of sp^3 hybridized carbon. Therefore, we must understand the relation between the structural properties of carbon grains and the optical behaviour at long wavelengths. Unfortunately, strong IR features due to C-H vibrations, which are of great diagnostic value for the structural characterization, have not been observed in the spectra of AGB stars (see, e.g., Chiar et al. 1998 and Schnaiter et al. 1998 for the discussion of possible reasons). Here, we should stress that the FIR/submm/mm optical properties of carbonaceous grains are not only influenced by their internal structure, but also by the morphology of the grains. It is a well-known fact that the clustering of carbon grains can strongly influence the spectral index β (Rouleau & Martin 1991; Henning et al. 1995b; Stognienko et al. 1995; Henning 1996).

The influence of the internal structure on the FIR behaviour of different carbon materials has been investigated by Rouleau & Martin (1991), Preibisch et al. (1993), Koike et al. (1994, 1995), Menella et al. (1995), Colangeli et al. (1995), and Zubko et al. (1996). These investigations are based on data from transmission spectroscopy. This method cannot exactly separate between structural and morphological effects on the FIR absorption. To overcome this problem, Jäger et al. (1998b) produced well-characterized bulk samples with different sp^2/sp^3 ratios and obtained optical constants from reflection spectroscopy in the wavelength range between 200 nm and 500 μm. Based on these fundamental data, an evaluation of the optical behaviour of spherical particles composed of such materials and a continuous distribution of ellipsoids (CDE) was possible (see Fig. 3). The β value for the spheres changed from 1.25 (sp^2 content=67%) to 2.28 (sp^2 content=80%). However, the CDE calculation demonstrated the importance of morphological effects for materials with high sp^2 content. Whereas the β value did not change for the first material, it moved to a value of 0.71 in the second material with the higher sp^2 content.

5. Conclusions

Observations of AGB stars with the ISO spectrometers have revolutionized our understanding of molecular and solid-state chemistry in circumstellar envelopes. However, this progress would not have been possible without high-quality experimental data whose production needs its own effort. Laboratory astrophysics will also contribute to the better understanding of the chemical and physical processess going on in the envelopes.

Acknowledgments. I acknowledge the support by my collaborators Drs. J. Dorschner, C. Jäger, H. Mutschke and M. Schnaiter and the funding by the Max Planck Society, the German Science Foundation, and the Thuringian Research Ministry. I thank the I.A.U. for the travel grant. I also thank D. Harker (Univ. of Wyoming) for critically reading the paper. Optical data of refractory solids can be retrieved from http://www.astro.uni-jena.de.

References

Aringer B., Kerschbaum F., Hron J., 1998, Ap&SS 255, 445

Bagnulo S., Doyle J.G., Andretta V., 1998, MNRAS 296, 545

Baron Y., de Muizon M., Papoular R., Pégourié B., 1987, A&A 186, 271

Begemann B., Dorschner J., Henning Th., Mutschke H., Thamm E., 1994, ApJ 423, L71

Begemann B., Dorschner J., Henning Th., Mutschke H., Gürtler J., Kömpe C., Nass R., 1997, ApJ 476, 199

Bohren C.F., Huffman D.R., 1983, Absorption and scattering of light by small particles, John Wiley & Sons, New York

Cami J., de Jong T., Justtanont K., Yamamura I., Waters L.B.F.M., 1998, Ap&SS 255, 339

Chan S.J., Kwok S., 1991, ApJ 383, 837

Cherchneff I., 1998, in The Molecular Astrophysics of Stars and Galaxies, T.W. Hartquist & D.A. Williams (eds.), Oxford University Press, Oxford, in press

Chiar J.E., Pendleton Y.J., Geballe T.R., Tielens A.G.G.M., 1998, ApJ 507, 281

Colangeli L., Mennella V., Palumbo P., Rotundi A., Bussoletti E., 1995, A&AS 113, 561.

Dorschner J., Henning Th., 1986, Ap&SS 128, 47

Dorschner J., Henning Th., 1996, Astron. Astrophys. Rev. 6, 271

Dorschner J., Begemann B., Henning Th., Jäger C., Mutschke H., 1995, A&A 300, 503

Edoh O., 1983, PhD Thesis, University of Arizona

Gail H.P., Sedlmayr E., 1998a, in The Molecular Astrophysics of Stars and Galaxies, T.W. Hartquist, D.A. Williams (eds.), Oxford University Press, Oxford, in press

Gail H.P., Sedlmayr E., 1998b, A&A, in press

Gail H.P., Sedlmayr E., 1998c, Faraday Discuss. 109, P. Sarre (ed.), p. 303

Goebel J.H., Bregman J.D., Witteborn F.C., 1994, ApJ 430, 317

Goebel J.H., Chesseman P., Gerbaut F., 1995, ApJ 449, 246

Habing H.J., 1996, Astron. Astrophys. Rev. 7, 97

Henning Th., 1983, Ap&SS 97, 405

Henning Th., 1996, in The Cosmic Dust Connection, J.M. Greenberg (ed.), Kluwer, Dordrecht, p. 399

Henning Th., 1997, in Molecules in Astrophysics: Probes and Processes, E. van Dishoeck (ed.), Kluwer, Dordrecht, p. 343

Henning Th., 1998a, Chem. Society Rev. 27, 315

Henning Th., 1998b, in Solid Interstellar Matter: The ISO Revolution, L. d'Hendecourt, C. Joblin and A. Jones (eds.), EDP Sciences, Les Ulis, in press

Henning Th., Salama F., 1998, Science 282, 2204

Henning Th., Schnaiter M., 1999, in Laboratory Astrophysics and Space Research, P. Ehrenfreund et al. (eds.), Kluwer, Dordrecht, p. 249

Henning Th., Gürtler J., Dorschner, J., 1983, Ap&SS 333, 349

Henning Th., Begemann B., Mutschke H., Dorschner J., 1995a, A&AS 112, 143

Henning Th., Michel B., Stognienko R., 1995b, Planet. Space Sci. 43, 1333

Jäger C., Mutschke H., Begemann B., Dorschner J., Henning Th., 1994, A&A 292, 641

Jäger C., Molster F.J., Dorschner J., Henning Th., Mutschke H., Waters L.B.F.M., 1998a, A&A 339, 904

Jäger C., Mutschke H., Henning Th., 1998b, A&A 332, 291

Jäger C., Henning Th., et al., 1999, in preparation

Jones T.W., Merrill T.M., 1977, ApJ 209, 509

Jura M., 1997, Ap&SS 251, 81

Justtanont K., Feuchtgruber H., de Jong T., Cami J., Waters L.B.F.M., Yamamura I., Onaka T., 1998, A&A 330, L17

Koike C., Shibai H., Tuchiyama A., 1993, MNRAS 264, 654.

Koike C., Kaito C., Shibai H., 1994, MNRAS 268, 321

Koike C., Kimura S., Kaito C., Suto H., Shibai H., Nagata T., Tanabe T., Saito Y., 1995, ApJ 446, 902

Kwok S., Chan S.J., 1993, AJ 106, 2140

LeVan P.D., Sloan G.C., Little-Marenin I.R., Grasdalen G.L. 1992, ApJ 392, 702

Little-Marenin I.R., 1986, ApJ 307, L15

Martin P.G., Rogers C., 1987, ApJ 322, 374
Mennella V., Colangeli L., Bussoletti E., 1995, A&A 295, 165
Michel B., Henning Th., Jäger C., Kreibig U., 1998, Carbon, in press
Mutschke H., Begemann B., Dorschner J., Gürtler J., Gustafson B.A.S., Henning Th., Stognienko R., 1998a, A&A 333, 188
Mutschke H., Andersen A.C., Clément D., Henning Th., Peiters G., 1998b, A&A, submitted
Nuth III J.A., 1996, in The Cosmic Dust Connection, J.M. Greenberg (ed.), Kluwer, Dordrecht, p. 205
Nuth III J.A., Hecht J.H., 1990, Ap&SS 163, 79
Olofsson H., 1997, Ap&SS 251, 31
Omont A., et al., 1995, ApJ 454, 819
Onaka T., de Jong T., Willems F.J., 1989, A&A 218, 169
Ossenkopf V., Henning Th., Mathis J.S., 1992, A&A 261, 567
Papoular R., Conard J., Guillois O., Nenner I., Reynaud C., Rouzaud J.-N., 1996, A&A 315, 222
Papoular R., Cauchetier M., Begin S., LeCaer G., 1998, A&A 329, 1035
Preibisch Th., Ossenkopf V., Yorke H.W., Henning Th., 1993, A&A 279, 577
Rouleau F., Martin P.G., 1991, ApJ 377, 526
Sandford S.A., 1996, Meteroritics & Planet. Sci. 31, 449
Schnaiter M., Mutschke H., Henning Th., Kohn B., Ehbrecht M., Huisken F., 1998, ApJ, submitted
Sloan G.C., LeVan P.D., Little-Marenin I.R., 1996, ApJ 463, 310
Sopka R.J., Hildebrand R.H., Jaffe D.T., Gatley I., Roellig T., Werner M., Jura M., Zuckerman B., 1985, ApJ 294, 242
Speck A.K., Barlow M.J., 1997, Ap&SS 251, 115
Speck A.K., Barlow M.J., Skinner C.J., 1997, MNRAS 288, 431
Stencel R.E., Nuth III J.A., Little-Marenin I.R., Little S.J., 1990, ApJ 350, L45
Stognienko R., Henning Th., Ossenkopf V., 1995, A&A 296, 797.
Tielens A.G.G.M., 1997, Ap&SS 251, 1
Tielens A.G.G.M., Waters L.B.F.M., Molster F.J., Justtanont K., 1998, Ap&SS 25 415
Yamamura I., de Jong T., Justtanont K., Cami J., Waters L.B.F.M., 1998, Ap&SS 255, 351
Waters L.B.F.M., et al., 1996, A&A 315, L361
Waters L.B.F.M., et al., 1998, Nature 391, 868
Willacy K., Millar T.J., 1997, A&A 324, 237
Willems F.C., de Jong T., 1988, A&A 196, 173
Zubko V.G., Menella V., Colangeli L., Bussoletti E., 1996, MNRAS 282, 1321

Asymptotic Giant Branch Stars
T. Le Bertre, A. Lèbre and C. Waelkens, eds.
I.A.U. Symposium 191, 1999

Dust formation in oxygen-rich circumstellar shells around long–period variables

Kyung Sook Jeong, Jan Martin Winters, Erwin Sedlmayr

Institut für Astronomie und Astrophysik, Technische Universität Berlin, Sekr. PN 8-1, Hardenbergstr. 36, D-10623 Berlin, Germany

Abstract. We present the first steps of our work aimed at a consistent time-dependent modeling of oxygen–rich circumstellar dust shells (CDS) around pulsating AGB stars. The nature of the most likely nucleation seeds is investigated for this situation and we find that TiO_2 is a most promising candidate to serve as the primary condensate, forming already at temperatures well above 1 000 K. These nuclei evolve to macroscopic dust grains by heterogeneous growth processes involving several chemical species. We investigate the varying chemical composition of the resulting dust grains as they evolve as a function of time in a fluid element moving through the CDS of an oxygen–rich long-period variable star (LPV).

1. Introduction

Circumstellar shells around cool pulsating late type stars are the major sources for the replenishment of the interstellar medium with processed material, in particular in the form of complex molecules and of solid dust particles (e.g. Gehrz 1989). Contrary to the carbon–rich situation, for which consistent dust shell models have been presented e.g. by Fleischer et al. (1992) and by Feuchtinger et al. (1993), a consistent physical description of CDS around oxygen–rich LPVs is still lacking. This is mainly due to the more complex problem of inorganic dust formation in an oxygen–rich environment. In contrast to the carbon–rich situation, where mainly *homogeneous* amorphous carbon grains and also PAHs are formed, in the case of an oxygen–rich composition of the gas almost all carbon is locked up in the chemically inert CO molecule. Therefore, grains only can be formed from molecules containing the remaining oxygen and those abundant heavy elements (Fe, Si, Mg, S, Al and Ti), which are contained in reactive gas phase species. In this case a *heterogeneous* composition of the emerging dust component has to be expected.

2. Basic considerations

In order to account for the observed high optical depths of the dust shells, formation of the major dust components requires the contribution of relatively **abundant elements**. Circumstellar dust is often observed at temperatures around 1 000 K or even higher. Therefore, the dust grains must consist of (a mixture of) **high temperature condensates**. Fig. 1 depicts the most promis-

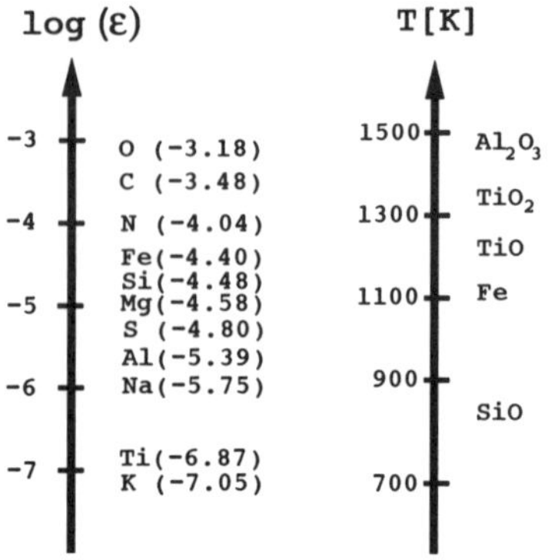

Figure 1. Solar element abundances and stability hierarchy of high temperature condensates (see text)

ing candidates for dust formation resulting from these criteria. The temperatures refer to a supersaturation ratio $S = 1$ for a mass density of $\rho = 10^{-12}\,\mathrm{g\,cm^{-3}}$. Corundum becomes stable around 1 500 K, titanium oxides around 1 300 K, iron around 1 100 K, and silicon monoxide around 900 K. However, this stability hierarchy does not necessarily mean that the respective compounds are actually formed in the circumstellar environment.

3. Nucleation seeds

Basically, the formation of dust grains can be conceived as a two step process: **i)** formation of a **critical cluster** which serves as a condensation seed and **ii) growth** of these clusters to macroscopic dust grains by addition of suitable molecules. Step one, the formation of the critical cluster requires high supersaturation ratios ($S \gg 1$), whereas the growth process, i.e. condensation of the growth molecules on preexisting nucleation seeds proceeds already if $S > 1$ for the respective growth species.

In order to determine the most likely primary condensation seed, nucleation rates J_* were calculated for the candidates shown in Fig. 1 in a range of the $p-T$ plane, relevant for the inner region of a dust forming circumstellar shell. Contour plots of the resulting nucleation rates are shown in Fig. 2 for various seed nuclei, which were calculated from kinetic nucleation theory (see Patzer et al. 1998b) assuming chemical equilibrium in the gas phase. This assumption of chemical equilibrium may be considered as highly unrealistic in pulsating situations, but due to the large numerical expenditure, no consistent non–equilibrium chemical structure exists so far for Miras and LPVs.

Since Al_2O_3 is already stable at very high temperatures, it might be expected to be the first condensate in the cooling outflows of oxygen–rich stars (e.g. Kozasa & Sogawa, this volume). However, using the thermodynamic functions resulting from quantum mechanical *ab initio* calculations for the structure of the Al_2O_3 molecule to calculate its equilibrium concentration shows that this molecule is virtually not present in the gas phase (see Chang et al. 1998, Patzer

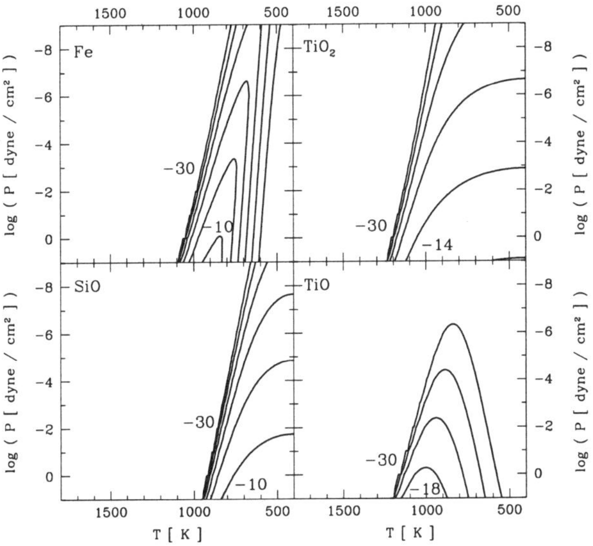

Figure 2. Contour lines of the nucleation rate for different nucleating species in the p–T plane. Contour levels are plotted in steps of $\Delta \log(J_*/n_{<\mathrm{H}>}) = 4$.

et al. 1998a).Therefore, homogeneous **nucleation** of molecular Al_2O_3 from the gas phase seems to be very unlikely in a circumstellar environment. However, the analysis of meteorites reveals that corundum–bearing grains in fact are formed in the circumstellar shells of oxygen–rich AGB stars (e.g. Zinner 1998). The formation of solid Al_2O_3 in principle might proceed via chemical reactions like $Al_2O + 2H_2O \rightarrow Al_2O_3 + 2H_2O$ (e.g. Gail & Sedlmayr 1998) or $2Al + 3H_2O \rightarrow Al_2O_3 + 3H_2$ (Kozasa & Hasegawa 1987) five body collision, respectively. Such a scheme seems to be realistic for the formation of corundum in a grain mantle, where the reaction partner can be adsorbed at the grain surface for a certain time. For the formation of an **isolated** $(Al_2O_3)_N$ cluster by this kind of chemical reactions, at least N (simultaneous) three body collisions in the gas phase are required, the effective occurrence of which is a delicate question considering the involved time scales. In the following we therefore consider corundum only as a growth species and are left with SiO, Fe, TiO and TiO_2 for the nucleation process (see Fig. 2).

Although the SiO molecule is the most abundant candidate for nucleation in oxygen–rich circumstellar shells, its nucleation rate only peaks around 600 K, which is significantly lower than the typical temperatures at the inner edge of the dust shells as derived e.g. from infrared interferometry (Danchi et al. 1994). Fe nucleation occurs only at temperatures well below 1 000 K. Only the titanium oxides show relevant nucleation rates at temperatures above 1 000 K where TiO_2 nucleation is by far more efficient, due to its higher molecular concentration in the relevant region of the $p - T$ plane.

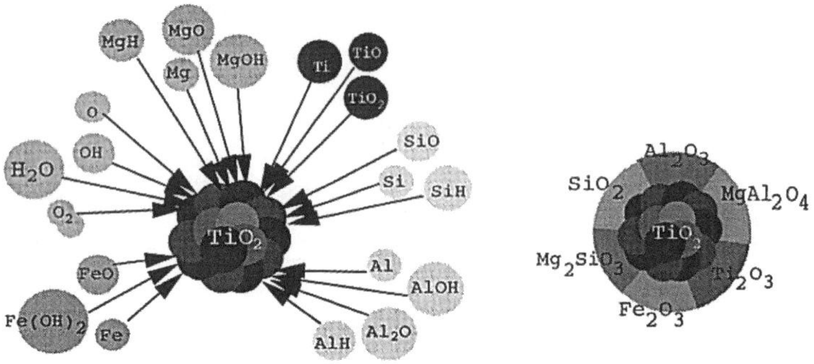

Figure 3. Heterogeneous growth on the seed material TiO_2. **l.h.s.**: molecules contributing to grain growth; **r.h.s.**: sketch of some compounds constituting the resulting grain mantle

4. Heterogeneous growth on TiO_2 grain nuclei

The TiO_2 seed nuclei grow by accretion of those molecular species which can form a stable condensate under the thermodynamic conditions prevailing in the circumstellar environment. The respective monomer of these condensates does not necessarily have to exist in the gas phase, but may be formed by chemical reactions on the grain surface from the atoms and molecules present in the gas phase (see Fig. 3).

For example, Mg_2SiO_4 can be formed by addition of the atoms Mg, Si and O, and the molecules O_2, MgO, SiO, SiO_2, etc. Actual formation of such a solid compound requires that all the generalized supersaturation ratios of the contributing atomic and molecular species are larger than unity (Dominik et al. 1993). Their method is applied here to describe this heterogeneous growth process. Generally, several compounds are simultaneously stable at a given pressure and temperature and there is a hierarchy of dust materials which successively become stable with decreasing temperature. The first condensing compounds are aluminum–bearing species like $Al_6Si_2O_{13}$, Al_2SiO_5 and Al_2O_3, followed by Mg_2SiO_4, $MgSiO_3$, SiO_2 and by the iron oxides and MgO, while sulphur–bearing compounds like $MgSO_4$ and SiS_2 only become stable at temperatures around 600 K. This stability hierarchy (shown in Fig. 4) might depend on the total gas pressure (which affects the partial pressure of the contributing species in a non–linear way) as can be seen e.g. from the stability limit of $MgAl_2O_4$ depicted in Fig. 4 (see also Gail & Sedlmayr 1998).

5. First results of a consistent model calculation

The consistent description of a dust forming circumstellar shell constitutes a non–linearly coupled problem comprising hydrodynamics, thermodynamics, chemistry, dust formation, growth and evaporation and the radiative transfer problem. Therefore, a simultaneous solution of the complete problem describing the dust forming system is necessary for a reliable description of a circumstellar dust shell. Applying the numerical method described in Fleischer et al. (1992), we

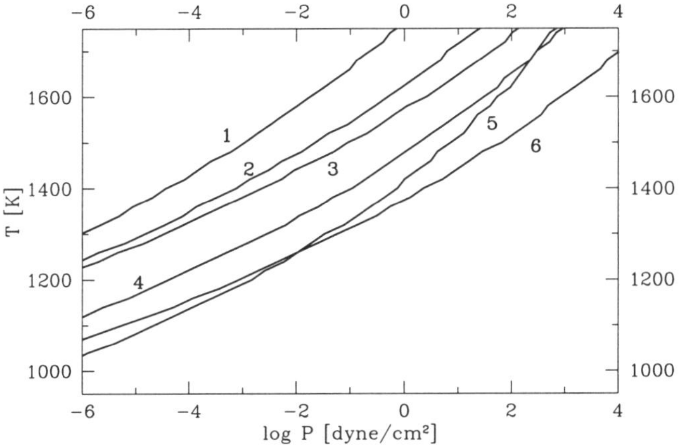

Figure 4. Stability limits ($S = 1$) for some high temperature condensates in the p–T plane. 1: Al_2O_3, 2: $Al_6Si_2O_{13}$, 3: Ti_2O_3, 4: $MgSiO_3$, 5: $MgAl_2O_4$, 6: $MgTiO_3$

calculate a consistent dust shell model including a proper treatment of heterogeneous dust formation as outlined in the preceding sections. The solution of the equation system is completely determined by prescribing the values of 6 parameters for which we choose $L_* = 2 \cdot 10^4\ L_\odot$, $T_* = 2\,000$ K, $M_* = 1\ M_\odot$, velocity amplitude at the inner boundary $\Delta u_p = 5\ \mathrm{kms}^{-1}$, P = 300 days, $\epsilon_C/\epsilon_O = 0.5$, and solar abundances for the other elements. This model produces a mass loss rate of $\dot{M} \approx 5 \cdot 10^{-5}\,M_\odot\,\mathrm{yr}^{-1}$ and an outflow velocity of $v_\infty \approx 15\,\mathrm{km\,s}^{-1}$. The radii of most of the dust grains range between 0.01 and 0.1 μm, but there still exist a few grains larger than 0.5 μm. Fig. 5 displays the temperature and velocity structure of the dust shell model at a fixed instant of time and a sketch of the chemical composition of the dust grains at different locations in the atmosphere. At $R \approx 1.9\ R_*$ the temperature is around 1 200 K and most of the dust volume consists of $Al_6Si_2O_{13}$. Titanium–bearing compounds like $MgTiO_3$ and titanium oxides like Ti_2O_3 and Ti_3O_5 are also included just after the onset of condensation, but their volume fraction is almost negligible. Already at slightly lower temperatures ($T \approx 1\,100$ K at $R \approx 2.2\ R_*$), Mg_2SiO_4, $MgSiO_3$ and SiO_2 are the dominant compounds contained in the dust grains. Just below 1 000 K the iron oxides are formed, which finally contribute about 30% to the dust volume. All sulphur–bearing compounds become only stable around 600 K and their volume fraction remains well below 10%.

This investigation about the composition of dust grains, together with the laboratory work (e.g. Jäger et al. 1994, Koike et al. 1993), will enable us to get a more precise insight about the "astrophysical silicates" which are considered as being responsible not only for the well known amorphous silicate bands, but also for several emission features recently detected at wavelengths between 12 and 70 μm (e.g. Waters & Molster, this volume).

Acknowledgments. This work has been supported by the FAZIT Stiftung and by the BMBF (grant 05 3BT13A 6).

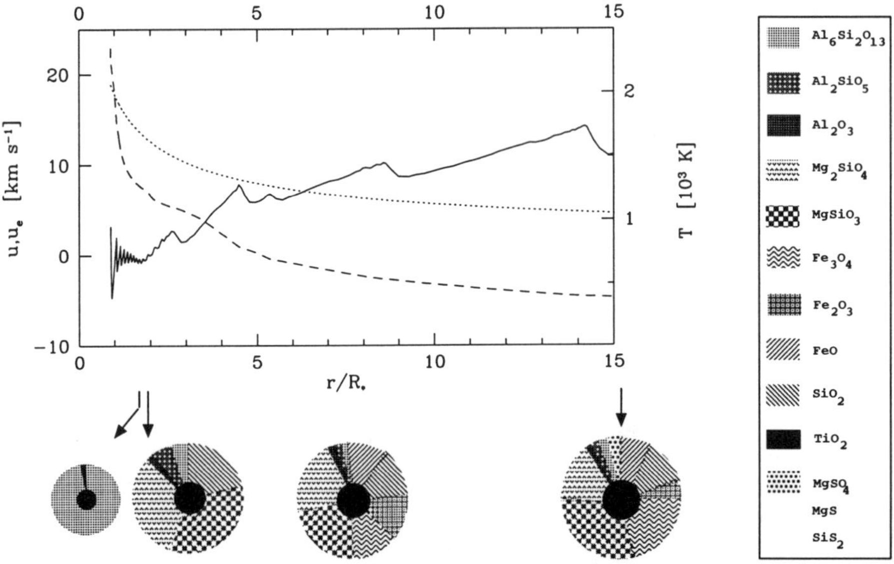

Figure 5. Upper panel: velocity (solid line), escape velocity (dotted line), and temperature structure (dashed line) of the dust shell model. Lower panel: composition of the dust component at different distances from the star.

References

Chang C., Patzer A.B.C., Sedlmayr E., Sülzle D., 1998, Eur. Phys. J. D 2, 57

Danchi W.C., Bester M., Degiacomi C.G., Greenhill L.J., Townes C.H., 1994, AJ 107, 1469

Dominik C., Sedlmayr E., Gail H.-P., 1993, A&A 277, 578

Feuchtinger M.U., Dorfi E.A., Höfner S., 1993, A&A 273, 513

Fleischer A.J., Gauger A., Sedlmayr E., 1992, A&A 266, 321

Gail H.-P., Sedlmayr E., 1998, in *Chemistry and physics of molecules and grains in space*, Faraday Discussion no. 109, P. Sarre (ed.), London, p. 303

Gehrz R.D., 1989, in *Interstellar Dust*, L.J. Allamandola and A.G.G.M. Tielens (eds.), Kluwer Academic Publishers, Dordrecht, p. 445

Jäger C., Mutschke H., Begemann B., Dorschner J., Henning T., 1994, A&A 292, 641

Koike C., Shibai H., Tuchiyama A., 1993, MNRAS 264, 654

Kozasa T., Hasegawa H., 1987, Prog. Theor. Phys 77, 1402

Patzer A.B.C., Chang C., Sedlmayr E., Sülzle D., 1998a, Eur. Phys. J. D, submitted

Patzer A.B.C., Gauger A., Sedlmayr E., 1998b, A&A 337, 847

Zinner E., 1998, Ann. Rev. Earth Planet. Sci 26, 147

Asymptotic Giant Branch Stars
T. Le Bertre, A. Lèbre and C. Waelkens, eds.
I.A.U. Symposium 191, 1999

Formation of crystalline silicate around oxygen–rich AGB stars

Takashi Kozasa and Hisato Sogawa

Department of Earth and Planetary Sciences, Kobe University, Kobe 657–8501, Japan

Abstract. Crystallization of silicate has been investigated within the framework of dust formation in steady state gas outflows around oxygen–rich AGB stars, where silicates are locked not only into homogeneous silicate grains but also into the mantles of heterogeneous grains. Based on the thermal history of dust grains after their formation, the crystallization calculation results in no crystalline silicate for the mass loss rate $\dot{M} \leq 2 \times 10^{-5} M_{\odot}$ yr^{-1}. Only silicate in the mantles of heterogeneous grains can be crystallized for $\dot{M} \geq 3 \times 10^{-5} M_{\odot}$ yr^{-1}, while homogeneous silicate grains remain amorphous. The mass fraction of crystalline silicate increases with increasing $\dot{M}$. The radiation transfer calculations confirm the appearance of an emission feature around 33.5 μm, taking olivine as a representative of crystalline silicates. On the other hand, the 10μm feature appears in absorption, being dominated by homogeneous silicate grains. These trends are consistent with the observations. Thus the crystalline silicate is a diagnostics of high mass loss rate at the late stage of AGB stellar evolution, reflecting the formation process of dust grains.

1. Introduction

The ISO–SWS observations revealed the presence of crystalline silicate in circumstellar envelopes of oxygen–rich evolved stars from emission features between 25 and 45 μm (Waters et al. 1996). Cami et al. (1998) have shown that oxygen–rich AGB stars with high mass loss rate exhibit the prominent emission features of crystalline silicates with absorption features around 10 and 18μm; otherwise no crystalline silicate feature is seen in AGB stars with low mass loss rate. The 10μm absorption features analyzed so far show no obvious structure due to crystalline silicates, being dominated by amorphous silicates (Tielens et al. 1998). The peak positions of the emission features suggest that crystalline silicates are Mg–rich olivine and pyroxene, although amorphous silicates have been considered to be Fe–rich (Tielens et al. 1998, and references therein).

The observed characteristics of silicate features seem to be closely related to the formation process of dust grains through the mass loss rate of AGB stars. Although Tielens et al. (1998) addressed the formation scenario of crystalline silicate, the details are still uncertain. In this paper we explore the origin and formation process of crystalline silicate within the framework of dust formation around oxygen–rich AGB stars. Summarizing the model of dust formation and

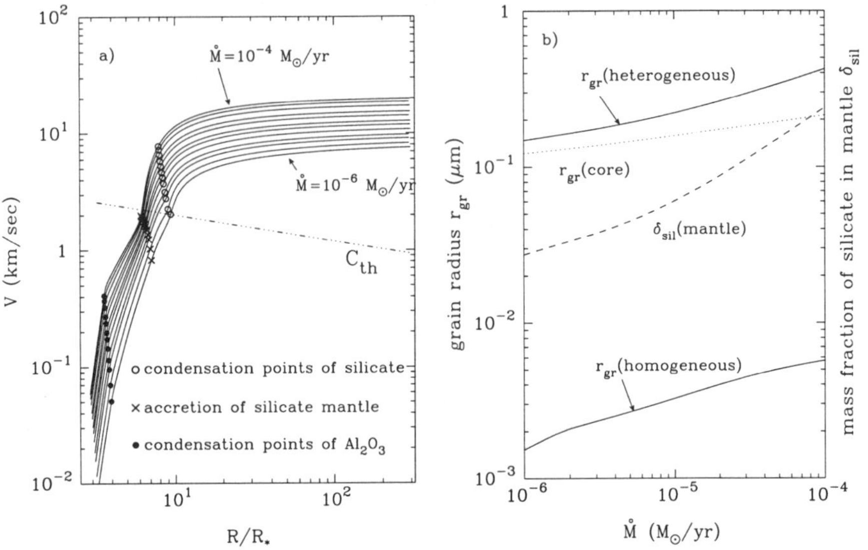

Figure 1. a) The gas outflow solutions and the locations of dust condensation for $\dot{M} = 10^{-6}$ to $10^{-4}\,M_\odot$/yr. b) The final average radii of dust grains (solid and dotted lines) and the mass fraction of silicate locked into the mantle of heterogeneous grains (dashed line) versus $\dot{M}$.

introducing a kinetic theory of crystallization, we shall show that crystalline silicate is a natural product when the mass loss rate is large enough.

2. Dust formation and the thermal history of dust grains

We have developed a model of dust formation in steady state dust driven winds around oxygen–rich AGB stars (see Kozasa & Sogawa 1997, 1998). The model considers the formation of not only homogeneous corundum (Al_2O_3) and silicate grains but also heterogeneous grains consisting of an Al_2O_3 core and a silicate mantle simultaneously. The radial dependence of the gas temperature is assumed to be given by $T(R) = T_*\,(R/R_*)^{-\beta}$ where R_* is the stellar radius and T_* the effective temperature. In the present calculations, the stellar parameters are $M_* = 1.0\,M_\odot$, $L_* = 2\times 10^4\,L_\odot$ and $T_* = 2\,500$ K. The mass loss rate $\dot{M}$ ranges from 10^{-6} to $10^{-4}\,M_\odot$ yr^{-1}, and $\beta = 0.45$.

The locations of condensation of dust grains and the resulting gas outflows are depicted in Fig. 1a, where the condensation point means the location at which the nucleation rate reaches its maximum. The accretion of silicate on pre–condensed Al_2O_3 grains starting slightly inside the sonic point drives the gas outflow into the supersonic region. Beyond the sonic point, homogeneous silicate grains condense. The final average radii of heterogeneous grains range from 0.15 to 0.4 μm (see Fig. 1b). Homogeneous silicate grains are very small: $r_{gr} = 15$ to 60 Å. The mass fraction of silicate locked into the mantles of heterogeneous grains δ_{sil} increases with increasing $\dot{M}$, and reaches 0.1 for $\dot{M} = 3\times10^{-5} M_\odot$ yr^{-1}.

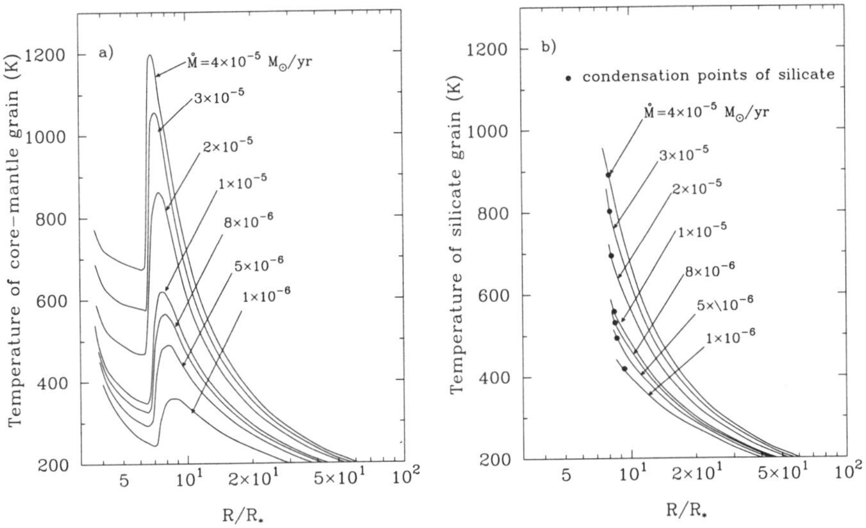

Figure 2. Temperatures of dust grains after their formation for the given mass loss rates; a) for Al_2O_3 and heterogeneous grains, and b) for homogeneous silicate grains

Fig. 2 shows the temperatures of dust grains after their formation for $\dot{M} \leq 4 \times 10^{-5}\, M_{\odot}\,\mathrm{yr}^{-1}$. They are calculated by a radiation transfer code using a Monte-Carlo method (Fig. 2a: Al_2O_3 and heterogeneous grains, Fig. 2b: homogeneous silicate grains). The temperatures of Al_2O_3 grains are very low, reflecting the transparency of Al_2O_3 in the visible and near infrared region. As an Al_2O_3 grain is coated with an absorptive silicate mantle, the temperature of heterogeneous grain increases steeply, reaches a maximum and then decreases with increasing R. The maximum temperature is higher than 1 000 K for $\dot{M} \geq 3 \times 10^{-5}\, M_{\odot}\,\mathrm{yr}^{-1}$. On the other hand, with the temperature $\leq$ 900 K at the condensation point, the temperatures of homogeneous silicate grains monotonically decrease.

At the time of accretion of silicate, the temperatures of Al_2O_3 cores are much lower than the glass temperature of silicate ($\sim$ 1 050 K, see Tielens et al. 1998). Also the condensation temperature of homogeneous silicate is lower than 1 000 K (Kozasa & Sogawa 1997, 1998). Thus newly formed silicates are to be amorphous. The thermal history of dust grains after their formation is crucial in determining the crystallinity of silicate as discussed in the following sections.

3. A kinetic theory of crystallization of amorphous materials

Crystallization of amorphous solid proceeds through the spontaneous nucleation of crystalline nuclei and the following growth during heating and/or cooling processes. The diffusion of molecules controls the kinetics. The steady state nucleation rate J and the growth rate w for formation of spherical crystal are given by

$$J(t) = \frac{2\pi}{\Omega}\left(\frac{\sigma}{kT}\right)^{1/2}\frac{D_0}{\Omega^{2/3}}\exp\left[-\frac{1}{kT}\left(E_a + \frac{16\pi}{3}\frac{\sigma^3}{\Delta G_v^2}\right)\right] \tag{1}$$

$$\omega(t) = \frac{D_0}{\Omega^{1/3}}\exp\left(-\frac{E_a}{kT}\right)\left[1 - \exp\left(-\frac{\Delta G_v}{kT}\right)\right] \tag{2}$$

where Ω is the molecular volume, σ the interfacial energy and ΔG_v the Gibbs free energy change per molecule between amorphous and crystalline phases (see Kouchi et al. 1994 for the details). The diffusion coefficient is written by $D = D_0 \exp(-E_a/kT)$ with E_a the activation energy. Then the temporal evolution of crystallized volume fraction $\theta(t)$ during heating and/or cooling processes is expressed by the standard and widely used "Avrami equation" (Avrami 1939)

$$1 - \theta(t) = \exp\left[-\frac{4\pi}{3}\int_0^t J(t')\left(\int_{t'}^t \omega(t'')dt''\right)^3 dt'\right] \tag{3}$$

In principle, applying the above equation for thermal history of dust grains after their formation, we can calculate the crystallized volume fraction of dust grains.

At present time the basic data such as D_0, E_a and ΔG_v are not available for crystallization of silicates. However, it is possible to estimate the necessary data by referring to the annealing experiments of "magnesium silicate smokes" by Hallenbeck et al. (1998). Monitoring the infrared spectrum during isothermal annealing, they obtained the Arrhenius plot for the stall time t_{stall} at which the change of the spectral features stalls for a while (see Figs. 1 and 8 in their paper). The stall spectrum allows us to estimate $\theta \sim 0.2$ at the stall time by comparison with the calculated spectrum for mixtures of amorphous and crystalline silicates. For isothermal annealing, equation (3) is reduced to $1-\theta = \exp\left[-(t/\tau_{cr})^4\right]$ with the characteristic time scale of crystallization τ_{cr} defined by

$$\tau_{cr} = \left(\frac{3}{2\pi^2}\right)^{1/4}\left(\frac{kT}{\sigma}\right)^{1/8}\frac{\Omega^{2/3}}{D_0}\exp\left[\frac{1}{kT}\left(E_a + \frac{4\pi}{3}\frac{\sigma^3}{\Delta G_v^2}\right)\right] \tag{4}$$

The estimated $\theta \sim 0.2$ at the stall time corresponds to $t_{stall} = 0.7\tau_{cr}$. Fitting the Arrhenius plot with Eq. (4) leads to the basic data: $\ln D_0 = 1.47 \times 10^2$ (in cgs units), $E_a/k = 1.72 \times 10^5$ K, $\sigma/\Omega^{2/3} = 731$ erg cm^{-2}, $\Delta G_v/k = 1.85 \times 10^4 (1 - T/1830)$ K, where we used $\Omega = 5.22 \times 10^{-23}$ cm^3. We shall adopt these derived values for the following calculations, although the "magnesium silicate smokes" may not be necessarily representative of "astronomical silicate".

4. Crystallization of silicate and the resulting spectral feature

We apply the theory presented in § 3 to investigate the crystallization of silicate. The characteristic time scale of crystallization τ_{cr} of silicate is very sensitive to the grain temperature because of the large activation energy E_a; for example, τ_{cr} = 20 sec., 30 days and $\sim 10^6$ years for $T = 1\,100$, $1\,000$, and 900 K, respectively. If the maximum temperature experienced by dust grains is less than 900 K, the crystallization of silicate is impossible during the lifetime of AGB stars. Thus, homogeneous silicate grains as well as silicate in the mantles of heterogeneous

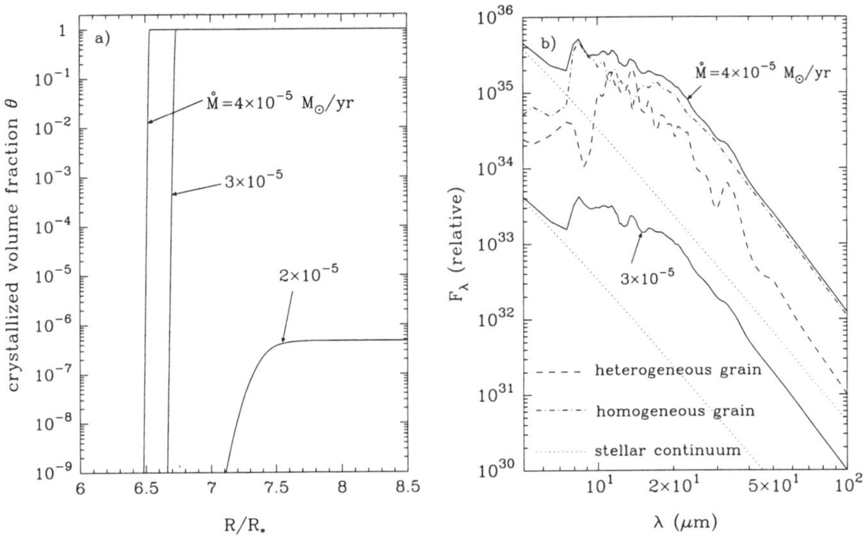

Figure 3. a) The spatial evolution of θ for silicate in the mantles of heterogeneous grains. b) The spectral energy distribution for the given mass loss rates. The contributions from the heterogeneous grains (dashed line) and the homogeneous silicate grains (dash–dotted line) to the total flux are indicated for $\dot{M} = 4 \times 10^{-5} M_{\odot}/\mathrm{yr}$

grains can not be crystallized for $\dot{M} \leq 2 \times 10^{-5}\, M_{\odot}\ \mathrm{yr}^{-1}$ (see Fig. 2). Even if the temperature is higher than 900 K, small homogeneous silicate grains can escape from crystallization due to the low probability of nucleation as discussed by Seki & Hasegawa (1981).

Figure 3a shows the spatial evolution of the crystallized volume fraction of silicate in the mantles of heterogeneous grains calculated by Eq. (3) using the evolution of dust temperatures given in Fig. 2a. For $\dot{M} = 2 \times 10^{-5}\, M_{\odot}\mathrm{yr}^{-1}$, crystallization is negligible ($\theta \leq 10^{-6}$) as discussed above. On the other hand, silicate in the mantle is completely crystallized for $\dot{M} \geq 3 \times 10^{-5}\, M_{\odot}\mathrm{yr}^{-1}$ as the temperature steeply increases above 1 000 K. Thus only silicate in the mantles of heterogeneous grains can be crystallized for $\dot{M} \geq 3 \times 10^{-5}\, M_{\odot}\mathrm{yr}^{-1}$, while homogeneous silicate grains stay amorphous.

The spectral energy distributions calculated on the basis of the results of condensation and crystallization calculations are shown in Fig. 3b, where we take olivine as a representative of crystalline silicates and use the optical constants of $(Mg_{0.9}, Fe_{0.1})_2SiO_4$ (Koike 1998). First it should be noted that the continuum beyond 20 μm is dominated by the thermal radiation from amorphous homogeneous–silicate grains as indicated for $\dot{M} = 4 \times 10^{-5}\, M_{\odot}\ \mathrm{yr}^{-1}$. The distinguishable olivine feature appears around 33.5 μm in emission over the continuum, while the 10 μm feature appears in absorption and is dominated by homogeneous silicate grains. The emission feature of crystalline silicate becomes prominent with increasing $\dot{M}$ because δ_{sil} increases as presented in Fig. 2b, though the present calculation is restricted to $\dot{M} \leq 4 \times 10^{-5}\, M_{\odot}\mathrm{yr}^{-1}$. These trends of the appearance of silicate features are consistent with the observational results.

5. Concluding remarks

The origin and formation process of crystalline silicate were explored in the framework of dust formation in circumstellar envelopes of oxygen–rich AGB stars. The results of the calculations show that the carriers of crystalline silicate are heterogeneous grains consisting of an Al_2O_3 core and a silicate mantle. When the mass loss rate $\dot{M}$ is larger than $3 \times 10^{-5}\ M_\odot\ \mathrm{yr}^{-1}$, the crystallization occurs through the thermal processing due to heating caused by the accretion of silicate on Al_2O_3 grains starting slightly inside the sonic point. On the other hand, small homogeneous silicate grains condensing beyond the sonic point remain amorphous.

The behavior of calculated silicate features is consistent with that revealed by the observations. The derived lower limit of the mass fraction of crystalline silicate is 0.1, which is larger than the value ~ 0.05 derived from the observations (Tielens et al. 1998). It should be noted that the appearance of emission feature of crystalline silicate could be sensitive to not only the mass fraction of crystalline silicate but also the spatial distribution of the temperatures of the carrier of crystalline silicate as well as the amorphous silicate grain.

We conclude that the crystallization of silicate is a natural consequence of the formation process of dust grains prevailing in circumstellar envelopes of oxygen–rich AGB stars. The crystalline silicate is a diagnostics of high mass loss rate at the late stage of AGB stellar evolution. The problem of Fe content in the crystalline silicate suggested by the observations (Tielens et al. 1998) has been left for the future; we may speculate that the thermal processing at the temperature above 1 000 K could make crystalline silicate Mg–rich. The crystalline silicate could be less active for reaction with Fe than small cluster of homogeneous silicate, which causes homogeneous silicate grains to be Fe-rich.

Acknowledgments. We are grateful to Dr. C. Koike for providing us the unpublished data of optical constants of crystalline olivine.

References

Avrami M., 1939, J. Chem. Phys. 7, 1103
Cami J., de Jong T., Justtanont K., et al., 1998, Ap&SS 255, 339
Hallenbeck S. L., Nuth III J.A., Daukantas P., 1998, Icarus 131, 198
Koike C., 1998, private communication
Kouchi A., Yamamoto T., Kozasa T., et al., 1994, A&A 290, 1009
Kozasa T., Sogawa H., 1997, Ap&SS 251, 165
Kozasa T., Sogawa H., 1998, Ap&SS 255, 437
Seki J., Hasegawa H., 1981, Prog. Theor. Phys. 66, 903
Tielens A.G.G.M., Waters L.B.F.M., et al., 1998, Ap&SS 255, 415
Waters L.B.F.M., Molster F.J., de Jong T., et al., 1996, A&A 315, L361

Asymptotic Giant Branch Stars
T. Le Bertre, A. Lèbre and C. Waelkens, eds.
I.A.U. Symposium 191, 1999

ISOSWS Spectral Variations of Oxygen-Rich Miras

M. J. Creech-Eakman
California Institute of Technology, MS 150-21, Pasadena, CA 91125, USA
University of Denver, Dept. of Physics and Astronomy, Denver, CO 80208, USA

R. E. Stencel
University of Denver, Dept. of Physics and Astronomy, Denver, CO 80208, USA

Abstract. In conjunction with an extensive ground-based monitoring program of 32 oxygen-rich Mira variables, a subset of this sample has been monitored with phase using ISO's Short Wavelength Spectrometer (ISOSWS) and ground-based mid-infrared spectrometers. Some of the six sources will be presented here in 2.5–45 μm ISOSWS spectra and 7.5–13.5 μm ground-based spectra using CGS3. Discussion of the spectral features of SiO, CO, H_2O, and silicates will be presented. These data are considered in the context of recent demonstrations of variations in the spectral features with phase of the Miras.

1. Introduction and background

This study was initiated based on classifications of silicate features of oxygen-rich Mira variables from the IRAS LRS database first assembled by Little-Marenin & Little (1990, hereafter LML). Sloan & Price (1995) later reclassified the LML features using statistical fits of black body curves modified to account for SiO absorption near 8 μm. They demonstrated that a sequence of color of the 10 μm features was evident by examining the colors of the black-body subtracted IRAS LRS spectra. Stencel et al. (1990), based on a paper by Lewis (1989), concluded that the silicate emission features when taken in conjunction with SiO, OH and H_2O masers implied an evolutionary sequence for M Miras. Based on this information, we undertook a photometric campaign of the 5 to 20 μm region to study the silicate features of the Miras to determine if these features changed with phase, and if more information on an evolutionary sequence could be obtained (Creech-Eakman 1997).

As part of the research, we proposed to study a sub-group of the sources in our photometric study with ISO's Short Wavelength Spectrometer (ISOSWS). Three epochs of each source were to be taken over the course of ISO's initial mission. The sources were chosen to span the LML classification scheme and to complement the Central Programme. Unfortunately, not all the allocated time was ultimately scheduled on the sources. Thus only two of the six sources had more than one epoch of ISOSWS data obtained. Concurrently, ground-based observations were obtained with CGS3 as part of the UKIRT Service Program on this same group of sources, so that between three and seven spectra exist for each of the sources in the 7–14 μm region over about 3 years.

2. Data reduction

2.1. ISOSWS data

ISOSWS AOT01 data were taken for six stars, two at more than one epoch (Table 1). The data were reduced using the SIA interactive software written by scientists at ESA and IPAC. Each scan was reduced from its ERD using the pipeline 7 reduction and was interactively dark subtracted, flat fielded, masked and corrected for anomalies in the up and down scans before being converted to flux values. Each scan was further reduced in ISAP by 2.0 sigma clipping and Gaussian smoothing at half the native resolution for each band. Finally, the data bands were merged together, rebinned onto a 0.005 μm grid and Gaussian smoothed at 0.02 μm. Some of the data are of low S/N and need more work; hence not all data will be shown here.

Star	IRAS (Jy)			Obs. Date	Phase	Sil Class
	12 μm	25 μm	60 μm			
TW Cyg	10.96	4.27	0.79	03/12/96	0.73	No Feature
ER Cyg	14.20	5.59	1.31	05/12/96	0.38*	Broad
U Dra	18.51	7.24	1.27	20/12/96	0.01	Broad
				23/04/97	0.40	
				10/10/97	0.94	
SV And	12.67	4.89	0.79	17/01/97	0.15	S Feature
				31/01/98	1.35	
RY And	23.97	11.55	2.01	03/12/96	0.26	Sil +
AU Her	17.31	10.41	2.53	12/02/97	0.61*	Sil

Table 1. Sources observed in the mcreeche ISOSWS program. Included here are the color-corrected IRAS fluxes assuming a 2500 K color-correction. Epochs and periods were taken from AAVSO data. Those marked by an asterisk were taken from the GCVS.

2.2. CGS3 data

Through the UKIRT Service Observing program Cooled Grating Spectrometer 3 (CGS3) has been used to conduct ground-based observations of the sources in Table 1 prior to and throughout the ISO mission. These observations were taken at a variety of variable star phases in the 10 μm and, when weather permitted, 20 μm windows at a resolving power of about 50. The observations were taken by J. K. Davies and standards from Cohen & Davies (1995) taken close in airmass and time were used for all the data reduction. We were supplied with completely reduced spectra, the entire collection of which will be published in an upcoming paper.

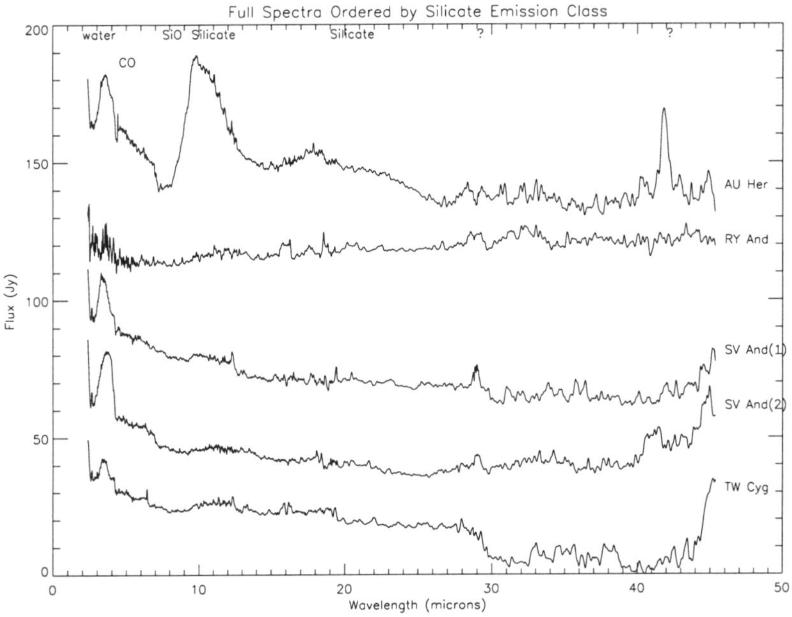

Figure 1. ISOSWS spectra of 4 M Miras - AU Her, RY And, SV And (2 epochs) and TW Cyg. Offsets of 120, 105, 35, 20 and 0 Jy have been applied to the data for plotting purposes.

3. Data and analysis

In Fig. 1 we show the complete ISOSWS spectrum for four different M Miras, AU Her, RY And, SV And (2 epochs) and TW Cyg, ordered from top to bottom by silicate class. It is clear that AU Her has pronounced 10 and 18 μm silicate features of both amorphous and crystalline nature. Based on the IRAS LRS spectrum of RY And, one would expect more pronounced silicates in this source. The ISOSWS spectrum is further evidence of the changing nature of the silicate features and general spectrum of M Miras with phase of the star (see also Little-Marenin et al. 1996, Creech-Eakman et al. 1997).

Also evident in these sources are a wealth of short wavelength features from CO and SiO gases and water present in these atmospheres. In some sources, there is a hint of the 13 μm feature attributable to Al_2O_3 (Begemann et al. 1997). While the spectra get considerably noisier past about 27 μm, there are two interesting long-wavelength features, at 29 μm (in all the sources) and 42 μm (in only AU Her). Based on the width of the features, it is likely they are related to crystalline solid state complexes of silicate dust (see for example Waters et al. 1998). It is possible, however, that the features near 30 μm are not real, caused by band mismatches when merging data from different detectors, so care must be taken in identifications at those wavelengths.

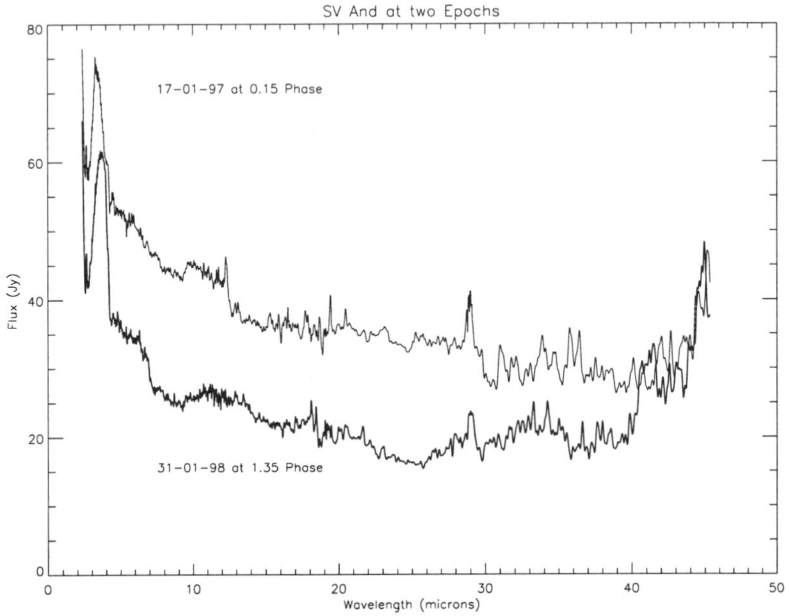

Figure 2. ISOSWS spectra of SV And at 2 epochs.

In Fig. 2 we show ISOSWS spectra of SV And at two epochs. While there is a change of nearly 20 Jy in overall level of the spectrum from the first to the second epoch, there are more pronounced changes in the individual spectra. In particular, the depth of the features near 3 μm, mainly attributable to H_2O and CO, are different by about 25%. Second, the silicate features at 10 and 18 μm and the SiO absorption at 7.5 μm combine to give an overall change in the appearance of the dust with phase. Finally the appearance of the 13 μm feature at only one epoch suggests dust formation processes on time scales of less than a pulsation period.

In Fig. 3 we show the 7.5–13.5 μm spectra from ISOSWS and CGS3 of SV And over a 3 year period. The top 2 spectra are from ISOSWS, while the bottom 3 are from CGS3. While it is possible that the CGS3 spectra are depressed with respect to the ISOSWS spectra due to telluric absorption, great care was taken in the calibration and reduction of this data to mitigate telluric effects. These data are presented to show that the character and flux level of the spectra change markedly with phase and do not precisely follow the visual light curve of the Mira. For instance, note the appearance of the spectral features from 9.5 to 10.5 μm and around 12 μm. Spectral features in these regions are most often associated with silicate dust. Changes in these regions are possibly indicative of short-time scale dust growth and destruction. Modeling of the

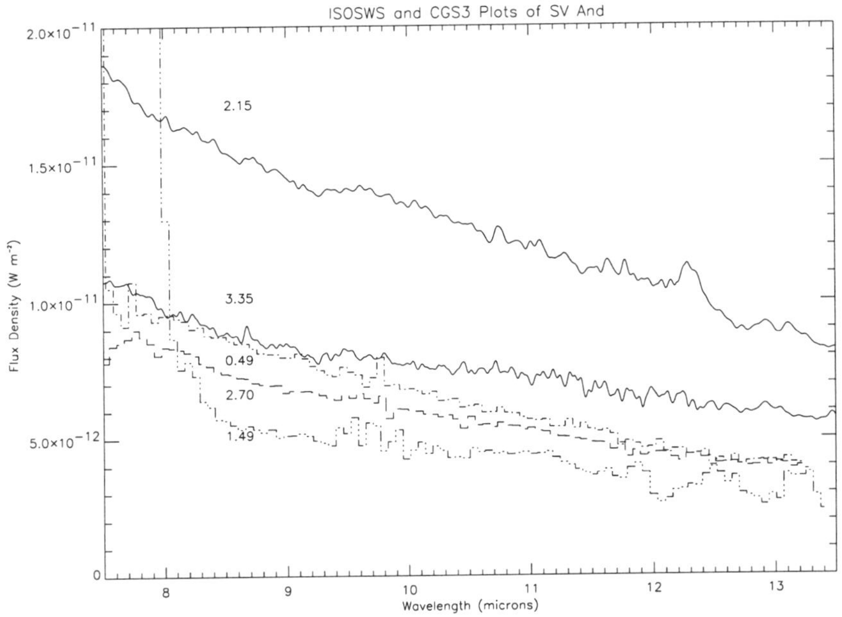

Figure 3. ISOSWS and CGS3 spectra of SV And over a 3 year time period demonstrate marked variability with phase. Epochs and phase were determined with concurrent AAVSO data.

major gas and dust constituents will help determine whether the changes seen are principally due to radiation and optical depth effects, or whether they indicate gross changes in the silicate dust.

4. Conclusions

In this paper, we have shown that marked variation exists among M Miras in the appearance of their spectral features throughout the 2.5–45 μm region. Along with differences among the various silicate emission features classes, individual stars show differences from one phase to the next, as is demonstrated by SV And. It is likely that the changes seen in the silicate features arise from formation, annealing and destruction of the dust during the stellar pulsation cycle (Creech-Eakman & Stencel 1998). Radiative hydrodynamical modeling with silicate dust formation will likely be needed to adequately reproduce the variations seen in these spectra.

Ongoing work in the form of both mid-IR photometry and near and mid-IR spectroscopy continues for some of these sources. In particular, we are currently trying to determine whether the short-term variations in M Miras reported by de Laverny et al. (1998) in the HIPPARCOS data are present at other wavelengths, in particular in the silicate dust and in the SiO and CO gas features. It is hoped that the new information presented here will encourage other groups to undertake dedicated photometric and spectroscopic monitoring of these sources.

Acknowledgments. We thank J. K. Davies and the UKIRT Service Program for the CGS3 data presented here. Also, we thank J. Mattei and the AAVSO for observations leading to the determinations of current epoch and phase for the Miras in this study. Finally, we thank ESA for ISO Observing time and NASA for funding support, and many folks at IPAC for assistance in data reduction. This work has made extensive use of the Simbad database operated by the CDS in France.

References

Begemann B., Dorschner J., Henning Th., Mutschke H., Guetler J., Koempe C., Nass R., 1997, ApJ 476, 199

Cohen M., Davies J.K., 1995, MNRAS 276, 715

Creech-Eakman M.J., 1997, PhD Dissertation, University of Denver

Creech-Eakman M.J., Stencel R.E., 1998, ApJ, submitted

Creech-Eakman M.J., Stencel R.E., Williams W.J., Klebe D.I., 1997, ApJ 477, 825

de Laverny P., Mennessier M.O., Mignard F., Mattei J., 1997, A&A 330, 169

Lewis B.M., 1989, ApJ 338, 234

Little-Marenin I.R., Little S.J., 1990, AJ 99, 1173

Little-Marenin I.R., Stencel R.E., Staley S.B., 1996, ApJ 467, 806

Sloan G.C., Price S.D., 1995, ApJ 451, 758

Stencel R.E., Nuth J.A.I., Little-Marenin I.R., Little S.J., 1990, ApJ 350, 45

Waters L.B.F.M., Beintema D.A., Zijlstra A.A., de Koter A., Molster F.J., Bouwman J., de Jong T., Pottasch S.R., de Graauw Th., 1998, A&A 331, L61

Asymptotic Giant Branch Stars
T. Le Bertre, A. Lèbre and C. Waelkens, eds.
I.A.U. Symposium 191, 1999

The chemistry of carbon dust formation

Isabelle Cherchneff & Piero Cau

Department of Physics, UMIST, PO Box 88, Manchester M60 1QD, United Kingdom

Abstract.
We shall review the various types of chemistry involved in the formation of carbonaceous material present in carbon-rich AGB stars, mainly amorphous carbon, silicon carbide and other metal carbides discovered in pre-solar stardust extracted from meteorites. The chemistry is discussed in the context of laboratory experiments and their application to circumstellar AGB winds. Emphasis is put on polycyclic aromatic hydrocarbons (PAHs), titanium carbide clusters and silicon carbide grains. Attempt to explain the condensation sequences derived from the study of pre-solar grains of meteoretical origin is made on the basis of physio-chemical models which describe the periodically shocked gas close to the photosphere of AGB stars.

1. Introduction

Carbon-rich stars ascending the Asymptotic Giant Branch (AGB) are known to form copious amount of dust at the base of their wind, as shown by the strong infrared (IR) excess they emit and the large mass-loss rate they develop. Except for the band seen at $\sim$ 11.3 microns in emission or absorption in most carbon stars and ascribed to silicon carbide (SiC) grains, the IR thermal emission of dust is usually featureless and can be well reproduced by a distribution of amorphous carbon (AC) particles (Groenewegen 1997). A new source of information on stardust comes from the study of pre-solar grains extracted from meteorites (Bernatowicz et al. 1991, 1996). These studies bring evidence for various sequences of nucleation and condensation in carbon stars, in particular the presence of transition metal carbides such as titanium or zirconium carbides embedded in the aromatic lattice, and the absence of silicon carbide clusters in graphite pre-solar grains.

Many studies have investigated the condensation of carbon grains in stationary or time-dependent winds (Gail & Sedlmayr 1988; Fleischer et al. 1994; Höfner et al. 1995) in order to explain ways of generating outflows and predict mass-loss rates. However, the nucleation and condensation of dust grains, which involve a full description of the chemistry describing the passage of the gas to the solid phase is not considered. Furthermore, the dust formation zone in AGB stars is characterised by the passage of strong periodic shocks induced by stellar pulsation implying physical conditions far from equilibrium. Therefore, it is important to bear in mind that studies of dust formation should consider a non-equibrium chemistry for the description of dust nucleation.

In this review, we discuss the various types of chemistry involved in the formation of AC grains, transition metal carbides and SiC clusters in the laboratory and the possible relevance of these mechanisms to stellar environments. We also present preliminary results on the formation of polycyclic aromatic hydrocarbons (PAHs) and molecular precursors to TiC clusters in the inner shocked wind of the carbon star IRC +10216 and draw potential explanations to the condensation sequences derived from meteoretical studies.

2. The chemistry of carbon grain formation in the laboratory

2.1. Amorphous Carbon

As already mentioned and investigated by several authors (Frenklach & Feigelson 1989; Frenklach et al. 1989; Cherchneff et al. 1992; Cadwell et al. 1994; Cherchneff et al. 1998), the chemical composition of the photospheres of carbon stars and some of the physical parameters characterising the inner stellar winds are very similar to conditions encountered in the combustion and/or the infrared pyrolysis of hydrocarbons in the laboratory. Indeed, acetylene, C_2H_2, is the dominant carbon-bearing molecule after carbon monoxide in the photospheres of carbon stars which have typical temperatures are 2500K. Such temperatures are also found in flames rich in acetylene which are efficient sources of amorphous carbon particles in the form of soot. Therefore, there exists a natural link between the two environments which permits the use of combustion chemistry to model the formation of AC grains in stars.

The major steps of soot particle nucleation and condensation are illustrated in Fig. 1. The first step referred to as *nucleation* consists of forming large aromatic molecules from the acetylene-rich gas phase. The second step is the *condensation* stage where a three-dimensional, small solid particle forms from large aromatic precursors. Formation of the first aromatic ring involves the isomerisation of acetylene and the formation of larger hydrocarbons via ter-molecular and bi-molecular reactions which may have relatively high activation energies. However, these are overcome due to the high gas temperature in the flame, shock-tube or stellar environment. The nucleation step includes the closure of the first aromatic ring (≡ benzene, C_6H_6) and the subsequent growth of PAHs. Once benzene or phenyl are formed, the growth to PAHs can be simply represented by a set of chemical reactions which consists of (Frenklach & Feigelson 1989; Cherchneff et al. 1992; Frenklach & Feigelson 1997):

1. Acetylene addition to aromatic radical

2. Atomic hydrogen extraction from the aromatic ring to form a radical site

3. Second acetylene addition and closure of the ring.

Because of the high gas densities, the reverse reactions for all processes have to be considered. This series of reactions occurs for a certain temperature window determined by the local gas number density and ring growth is favoured

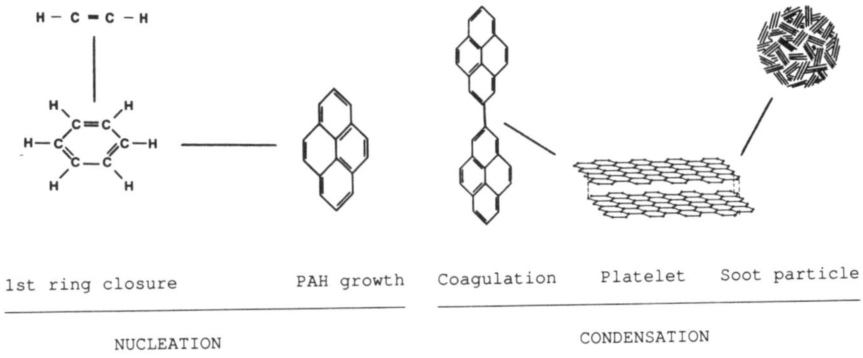

Figure 1. Dominant pathways to PAH growth (adapted from Cherchneff et al. 1992).

when Reaction 2 is at equilibrium but the C_2H_2 addition in Reaction 3 becomes irreversible. In flames, this temperature window is 900 – 1100 K, but in astrophysical environments, in particular in the inner shocked regions of carbon star winds, this window can vary greatly as shown in Sect. 3.

As for condensation, the various stages comprise a series of chemical processes which permit the growth of small solid condensates. Whereas nucleation treats the conversion of linear gas-phase species into planar, large aromatic molecules, condensation involves the growth of planar species into three-dimensional solid grains. Condensation is initiated with the formation of PAH dimers where the aromatics are linked by van der Waals forces. Condensation products form simultaneously and these differ from the dimers by the formation of a six-membered or five-membered ring as a link between the two aromatic rings. For example, both dimers and condensation products of pyrene are observed in high-temperature pyrolysis of pyrene and the growth to soot particles was suggested to take place through aromatic polymerisation and cyclodehydrogenation rather than through acetylene addition (Mukherjee et al. 1994). The growth then proceeds through coagulation of condensation products of PAH molecules, deposition of carbon via acetylene surface reactions and surface condensation of free PAH molecules on the grains.

2.2. Metal carbides

The reactions of early transition metals with hydrocarbons and the resulting formation of various metal-carbon clusters have been extensively studied in laser-induced plasma reactor experiments (Guo et al. 1992; Wei et al. 1992). The metal usually considered were titanium, vanadium and zirconium and the metal-carbon clusters observed spanned many molecular weights in the mass spectrometer. An important discovery was the identification of a dominant

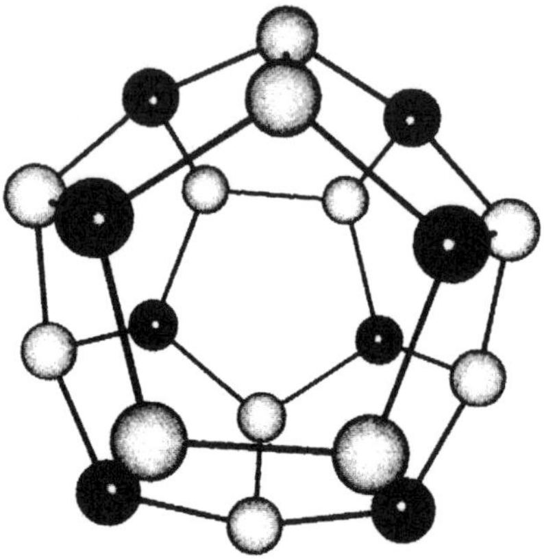

Figure 2. A possible T_h point group structure for Ti_8C_{12} (from Guo et al. 1992).

cluster among the cluster distribution peaking at 528 atomic mass units and due the metallo-carbohydrene M_8C_{12} (or met-car), where M is the transition metal under consideration. Due to its 'super-magic' nature and the fact that the cluster was always observed dominant whatever the hydrocarbon reactant used (methane, CH_4, ethylene, C_2H_4, and acetylene), it was suggested that the species had a cage structure to explain its stability and ubiquity (Guo et al. 1992; Reddy & Khanna 1993; Wei & Castleman 1994; Cartier et al. 1996). An idealised pentagonal dedocadedron structure for Ti_8C_{12} is illustrated in Fig. 2 and shows that each titanium atom can bond to three carbon atoms through Ti$-$C σ-bonds while each carbon atom is bonded to two titanium atoms and its adjacent carbon through a C$-$C σ-bond.

The chemical mechanisms leading to the formation of 'met-car' clusters have not been identified yet but there is evidence from the cluster size distribution for structural growth through addition of the MC_2 unit (Guo et al. 1992) and closure at M_8C_{12}. The chemistry of met-car molecular precursors is still unexplored but studies on organometallics and the gas-phase reactions of transition metals with various hydrocarbons have shown that bare transition metal cations (Ti^+, Zr^+, V^+, etc.) are very reactive chemicals that can break C$-$H and C$-$C bonds in alkanes and alkenes (Byrd et al. 1982; Tolbert & Beauchamp 1986). A typical reaction of Ti^+ with C_2H_4 gives

$$Ti^+ + C_2H_4 \rightarrow TiC_2H_2 + H_2 \qquad (1)$$

where two hydrogen atoms have been eliminated to form H_2.

The reactions of neutral transition metals with hydrocarbons have been investigated with less extent. Ritter & Weisshaar (1990) and Ritter et al. (1992)

have studied the kinetics of neutral transition metals in the gas phase and conclude that they are less reactive than their cation counterparts. In particular, they derive a low reaction rate for the bimolecular reaction between neutral titanium (and vanadium) with C_2H_4. The reaction of neutral titanium with acetylene remains unexplored at present time, although it is known from the study of Guo et al. (1992) that Ti^+ (and possibly neutral Ti) must react very efficiently with acetylene to explain the very large abundance of Ti_8C_{12} observed.

2.3. Silicon carbide

Silicon carbide powders have been extensively studied in the laboratory as a result of the pyrolysis of silane, SiH_4, and various hydrocarbons (Harano et al. 1990; Lirhmann & Cauchetier 1994). Although the end products and the powder are usually well characterized, the mechanisms explaining the formation of SiC cluster precursors and the condensation processes are still unclear. Fantoni et al. (1991) have investigated the by-products of SiC cluster formation and have identified with the help of infrared spectroscopy the presence of C_2 and SiC_2 as intermediates in the nucleation and condensation processes. Harano et al. (1990) in their laser-induced plasma experiment have brought evidence for the presence of several Si and SiC clusters in the final cluster distribution. In particular they found that SiC_n, Si_2C_n (with $n > 1$) and pure carbon clusters were dominant as the percentage of hydrocarbons was raised in the buffer gas. There is no 'super-magic' molecule such as met-cars in the cluster distribution indicating that no specific cage structure can act as condensation nucleus in the grain growth processes. This result may hint to the fact that SiC and metal carbides have very different fates in stellar winds, the former acting as a separate dust population whereas the latter is trapped in the aromatic lattice of the forming grain.

3. Condensation sequences and preliminary models

An important implication of the studies of meteorites and the extraction of isotopically anomalous pre-solar graphite spherules, which origin has been ascribed to carbon stars on the AGB, is the possibility to gain insight on dust condensation processes and to test theoretical models of grain formation. For example, the study of graphite spherules by Bernatowicz et al. (1996) points to various possible dust condensation sequences in carbon stars, as illustrated in Table 1.

Transmission electron microscopy and electron diffraction studies conducted by Bernatowicz et al. have provided evidence for the presence of three types of configuration for the spherules: (1) a core of small graphene sheets (size ~ 40 Å) and small PAHs (size < 10 Å) surrounded by a well graphitized carbon rim. About 80 % of the spherules are of this type; (2) a core of metal carbide nanocrystals surrounded by the same rim as in (1); (3) metal carbide nanocrystals randomly distributed in the graphite spherule. Such information can help understanding the condensation of AC grains in carbon stars, in particular, it strengthens the fact that silicon carbide clusters are not observed to act as condensation nuclei to AC particles as it has been proposed by Frenklach et al. (1989) and Cadwell et al. (1994).

Table 1. Condensation sequences observed in pre–solar graphite spherules (from Cherchneff 1998).

	Core	Mantle	Core size (Å)	Graphite rim size (μm)
1	Graphene + PAHs	Graphitic concentric sheets (GCS)	?	0.1 - 0.4
2	TiC, ZrC, MoC	GCSs	50-300	0.1 - 0.4
3	none	metal carbides spread in GCSs	–	0.1 - 0.4

In order to gain insight on condensation sequences in a mixture of dust grains, we have investigated the nucleation and early condensation of carbon grains using a formalism based on combustion chemistry simultaneously to the formation of molecular precursors to titanium carbide clusters, in particular the molecule TiC_2. We apply the chemistry to the dust formation region close to the stellar photosphere of the archetypical carbon star IRC +10216 and use the model for the inner, periodically shocked wind as described in Willacy & Cherchneff (1998) and Cherchneff et al. (1998). For AC grain precursors, we consider the formation of free PAHs up to coronene ($C_{24}H_{12}$) and their dimers. We do not include destruction processes for the latter as we assume that the dimers convert quickly to condensation products in which the two aromatic molecules are bridged by a pentagonal or an aromatic ring. These products are very stable and are observed to form at the expense of PAH dimers in aromatic pyrolysis experiments at high temperature (Mukherjee et al. 1994).

We define a PAH (dimer) yield as the ratio between the number of carbon atoms locked in PAHs (dimers) divided by the number of carbon atoms initially present in hydrocarbons (Cherchneff et al. 1992). The yields for benzene, PAHs and aromatic dimers are presented in Fig. 3 as a function of stellar radius. The yields at each radius correspond to the values calculated in the periodic shocks and in the gas excursions induced by the shocks over two pulsation periods (see Cherchneff et al. 1998, for more details). From inspection of Fig. 3, several conclusions can be drawn. First, the non-equilibrium chemistry favours the formation of aromatic species close to the photosphere and the shock formation radius (equal to 1.2 $R_\star$ in this model). Benzene starts to form as early as 1.4 $R_\star$ and represents the dominant aromatic hydrocarbon until 1.7 $R_\star$. At this radius, the conversion of single-ring aromatics to PAHs starts as the growth due to acetylene addition becomes irreversible, but it is important to notice that the temperature at which this growth starts is $\sim$ 1 700 K for the physical parameters of the gas excursion considered (typical gas number density of $\sim 10^{13} - 10^{14}$ cm^{-3}). This temperature is much higher than the temperature window derived by Frenklach & Feigelson (1989) and Cherchneff et al. (1992) as these previous studies considered stationary expanding outflows and lower gas number densities. PAH dimers form in large amount along with the growth of their aromatic progenitors and yields between 10^{-5} and 10^{-4} are obtained.

The conversion of atomic Ti to TiC_2 is illustrated in Fig. 4 along with the PAH formation yield. We have modelled the formation of TiC_2 assuming that

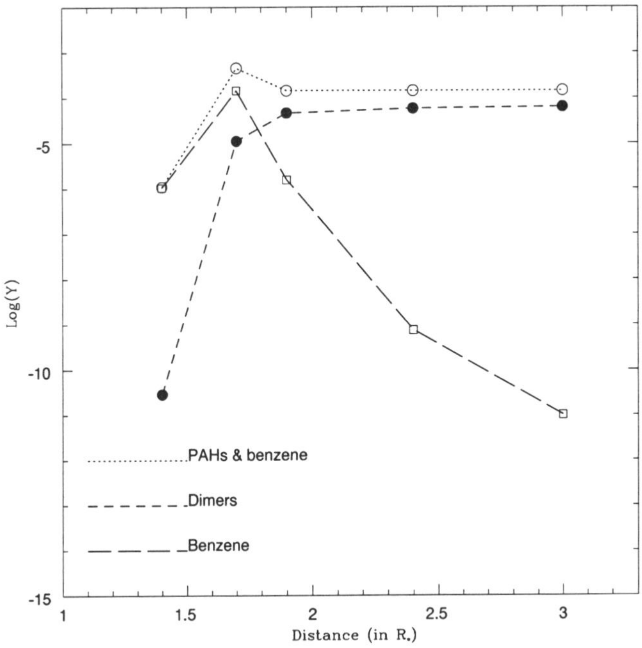

Figure 3. Benzene, PAH and PAH dimer formation yields as a function of stellar radius in the inner wind of IRC +10216.

atomic titanium reacts with acetylene and using the dehydrogenation property of neutral Ti. Considering the reaction

$$Ti + C_2H_2 \rightarrow TiC_2 + H_2 \tag{2}$$

we use a very low rate $k = 10^{-16}$ cm^3 s^{-1} as suggested by Ritter et al. (1992). It appears that all the phostospheric Ti can be converted very rapidly to TiC_2 in the first shock and its excursion and that the process is irreversible once this transformation takes place. This implies that precursors to TiC clusters can form at very high temperatures, and certainly before PAH growth.

4. Discussion

The formation of AC grains in carbon stars is far from beeing understood yet but a description of the various steps involved in the passage from gas to solid phase relying on non-equilibrium chemistry appears to give promising results. A first important point is that PAHs and their products form at higher gas temperatures and densities than previously thought resulting in more efficient aromatic growth. Further growth processes such as acetylene deposition and

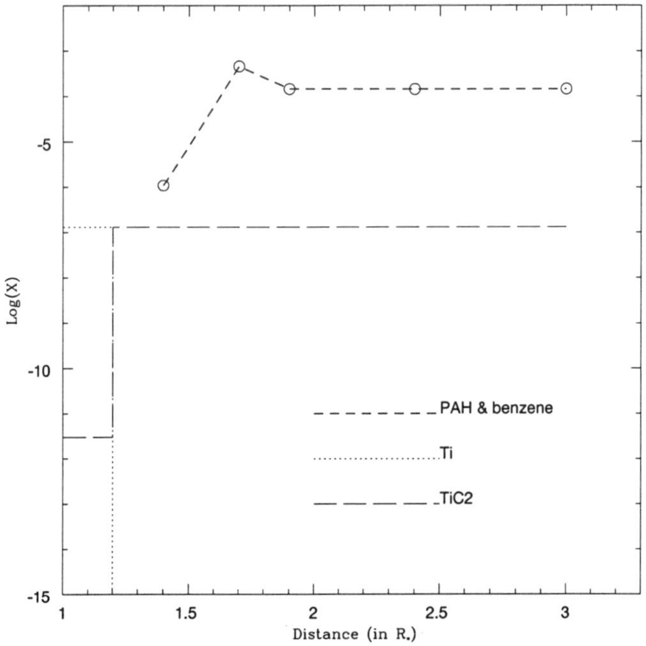

Figure 4. The conversion of atomic Titanium to TiC_2 along with the PAH formation yield as a function of stellar radius in the inner wind of IRC +10216.

condensation product coalescence must be investigated for the new conditions encountered in the inner wind.

Although there is evidence for formation of TiC precursors at high temperatures and certainly before the growth of aromatics, the conversion of TiC_2 into met-car molecules and larger clusters remains an open question. If this conversion is fast enough and occurs efficiently at high temperatures, then TiC clusters are expected to form before the aromatics resulting in a core-mantle structure for carbon grains as seen in some pre-solar grains. This scenario could occur in stars in which *strong shocks* develop or characterised by high effective temperature. In case the conversion to larger TiC clusters is slow, their formation would be delayed to the regions of aromatic growth, resulting in a mixed AC grain where TiC cristals are spread in the lattice. This scenario could also happen for the *slower shocks* or cooler stars. As for the pre-solar grains containing *only* an AC/PAH core, can they correspond to inhibited TiC cluster formation or to a case where Ti_8C_{12} clusters form but do not grow, being below the resolution limit to be detected but still acting as small condensation nuclei to AC dust grains ?

The case of silicon carbide is even more puzzling and much less well defined as there exist only a few measured rates for reactions involving silicon compounds. The fact that no 'super-magic' cluster as the met-car molecule has been detected as bi-products of SiC powder formation could restrict the efficiency of SiC formation in stellar environments. If the chemistry of SiC grain growth is slower than that of transition metal carbides and aromatics, the SiC dust formation will certainly be decoupled from other dust condensation mechanisms, resulting in a separate grain population. This question can only be answered by the combined efforts of the various scientific communities working on carbon dust in the laboratory and in space.

References

Bernatowicz T.J., Amari S., Zinner E.K., Lewis R.S., 1991, ApJ 373, L73

Bernatowicz T.J., Cowsik R., Gibbons P.C., et al., 1996, ApJ 472, 760

Byrd G.D., Burnier R.C., Freiser B.S., 1982, J. Am. Chem. Soc. 104, 3565

Cadwell B.J., Wang H., Feigelson E.D., Frenklach M., 1994, ApJ 429, 285

Cartier S.F., May B.D., Castleman A.W., 1996, J. Phys. Chem. 100, 8175

Cherchneff I., 1998, in "Abundance profiles: Diagnostic tools for Galaxy history", San Francisco, ASP Conf. Ser. 147, p. 179

Cherchneff I., Cau P., Tielens A.G.G.M., 1998, A&A, submitted

Cherchneff I., Barker J.R., Tielens A.G.G.M., 1992, ApJ 401, 269

Fantoni R., Bijen F., Djurić N., Piccirillo S., 1991, Appl. Phys. B 52, 176

Fleischer A.J., Gauger A., Sedlmayr E., 1994, A&A 297, 543

Frenklach M., Carmer C.S., Feigelson E.D., 1989, Nature 339, 196

Frenklach M., Feigelson E.D., 1989, ApJ 341, 372

Frenklach M., Feigelson E.D., 1997, in "From Stardust to Planetesimals", Y. Pendleton (ed.), ASP Conference Series 122, p. 107

Gail H.-P., Sedlmayr E., 1988, A&A 206, 153

Groenewegen M.A.T., 1997, A&A 317, 503

Guo B.C., Kerns K.P., Castleman A.W., 1992, Science 255, 1411

Harano A., Kinoshita J., Koda, Seiichiro, 1990, Chem. Phys. Letters 172, 219

Höfner S., Feuchtinger M.U., Dorfi E.A., 1995, A&A 297, 815

Lihrmann J.-M., Cauchetier M., 1994, J. European Ceramic Soc. 13, 41

Mukherjee J., Sarofim A., Longwell J.P., 1994, Combustion and Flame 96, 191

Reddy B.V., Khanna S.N., 1993, Chem. Phys. Letters 209, 104

Ritter D., Weissharr J.C., 1990, J. Am. Chem. Soc. 112, 6425

Ritter D., Carroll J.J., Weissharr J.C., 1992, J. Phys. Chem. 96, 10636

Tolbert M.A., Beauchamp J.L., 1986, J. Am. Chem. Soc. 108, 7509

Wei S., Guo B.C., Purnell J., Buzza S., Castleman A.W., 1992, J. Phys. Chem. 96, 4166

Wei S., Castleman A.W., 1994, Chem. Phys. Letters 227, 305

Willacy K., Cherchneff I., 1998, A&A 330, 676

Hans Olofsson and Issei Yamamura

Asymptotic Giant Branch Stars
T. Le Bertre, A. Lèbre and C. Waelkens, eds.
I.A.U. Symposium 191, 1999

Infrared appearance of dust forming LPVs

Jan Martin Winters

Institut für Astronomie & Astrophysik, Technische Universität Berlin, Sekr. PN 8-1, Hardenbergstr. 36, D-10623 Berlin, Germany

Thibaut Le Bertre

DEMIRM, Observatoire de Paris, 61, av. de l'Observatoire, F-75014 Paris, France

John J. Keady

Theoretical Division, Los Alamos National Laboratory, T4, MS-B212, Los Alamos, NM 87545, USA

Abstract. The consistent treatment of the physical problem describing a dust forming circumstellar shell yields, besides the time variation of the hydrodynamic and thermodynamic shell structure, also the temporal evolution of the grain–size distribution function. We discuss the influence of the dynamical behavior of corresponding dust shell models on the resulting infrared light curves, brightness distributions, and CO infrared line profiles. Some remarks on the grain–size distribution are also given.

1. Introduction

Only in recent years it has become possible to build consistent physical models of the atmospheres and the dust forming circumstellar shells around pulsating AGB stars (see Höfner, this volume and Fleischer et al., this volume). Corresponding model calculations reveal a complex, non-linear dynamical structure of the circumstellar shells, which is dominated by the strong mutual coupling between the emerging dust component and the physical conditions prevailing in the dust forming circumstellar environment. In this contribution we report on several pronounced observable implications resulting from such model calculations for carbon-rich circumstellar shells. We argue that these model calculations are already capable of a physical explanation of the typical characteristics observed in dust forming carbon–rich long–period variable stars.

2. Typical radial structure of a dust shell model

The radial structure of a carbon-rich dust shell model is depicted on the l.h.s. of Fig. 1 at a fixed instant of time, indicating the very close coupling among the dust distribution (represented by the degree of condensation f_c), the hydrodynamic, and the thermodynamic structure of the shell, which is only revealed by

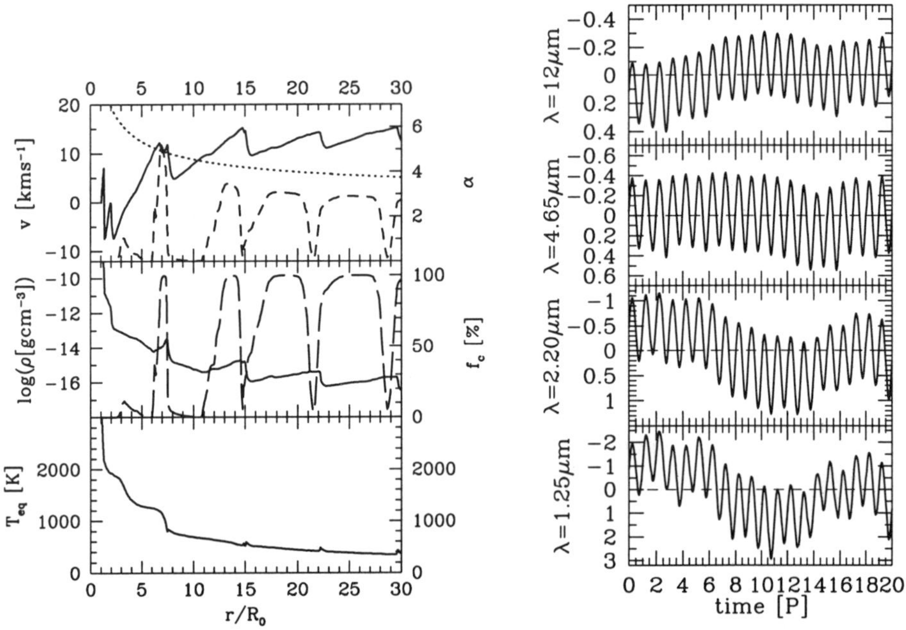

Figure 1. **l.h.s.**: Radial structure of the dust shell model (see text) at $t = 6.25\,P$ and **r.h.s**: light curves of this model

the consistent treatment of the complete physical problem. The parameters of this model are: stellar mass $M_* = 1\,M_\odot$, stellar luminosity $L_* = 10^4\,L_\odot$, stellar temperature $T_* = 2\,600\,\mathrm{K}$, carbon–to–oxygen ratio $\epsilon_\mathrm{C}/\epsilon_\mathrm{O} = 1.30$ (the other elements are assumed to have solar abundances), pulsation period $P = 640\,\mathrm{days}$, and velocity amplitude at the inner boundary $\Delta u_p = 5\,\mathrm{km\,s^{-1}}$. This model yields an outflow velocity of $v_\infty = 13.2\,\mathrm{km\,s^{-1}}$ and a mass loss rate of $\dot{M} = 3.5\,10^{-5}\,M_\odot\,\mathrm{yr^{-1}}$. Due to the self–regulating dust formation process, which is triggered by the pulsation of the star (providing high densities at sufficiently low temperatures), a layered structure of the dust shell results. Radiative acceleration ($\alpha = g^\mathrm{rad}/g^\mathrm{grav}$) of these dust layers produces regions of enhanced density and, caused by the backwarming effect, a step-like temperature stratification results. For details see Fleischer et al. (1992). Moreover, the process of dust formation introduces a new time scale into the system, which leads to the formation of new dust layers over time intervals, which are usually longer than the pulsation period of the star (cf. Fleischer et al. 1995, Höfner et al. 1995). Observable consequences of this *multiperiodic* formation of discrete dust layers are presented in the following sections.

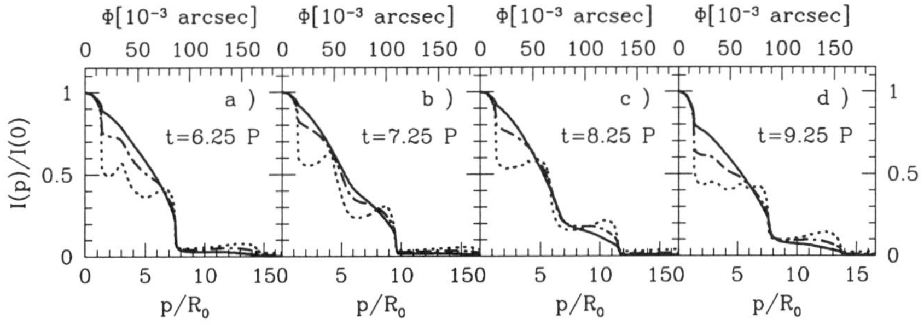

Figure 2. Brightness profiles of the dust shell model at $\lambda = 2.2\,\mu$m (solid), $3.8\,\mu$m (dash-dot), and $12\,\mu$m (dotted) at different instants of time corresponding to the light curves shown on the r.h.s. of Fig. 1

3. Observable properties

3.1. Light curves

The r.h.s. diagram in Fig. 1 shows the light curves of the model which produces a new dust layer on a time scale of about 5 pulsation cycles. More strictly, the dust optical depth increases during the time interval between $t \approx 6\,$P and $t \approx 11\,$P when a new dust layer is being formed, it reaches a maximum value of $\tau^{\rm d}_{1\mu} \approx 28$ around $t = 11\,$P and then decreases again, since the newly formed layer subsequently expands as it is driven outwards by radiation pressure. Hence, the material becomes diluted, i.e. the density and, thus, the optical depth decreases again ($\tau^{\rm d}_{1\mu} \approx 23$ around $t = 16\,$P). The increase of the dust optical depth results in an increasing absorption of the near infrared radiation from the star and, therefore, to a decreasing mean magnitude of the corresponding near infrared light curves. The absorbed photon energy is very efficiently redistributed over the internal degrees of freedom of the dust grains, and is re-radiated at mid- and far infrared wavelengths according to the internal temperature of the grains, leading to the simultaneous increase of the mean magnitude of the mid- and far infrared light curves. By this mechanism the flux observed at near-infrared wavelengths is dominated by absorption of the short wavelength stellar radiation field in the innermost shell region, whereas the mid- and far-infrared flux is mainly produced by thermal emission emanating from a rather extended radial region of the dust shell. This *multiperiodic* behavior and the inversion effect seen in the light curves is a direct consequence of the *multiperiodic* formation of discrete dust layers. For more details on the light curves see Winters et al. (1994).

3.2. Brightness profiles of the dust shell

In principle, the formation of new dust layers might be observed directly from the spatial intensity profiles emerging from the circumstellar shell. An example of synthetic profiles is shown in Fig. 2 in a time sequence, where the dynamics of the inner shell region is most obvious at mid-infrared wavelengths (e.g. at $\lambda = 12\,\mu$m). A detailed investigation of the synthetic brightness profiles is given

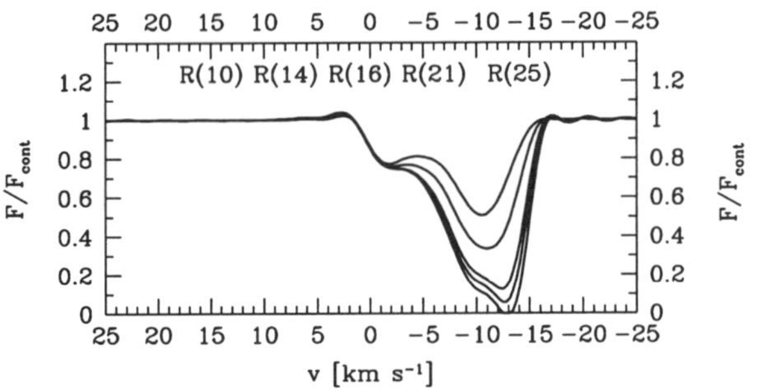

Figure 3. Synthetic CO (v = 2–0) overtone line profiles of the dust shell model at $t = 11.25\,P$

in Winters et al. (1995). However, a very high spatial resolution at the respective infrared wavelengths is required for this kind of direct observations. This can be seen from the upper abscissa in Fig. 2, which gives the angular scale for an assumed distance of 200 pc. The required spatial resolution in fact seems to be within reach in the very near future (see e.g. the contributions by Monnier, Tuthill et al. and Weigelt et al., in this volume, and Weigelt et al. 1998).

3.3. Infrared CO molecular line profiles

A very different, and simultaneously very powerful diagnostic tool for probing the structure and dynamical behavior of the inner dust shell region is provided by high-resolution molecular ro-vibration line profiles. The best suited species for this purpose is the CO molecule since i) it is already present in the stellar photosphere because of its high binding energy, ii) it is very abundant both in oxygen-rich and in carbon-rich environments due to its element composition, and iii) once the CO molecule is formed, it remains in the gas phase (at least in the inner part of the dust shell). It is neither destroyed by chemical reactions, nor is it depleted from the circumstellar gas by condensation.

The most useful diagnostic information is contained in the profiles of the CO first-overtone lines, which form close to the star and, with increasing rotational excitation, are excellent probes for the acceleration region of the wind. These lines show a pronounced multicomponent structure and a clear shift of the line core position towards lower outflow velocities with increasing rotational excitation. Fig. 3 displays calculated CO (v = 2–0) overtone line profiles for different rotational transitions. The strong $10\,\mathrm{km\,s^{-1}}$ absorption core arises from the shell region, where a new dust layer has just been formed, leading to a sharp increase of the density by sweeping up the gas in the corresponding dust induced shock (cf. Fig. 1). This type of line profiles has been observed e.g. for IRC +10216 (see Sada 1993). Since the formation of a new dust layer usually takes a time longer than the pulsation period of the star, the line profiles should display a temporal evolution on the corresponding time scale. In fact, such observations have been performed, again for IRC +10216, and show the evolution of a secondary feature (similar to the one seen at $2\,\mathrm{km\,s^{-1}}$ in Fig. 3), which considerably increases in strength and shifts to higher outflow velocities over a time interval of about 10 yr corresponding to about 6 pulsation cycles of the star. A direct

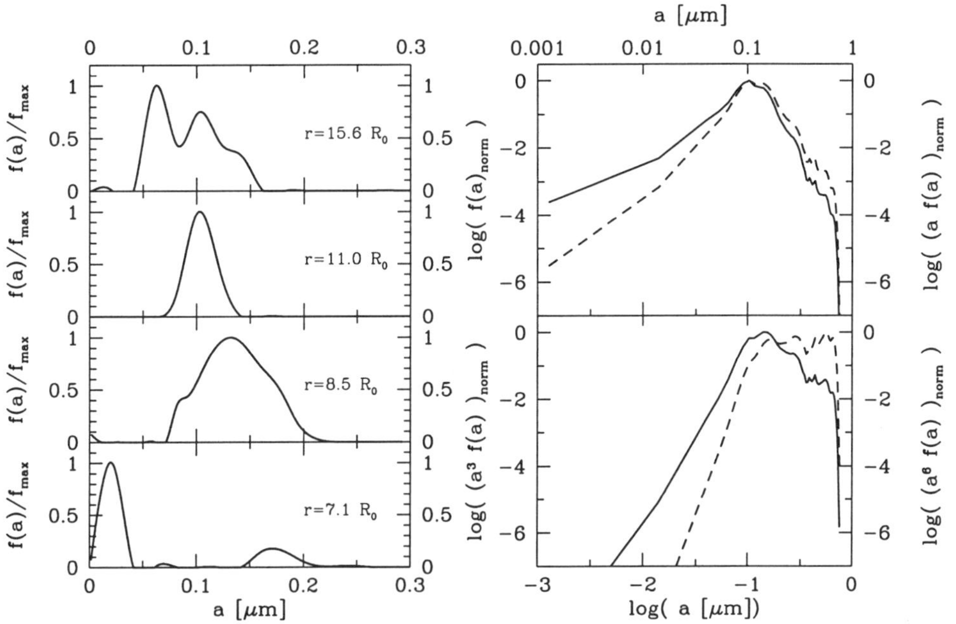

Figure 4. Grain–size distribution functions in different fluid elements at the same instant of time (l.h.s.) and logarithmic plot of the average distribution function, weighted by different powers of the particle size (r.h.s., solid lines refer to the l.h.s. ordinate)

comparison between this observation and a corresponding model calculation is given in Winters (1998).

4. Remarks on the grain–size distribution function

The observable results presented in the preceding sections rely on the assumption that the dust grains are small compared to the wavelength of the radiation field interacting with these particles. However, the consistent treatment of the dust formation process allows to calculate the size distribution function of the emerging dust component (cf. Dominik et al. 1989, Gauger et al. 1990). Since each fluid element moving through the circumstellar shell has its own history, different distribution functions evolve in different fluid elements (see l.h.s. of Fig. 4). In particular, multiple–peaked size distributions are formed in those elements, which experience favorable nucleation conditions several times during their evolution.

An average distribution function expected from the calculation to be injected into the ISM is shown in the r.h.s. of Fig. 4. Although this size distribution is sharply peaked around a dominant grain size of $a \approx 0.1\mu\text{m}$, considerably larger particles are also formed in a quantity sufficient to influence the transport coefficients of the dust component (see e.g. the quantity $a^6 f(a)$ which is relevant for the scattering coefficient, and the discussion in Winters et al. 1997). In fact, the large particle tail of this average distribution function can roughly be approximated by a power–law having a spectral index of about −4.5. Calculating the extinction and scattering coefficients from this average distribution function by applying Mie theory for arbitrary particle sizes yields e.g. a scattering albedo of about 30% at a wavelength of $\lambda = 2.2\ \mu$m (10% at 5 μm, and 1% at 11 μm, respectively). The next step to improve the model calculations, therefore will be the consistent consideration of the size distribution function in calculating the transport coefficients of the dust component.

5. Conclusion

By the consistent treatment of the physical problem describing a dust forming circumstellar shell around LPVs, we find a strong interplay among the involved processes, which in particular leads to a discrete internal structure of the dust shell, evolving on time scales usually longer than the pulsation period of the star. These structures influence the observable appearance of these models, in particular the light curves, the brightness profiles and the CO infrared line profiles. We find, that the formation of discrete dust layers on a time scale different from the pulsation period of the star leads to rather distinct observable (and observed!) characteristics in a natural way, i.e. without introducing any ad-hoc assumptions in order to fit those observations.

Acknowledgments. This work has been supported by the BMBF (grant 05 3BT13A 6). Research at Los Alamos National Laboratory was performed under the auspices of the U.S. Department of Energy.

References

Dominik C., Gail H.-P., Sedlmayr E., 1989, A&A 223, 227

Fleischer A.J., Gauger A., Sedlmayr E., 1992, A&A 266, 321

Fleischer A.J., Gauger A., Sedlmayr E., 1995, A&A 297, 543

Gauger A., Gail H.-P., Sedlmayr E., 1990, A&A 235, 345

Höfner S., Feuchtinger M.U., Dorfi E.A., 1995, A&A 297, 815

Sada P.A.V., 1993, Temporal variations in the circumstellar shell IRC +10216, PhD thesis, New Mexico State University, Las Cruces, New Mexico

Weigelt G., Balega Y., Blöcker T., Fleischer A.J., Osterbart R., Winters J.M., 1998, A&A 333, L51

Winters J.M., 1998, Ap&SS 255, 257

Winters J.M., Fleischer A.J., Gauger A., Sedlmayr E., 1994, A&A 290, 623

Winters J.M., Fleischer A.J., Gauger A., Sedlmayr E., 1995, A&A 302, 483

Winters J.M., Fleischer A.J., Le Bertre T., Sedlmayr E., 1997, A&A 326, 305

Asymptotic Giant Branch Stars
T. Le Bertre, A. Lèbre and C. Waelkens, eds.
I.A.U. Symposium 191, 1999

The 14 μm Band of Carbon Stars[1]

I. Yamamura, T. de Jong[2], L.B.F.M. Waters[3], J. Cami[3]

Astronomical Institute 'Anton Pannekoek', University of Amsterdam, Kruislaan 403, 1098 SJ, The Netherlands

K. Justtanont

Stockholm Observatory, 13336 Saltsjöbaden, Sweden

Abstract. We have studied the absorption bands around 14 μm in the spectra of 11 carbon stars with mass-loss rates ranging from 10^{-8} to 10^{-4} $M_\odot\,yr^{-1}$, based on data obtained with the Short Wavelength Spectrometer (SWS) on board the Infrared Space Observatory (ISO). All stars clearly show a C_2H_2 absorption band peaking at 13.7 μm, while the contribution from HCN molecules is small in this wavelength region. A simple plane-parallel LTE model with two layers at different temperatures is used to derive the C_2H_2 abundances in the outer photosphere and in the circumstellar envelope. We find that (1) the column density of the hot-layer, placed at about $3R_\star$ with a temperature of 1400 K is roughly the same for all stars regardless of the mass-loss rate, and (2) the contribution of cool molecules in the circumstellar envelope increases with the dust mass-loss rate, (3) the abundance of C_2H_2 in the two layers is about the same, i.e. no obvious depletion of C_2H_2 molecules seems to occur in the circumstellar envelope.

1. Introduction

Acetylene (C_2H_2) is the most abundant molecule in the atmosphere of carbon stars after CO and HCN (Tsuji 1964). The molecule probably plays an essential role in the atmosphere of these stars, not only in the energy budget through their ro-vibrational transitions but also as a seed molecule for dust formation. C_2H_2 has been little studied, however, since the linear and symmetric structure of C_2H_2 prevents us to observe the molecule directly at radio frequencies. The abundance of the molecule has been estimated from its photodissociation products such as C_2H (Keady & Hinkle 1988; Fuente et al. 1998). The ro-vibrational

[1]Based on observations with ISO, an ESA project with instruments funded by ESA Member States (especially the PI countries: France, Germany, the Netherlands and the United Kingdom) with the participation of ISAS and NASA. The SWS is a joint project of SRON and MPE.

[2]also SRON-Utrecht, Sorbonnelaan 2, 3584 CA, Utrecht, The Netherlands

[3]also SRON-Groningen, Landleven 12, 9700 AV, Groningen, The Netherlands

transitions in the infrared are also not easy to measure. The 3.05 μm absorption band is accessible from the ground, but very high resolution is required in order to avoid confusion with other molecules such as HCN. The band centers of other major transitions at 7.5 and 13.7 μm are inaccessible from the ground, and have only been partially observed (Keady & Ridgway 1993).

The ISO/SWS (Kessler et al. 1996; de Graauw et al. 1996) provides infrared spectra in a broad wavelength range with moderately high resolution. As a part of the ISO guaranteed time program AGBSTAR, we observed 11 carbon stars with various mass-loss rates ranging from 10^{-8} and 10^{-4} $M_{\odot}\,yr^{-1}$. Based on the infrared spectra of these stars we concluded that not only the energy distribution and the broad dust features but also several narrow molecular features drastically change with the mass-loss rate (Yamamura et al. 1998). In this article, we focus on the C_2H_2 band at 13.7 μm, and discuss the C_2H_2 distribution around the stars. This band has a sharp Q-branch, which makes it easy to identify and distinguish from other molecular features.

2. Observations and results

The observations were carried out using the SWS AOT01 full-grating scan mode. All stars were observed with a resolution of about 300, except IRC +10216 which was observed at a resolution of about 1 000. The data were reduced by the SWS Interactive Analysis package developed by the SWS instrument dedicated team, using the most recent set of calibration parameters (as of April 1998).

The carbon star sample is divided into four groups (II–V) based on their near-infrared color temperature, $T_{\rm NIR}$, which is supposed to be a good indicator of mass-loss rate (Groenewegen et al. 1992). The spectra of our carbon stars between 13 and 15 μm are shown in Fig. 1, labeled with their name and group number. The spectra are normalized to the *local continuum.* The spectra are also shifted slightly with respect to the model spectra in order to correct for the ambiguity in the absolute continuum level. The uncertainty in the continuum determination will be discussed in Sect. 4.1.

The C_2H_2 band at 13.7 μm is clearly detected in *absorption* in all stars, regardless of the mass-loss rate. The band is actually a mixture of many Q-branches including the ν_5 fundamental transition. The band is broader in stars with smaller mass-loss rate, indicating that the absorption is due to hot gas near the photosphere. On the other hand, the HCN band at 14.0 μm is hardly recognized in any of the spectra in Fig. 1. Indeed, the feature is much weaker than expected from the far-infrared spectrum of IRC +10216 (Cernicharo et al. 1996) or from the 3.6 μm feature in the same SWS spectra. This may be explained in terms of the distribution of the molecule as well as the excitation mechanism, but it is a subject for future studies.

The spectra are in contrast with those of oxygen-rich stars, which exhibit CO_2 emission bands in the same wavelength range (Justtanont et al. 1998; Cami et al. 1997). This indicates that there is always a continuum source behind the C_2H_2 molecule.

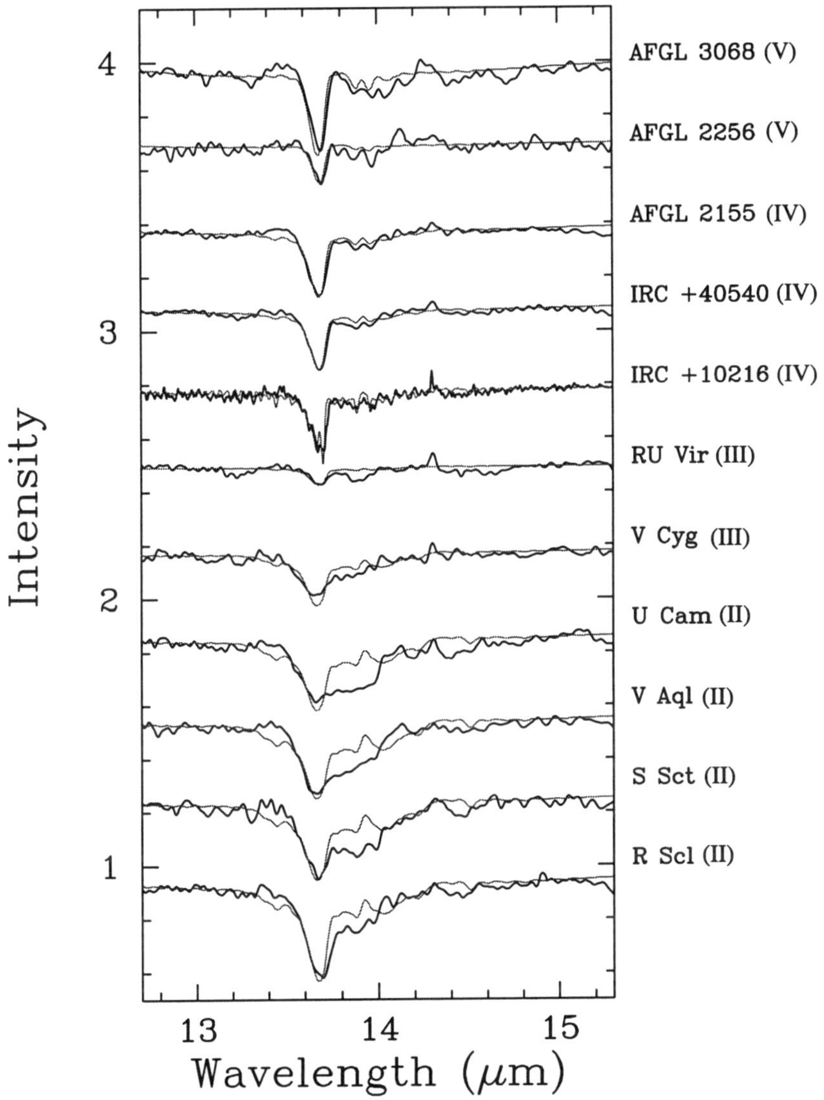

Figure 1. The 14 μm band of carbon stars (*solid line*) is shown with the model spectra (*gray line*). The spectra are ordered by decreasing near-infrared color temperature ($T_{\rm NIR}$) from bottom to top. The group number of each object is indicated after the object name. The spectra are normalized to the *local continuum*. In addition, they are shifted by at most 7 % when compared with the models, in order to correct for the uncertainty in the absolute continuum levels. The fit is not so good for group II stars especially between 13.8 and 14.0 μm due to insufficient molecular line data.

3. Modeling

Quantitative analysis of the spectra is attempted with a simple plane-parallel model. In order to simplify the problem as much as possible, we try to explain the observed spectra by a combination of two components: a *hot-component* representing the photosphere and a *cool-component* corresponding to the circumstellar envelope. Each layer is characterized by three parameters, the column density (N), the excitation temperature (T), and the turbulent velocity ($v_{\rm turb}$). In the present calculation, $v_{\rm turb}$ is fixed to 3 km s^{-1} in both layers. The results are insensitive to the value of $v_{\rm turb}$. For instance, for $v_{\rm turb} = 10$ km s^{-1} the column density would only decreases by 0.1 dex. Only C_2H_2 is considered. The line data are based on Kabbadj et al. (1991), Rinsland et al. (1982), Palmer et al. (1972), and Weber et al. (1994). About 20 000 lines are included in the wavelength region in Fig. 1. From a comparison with laboratory spectra (Kawaguchi, private communication), it is known that the line data are incomplete, especially between 13.8 and 14.0 μm. We ignored that part in the fitting.

First, the temperatures of the two layers are determined. We assume that group II and group V stars represent the hot- and cool-component, respectively, and we fit their spectra by a single layer model. For every combination in the parameter space of $10^{15} \leq N \leq 10^{19}$ cm^{-2} with 0.1 dex step and $300 \leq T \leq 1800$ K with 50 K interval, the model spectrum is characterized by the *rms* of the residual from the observed spectrum, and the best fit is the one which minimizes the *rms*. We obtain $T_{\rm hot} = 1400$ K and $T_{\rm cool} = 450$ K. In the second step, we fix the temperatures at these values and fit all stars with a two-layers model by the same procedure in ($N_{\rm hot}$, $N_{\rm cool}$) space. The best fit spectra are shown in Fig. 1 by gray lines, and the derived column densities for the spectra are plotted in Fig. 2 as a function of $T_{\rm NIR}$.

The emission from the dust envelope fills in and weakens the absorption in the photosphere. The correction of this effect is attempted as follows: A model spectrum with $N_{\rm hot}$ and $T_{\rm hot}$ is calculated. The spectrum is then inserted in a spherical dust envelope model with a certain mass-loss rate, and a synthesized spectrum is calculated. A star with a blackbody temperature of 2 500 K and radius 3×10^{13} cm is used when the dust temperature distribution is calculated. The envelope is assumed to extend from $10\,R_\star$ to $10^4\,R_\star$. The weakened molecular feature in the output spectrum may have an intensity similar to that with a lower column density and no dust effect. The difference between the original and weakened spectrum in terms of the column density is the correction factor corresponding to the dust mass-loss rate. The dust mass-loss rate of each star is determined by adopting the same model to their SWS spectra. The corrected values of $N_{\rm hot}$ are also plotted in Fig. 2.

4. Discussion

4.1. Uncertainty in the result

C_2H_2 molecules can be as abundant as 10^{-4} in the photosphere when the temperature is below about 2 000 K (Tsuji 1964). We find that when the column density exceeds about 10^{19} cm^{-2} and the temperature is high, the sharp feature at 13.7 μm disappears and the C_2H_2 band becomes a rather smooth and deep

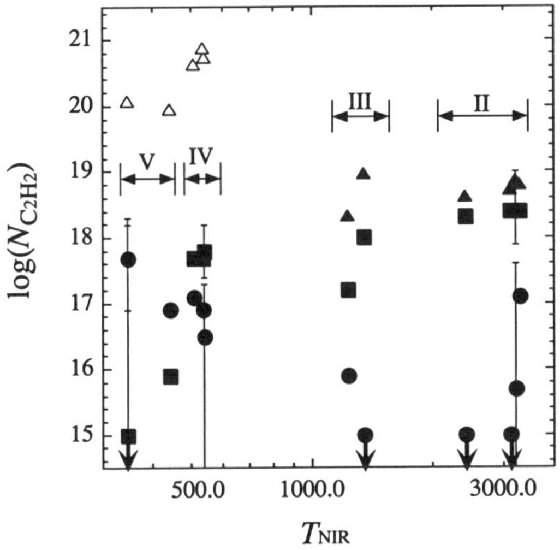

Figure 2. The column densities of C_2H_2 molecules in hot (*squares*) and cool (*circles*) layers are plotted against the near-infrared temperature. The hot-layer column densities after correction for dust emission effect are also plotted (*triangle*). The correction may be proper only for the low mass-loss rate stars (*filled triangles*). The error bars given in some data indicate the approximate uncertainty of the results in general. Symbols with arrows are upperlimits.

absorption from 10 μm to over 20 μm due to heavy saturation effects. Presumably, the *continuum* we draw in the present analysis is actually this continuous absorption, and the features we call *hot-component* arise from a layer just above this hottest, innermost part of the atmosphere.

The dust envelope also affects the results of the cool layer. Especially when the envelope has a moderate optical depth, the present plane parallel model potentially underestimates the column density of the cool layer, since it does not consider the size difference between two layers. This effect is small even in the most optically thick envelopes (group V), in which the circumstellar dust emission is the only continuum source.

4.2. Column densities of two layers

The column density in the hot-layer is almost constant at $10^{18.7}$ cm^{-2} in group II and III stars. The value increases considerably in groups IV and V, but this may be an artifact because the dust mass-loss rate of these stars is too high for this correction method to be applicable.

N_{cool} increases as T_{NIR} decreases. Since T_{NIR} refers to the dust mass-loss rate of the stars, the good correlation between the two quantities indicates that the C_2H_2 molecule is associated with the dust envelope. One interesting exception is R Scl, which shows a significant N_{cool} although its T_{NIR} is the highest in the sample. This star is known to have a rather thick, probably detached circumstellar shell (Olofsson et al. 1990) and a relation between the cool C_2H_2 and such an envelope is probable.

4.3. C_2H_2 abundance

Despite the uncertainties described above, it is still worth to make estimates of the abundance of C_2H_2 in the two layers. According to the dynamic model atmosphere by Höfner & Dorfi (1997), the temperature of the hot layer, 1 400 K, can be achieved at a distance of about $3\,R_\star$ from the central star. The density of the region is about 3×10^9 H_2 cm^{-3}. Assuming that the molecule is distributed between the surface ($1\,R_\star$) and $3\,R_\star$, the column density $10^{18.7}$ cm^{-2} is converted to an abundance of 2×10^{-5}. In the circumstellar envelope, we assume that the density decreases proportionally to the square of radius. We take AFGL 3068 as an example. The expansion velocity and the mass-loss rate measured by CO observations is 14 km s^{-1} and 2×10^{-5} $M_\odot$ yr^{-1} (Loup et al. 1993), and the column density derived by the model analysis is $10^{17.7}$ cm^{-2}. Assuming the inner radius of the envelope as 3×10^{14} cm, we obtain an abundance of 7×10^{-6}.

These values are in good agreement with previous estimates of the C_2H_2 abundance (e.g., Keady & Hinkle 1988, Fuente et al. 1998). Our results indicate that the abundance does not change drastically in the regions below and above the dust forming region, i.e., no significant depletion of C_2H_2 onto grains seems to take place in the dust forming region.

Acknowledgments. The authors are grateful to Prof. K. Kawaguchi for providing the laboratory spectra of C_2H_2. I. Y. and L.B.F.M. W. acknowledge financial support from a NWO PIONIER grant.

References

Cami J., Justtanont K., de Jong T., et al., 1997, First ISO Workshop on Analytical Spectroscopy, A.M. Heras et al. (eds.), ESA SP-419, p. 159

Cernicharo J., Barlow M.J., González-Alfonso E., et al., 1996, A&A 315, L201

de Graauw Th., Haser L.N., Beintema D.A., et al., 1996, A&A 315, L49

Fuente A., Cernicharo J., Omont A., 1998, A&A 330, 232

Groenewegen M.A.T., de Jong T., van der Bliek N.S., Slijkhuis S., Willems F.J., 1992, A&A 253, 150

Höfner S., Dorfi E.A., 1997, A&A 319, 648

Justtanont K., Feuchtgruber H., de Jong T., et al., 1998, A&A 330, L17

Kabbadj Y., Herman M., Di Lonardo, et al., 1991, J. Mol. Spec. 150, 535

Keady J.J., Hinkle K.H., 1988, ApJ 331, 539

Keady J.J., Ridgway S.T., 1993, ApJ 406, 199

Kessler M., Steinz J.A., Anderegg M.E., et al., 1996, A&A 315, L27

Loup C., Forveille T., Omont A., Paul J.F., 1993, A&AS 99, 291

Olofsson H., Carlström U., Eriksson K., et al., 1990, A&A 230, L13

Palmer K.F., Mickelson M.E., Rao K.N., 1972, J. Mol. Spec. 44, 131

Rinsland C.P., Baldacci A., Rao K.N., 1982, ApJS 49, 487

Tsuji T., 1964, Ann. Tokyo Astr. Obs. 2nd Ser. 9, 1

Weber M., Blass W.E., Halsey G.W., et al., 1994, J. Mol. Spec. 165, 107

Yamamura I., de Jong T., Justtanont K., et al., 1998, Ap&SS 255, 351

Asymptotic Giant Branch Stars
T. Le Bertre, A. Lèbre and C. Waelkens, eds.
I.A.U. Symposium 191, 1999

Diffraction-limited 76 mas speckle-masking interferometry of the carbon star IRC +10 216 and related AGB objects with the SAO 6 m telescope

G. Weigelt, T. Blöcker, K.-H. Hofmann, and R. Osterbart

Max-Planck-Institut für Radioastronomie, Auf dem Hügel 69, 53121 Bonn, Germany

Y.Y. Balega

Special Astrophysical Observatory, Nizhnij Arkhyz, Karachaevo-Cherkesia, 357147, Russia

A. J. Fleischer and J. M. Winters

Technische Universität Berlin, Institut für Astronomie und Astrophysik, 10623 Berlin, Germany

Abstract. We present high-resolution J-, H-, and K-band observations of the carbon star IRC +10 216. The images were reconstructed from 6 m telescope speckle interferograms using the speckle masking bispectrum method. The H image has the unprecedented resolution of 70 mas. The H and K images consist of at least five dominant components within a 0.21 arcsec radius and a fainter asymmetric nebula. The J–, H–, and K–band images seem to have an X-shaped bipolar structure. A comparison of our images from 1995, 1996, 1997, and 1998 shows that the separation of the two brightest components A and B increased from $\sim$193 mas in 1995 to $\sim$246 mas in 1998.

The cometary shapes of component A in the H and J images and the 0.79 μm and 1.06 μm HST images suggest that the core of A is not the central star, but the southern (nearer) lobe of the bipolar structure. The position of the central star is probably at or near the position of component B, where the H–K color has its largest value of H–K = 4.2.

If the star is located at or near B, then the components A, C, and D are located close to the inner boundary of the dust shell at separations of $\sim$200 mas $\sim$30 AU (projected distance) $\sim$6 stellar radii for a distance of $\sim$150 pc, in agreement with our 2-dimensional radiative transfer modelling.

In addition to IRC +10 216 we studied the stellar disks and the dust shells of several related objects. Angular resolutions of 24 mas at 700 nm or 57 mas 1.6 μm were achieved.

1. Speckle masking observations of IRC +10 216

IRC +10 216 (CW Leo) is the nearest, brightest and best–studied carbon star. The central star of IRC +10 216 is a long–period variable with a period of about 650 days. Distance estimates between 110 pc and 170 pc were reported (see Crosas & Menten 1997 and references therein). High-angular-resolution IR observations of IRC +10 216 and of its circumstellar dust shell were reported by Dyck et al. (1991), Danchi et al. (1994), McCarthy et al. (1990), Christou et al. (1990), Osterbart et al. (1997), Weigelt et al. (1997, 1998), and Haniff & Buscher (1998). Recent detailed radiative transfer calculations for IRC +10 216 were presented by Groenewegen (1997).

The IRC +10 216 speckle interferograms were obtained with the 6 m telescope at the Special Astrophysical Observatory in Russia and our NICMOS 3 camera at four different epochs. The high–resolution J, H, and K images (Figs. 1 to 3) were reconstructed from the speckle interferograms using the speckle masking bispectrum method (Weigelt 1977, Lohmann et al. 1983, Weigelt 1991). The resolution of the J (April 2, 1996), H (January 23, 1997), and K (January 23, 1997) images are 149 mas, 70 mas, and 87 mas, respectively. We denote the resolved components in the H and K images as A, B, C, and D (in the order of decreasing peak intensity; Weigelt et al. 1998).

The separation of the components A and B was measured for four different epochs. The four obtained separations are: 193 mas (October 1995, phase 0.88), 201 mas (April 1996, phase 0.15), 212 mas (January 1997, phase 0.62), and 246 mas (June 98, phase 0.39). The linear regression fit gives a velocity for this motion of $\sim$20 mas/yr. Assuming a distance of 150 pc (Crosas & Menten 1997) for IRC +10 216, this tangential velocity transforms to 14 km/s. This velocity is of the same order as the value of the radial expansion velocity of the circumstellar matter, 15 km/s (see Gensheimer & Snyder 1997 and references therein). From the present data there is no evidence that the motion inside the nebula is correlated with the stellar pulsation cycle.

2. Discussion

The core structure and the faint nebula in the H and K images. The multi–component structure of the IRC +10 216 K–band image (of the innermost 300 mas $\times$ 300 mas) has already been reported by Weigelt et al. (1998) and Haniff & Buscher (1998). This bright inner region is surrounded by a larger faint bipolar nebula (with $\sim$1% of the peak brightness of A). On the northern side of this faint nebula two arms of the nebula can be seen at position angles of roughly 30°(NE) and 340°(NW) with respect to component A. On the southern side one arm is present at about 210°(SW). At 160°(counterside of the NW-arm) the nebula is much fainter. It may, therefore, be adequate to take the direction from component A to component B (about 20° position angle) as the direction of the main axis (see also Kastner & Weintraub 1994).

The bipolar J–band image. The J–band image (Fig. 1) and the 0.79 μm and 1.06 μm HST images (Haniff & Buscher 1998) show a bipolar structure of the nebula. The southern lobe has a cometary shape whereas the northern region of

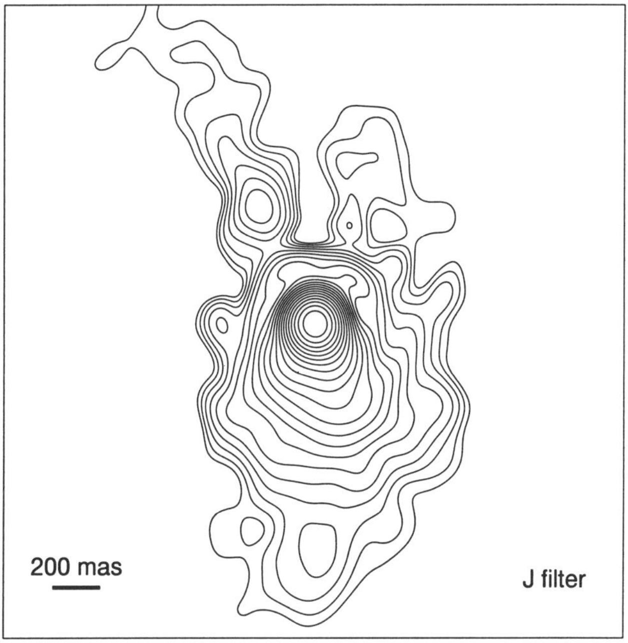

Figure 1. J–band image of IRC +10 216 with 149 mas resolution. In all images north is up and east to the left and the contour spacing is 0.25 mag.

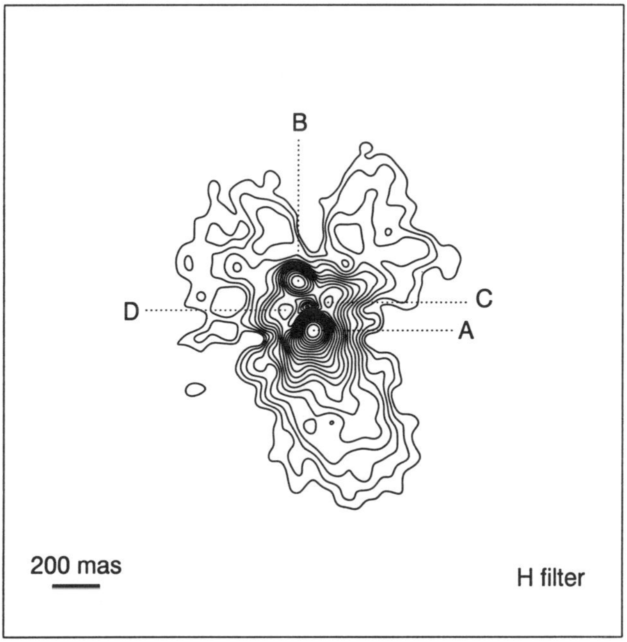

Figure 2. H–band image of IRC +10 216 with 70 mas resolution.

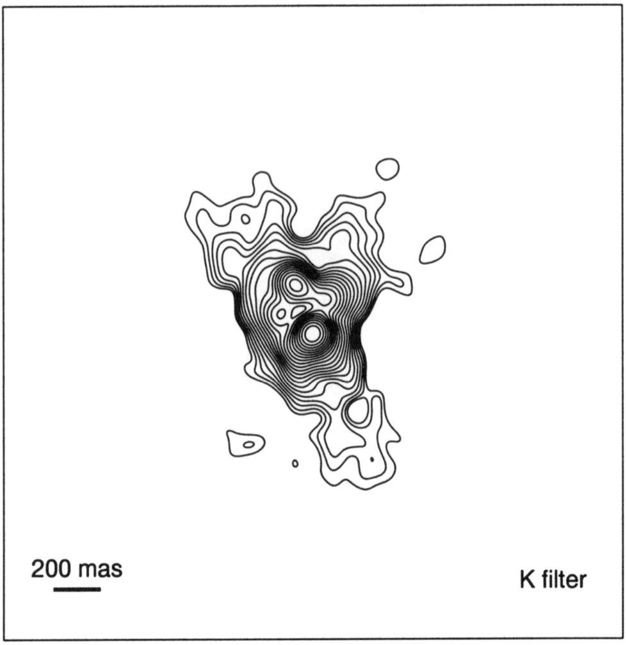

Figure 3. K–band image of IRC +10 216 with 87 mas resolution.

the images shows two arms reminiscent to (but much weaker than) the X–shaped structure of the Red Rectangle (see Men'shchikov et al. 1998).

H–K color image. In a square aperture of 1.6 arcsec the H and K magnitudes were determined to be K = 2.5 and H = 5.7. The integral color in this field is thus H–K = 3.2 ± 0.2. Our high–resolution H–K color image (Osterbart et al. 1999) shows that the components B, C, and D are rather red (H–K ≈ 4.2) in comparison with the integral color. The brightest component A (H–K = 3.2) as well as the *cometary* southern tails (H–K = 2 to 3) in the H and J image are bluer. These tails are a striking feature best seen in the high–resolution H–band image (Fig. 2). This structure and its relatively blue color suggest that component A is produced to a significant fraction by scattering of stellar light. This morphology can easily be understood under the assumption that the density of the dust shell is not sperically symmetric. The polar axis probably points with its southern side towards us. Component B may then be at or near the position of the star, strongly obscured and reddened by its environment. This suggests that the main axis of the nebula has a position angle of ~ 20°.

Relative motion of A and B. The relative motion of the nebula components is clearly not related to the stellar variability which has a period of about 650 days. It may thus be related to either an overall expansion or a variability of the dust shell with a period significantly larger than the stellar pulsation period (cf. Winters et al. 1995).

Bipolar structure, clumpiness, and position of the central star. Theoretical models treating the dust formation mechanism in the envelope of long–period variable carbon stars (Winters et al. 1994a, b) predict that dust formation does not occur permanently but ceases regularly. Periods of this mechanism may be significantly different from the stellar pulsation period. This leads us to the conclusion that in our IRC +10 216 images we see the inner boundary of the dust shell moving outwards with approximately 15 km/s. The components in the central region of IRC +10 216 have separations of the order of only a few stellar radii and the travel time of newly formed dust over these distances is only a few years. The components A to D may be caused, for instance, by large convection cells in the stellar atmosphere leading to asymmetric and stochastic fluctuations of the dust formation and thus to disturbances of the overall bipolar geometry.

Consistent with preliminary results from our radiative transfer calculations (Men'shchikov et al. 1999, in preparation), the cometary shapes of A in the H and the J images and the 0.79 μm and 1.06 μm HST images (Haniff & Buscher 1998) suggest that the core of A is *not* the central star, but the central star is at or near B, between the northern and southern J–band lobes separated by $\sim$500 mas. B is very red (H–K$\approx$4.2) and, therefore, it is not visible in the J–band image. If the star is now at or near B, then the components A, C, D and the northern components in the J image are probably located at the inner boundary of the dust shell. The radiative transfer models imply that the polar axis is inclined by intermediate angles (roughly 50°) with respect to the plane of the sky.

3. Diffraction-limited speckle masking studies of R Cas, AFGL 2290, and the Red Rectangle

In addition to IRC +10 216 we have studied the wavelength dependence of the size and the shape of the Mira stars R Cas (Weigelt et al. 1996) and R Leo, the aspherical dust shell of the oxygen-rich AGB star AFGL 2290 (Gauger et al. 1999), and the bipolar structure of the Red Rectangle (Men'shchikov et al. 1998). One-dimensional radiative transfer modelling of AFGL 2290 and two-dimensional radiative transfer modelling of the Red Rectangle was used for the interpretation of the observations.

Acknowledgments. This research has made use of the SIMBAD database, operated at CDS, Strasbourg, France.

References

Christou J.C., Ridgway S.T., Buscher D.F., Haniff C.A., McCarthy D.W., 1990, in Astrophysics with infrared arrays, R. Elston (ed.), ASP Conf. Series 14, p. 133

Crosas M., Menten K.M., 1997, ApJ 483, 913

Danchi W.C., Bester M., Degiacomi C.G., Greenhill L.J., Townes C.H., 1994, AJ 107, 1469

Dyck H.M., Benson J.A., Howell R.R., Joyce R.R., Leinert C., 1991, AJ 102, 200

Gauger A., Balega Y.Y., Irrgang P., Osterbart R., Weigelt G., 1999, High–resolution speckle masking interferometry and radiative transfer modeling of the oxygen–rich AGB star AFGL 2290, A&A, submitted

Gensheimer P.D., Snyder L.E., 1997, ApJ 490, 819

Groenewegen M.A.T., 1997, A&A 317, 503

Haniff C.A., Buscher D.F., 1998, A&A 334, L5

Kastner J.H., Weintraub D.A., 1994, ApJ 434, 719

Lohmann A.W., Weigelt G., Wirnitzer B., 1983, Appl. Opt. 22, 4028

McCarthy D.W., McLeod B.A., Barlow D., 1990, in Astrophysics with infrared arrays, R. Elston (ed.), ASP Conf. Series No. 14, p. 139

Osterbart R., Balega Y.Y., Weigelt G., Langer N., 1997, in Planetary Nebulae, H.J. Habing and H.J.G.L.M. Lamers (eds.), IAU Symposium 180, Kluwer Academic Press, Dordrecht, p. 362

Osterbart R., Balega Y.Y., Blöcker T., Men'shchikov A.B., Weigelt G., 1999, High–resolution bispectrum speckle interferometry of the bipolar dust shell around the carbon star IRC +10 216, in preparation

Men'shchikov A.B., Balega Y.Y., Osterbart R., Weigelt G., 1998, New Ast. 3, 601

Men'shchikov A.B., Balega Y.Y., Blöcker T., Osterbart R., Weigelt G., 1999, in preparation

Weigelt G., 1977, Optics Commun. 21, 55

Weigelt G., 1991, in Progress in Optics Vol. 29, E. Wolf (ed.), North Holland, Amsterdam, p. 293

Weigelt G., Balega Y.Y., Hofmann K.-H., Scholz M., 1996, A&A 316, L21

Weigelt G., Balega Y.Y., Hofmann K.-H., Langer N., Osterbart R., 1997, in Science with the VLT Interferometer, F. Paresce (ed.), Springer, Berlin, p. 206

Weigelt G., Balega Y.Y., Blöcker T., Fleischer A.J., Osterbart R., Winters J.M., 1998, A&A 333, L51

Winters J.M., Dominik C., Sedlmayr E., 1994a, A&A 288, 255

Winters J.M., Fleischer A.J., Gauger A., Sedlmayr E., 1994b, A&A 290, 623

Winters J.M., Fleischer A.J., Gauger A., Sedlmayr E., 1995, A&A 302, 483

Asymptotic Giant Branch Stars
T. Le Bertre, A. Lèbre and C. Waelkens, eds.
I.A.U. Symposium 191, 1999

Condensation Chemistry of Circumstellar Grains

Katharina Lodders and Bruce Fegley, Jr.

Washington University, Campus Box 1169, St. Louis, MO 63130, USA

Abstract. Thermochemical equilibrium calculations are successful in predicting the mineralogy as well as the major and trace element chemistry of circumstellar grains found in meteorites. The calculations also explain observations of dust close to AGB stars (within 1-3 stellar radii). The trace element chemistry in circumstellar graphite, SiC, and other refractory carbide grains agrees with equilibrium condensation calculations for circumstellar shells of carbon stars. Observed trace element abundance patterns in N stars are complementary to those found in SiC grains indicating fractional condensation in circumstellar shells. Condensation temperatures depend upon total pressure, C/O ratio, nitrogen abundances, and overall metallicity. Therefore for condensation temperatures to be meaningful, the total pressure and elemental abundances (i.e., C/O ratio, metallicity) must be specified.

1. Introduction

Dust formation in cool circumstellar shells (CS) of AGB stars drives stellar winds and mass-loss leading to enrichment of the interstellar medium with heavy elements. Most of the stardust is composed of the abundant elements such as O, C, Mg, Si, Fe, S, Al, Ca, and Ti. Thermochemical equilibrium calculations have long been successful in modeling the gas chemistry in giant stars (Tsuji 1973) and they are successful as well in predicting the observed types and the trace element chemistry of dust. Many topics discussed in this review have already been described in detail by Lodders & Fegley (1997a) and this review focuses on the following questions:

1. What types of dust are expected to form in CS of M, S, and C stars? What types of dust are known from astronomical observations and studies of circumstellar dust in meteorites?

2. What is the condensation sequence? How is the condensation sequence influenced by the C/O ratio and the nitrogen abundance?

3. At which distance from the star do condensates form? What is the influence of stellar variability?

4. What can be learned about elemental fractionation processes from trace element abundances in stellar atmospheres and circumstellar SiC grains?

5. What future observational and laboratory work is indicated?

2. Applicability of thermochemical equilibrium calculations to circumstellar shells

Several different arguments indicate that thermochemical equilibrium calculations are applicable to gas and grain chemistry in the circumstellar shells of AGB stars. First, during the past forty years, thermochemical equilibrium (i.e., condensation) calculations have been used to predict the chemistry and mineralogy of dust grains in the solar nebula. The reviews by Larimer (1988), Lewis (1995), and Fegley (1997) show that these predictions generally agree well with the observed chemistry and mineralogy of chondritic meteorites and with the bulk compositions of planets and satellites. This agreement is significant because model pressures and temperatures in the solar nebula are similar to model pressures and temperatures in grain-forming regions of CS of cool stars.

Second, as discussed below and in more detail by Lodders & Fegley (1997a), thermochemical equilibrium calculations of grain formation from an otherwise solar gas with a C/O ratio >1 (i.e., around carbon stars) predict several features of circumstellar grain chemistry that are observed in presolar grains in meteorites. These features include the incorporation of TiC-bearing grains inside graphite grains, the formation of solid solutions of Mo- and Zr carbides with TiC, and the trace element patterns observed in SiC grains.

Third, consideration of the characteristic time scales for chemical reactions (t_{chem}) and of the characteristic time scales for expansion or cooling of gas in circumstellar shells (t_{cool}) indicate that circumstellar shells can be divided into three regions. In the inner regions where temperatures are high and grain formation occurs, $t_{chem} < t_{cool}$ and chemical equilibrium is reached. Conversely, in the outer regions where temperatures are low, $t_{chem} > t_{cool}$ and chemical equilibrium is not reached. In between these two regions is an intermediate zone where $t_{chem} = t_{cool}$ and chemical equilibrium is quenched. Finally, calculations of gas-grain chemical reaction rates using the simple collision theory of Fegley (1988) show that gas-grain chemistry proceeds very rapidly at high temperatures at which circumstellar grains such as TiC, graphite, and SiC form.

3. Major element condensates in circumstellar shells

The gas and dust chemistry in red giant atmospheres depends on how much CNO- and triple alpha-processed material is dredged-up to the surface. An increase of C in an otherwise approximately solar composition gas increases the C/O ratio and is responsible for the different chemistry seen in giants of type M (C/O from solar to < 1), type S (C/O ~ 1) and type N (normal carbon stars, C/O > 1). The specific types of condensates depend on temperature (T), pressure (P), and bulk elemental composition and are predicted by thermochemical equilibrium calculations. Descriptions of the computational methods for these calculations are given by Larimer (1988) and Fegley & Lodders (1994). In oxidizing environments (C/O $\sim$ solar) refractory oxides and silicates form while under reducing conditions (C/O > 1), refractory carbides, nitrides and sulfides form. Table 1 gives a list of the condensates (in order of appearance) of the major rock-forming elements for each element. Interestingly, Mg and Si form silicates under oxidizing and reducing conditions, although the condensation

Table 1. Major Element Condensates Expected in M- and C-stars *

Element	Abundance§	M-stars		C-stars	
		Formula	Mineral Name	Formula	Mineral Name
O	2.09×10^7	oxides and silicates		silicates	
C	1.00×10^7	—	—	TiC	titanium carbide M
				C	graphite CS, M
				SiC	silicon carbide CS, M
N	2.63×10^6	—	—	TiN	osbornite
				AlN	aluminum nitride
Mg	1.02×10^6	$MgAl_2O_4$	spinel M	MgS	niningerite CS
		Mg_2SiO_4	forsterite CS	$MgAl_2O_4$	spinel M
		$MgSiO_3$	enstatite CS	Mg_2SiO_4	forsterite CS
				$MgSiO_3$	enstatite CS
Si	1.00×10^6	$Ca_2Al_2SiO_7$	gehlenite	SiC	silicon carbide CS, M
		Mg_2SiO_4	forsterite CS	FeSi	iron silicide
		$MgSiO_3$	enstatite CS	Mg_2SiO_4	forsterite CS
				$MgSiO_3$	enstatite CS
Fe	8.91×10^5	Fe	iron metal	FeSi	iron silicide
		$(Fe,Ni)_3P$	schreibersite	Fe	iron metal
		FeS	troilite	FeS	troilite
S	4.47×10^5	FeS	troilite	CaS	oldhamite
				MgS	niningerite CS
				FeS	troilite
Al	8.51×10^4	Al_2O_3	corundum CS, M	AlN	aluminum nitride
		$CaAl_{12}O_{19}$	hibonite	Al_2O_3	corundum CS, M
		$CaAl_4O_7$	grossite	$MgAl_2O_4$	spinel M
		$Ca_2Al_2SiO_7$	gehlenite	$CaAl_2Si_2O_8$	anorthite
		$MgAl_2O_4$	spinel M		
		$CaAl_2Si_2O_8$	anorthite		
Ca	6.46×10^4	$CaTiO_3$	perovskite	CaS	oldhamite
		$CaAl_{12}O_{19}$	hibonite	$CaAl_2Si_2O_8$	anorthite
		$CaAl_4O_7$	grossite		
		$Ca_2Al_2SiO_7$	gehlenite		
		$CaAl_2Si_2O_8$	anorthite		
Na	5.75×10^4	$NaAlSi_3O_8$	albite	$NaAlSi_3O_8$	albite
Ni	5.01×10^4	FeNi	kamacite & taenite	FeNi	kamacite & taenite
		$(Fe,Ni)_3P$	schreibersite	$(Fe,Ni)_3P$	schreibersite
Cr	1.35×10^4	Cr in FeNi	alloy	$FeCr_2S_4$	daubréelite
Mn	9.33×10^3	Mn_2SiO_4	rhodonite in olivine	(Mn,Fe)S	alabandite
P	8.13×10^3	$(Fe,Ni)_3P$	schreibersite	$(Fe,Ni)_3P$	schreibersite
Cl	5.25×10^3	$Na_4[AlSiO_4]_3Cl$	sodalite	NaCl (?)	halite (?)
K	3.72×10^3	$KAlSi_3O_8$	orthoclase	$KAlSi_3O_8$	orthoclase
Ti	2.40×10^3	$CaTiO_3$	perovskite	TiC	titanium carbide M
				TiN	osbornite

* condensates are listed in order of appearance for each element
§ solar abundances on a scale Si = 1×10^6 atoms (Lodders & Fegley, 1998)
CS: dust observed in circumstellar shells of AGB stars.
M: circumstellar grains found in meteorites.

temperatures and absolute amount of silicates formed will vary with different redox state. The condensation temperatures of major compounds as a function of pressure are shown in Figures 1 & 2 for a solar composition gas and for a gas relevant for C stars. Graphs showing the percentage distribution of an element between different gases and condensates as a function of temperature can also be calculated, and some examples are given in Figure 3 of Glassgold (1996).

Condensates observed by infrared spectroscopy of CS and by analytical studies of circumstellar dust in meteorites are identified by the abbreviations "CS" and "M" in Table 1. Many of the condensates observed around cool stars are found as presolar (i.e., circumstellar) grains in meteorites. For example, silicate (olivine, pyroxene; 9.7 μm emission; Woolf & Ney 1969) and corundum (13 μm emission, Onaka et al. 1989) dust is observed in CS of M giants and circumstellar corundum and spinel grains are found in acid-resistant residues from meteorites (see Zinner 1997). Silicon carbide (11.3 μm emission; Gilra 1971; Gilra & Code 1971; Forrest et al. 1975), carbon (7.5 & 23 μm emission; Omont 1991, Martin & Rogers 1987), and MgS (30 μm emission; Goebel & Moseley 1985) are observed in infrared spectra of N stars; and circumstellar graphite and SiC are found in meteorites. Titanium carbide (and other refractory carbide) inclusions in meteoritic circumstellar graphite indicates the condensation sequence TiC, graphite, SiC around carbon stars.

Some of the condensates listed in Table 1 are not yet observed around cool stars. For example, Fe is about as abundant as Si and Mg and it is implausible that Fe-bearing condensates will not form. As shown in Table 1, equilibrium calculations predict that Fe alloy should form around M stars and some of it may react with H_2S gas to form FeS (troilite) at lower temperatures. In the more reducing environment of N stars, FeSi should form. If equilibrium persists down to very low temperatures <500 K, which may be unlikely (cf. Palme & Fegley 1990), appreciable amounts of Fe can be incorporated into olivine as Fe_2SiO_4 and into pyroxene as $FeSiO_3$. Thus the failure to observe Fe-bearing condensates such as Fe alloy, FeSi, and FeS must be due to other causes. Two possible causes are an absence of distinctive IR emission features at observable wavelengths or a lack of laboratory data on the IR spectra of predicted phases. Condensates such as TiC, CaS, and AlN have not yet been identified around carbon stars, but the abundances of these condensates are limited by the elemental abundances of Ti, Ca, and Al, which are much lower than the elemental abundance of Si. Finally, some condensates have not yet been found as circumstellar grains in meteorites because they are sensitive to water (CaS, MgS, AlN) and are also less resistant than SiC or graphite to the corrosive acids used to dissolve away the bulk of the meteorite. Sensitive in situ searches can be used to detect these condensates.

4. Influence of total pressure and C/O ratio

Condensation temperatures generally decrease with decreasing total pressure. Fig. 1 shows the condensation temperatures for major element condensates as a function of total pressure for a solar composition gas appropriate for M stars. Also shown are condensation temperatures for very refractory trace element oxides (HfO_2, ZrO_2), which condense at even higher temperatures than corundum. The graph also indicates that the initial condensates can trans-

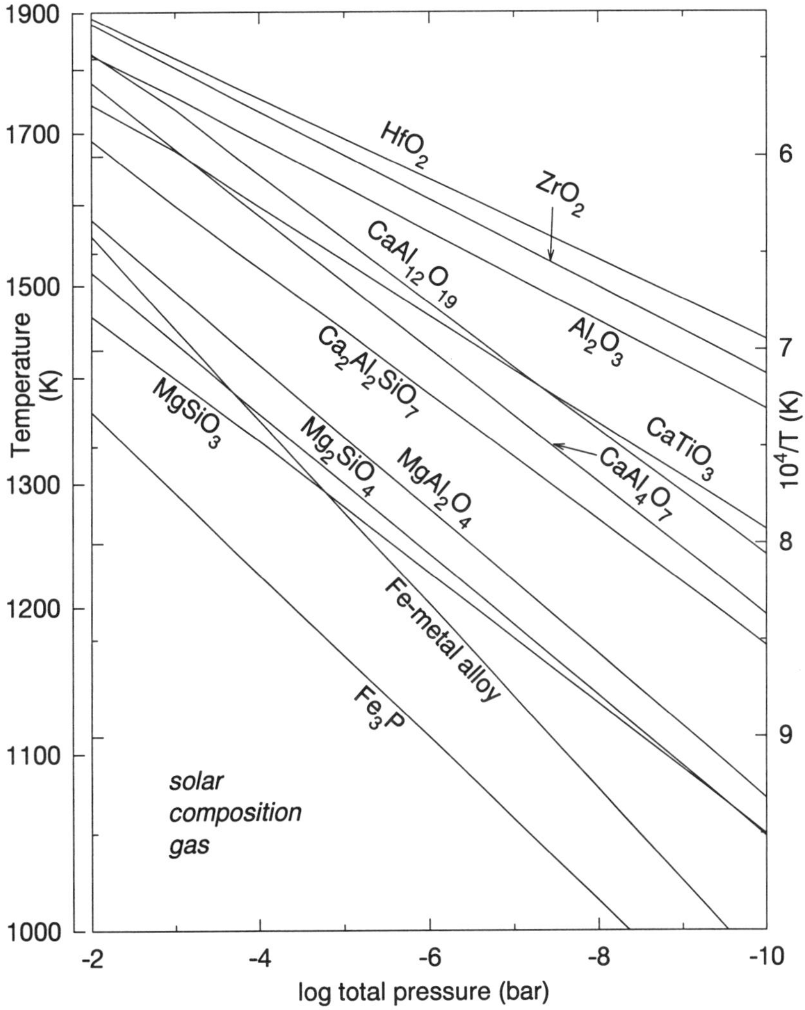

Figure 1. Major element condensation temperatures as a function of total pressure for a solar composition gas (C/O = 0.48). Note that the y-axis is linear with 1/T.

form into other phases. For example, corundum reacts with Ca (g) to hibonite, which changes into grossite and eventually into gehlenite at lower temperatures. Gehlenite is the first Si-bearing condensate but because Ca and Al are much less abundant than Si, the amount of total Si tied up in gehlenite is very small. Similarly, some corundum reacts with Mg (g) + H_2O (g) to form spinel but only a small amount of Mg is in spinel, because the spinel abundance is limited by the Al abundance. The major Mg- and Si-bearing condensate is forsterite (Mg_2SiO_4, the Mg-endmember of the olivine $(Mg,Fe)_2SiO_4$ solid solution). At lower temperatures, forsterite reacts with SiO (g) to form enstatite ($MgSiO_3$, the Mg-endmember of the pyroxene $(Mg,Fe)SiO_3$ solid-solution).

In addition to the pressure sensitivity of condensation temperatures, the C/O ratio influences the condensation temperatures (see detailed discussion by Lodders & Fegley 1995, 1997a). An increase of the C/O ratio from solar (0.48) to unity causes a drop in oxide and silicate condensation temperatures because more oxygen is tied up in CO and less H_2O is available for condensation reactions at higher temperatures. This effect becomes important in atmospheres of S stars, which have elevated C/O ratios. Thus, we expect dust formation in S stars to occur at lower temperatures (i.e., farther out in the CS) than in M stars (assuming similar pressure-temperature conditions in atmospheres of M and S stars). As a consequence, dust-driven winds in S stars may become less effective in increasing mass-loss rates, which could explain the generally lower mass-loss rates in S stars than in M- or C stars (e.g., Bieging & Latter 1994, Jura 1988).

Near a C/O ratio of unity, the types of condensates change from oxides and silicates to carbides, sulfides, and nitrides (e.g., Larimer 1975, Larimer & Bartholomay 1979, Lodders & Fegley 1995, 1997a). Condensation temperatures for wide P-T and C/O ranges are listed by Lodders & Fegley (1995). Fig. 2 shows condensation temperatures for major element condensates in a N star composition gas with C/O = 1.1 (CNO abundances from Lambert et al. 1986, otherwise solar composition). In addition to the major elements, ZrC, the most refractory trace element carbide, is also shown. As is the case for condensates under more oxidizing conditions, condensation temperatures drop with decreasing total pressure. The exception is graphite, whose condensation temperature is not very pressure sensitive. As a consequence, the condensation sequence of TiC, graphite and SiC is very sensitive to the total pressure. At $P > 10^{-3.5}$ bar, the condensation sequence is TiC, SiC, C; from $10^{-3.5}$ bar to $10^{-5.5}$ bar the sequence is TiC, C, SiC; and at low pressures $P < 10^{-5.5}$ bar, the sequence is C, TiC, SiC. These P-dependent condensation sequences are only valid for C/O = 1.1 because the condensation sequence depends on total pressure and C/O ratio. Higher C/O ratios at a given P increase the condensation temperature of carbon thus shifting the pressure range where the condensation sequence is TiC, C, SiC to higher pressure values.

The condensation sequence TiC, C, SiC is recorded in circumstellar graphite grains where TiC as well as ZrC-MoC-TiC carbide alloys are often embedded in the center of graphite, indicating that these refractory carbides formed prior to graphite and served as nucleation seeds (see Bernatowicz et al. 1996 for details).

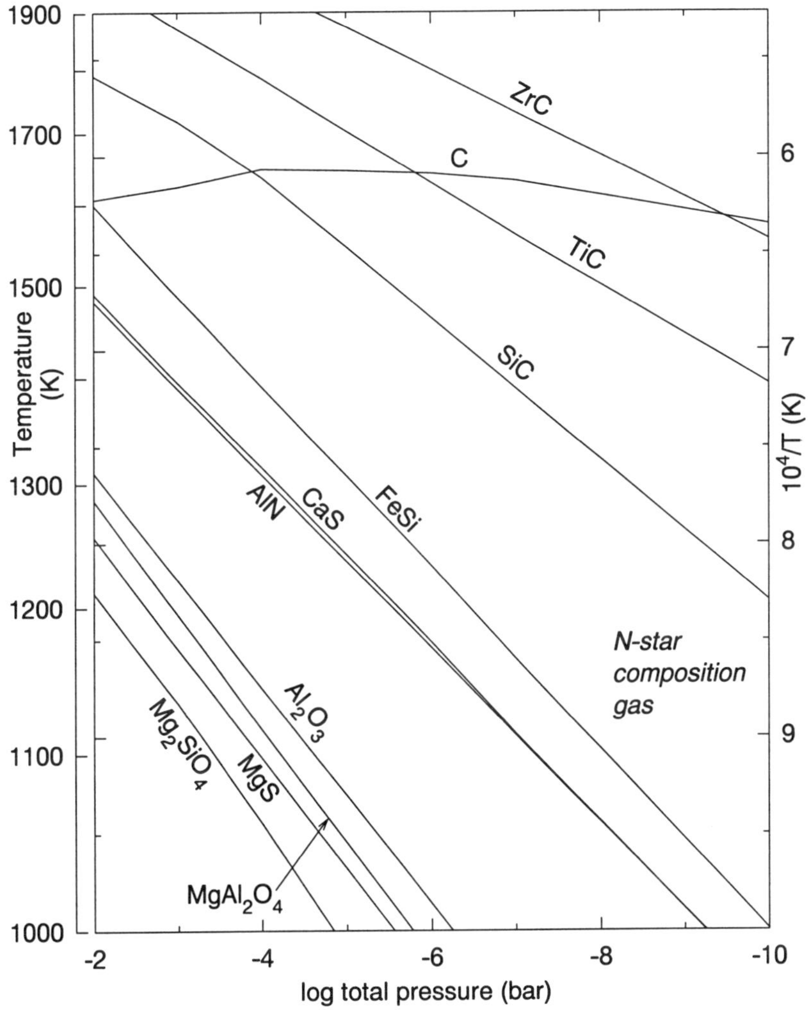

Figure 2. Major element condensation temperatures as a function of total pressure for a gas with average N star abundances of C, N, O (C/O = 1.1; Lambert et al. 1986) and otherwise solar composition.

5. Influence of nitrogen abundance

Another important influence on condensation temperatures is the nitrogen abundance. The relative nitrogen abundances increase from M- to S stars because of CNO-cycle processing and dredge-up; and enhancements of nitrogen may also be present in C stars. This could be the case for some C stars with enhanced HCN emissions (Jura 1991). Higher than solar abundances of N are known for R stars (Dominy 1984) and although Lambert et al. (1986) find approximately solar N-abundances, further analyses of N- and J carbon stars are needed to clarify the N abundances in these stars. An increase in N abundances does not influence condensation temperatures of oxides and silicates (e.g., M- and S stars) because N does not affect the gas chemistry relevant for oxide condensation. However, an increase in N abundances in C stars locks more carbon into CN (g) and HCN (g) which causes a decrease in condensation temperatures of C-bearing condensates. For example, an increase of the solar nitrogen abundance by a factor of 100 lowers graphite condensation temperatures by about 100 - 200 K and carbide condensation temperatures by 20 - 50 K (depending on total P and C/O ratio). Such an increase in N abundance also causes an increase in nitride condensation temperatures (e.g., AlN condenses 50 K higher) but does not lead to formation of trace element nitrides instead of carbides as initial condensates.

6. Mythical condensates

There are also several compounds (e.g., pyrite, FeS_2; wüstite, FeO; hematite, Fe_2O_3; silica, SiO_2; silicon disulfide, SiS_2; calcium dicarbide, CaC_2) often discussed as potential condensates in the literature but none of these compounds can form at equilibrium in CS. For example, formation of pyrite from Fe-metal proceeds via Fe (metal) + H_2S = FeS + H_2 and FeS + H_2S = FeS_2 + H_2, which implies that all Fe has to be converted to FeS first before sulfurization to FeS_2 proceeds. This requires very large sulfur partial pressures which are inconsistent with the solar atomic S/Fe ratio of ~ 0.5. Similarly, the formation of SiS_2 is unreasonable. Silicon, together with Mg, forms more stable forsterite and enstatite instead of SiS_2 and SiO_2 (see Figures 1 & 2). Hematite formation requires a very oxidizing gas and proceeds via Fe (metal) to magnetite to hematite. However, low temperature oxidizing conditions in a solar composition gas are only sufficient to allow magnetite formation at low T, so stars don't rust.

7. Dust formation locations and nucleation

As discussed in the previous paragraphs, modeling dust forming locations in CS depends on the atmospheric P-T conditions and bulk elemental abundances. Although both these parameters need refinement, using available P-T models and abundances, dust formation locations predicted at thermochemical equilibrium (i.e., supersaturation ratios S = 1) concur with astronomical observations that dust forms close to the star (3-10 stellar radii) in M as well as N stars (e.g., Anandarao et al. 1993; Bergeat et al. 1976; Danchi & Bester 1995). This suggests that nucleation does not require high supersaturation ratios and proceeds relatively rapidly. In this respect it is also very important to consider stellar variability. During minimum dust formation is enhanced and dust must nucleate and move outward from the CS before the variability cycle moves to maximum, where dust evaporation can occur. Such a scenario has support from the observations of the carbon Mira R For, of which the central star shows increased obscuration during minimum phase (Le Bertre 1988). Recently, variations of the silicate emission as a function of light curve have been reported for oxygen-rich long-period variables (Creech-Eakman et al. 1997).

This scenario can be tested by making observations of the C/O ratio, SiC and C dust abundances and trace element abundances. For example, during minimum, formation of carbon and SiC dust removes carbon from the gas and the C/O ratio in the gas decreases toward unity (Fix 1969; Lodders & Fegley 1997a). As noted by Fix (1969), grains formed during minimum will evaporate again and raise the C/O ratio back to the original value. However, if dust is removed from the CS by radiation pressure, the gas remains at a lower C/O ratio. Gas fall-back onto the star during contraction then causes a "dilution" of the photospheric gas, which then is observed to have a lower C/O ratio. This could be one reason why the C/O ratios in N stars determined from the gas phase are close to unity (Lambert et al. 1986) while the hotter and dust-free R stars have higher C/O ratios (Dominy 1984). If condensation and removal of carbonaceous dust takes place in C stars, the C/O ratio determined from the gas cannot be representative of the bulk C/O ratio, which then needs to be determined from $(C_{dust}+C_{gas})/O$. Nevertheless, the determination of the C/O ratio as a function of variability may prove a sensitive test of dust formation.

Similarly, the trace element chemistry in circumstellar grains in meteorites and stellar atmospheres indicates that fractional condensation takes place in CS (see detailed discussion by Lodders & Fegley 1997a, b). Fractional condensation into SiC depletes the gas in refractory elements such as Ti, Zr, REE (Rare Earth Elements), while the more volatile elements such as Sr and Ba require much lower temperatures to be scavenged into SiC. The analyses of circumstellar SiC grains in meteorites revealed that many SiC grains are depleted in Sr and Ba implying that the SiC grains were removed from equilibrium with the gas at relatively high temperatures in the CS. This means that the grains formed relatively close to the star. The trace element abundances also indicate that grain removal from the CS must have taken place at different distances in the CS to explain the different types of trace element abundance patterns seen in the grains. As in the case above for carbon dust removal, we expect to see an effect in the remaining gas abundances in the stellar atmospheres. Since Sr and Ba are not removed into grains, the remaining gas becomes relatively enriched in Sr and Ba. If grain formation and removal from the CS takes place preferentially during minimum, the gas enriched in Sr and Ba falls back onto the photosphere (and thus line-forming region) during maximum phase and relative enrichments of these elements should become visible. In other words, the trace element abundances in the SiC grains and the stellar atmosphere gas become complementary and should mirror each other. Such complementary abundances are observed, supporting the fractional equilibrium condensation model.

Thus, the variability cycle puts constraints on nucleation timescales. If nucleation timescales were longer than the time between minimum (enhanced condensation) and maximum (evaporation of dust), we would not expect much dust at all in CS. The observation of dust in CS and the existence of circumstellar grains in meteorites means that dust nucleation and expulsion to the ISM proceeds within timescales of half the variability periods (e.g., typically 100/2 to 500/2 days for SR and Mira variables). An observational constraint on nucleation timescales may also come from optical and IR observations as a function of light curve. As noted by Creech-Eakman et al. (1997), there seems to be a quarter cycle phase lag between the IR and optical light curves for the oxygen Mira AU Cyg (period 435 days). This suggests that appreciable amounts of dust form within about 110 days. The nucleation timescales are also reflected by the correlation of variability period with mass-loss rate, which depends on the amount of dust present.

8. Required observations and laboratory work

Better and more detailed abundance determinations in giant atmospheres are needed. Observations require refinements of C/O ratios, nitrogen abundances and determination of metallicity and s-process element abundances. Of particular interest are studies of abundances as a function of variability cycle. For example, the condensation calculations suggest that the C/O ratio (determined in the gas) is lower at minimum because of condensation of carbon into dust. Similarly, SiC band strength may increase at minimum due to increasing SiC condensation. The trace elements are diagnostic of gas-solid fractionation processes and volatility related fractionation processes may lead to enhancements of the more volatile trace elements in the gas phase at minimum.

In addition to stellar observations, more analytical data on individual submicron circumstellar grains in meteorites are needed. Their trace element abundances can reveal much about the formation conditions in CS.

Condensation calculations predict many yet unidentified major element condensates in CS. In O-rich CS, we expect Ca-Al-bearing phases (hibonite, grossite, gehlenite), spinel, and metal-alloy. Not yet identified minerals expected to be present in C-rich CS are spinel, AlN, CaS, FeSi. The infrared spectra for many of these minerals are poorly known and need to be investigated to aid identification of these phases in stellar spectra. There are also predicted refractory trace element condensates which may be present in circumstellar grains, such as HfO_2 and ZrO_2 from M stars. Refractory carbides such as HfC, NbC, TaC are expected to be present in the TiC-ZrC-MoC alloys observed in circumstellar graphite grains. However, Hf, Nb, and Ta are much less abundant than Zr or Mo and cannot be detected until advanced instrumentation becomes available.

Acknowledgments. K. L. thanks the IAU Symposium organizing committee and the Université de Montpellier for the invitation to the enjoyable meeting and for travel support. Work supported by the NASA Origins Program (NAG5-4323).

References

Anandarao B.G., Pottasch S.R., Vaidya D.B., 1993, A&A 273, 570
Bergeat J., Lefèvre J., Kandel R., Lunel M., Sibille F., 1976, A&A 52, 245

Bernatowicz T.J., Cowsik R., Gibbons P.C., et al., 1996, ApJ 472, 760

Bieging J.H., Latter W.B., 1994, ApJ 422, 765

Creech-Eakman M.J., Stencel R.E., Williams W.J., Klebe D.I., 1997, ApJ 477, 825

Danchi W.C., Bester M., 1995, Ap&SS 224, 339

Dominy J.F., 1984, ApJS 55, 27

Fegley B., 1988, in Workshop on the Origins of Solar Systems, eds. J.A. Nuth and P. Sylvester, LPI Technical Report 88-04, Houston TX, Lunar and Planetary Institute, p. 51

Fegley B., 1997, in Encyclopedia of Planetary Sciences, eds. J.H. Shirley and R.W. Fairbridge, London, Chapman & Hall, p. 169

Fegley B., Lodders K., 1994, Icarus 110, 117

Fix J.D., 1969, MNRAS 146, 37

Forrest W.J., Gillett F.C., Stein W.A., 1975, ApJ 195, 423

Gilra D.P., 1971, Nature 229, 237

Gilra D.P., Code A.D., 1971, BAAS 3, 379

Glassgold A.E., 1996, ARA&A 34, 241

Goebel J.H., Moseley S.H., 1995, ApJ 290, L35

Jura M., 1988, ApJS 66, 33

Lambert D.L., Gustafsson B., Eriksson K., Hinkle K.H., 1986, ApJS 62, 373

Larimer J.W., 1975, Geochim. Cosmochim. Acta 39, 389

Larimer J.W., 1988, in Meteorites and the Early Solar System, eds. J.F. Kerridge and M.S. Matthews, Tucson AZ, Univ. of Arizona Press, p. 375

Larimer J.W., Bartholomay M., 1979, Geochim. Cosmochim. Acta 43, 1455

Le Bertre T., 1988, A&A 190, 179

Lewis J.S., 1995, Physics and Chemistry of the Solar System, Academic Press, New York

Lodders K., Fegley B., 1995, Meteoritics 30, 661

Lodders K., Fegley B., 1997a, in Astrophysical Implications of the Laboratory Studies of Presolar Materials, eds. T. J. Bernatowicz and E. Zinner, Woodbury NY, Amer. Institute of Physics, AIP Conf. Proc. 402, p. 391

Lodders K., Fegley B., 1997b, ApJ 484, L71

Lodders K., Fegley B., 1998, The Planetary Scientist's Companion, New York, Oxford University Press

Martin P.G., Rogers C., 1987, ApJ 322, 374

Omont A., 1991, in Chemistry in space, eds. J. M. Greenberg and V. Pirronello, Kluwer Acad. Press, Dordrecht, p. 171

Onaka T., de Jong T., Willems F.J., 1989, A&A 218, 169

Palme H., Fegley B., 1990, Earth Planet. Sci. Lett. 101, 180

Tsuji T., 1973, A&A 23, 411

Woolf N.J., Ney E.P., 1969, ApJ 155, L181

Zinner E., 1997, in Astrophysical Implications of the Laboratory Studies of Presolar Materials, eds. T.J. Bernatowicz and E. Zinner, Woodbury NY, AIP Conf. Proc. 402, p. 3

The audience

Asymptotic Giant Branch Stars
T. Le Bertre, A. Lèbre and C. Waelkens, eds.
I.A.U. Symposium 191, 1999

Silicate and ice emission bands in the *ISO* spectrum of the PAH-emitting carbon-rich planetary nebula CPD −56°8032

Martin Cohen

Radio Astronomy Laboratory, University of California, Berkeley, CA 94720, U.S.A.

M. J. Barlow, R. J. Sylvester, X.-W. Liu

Dept. of Physics and Astronomy, University College London, London WC1E 6BT, U.K.

P. Cox

Institut d'Astrophysique Spatiale, Université de Paris XI, F-91405 Orsay Cedex, France

T. Lim

Rutherford Appleton Laboratory, Chilton, Didcot, Oxon OX11 0QX, UK

B. Schmitt

Laboratoire de Glaciologie et Géophysique de l'Environnement, CNRS, F-38041 Grenoble/Saint Martin d'Hères, France

A. K. Speck

Dept. of Physics and Astronomy, University College London, London WC1E 6BT, U.K.

Abstract. Combined ISO SWS and LWS spectroscopy is presented of the late WC-type planetary nebula nucleus CPD −56°8032 and its carbon-rich nebula. The extremely broad coverage (2.4–197 μm) enables us to recognize the clear and simultaneous presence of emission features from both oxygen- and carbon-rich circumstellar materials. Removing a smooth continuum highlights bright emission bands characteristic of polycyclic aromatic hydrocarbons (hereafter PAHs) in the 3–15 μm region, bands from crystalline silicates longwards of 18 μm, and the 43- and 62-μm bands of crystalline water ice. We discuss the probable evolutionary state and history of this unusual object in terms of (a) a recent transition from an O-rich to a C-rich outflow following a helium shell flash; or (b) a carbon-rich nebular outflow encountering an O-rich comet cloud orbiting in a Kuiper-belt-like distribution.

1. Introduction

CPD −56°8032 (hereafter CPD) belongs to the rare class of late WC-type ([WCL]) nuclei of planetary nebulae (PNNs). These objects may result when a helium shell-flash in a low- or intermediate-mass star on the Asymptotic Giant Branch (AGB) either ingests or ejects all the remaining hydrogen-rich outer envelope, leaving a H-poor, carbon-rich star spectrally like late WC-type population I Wolf-Rayets.

CPD's large infrared excess has long been attributed to dust emission (Webster & Glass 1974; Cohen & Barlow 1980). PAH bands also appear in emission (Aitken et al. 1980; Cohen, Tielens & Allamandola 1985; Cohen et al. 1989). CPD has the highest measured IR luminosity fraction in the 7.7-μm band of any object (Cohen et al. 1989), consistent with its very large nebular gas-phase C/O number ratio of 13 (De Marco, Barlow & Storey 1997; hereafter DMBS).

We present Infrared Space Observatory (ISO) Long Wavelength Spectrometer (LWS; Clegg et al. 1996; Swinyard et al. 1996) 43–197 μm full grating spectra of CPD, combined with Short Wavelength Spectrometer (SWS; de Graauw et al. 1996) 2.4–45 μm grating spectra of this object, obtained in the LWS Guaranteed Time program. Barlow (1998) has summarized our preliminary ISO results on CPD.

2. The ISO spectrum of CPD -56°8032

Full wavelength coverage grating mode LWS01 spectra of CPD were secured in eight fast scans, each comprising a 0.5-s integration ramp at each grating position, sampled at 1/4 of a spectral resolution element. Our low-resolution 2.4–45 μm SWS grating spectrum of CPD was taken at Speed 1, yielding a mean spectral resolving power of ~250. Standard pipeline processing (LWS OLP6.0 and SWS OLP6.1) was used to extract, reduce and calibrate the separate subspectra. The ISAP and SIA packages provided the capability to examine the spectral fragments in detail.

Fig. 1 presents our complete wavelength coverage of CPD. We first spliced all the SWS and LWS subspectra separately, following the methods described by Cohen, Walker & Witteborn (1992; CWW), then joined the composite SWS and LWS portions, which required scaling the LWS spectrum by 0.99±0.01 to register it with the SWS spectrum. The resultant 2.4–197 μm spectrum was normalized to the Point Source Catalog (PSC) photometry (see CWW) in all 4 IRAS bands, necessitating a further rescaling of the total spectrum by 1.05±0.04.

The ISO spectrum of CPD in Fig. 1 exhibits the canonical spectrum of PAH emission bands but the most striking aspect is that, despite the carbon-dominated stellar and nebular chemistry, the spectrum longwards of 15 μm is dominated by emission features usually associated with the circumstellar envelopes of O-rich stars (Glaccum 1990; Waters et al. 1996). Waters et al. (1998a, b) have presented SWS spectra of several other C-rich post-AGB objects, also with both PAH and crystalline silicate emission features. The ISO spectrum of CPD has an additional remarkable property in that crystalline water ice features are present in emission at 43 and 62 μm (see below).

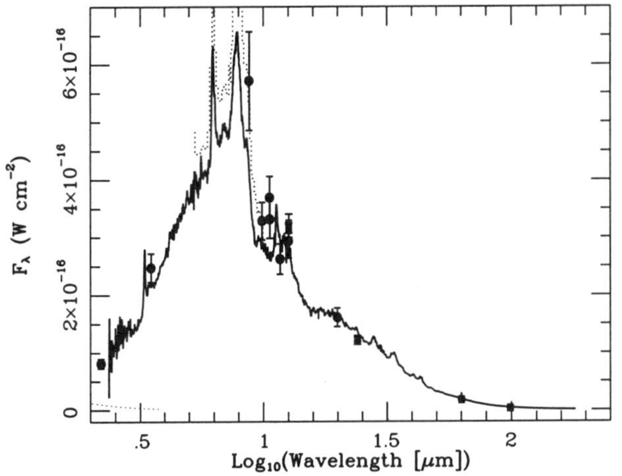

Figure 1. Full SWS+LWS spectrum of CPD–56°8032 (heavy solid line). Also shown are IRAS PSC fluxes (filled squares, plotted at their isophotal wavelengths), the 2–20 μm photometry of Cohen & Barlow (1980, filled circles), and (as dotted lines) the 5–8 μm KAO spectrum of Cohen et al. (1989) and the 7.7–22.7 μm IRAS LRS spectrum.

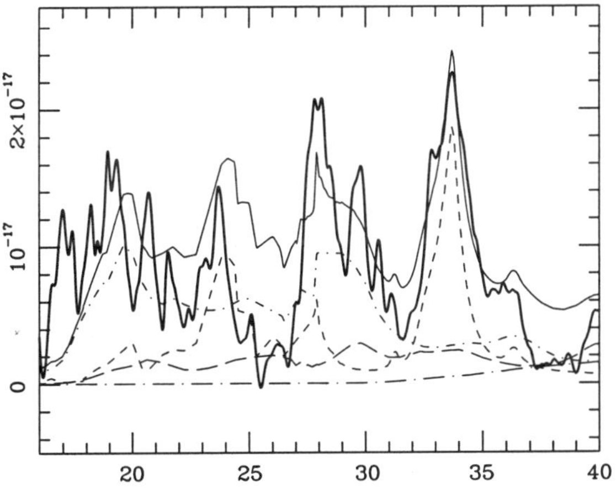

Figure 2. The emission from CPD–56°8032 in excess of the two-blackbody continuum, from 16–40 μm. Heavy solid line: observed excess emission; short-dashed line, forsterite; long-dashed line, clinopyroxene; short-dash dotted line, orthopyroxene; long-dash dotted line, crystalline ice; light solid line, the sum of all these components. The units of excess flux are W cm^{-2} μm^{-1}.

To amplify the intensities of the many emission bands, we subtracted a continuum as the sum of two blackbodies (temperatures and solid angles optimized by least-squares fit to the lower envelope of the observed spectrum). The best fit is for temperatures of 470±5K and 135±5K. Fig. 2 illustrates a portion of CPD's "excess" spectrum. Between 16 and 40 μm (Fig. 2), emission features characteristic of crystalline silicates are seen prominently at 19, 24, 28, and 33 μm. The longer wavelength region (not shown) exhibits emission bands at 41, 43, 48, and 69 μm, along with a very broad, low-level, emission hump centered near 62 μm, under the [O I] 63-μm line. We identify the 43- and 62-μm bands with crystalline water ice, as first detected by Omont et al. (1990) in the KAO spectrum of the Frosty Leo nebula.

We have attempted to identify the emission features seen above the continuum of CPD using data for clinopyroxene, orthopyroxene and 100% forsterite (Koike et al. 1993) and optical constants for both amorphous and crystalline water ice recently measured in the laboratory (Trotta 1996; Schmitt et al. 1998). Only 100% pure forsterite shows the 69 μm feature (Koike et al. 1993). Fig. 2 distinguishes the separate contributions of these four materials, and their total. We modeled the optically thin case, for small spherical grains (0.1 or 1.0 μm radius) and, by matching relative band strengths, we found plausible temperatures (forsterite, clinopyroxene, and crystalline ice: 65 K; orthopyroxene: 90 K), suggestive of a common physical location. The cold silicates produce no measurable emission below the 19-μm band. Without combined SWS+LWS coverage, one cannot constrain these temperatures.

Modelling indicates that: forsterite causes the 24- and 33-μm bands; crystalline silicates such as orthopyroxene the 19- and 28-μm bands; crystalline ice the 43- and 62-μm features; and clinopyroxene the prominent 41-μm band and part of the broad 66-μm emission feature. It is striking that CPD displays both carbon- and oxygen-rich materials and that its materials are so much more ordered (crystalline) than the laboratory data.

3. Discussion

From the luminosity and distance of CPD (1.5 kpc: DMBS), we can deduce the angular extent of each of the emitting dust components, summarized in Table 1. The diameter of the 3.3-μm PAH emission is from Roche, Allen & Bailey (1986) and the ionized zone from HST Hβ imaging (see DMBS).

The black grains are probably carbon-rich, like the dust that forms in the outflows from Population-I WC9 stars (e.g., Cohen, Barlow & Kuhi 1975). The crystalline materials seem to exist just outside the ionized zone while the PAHs appear within the ionized region. Other [WCL] PNe show the signatures of both C-rich and O-rich material, e.g. the [WC10] PNN IRAS 07027–7934 (Menzies & Wolstencroft 1990) with PAH features and a 1612 MHz OH maser (Zijlstra et al. 1991); the Type I bipolar PN NGC 6302 with weak 8.7- and 11.3-μm PAH bands (Roche & Aitken 1986), a weak OH maser (Payne, Phillips & Terzian 1988), and ice and crystalline silicate features (Barlow 1998; Waters et al. 1996); and the [WCL] PNNs BD +30°3639 and He 2-113 (Waters et al. 1998a, b).

If a recent thermal pulse converted an O-rich mass loss outflow to a C-rich one, then why do we observe over 6 [WCL] PNNs in the same evolutionary state

Table 1. Cross-section through CPD's different zones.

Material	Temperature K	Mass $10^{-4}M_{\odot}$	Ang. diam. arcsec	Diameter AU
Hot grains	1600	–	?	~few
BB	470	–	0.06	80
BB	135	–	0.7	1 000
3.3 μm PAHs	–	–	1.3	2 000
ionized gas	–	–	1.6×2.1	2 400×3 000
forsterite	65	1.6	2	3 000
clinopyroxene	65	1.3	2	3 000
orthopyroxene	90	0.3	2	3 000
crystalline ice	65	0.6	2	3 000

when such a transition should occur only once during the lifetime of a star? This seems even less probable when one estimates that the O-rich grains must have been ejected only 250 yr ago by CPD (the nebular expansion velocity is 30 km s^{-1} (DMBS) and this material is now 1" from the star), such a tiny fraction of the 6×10^4 yr interpulse interval (for a 0.62 $M_{\odot}$ core: Boothroyd & Sackmann 1988). Waters et al. (1998a) suggest that such stars are somehow particularly susceptible to a thermal pulse during their immediate post-AGB phase.

An alternative possibility is that the crystalline ice and silicates are pre-existing, and residing in a Kuiper belt or inner Oort comet cloud around CPD. The radius of 1500 AU is comparable to current estimates for the outer edge of the Kuiper belt around our own Sun (Weissman 1995). The interaction of cometary nuclei in such a belt with CPD's mass outflow and ionization front could provide the conditions needed to liberate the small particles ($< 3 - 10$ μm radius) that are required in order to explain the observed far-infrared silicate and ice bands. It is, therefore, of particular interest to note the presence of the same bands in the ISO spectrum of Comet Hale-Bopp (Crovisier et al. 1997). The annealing and recrystallization of silicates and ice grains liberated from comets, leading to the required highly ordered structures with correspondingly "sharp" emission features, may have resulted from the sudden increase in the UV photon flux from CPD during its post-AGB evolution.

References

Aitken D.K., Barlow M.J., Roche P.F., Spenser P.M., 1980, MNRAS 192, 679

Barlow M.J., 1998, in ISO's View on Stellar Evolution, eds. L.B.F.M. Waters, C. Waelkens, K.A. van der Hucht & P.A. Zaal, Kluwer, Dordrecht, Ap&SS 255, 315

Boothroyd A.I., Sackmann I.-J., 1988, ApJ 328, 653

Clegg P.E., et al., 1996, A&A 315, L38

Cohen M., Barlow M.J., 1980, ApJ 238, 585

Cohen M., Barlow M.J., Kuhi L.V., 1975, A&A 40, 291
Cohen M., Tielens A.G.G.M., Allamandola L.J., 1985, ApJ 299, L93
Cohen M., et al., 1989, ApJ 341, 246
Cohen M., Walker R.G., Witteborn F.C., 1992, AJ 104, 2030
Crovisier J., et al., 1997, Science 275, 1904
de Graauw Th., et al., 1996, A&A 315, L49
De Marco O., Barlow M.J., Storey P.J., 1997, MNRAS 292, 86 (DMBS)
Glaccum W., 1990, Ph.D. dissertation, Univ. of Chicago
Koike C., Shibai H., Tuchiyama A., 1993, MNRAS 264, 654
Menzies J.W., Wolstencroft R.D., 1990, MNRAS 247, 177
Omont A., et al., 1990, ApJ 355, L27
Payne H.E., Phillips J.A., Terzian Y., 1988, ApJ 326, 368
Roche P.F., Aitken D.K., 1986, MNRAS 221, 63
Roche P.F., Allen D.A., Bailey J.A., 1986, MNRAS 220, 7P
Schmitt B., Quirico E., Trotta F., Grundy W.M., 1998, in Solar System Ices, B. Schmitt, C. de Bergh & M. Festou (eds.), Kluwer, Dordrecht, Astrophys. Space Sci. Lib. 227, 199
Swinyard B.M., et al., 1996, A&A 315, L43
Trotta F., 1996, Ph.D thesis, Univ. of Grenoble, France
Waters L.B.F.M., et al., 1996, A&A 315, L361
Waters L.B.F.M., et al., 1998a, A&A 331, L61
Waters L.B.F.M., et al., 1998b, Nature 391, 868
Webster B.L., Glass, I.S., 1974, MNRAS 166, 491
Weissman P.R., 1995, ARA&A 33, 327
Zijlstra A.A., Gaylard M.J., te Lintel Hekkert P., Menzies J., Nyman L.-Å., Schwarz H.E., 1991, A&A 243, L9

Asymptotic Giant Branch Stars
T. Le Bertre, A. Lèbre and C. Waelkens, eds.
I.A.U. Symposium 191, 1999

On the Origin of the 21 Micron Feature in Post-AGB Stars

Sun Kwok
Department of Physics & Astronomy, University of Calgary, Calgary, Canada

Kevin Volk
Department of Physics & Astronomy, University of Calgary, Calgary, Canada

Bruce J. Hrivnak
Department of Physics & Astronomy, Valparaiso University, Valparaiso, Indiana, USA

Abstract. The unidentified emission feature at 21 μm is now observed in 12 sources, all being objects in transition between the asymptotic giant branch and planetary nebulae phases. The relations between the 21 μm and other emission features, such as the PAH features and the broad 30 μm feature, and the possible origins of the 21 μm feature are discussed.

1. Introduction

Stars on the asymptotic giant branch (AGB) are classified as oxygen (M) or carbon (C) rich based on their photospheric abundances. Over 4 000 and 700 stars have been found to show the silicate and the SiC features, respectively, in the IRAS Low Resolution Spectrometer (LRS) database and these stars are generally associated with M and C stars, respectively (Kwok, Volk & Bidelman 1997). In addition to these features, strong emission features attributed to the PAH molecules are also found in planetary nebulae (PN). These molecules are believed to be synthesized in the AGB envelopes but are only excited by the UV radiation field during the PN stage. It therefore came as a surprise that proto-planetary nebulae (PPN), objects in transition between the AGB and PN phases, were found to show an entirely new, strong emission feature at 21 μm, a feature which is not seen in AGB stars or PN.

The unidentified emission feature at 21 μm was first discovered in the IRAS LRS spectra of four carbon-rich post-AGB stars (Kwok, Volk & Hrivnak 1989). Since the initial discovery, this feature has been detected by both airborne (Omont et al. 1995) and ground-based observations (Kwok, Hrivnak & Geballe 1995; Justtanont et al. 1996).

To date, twelve 21 μm sources have been identified. Their properties are summarized in Table 1. All these sources have relatively uniform properties. They are mostly carbon-rich F and G supergiants with strong infrared excesses.

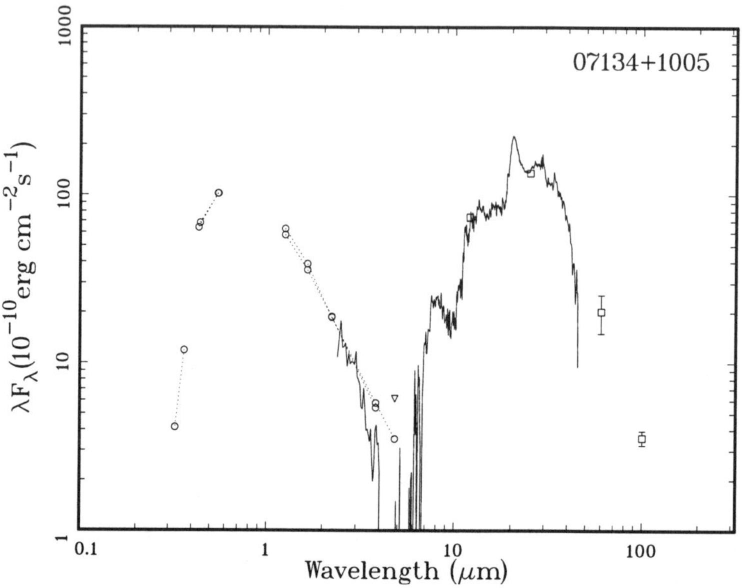

Figure 1. The spectral energy distribution of a typical 21 μm source. The high temperature component is the reddened photosphere and the low temperature component is the remnant dust envelope from the AGB phase.

Table 1. List of 21 μm sources and their common properties

Name	21μm	30 μm	Sp. Ty.	Optical	PAH features	Mol. lines
02229+6208	medium	...	G8-K0 0-Ia	C_2, C_3	...	CO
04296+3429	strong	...	G0 Ia	C_2, C_3, CN	3.3, 3.4–3.5, 7.7,11.3	CO, HCN
05113+1347	medium	...	G8 Ia	C_2, C_3, CN	3.3, 11.3	CO
07134+1005	v. strong	medium	F5 I	C_2, CN	3.3, 6.9	CO, HCN
16594−4656	strong	...	...	...	...	CO
19500−1709	v. weak	...	F3 I	...	...	CO, HCN
20000+3239	weak	v. strong	G8 Ia	C_2, CN	7.7, 11.3	CO, HCN
AFGL 2688	weak	medium	F5 Iae	C_2, C_3, CN	3.3, 3.4–3.5	CO, HCN
22223+4327	medium	...	G0 Ia	C_2, C_3, CN	...	CO, HCN
22272+5435	strong	v. strong	G5 Ia	C_2, C_3, CN	3.3,3.4–3.5, 6.9, 7.7, 11.3	CO, HCN, CS
22574+6609	medium	...	...	...	7.7, 11.3	CO
23304+6147	v. strong	v. strong	G2 Ia	C_2, C_3, CN	7.7, 11.3	CO, HCN

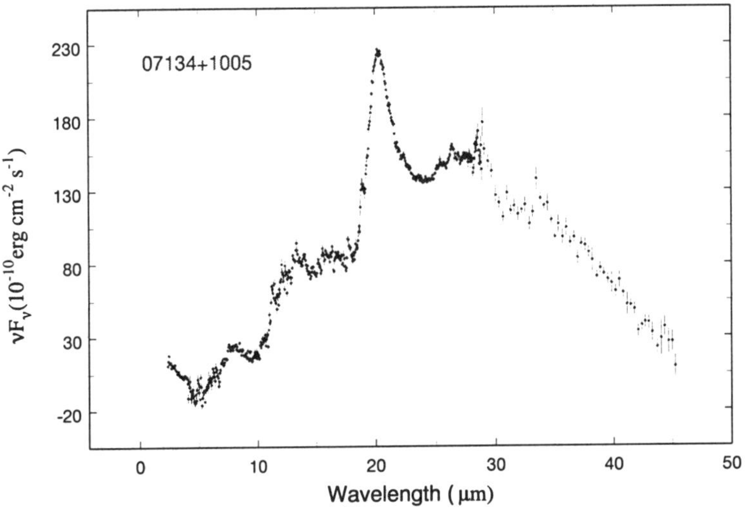

Figure 2. ISO SWS01 spectrum of IRAS 07134+1005.

2. The post-AGB nature of 21 μm sources

The post-AGB nature of the 21 μm sources is evident from (1) their intermediate spectral types (F-G); (2) the red IRAS colors (intermediate between the reddest AGB stars and PN); (3) the presence of C_2 and CN (Hrivnak 1995; Bakker et al. 1997) absorption bands consistent with large carbon enrichment through the 3rd dredge-up process; (4) excess of s-process elements (Van Winckel, this volume); and (5) their "double-peaked" spectral energy distributions, characteristic of dust shells detached from the stellar photospheres, implying the termination of mass loss sometime in the past (Fig. 1). These properties make the 21 μm sources the most unambiguous PPN objects.

3. ISO observations of 21 μm sources

Fig. 2 shows a typical infrared spectrum of a 21 μm source. Other than the strong, broad emission feature at 21 μm, the spectrum shows an almost flat continuum between 12 and 18 μm, and another broad feature between 11 and 13 μm. The sharp feature at 11.3 μm is likely to be associated with the C-H out-of-plane bend of PAH molecules. An unidentified, very broad emission feature at 30 μm is also often found in 21 μm sources.

The 21 μm features have almost exactly the same shape in each source after continuum subtraction. Fig. 3 shows the profiles of the 21 μm feature of 3 different sources at the higher resolution SWS06 mode of ISO ($\Delta\lambda/\lambda \sim 500$). The feature spans 19 to 23 μm with no obvious narrow substructures.

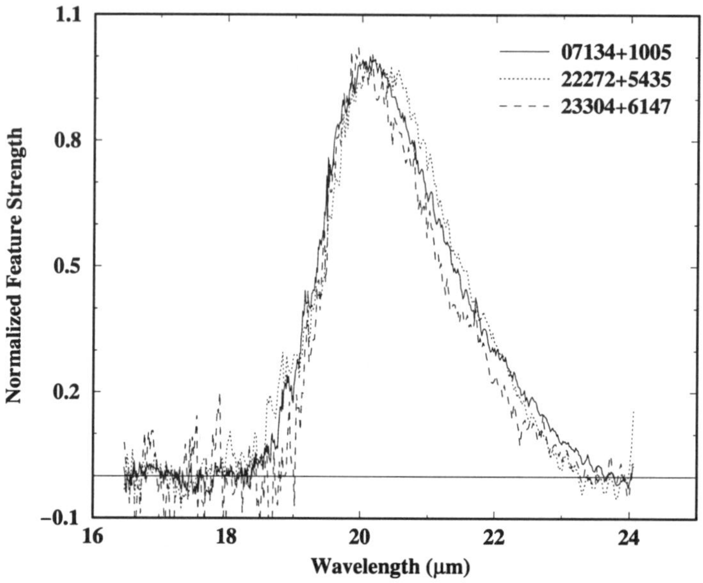

Figure 3. Overlay of the ISO SWS06 spectra of three 21 μm features.

4. Relationship with the 30 μm feature

The detection of a 30 μm emission feature in four 21 μm sources (07134+1005, 20000+3239, 22272+5435 and 23304+6147; Omont et al. 1995) suggests that these two features may share a common origin. However, the 30 μm feature has been seen in AGB stars (e.g. IRC +10216) and PN (e.g. IC 418) as well as in PPN (Cox 1993), while the 21 μm feature thus far appears to be restricted to PPN.

5. Relationship with the PAH molecule

There is strong evidence that the 21 μm feature is related to the PAH molecules. The 3.4-3.5 μm emission is particularly strong in 21 μm sources (Geballe et al. 1992). However, the 3.4-3.5 μm feature is absent in the strongest 21 μm emission source (IRAS 07134+1005; Kwok et al. 1990), so the correlation is not perfect.

The 3.4-3.5 μm feature has been suggested to be due to the hot bands of the fundamental 3.3 μm (CH stretch) feature (Barker et al. 1987), although this cannot be the case in PPN because the 3.4-3.5 μm feature is almost as strong as the 3.3 μm feature. More possibly, the 3.4-3.5 μm feature is due to aliphatic sidegroups (-CH_2, -CH_3) attached to the PAH molecules (Jourdain de Muizon et al. 1989).

The 6.9 μm feature is not seen in PN but is strong in 21 μm sources. It is therefore possible that the 3.3-3.4, 6.9, and 21 μm sources are related. However, IRAS 07134+1005 has a strong 6.9 μm feature but not the 3.4 μm feature.

6. Origin of the 21 μm feature

Many candidates have been proposed as possible carriers of the 21 μm feature, including:

- amides: urea or thiourea (Sourisseau et al. 1992)
- hydrogenated amorphous carbon (HAC; Buss et al. 1990)
- hydrogenated fullerenes ($C_{60}H_m$, m = 1, 2, ...60, or fulleranes; Webster 1995)
- maghemite (Fe_2O_3) or magnetite (Fe_3O_4; Cox 1990)
- solid SiS_2 (Nuth et al. 1985; Goebel 1993; Begemann et al. 1996)

In view of the fact that the 21 μm feature is seen exclusively in carbon-rich objects, it seems unlikely that it is due to inorganic materials such as SiS_2 or Fe_2O_3. The correlation of the 21 μm feature with the PAH features suggests that its carrier is a large carbon-based molecule. Most PAH molecules have ring deformation vibration modes around 20 μm, and mid-infrared laboratory emission spectroscopy of simple PAH molecules such as naphthalene, pyrene, and chrysene shows broad emission features in that region (P. Bernath, private communication).

Considering that fullerenes (C_{60}) are the 3-D analogues of the 2-D PAH molecules, their existence in carbon-rich objects is not unexpected. Emission spectra of gas-phase C_{60} show strong features at 7.1, 8.6, 17.5, and 19 μm (Frum et al. 1991), whereas the lowest vibrational mode of $C_{60}H_{60}$ is calculated to be at 23 μm (Webster 1995). For the other unsaturated hydrides ($C_{60}H_m$, m = 1 to 59), the perfect symmetry of the molecule is broken and the mode is no longer a narrow feature but bands peaking at wavelengths from 19 to 23 μm as m changes from 0 to 60. Webster (1995) suggests that the observed broad 21 μm feature is the result of a mixed population of fulleranes.

7. Evolutionary scenario

As a star evolves from the AGB to PN, its circumstellar environment (density, temperature, and radiation background) undergoes drastic changes. The dominant infrared emission features will likely change as the conditions for molecular formation, destruction, and excitation change. Acetylene (C_2H_2), widely considered to be the first building block of PAH molecules, has recently been detected in the atmospheres of carbon stars (Yamamura et al., this volume). Thus it is possible that PAH molecules are synthesized in AGB stars, but that the molecules are not excited because of the low temperatures of the central stars. HAC grains, made up of aromatic rings of various sizes bonded peripherally to polymeric or hydrocarbon species (Duley & Willams 1986), can also be made during the late AGB phase. Extreme carbon stars on the AGB have featureless dust continua (Volk et al. 1992) and these could be due to HAC grains. During the transition from the AGB to PN, these grains may be gradually shattered into smaller PAH molecules, possibly by high-velocity outflows now observed in

PPN. Then as the UV radiation background increases as the central star evolves to higher temperatures, the strengths of the PAH features increase. If 21 μm features are due to HAC grains, then the destruction of HAC grains will also imply the disappearance of the 21 μm feature beyond the PPN phase.

If the 21 μm feature is related to fulleranes and the disappearence of the 21 μm feature is the result of the loss of H atoms, then fullerenes should be detected in PN if they can be excited by UV photons or collisions. The infrared features of C_{60} have been searched in R CrB and IRC +10216 without success (Clayton et al. 1995), but a more extensive search in PN may be fruitful.

References

Bakker E.J., van Dishoeck E.F., Waters L.B.F.M., Schoenmaker T., 1997, A&A 323, 469

Barker J.R., Allamandola L.J., Tielens A.G.G.M., 1987, ApJ 315, L61

Begemann B., Dorschner J., Henning Th., Mutschke H., 1996, ApJ 464, L195

Buss R.H., Cohen M., Tielens A.G.G.M., Werner M.W., Bregman J.D., Rank D., Witteborn F.C., Sandford S., 1990, ApJ 365, L23

Clayton G.C., Kelly D.M., Lacy J.H., Little-Marenin I.R., Feldman P.A., Bernath P.F., 1995, AJ 109, 2096

Cox P., 1990, A&A 236, L29

Cox P., 1993, in *Astronomical Infrared Spectroscopy*, S. Kwok (ed.), ASP Conf. Ser. 41, p. 163

Duley W., Willams D.A., 1986, MNRAS 219, 859

Frum C.I., Engleman R., Hedderich H.G., Bernath P.F., Lamb L.D., Huffman D.R., 1991, Chem. Phys. Let. 176, 504

Geballe T.R., Tielens A.G.G.M., Kwok S., Hrivnak B.J., 1992, ApJ 387, L89

Goebel J.H., 1993, A&A 278, 226

Hrivnak B.J., 1995, ApJ 438, 341

Jourdain de Muizon M., Cox P., Lequeux J., 1989, A&AS 83, 337

Justtanont K., Barlow M.J., Skinner C.J., Roche P.F., Aitken D.K., Smith C.H., 1996, A&A 309, 612

Kwok S., Hrivnak B.J., Geballe T.R., 1990, ApJ 360, L23

Kwok S., Hrivnak B.J., Geballe T.R., 1995, ApJ 454, 394

Kwok S., Volk K., Bidelman W.P., 1997, ApJS 112, 557

Kwok S., Volk K., Hrivnak B.J., 1989, ApJ 345, L51

Nuth J.A., Moseley S.H., Silverberg R.F., Goebel J.H., Moore W.J., 1985, ApJ 290, L41

Omont A., et al., 1995 ApJ 454, 819

Sourisseau C., Coddens G., Papoular R., 1992, A&A 254, L1

Volk K., Kwok S., Langill P.P., 1992, ApJ 391, 285

Webster A., 1995, MNRAS 277, 1555

Part 4. Circumstellar Envelopes

Alfred Glassgold and Robert Lucas probing organic molecules

Asymptotic Giant Branch Stars
T. Le Bertre, A. Lèbre and C. Waelkens, eds.
I.A.U. Symposium 191, 1999

Millimeter-wave Interferometry of Circumstellar Envelopes

Robert Lucas and Michel Guélin
Institut de RadioAstronomie Millimétrique, 300, rue de la Piscine, 38406 Saint Martin d'Hères, France

Abstract. We present some of the most significant results obtained in the last few years with the IRAM Plateau de Bure interferometer and with other instruments on the detailed mapping of circumstellar envelopes.

1. Introduction

In the last few years millimeter-wave observations of circumstellar envelopes have considerably improved in both angular resolution and sensitivity. This is due to progress in receiver technology, to the increasing collecting area available at all millimeter-wave interferometers, and to the possibility to use these instruments down to 1mm wavelength. Sub-arcsecond resolution maps may now be obtained for many objects. The spectacular images contain information on the detailed morphology of expanding circumstellar envelopes, both for standard AGB stars and more evolved proto-planetary envelopes; they allow to conduct studies on the chemistry in the innermost region of the outflows, as well as to map the remnants of old high mass-loss episodes.

The millimeter-wave interferometers currently in operation are Berkeley Illinois Maryland Array and the Owens Valley Radio Observatory (both in California, USA), the Nobeyama Millimeter Array (Nobeyama, Japan), and the Plateau de Bure Interferometer of IRAM (France). They operate in the 3 mm (80-115 GHz) and 1.3 mm (210-250 GHz) windows, and reach typical resolutions of $\sim 2''$ (at 3 mm) or $0.5''$ (at 1.3 mm), with fields of view of 20-120″, which may be extended by combining several fields (mosaics).

2. A CO survey of circumstellar envelopes

A CO survey of circumstellar envelopes has been undertaken at the Plateau de Bure interferometer by Neri et al. (1998). Forty-six AGB and post-AGB stars were mapped in the CO J=1–0 line in the snapshot mode. The IRAM 30 m telescope was used to obtain the J=1–0 short spacings and CO J=2–1 maps. The derived properties include positions (with an accuracy of 0.5″), fluxes, and radii using both model fits to the visibilities and single dish maps. Crude interferometer maps are also presented, despite the limited *uv* coverage. The observed line shapes are mostly characteristic of thick envelopes, while some objects may be two-component (detached) shells. The morphology, size and asymmetries were analysed. Most envelopes have a spherical appearance ($\sim 70\%$), but some shells present a significant asymmetry (RS Cnc, M1-92, R Cas), with a pronounced bipolar kinematical signature. Mass loss rates and ^{12}CO photodissociation radii are calculated for 38 stars, using the method of Loup et al. (1993). An empirical relation is found between the ^{12}CO photodissociation radii and the observed radii, indicating that the CO extents are limited by photodissociation by interstellar UV radiation.

3. Observations of IRC +10216

IRC +10216 (CW Leo) is the nearest carbon star; its distance is of the order of 150 pc. It has a massive envelope (mass loss rate $\sim 2\ 10^{-5}$ $M_\odot$), expanding at 14.5 km/s. The stellar radius is $R_* \sim 7\ 10^{13}$ cm. The optical nebula is elongated along a position angle of 20°. In the inner envelope, one detects photospheric species, affected by neutral reactions and interaction with dust in the zone of dust formation and acceleration. The outer envelope extends in CO out to 200″ radius. It has a rich carbon chemistry, with about 50 different molecules detected out to 30″ radius, which appears dominated by photodissociation and reactions with photodissociation products, as well as, possibly, photodesorption from dust grains (see the review by Glassgold in this volume).

3.1. The inner envelope

Continuum emission at millimeter wavelengths The continuum emission of IRC +10216 is easily detected with the IRAM interferometer at 3 mm wavelength (Lucas & Guilloteau 1992) and at 1.3 mm (Lucas 1997). At the highest frequencies the source emission may be split into a point source and extended emission (Lucas et al., in preparation). Table 1 gives a summary of the source properties.

Table 1. IRC +10216 continuum emission at millimeter wavelengths

Frequency (GHz)	Point source (mJy)	Extended emission (mJy)	size (″)
89.	65 ± 1	8 ± 1	2.4 ± 0.1
216.	316 ± 2	178 ± 5	1.3 ± 0.1
235.	407 ± 6	280 ± 42	1.8 ± 0.4
242.	486 ± 7	270 ± 22	1.6 ± 0.3

The point source is identified to the photospheric emission, while the extended part is due to thermal emission from dust in the inner envelope. Groenewegen et al. (1997) have observed the continuum emission at 1 mmusing a bolometer at the 30-m telescope. They find a peak flux of 0.7 Jy, after subtracting a contribution of similar magnitude due to line contamination in the bolometer passband. The integrated single dish flux in the primary beam of the 15m Plateau de Bure antennas, estimated from the bolometer map, is 1.6 Jy, which could mean, if the line contamination in the 30m data has been correctly estimated, that half of the emission from dust is resolved out at PdBI.

Vibrationally excited lines Vibrationally excited rotational lines of SiS and CS may be used to investigate the inner envelope of IRC +10216: their lower levels lie in energy about 10^3K above the ground state, and thus they can only be excited at a few stellar radii from the star in the presence of a high enough kinetic temperature or of a strong IR field. We have observed the $v = 1, J = 13 - 11$ transition of SiS and the $v = 1, J = 5 - 4$ transition of CS in IRC +10216, with the Plateau de Bure interferometer. Both transitions are in the 1.3 mmwindow, where the available resolution can reach $0.5''$ in the most extended configuration.

Both emissions are spatially resolved, and no departure from spherical symmetry is seen (Fig. 1). The source size, derived by a Gaussian fit to the visibilities, is $\sim 0.35''$ or 7 10^{14}cm, which represents $\sim$ 10 stellar radii, for both lines. The brightness temperatures are 1200 K for SiS and 900 K for CS, which means that both lines are close to be optically thick, given the kinetic temperatures expected at a few stellar radii (1000-1500 K). We have tentatively modelled the SiS emission, using source model parameters from Groenewegen (1997); the rotational emission from SiS in the excited vibrational state is due to collisional and radiative excitations (the source size at $\lambda = 10\mu$m is $0.9''$ in diameter); the emerging line profile and the spatial distribution are well reproduced with an abundance of SiS decreasing from 10^{-5} at 3 stellar radii to 7×10^{-6} at 12 stellar radii, in agreement with infrared absorption measurements (Boyle et al. 1994). The predicted size of the SiS $v = 1$ emission is mainly determined by the size of the 10 μm emitting region. We thus mainly check that the dust forming region actually coincides with the region where the AGB wind is accelerated. Measurements of this kind at higher angular resolution should in the next decade or so provide detailed information on the kinematics of this important part of the circumstellar envelope. Radiative transitions within the lowest bending states of HC_3N, C_3H, C_4H and MgNC, a few 10^2 K high in energy, are also detected at $12''$ radius.

Ground state lines The emission in the inner envelope is dominated by lines of CO, HCN, SiO, CS, SiS, SiC_2, etc. Some of their lines have been mapped with interferometers by Bieging et al. (1984), Bieging & Nguyen-Q-Rieu (1989), Bieging & Tafalla (1993), Dayal & Bieging (1995), Gensheimer et al. (1995); maps of SiS, SiC_2, CS have been presented by Lucas et al. (1995). Those maps lack the sub-arcsecond resolution needed to study the inner shell.

Recent, unpublished CO $J = 2-1$ observations at Plateau de Bure show no obvious departure from spherical symmetry at angular radii larger than $\sim 2''$. On the opposite, maps of SiS ($J = 12-11$) at 217 GHz show enhanced emission in the southern half on the envelope; apparently two inner shells are present at radii $\sim 4''$ and $\sim 8''$. They are also seen in AlF emission.

3.2. The outer envelope

Many maps have been made, using the Hat Creek interferometer (Bieging et al. 1984; Bieging & Nguyen-Q-Rieu 1988, 1989; Bieging & Tafalla 1993; Dayal & Bieging 1993, 1995; Gensheimer et al. 1995), which support a chemistry

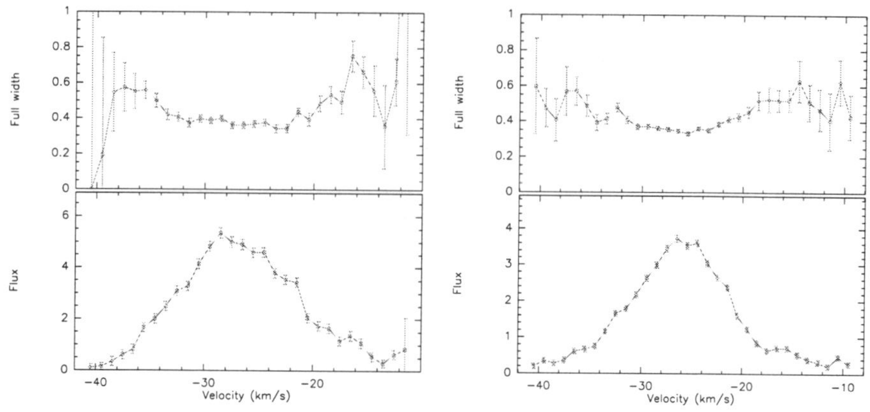

Figure 1. SiS $v = 1, J = 13 - 11$ (left) and CS $v = 1, J = 5 - 4$ (right) emission profiles from IRC +10216. We have plotted the result of Gaussian source model fits to the visibilities: the upper boxes give the size at half-power, in arcseconds, and the lower boxes the integrated fluxes in Jy.

based on photochemical and radical reactions (see the review by Glassgold in this volume). More detailed maps are now available from the Plateau de Bure interferometer (Guélin et al. 1993a, 1993b, 1996; Lucas et al. 1995). They are obtained at better angular resolution (3″) and give more insights on the detailed structure of the outer envelope. More species are mapped, resulting from the HCN chemistry (HCN, CN, HNC, C_3N, $HC^{13}CCN$, HC_5N), from the C_2H_2 chemistry (C_2H, C_3H, C_4H, C_6H), but also species containing metals, like NaCl, NaCN, AlF, MgNC (Guélin et al. 1993a). A recent map of the ion HCO^+ is shown in Fig. 2 (see Glassgold's review for a discussion of this result).

A comparison of the spatial distributions of four species is given in Fig. 3. Both HNC and HC_3N are less extended than CN; emissions of similar products with different number of carbon atoms tend to be coincident, e.g. C_3H and C_4H, or HC_5N and HC_3N. The latter is not really expected from current theory, where chains are built step by step through gas phase reactions, which would predict a larger radial extent for the longer carbon chains. In fact these models are assuming constant and isotropic mass loss, while the maps show considerable structure, both radially and azimuthally, in most species. For instance the CN and HNC emissions show clearly a double shell structure, especially in the NW region.

In Fig. 4 we have plotted the location of the emission maxima ('clumps') in all maps obtained so far. A general picture emerges where a series of concentric shells are present, at approximate radii of 4, 8, 15, and 25″. The shell at 15″ is the one where the the largest number of different species are found; it appears smaller in CS and SiC_2 than in HNC and CN. At this point it is not yet clear if we are seeing substructure in this shell or only chemical segregation.

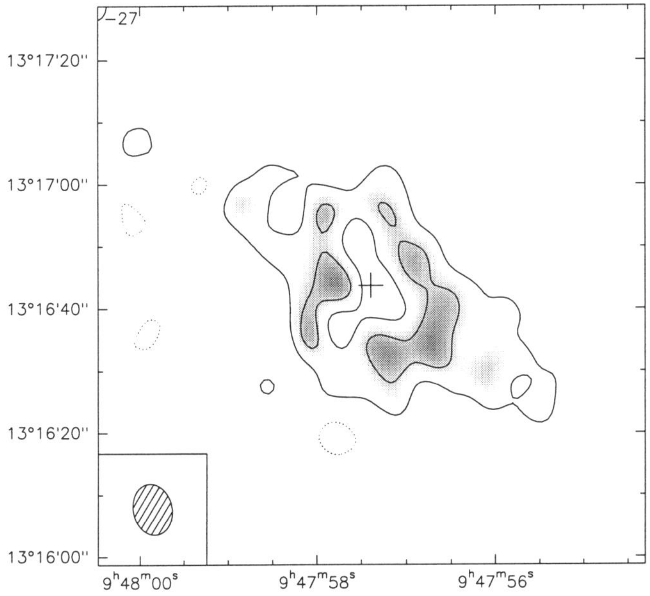

Figure 2. The IRC +10216 emission of HCO^+ mapped with the Plateau de Bure interferometer. The central 10 km/s of the emission profile are integrated. This is a very weak line: the contour spacing is 5 mJy/beam.

4. Detached envelopes

The rate at which AGB stars lose mass is known to evolve with time, very probably both on long time scales and on short time scales. Thermal pulses and mechanical oscillations are believed to modulate on short time scales. The most direct indications for actual changes of mass loss rates came from the detection of geometrically thin CO shells around a few carbon stars: R Scl, U Ant, S Sct, and TT Cyg (Olofsson et al. 1990, 1996). Two other carbon stars have large, detached dust shells (Waters et al. 1994; Izumiura et al. 1996). The CO shells contain a few hundredths of a solar mass. The shell widths, however, cannot be resolved by single dish observations. Only two such objects have been so far studied with interferometers, U Cam and TT Cyg.

The shell around U Cam was observed in HCN and CN at Plateau de Bure by Lindqvist et al. (1996). They separated an outer extended shell of radius 7 10^{16}cm, expanding at 25 km/s, and an inner envelope, of radius 6 10^{15}cm, expanding at 13 km/s. A preliminary ^{12}CO $J = 2-1$ map by Linqvist et al. (in prep.) was obtained at Plateau de Bure, which unambiguously shows the detached nature of the outer shell.

The shell around TT Cyg was partially mapped in ^{12}CO $J = 1-0$ and $2-1$ by Olofsson et al. (1996, 1998). They found a remarkably thin shell of CO gas (radius 34$''$, thickness 1.3$''$), containing 0.02 solar masses of gas. The shell shows only small deviations from spherical symmetry. Mass loss is currently occurring

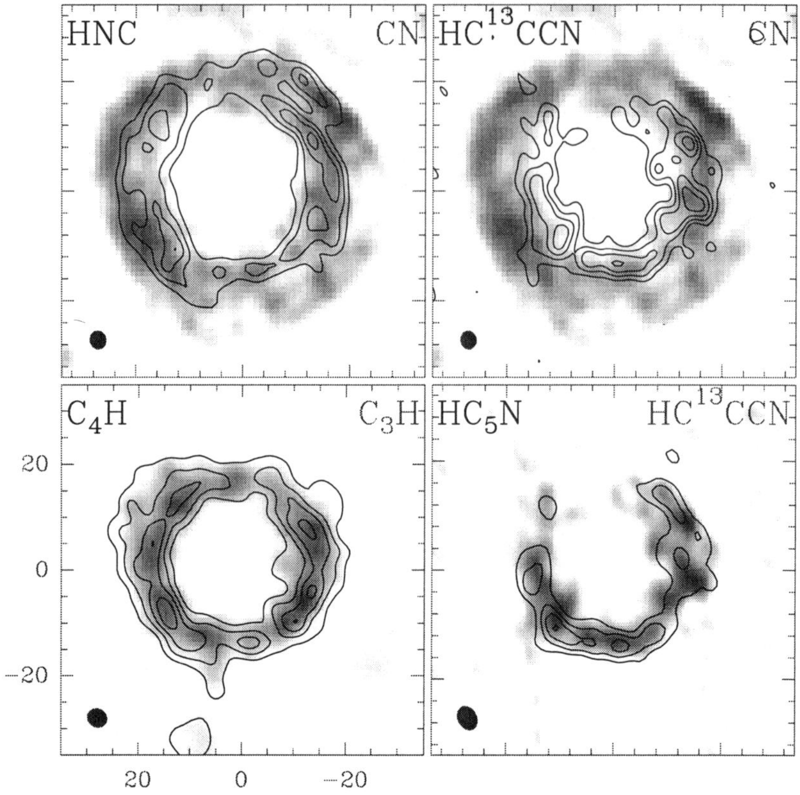

Figure 3. Comparison of the extents of various molecular emissions in the outer envelope of IRC +10216. All plots represent the emission at the systemic velocity (-27 km/s) and are at the same angular scale. Grey scale correspond to the species noted in the upper right corners, contours to those noted in the upper left. Except for CN, no correction has been made for primary beam attenuation. Beam widths are drawn in the lower left corner of each box.

at a rate of $\sim 10^{-8}$ $M_\odot$/yr. A preliminary map of the complete shell is shown in Fig. 5.

5. The multiple molecular jets in AFGL 2688

AFGL 2688, the Egg Nebula, is one of the few objects in the transition phase from the AGB to the planetary nebula phase. It left the AGB about a hundred years ago. Its optical appearance in the visible and in the infrared continuum is that of a bipolar nebula (Sahai et al. 1998a, 1998b), scattering the light of the cool central star, which is surrounded by a cold and dense circumstellar envelope shocked by a warm fast wind (Young et al. 1992). Observations of H_2 in the near-infrared have revealed that high velocity gas is also present in the east-west

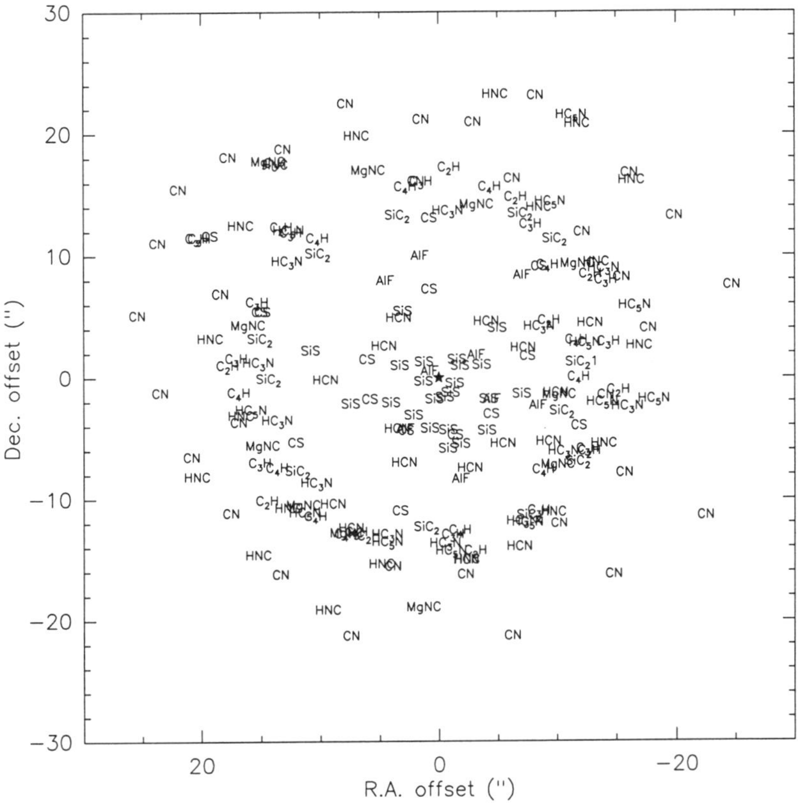

Figure 4. The location of clumps in the IRC +10216 envelope. The relative maxima in the emission map of each species are plotted on the same map and labelled with the name of the molecule.

direction, suggesting a quadrupolar outflow (e.g.: Cox et al. 1997; Sahai et al. 1998b).

Previous interferometric images in the molecular lines (Yamamura et al. 1995; Bieging & Nguyen-Q-Rieu 1996; Kasuga et al. 1997) lacked the resolution to investigate the detailed kinematics of this source. We have made ^{12}CO $J = 2-1$ observations of this source at Plateau de Bure (Cox et al., in preparation), with an angular resolution of $1.0 \times 1.6''$. Four outflows are observed, which are obviously linked to the four H_2 emission regions, and a strong 1 mm continuum (peak flux 200 mJy/beam) is coincident with the obscuring dust in the IR maps (Fig. 6).

6. Prospects

On the long term, considerable progress in this field is expected from the availability of a new generation of interferometers during the next decade. Large arrays are being planned, such as the MMA in the US, the LSA in Europe,

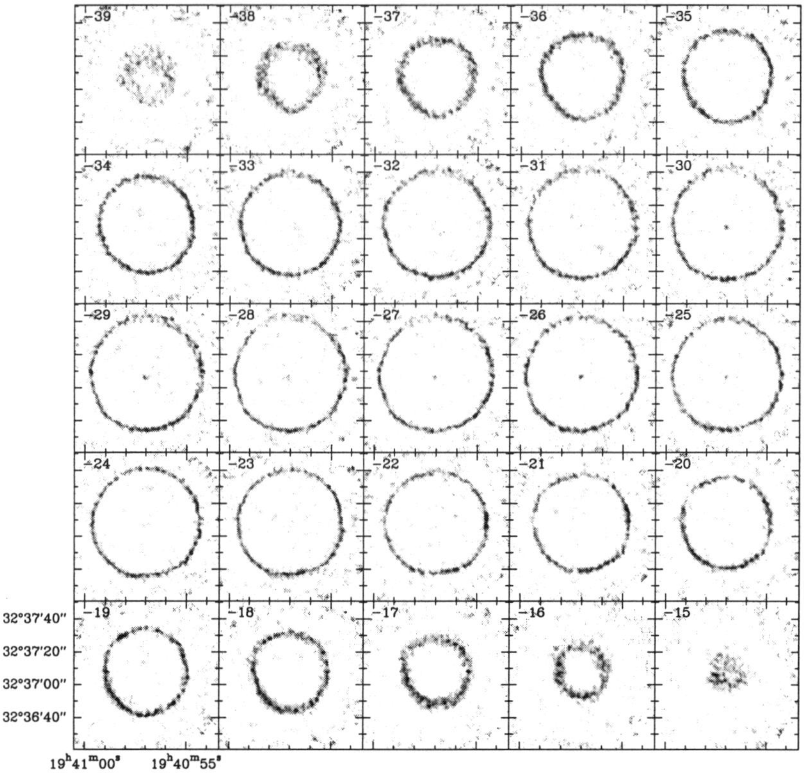

Figure 5. TT Cyg ^{12}CO $J = 1-0$ emission from Olofsson et al. (in preparation). The channel maps are labelled by their LSR velocity in km/s.

the LMSA in Japan. It is expected that at least two of these projects will be merged into a single, even more powerful instrument. Collecting areas of several thousand square meters are foreseen. Those instruments should reach $\sim 0.1''$ resolutions, enabling much more detailed studies of the prototypal objects, down to a few stellar radii, and to investigate the problems of dust formation and wind acceleration. The frequency coverages will be extended to the sub-millimeter range, where dust emission is stronger, and other chemical species (hydrides) may be observed. The ten-fold improved sensitivity will enable to map circumstellar envelopes of AGB stars up to the center of the Galaxy.

References

Bieging J.H., Chapman B., Welch W.J., 1984, ApJ 285, 656

Bieging J.H., Nguyen-Q-Rieu, 1989, ApJ 343, L25

Bieging J.H., Nguyen-Q-Rieu, 1996, AJ 112, 706

Bieging J.H., Nguyen-Q-Rieu, 1988, ApJ 329, L107

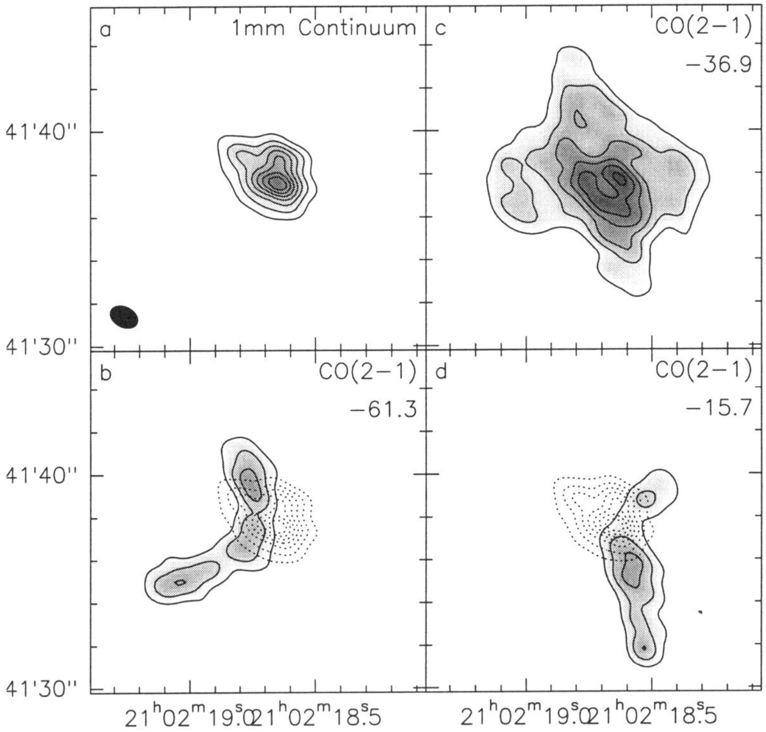

Figure 6. AFGL 2688: CO emission in the central channel; CO emission in two outer channels overlaid with the H_2 emission.

Bieging J.H., Tafalla M., 1993, AJ 105, 576

Boyle R.J., Keady J.J., Jennings D.E., Hirsch K.L., Wiedemann G.R., 1994, ApJ 420, 863

Cox P., Maillard J.-P., Huggins P., Forveille T., Simons D., Guilloteau S., Rigaut F., Bachiller R., Omont A., 1997, A&A 321, 907

Dayal A., Bieging J.H., 1993, ApJ 407, L37

Dayal A., Bieging J.H., 1995, ApJ 439, 996

Gensheimer P.D., Likkel L., Snyder L.E., 1995, ApJ 439, 445

Groenewegen M.A.T., 1997, A&A 317, 503

Groenewegen M.A.T., van der Veen W.E.C.J., Lefloch B., Omont A., 1997, A&A 322, L21

Guélin M., Lucas R., Cernicharo J., 1993a, A&A 280, L19

Guélin M., Lucas R., Neri R., 1993b, in *34th Herstmonceux Conference*, ed. R. Clegg, Cambridge

Guélin M., Lucas R., Neri R., 1996, in Science with large millimeter arrays, P.A. Shaver (ed.), Springer, p. 276

Izumiura H., Hashimoto O., Kawara K., Yamamura I., Waters L.B.F.M., 1996, A&A 315, L221

Kasuga T., Yamamura I., Deguchi S., 1997, A&A 320, 575

Lindqvist M., Lucas R., Olofsson H., Omont A., Eriksson K., Gustafsson B., 1996, A&A 305, L57

Loup C., Forveille T., Omont A., Paul J.F., 1993, A&AS 99, 291

Lucas R., 1997, Ap&SS 251, 247

Lucas R., Guélin M., Kahane C., Audinos P., Cernicharo J., 1995, Ap&SS 224, 293

Lucas R., Guilloteau S., 1992, A&A 259, L23

Neri R., Kahane C., Lucas R., Bujarrabal V., Loup C., 1998, A&AS 130, 1

Olofsson H., Bergman P., Eriksson K., Gustafsson B., 1996, A&A 311, 587

Olofsson H., Bergman P., Lucas R., et al., 1998, A&A 330, L1

Olofsson H., Carlstrom U., Eriksson K., Gustafsson B., Willson L.A., 1990, A&A 230, L13

Sahai R., Hines D.C., Kastner J.H., Weintraub D.A., Trauger J.T., Rieke M.J., Thompson R.I., Schneider G., 1998a, ApJ 492, L163

Sahai R., Trauger J.T., Watson A.M., et al., 1998b, ApJ 493, 301

Waters L.B.F.M., Loup C., Kester D.J.M., Bontekoe T.R., de Jong T., 1994, A&A 281, L1

Yamamura I., Onaka T., Kamijo F., Deguchi S., Ukita N., 1995, ApJ 439, L13

Young K., Serabyn G., Phillips T.G., Knapp G.R., Guesten R., Schulz A., 1992, ApJ 385, 265

Asymptotic Giant Branch Stars
T. Le Bertre, A. Lèbre and C. Waelkens, eds.
I.A.U. Symposium 191, 1999

A MERLIN movie of mass-loss from RT Vir

A. M. S. Richards, R. J. Cohen, I. Bains

N.R.A.L., Jodrell Bank, University of Manchester, U.K.

J. A. Yates

Department of Physical Sciences, University of Hertfordshire, U.K.

Abstract. We used MERLIN to observe RT Vir at 22 GHz at six epochs during 10 weeks. The water maser emission comes from a thick expanding shell with an elliptical velocity field. MERLIN has a velocity resolution of 0.1 km s^{-1} and milli-arcsecond angular resolution, revealing details within the individual maser clouds, typically 12 mas in diameter spanning 15 velocity channels. The brightest peak doubles in intensity to 800 Jy/beam. Features at velocities close to the stellar velocity show the largest proper motions of $\sim$ 3 mas away from the centre of emission. Some features are seen near the outer limits to the maser shell at early epochs only, but new masers appear close to the inner rim. The variability of individual maser features is not a simple function of the stellar luminosity.

1. Observations of RT Vir

RT Vir is a semi-regular variable star (SRb) (Kholopov et al. 1985) which may be in a pre- or post- Mira phase (Szymczak et al. 1998), or an AGB star from a distinct population (Whitelock et al. 1994). Optical data provided by the AAVSO and AFOEV[1] shows irregular variations between mag. 9.3 – 8.2. Fig. 1a shows its state during our radio observations. It is classed as an unsolved variable by HIPPARCOS; the parallax gives a distance of 140 ± 20 pc. The mass loss rate estimated from CO observations is 3×10^{-6} $M_{\odot}$ yr^{-1} (Bujarrabal et al. 1989). The stellar velocity V_* is 18.2 km s^{-1} (Nyman et al. 1986) which is close to the midpoint of its strong H_2O maser emission, weak OH mainline masers, and CO (e.g. Mendoza-Torres et al. 1997; Kirrane 1986; Stanek et al. 1995).

RT Vir was observed using MERLIN[2] at 22 GHz at epochs 1 – 6 during April – June 1996. Details are given in Table 1 along with a 1994 MERLIN observation (Bains 1995). The maximum baseline is 217 km giving $\lambda/d \approx 12$ mas. A bandpass of 2 MHz was divided into 256 channels providing a velocity

---

[1]This research has made use of the databases of these organisations of variable star observers, which we acknowledge with thanks.

[2]The Multi Element Radio Linked Interferometer Network, operated by the University of Manchester on behalf of PPARC `http://www.jb.man.ac.uk/merlin/`

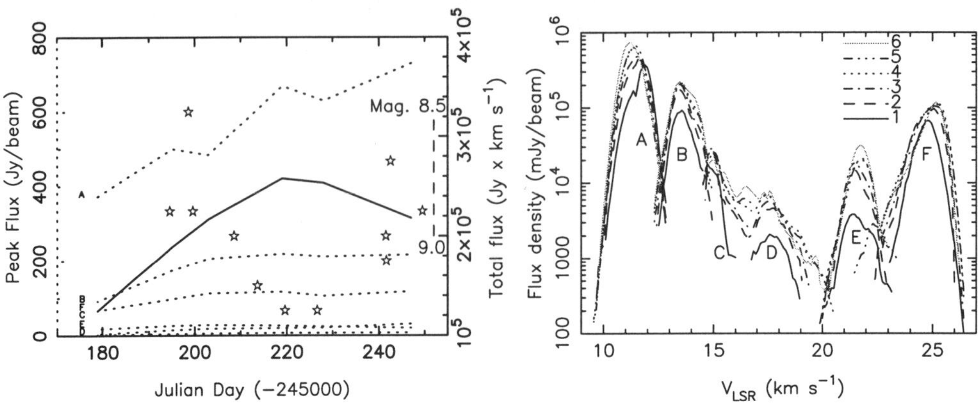

Figure 1. *left* (a) RT Vir total 22 GHz flux density (solid line) and selected peaks of features A – F (dotted lines) at six epochs. The stars show the optical magnitude, scale is dashed line. *right* (b) Velocity profiles of features A – F at epochs 1 – 6.

resolution of 0.105 km s^{-1}. 3C273 (flux density 31 Jy, Stevens et al. 1998) was used to establish the flux scale and correct the spectral bandpass at each epoch.

Table 1. MERLIN 22 GHz observations of RT Vir since 1994

Epoch	1994	1	2	3	4	5	6
JD 2400000+	49459	50179	50195	50203	50219	50228	50247
Duration (hr)	11.7	4.5	6.3	11.5	9.5	7.7	8.4

At every epoch a map of the brightest channel of RT Vir was used to derive phase and amplitude corrections by self-calibration and the solutions were applied to all channels. The data cubes were mapped and CLEANed using a 12 mas diameter restoring beam. The off-source noise level was $\sigma_{\rm rms} \approx 20$ mJy/beam. 2D Gaussian components were fitted to each patch of flux in each channel above the greater of $5\sigma_{\rm rms}$ or 2 – 4% of the peak. The relative position accuracy is proportional to the signal to noise ratio. Under these conditions it is typically 1 mas for a 1 Jy component (Richards et al. 1998b; Condon 1997).

2. Results

2.1. The RT Vir H_2O maser shell

Emission at 22 GHz was detected between 9 and 27 km s^{-1}. The brightest peak occurs between 11.27 and 11.52 km s^{-1} during these observations. It was used to align the epochs for this preliminary analysis. Fig. 1b shows the velocity profiles of selected features at all epochs and Fig. 2a shows emission integrated across all channels. Fig. 2b shows the the flux density and velocity of all maser components at epoch (1). These Figures show that the brightest maser emission

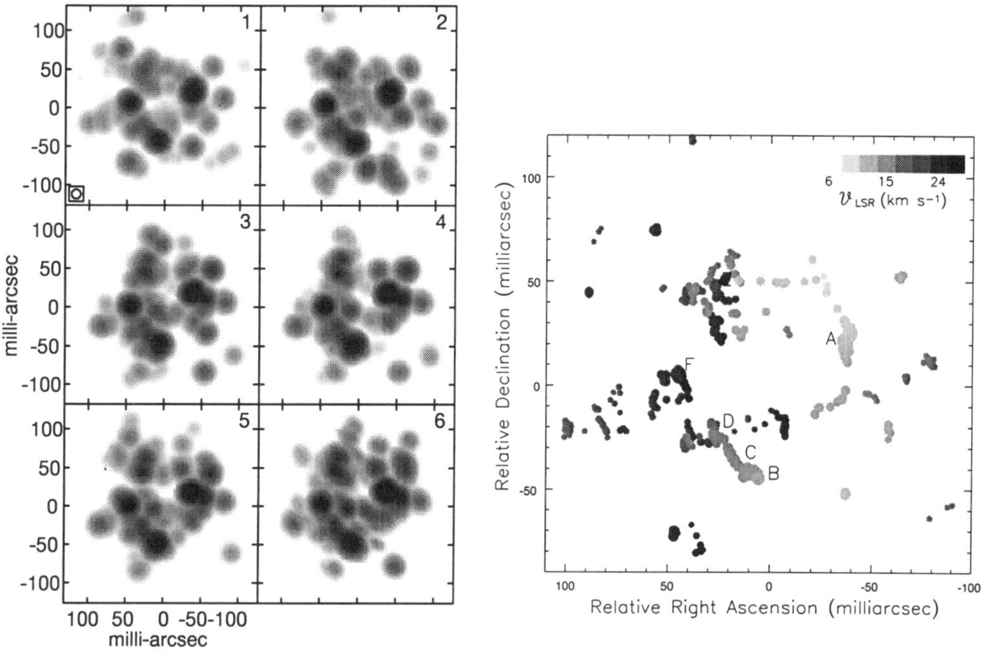

Figure 2. *left* (a) Maps of total 22 GHz flux from RT Vir at epochs 1 – 6. *right* (b) Individual maser components at epoch 1. Symbol diameter is proportional to log(flux density) and the shade shows velocity.

forms a ring typical of tangential beaming due to an outflow accelerating radially away from the star (Chapman & Cohen 1985). The centre of emission was found by assuming that the maser distribution forms a shell which is point-symmetric in space and velocity, this is at (0, 0) in Fig. 2.

Some red- and blue-shifted emission is seen in most parts of the shell, and emission close to V_* is seen at large angular separations at all points of the compass. The simplest model is a thick spherical shell with inner and outer radii and expansion velocities marked in Fig. 3. These are respectively $r_{\rm i} = 6.9 \times 10^{11}$ m; $r_{\rm o} = 28.8 \times 10^{11}$ m; $v_{\rm i} = 3.5\,{\rm km\,s^{-1}}$; $v_{\rm o} = 11.5\,{\rm km\,s^{-1}}$. This gives a logarithmic velocity gradient $\epsilon \sim 0.8$.

However the outflow deviates from spherical symmetry. There is a SE – NW offset between much of the red-shifted emission and much of the blue-shifted emission. There is no evidence for any systematic rotation (Imai et al. 1997). Comparison of the maps and spectral profile of RT Vir with the models of Bowers (1991) shows that radial outflow, in which the velocity depends on direction alone, can produce an oblate ellipsoid (his figure 2, $i = -45°$, rotated by 45° clockwise with respect to the observer). The appearance of RT Vir implies some modifications to this model. The most extreme red- and blue- shifted features have a large angular separation SE – NW but the whole shell has no obvious spatial asphericity despite a greater flux density in the SE and NW. It is possible that the star loses mass asymmetrically. A higher density in the

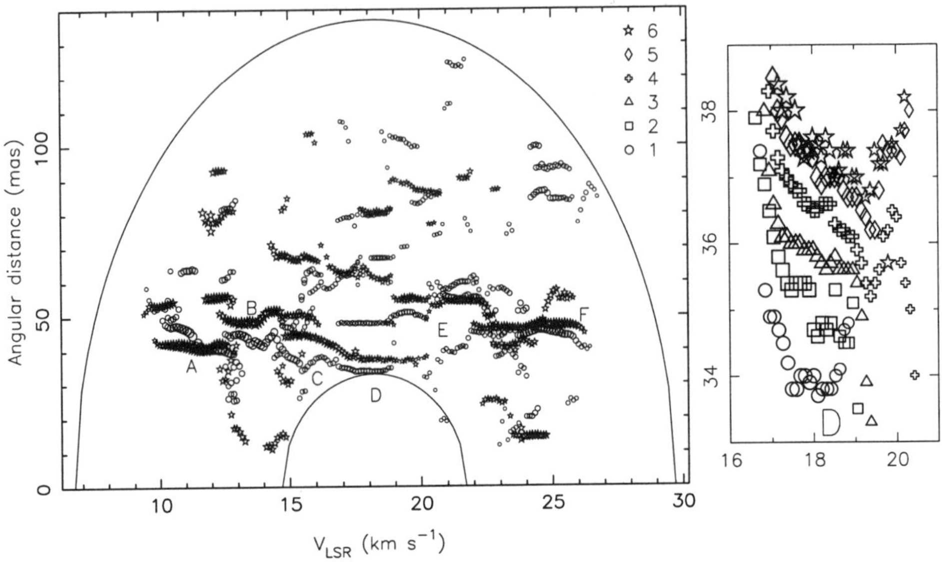

Figure 3. The angular separation of maser components from the assumed stellar position *a* as a function of velocity; $V_* = 18.2$ km s^{-1}. Epochs 1 and 6 are shown on the left; all epochs are shown for feature D on the right.

equatorial plane would be associated with more effective dust-gas momentum coupling, more acceleration and brighter emission in these directions.

2.2. Maser clouds

Fig. 2b shows that in most cases the positions of components making up each feature shifts systematically with velocity. The average angular size of a feature is 11 ± 5 mas, corresponding to $2.3 \pm 1.0 \times 10^{11}$ m. This represents the unbeamed size of maser clouds. Their radius is $\sim$ 6% of the mean radius of the RT Vir H_2O maser shell. The measured radius of H_2O clouds in the supergiant S Per (Richards et al. 1998b) is the same fraction of its average shell radius although the scale is larger. 40 – 60 features were identified in RT Vir at each epoch. The inner limit to 22 GHz maser emission is determined by the maser quenching density, 5×10^{15} m^{-3} at 1000 K (Cooke & Elitzur 1985; Yates et al. 1997). However the number density at $r_{\rm i}$ derived from the CO mass loss rate (Bujarrabal et al. 1989, adjusted to 140 pc) is only $\sim 10^{14}$ m^{-3} , suggesting that the H_2O masing clouds are 50 times denser than their surroundings, consistent with the findings for other late-type stars (Richards et al. 1996, 1998a, 1998b).

2.3. Time variations

The maps of the six epochs were made into a movie which shows small-scale shifts and brightening of the peaks, and the winking on and off of faint extended clouds. In Fig. 3 at least one new feature near $r_{\rm i}$ is only seen at later epochs;

conversely some features near r_o are only seen at earlier epochs (Figs. 1a and 3), suggesting that new clouds enter the region suitable for masing as old ones leave.

Fig. 1a shows the variation with epoch of total 22 GHz flux density. There is no obvious correlation with optical brightness. The flux density of the brightest features increases unsteadily throughout; the total flux peaks 20 days after the optical peak. The light travel time from RT Vir through the H_2O maser shell is $\leq$ 3 hr. The dust is directly heated by absorbing stellar radiation, which also provides radial acceleration. The gas is in turn heated and accelerated by collisions. If this involved a delay of several weeks then the denser regions nearest the star would brighten first. There is no clear evidence for this, and it is likely that the propagation of shocks and changes in the velocity gradient are important.

Some preliminary quantitative results were obtained by comparing six features which are clearly defined at each epoch. These are labelled A – F in the figures. Table 2 lists the flux-weighted mean Doppler velocity of each feature ($u = V_{\rm LSR} - V_*$); the error-weighted mean angular separation from the assumed stellar position (a) and the peak flux (P); the errors are less than a few percent. An overline denotes the mean value for six epochs and prefix δ is the greatest change. An asterisk shows that this is not between the first and last epochs. $\delta P\%$ is $\delta P/P$(epoch 1). The mean angular sizes and velocity width of the features ($\overline{L}$ and $\overline{\Delta V}$), do not show any systematic variation with time.

Table 2. Six H_2O maser features in RT Vir (see text for definitions).

Feature	$\overline{u}$ (km s^{-1})	δu	$\overline{a}$ (mas)	δa	$\overline{P}$ (Jy)	δP	$\delta P\%$ %	$\overline{L}$ (mas)	$\overline{\Delta v}$ (km s^{-1})
A	−6.8	−0.43	42	−1	550	+357	96%	18 ± 5	3.1 ± 0.2
B	−4.6	+0.07	48	+6	150	+125	136%	11 ± 3	2.4 ± 0.2
C	−2.9	+0.68	40	+6	24	*11	*62%	15 ± 3	2.0 ± 0.3
D	−0.4	*+0.27	36	+3	5	+6	288%	12 ± 2	2.9 ± 0.5
E	+3.5	*+0.16	52	*±3	20	+27	715%	13 ± 7	3.0 ± 0.3
F	+6.6	+0.17	45	*+3	93	+51	76%	16 ± 5	3.9 ± 0.6

The mean increase in separation between all pairs of the six features is not affected by alignment uncertainties and is 3.0 ± 0.5 mas in 68 days, corresponding to $11 \pm 2\,{\rm km\,s^{-1}}$ if RT Vir is at 140 pc. The features lie 36 – 52 mas from RT Vir, so for uniform acceleration the expansion velocity in this region should be $\leq 5\,{\rm km\,s^{-1}}$. Fig. 1b shows that for features with larger values of $|u|$ the lines as a whole move towards higher outflow velocities even though δu is small. These bright features may be denser, dustier and more effectively accelerated than their surroundings. Alternatively the apparent outward proper motion of the brightest emission may be a pattern motion of optimum masing conditions (e.g. the passage of shocks).

RT Vir was observed by MERLIN under similar conditions in 1994 (Bains 1995) and showed a similar structure in the CSE as a whole, with brighter blueshifted emission and the SE – NW velocity asymmetry; however the brightest features were in different locations and if any features had survived their internal structure and brightness were very different. There are even greater differences with earlier maps by Yates & Cohen (1994) and Bowers et al. (1993).

3. Conclusions

The 22 GHz H_2O emission from around RT Vir forms an expanding shell with a velocity gradient $\epsilon \approx 0.8$. Emission is brightest in the SE (red-shifted) and NW (blue-shifted) suggesting an elliptical velocity field. Individual masing clouds are typically $\sim 2.3 \times 10^{11}$ m in diameter and span $1 - 3\,\mathrm{km\,s^{-1}}$. Many individual features are recognisable for at least 10 weeks but their flux changes severalfold. There is no obvious correlation with stellar optical magnitude. The brightest show proper motions 1 or 2 times the uniform acceleration model shell expansion velocity. Further analysis is underway to establish the physical conditions in the masing regions and to investigate the rôle of the stellar brightness, pulsations, uneven mass loss, and so on, in determining the appearance of the masers.

References

Bains I., 1995, Master's Thesis, University of Manchester

Bowers P.F., 1991, ApJS 76, 1099

Bowers P.F., Claussen M.J., Johnston K.L., 1993, AJ 105, 284

Bujarrabal V., Gómez-González J., Planesas P., 1989, A&A 219, 256

Chapman J.M., Cohen R.J., 1985, MNRAS 212, 375

Condon J.J., 1997, PASP 109, 166

Cooke B., Elitzur M., 1985, ApJ 295, 175

Imai H., Shibata K.M., Sasao T., et al., 1997, A&A 319, L1

Kholopov P.N., et al., 1985, GCVS, Nauka

Kirrane M.-T., 1986, Master's Thesis, University of Manchester

Mendoza-Torres J.E., Lekht E.E., Berulis I.I., Pashchenko M.I., 1997, A&AS 126, 257

Nyman L.-Å., Johansson L.E.B., Booth R.S., 1986, A&A 160, 352

Richards A.M.S., Yates J.A., Cohen R.J., 1996, MNRAS 282, 665

Richards A.M.S., Yates J.A., Cohen R.J., 1998a, MNRAS 299, 319

Richards A.M.S., Yates J.A., Cohen R.J., 1998b, MNRAS, submitted

Stanek K.Z., Knapp G.R., Young K., Phillips T.G., 1995, ApJS 100, 169

Stevens J.A., Robson E.I., Gear W.K., et al., 1998, ApJ 502, 182

Szymczak M., Le Squeren A.M., 1998, A&A, submitted

Whitelock P.A., Menzies J., Feast M., et al., 1994, MNRAS 267, 711

Yates J.A., Cohen R.J., 1994, MNRAS 270, 958

Yates J.A., Field D., Gray M.D., 1997, MNRAS 285, 383

Asymptotic Giant Branch Stars
T. Le Bertre, A. Lèbre and C. Waelkens, eds.
I.A.U. Symposium 191, 1999

Infrared interferometry of circumstellar envelopes

John D. Monnier

Physics Dept, University of California, Berkeley, CA 94720-7450, USA

Abstract. This paper will review the technical progress of interferometric infrared observing techniques from the first 2-element interferometer 25 years ago to the 3+ element arrays now coming into service. To date, only the Infrared Spatial Interferometer (ISI) has published separate-element interferometric data on circumstellar dust shells in the infrared and many of these scientific results will be discussed. Speckle interferometry has also evolved significantly over the last few decades as slit-scanning techniques over single-pixel detectors have largely been replaced by fast-readout of large format detector arrays. Important near-infrared and mid-infrared results derived from speckle data will also be reviewed.

Until recently, two-dimensional information about circumstellar dust distributions has been sorely lacking, hence detections of dust shell asymmetries have been difficult and uncertain. New speckle observations using modern, 10-m class telescopes have yielded surprising results, demonstrating the importance of accurate closure phase information in interpreting interferometric data. These discoveries hopefully precursor those to be made from closure-phase imaging with the new generation of separate-element, interferometric arrays.

1. Introduction

Infrared interferometry arises from the desire to image the heavens in ever increasing spatial resolution. Other techniques exist with the same goal in mind, but the limited space afforded to this review necessarily restricts the scope to exclude lunar occultation (e.g., Ragland, Chandrasekhar & Ashok 1997; Käufl et al. 1998) and adaptive optics works (e.g., Cruzalèbes et al. 1998; Monnier et al. 1999). This review will also ignore the rapidly increasing interferometric successes at observing the stellar surfaces themselves (Dyck et al. 1996; Wilson, Dhillon & Haniff 1997; Burns et al. 1998; Van Belle et al. 1998; Tuthill, Monnier & Danchi 1999; Young et al. 1999), but rather will concentrate on the circumstellar envelopes of late-type stars.

The objects under scrutiny here are mass-losing asymptotic giant branch (AGB) stars; many of these stars are Mira variables, both with oxygen- and carbon-rich shells. This volume is devoted to the study of these objects and I invite you to look through the other review articles in order to gain a greater understanding of these interesting astrophysical laboratories. That said, what do we hope to measure with interferometers and then hopefully will learn about the astrophysics of the AGB? Interferometrists strive to get the most detailed information possible concerning the morphologies of circumstellar dust shells. This requires observations at various wavelengths as well as at various times. These measurements then can be used to directly test theories of dust production and mass-loss mechanisms in general, which are important in guiding our understanding of stellar evolution along the AGB. When coupled with spectral line observations, high angular resolution studies can directly measure gas densities, velocities, and temperatures around these stars, as well as probe the chemistry of molecular formation.

2. Introduction to interferometry

Although images, or maps, can be made from interferometry data (a process known as aperture synthesis), an interferometer actually measures a quantity called the *visibility*. When light beams from two separate telescopes are brought together to produce a fringe signal, the visibility is proportional to the strength of the detected fringes. The visibility, which is a function of the projected telescope separation (or baseline), can be shown to be equivalent to one of the Fourier component amplitudes of the object brightness distribution. Nice discussions of interferometry can be found in Thompson, Moran & Swenson (1986) and Born & Wolf (1964).

For reconstructing reliable images of complex extended objects, one also requires accurate estimates of the Fourier phases, not just the amplitudes. While atmospheric turbulence scrambles the phase of individual Fourier components, a quantity called the *closure phase* retains its integrity. By summing the fringe phases for baselines in a closed triangle, the closure phase is formed and allows a majority of the intrinsic phase information to be recovered for interferometric arrays with three or more telescopes (e.g., Jennison 1958; Readhead et al. 1988). This quantity is critically important for accurate reconstructions of highly asymmetric structures in images.

Since any image can be alternatively represented by its Fourier components, the collection of all "interesting" components allows the interferometric data to be inverted, thus producing an image. The Caltech VLBI and the VLBMEM packages can be used to estimate the original object brightness distribution using Fourier inversion algorithms such as CLEAN (with self-calibration) and the Maximum Entropy Method (e.g., Högbom 1974; Cornwell & Wilkinson 1981; Sivia 1987; Haniff et al. 1989). Speckle interferometry data (more on this later) is often reduced using a bispectral analysis code which performs a direct inversion from the estimated Fourier amplitudes and phases. However, modifications of this technique lead to more robust reconstructions in the presence of noise (e.g., Hofman & Weigelt 1993).

3. First generation of infrared Michelson interferometers

The first interferometric observations of circumstellar dust shells were performed by the Arizona group (McCarthy & Low 1975). The interferometer setup was mounted at the Cassegrain focus of a large telescope, beams from two subapertures interfering together onto a single-pixel InSb detector (later upgraded to a Ge bolometer). Additional observations were presented in McCarthy, Low &

Howell (1977), McCarthy, Howell & Low (1978, 1980), and McCarthy (1979). These measurements were sensitive in the mid-infrared, usually between 5 μm and 12.5 μm.

Meanwhile C. H. Townes was leading the U.C. Berkeley group in building a separate-element heterodyne interferometer. Infrared light incident on each of the Auxiliary McMath Solar Telescopes at Kitt Peak (5.5 m separation) were combined with a CO_2 laser operating near 10 μm to produce a down-converted RF signal from a Ge:Cu detector (later HgCdTe). The RF signals (original bandwidth $\approx \pm 1200$ MHz) from the two telescopes were sent through a delay line and subsequently correlated, acting much like a typical radio or mm interferometer. This system detected fringes on the limb of Mercury as early as 1974 (Johnson 1975), and observations of circumstellar envelopes soon followed (Sutton et al. 1977, 1978, 1979).

These two groups observed many of the same stars and reached the same general conclusions. A diversity of dust shell types were detected, both large and small (in terms of stellar size). Both groups measured changes in the dust shell around the Mira variable *o* Cet, and attributed these to the large amplitude pulsation. These first set of measurements also indicated asymmetries for some of the thickest (i.e. most resolved) dust shells (McCarthy 1979).

4. One-dimensional speckle interferometry

The promise of speckle interferometry was first realized by Antoine Labeyrie (Labeyrie 1970). In the original formulation, short exposures of an astrophysical object are made to freeze the "speckling" induced by the turbulent atmosphere. The amount of high-resolution structure in the speckle pattern, as quantified by its power spectrum, is a measure of two things: 1) the quality of the atmospheric seeing, and 2) the high resolution structure in the object of interest. Observing a nearby point-source star allows the calibration of the seeing contribution and thus the extraction of interferometric visibility measurements out to the diffraction limit of the telescope. It was only later that the Fourier phases were estimated from such data (e.g., Knox & Thompson 1974; Weigelt 1977).

Although speckle interferometry was applied to measurements of visible star diameters and binary star orbits in the early- to mid-1970's, it was not until 1979 that the first infrared application to AGB circumstellar envelopes emerged. A group from France published a number of papers on the subject, resolving a few dust shells for the first time in the near-infrared (Foy et al. 1979; Sibille, Chelli & Léna 1979; Mariotti et al. 1983). Because of the lack of two-dimensional array cameras, a single pixel detector (typically InSb) was used to sample the speckle cloud. In most setups, a slit was scanned across the speckled image (or the image was passed over the slit). Power spectra for all scans were averaged together to estimate the visibility, after calibrating the mean atmospheric (and telescope) transfer function.

While many groups made speckle observations in the 1980's, I want to discuss the few representing significant advances. Dyck et al. (1984) contains the largest sample of resolved dust shells to date, observing 16 late-type stars and proto-planetary nebula (PPN) between 1.65 μm and 4.8 μm. This work was able to confirm, in a general sense, that dust condenses at a few stellar radii from the stellar surface, and the existence of dust shell asymmetries for a number of the most resolved sources.

Another significant advance occurring was the use of radiative transfer models for interpreting visibility curves. Because spatial resolution is often inadequate to unambiguously interpret the data, radiative transfer codes allow the inner radius of dust formation to be estimated and can utilize multi-wavelength data in a natural way. Two illustrations can be found in Ridgway et al. (1986) and Ridgway & Keady (1988), which estimated the dust shell optical depths for IRC +10216, NML Cyg, and IRC +10420.

A number of other firsts occurred during this very productive time. Dyck, Beckwith & Zuckerman (1983) published the first (and only?) infrared speckle observations on spectral lines, five fundamental vibration-rotation lines of CO around IRC +10216. Phase retrieval methods were first applied to reconstructing 1-D profiles of asymmetric dust shells (Chelli, Perrier & Biraud 1983; Leinert & Haas 1989; Navarro, Benitez & Fuentes 1990), and speckle techniques were applied in the mid-infrared (Cobb & Fix 1987; Benson, Turner & Dyck 1989). Dyck et al. (1991) compared the existing IRC +10216 speckle data to show that the circumstellar emission changed significantly during the 1980's. There were a number of other interesting near-infrared (e.g., Dainty et al. 1985; Dyck et al. 1987; Mariotti et al. 1992; Lopez et al. 1993) and mid-infrared speckle results (Fix & Cobb 1988; Dyck & Benson 1992) published during this time.

5. The Infrared Spatial Interferometer

After the successful interferometry experiment at Kitt Peak, the UC Berkeley group began constructing a pair of movable telescopes to be placed on Mt. Wilson, California, based on the same heterodyne detection design (present bandwidth $\approx \pm 2700$ MHz). The Infrared Spatial Interferometer (ISI) published its first results on IRC +10216 in 1990 (Danchi et al. 1990; Bester et al. 1991) and showed the dust shell size changes as a function of the stellar pulsational phase.

The most significant single paper was a compilation of dust shell sizes for 13 late-type stars (Danchi et al. 1994), which established two classes of dusty stars, those with hot dust close to the star and those with detached, relatively cool shells. In addition, the variation of inner radii indicated an episodic emission time-scale of 20-50 years. The ISI has monitored a number of AGB stars for a 10-year period now, discovering complex shell geometries and time variation around multiple sources (Greenhill et al. 1995; Bester et al. 1996; Monnier et al. 1997; Hale et al. 1997; Lopez et al. 1997). High spatial resolution observations with $\lambda/\Delta\lambda \approx 10^5$ presently underway should reveal the location of NH_3 and SiH_4 molecules, and help explain their anomalously high observed abundances around AGB stars (e.g., Betz, McLaren & Spears 1979; Goldhaber 1988).

6. Two-dimensional speckle interferometry

Two major technical advances in the 1990's have ushered in a renaissance in speckle observations. The development of fast-readout, infrared array cameras allows the entire speckle cloud to be recorded and thousands of images to be collected in a short period of time. Secondly, a new generation of large aperture telescopes has come into service, allowing 2-3 times higher spatial resolution.

Most present results are coming from two groups using different techniques and telescopes, although other groups are making significant contributions as well (e.g., Haniff & Buscher 1998; Buscher, Haniff & Oudmaijer 1998). A team headed by G. Weigelt has been using so-called "speckle masking," a modified bispectral analysis, to analyze speckle data collected at the 6m Special Astrophysical Observatory telescope and the 3.5m+ ESO telescopes on La Silla, Chile. Very nice results have recently been reported (Osterbart et al. 1997; Weigelt et al. 1998; Osterbart et al. 1998; Irrgang et al. 1998; Weigelt et al. 1999). P. Tuthill leads the other group, which uses aperture masking techniques on the Keck-I telescope with maximum baselines of 11 m. Results can be found in this volume (Tuthill et al. 1999) and in other publications (Tuthill, Monnier & Danchi 1998; Monnier et al. 1999).

The aperture masking experiment at Keck has succeeded in obtaining good visibility and Fourier phase information down to K-band magnitude of 5. Probably, standard speckle techniques will need to be used to go much dimmer. Using "speckle masking" (not to be confused with "aperture masking"), Weigelt's group has been able to go down to $K \sim 9$.

Both groups have succeeded in constructing very detailed images at the telescope diffraction limit. The resolution achieved equals about one stellar radius for the closest objects (e.g., IRC +10216), allowing the dust formation process to be studied in detail. Three different groups presented images of IRC +10216 at 2.2 μm, and all images showed the same asymmetric and clumpy dust shell (Weigelt et al. 1998; Haniff & Buscher 1998; Tuthill et al. 1999). In addition, proper motions of the dust clouds have been detected, which promises to yield important dynamical and three-dimensional information on the dust shell geometry. These observations are showing that mass-loss processes can be very inhomogeneous in the inner dust shell regions, for reasons not entirely clear.

7. Conclusions

I have summarized the progress of infrared interferometry on circumstellar envelopes in Fig. 1 and Table 1. Fig. 1 shows the spatial resolution and wavelength coverage attainable with present instruments, while Table 1 lists all dust shells resolved by papers referenced here (and is a fairly complete list of published results). This list will surely grow by a dozen or so in the coming years just from analyses of existing data, especially from the speckle imaging groups. While many measurements do exist, a lack of coordinated observations at multiple wavelengths has hindered interpretation of these data because of the time variable nature of these sources. Also, combining speckle observations with polarization and spectral line work can make unique and valuable contributions to the study of AGB stars, but have only rarely been performed. In addition, photometric monitoring of these sources in the infrared can be very valuable when trying to understand temporal changes in the dust shell properties, but is also rarely done, despite their high flux levels.

The inhomogeneous, asymmetric and time-variable dust shells observed around many stars indicate that mass-loss on the AGB may not be understood

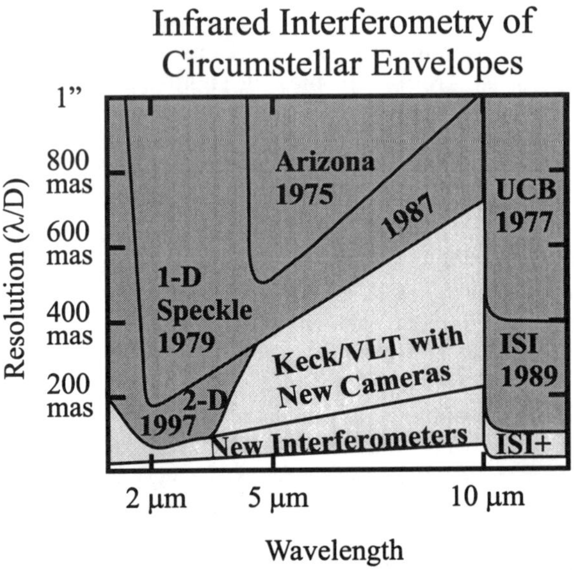

Figure 1. Spatial resolution vs. wavelength for past, present and future interferometric observations. Parameter space to be covered in the next few years is shaded in light grey.

quite as well as previously thought. More theoretical work and modeling (two- or even three-dimensional radiative transfer) are needed to make full use of the new high resolution data. Understanding of these processes should continue to improve as a number of infrared interferometer arrays are being constructed or upgraded (IOTA, COAST, PTI, CHARA, Keck, VLTI, MIRA, ISI+, more), leading to spatial resolutions (at some wavelengths) of less than 1 mas in the near future.

Acknowledgments. I am thankful to A. P. Zemgulys for enlightening discussions.

References

Benson J.A., Turner N.H., Dyck H.M., 1989, ApJ 97, 1763

Bester M., Danchi W.C., Degiacomi C.G., Townes C.H., Geballe T.R., 1991, ApJ 367, L27

Bester M., Danchi W.C., Hale D., et al., 1996, ApJ 463, 336

Betz A.L., McLaren R.A., Spears D.L., 1979, ApJ 229, 97

Born M., Wolf E., 1964, Principles of Optics, Macmillan, New York

Burns D., et al., 1998, MNRAS 298, 697

Buscher D.F., Haniff C.A., Oudmaijer R.D., 1998, poster contribution, this conference

Table 1. Circumstellar shells resolved by infrared interferometry

Star name	Type	Wavelengths (μm)	Other comments
NML Cyg	Red Supergiant	2.2, 2.4, 3.5, 3.6, 3.8, 4.8, 5.0, 7.9, 8.7, 10, 10.4, 11.4, 12.6	Asymmetric
α Ori	Red Supergiant	7.9, 8.7, 10,10.2, 10.4, 11.1, 11.4	Time variable
VY CMa	Red Supergiant	1.65, 2.2, 3.1, 3.8, 4.8, 8.3, 10.2, 11.1	Asymmetric 2-D images
S Per	Red Supergiant	7.9, 8.7, 10, 10.4, 11.4, 12.6	
α Sco	Red Supergiant	10.0, 11.1	Asymmetric, 1-D profile
VX Sgr	Red Supergiant	2.2, 3.8, 11.15	
μ Cep	Red Supergiant	10	
α Her	Red Supergiant	11.15	
IRC +10216	Carbon Star	2.2, 3.1, 3.5, 8.3, 10, 10.2, 11.1, 12.5	Asymmetric, Time variable, Polarization data, 1-D profiles, Spectral Line Work (CO, NH_3, SiH_4), 2-D images, Proper motions detected
CIT 6	Carbon Star	2.2, 3.1, 3.8, 4.8, 10, 11.5	Asymmetric, 2-D images
IRAS 15194-5115	Carbon Star	2.2, 3.6	
CIT 5	Carbon Star	4.8	
χ Cyg	Mira	2.2, 4.6, 4.8, 10, 11.15	Asymmetric
IK Tau	Mira	3.8, 11.15	Time variable
o Cet	Mira	10.2	Asymmetric, Time Variable
R Leo	Mira	11.1	
R Cas	Mira	7.9, 8.7, 10, 10.4, 11.4, 12.6	
CIT 3	Mira	1.65, 2.2, 11.15	Asymmetric
AFGL 2290	Mira	2.2	Asymmetric
T Cas	Mira	10	
U Her	Mira	10	
U Ori	Mira	11.15	
W Aql	Mira	11.15	
R Aqr	Mira	11.15	
BC Cyg	Mira?	7.9, 8.7, 10, 10.4, 11.4, 12.6	
BI Cyg	Mira?	7.9, 8.7, 10, 10.4, 11.4, 12.6	
RX Boo	Semiregular	7.9, 8.7, 10, 10.4, 11.4, 12.6	
SW Vir	Semiregular	2.4, 10	
X Her	Semiregular	10	
OH 26.5+0.6	OH/IR	3.8, 4.6, 4.8, 5, 10	Asymmetric
OH 354.9-0.5	OH/IR	10	Asymmetric
OH 1.1-0.8	OH/IR	10	Asymmetric
Red Rectangle	PPN	1.25, 1.6, 2.2, 3.1, 3.3, 3.8, 4.8	Asymmetric, 1-D Profile, Polarization data, 2-D images
OH 231.8+4.2	PPN	3.8	Asymmetric
AFGL 618	PPN	3.8	
M1-9	PPN	3.8, 4.8	Asymmetric
M2-9	PPN	3.8	Asymmetric
IRC +10420	F8-G0 I	2.2, 3.5, 3.8, 4.8, 5, 10	Asymmetric
η Car	LBV	4.6	1-D profiles, 2-D images

Chelli A., Perrier C., Biraud Y.G., 1983, A&A 117, 199

Cobb M.L., Fix J.D., 1987, ApJ 315, 325

Cornwell T.J., Wilkinson P.N., 1981, MNRAS 196, 1067

Cruzalebes P., Lopez B., Bester M., Gendron E., Sams B., 1998, A&A 338, 132

Dainty J.D., Pipher J.L, Lacasse M.G., Ridgway S.T., 1985, ApJ 293, 530

Danchi W.C., Bester M., Degiacomi C.G., Greenhill L.J., Townes C.H., 1994, AJ 107, 1469

Danchi W.C., Bester M., Degiacomi C.G., McCullough P.R., Townes C.H., 1990, ApJ 359, 59

Dyck H.M., Beckwith S., Zuckerman B., 1983, ApJ 271, L79

Dyck H.M., Zuckerman B., Leinert Ch., Beckwith S., 1984, ApJ 287, 801

Dyck H.M., Zuckerman B., Howell R.R., Beckwith S., 1987, PASP 99, 99

Dyck H.M., Benson J.A., Howell R.R., Joyce R.R., Leinert C., 1991, AJ 102, 200

Dyck H.M., Benson J.A., 1992, AJ 104, 377

Dyck H.M., Benson J.A., Van Belle G.T., Ridgway S.T., 1996, AJ 111, 1705

Fix J.D., Cobb M.L., 1988, ApJ 329, 290

Foy R., Chelli A., Sibille F., Léna P., 1979, A&A 79, L5

Goldhaber D.M., 1988, University of California at Berkeley, PhD Dissertation

Greenhill L.J., Colomer F., Moran J.M., et al., 1995, ApJ 449, 365

Hale D.D.S., Bester M., Danchi W.C., et al., 1997, ApJ 490, 407

Haniff C.A., Buscher D.F., Christou J.C., Ridgway S.T., 1989, MNRAS 241, 51

Haniff C.A., Buscher D.F., 1998, A&A 334, L5

Hofman K.-H., Weigelt G., 1993, A&A 278, 328

Högbom J., 1974, ApJS 15, 417

Irrgang P., Balega Y.Y., Gauger A., Osterbart R., Schniggenberg G., Weigelt G., 1998, poster contribution, this conference

Jennison R.C., 1958, MNRAS 118, 276

Johnson M.A., 1975, University of California at Berkeley, PhD Dissertation

Käufl H.U., Stecklum B., Richichi A., Richter S., 1998, poster contribution, this conference

Knox K.T., Thompson B.J., 1974, ApJ 193, L45

Labeyrie A., 1970, A&A 6, 85

Leinert C., Haas M., 1989, A&A 221, 110

Lopez B., Perrier C., Mekarnia D., Lefèvre J., Gay J., 1993, A&A 270, 462

Lopez B., Danchi W.C., Bester M., et al., 1997, ApJ 488, 807

Mariotti J.-M., Chelli A., Foy R., et al., 1983, A&A 120, 237

Mariotti J.-M., Monin J.-L., Ghez P., Perrier C., Zadrozny A., 1992, A&A 255, 462

McCarthy D.W., Low F.J., 1975, ApJ 202, L37

McCarthy D.W., Low F.J., Howell R., 1977, ApJ 214, L85

McCarthy D.W., Howell R., Low F.J., 1978, ApJ 223, L113

McCarthy D.W., 1979, in High Angular Resolution Stellar Interferometry, IAU Colloq. 50, p. 18

McCarthy D.W., Howell R., Low F.J., 1980, ApJ 235, L27

Monnier J.D., Tuthill P.G., Lopez B., et al., 1999, ApJ 512, in press

Monnier J.D., Bester M., Danchi W.C., et al., 1997, ApJ 481, 420

Navarro R., Benitez D., Fuentes F.J., 1990, J. Opt. Soc. Am. A 7, 459

Osterbart R., Langer N., Weigelt G., 1997, A&A 325, 609

Osterbart R., Balega Y.Y., Langer N., Man'shchikov A.B., Weigelt G., 1998, poster contribution, this conference

Ragland S., Chandrasekhar T., Ashok N.M., 1997, MNRAS 287, 681

Readhead A.C.S., Nakajima T.S., Pearson T.J., Neugebauer G., Oke J.B., Sargent W.L.W., 1988, AJ 95, 1278

Ridgway S.T., Joyce R.R., Connors D., Pipher J.L., Dainty C., 1986, ApJ 302, 662

Ridgway S.T., Keady J.J., 1988, ApJ 326, 843

Sibille F., Chelli A., Léna P., 1979, A&A 79, 315

Sivia D.S., 1987, Cambridge University, PhD Dissertation

Sutton E.C., Betz A.L., Storey J.W.V., Spears D.L., 1979, ApJ 230, L105

Sutton E.C., Storey J.W.V., Betz A.L., Townes C.H., Spears D.L., 1977, ApJ 217, L97

Sutton E.C., Storey J.W.V., Townes C.H., Spears D.L., 1978, ApJ 224, L123

Thompson R., Moran J.M., Swenson G.W., 1986, Interferometry and Aperture Synthesis in Radio Astronomy, John Wiley & Sons

Tuthill P.G., Monnier J.D., Danchi W.C., 1998, A Half Century of Stellar Pulsation Interpretations, p. 322

Tuthill P.G., Monnier J.D., Danchi W.C., 1999, this volume

Van Belle G.T., PTI Collaboration, Thompson R., 1998, poster contribution, this conference

Weigelt G., 1977, Opt. Commun. 21, 55

Weigelt G., Balega Y., Blöcker T., Fleischer A.J., Osterbart R., Winters J.M., 1998, A&A 333, L51

Weigelt G., Balega Y.Y., Blöcker T., Fleischer A.J., Hofmann K.-H., Osterbart R., Winters J.M., 1999, this volume

Wilson P.W., Dhillon V.S., Haniff C.A., 1997, MNRAS 291, 819

Young J.S., et al., 1999, this volume

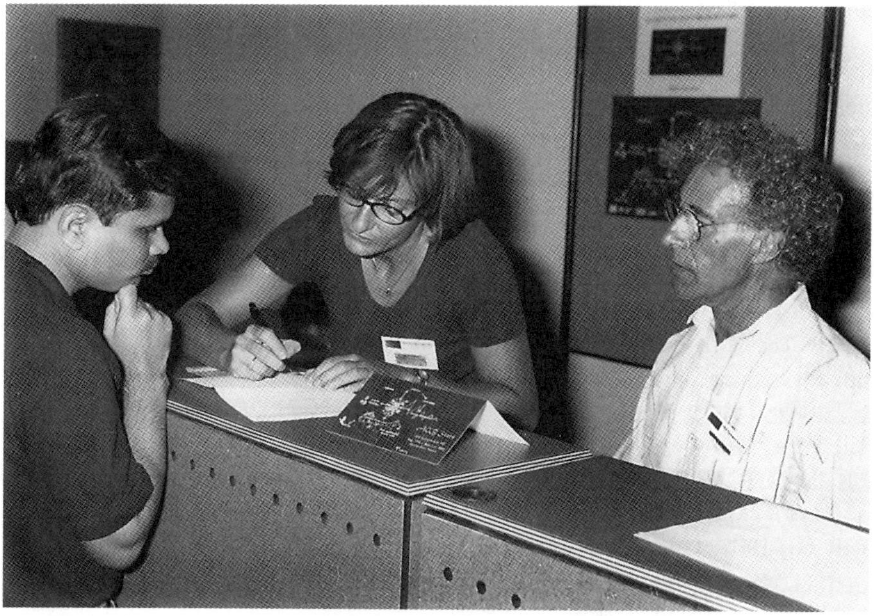

Busy L.O.C. (Agnès Lèbre with Christian Magnan) and Devendra Ojha

Asymptotic Giant Branch Stars
T. Le Bertre, A. Lèbre and C. Waelkens, eds.
I.A.U. Symposium 191, 1999

Imaging of Stellar Disks and Mass Loss Envelopes in Evolved Stars

P.G. Tuthill, J.D. Monnier & W.C. Danchi

Space Science Laboratory, University of California, Berkeley, CA 94720, USA

Abstract. Interferometric imaging of evolved stars has been performed in the near-IR at the Keck Observatory revealing diffraction-limited structure at unprecedented spatial resolution. By studying the hottest regions of the circumstellar dust shells of a number of bright prototypical objects, we have revealed complex and anisotropic structures in the inner regions where dust nucleation occurs. We find that dramatic dust plumes and clumps dominate the morphology of the circumstellar environment at a scale of a few stellar radii for extreme carbon stars such as IRC +10216 and CIT 6. Furthermore our observations have yielded near-IR diameters and images of the stellar photospheres of nearby red giant and supergiant stars. With the additional benefit of multi-epoch observations, we have begun to investigate the time-evolution of our enshrouded objects as dust is created and accelerated away from the nucleation zone. Armed with such a powerful new set of observational constraints, we are able to address contemporary problems such as the origin of structure in PNe, the physics of mass loss, and the dynamical properties of Mira pulsation.

1. Introduction

Interferometry in the infrared is providing a new observational window on the environments of late type stars from the stellar photospheres out to the extended circumstellar dust shells. The resolution attainable with the latest generation of 10 m class telescopes is well suited to the study of the nearest and brightest dust-enshrouded stars. This paper describes a program running at the Keck observatory over the last few years obtaining images of bright evolved stars at the diffraction limit of the Keck aperture. For these types of objects, studies in the near-IR are particularly profitable. This is due to the low dust opacity, the low contamination from molecular lines, and the high luminosity of the sources. Coupled with this, the hottest dust in the inner regions of the dust shells becomes bright enough so that the blackbody emission can rival or even outshine the radiation coming from the parent star, enabling us to make images of the flux coming from this circumstellar material.

Using multi-wavelength images spanning the infrared JHK and L bands, we have begun to address some of the most pressing questions related to the latest stages of stellar evolution. Our observations of a handful of bright evolved LPV's, supergiants and enshrouded stars have allowed us to explore the processes underlying mass loss, a key feature of such evolved systems. Although theoretical models usually assume spherical symmetry, there has been a growing body of evidence (e.g. from molecular and polarization data, spectrophotometry, and observations of planetary nebulae) that mass loss is often anisotropic. Our high-resolution images have provided some of the newest insights into this phenomenon.

2. Aperture masking interferometry

Our observations have utilized the technique of aperture masking, in which a mask is placed over the telescope pupil so as to only pass light from selected regions, thus transforming the telescope into an array of small subapertures. When the light from each of these separate subapertures is combined, interference fringes are formed which encode information on the spatial structure of the source object. The masking of the Keck 1 primary was achieved by mounting custom-made plates in front of the f/25 secondary mirror. An example showing the hole-pattern of our 15-hole mask, as projected onto the Keck primary mirror, is given in Fig. 1. There are many constraints in arriving at such pupil geometries, with the primary concern being to sample the Fourier plane evenly and densely out to the cut-off of the telescope, in this instance with a *non-redundant* array – a sampling in which no vector baseline occurs more than once.

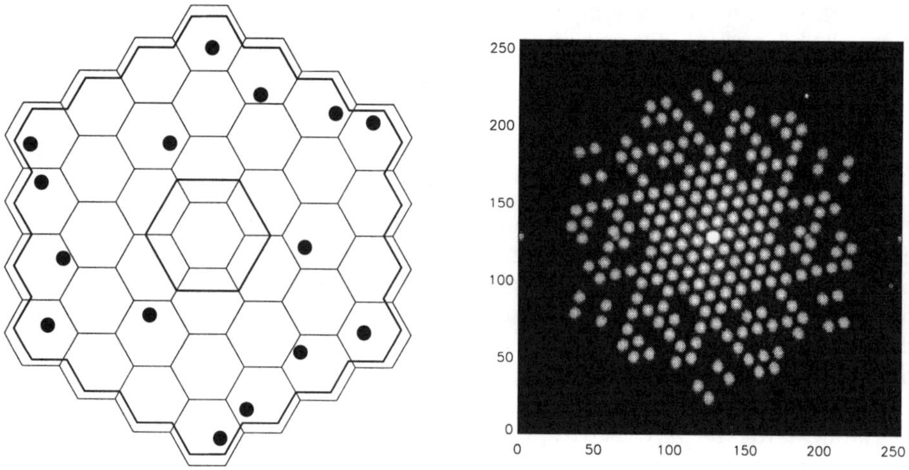

Figure 1. A 15-hole non-redundant array as it appears projected onto the Keck primary mirror (left) and the corresponding power spectrum obtained by Fourier analysis of the resultant fringe pattern (right).

After passing through the mask, starlight was brought to focus in the Near Infra-Red Camera (NIRC), a cryogenically cooled 256×256 pixel InSb array

(Matthews & Soifer 1994). Using the NIRC magnifying optics (Matthews et al. 1996) gave a pixel scale of 20.57 milli-arcsec/pixel – sufficient to Nyquist sample fringes on the longest baselines in K-band. With the aperture mask in place, the image formed at the camera focus is an Airy disk crossed by multiple sets of fringes corresponding to each pair of holes in the mask. In order to preserve as much high-resolution information as possible in this rapidly-evolving pattern, the array was clocked out in a 'speckle' readout mode as rapidly as possible resulting in data cubes of 100 consecutive frames each with 140 msec exposure time.

Data reduction proceeded by extracting the visibilities and phases from each exposure which are related to the brightness distribution of the science target by a Fourier transform (e.g. Born & Wolf 1964). Fourier amplitudes may be conveniently extracted from the power spectrum, an example of which is given in Fig. 1. Although the Fourier phases were corrupted by atmospheric fluctuations, an observable known as the *closure phase* – the sum of baseline phases around closed triangles – has been developed in radio astronomy to overcome this problem. These techniques are now well established at optical wavelengths (Haniff et al. 1987; Roddier 1988).

Raw visibilities were calibrated for the mean telescope–atmosphere transfer function by utilizing nearly contemporaneous observations of point-source stars. After such calibration, Fourier data were inverted to yield images with the help of reconstruction algorithms such as the "Maximum Entropy Method" (Gull & Skilling 1984; Sivia 1987) or CLEAN algorithm (Högbom 1974) – both of which have been used extensively in radio astronomy.

Both theoretical and experimental studies have highlighted the gains in signal-to-noise ratio (SNR) made available by using sparse rather than filled apertures in interferometric observations of bright sources (Roddier 1988; Haniff & Buscher 1992). Our own results have verified these predictions, and enabled us to reconstruct complex images at diffraction-limited angular resolution in the near-infrared.

3. Results

This program has extensive data on a wide variety of Supergiants, Miras and enshrouded evolved stars. The supergiant VY CMa has been found to exhibit a one-sided extension at high resolution in the near-IR which appears to be linked to a curved plume outflow in the extended circumstellar environment. Further details on this fascinating system can be found in Monnier et al. (1999). Rather than trying to summarize observations of dozens of stars, and to give a flavor of this work within the constraints of this report, we focus here on two famous bright objects: *o* Ceti and IRC +10216.

Our diffraction-limited observations with the Keck telescope have enabled us to measure photospheric stellar diameters in a number of near-IR wavebands, as is illustrated in Fig. 2 for *o* Ceti, the prototype Mira variable. Such high-quality visibility data allows us to address a number of important topics in the study of evolved stars.

The angular diameter is amongst the most fundamental stellar observables; from it may be computed the effective temperature and the luminosity. Fur-

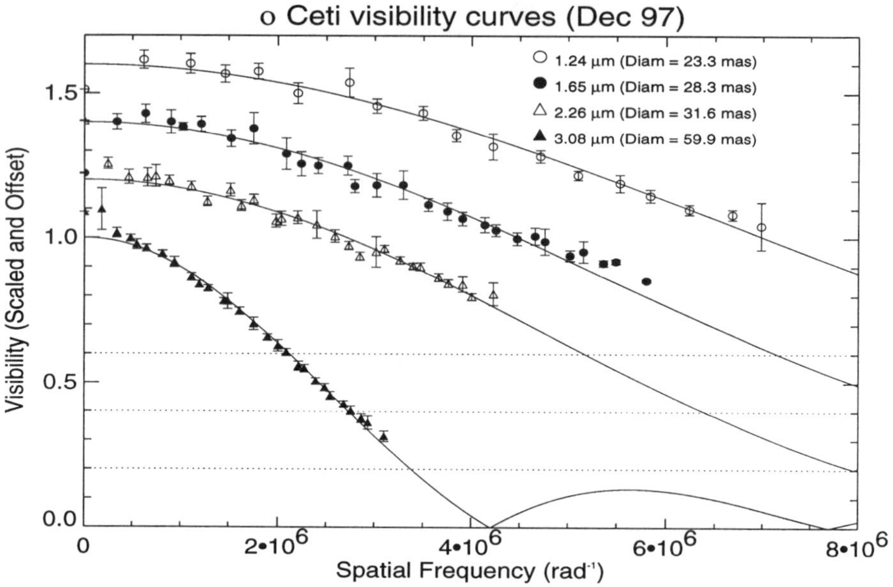

Figure 2. Azimuthally averaged visibility curves produced for *o* Ceti from multi-wavelength interferometer data (vis offsets have been added to avoid confusion on the plot). Diameters given correspond to the best-fit uniform-disk models. The larger apparent diameter recorded at 3.08 μm may be due to molecular blanketing around the star.

thermore, in cases where stars are known to be pulsating, the diameter can be used for dynamical studies such as those of Haniff et al. (1995) to determine the pulsation mode – a question which remains open despite extensive debate. Additional data forming a time-series record of stellar diameters through the pulsation cycle should be able to directly observe changes in photospheric radius, greatly clarifying observational aspects of stellar pulsation in LPV's.

Our quasi-monochromatic diameters at a range of different wavelengths probe the atmospheric structure, helping to discriminate emission from the photosphere, molecular atmosphere and circumstellar dust, thereby providing valuable experimental testing of numerical model atmospheres. We report here a particularly marked enlargement in the apparent size of *o* Ceti in the 3.08 μm filter, an effect we believe is due to absorption of the stellar flux by some molecular species, probably water. Clearly, such data is best interpreted in the context of a numerical simulation involving radiative transfer through a stellar atmosphere with appropriate abundances of molecular and atomic species.

As a final illustration of the results from the Keck aperture masking project, we show in Figure 3 diffraction limited maps of the enshrouded carbon star IRC +10216 taken in January 1997 at 2.26 μm and 3.08 μm. First published in Danchi et al. (1997), these maps have the highest angular resolution to date, and furthermore show beautiful agreement with recent work of Weigelt et al. (1998) and Haniff & Buscher (1998).

The maps show dramatic departure from the spherical symmetric outflow scenarios common in the literature, with pronounced plumes and clumps dominating the morphology. We believe that IRC +10216 has an underlying bipolar morphology with an axis of symmetry running at a position angle of about 20° – an interpretation which is reinforced by the bipolar nature of the visible scattering nebula (Haniff & Buscher 1998).

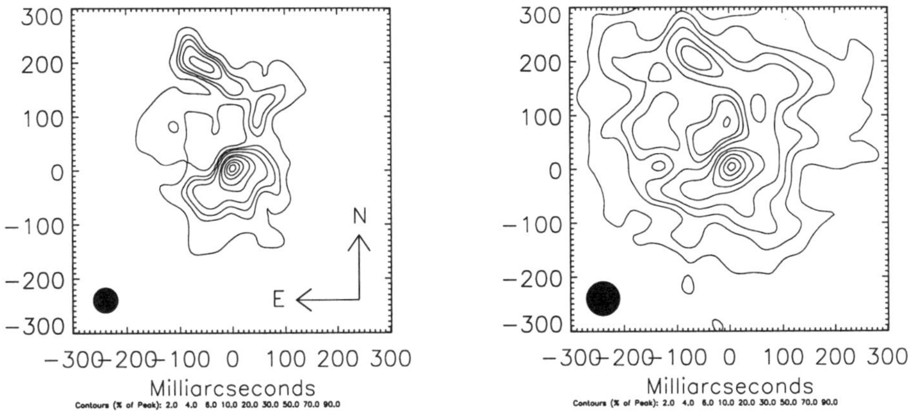

Figure 3. Contour maps of the enshrouded carbon star IRC +10216 from interferograms recorded in January 1997 at 2.26 μm (left) and 3.08 μm (right).

It is not certain which, if any, of the features visible in the maps corresponds to the central star. However if we make the tentative identification that the brightest (southern) component is the star itself, then we can derive an apparent diameter of about 50 mas for this object. This is in reasonable agreement with expectations based upon effective temperature and luminosity arguments.

Our study has also produced tentative evidence for time-evolution of the circumstellar environment of IRC +10216 by examining maps taken over a period from January 1997 through June 1998. The displacement between the bright southern feature (which we assume for the present to be the star itself) and the northern arm was seen to exhibit a motion of some 10 mas over a one year timescale. This is in agreement with expectations: if we take an outflow velocity of 14 km/sec (Goldhaber 1988) and a distance to IRC +10216 of 290 pc (Herbig & Zappala 1970) we arrive at a linear motion of 10 mas per year. If, however, IRC +10216 is taken to lie at 150 pc (Crosas & Menten 1997) then we must consider the possibility that the dust we see has a significant projected velocity on our line of sight (we only measure the tangential component in our proper motion study), or alternatively that the dust in the northern arm is still in a region where it is being accelerated by the radiation pressure and has not yet attained its terminal velocity. With continued monitoring, we should be able to answer these questions as to the detailed behavior of the outflow around this and a handful of other bright evolved stars.

References

Born M., Wolf E., 1964, *Principles of Optics*, Macmillan, New York

Crosas M., Menten K.M., 1997, ApJ 483, 913

Danchi W.C., Tuthill P.G., Bester M., Lipman E.A., Monnier J.D., Townes C.H., 1997, in *The Tenth Cambridge Workshop on Cool Stars, Stellar Systems, and the Sun*, eds. R.A. Donahue & J.A. Bookbinder, ASP Conf. Ser. 155, 361

Goldhaber D.M., 1988, *Ph.D. Dissertation,* University of California, Berkeley

Gull S.F., Skilling J., 1984, IEEE Proceedings 131, Pt. F, No. 6

Haniff C.A., Mackay C.D., Titterington D.J., Sivia D., Baldwin J.E., Warner P.J., 1987, Nature 328, 694

Haniff C.A., Buscher D.F., 1992, J. Opt. Soc. Am. A 9, 203

Haniff C.A., Scholz M., Tuthill P.G., 1995, MNRAS 266, 745

Haniff C.A., Buscher D.F., 1998, A&A 334, L5

Herbig G.H., Zappala R.R., 1970, ApJ 162, L15

Högbom J., 1974, ApJS 15, 417

Matthews K., Soifer B.T., 1994, Ap&SS 190, 239

Matthews K., Ghez A.M., Weinberger A.J., Neugebauer G., 1996, PASP 108, 615

Monnier J.D., Tuthill P.G., Lopez B., Cruzalèbes P., Danchi W.C., Haniff C.A., 1999, ApJ 512, 351

Roddier F., 1988, Physics Reports 170, 97

Sivia D.S., 1987, *Ph.D. Dissertation*, Cambridge University

Tuthill P.G., Haniff C.A., Baldwin J.E., 1997, MNRAS 285, 529

Weigelt G., Balega Y., Fleischer A.J., Osterbart R., Winters J.M., 1998, A&A 333, L51

Asymptotic Giant Branch Stars
T. Le Bertre, A. Lèbre and C. Waelkens, eds.
I.A.U. Symposium 191, 1999

Circumstellar Chemistry of AGB Winds

A. E. Glassgold

Physics Department, New York University, 4 Washington Place, New York, NY 10003, USA

Abstract.
Carbon-rich AGB stars are an excellent place to study circumstellar chemistry. The best case is IRC +10216, where observations at optical, infrared, and radio wavelengths have led to the discovery of more than 50 molecules. When the detected species are grouped in chemically related families, the observed spatial distributions and abundances can be understood with a photochemical model. Most of the species are either progenitors that arise in the photosphere or inner envelope, or are unsaturated radicals and hydrocarbon chains photochemically produced in the outer envelope. The few saturated molecules with a single heavy atom are probably produced on grain surfaces. Inside the ionization radius of atomic carbon, the circumstellar envelope of IRC +10216 is weakly ionized by suprathermal particles. The HCO^+ detected and mapped in the $J = 1 - 0$ line by the *IRAM* telescopes can be interpreted as arising from either partially-excluded galactic cosmic rays or β-particles emitted by radioactive nuclei dredged up from the H-burning zone into the wind. An estimate of the ionization rate due to ^{26}Al based on the $^{26}Al/^{27}Al$ ratio of "mainstream" pre-solar grains is the same order of magnitude as determined from the measured abundance of HCO^+ in IRC +10216.

1. Introduction

High-density winds emerge from stars at various stages of evolution. For low and intermediate mass stars, this occurs when they form and just before they die, e.g., as AGB stars with mass-loss rates in the range $10^{-4} - 10^{-5}\, M_{\odot}\,\mathrm{yr}^{-1}$. The rapid decrease with distance of the density of these winds is accompanied by strong adiabatic and line cooling[1]. The combination of decreasing wind density and temperature means that the abundances close to the star tend to get frozen into the wind, to be carried out to large distances where they can be observed by radio astronomy. This conclusion is dramatically altered when the effects of external agents such as interstellar UV radiation are considered. A main contention of this review is that most of the molecular complexity observed in AGB winds is induced by the interstellar radiation field. We also show that

[1]In the absence of heating, the temperature would fall below that of the cosmic background radiation.

live radionuclides dredged up from the interior of an AGB star can affect the ionization of the outer circumstellar envelope (CSE) and be observable by radio observations of molecular ions. An earlier but more complete review of wind chemistry is also available (Glassgold 1996a).

Nuclear burning in an AGB star occurs in a double shell outside of a burnt-out core. Convection currents dredge up newly processed material through the huge stellar envelope to the photosphere. Stellar dynamical processes, possibly pulsational shocks (see the review by Willson et al. 1997), drive the outer layers of the star and distend the atmosphere, producing physical conditions favorable for the formation of dust. Once formed, the dust can be driven out by radiation pressure and drag the gas along to generate the wind. The complex and poorly understood processes that form dust and produce the wind occur in a "transition region" between radial distances $\sim 10^{14} - 10^{15}$ cm. A detailed knowledge of the dynamics of this region is currently unavailable but is crucial for determining its physical and chemical properties. Although interferometry is now beginning to provide images on this scale, most of the chemical information that we have on the transition region is based on high spectral resolution infrared spectroscopy (e.g., Keady & Ridgway 1993).

Once dust formation and wind generation have ceased, the dust and gas (with depleted abundances) expand at constant velocities. At first the dust and gas interact but they eventually become uncoupled as the density and the temperature decrease. The gas-phase abundances become fixed at a new quasi-LTE equilibrium (see Sect. 2.5). If nothing else occurred, there would be no further chemical evolution. What happens, however, is that the wind becomes exposed to interstellar UV radiation, and all of the molecular species are broken down by photodissociation into atoms, which are then photoionized (if their ionization potentials $IP < 13.6$ eV). Before being destroyed, however, the radicals produced by molecular photodissociation can synthesize other (possibly more complex) molecules, especially in the case of C-rich CSEs. The only complexes that survive from the inner part of the CSE are solids that serve as the seeds for the development of interstellar dust.

Thus the circumstellar matter that leaves the cool surface of an AGB star is profoundly altered by its passage through the inner transition region, where dust is formed, and by photodissociation at intermediate distances. The latter exposure involves both molecular synthesis and destruction, and produces shells of radicals and other compounds, including molecular ions. The physical conditions in the shells are similar to those in interstellar molecular cloud cores, and are observable in a wide range of molecular rotational lines. In the next section we review the extensive observations of the nearby C-rich AGB star, IRC +10216. Focusing on this well-studied object allows detailed theoretical interpretations to be made. This envelope is approximately spherical at intermediate to large distances, as is the case for most AGB stars (Guélin & Lucas, personal communication). However, the detection of clumps and other substructure in the photochemical shells clearly indicates that the envelope is inhomogeneous. Nonetheless, the application of a very simple photochemical model based on a spherical and homogeneous CSE proves to be quite illuminating.

2. Chemical observations of IRC +10216

2.1. Astrophysical properties

High-resolution radio line spectra of IRC +10216 conform closely to expectations for an expanding envelope and permit an accurate determination of the maximum expansion velocity, $V = 14\,\mathrm{km\ s^{-1}}$. The measurement and mapping of CO lines, together with the distance d, yield the hydrogen nucleus mass-loss rate $\dot{M}$ and the CO abundance $x(\mathrm{CO})$. For $d = 200\,\mathrm{pc}$, values accurate to 50% are $\dot{M}_\mathrm{H} = 2 \times 10^{-5} M_\odot\ \mathrm{yr}^{-1}$ and $x(\mathrm{CO}) = 4.5 \times 10^{-4}$. Considerations on the bolometric luminosity (Zuckerman et al. 1986) indicate a smaller distance, possibly $d \approx 150\,\mathrm{pc}$ (see, e.g., Groenewegen et al. 1998).

The abundances of the two next most abundant species after CO, C_2H_2 and HCN, can be determined directly by IR absorption and indirectly from the mm emission of the photodissociation products, C_2H and CN. The two methods agree to within the errors (probably no better than a factor of 2), and yield $x(C_2H_2) = 4.0 \times 10^{-5}$ and $x(\mathrm{HCN}) = 2.0 \times 10^{-5}$. All of the abundances quoted so far are relative to total hydrogen nuclei and refer to the part of the CSE that is outside of the region where dust is formed and inside the photodissociation zone. These abundances should be smaller than photospheric due to inclusion of certain elements into or onto dust grains. The most likely example in a C-rich environment is the C in C_2H_2, generally considered the molecular progenitor of carbonaceous dust. Since IR spectroscopy measures C_2H_2 just above the photosphere, and radio astronomy measures C_2H in the outer envelope, a comparison between the two could yield the amount of C in dust. However, the observations are not yet accurate enough for this purpose.

2.2. The photochemical model

The underlying ideas of the model are closely related to the Haser-type model for cometary comae and to photodissociation models that were developed for interstellar clouds in the 60's and 70's. The early development of the photochemical model for CSEs is reviewed by Glassgold (1996a) with full literature references.

The essence of the photochemical model is the processing by UV radiation of species that enter the inner envelope from the transition region. The main processing of carbon can be represented by:

$$C_2H_2 \ \&\ \mathrm{HCN} \rightarrow C_2H \ \&\ \mathrm{CN} \rightarrow \mathrm{hydrocarbons} \rightarrow \mathrm{C} \rightarrow \mathrm{C}^+,$$

where the first and last stages are radiative. The hydrocarbon synthesis is mainly accomplished by neutral reactions starting with the radicals C_2H & CN (see Sect. 2.4). Fig. 2 of Glassgold (1996a) shows the spatial distribution of the most abundant C species, particularly how the radicals and their products are distributed in shells. This prediction of the first photochemical calculations has been confirmed by maps made with mm-wave interferometers (e.g., Lucas 1994, Lucas & Guélin 1998). We are thus led to a simple classification of circumstellar molecules according to their spatial distribution as (1) centrally peaked species coming out of the inner transition region or (2) shell molecules that are synthesized in the outer CSE. Of course there are some exceptions, such as molecular ions and others to be mentioned in the next section.

2.3. Observed species

In Table 1, the more than 50 species detected in IRC +10216 are grouped into families (columns) and listed according to approximate measured abundances. The last column is a mixed bag of heavy-element molecules (excluding S and Si, columns 7 & 8) that may not be all related. Except for these heavy elements, whose study is really only getting started, we have a reasonably good understanding of each of the families in Table 1.

Table 1. Species detected in IRC +10216[1,2]

10^{-3}								
	CO							
10^{-4}								
			$C_2H_2^*$					
	C				HCN			
10^{-5}								
			C_2H		CN			
		CH_4^*	C_4H			SiS	CS	
10^{-6}					HC_3N			
			C_3^*		C_3N			
					HC_5N	SiC_2		$AlCl$
10^{-7}		SiH_4^*	C_6H		HNC	SiO		AlF
			C_5^*					
		NH_3	lC_3H		HC_7N	SiC		$NaCN$
			cC_3H_2		HC_9N	SiN		$MgNC$
10^{-8}			C_5H	H_2C_4	HC_2NC		C_2S	CP
			C_8H	HC_2N	CH_3CN		C_3S	$NaCl$
			C_7H	H_2C_3	$[C_5N]$	SiC_4		$MgCN$
10^{-9}			cC_3H				H_2S	KCl
	HCO^+							
10^{-10}								

1. C_2 has been detected but its abundance has not been determined.
2. The species marked * are detected only in the infrared.

The second column of Table 1 lists CO and its observed C-bearing descendants, atomic C and HCO^+. Attempts to detect the 158 μm line of C^+ with ISO have not been successful. The CO family is not important chemically because CO photodissociation occurs at rather large distances due to line self-shielding (Mamon et al. 1988) where the density and the abundances of potential reactants are low. We will discuss the very weak production of HCO^+ in Sect. 3. The 4th and 6th columns contain the combined effects of the photodissociation and radical chemistry associated with C_2H_2 and HCN. The 5th column contains some of the isomers (carbenes) obtained in the process. These three families illustrate the high level of molecular complexity observed in IRC +10216, characterized by unsaturated carbon chains. The 3rd column contains the saturated molecules, CH_4, NH_3, and SiH_4. Near-IR spectroscopy indicates that they are *not* produced in the photosphere, but further out in the inner CSE, where the timescale for gas atoms to interact with dust is still short compared to the expansion time, r/V. These molecules are probably desorbed by the energy released in formation.

The 7th and 8th columns list the 10 Si- and S-bearing molecules that have been observed in the radio. The LTE calculations of Lodders and Fegley (1995) suggest that SiS and atomic Si are the main progenitors, and this is supported by the IRAM maps of SiS and CS, which are concentrated towards the star. The SiS distribution actually shows a rapid decrease very close in, which has been interpreted in terms of depletion of SiS onto grains (e.g., Glassgold & Mamon 1993). In any case, substantial amounts of SiS and CS reach the photodissociation region, where they are transformed into Si and S atoms and ions. The latter then engage in ion-molecule and neutral reactions to produce the rest of the species listed in columns 7 & 8. Maps of SiC_2 (Gensheimer et al. 1995) indicate that, although its abundance peaks in the outer envelope, some of it is produced in the inner envelope. This supports the idea that SiC_2 is also a progenitor molecule (Howe & Millar 1990). It would be particularly interesting to measure SiC_2 in the near-IR to learn more about its distribution near the star. MacKay & Charnley (1998) have suggested that it is also produced on grains.

In summary, the molecular observations of IRC +10216 can be understood qualitatively in terms of the photochemical model. We do not attempt to present a full case for this conclusion in this short report. However, it is important to emphasize that mapping a species is crucial in determining whether it is centrally peaked or peaked in a shell far from the star. For theoretical modeling, we refer the reader to the literature, e.g., Cherchneff at al. (1993a, henceforth CGM), Cherchneff at al. (1993b), Millar & Herbst (1994), and Glassgold (1996a), as well as to the discussion of modeling assumptions in Sect. 2.5.

2.4. Dominance of neutral reactions

In light of the qualitative success of the photochemical model, it is important to identify the main mechanisms for synthesis in the outer CSE. We use the simplest polyacetylene, HC_3N, as an example and discuss the relative importance of ion-molecule and neutral radical reactions.

Synthesis by ion-molecule reactions

$$h\nu + C_2H_2 \rightarrow C_2H_2^+ + e$$
$$C_2H_2^+ + HCN \rightarrow H_2C_3N^+ + H,$$

followed by $e + H_2C_3N^+ \rightarrow HC_3N + H$.

Synthesis by neutral reactions

$$h\nu + HCN \rightarrow CN + H$$
$$CN + C_2H_2 \rightarrow HC_3N + H$$

In both cases, the first step is a photoreaction that produces an ion or radical. The measured reaction rate coefficients for the next steps, which produce the protonated or the neutral form of HC_3N, are about the same, $\approx 5\times10^{-10}\,cm^3\,s^{-1}$ at 30 K (the temperature in the photochemical shell). Recombination of protonated HC_3N is very fast. *Thus neutral reactions dominate because the abundance of CN is between 100 and 1000 times larger than that of* $C_2H_2^+$.

2.5. Modeling issues

Although we omit detailed discussion of the theory, the qualitative success of the photochemical model suggests that we examine the basic assumptions of the simple version used so far.

Dynamical assumptions The wind density is almost always prescribed by assuming that the flow is steady, spherically symmetric, and locally homogeneous (unclumped). Although there is observational evidence to the contrary, CO observations that sample the average density of the wind in IRC +10216 indicate approximate spherical symmetry in the region where most molecules are observed.

Chemistry In accord with the observations summarized in Table 1, the modeling of an AGB wind requires the chemistry of a diverse array of elements: C, N, O, Si, S, Mg, and other heavy elements. Both neutral and ion-molecule reactions have to be specified, generally at low temperatures, e.g., $T \approx 30\,K$ in the photochemical shell. Most of the required reaction rates have not been measured, and very few at the appropriate temperature. Similarly most of the photorates have not been measured or calculated theoretically, and very little is known about dust surface chemistry. This situation obviously requires considerable discretion on the part of both modelers and those who use the theory to interpret observations.

Initial conditions The initial abundance of the main progenitors entering the outer CSE from the transition region were originally considered as unknown phenomenological parameters. In time, the initial abundances of CO, C_2H_2, and HCN determined by the analysis of mm observations of the outer envelope have come into reasonable agreement with those obtained from near infrared spectroscopy (Keady & Ridgway 1993). The accuracy of both the radio and infrared measurements are not yet good enough to definitively show any significant difference between the inner and the outer envelope. The envelope abundances of the major species probably do not differ much from photospheric abundances. The calculations of Willacy & Cherchneff (1997) support this position. They show that, when such molecules are destroyed by pulsational shocks, they are reformed as the gas cools to roughly the original abundances. The calculations of Lodders & Fegley (1995) also indicate that an effective LTE holds in the inner regions of AGB envelopes.

3. Molecular ions

Despite the dominance of neutral reactions in the synthesis of complex hydrocarbons, ion-molecule reactions and molecular ions are still important in circumstellar chemistry (Glassgold, Lucas & Omont 1986). Dissociative recombination of molecular ions can produce isomeric as well as normal molecular forms, e.g., e + $HCNH^+ \rightarrow$ HNC as well as HCN and H + CN, as discussed further by CGM. Another example is the production of the ring molecule C_3H_2 in the dissociative recombination of $C_3H_3^+$. Of course, the most basic goal of observing molecular ions is to measure the electron fraction.

Table 2. Most Abundant Molecular Ions

Molecule	*IP* (eV)	*pa* (eV)	Ion	Production
H_2	15.4	4.39	H_3^+	CR
N_2	15.6	5.13	N_2H^+	CR
CO	14.0	6.15	HCO^+ [1]	CR
C_2H_2	11.4	6.65	$C_2H_3^+$	UV, CR
HCN	13.7	7.43	$HCNH^+$	UV, CR
H_2O	12.6	7.22	H_3O^+ [2]	UV,CR

1. Detected in many post-AGB shells.
2. Possible detection in VY CMa.

Table 2 gives the most abundant molecular ions in CSEs in order of increasing proton affinity or *pa*, the binding energy of the proton. The entry for H_2O is relevant for O-rich CSEs, where it replaces C_2H_2 and HCN as the most abundant molecule after CO. The symbol CR indicates that the molecular ion is produced by collisional ionization of H_2 by a suprathermal particle, usually thought to be a galactic cosmic ray. Ions of smaller *pa* rapidly lose their proton to ions of larger *pa* by "proton-transfer" reactions, e.g., $H_3^+ + CO \rightarrow HCO^+ + H_2$. Molecules with $IP < 13.6$ eV can be photoionized by the interstellar UV radiation field, and they produce molecular ions by "abstracting" a H atom from H_2, e.g., $C_2H_2^+ + H_2 \rightarrow C_2H_3^+ + H$. We see from the table that the ions H_3^+, N_2H^+, and $C_2H_3^+$ can only be produced from abundant neutrals by suprathermal particles, whereas $C_2H_3^+$ and $HCNH^+$ can also be produced by UV. However, in the neighborhood of the peak abundances of the latter two ions, UV is the main mechanism. Because H_2 is the most abundant species in AGB winds with large mass-loss rates, the suprathermal particles mainly create H_2^+ ions that are rapidly transformed into H_3^+ ions by proton transfer to another H_2. The H_3^+ is then transformed into heavier ions, especially $HCNH^+$ and $C_2H_3^+$. However, these more abundant molecular ions are also difficult to detect (e.g., see Glassgold et al. 1992 for the case of $C_2H_3^+$), and most efforts to discover molecular ions have focused on HCO^+. In addition to its large dipole moment, this ion has the advantage that it directly measures the suprathermal ionization rate.

Ten years ago Lucas & Guélin (1988) reported the detection of HCO^+ with the *IRAM* 30-m telescope. Lucas & Guélin (1998) has reported more definitive results at this symposium based on sensitive observations with the *IRAM* interferometer (Plateau de Bure) and the 30m. The measurements show that the $J = 1-0$ transition has the lineshape expected for an optically-thin, expanding-shell and that the observed and laboratory frequencies agree to one part in a million. The emission has also been mapped with the interferometer and is consistent with the partially filled-in shell expected from theory. The line is rather weak, with an average brightness over the line of slightly more than 20 mK. The peak abundance of HCO^+ in IRC +10216 is $\sim 10^{-10}$, and corresponds to a suprathermal ionization rate of $\zeta \approx 2 \times 10^{-18}\,s^{-1}$. This estimate of the ionization rate is one order of magnitude smaller than the ionization rate in the solar neighborhood due to (mainly low-energy) galactic cosmic rays, $\zeta_{\rm CR} \approx 2 \times 10^{-17}\,{\rm s}^{-1}$.

One explanation of this discrepancy is that the wind of IRC +10216 is magnetized, and that it excludes low-energy cosmic rays in roughly the same way that the solar wind does. [2] If one uses the modulated cosmic-ray ionization rate measured in the solar neighborhood, then the observed ionization rate in IRC +10216 requires the exclusion of cosmic rays with energies per nucleon $E/A > 400\,{\rm MeV}$. If this is the right explanation, one should be able to explain the numerical value of this cutoff. Although the cosmic ray transport problem is formally solvable, it requires a knowledge of the (unknown) magnetic scattering properties of AGB winds. Alternatively, one could deduce these properties for the wind of IRC +10216 from a wind exclusion calculation.

It is also conceivable that the powerful AGB wind is effective in excluding even higher-energy cosmic rays, in which case the ionization indicated by the presence of HCO^+ must be generated within the envelope. One possible source is radioactive nuclides dredged up from the stellar interior and carried out into the wind which they ionize by fast β particles (Glassgold 1995, 1996b). Two interesting radionuclides synthesized by AGB stars are ^{26}Al and ^{14}C, with mean lifetimes $\tau = 1.0 \times 10^6\,{\rm yr}$ and $\tau = 8100$ yr, respectively. The decay of ^{26}Al is accompanied by the emission of γ rays, with 82% producing a positron ($^{26}{\rm Al} \to {}^{26}{\rm Mg}^* + {\rm e}^+ + \nu$) and 18% proceeding by electron capture (${\rm e}^- + {}^{26}{\rm Al} \to {}^{26}{\rm Mg}^* + \overline{\nu}$). The latter channel gives a few Auger electrons and can be neglected, whereas the positron with mean energy $E_\beta(^{26}{\rm Al}) = 0.66$ MeV can produce about 20 000 ionizations. AGB winds are too thin to stop the γ rays (they interact with the Compton cross-section), and they escape. The decay of ^{14}C is simpler, ($^{14}{\rm C} \to {}^{14}{\rm N} + {\rm e}^- + \overline{\nu}$), with 100% going to the ground state of ^{14}N with $E_\beta(^{14}{\rm C}) = 0.047\,{\rm MeV}$. The two β decays heat and ionize the wind in a similar fashion since the physics of the stopping of electrons and positrons is the same for energies above 5 keV. However, the depletions of the parent nuclei are different. Whereas about half of the C is in circumstellar dust particles, much of the Al may be in solids. A qualitative analysis of the escape of ^{26}Al positrons from dust grains indicates, however, that most of them get into the gas.

The ionization rate ζ per H nucleus due to a radioactive decay is (Glassgold 1995)

$$\zeta = b_\beta \, \frac{1}{\tau} \, \mathcal{R} \, x \, \frac{E_\beta}{e_{\rm ion}} \, \exp(-r/V\tau),$$

where b_β is the branching ratio for β-emission (0.82 for ^{26}Al and 1.00 for ^{14}C), $e_{\rm ion} = 35\,{\rm eV}$ is the mean energy to create an ion pair in a predominantly H_2/He gas, $\mathcal{R}$ is the abundance ratio of a radioactive to a stable nuclide, the latter with absolute abundance x. For the cases of interest here, $r < V\tau$, and the rates are:

$$\zeta_{26} \approx 1.4 \times 10^{-15}\,\mathcal{R}_{26}\,{\rm s}^{-1} \qquad \zeta_{14} \approx 1.0 \times 10^{-12}\,\mathcal{R}_{14}\,{\rm s}^{-1}, \qquad (1)$$

with $x(^{27}{\rm Al}) = 3 \times 10^{-6}$ (solar) and $x(^{12}{\rm C}) = 4.5 \times 10^{-4}$.

A small upper limit to the abundance of ^{14}C in IRC +10216 has been obtained by a sensitive search for ^{14}CO (Forestini et al. 1997), with the result

[2] No one has ever solved the cosmic ray transport problem for any wind other than the Sun's.

$\mathcal{R}_{14} < 1.5\times10^{-5}$. Using Eq. (1), this yields $\zeta_{14} < 1.5\times10^{-17}\,s^{-1}$ for IRC +10216. This upper limit is significantly larger than the ionization rate determined from the observations of HCO^+. Forestini et al. (1997) consider the prospects for detecting ^{14}CO to be dim. The attempts by Guélin et al. (1995) to detect ^{26}Al molecules in the CSE of IRC +10216 have not been definitive because of confusion and blending with other lines. They quote an upper limit, $\mathcal{R}_{26} < 0.04$, based on observations of the $J = 7-6$ line of ^{26}AlF (assuming that all Al is in the gas phase). The most recent nucleosynthetic calculations for an intermediate-mass AGB star ($M = 4.5M_\odot$) give $\mathcal{R}_{26} \sim 10^{-3}$ (Lattanzio & Forestini 1998). If we substitute this estimate into Eq. (1), we find $\zeta_{26} \sim 1.4 \times 10^{-18}\,s^{-1}$, the same order as the value deduced from the observations of HCO^+ in IRC +10216. Amari et al. (1998) report measurements of $\mathcal{R}_{26}$ for "mainstream" presolar SiC grains ($^{12}C/^{13}C = 30 - 80$) that are in the range $10^{-4} - 10^{-3}$. The upper part of this range is the same order of magnitude as predicted by nucleosynthesis; it is also consistent with the ionization rate determined from HCO^+.

We therefore regard live ^{26}Al as a viable explanation for the level of suprathermal ionization observed in IRC +10216. It would be reassuring to detect ^{26}Al molecules, as attempted by Guélin et al. (1995), and to detect other molecular ions such as $C_2H_3^+$ and $HCNH^+$.

4. Discussion

Although the photochemical model is successful in providing a qualitative understanding of many aspects of AGB wind chemistry, there are problems in making it more quantitative. Some of these issues have been mentioned in Sect. 2.5, and here we address two important observational issues, the clumpiness and the size of the photochemical shells.

The IRAM observations (e.g., Lucas & Guélin 1998) show that the outer shell distributions are fragmented into what appears to be large clumps, (typically $5'' \times 10''$) in IRC +10216. The fact that they are common to several species implies that they are density enhancements. If so, they potentially can affect the dynamics, radiative transfer, and physical processes, including the chemistry. The origin as well as the detailed nature of these clumps are unknown, but they could be studied directly by mm interferometry. For example, it would be of considerable interest to map the clumps in a density-sensitive line such as ^{13}CO. To obtain a deeper theoretical understanding, one needs more sophisticated modeling that incorporates a richer dynamical and physical treatment than attempted so far.

The IRAM observations (e.g., Lucas & Guélin 1998) also show that many of the shells have same radii ($\approx 15''$, or 3000 AU at 200 pc). There are some exceptions that result from interior production, e.g., SiC_2, HNC, HCO^+, or from weak photo-destruction that yields large shells, e.g., CN. In the current version of the photochemical model, as described in Sect. 1 and Sect. 2, the shell separation arises because the dynamical (expansion) and chemical timescales are about the same in the shells, ≈ 300 yr. Thus one simple way of reducing the differences in shell radii is to reduce the chemical time by increasing the density. This idea is consistent with clumping. Like the treatment of clumps, however, a sound understanding of this issue requires a new dynamical chemistry of CSEs.

Acknowledgments. The research on which this survey is based is funded in part by the National Science Foundation.

References

Amari S., Nittler L.R., Zinner E., Lewis R.S., 1998, poster contribution, this conference

Cherchneff I., Glassgold A.E., Mamon G.A., 1993a, ApJ 410, 188 (CGM)

Cherchneff I., Glassgold A.E., Mamon G.A., 1993b, ApJ 419, L41

Forestini M., Guélin M., Cernicharo J., 1997, A&A 317, 883

Gensheimer P.D., Likkel L., Snyder L.E., 1995, ApJ 439, 445

Glassgold A.E., 1995, ApJ 438, L111

Glassgold A.E., 1996a, ARA&A 34, 241

Glassgold A.E., 1996b, in Amazing Light, R. Chiao (ed.), Springer, New York, p. 277

Glassgold A.E., Lucas R., Omont A., 1986, A&A 157, 35

Glassgold A.E., Omont A., Guélin M., 1992, ApJ 396, 115

Glassgold A.E., Mamon G.A., 1993, in Chemistry and Spectroscopy of Interstellar Molecules, D.K. Bohme, E. Herbst, N. Kaifu and S. Saito (eds.), Tokyo, p. 263

Groenewegen M.A.T., van der Veen W.E.C.J., Matthews H.E., 1998, A&A 338, 491

Guélin M., Forestini M., Valiron P., Ziurys L.M., Anderson M.A., Cernicharo J., Kahane C., 1995, A&A 297, 183

Howe D.A., Millar T.J., 1990, MNRAS 244, 444

Keady J.J., Ridgway S., 1993, ApJ 406, 199

Lattanzio J., Forestini M., 1998, this volume

Lodders K., Fegley B., 1995, Meteoritics 30, 661

Lucas R., 1994, in Astrometry with Millimeter & Submillimeter Interferometry, M. Ishiguro & W.J. Welsh (eds.), ASP, San Francisco, p. 135

Lucas R., Guélin M., 1988, in Submillimeter and Millimeter Astronomy, W.B. Burton & R.D. Wolstencroft (eds.), Reidel, Dordrecht, p. 97

Lucas R., Guélin M., 1988, this volume

MacKay D.D.S., Charnley S.B., 1998, MNRAS in press

Mamon G.A., Glassgold A.E., Huggins P.J., 1988, ApJ 328, 797

Millar T.J., Herbst E., 1994, A&A 288, 561

Willacy K., Cherchneff I., 1997, A&A 330, 676

Willson L.A., Struck C., Bowen G.H., 1997, in Cosmic Winds & The Heliosphere, J.R. Jokipii, P.C. Sonnett and M.S. Giampapa (eds.), Tucson, p. 155

Zuckerman B., Dyck H.M., Claussen M., 1986, ApJ 304, 401

Asymptotic Giant Branch Stars
T. Le Bertre, A. Lèbre and C. Waelkens, eds.
I.A.U. Symposium 191, 1999

OH 231.8+4.2: its energetic bipolar outflow and rich chemistry

C. Sánchez Contreras[1], V. Bujarrabal and J. Alcolea

Observatorio Astronómico Nacional (IGN), Ap. 1143, E-28800 Alcalá de Henares, Spain

Luis F. Miranda

Instituto de Astrofísica de Andalucía, CSIC, Ap. 3004, C/ Sancho Panza s/n, E-18080 Granada, Spain

J. Zweigle

German Aerospace Center (DLR), Institute of Optoelectronics, 82230 Wessling, Germany

Abstract. We report results from observations taken at different wavelengths (optical, radio and NIR) of the bipolar Protoplanetary Nebula OH 231.8+4.2. Radio interferometry with high spatial resolution has been particularly revealing. We study the complex structure and dynamics of this object as well as its rich chemistry.

1. Introduction

OH 231.8+4.2 (hereafter OH 231.8) is a well studied source presumably in transition between the AGB and the planetary nebula (PN) phase. This object is unique, due to its large angular extent, for studying the complex dynamics present in this kind of objects. Like most protoplanetary nebulae (PPNe), OH 231.8 has a clear bipolar morphology. In the visible it shows two lobes emerging from an obscured region where the central star lies (Reipurth 1987). Herbig-Haro-like objects have been observed beyond the lobes (Cohen et al. 1985), which are attributed to the post-AGB wind interaction phenomenon. OH 231.8 is classified as an O-rich source, in view of its SiO, OH, and H_2O maser emission (Bowers & Morris 1984, and references therein). Also, emission from a large variety of C-bearing molecules has been detected in this source. The presence of a Mira variable in the heart of OH 231.8 has been recently confirmed by direct imaging in the NIR (Kastner et al. 1998), the central star being classified as M9 III (Cohen et al. 1981). However, the detection of a blue excess in the optical continuum could suggest the presence of a hotter companion (Cohen et al. 1985).

[1]Also Departamento de Astrofísica, F. C. Físicas, U. Complutense, E-28040 Madrid, Spain

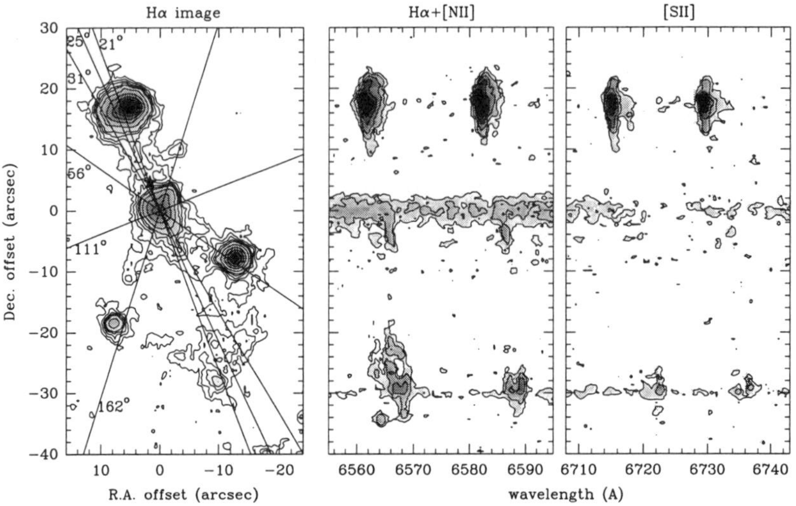

Figure 1. Hα image and long-slit Hα, [NII] and [SII] spectra along the symmetry axis of OH 231.8+4.2

2. Optical long-slit spectroscopy of OH 231.8+4.2

We have obtained long-slit spectra of OH 231.8 using the IDS spectrograph of the 2.5 m Isaac Newton Telescope of the Roque de los Muchachos Observatory (Canary Islands, Spain). Six slit positions were observed at position angles 21° (the symmetry axis of the object), 25°, 31°, 56°, 111° and 162°, (see Fig. 1). We have detected the main nebular lines (Hα, [NII] 6548+6583, [SII] 6716+6731, [OI] 6300+6363, [OII] 3726+3729, [OIII] 4959+5007 and Hβ).

In Fig. 1 we show the Hα, [NII] and [SII] spectra for the 21° slit (symmetry axis). Note the wide profiles of the lines, with a total projected velocity range of $\sim$ 500 km s^{-1}. The emission from the North/South lobe is blue/red-shifted. This fact indicates that the gas is flowing outwards, since the North lobe is the nearest of the two. The axial velocity gradient present in OH 231.8 is $\sim$5 km s^{-1} arcsec^{-1}. This value provides a direct measurement of the kinematic age of the lobes, of about 1 000 yr, in agreement with previous estimations of the time spent in the PPN phase. We have assumed an inclination of the nebula with respect to the plane of the sky of 40°, and a distance of 1 500 pc (Bowers & Morris 1984; Kastner et al 1992). The complexity of the kinematics and of the structure of the lobes is remarkable. Particularly, part of the gas in the South lobe seems to expand like a bubble (see the nearly elliptical spectral feature at the end of the South lobe on the Hα spectrum, Fig. 1). We suggest that such a bubble-like structure is originated by the shock-accelerated gas moving outward and its backward countershock. On the other hand, we have found intense emission from forbidden lines of [NII], [SII] and [OIII] which demonstrates the shocked nature of the line emitting gas in the lobes of OH 231.8. In the long-slit spectra in Fig. 1, we can also see the stellar continuum reflected by the nebula. The detailed study of such a spectrum confirms the M9 III type of the central star. From the [SII] λ6716/[SII] λ6731 lines ratio we have estimated the average electron density in the lobes of OH 231.8, which ranges between 400 cm^{-3} (North Lobe) and $\lesssim$ 100 cm^{-3} (South Lobe). We find that the total ionized mass of the nebula is of the order of 10^{-4} $M_{\odot}$.

3. Structure and kinematics of the neutral envelope of OH 231.8+4.2

In order to study the neutral envelope of OH 231.8, and not only the relatively highly excited component traced by the previous optical observations, we have carried out single-dish and interferometric maps at millimeter wavelengths of different molecular lines. We have used the IRAM 30 m radiotelescope (Granada, Spain) and the Plateau de Bure interferometer (French Alps). The observed transitions are: ^{12}CO and ^{13}CO (J=1–0 and J=2–1), HCO^+ (J=1–0), HCN (J=1–0), HNC (J=1–0), SO ($J=2_2$-1_1), SO_2 ($J=10_{0,10}$-$9_{1,9}$), CS (J=5–4), $H^{13}CN$ (J=1–0), SiO (J=5–4), SiO maser (v=1 J=2–1) and NS ($^2\Pi_{1/2}$, J=5/2–3/2). For ^{12}CO, the best tracer of the total mass, we have performed 3-point mosaicing, interferometric observations to properly study the mass distribution in the whole nebula.

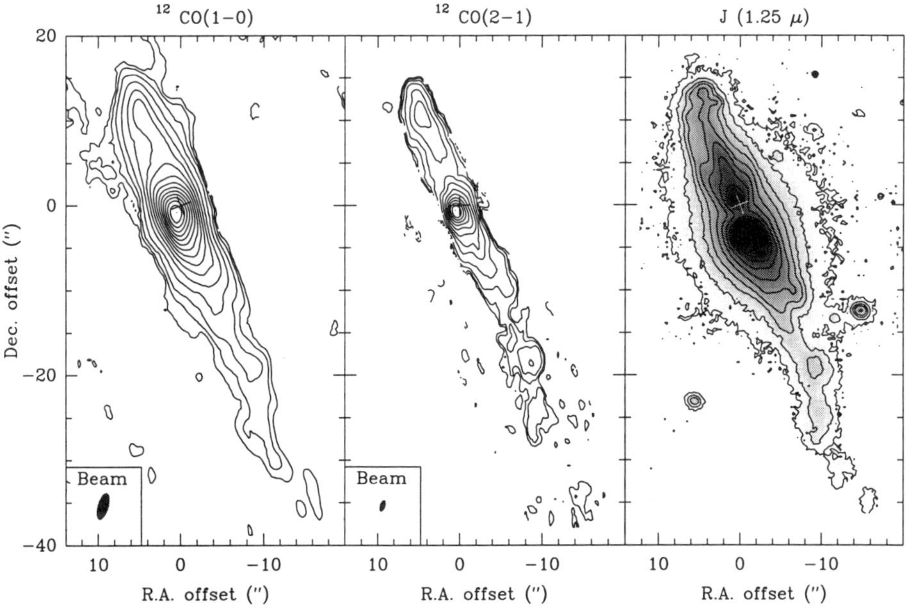

Figure 2. Integrated intensity ^{12}CO maps and J image of OH 231.8+4.2. Levels are: 1, 2.5, 5, 10 to 100 by 10% for the CO maps, and 0.2, 0.4, 0.75, 1.25, 2.5, 5, 10 to 100 by 10% for the J image

In Fig. 2 we show the integrated intensity maps of ^{12}CO. We find that the emission is very extended and narrow ($\sim 55'' \times 4''$). Note also the asymmetry with respect to the equator and the clumpiness of the nebula, which, in fact, consists of different mass condensations. Another good tracer of the mass, independently of the excitation, is the NIR emission. Such an emission arises from reflected light by dust grains and outlines the mass distribution, except for the center where there is a large extinction. We have obtained maps at different NIR bands (J, H and K') using the IRAC-2b camera of the 2.2 m ESO Telescope (La Silla, Chile). In Fig. 2 we show the J image. Note the agreement between both the CO and J maps. The total axial extent is very similar in both maps, the J map also showing an undulating tail of material toward the South and a clear clumpy structure. The different distributions of the neutral and of the high-excitation components are remarkable (Figs. 1 and 2). In the direction perpendicular to the nebula axis, the atomic gas is much more extended than the CO gas which seems to be confined to the axis. The extent in the axial direction is very similar in both cases.

The very wide CO profiles ($\sim$ 330 km s^{-1}, projected) indicate the presence of a high-velocity molecular flow in OH 231.8. We find a displacement of the emitting regions along the axis for the different velocity intervals, so that the red-and blueshifted emissions arise from the South and North lobe, respectively. In Fig. 3 we show the interferometric velocity-position diagrams along the symmetry axis for ^{12}CO and other two representative molecules. For all the molecular lines we find a similar emission distribution (that from CO being the most extended) and kinematics, except for HCO$^+$ (see below). We have found a strong velocity gradient along the nebula axis, which is in general compatible with that of the high-excitation component. We stress on the high velocities reached by the neutral material at the end of the lobes of OH 231.8. We think that the acceleration of the lobes must be due to the interaction between the fast, post-AGB ejections and the slow, circumstellar envelope originated during the red giant phase. In all the molecules, we find a relatively narrow component at the systemic velocity located at the nebula center, so it could be the remnant of the AGB envelope which has not been accelerated by the fast post-AGB ejections.

4. OH 231.8+4.2: mass, momentum and abundances

We have calculated the mass, momentum and abundances of the different clumps of OH 231.8. The method followed is described in detail in Sánchez Contreras et al. (1997). For simplicity, we have chosen six different clumps related to six velocity intervals (indicated in the tables) corresponding to the different features found in both our line profiles and spatial intensity distributions.

In the estimation of the mass from the ^{13}CO(1-0) line, we have assumed that it is optically thin, and a standard value for the ^{13}CO abundance in AGB envelopes. The excitation state has been estimated from the ^{12}CO(2-1)/^{12}CO(1-0) brightness temperature ratio. We have found such a ratio to be almost constant along the axis, giving a value for the rotational temperature of about 10 K. Note that such a low temperature suggests a strong cooling after the post-AGB shock.

In Table 1 we show the mass, the average (deprojected) velocity with respect to the systemic velocity, and the momentum associated to the different velocity intervals or clumps. We have found a total mass $\sim$ 0.6 $M_\odot$, of which at least 0.2 $M_\odot$ are flowing at high velocity. These values can be underestimated, probably not more than a factor 2, but not significantly overestimated (see discussion in Sánchez Contreras et al. 1997). Our mass estimations are compatible

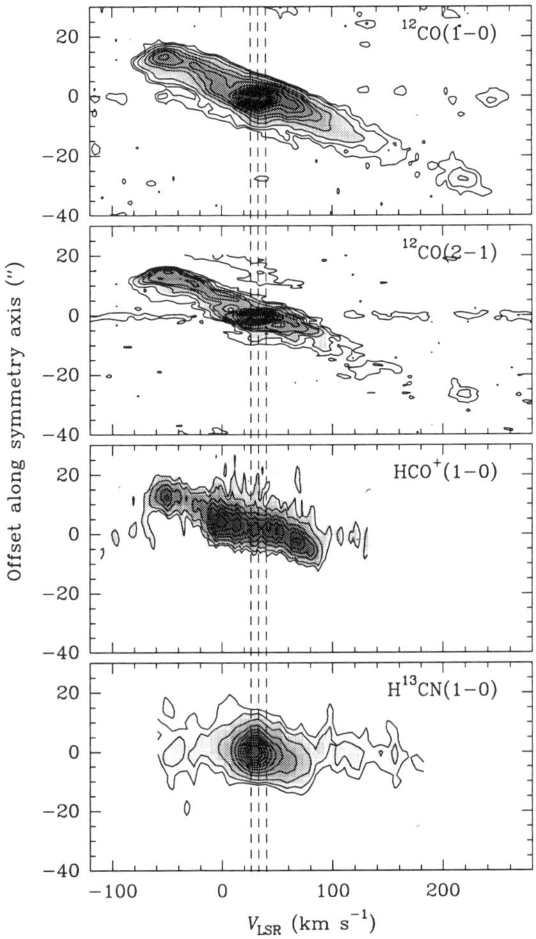

Figure 3. IRAM Plateau de Bure velocity-position maps along the symmetry axis of OH 231.8+4.2 in some of the observed transitions

with the total dust mass in OH 231.8, $\sim 0.01\ M_\odot$ (Kastner et al. 1995), which leads to a molecular mass of about 1 $M_\odot$ if we assume a standard value for the gas-to-dust ratio. We have also estimated the mass-loss during the last AGB phases ($\sim 10^{-4}\ M_\odot\ \mathrm{yr}^{-1}$). We have found very high values for the momentum carried by the gas in the lobes, in both of about 10 $M_\odot\ \mathrm{km\,s}^{-1}$. If the radiation pressure was the only mechanism responsible from the momentum increase, then the interaction time should have been $\sim$50 000 yr. This time is, however, extremely large compared with the expected value, which must be a small fraction of the PPN stage duration ($\sim$1 000 yr). Even if we consider a strong effect of multiple scattering, we need another additional mechanism to produce such high momentum.

In Table 2 we show the derived abundances of the different species observed in the different clumps of OH 231.8. We assumed that all lines are optically thin

Table 1. Mass and momentum of the clumps

	interval km s^{-1} (LSR)	mass ($M_\odot$)	V_i (km s^{-1})	momentum ($M_\odot$ km s^{-1})
I1	−80:−30	0.04	137	5.5
I2	−30:+10	0.11	67	7.4
I3	+10:+55	0.29		
I4	+55:+80	0.08	54	4.3
I5	+80:+150	0.03	128	3.8
I6	+150:+250	0.01	260	2.6

Table 2. Molecular abundances for the clumps

interval km s^{-1} (LSR)	−80:−30	−30:+10	+10:+55	+55:+80	+80:+150	+150:+250
X(HCO^+)	1.3 10^{-8}	1.3 10^{-8}	3.5 10^{-9}	1.0 10^{-8}	1.4 10^{-8}	$<$ 2.5 10^{-8}
X(SiO)	0.9 10^{-8}	1.8 10^{-8}	1.7 10^{-8}	1.2 10^{-8}	1.1 10^{-8}	$\lesssim$ 1.0 10^{-8}
X(HNC)	0.6 10^{-8}	1.0 10^{-8}	2.1 10^{-8}	0.7 10^{-8}	1.0 10^{-8}	
X($H^{12}CN$)	4.0 10^{-8}	3.6 10^{-8}	6.9 10^{-8}	5.2 10^{-8}	2.4 10^{-8}	
X($H^{13}CN$)		3.6 10^{-9}	1.3 10^{-8}	6.1 10^{-9}		
X(CS)	0.6 10^{-7}	0.3 10^{-7}	0.8 10^{-7}	0.2 10^{-7}	0.4 10^{-7}	$<$ 1.0 10^{-7}
X(SO_2)		0.2 10^{-5}	1.4 10^{-5}	1.1 10^{-5}	1.3 10^{-5}	
X(SO)			2.4 10^{-6}			

and the masses of the clumps previously estimated. We also assumed the same rotational temperature (10 K) in all cases. The most interesting result from these calculations is the very strong decrease in the central clump of the HCO^+ abundance, that is about 3 or 4 times lower than in the outer parts of the nebula. Moreover, the HCO^+ abundance is relatively high. Our results suggest that the efficient formation of this molecule in OH 231.8 is due to shock chemistry, since the HCO^+ emission appears particularly intense in the shock-accelerated lobes, and not in the central clump where one expects the maximum photoionization. The high abundance of SiO and SO_2 at relatively large distances from the center can also be due to the extraction of SiO and S from dust grains by shocks.

Acknowledgments. C.S.C. is grateful to the IAU for a travel grant. We thank A. Gil de Paz and J. Zamorano for the Hα image of OH 231.8. Part of this work was supported by the DGES, project numbers PB96-0104 and PB95-0066.

References

Bowers P.F., Morris M., 1984, ApJ 276, 646
Cohen M., 1981, PASP 93, 288
Cohen M., Dopita M.A., Schwartz R.D., Tielens A.G.G.M., 1985, ApJ 297, 702
Kastner J.H., Weintraub D.A., Zuckerman B., et al., 1992, ApJ 398, 552
Kastner J.H., Weintraub D.A., 1995, AJ 109, 1211
Kastner J.H., Weintraub D.A., Merril K.M., Gatley I., 1998, AJ 116, 1412
Reipurth B., 1987, Nature 325, 787
Sánchez Contreras C., Bujarrabal V., Alcolea J., 1997, A&A 327, 689

Asymptotic Giant Branch Stars
T. Le Bertre, A. Lèbre and C. Waelkens, eds.
I.A.U. Symposium 191, 1999

ISO results on circumstellar envelopes

M. J. Barlow

Dept. of Physics & Astronomy, University College London, Gower St., London WC1E 6BT, U.K.

Abstract.
This review covers spectroscopic results obtained with ISO on molecular and atomic line emission from the circumstellar envelopes of oxygen-rich AGB stars and post-AGB objects and from cool supergiant stars. As well as the SWS detection of the 4.27-μm CO_2 band in absorption and emission from cool star envelopes, several new emission bands which have been detected in SWS spectra between 13.9 and 16.2 μm have been identified with CO_2 bands. Strong H_2O line emission has been observed in the LWS and SWS spectra of high mass loss rate O-rich AGB and M supergiant stars, confirming the predicted importance of this molecule as a coolant in their outflows. The proposed infrared radiative pump mechanism for circumstellar OH masers has been directly confirmed for the first time via measurements for several stars of the OH infrared pumping lines between 34.6 μm and 163 μm. Water vapour emission lines are not present in the spectra of O-rich post-AGB objects, having been replaced by broad emission features at 43 and 62 μm due to crystalline water ice that has condensed in the very cool outflows.

1. Introduction

In this review I will concentrate on ISO observations of atomic and molecular species around oxygen-rich cool evolved stars, although some discussion is included of observations of water ice emission bands. Space limitations preclude more than a passing mention of ISO results for carbon-rich objects – reviews of results on C-rich objects have been presented by Barlow (1998a) and by Cernicharo (1998). A companion review (Waters & Molster, this volume) describes ISO observations of solid-state dust features from evolved stars.

The plan of this review will be to concentrate first on results pertaining to smaller circumstellar radii (i.e. from warm material, absorbing or emitting at SWS infrared wavelengths), followed by a discussion of observations of species located at larger circumstellar radii (i.e. cooler material, emitting at longer, LWS, wavelengths).

2. ISO SWS observations of molecules in cool star envelopes

Prior to the launch of ISO, telluric absorption lines arising high in our own atmosphere had prevented observations of many of the bands of H_2O and CO_2 that had been predicted to be strong in cool star spectra. Justtanont et al. (1996) presented the 2.4–45-μm SWS grating spectrum of the highly evolved M supergiant NML Cyg, obtained with a resolving power (depending on wavelength) of between 700 and 1400. The 6.2-μm ν_2 (bending mode) ro-vibrational band of H_2O was detected in absorption, together with the 2.7-μm ν_1 and ν_3 stretching bands. The strength of the higher-excitation 2.7-μm absorption band implied that it must originate from a hotter region (with T > 500 K) interior to where the 6.2-μm band absorption originates. Justtanont et al. (1996) also detected for the first time absorption due to the 4.27-μm stretching band of CO_2. The 15.2-μm bending mode band of CO_2 was not detected however. Since they also detected strong absorption in the CO fundamental ro-vibration band at 4.6 μm, Justtanont et al. were able to derive an overall CO/CO_2 column density ratio of 5. From fits to the 2.7-μm H_2O band, they obtained a surprisingly high $^{12}C/^{13}C$ ratio of > 100.

Tsuji et al. (1997) presented similar SWS observations of these bands for the M2-M7 giants β Peg, 30g Her, SW Vir and RT Vir, and for the Mira variable S Vir. Their observations of the 2.7-μm H_2O absorption band could be fitted by a H_2O layer located at about two stellar radii, with an excitation temperature of 1250 K. They also found broad excess absorption at 4.27-μm in the spectra of the M7III stars SW Vir and RT Vir, and in the spectrum of the Mira variable S Vir. This absorption could be fitted by CO_2 columns with excitation temperatures of 750–1250 K, located at a similar radius ($\sim 2R_*$) to that deduced for the water vapour responsible for the 2.7-μm H_2O band.

Justtanont et al. (1998) reported the identification of four new emission lines discovered in the mid-infrared SWS spectra of five AGB stars. The detective story started with the observation of emission features at 13.87 and 16.18 μm, together with a feature at 14.97 μm which could appear in emission or absorption, in low-resolution (R $\sim$ 250) SWS grating spectra of several AGB stars having prominent 13-μm broad dust emission features. At first it was suspected that the new features might be linked to the 13-μm dust band carrier, but further analysis, together with the subsequent acquisition of higher resolution grating and Fabry-Perot SWS spectra of the stars, led to their identification with Q-branch ro-vibrational bands of CO_2. The 14.97-μm band is the fundamental (ground-state) ro-vibrational ν_2 band of CO_2 – it appears in absorption in the spectra of two of the AGB stars (R Cas and W Hya) and in emission in the spectra of RX Boo and EP Aqr. Its appearance in absorption was attributed to the additional presence of hot ν_2 bands of CO_2, with the ensemble of transitions, estimated to arise from material at $\sim$ 1250 K, shifting the absorption peak to a slightly shorter wavelength than for the fundamental band alone. The transitions at 13.87, 15.40 and 16.18 μm (all seen in emission), originate from excited ro-vibrational transitions from levels in the ladder above that responsible for the fundamental 14.97-μm transition. Justtanont et al. estimated typical excitation temperatures of $\sim$ 650 K from the observed relative intensities and estimated that the lines are formed a few stellar radii above the photosphere. These physical parameters are very similar to those derived by Tsuji et al. (1997) for the material inferred to be responsible for the 2.7-μm H_2O and 4.27-μm CO_2 absorption bands observed in the spectrum of the AGB star S Vir.

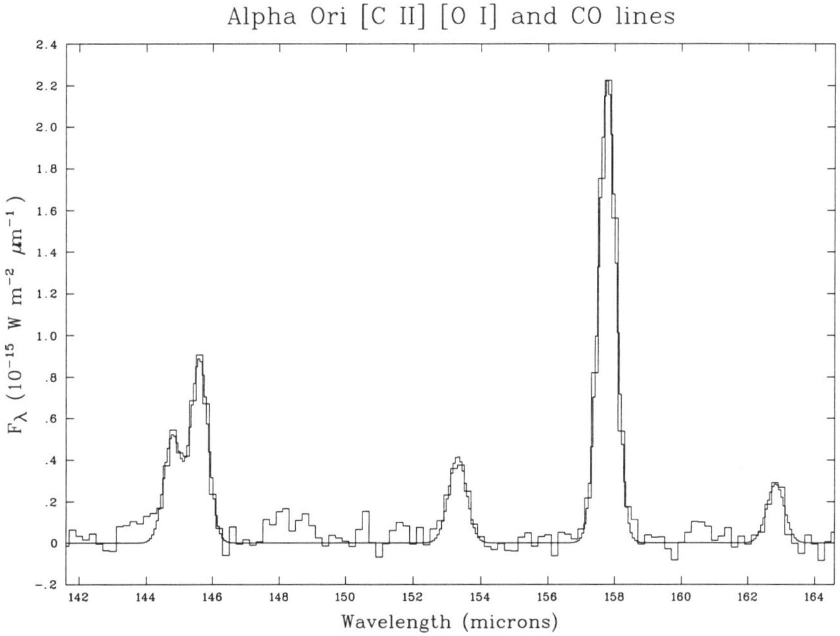

Figure 1. A portion of the background-subtracted and continuum-subtracted LWS spectrum of α Ori. The smooth curve shows Gaussian fits to the observed lines of [O I] and [C II] at 145.5 and 157.7 μm and to the J = 18-17, 17-16 and 16-15 rotational lines of CO at 144.8, 153.3 and 162.8 μm.

3. Infrared fine structure emission lines in the spectra of cool stars

Glassgold & Huggins (1986) and Rodgers & Glassgold (1991) have calculated the ionization structure and energy balance for the wind of the M2Iab supergiant α Ori. Due to the presence of a chromosphere, the recombining wind beyond the chromosphere is predicted to be mainly atomic, with the small amount of carbon that associates into CO eventually being dissociated further out in the wind by the interstellar radiation field. The cooling of the wind was predicted to be dominated by atomic fine structure lines. These predictions were confirmed by the KAO observations of α Ori by Haas & Glassgold (1993), who detected strong [O I] 63.2-μm and moderate [Si II] 34.8-μm emission. Haas et al. (1995) extended these KAO observations to include detections of the same lines from α Sco and α Her as well. Both these M supergiants have B-star companions which are expected to irradiate the cool star winds with photodissociating flux levels much higher than the interstellar value. Justtanont et al. (1998, poster contribution, this conference) have obtained SWS grating spectra of both α Ori and α Sco. In addition to the previously detected 34.8-μm line of [Si II], they also detected the 26.0 and 35.3-μm fine structure lines of [Fe II] from both stars and used the [Fe II] flux ratios to estimate an excitation temperature of 1230 K for the emitting region around α Ori, and 1785 K for α Sco.

The LWS grating spectrum of α Ori (part of which is shown in Fig. 1), obtained by the LWS Consortium just two days before the end of the ISO mission, shows the expected emission from the [O I] 63- and 146-μm lines and the [C II] 158-μm line (after subtraction of the background interstellar component in the latter case), as well as the detection of CO rotational lines extending from 174 μm (J = 15-14) to 130 μm (J= 20-19). Although not as strong as the fine structure lines, the integrated flux of these CO lines means that CO cannot be neglected as a wind coolant. The LWS spectrum of α Sco shows the CO lines to be absent and the [O I] and [C II] lines to have line-to-continuum ratios that are four times higher than in the case of α Ori. These two effects, as well as the higher [Fe II] excitation temperature found by Justtanont et al. (1998, poster contribution, this conference), can be attributed to the additional photodissociation and heating of α Sco's wind by its B-star companion.

Aoki et al. (1998) have reported the detection of [Fe II] 26.0 and 35.3 μm and [Si II] 34.8 μm line emission from the wind of the M6III giant 30g Her. They determined column densities for the emitting Fe^+ and Si^+ ions and, with the assumption of standard solar abundance ratios relative to hydrogen, estimated an emitting region mass of 2–3$\times10^{-6}$ $M_\odot$. Aoki et al. (1998a) also detected fine structure lines in their SWS spectra of two carbon stars, with one of them, the C7,2 star TX Psc, showing emission in the [Fe I] 24.0 μm, [S I] 25.2 μm and [Fe II] 26.0 μm lines. The neutral and singly ionized lines of Fe yielded an emitting region mass of 6$\times10^{-6}$ $M_\odot$, while the neutral S line yielded an emitting mass estimate of 7$\times10^{-6}$ $M_\odot$. The close agreement between these two mass estimates indicates a solar Fe/S ratio, i.e. there seems to be no evidence for depletion of Fe onto dust grains in this carbon star's outflow.

4. UIR band emission from M supergiants

Sylvester et al. (1994) discovered the well-known 8.6- and 11.3-μm UIR bands, usually attributed to aromatic hydrocarbons such as PAHs or HAC complexes, superposed on the broad 10-μm silicate emission features of half a dozen M supergiants in the double cluster h & χ Per. Not only were these the coolest objects from which UIR-band emission had been observed, but M supergiants are oxygen-rich, for which equilibrium condensation theory predicts that all available carbon atoms should be locked up in CO molecules, allowing only oxygen-rich molecules and particles to form. Sylvester et al. interpreted the observed presence of carbon-rich species in terms of a non-equilibrium chemistry model by Beck et al. (1992), whereby chromospheric UV radiation (a) liberates atomic carbon by photodissociating CO molecules, enabling the formation of C-rich species as well as the usual silicates; and (b) provides the photons needed to excite the observed UIR-band emission.

While the 8.6-μm band has only ever been seen in the company of other members of the ensemble of PAH-type UIR bands, the UIR band at 11.3 μm is susceptible to confusion with a narrow crystalline olivine band appearing at the same wavelength. It was therefore desirable to obtain further confirmation of the presence of PAH-type hydrocarbons around these M supergiants by searching for the presence of the well-known PAH bands at 6.2 and 7.7 μm in the SWS spectra of these stars. Tsuji et al. (1998) confirmed the presence of these UIR

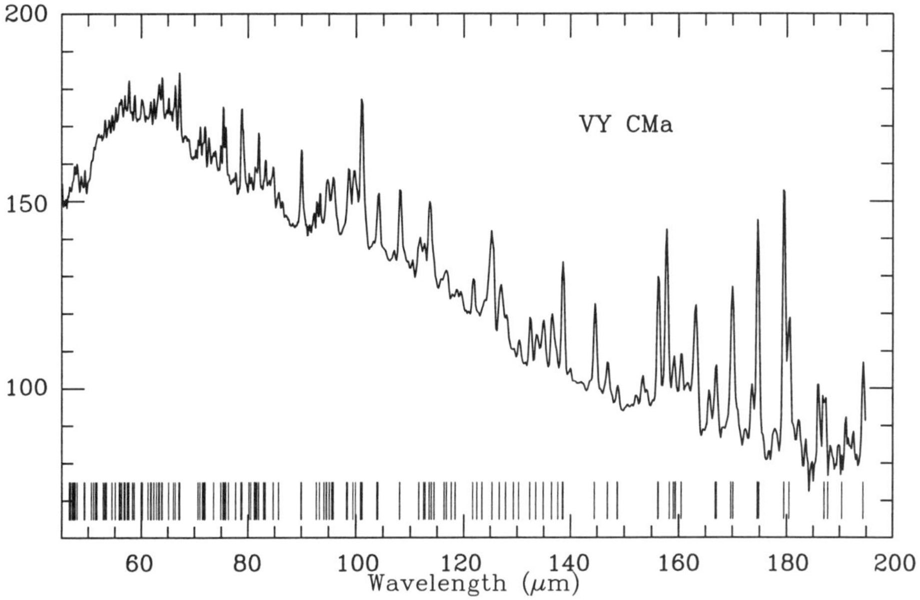

Figure 2. The ISO LWS spectrum of the red supergiant VY CMa. The observed flux (F_λ) has been multiplied by λ^4 to reduce the dynamic range of the vertical axis. Over 100 water lines are detected above the strong dust continuum. The vertical bars below the spectrum indicate the positions of some of the ortho- and para-H_2O rotational lines occurring in the LWS wavelength range.

bands in their SWS spectra of the h & χ Per M supergiants KK Per, FZ Per and HD 14580, while Sylvester (1998) has confirmed their presence in the SWS spectra of several other h & χ Per supergiants, including AD Per.

5. Water vapour line emission and the OH maser pump for cool evolved stars

Because of the huge opacity of the Earth's atmosphere in the far-infrared rotational lines of H_2O, these same lines had never been observed in the spectra of astronomical sources prior to the launch of ISO. Predictions of the importance of these lines as coolants in a variety of astrophysical environments (e.g. Deguchi & Nguyen-Q-Rieu 1990, Neufeld & Kaufman 1993) were quickly confirmed by their widespread detection by ISO. Fig. 2 shows the 43–190-μm LWS grating spectrum of the high luminosity M supergiant VY CMa, whose high mass loss rate wind exhibits a particularly rich water line spectrum, superposed on the strong dust continuum emission from the wind.

Numerous water lines were detected in the 43–190-μm LWS spectrum of the M8e semi-regular variable W Hya by Barlow et al. (1996), and modelled with Sobolev escape probability techniques. While the output spectrum was found to be sensitive to the mass loss rate and H_2O/H_2 abundance ratio, it was only weakly dependent on the wind temperature gradient and velocity field. Although radiative excitation processes were included in the modelling, the excitation of H_2O was found to be dominated by collisions with H_2. All of the lines observed are optically thick, explaining why so many transitions from high rotational levels are easily detectable. The best fit to the observed water line spectrum was obtained for a mass loss rate of 6×10^{-7} $M_\odot$ yr^{-1}, with H_2O/H_2 ratios of 8×10^{-4} and 3×10^{-4} inside and outside 16 stellar radii, respectively. A distance of 130 pc was adopted by Barlow et al. (1996) – use of the subsequently published HIPPARCOS distance for W Hya of 114 pc would have reduced the derived mass loss rate slightly.

Neufeld et al. (1996) used the SWS Fabry-Perot (R = 30 000) to measure 3 high-J rotational emission lines of water in the 29.8–40.7-μm region of W Hya's spectrum, and also detected 2 further water line features, one in emission and the other in absorption, using the highest resolution grating mode of the SWS (R = 2000). Their modelling of the fluxes observed for these high-J water lines led to a very much larger derived mass loss rate, $(0.5\text{–}3)\times10^{-5}$ $M_\odot$ yr^{-1}, than obtained by Barlow et al. (1996) from their modelling of the lower-J lines observed at longer wavelengths. As pointed out by Neufeld et al., the high mass loss rate which they derived for W Hya is at least a factor of 30 above previous determinations based on mm-wave CO observations. One possible explanation for the discrepancy is that in the innermost regions of the wind, radiative pumping of the ν_2 vibrational level of H_2O from the $\nu = 0$ ground vibrational state, via the absorption of photons in the 6-μm region, may lead to significant overpopulation of high-J rotational levels following the radiative decay back down to the $\nu = 0$ state, thereby causing enhanced emission in the ($\nu = 0$) high-J rotational transitions that are observed in the SWS spectral region.

The many water lines detected in the 43–197 μm LWS grating spectrum of the Mira variable R Cas have been fitted by Truong-Bach et al. (1998) using a circumstellar model that treats in a self-consistent manner the radiative transfer, photodissociation and chemical exchange reactions, together with the heating and cooling processes. Adopting the HIPPARCOS distance of 107 pc and the corresponding stellar radius (1.3 AU) and effective temperature (2400 K) measured by Haniff et al. (1995), Truong-Bach et al. obtained a mass loss rate of 5×10^{-7} $M_\odot$ yr^{-1} from their fit to the observed water line spectrum of R Cas, very similar to the mass loss rate derived for W Hya by Barlow et al. (1996). The H_2O/H_2 abundance ratio was again found to be constant out to a certain radius (20 R_* in this case), beyond which it falls by a factor of three.

Photochemical models (e.g. Huggins & Glassgold 1982) predict that H_2O should be photodissociated by the interstellar radiation field into OH + H at large circumstellar radii. Elitzur, Goldreich & Scoville (1976; EGS) worked out a detailed radiative pump mechanism for OH/IR stars whose maser emission is dominated by the 1612 MHz satellite transition within the J = 3/2 $^2\Pi_{3/2}$ ground state of OH. Their predicted excitation pathway is shown in Fig. 3. The ground $^2\Pi_{3/2}$ J=3/2 state is excited to the $^2\Pi_{1/2}$ J=5/2 state by 34.6-μm line absorption – the preferred decay from there is to the $^2\Pi_{1/2}$ J=3/2 state, with emission at 98.7 μm. The next preferred stage is decay to the $^2\Pi_{1/2}$ J=1/2 state, with emission at 163.2 μm, and finally there is a decay back to

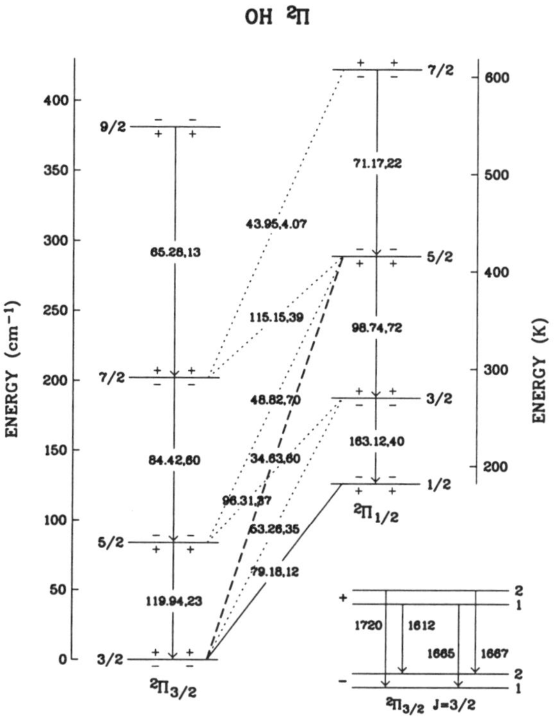

Figure 3. The low-lying energy levels of OH, showing the transitions involved in the maser pumping scheme. The maser action arises from transitions within the J = 3/2 $^2\Pi_{3/2}$ ground state.

the $^2\Pi_{3/2}$ J=3/2 ground state, with emission at 79.2 μm. Excitation to the $^2\Pi_{1/2}$ J=3/2 state by 53.3-μm photons can also contribute to the pump. If the transitions to and from the ground state (34.6, 53.3 and 79.2 μm) are optically thick, while those between the excited levels are optically thin, this scheme causes an inversion of the F=2/F=1 levels of the ground state, leading to 1612 MHz maser action. Another requirement is that there be sufficient 34.6- and/or 53.3-μm flux available: this is provided by continuum emission from dust in the outflow. Elitzur et al.'s proposed pumping mechanism had not been directly confirmed before the launch of ISO, because the OH rotational lines involved in this pumping scheme all occur at far-infrared wavelengths, between 34.6 and 163 μm, unobservable from the ground but accessible to ISO's spectrometers.

Sylvester et al. (1997) directly observed the OH pump line strengths in the ISO spectrun of the A-F hypergiant source IRC+10°420. The OH 34.6-μm doublet was confirmed to be in absorption in their SWS grating spectrum, while the lines in the resulting downward cascade (98.7, 163.2 and 79.2 μm; see Fig. 3) were all clearly detected in emission in the LWS spectrum, in agreement with predictions. Comparison of the observed photon absorption rate in the 34.6-μm doublet of 83×10^4 m^{-2} s^{-1} with observations made with the Nançay radiotelescope at the same epoch of a 1612 MHz maser line photon flux of

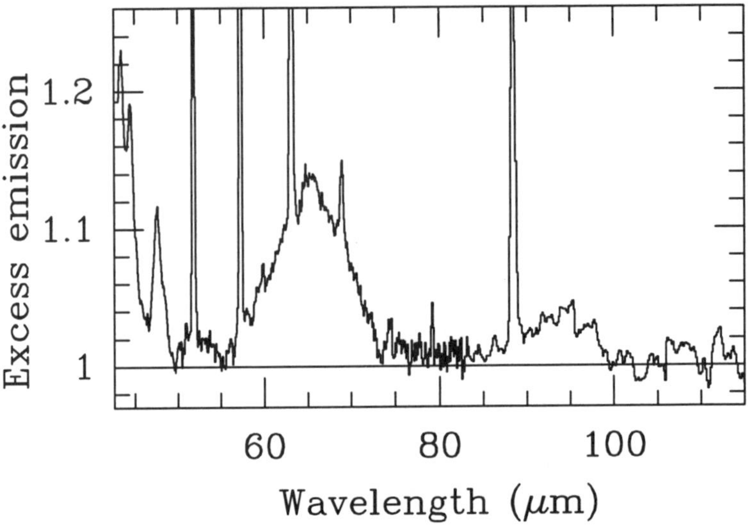

Figure 4. The LWS grating spectrum of NGC 6302 after background subtraction and division by a modified blackbody fit (45 K, $\lambda^{-1.2}$ emissivity) to the underlying continuum.

6.2 photons m^{-2} s^{-1}, implies a corresponding pumping efficiency of 7.5%. Since the 1612 MHz maser can also be pumped by radiative absorption by the 53.3 μm transition, as well as by radiative absorption and collisional excitation in the 79.2 μm line (see Fig. 3), a more appropriate estimate for the pumping efficiency comes from ratioing the emission rate in the 1612 MHz maser transition to that in the final downward pumping cascade line at 79.2 μm. The latter rate is measured to be $(200 \pm 30) \times 10^4$ m^{-2} s^{-1}, corresponding to an overall pumping efficiency for the 1612 MHz transition of IRC+10°420 of 3%. Similar observations of the M supergiant OH/IR source NML Cyg by the LWS consortium yield an overall pumping efficiency of 7%. The two measurements show good agreement with the predictions of Elitzur et al. (1976). Tung et al. (1998) have presented detailed modelling of the OH pumping conditions and maser radiative transfer for IRC+10°420, finding agreement with the observations of Sylvester et al.

6. Water ice emission bands in the IR spectra of cool evolved objects

The stretching-mode absorption band of water ice at 3.1 μm has been studied in the spectra of astronomical sources since the 1970's but, until the advent of ISO, the far-infrared ice bands had only been seen in the KAO spectra of a small number of sources (Omont et al. 1990), the most well known of which is IRAS 09371+1212 (the Frosty Leo Nebula; Forveille et al. 1987). The 3.1 μm band is only ever seen in absorption, because ice would evaporate before it could get hot enough to emit at 3 μm. Thus only water ice column densities in front of background infrared sources can be derived from observations of the optical depth in the 3.1 μm band. However, the lattice translational mode far-infrared bands of water ice can be observed in *emission*, so that the total emitting mass of ice can in principle be derived. Amorphous ice only exhibits a broad 44-μm band (transverse optical mode), whereas crystalline ice exhibits a narrow 44-μm band and a broader 62-μm peak (longitudinal acoustic mode). Thus the detection of the 62-μm peak requires the presence of crystalline ice (Omont et al. 1990; Moore et al. 1994).

Fig. 4 shows the LWS 43–197 μm grating spectrum of the Type I bipolar planetary nebula NGC 6302, after background subtraction and division by a modified blackbody fit to the underlying dust continuum (Lim et al., in preparation). A strong narrow emission feature is visible at about 43–44 μm. This was also seen at the long wavelength end of the SWS spectrum of NGC 6302 shown by Waters et al. (1996) and attributed by them to water ice and crystalline silicates. As noted above, this band can in principle arise from either amorphous or crystalline water ice. A strong, very broad emission feature (FWZI $\sim$ 15 μm) peaks at about 65 μm in the LWS spectrum of NGC 6302 (Fig. 4) and its presence implies that the ice is crystalline (there is also likely to be an underlying contribution by orthopyroxene to this feature). Laboratory measurements (e.g. Smith et al. 1994) show that the peak intrinsic strength of the 44-μm ice band is usually 2–3 times larger than that of the 62-μm band, whereas, from Fig. 4, it is clear that for NGC 6302 the peak strength of the 44-μm feature is less than 50% larger than that of the 65-μm feature. This can largely be explained by the fact that NGC 6302's low characteristic temperature of emission in this region ($\sim$45 K; see Fig. 4) shifts more emission into the longer wavelength band.

In addition to the two ice bands, the LWS spectrum of NGC 6302 in Fig. 4 exhibits several other emission features that are broader than the instrumental resolution of 0.3 μm. There is a band at 47.5 μm; a possible band at 52 μm (underlying the narrow [O III] 52-μm nebular emission line; this feature may be due to the longitudinal optical branch of crystalline ice; Bertie & Whalley 1967); and one at 69 μm, which is only slightly broader than the instrumental resolution, may be attributable to crystalline forsterite, Mg_2SiO_4 (see Waters & Molster, this volume). Finally, there is a very broad feature extending from 88–98 μm.

Barlow (1998b) has presented LWS spectra of a number of other sources that exhibit prominent 44- and 62-μm ice emission features, including those of the OH/IR star GL 5379, the O-rich post-AGB stars HD 101584 and HD 161796 and the young planetary nebula CPD–56°8032 (see Cohen et al., this volume, for a fuller discussion of the latter object). None of the O-rich AGB stars and M supergiants that show strong far-IR rotational line emission from H_2O have been observed to exhibit water ice bands, while none of the post-AGB objects that *do* show water ice band emission have been found to exhibit H_2O rotational lines. This is consistent with the scenario proposed by Omont et al. (1990) for the formation of crystalline water ice particles in the Frosty Leo Nebula (an O-rich post-AGB object), namely that water ice is only deposited onto silicate dust cores once the outflowing material has moved far enough out from the star for ambient temperatures to have dropped low enough for ice condensation to occur. While this process might well occur around AGB stars, warmer dust grains in the inner parts of their flows will dominate their overall spectra, whereas once strong mass loss has effectively ceased, after departure from the AGB, the absence of warmer dust particles will render ice band emission more prominent, particularly if the final mass loss on the AGB had ratcheted up to superwind levels.

Acknowledgments. I would like to thank Drs. X.-W. Liu and R.J. Sylvester for their help in preparing several of the figures used here.

References

Aoki W., Tsuji T., Ohnaka K., 1998, A&A 333, L19

Barlow M.J., 1998a, Ap&SS 251, 15

Barlow M.J., 1998b, Ap&SS 255, 315

Barlow M.J., Nguyen-Q-Rieu, et al., 1996, A&A 315, L241

Bertie J.E., Whalley E., 1967, J. Chem. Phys. 46, 1271

Cernicharo J., 1998, Ap&SS 255, 303

Chen W., Neufeld D., 1995, ApJ 453, L99

Deguchi S., Nguyen-Q-Rieu, 1990, ApJ 360, L27

Elitzur M., Goldreich P., Scoville N., 1976, ApJ 205, 384

Forveille T., Morris M., Omont A., Likkel L., 1987, A&A 176, L13

Glassgold A.E., Huggins P.J., 1986, ApJ 306, 605

Haas M.R., Glassgold A.E., 1993, ApJ 410, L111

Haas M.R., Glassgold A.E., Tielens A.G.G.M., 1995, ASP Conf. Ser. 73, 397

Haniff C.A., Scholz M., Tuthill P.G., 1995, MNRAS 276, 640

Huggins P.J., Glassgold A.E., 1982, AJ 87, 1828

Justtanont K., de Jong T., et al., 1996, A&A 315, L217

Justtanont K., Feuchtgruber H., de Jong T., Cami J., Waters L.B.F.M., Yamamura I., Onaka T., 1998, A&A 330, L17

Moore H.H., Ferrante R.F., Hudson R.L., Nuth J.A., Donn B., 1994, ApJ 428, L81

Neufeld D.A., Chen W., et al., 1996 A&A 315, L237

Omont A., et al., 1990, ApJ 355, L27

Rodgers B, Glassgold A.E., 1991, ApJ 382, 606

Smith R.G., Robinson G., Hyland A.R., Carpenter G.L., 1994, MNRAS 271, 481

Sylvester R.J., 1998, in 'Solid Interstellar Matter - the ISO Revolution', ed. A.P. Jones, C. Joblin and L. d'Hendecourt, Editions Frontières, in press

Sylvester R.J., Barlow M.J., Skinner C.J., 1994, MNRAS 266, 640

Sylvester R.J., Barlow M.J., et al., 1997, MNRAS 291, L42

Thai-Q-Tung, Dinh-V-Trung, Nguyen-Q-Rieu, Bujarrabal V., Le Bertre T., Gérard E., 1998, A&A 331, 317

Truong-Bach, et al., Ap&SS 255, 325

Tsuji T., Ohnaka K., Aoki W., Yamamura I., 1997, A&A 320, L1

Tsuji T., Ohnaka K., Aoki W., Yamamura I., 1998, Ap&SS 255, 293

Waters L.B.F.M., et al., 1996, A&A 315, L361

Asymptotic Giant Branch Stars
T. Le Bertre, A. Lèbre and C. Waelkens, eds.
I.A.U. Symposium 191, 1999

AGB circumstellar envelopes: molecular observations

V. Bujarrabal

Observatorio Astronómico Nacional, Apartado 1143, E-28800 Alcalá de Henares, Spain

Abstract. Due to the low excitation requirements and easy observation from the ground, molecular line observations probably constitute our main source of empirical knowledge on circumstellar envelopes (CSEs) around AGB stars. The CO rotational transitions, the most intense 'thermal' lines, are efficiently used to determine the total gas content and its spatial distribution in wide samples of objects. Thermal emission from other molecules is mainly useful in order to study their abundances and the chemical reactions taking place in CSEs. Maser lines are easily observed due to their high intensity and the flux distribution in very compact spatial spots and narrow profile spikes, characteristic of the exponential amplification; however the data interpretation is difficult due to the intricate pumping processes. The most important maser lines (of SiO, H_2O and OH) arise from very different regions, which allows the study of various components of the CSEs. We will focus here on SiO masers.

1. Excitation of molecular transitions in AGB circumstellar envelopes

Molecular lines provide us with data on the total amount of gas in circumstellar envelopes (CSEs) around AGB stars, as well as on its structure, the physical conditions in them, and the abundances of more than 50 species. Thousands of stars have been observed. Such observations are eased by a) the very reliable, ground based observational techniques applied (mostly at mm wavelengths), and b) the low excitation present in CSEs, which allows that most of the gas is molecular, but is high enough to excite the molecular rotational transitions.

Molecular lines are useful to study the circumstellar gas except for very small, inner regions ($< 10^{15}$ cm), that are not specifically probed by these low-excitation lines, and the very outer regions ($\gtrsim 10^{18}$ cm), where molecules are photodissociated by the interstellar UV radiation field. The photodissociation radii are thought to vary between about 10^{16} cm and 10^{18} cm; see Mamon et al. (1988) for the case of CO, and e.g. Lafont et al. (1982) and the review by A. Glassgold (in this volume) for other molecules. The largest radius values hold for the densest envelopes and most abundant molecules, due to shielding; for CO, we often expect values for the photodissociation radii $R_{CO} \sim 10^{17}$ cm.

The temperature of the circumstellar gas around AGB stars ranges between about 1 000 K in the innermost region, ~ 100 K at $\sim 10^{16}$ cm, and about 20 K in outer layers ($\sim 10^{17}$ cm). Variations of about a factor 3 are expected from object to object, the less dense envelopes being predicted to be the hotter ones. From comparison between the different models for circumstellar thermodynamics (the predictions of which have been by themselves exhaustively checked with observations), the above figures seem relatively well established, with uncertainties also of about a factor 3. Only in the very inner and outer regions the uncertainties can be larger, due to the difficult comparison with observations. For more details on the circumstellar thermodynamics see the seminal paper by Goldreich & Scoville (1976), Justtanont et al. (1994, 1996) for the case of O-rich stars, Kwan & Linke (1982) and Groenewegen et al. (1994, 1998) for the well known C-rich star IRC +10216, and Kastner (1992) for several other cases.

With these temperatures, the circumstellar gas is expected to be always molecular, except for the outer regions, as we have mentioned. These low temperatures can almost only excite rotational transitions in the ground vibrational state, since the relevant species have vibrational energies $\gtrsim 2\,000$ K. For instance the first rotational level of CO is placed 5.5 K over the ground, but its first vibrational state is at more than 3 000 K. (In some cases, the absorption of stellar photons can yield a significant vibrational excitation, as we will see later.)

In this review we will mostly deal with rotational molecular lines in CSEs, discussing both maser and 'thermal' (i.e. non maser) lines at millimeter wavelengths. High-J and vibrational transitions, observed in the IR, are discussed in reviews by M. Barlow and S. Höfner in this volume.

2. CO 'thermal' emission

CO is the most abundant molecule in CSEs (after H_2), its lowest rotational transitions being the most intense molecular lines. They have been detected in about five hundred objects. The rare isotopic substitutions (mainly ^{13}CO) can also be observed. The chemistry of CO is particularly well known. Its high abundance is due to its very high dissociation energy, and almost all the available C/O atoms are locked in CO, up to R_{CO}, see 3.1.

The excitation of the low-J transitions of this molecule is also easy to understand. The level structure is simple, consisting of a single rotational ladder with transitions at 2.6 mm (J=1–0), 1.3 mm (J=2–1), 0.9 mm (J=3–2), etc. Due to the anomalously low value of the Einstein coefficients (for instance, the A-coefficient of the CO J=1–0 transition is only 7 10^{-8} s^{-1}), the population by collisions of the low-energy levels is very efficient.

If we assume that a CO transition is optically thin and the rotational excitation temperature T_{rot} is much larger than the energy of the levels joined by the transition (and than the background radiation temperature), it is easy to show that the net photon production rate of this transition per unit time and volume is: $n_{ph} \sim A n_u \underset{\sim}{\propto} n_{CO}/F_{par} \underset{\sim}{\propto} n_{CO}/T_{rot}$. Where A is the Einstein coefficient of the transition, n_u is the population of its upper level, n_{CO} is the total number of CO molecules per cubic centimeter, and F_{par} is the partition function. Note that only the rotational temperature (i.e. an average of the excitation temperatures of the rotational transitions) appears, through the partition function, and not the excitation temperature of our transition. The moderate dependence on the inverse rotational temperature is remarkable. Since the radiative transition probabilities are low for CO, T_{rot} is often close to be thermalized, i.e. close to the kinetic temperature T_k. Detailed calculations show indeed that the overexcitation (relatively large T_{rot}) and approximate thermalization are widely satisfied in relevant cases (closer than 10^{16} - 10^{17} cm); however, T_{rot} is often somewhat lower than T_k, since the population of the high-J levels is difficult. This simple formulation is only illustrative of how the CO emission behaves in the case of overexcitation, but it shows that the dependence of the line intensity on the physical conditions is moderate in our case, which means that the model calculations must be robust and give reliable results.

However, when the overexcitation condition is not satisfied the line excitation becomes much more difficult to treat. In the limit, we are dealing with a differential population effect that is very elusive to accurate calculations. Such a case appears in the outer regions of the envelope, where the kinetic temperature and density are low. Since the volume of these regions can be enormous, the total emission from them should be dominant, in spite of underexcitation. Note that the excitation via the absorption of stellar IR photons by the vibrational transitions becomes important precisely in these outer regions; such a phenomenon is also difficult to treat in the models due to the non-local nature of the strong selfabsorption effects in the rovibrational components. We also note that the detection of these very extended, low-brightness shells is difficult, particularly in interferometric observations. (The fact that the total emission of the outer shells is significant is not in contradiction with the brightness drop with the radius in the images.)

The situation is significantly improved since, as we have seen, the photodissociation radius tends to be comparable to the distance at which underexcitation appears, at least for the two most often observed CO lines, J=1–0 and J=2–1. The contribution of the outer layers, beyond R_{CO}, can be neglected, just because there is no CO in them. This helps in the calculation of the CO emissivities. But, since the emission of the outer layers is important if CO is abundant in them, we find a strong dependence of the total calculated emission on the assumed value for R_{CO} (this dependence is approximately linear when the emissivity is constant with the radius).

Finally we must mention that the ^{12}CO lines can be optically thick; but rarely they are very optically thick in all the points of the envelope and in all the directions, so, the 'growth curve' is never very flat and such an effect can be treated in the models. In this respect, observations of rare isotopes, mainly ^{13}CO, help in the analysis.

Examples of the theoretical analysis of the CO line emission can be found in Justtanont et al. (1994), Groenewegen et al.(1998), Loup et al. (1993), and references therein.

2.1. Observational results: mass loss rates

The main expected result from the comparison of the CO observations and calculations is the determination of the mass loss rates ($\dot{M}$). For this task, we must know the CO relative abundance and photodissociation radius. The CO abundance is in fact determined in a selfconsistent way from models for the emission of several lines in well known sources: values between 3 and 8 10^{-4} are found. R_{CO} is expected to depend on the mass loss rate, so both parameters must be calculated at the same time, by means of some kind of iterative procedure. In any case, the favorable CO excitation ensures that it is possible to calculate the mass loss rates from fitting of the observed intensities by means of models, or using relatively simple formulae (often obtained from parametrization of model grids). Comparison between different calculations indicate that the errors in the mass loss determinations are small, not exceeding a factor 3. Those calculations, performed for hundreds of stars, constitute the most important element of knowledge on CSEs around AGB stars.

The mass loss rates determined by these means range between a few 10^{-8} and a few 10^{-5} $M_\odot$ yr^{-1}, being in general well correlated with the FIR intensity and the dust mass loss rates determined from it (the gas to dust mass ratio ranges between 150 and 500).

For more details, see the very useful and thoughtful compilation by Loup et al. (1993), and the surveys by Knapp and Morris (1985; a pioneering, important work), Margulis et al. (1990; on CO in nearby stars), Olofsson et al. (1993a; on diffuse C-rich envelopes), Kerschbaum et al. (1996, 1998; on emission from semiregular and irregular variables), Young (1995; containing J=3–2 and J=4–3 data), Omont et al. (1993a; on very thick and cold envelopes), etc.

Both of the above limits are interesting. Except for a few objects (that are probably peculiar from the chemical point of view, see Kerschbaum et al. 1996, 1998), AGB stars show mass loss rates larger than 10^{-8} $M_\odot$ yr^{-1}. It is of course difficult to work with these lower limits, because of the uncertain interpretation of the sample meaning and nondetections in the different papers. But it is remarkable that, even in surveys of diffuse envelopes, the detection rate is very high. Olofsson et al. (1993a) observed a 'relatively complete' sample of C-rich AGB stars up to a distance of 900 pc, up to 600 pc all the objects were detected and show values of $\dot{M}$ over the limit given above. Other surveys, more or less biased towards dense envelopes, give similar results. No definitive conclusion can be drawn on this important subject, but it seems that at least an important subsample of the AGB stars (perhaps limited by the stellar mass) are loosing mass with $\dot{M} > 10^{-8}$ $M_\odot$ yr^{-1}.

The upper limit is also interesting. It is hard to find AGB stars with $\dot{M}$ higher than 10^{-5} $M_\odot$ yr^{-1}. For instance, the largest value listed by Loup et al. (1993) for an AGB star is 4 10^{-5} $M_\odot$ yr^{-1} (but see Justtanont et al. 1994), and the famous, dense envelope of IRC +10216 has $\dot{M} \sim 2\ 10^{-5}$ $M_\odot$ yr^{-1} (Kastner 1992, Groenewegen et al. 1998). However, the mass loss rates measured in protoplanetary (post-AGB) objects, determined from the values of the total nebular mass ($\sim$ 0.1 - 1 M$_\odot$) and the life time ($\sim$ 500 - 1 000 yr), are larger than or about 10^{-4} $M_\odot$ yr^{-1} in most well studied cases (as for example in CRL 618, CRL 2343, CRL 2688, IRC +10420, NGC 7027, OH 231.8+4.2, Roberts 22, the Boomerang Nebula, etc). This means that the long-lasting mass loss process detected in AGB stars is not responsible for the ejection of the planetary nebula. Instead, we should invoke a discrete phenomenon: a copious, quasi-explosive ejection of matter, that also triggers the end of the AGB phase.

2.2. Observational results: line profiles and envelope structure

We will only marginally deal, for the sake of brevity, on the theoretical and observational characteristics of the molecular line profiles in AGB stars. It is well known (e.g. Morris 1975, Olofsson et al. 1982) that the line shape depends on the opacity and degree of spatial resolution attained by the used telescope.

Two-peak profiles are observed in optically thin lines and (at least) partially resolved envelopes. Parabolic shapes appear in optically thick lines. A variety of intermediate cases are possible, with also some exceptions and deviations from the standard scheme (e.g. Bujarrabal et al. 1989, Knapp et al. 1998). The main parameters determined from the line profiles are the systemic and circumstellar expansion velocities, that correspond respectively to the line centroid and to half the observed line width; these parameters have been accurately measured in hundreds of stars (see data in Loup et al. 1993).

Our knowledge on the spatial structure of CO envelopes has recently received an important boost thanks to the publication of the IRAM atlas of CSEs (Neri et al. 1998). The AGB envelopes systematically show a circular CO image, again in opposition to the case of protoplanetary nebulae (that are mostly bipolar), as already shown by previous works (e.g. Stanek et al. 1995). In the accurate measurements by Neri et al., less than 20% of the CO images of AGB objects show a significant deviation from circular symmetry. In general, the measured radial decrease of the intensity is quite in agreement with the theoretical predictions. The measured outer radii range between 10^{16} and a few times 10^{17} cm, increasing for objects with large mass loss rates. However, it has been impossible in practice to check the theoretical values of the photodissociation radius. There are good indications that the actual total CO extent is comparable to the theoretical R_{CO} (Neri et al.), but the measured outer radii (for instance, the last significant contour in maps) remain somewhat smaller than the theoretical values. The reason is simply that the dynamic range of the existing interferometric maps is not high enough to reach the low brightness expected close to R_{CO}. (Single dish data seem to reach the required sensitivity in some published maps, but the poor spatial resolution and/or small map extent strongly hamper the outer radius measurement.) In any case, the present instrumental performances should allow observations accurate enough, and we can hope to get soon direct information on this important question.

The agreement between the radial variation of the CO brightness and the predictions by simple models suggests that in many objects, about 70% of the sample by Neri et al. (1998), the mass loss rate has been practically constant in the last few thousand years. But it is clear that there are sudden changes in the brightness in a nonnegligible number of objects, that probably correspond to changes in the ejection rate (see also Justtanont et al. 1996, Knapp et al. 1998). The case of the detached envelopes is particularly interesting. In these objects, a very thin molecular shell surrounds an AGB star, being clearly detached from it. In general there is also a more compact component very close to the star, characterized by lower mass loss rate and expansion velocity. Probably the CO image of the detached shell is limited by the ejection duration, though that of the compact component is limited by photodissociation. There is an interesting poster in this Symposium on this topic (Olofsson et al.), see also the published work by Olofsson and coworkers (1998a, and references therein).

3. Other molecules

The main purpose of the observation of the thermal emission from other molecules, less abundant than CO, is the determination of their fractional abundance, the main empirical input to understand the very rich circumstellar chemistry.

The theoretical understanding of the emission from these molecules is much worse than for CO. There are two reasons for that : a) The excitation of these molecules is often more difficult to model than for CO, because in most cases the level structures are more complex and the emission takes place under a regime of 'underexcitation' (as defined above). b) It is impossible that the big effort devoted to understand the CO line excitation can be also devoted to each of the many molecules found in AGB envelopes. So, in most cases only very simple calculations of the molecular abundances are performed, often assuming optically thin emission (not justified for some molecules, like SiO and HCN), an outer radius (usually of the order of 10^{16} cm) and a typical rotational temperature (derived from the observation of several transitions of the molecule). Under these conditions, the results only have a statistical meaning and differences as high as a factor ten can appear between different works.

Only in the case of the very well studied source IRC +10216 and/or for a few interesting molecules, the abundance determination is more sophisticated. In these cases, the studies are of course helped by a large observational information, including maps with high spatial resolution and multiline measurements. These well studied cases are discussed in the reviews by Lucas & Guélin and Glassgold in this volume. A. Glassgold has also explained the main features of the chemistry models. For detailed analysis of molecule abundances, see the papers by Omont et al. (1993b) and Sahai & Wannier (1992) on S-bearing molecules, by Bujarrabal et al. (1989) and Lucas et al. (1992) on SiO, by Nercessian et al. (1989), Fuente et al. (1998) and Olofsson et al. (1993b) on HCN and CN, etc. For the search of more exotic molecules, see for instance Cernicharo & Guélin (1987) and Guélin et al. (1995). Isotopic abundance ratios have been studied in depth by Kahane et al. (1992).

Probably the most complete set of molecular abundances in AGB stars is still that calculated by Nyman et al. (1993b) for two C-rich stars, IRC +10216 and IRAS 15194-5115. (IRAS 15194-5115 is a southern star with a mass loss rate comparable to that of IRC +10216, but placed much farther, at a distance of about 1200 pc.) These authors derived abundances in both sources for 18 species, including some isotopic substitutions. Even relatively heavy molecules, like HC_3N, C_3N and SiC_2, show high abundances, $> 10^{-7}$. Both stars show a very similar chemistry, with abundances coincident within a factor 3. We note that IRC +10216 and IRAS 15194-5115 are quite similar from other points of view, and that we lack for information on many of these species for objects with significantly different physical conditions and O/C abundance.

3.1. Dependence on the O/C abundance ratio

Due to the efficient formation of CO, the O/C relative abundance plays a very important role in the chemistry of the stellar atmosphere and the circumstellar envelope. In carbon rich objects, almost all O nuclei are locked in CO; so, the O-bearing molecules are rare, while the remaining C serves to form the (many) possible carbon-bearing molecules. The opposite situation occurs in O-rich envelopes. The first systematic, quantitative determination of the different molecular emission properties of O and C-rich envelopes has been performed in Bujarrabal et al. (1994a,b), from observations of a sample of 47 AGB stars. As expected, the lines of C(O)-bearing molecules are systematically much more intense in C(O)-rich stars. For instance the SiO(J=2–1)/HCN(J=1–0) line intensity ratio was found to be always larger than 2.3 in O-rich stars, while this ratio is smaller than 0.23 in all the studied carbon rich objects (note that these figures are obtained from main-beam temperatures and slightly depend on the telescope used). Similar quantitative conclusions were reached for other molecules, namely SO, HNC, HC_3N, SiS (that behaves as the C-bearing molecules), and CS. CO was also observed, its behavior being intermediate (as expected) between that of C- and O-bearing molecules. These results were recently confirmed by Olofsson et al. (1998b), for a larger sample of stars. For recent studies on S-stars, see Bieging et al. (1998).

The empirical discrimination is so clear that it can be applied to determine the O/C-rich nature of envelopes of uncertain classification by other means. It is remarkable that S-type stars are intermediate in this respect between the O- and C-rich groups, which shows that their envelopes are also S-type.

These works also allowed statistical studies of the molecule abundances for the different types of stars. The variations of the abundances (averaged for the detections) between O- and C-rich stars are in general of about a factor ten (except for CO). It is in particular noticeable that the abundance of C(O)-bearing molecules in O(C)-rich stars is in some cases not negligible. For instance, the average abundance of HCN in O-rich envelopes was found to be $\sim 4\ 10^{-7}$, with a big dispersion, being its J=1–0 transition detected in the majority of the O-rich objects observed. On the other hand, HC_3N was not detected in the 8 O-rich stars observed in its J=10–9 transition, with upper limits to the abundance $\sim 10^{-7}$, 20 times smaller than the average abundance in C-rich stars.

4. Maser transitions

Maser effect appears in a transition when the populations of the joined levels are 'inverted', or, equivalently, when the excitation temperature is negative and the (also negative) absorption coefficient becomes an amplification coefficient. For large absolute values of the opacity, the exponential amplification yields very intense lines and strong contrasts in the spatial structure and line shape. For these reasons, masers are easily observed. There are data on about 2 000 stars, including VLBI experiments with spatial resolution as high as 1 mas.

The key to understand maser emission is the pumping: the mechanism that maintains the population inversion. Very fine work has been published on the relevant microprocesses, that are interesting by themselves, but the pumping is often so complex that it is very difficult to derive physical or chemical conditions from comparison between models and observations. In fact, the maser observations are more efficient to determine the spatial and kinetic structure of the emitting regions. OH masers have been also useful to study the AGB stellar populations, including the detection and study of the OH/IR stars (very red, optically unidentified AGB stars).

Several molecules show maser emission in AGB CSEs, the most intense masers being those of OH, H_2O, and SiO. The different lines come from different regions of the envelope, so they can be used to study the various circumstellar components. SiO mm-wave masers arise in the very inner shells (at a few 10^{14} cm from the star), OH 18 cm emission comes from much outer regions ($\gtrsim 10^{16}$ cm) and the H_2O 1.3 cm line comes from intermediate layers (2 - 10 10^{14} cm). We can only discuss in this paper the case of SiO masers; for a quite complete review on circumstellar masers, and in general on AGB CSEs, see Habing (1996).

4.1. SiO masers

SiO shows maser emission in rotational transitions within **excited** vibrational states. Transitions from J=1–0 up to J=8–7 in the states v=1, 2, 3, 4 have been observed, the total star sample including more than 500 stars. The J=1–0 transitions of v=1,2 and the J=2–1 transition of v=1 show the most intense masers. The observed line profiles are often composed of several spikes spread over a few km s^{-1}, and are centered on the stellar velocity. Weaker masers in ^{29}SiO and ^{30}SiO are also found, in this case also in the v=0 state. See for surveys of SiO masers, including observations of different lines: Jewell et al. (1987), Bujarrabal et al. (1987), Cernicharo et al. (1992, 1993), Nyman et al. (1993a), Izumiura et al. (1995), and references therein.

SiO masers are mostly detected in Mira-type variables, but they are also found in semiregular and irregular giants and some red supergiants (as well as in a few starforming regions and protoplanetary nebulae); SiO emitters are always O-rich, except for a few S-type stars. Miras show a very good correlation between SiO and IR absolute fluxes (Bujarrabal et al. 1987). The correlation is particularly good for the IR flux at 8 μm, which is the wavelength of the Δv=1 vibrational transition; this transition is thought to be responsible for the maser pumping in the radiative mechanisms (see below). Semiregular and irregular variables (including supergiants) are often weaker, i.e. they show a lower SiO/IR flux ratio. Only objects of these types showing a variability amplitude in the visible larger than 2.5 mag are comparable to Miras in SiO intensity, see Alcolea et al. (1990), Hall et al. (1990). (Remember that Miras are regular variable stars satisfying this requirement.)

SiO masers are found to be linearly polarized, with a degree that can reach almost 100% in some spikes, and in the average is $\sim$ 30%; see e.g. Clark et al. (1985). Circular polarization is also detected, but not exceeding $\sim$ 10%.

VLBI data show that SiO masers are placed in a ring-like structure surrounding the star at a distance of few stellar radii, see details in the contribution by Diamond & Kemball in this volume. Other VLBI observations have also yielded interesting results, like the very small size of the emitting spots ($\sim 10^{12}$ cm, Colomer et al. 1992) and the partial coincidence between the clumps found in the J=1–0 v=1 and v=2 transitions (Miyoshi et al. 1994).

Finally, monitoring observations have shown that the SiO masers vary with the same period as the central star. There is a remarkable coincidence in phase between the variability curves of the SiO maser emission and of the IR; but this is not the case of the optical curve, since the SiO and IR maxima appear at phase 0.1 - 0.2 with respect to the optical maxima. See the Yebes monitoring, the most complete one performed to date, in Alcolea et al. (1999).

The basic mechanism in the inversion of ^{28}SiO masers was discovered by Kwan & Scoville (1974). In this process, the rotational lines for $v > 0$ are inverted due to photon trapping in the $v \rightarrow v-1$ radiative deexcitation. The modified transition probability favors the vibrational components from the lowest J-levels, and so systematically overpopulates levels with high J in v. What is still under discussion is the way in which the $v > 0$ states are significantly populated in AGB envelopes. There are two basic pumping mechanisms, collisional (Lockett & Elitzur 1992; Doel et al. 1995) and radiative (Bujarrabal 1994, and references therein).

It has been shown that collisions up to an excited vibrational state do not show any significant selection of the rotational levels, so the subsequent radiative decay can yield the observed population inversion. From the theoretical point of view, the advantage of the collisional pumping is that it only requires the existence, close to the star, of temperatures $\gtrsim 1500$ K and a high SiO abundance; although, on the other hand, its efficiency is low. The collisional mechanism has problems to explain some observational results, like the relations between the SiO and IR intensities shown above.

In the radiative pumping, molecules are vibrationally excited by means of the absorption of stellar photons at 8μm. The mechanism needs that the high opacity applied to the decay does not hold for the stellar photons traversing the maser region, which occurs in regions with high velocity gradients or shell-like shape. This requirement is the main drawback of the radiative pumping, though it is compatible with the observations. The radiative pumping naturally explains the correlation between the maser and IR intensities and the linear polarization, that in this case would appear not as a consequence of a magnetic field, but just as due to the anisotropic excitation. On the other hand, the coincidence between spots of different lines (see above) is more convincingly explained by the collisional pumping models.

Acknowledgments. This work has been supported by the Spanish DGES, under project PB96-0104, and by a grant from the IAU Symposium n° 191 Local Organizing Committee. I am also grateful to J. Alcolea and T. Le Bertre for critical reading of the manuscript.

References

Alcolea J., Bujarrabal V., Gómez-González J., 1990, A&A 231, 431
Alcolea J., Pardo J.R., Bujarrabal V., et al., 1999, A&AS, submitted
Bieging J.H., Knee L.B.G., Latter W.B., Olofsson H., 1998, A&A 339, 811
Bujarrabal V., 1994, A&A 285, 953
Bujarrabal V., Fuente A., Omont A., 1994a, A&A 285, 247
Bujarrabal V., Fuente A., Omont A., 1994b, ApJ 421, L47
Bujarrabal V., Planesas P., Del Romero A., 1987, A&A 175, 164
Bujarrabal V., Gómez-González J., Planesas P., 1989, A&A 219, 256
Cernicharo J., Bujarrabal V., 1992, ApJ 401, L109
Cernicharo J., Bujarrabal V., Santarén J.L., 1993, ApJ 407, L33
Cernicharo J., Guélin M., 1987, A&A 183, L10
Clark F.O, Troland T.H., Miller J.S., 1985, ApJ 289, 756
Colomer F., Graham D.A., Krichbaum T.P., et al., 1992, A&A 254, L17
Doel R.C., Gray M.D., Humphreys E.M.L., et al., 1995, A&A 302, 797
Fuente A., Cernicharo J., Omont A., 1998, A&A 330, 232

Goldreich P., Scoville N., 1976, ApJ 205, 144
Groenewegen M.A.T., 1994, A&A 290, 531
Groenewegen M.A.T., van der Veen W.E.C.J., Matthews H.E., 1998, A&A 338, 491
Guélin M., Forestini M., Valiron P., et al., 1995, A&A 297, 183
Habing H.J., 1996, A&A Rev 7, 97
Hall P.J., Allen D.A., Troup D.R., et al., 1990, MNRAS 243, 480
Izumiura H., Deguchi T., Hashimoto O, et al., 1995, ApJ 453, 837
Jewell P.R., Dickinson D.F., Snyder L.E., Clemens D.P., 1987, ApJ 323, 749
Justtanont K., Skinner C.J., Tielens A.G.G.M., 1994, ApJ 435, 852
Justtanont K., Skinner C.J., Tielens A.G.G.M., et al., 1996, ApJ 456, 337
Kahane C., Cernicharo J., Gómez-González J., Guélin M., 1992, A&A 256, 235
Kastner J.H., 1992, ApJ 401, 337
Kerschbaum F., Olofsson H., Hron J., 1996, A&A 311, 273
Kerschbaum F., Olofsson H., 1998, A&A 336, 654
Knapp G.R., Morris M., 1985, ApJ 292, 640
Knapp G.R., Young K., Lee E., Jorissen A., 1998, ApJS 117, 209
Kwan J., Linke R.A., 1982, ApJ 254, 587
Kwan J., Scoville N., 1974, ApJ 194, L97
Lafont S., Lucas R., Omont A., 1982, A&A 106, 201
Lockett P., Elitzur M., 1992, ApJ 399, 704
Loup C., Forveille T., Omont A., Paul J.F., 1993, A&AS 99, 291
Lucas R., Bujarrabal V., Guilloteau S., et al., 1992, A&A 262, 491
Mamon G.A., Glassgold A.E., Huggins P.J., 1988, ApJ 328, 797
Margulis M., Van Blerkom D.J., Snell R.L., Kleinmann S.G., 1990, ApJ 361, 673
Miyoshi M., Matsumoto K., Kameno S., et al., 1994, Nature 371, 395
Morris M., 1975, ApJ 197, 603
Nercessian E. Guilloteau S., Omont A., Benayoun J.J., 1989, A&A 210, 225
Neri R., Kahane C., Lucas R., Bujarrabal V., Loup C., 1998, A&AS 130, 1
Nyman L.-Å., Hall P.J., Le Bertre T., 1993a, A&A 280, 551
Nyman L.-Å., Olofsson H., Johansson L.E.B., et al., 1993b, A&A 269, 377
Olofsson H., Bergman P., Lucas R., et al., 1998a, A&A 330, L1
Olofsson H., Eriksson K., Gustafsson B., Carlström U., 1993a, ApJS 87, 267
Olofsson H., Eriksson K., Gustafsson B., Carlström U., 1993b, ApJS 87, 305
Olofsson H., Johansson L.E.B., Hjalmarson Å, Nguyen-Q-Rieu, 1982, A&A 107, 128
Olofsson H., Lindqvist M., Nyman L.-Å., Winnberg A., 1998b, A&A 329, 1059
Omont A., Loup C., Forveille T., et al., 1993a, A&A 267, 515
Omont A., Lucas R., Morris M., Guilloteau S., 1993b, A&A 267, 490
Sahai R., Wannier P.G., 1992, ApJ 394, 320
Stanek K.Z., Knapp G.R., Young K., Phillips T.G., 1995, ApJS 100, 169
Young K., 1995, ApJ 445, 872

Asymptotic Giant Branch Stars
T. Le Bertre, A. Lèbre and C. Waelkens, eds.
I.A.U. Symposium 191, 1999

Variability of 22 GHz H_2O masers in circumstellar shells

D. Engels

Hamburger Sternwarte, Gojenbergsweg 112, D-21029 Hamburg, Germany

A. Winnberg

Onsala Space Observatory, S-43992 Onsala, Sweden

J. Brand

Istituto di Radioastronomia, C.N.R., Via Gobetti 101, I-40129 Bologna, Italy

C.M. Walmsley

Osservatorio Astrofisico di Arcetri, Largo E. Fermi 5, I-50125 Firenze, Italy

Abstract. We have monitored different types of late-type stars in the 22 GHz maser line of water vapor with single-dish telescopes and the Very Large Array over several years. We find that the emission pattern in the circumstellar shells of Semi-Regular (SR) and Mira variables mapped by the VLA remain stable on the scale of several years while the maser profiles from the single-dish telescopes vary strongly. Thus the strong variability is mainly due to incoherent intensity fluctuations of the individual maser lines, which have lifetimes less than a year.

In contrast to this, the variability in OH/IR stars and M-Supergiants is more regular and is mainly a response to the long-period variations of the central star. H_2O maser in many OH/IR stars are "extinguished" most of the time, as the excitation temperature is only high enough during the phases close to the maximum of the variability.

The strength of the profile variability is decreasing as a function of the radial distance of the maser shell from the star. This gives a natural explanation for the increasing regularity of the maser variations along the sequence SR-, Mira variables, OH/IR stars, M-Supergiants.

1. Introduction

Ever since the first discovery of an H_2O maser at 1.35 cm in the circumstellar shell of VY CMa the strong variability was mentioned, although the variations in this supergiant star are small compared to those in star forming regions (Knowles et al. 1969; Buhl et al. 1969). Monitoring programs were immediately started (Schwartz et al. 1974) and evidence was found that the masers varied in peak and integrated flux, in line profiles and possibly radial velocity (Dickinson et al. 1973). These variations were found to be correlated with the variations in the optical and in the infrared (Schwartz et al. 1974). Later observations showed the variations to be less regular (Engels et al. 1988; Cohen 1989). The first interferometric maps of the Miras R Aql and RR Aql taken at two epochs separated by ≈ 1 year also showed dramatic changes (Johnston et al. 1985). Monitoring of OH/IR stars had not been attempted, but the available observations indicated more regular variations than in SR and Mira variables (Engels et al. 1986). To improve the understanding of the properties of maser variability for the different types of late-type stars, we have monitored since 1990 several of them with single-dish telescopes and the VLA.

2. Observations

Single dish observations have been made with the Effelsberg 100m and Medicina 32m telescopes at typical intervals of a few months for the years 1990-1997. The spectra have a velocity resolution of 0.1-0.3 $km\,s^{-1}$ with the noise varying between 0.1 and 5 Jy. Interferometric maps were obtained with the VLA in its A-configuration for a few representative stars at several epochs. These stars were the SR variable RX Boo, the Mira U Her, the OH/IR stars OH 39.7+1.5 and OH 83.4-0.9, and the M-Supergiant VX Sgr. The smoothed spectral resolution of the VLA data is 1.32 $km\,s^{-1}$, the angular resolution ≈80 mas, and the rms sensitivity ≤30 mJy.

In the following we discuss the results for U Her and OH 83.4-0.9 as representative cases. First results on RX Boo were given by Engels et al. (1992), and on OH 39.7+1.5 by Engels et al. (1997).

3. U Her

U Her is a well-observed Mira variable with a period of 406 days. It exhibits strong maser emission not only in H_2O, but also in OH and SiO. Maps in the 1.35 cm line were obtained in 1990 February and June, in 1991 October and 1992 December. In 1991 the VLA was in BnA-configuration (intermediate between A and B) which degraded the east-west resolution to ≈300 mas. Generally the maser spots were not resolved and individual spots were identified by decomposition of the channel maps with multiple gauss fits. The maps, containing all maser spots for a given epoch, and two typical single-dish spectra are shown in Fig. 1, and the light curve for the integrated flux in Fig. 2. The maps were shifted tentatively to a common origin, assuming that the masers lie in a ring indicated by the circle drawn in each map. We also tentatively assume that the star is located at the center of the ring. The size of the symbols is scaled with intensity and a few components are labeled by their velocities.

Maser spots were detected between -22 and -8 $km\,s^{-1}$, with two dominating components at ≈ -18 and ≈ -15 $km\,s^{-1}$. To distinguish between them in the maps all components with $v \leq -16$ $km\,s^{-1}$ were drawn as open symbols. The stellar radial velocity is at ≈ -14 $km\,s^{-1}$, so that almost all of the emission is blue shifted. In the period 1990-1997, except for about a year in 1991/92, the -15 $km\,s^{-1}$ component was always the strongest. The maps show most of

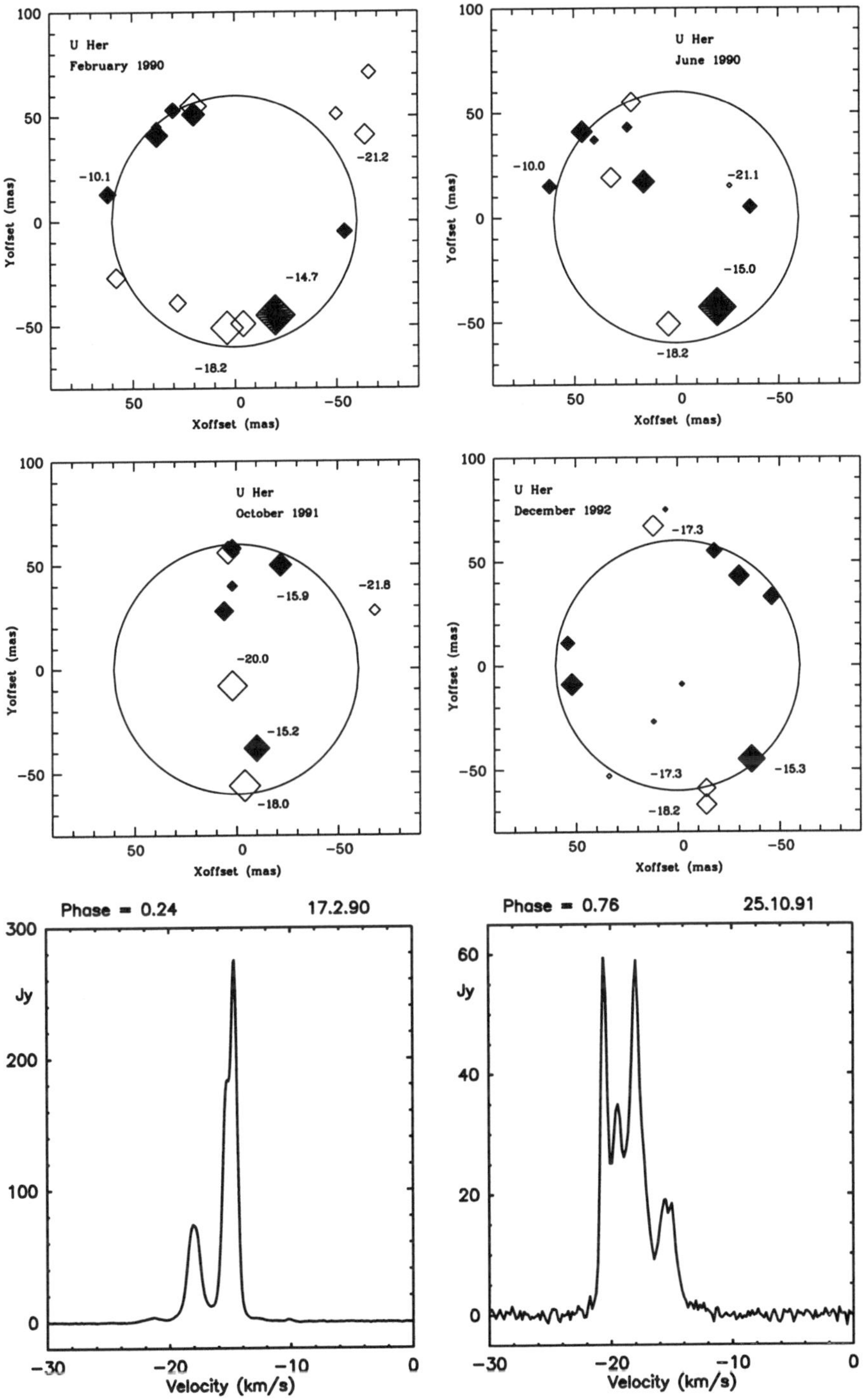

Figure 1. U Her. Maps and representative single-dish spectra. See text for the meaning of the symbols.

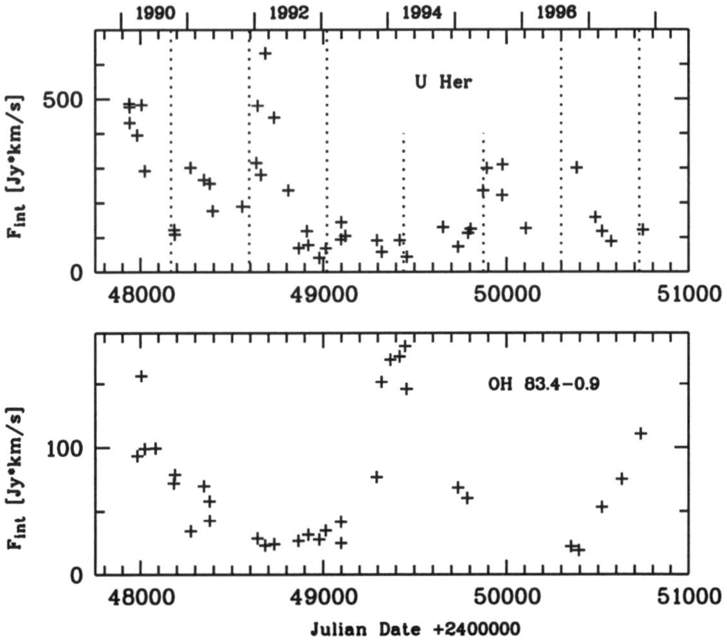

Figure 2. Light curves of the integrated maser flux. The vertical lines mark the maxima of the optical U Her lightcurve.

the individual maser spots to be located in a ring of ≈60 mas radius, which corresponds to 21 AU at the star distance of 350 pc. All four maps show the two dominating components to be located in the southern part of the ring. In the two maps obtained in 1990 within four months, the maser components detected are almost identical. The differences between the maps in Fig. 1 are due to blending of components and intensity variations. This leads to apparent positional changes (cf. the −21 km s^{-1} component) and to the presence of several spots in only one epoch. The 1991 map was taken while the emission was dominated by features with v≤-18 km s^{-1}. Because of the degraded resolution we have no information of the east-west extent of the shell. The very blue maser line came from about the center of the map and could be amplified light of the central star. The 1992 map shows again the ring-like structure as in 1990, but it is not possible to identify individual spots common to both epochs. Interestingly the strongest components still come from the southern part of the ring but shifts in the velocity indicate that these are also different masers from those in 1990.

This star was also observed with MERLIN in 1985 (Yates & Cohen 1994), with the VLA in 1983 and 1988 (Lane et al. 1987; Bowers & Johnston 1994), and with the VLBA in 1995 by Marvel (1997). Comparing our maps with these ones shows that at least between 1988 and 1995 the main feature, the location of the strongest maser in the south of the spatial distribution, was maintained. During the early VLA and MERLIN observations only emission over a rather small velocity range was detected, precluding conclusions on the shell structure. The U Her observations show that the individual maser spots have a lifetime of

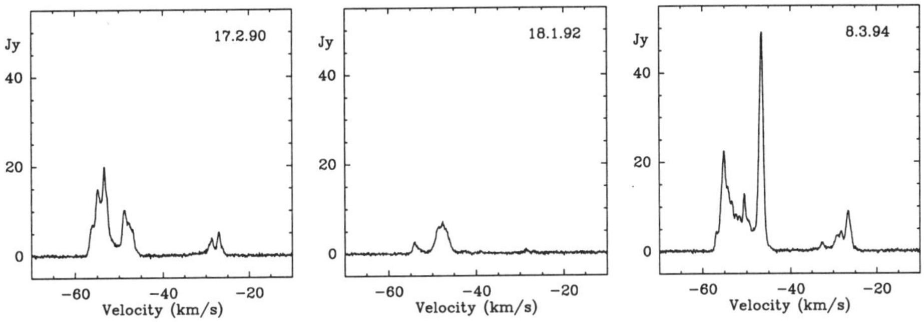

Figure 3. Spectra of OH 83.4-0.9 during the maxima in 1990 and 1994 and during the minimum in 1992.

at least several months, while the general structure of the maser shell remains stable over $\geq$7 years. The crossing time through the maser shell is of the order of 10 years and more. Thus we may expect to observe changes in the map structure in the future.

4. OH 83.4-0.9

This is an OH/IR star with a period of $\approx$1400 days. Examples of the double-peaked H_2O maser spectra, which are typical for OH/IR stars, are shown in Fig. 3. The light curve (Fig. 2) shows a regular variation with the period of the star. Intensity fluctuations of individual maser lines are less pronounced than in U Her. Two VLA maps taken in 1990 did not resolve the maser shell but a size of $\approx$30 mas can be inferred from the positional shifts with velocity. A new VLBA map that we made in 1998 confirms this estimate. The distance of OH 83.4-0.9 is 2-4 kpc yielding a shell radius of 60-120 AU. This compares well with the maser shell radius obtained by Engels et al (1997) for OH 39.7+1.5 and shows that the maser shells of OH/IR stars are $\approx$5 times larger than in SR and Mira variables.

5. Discussion

So far, the discussion of H_2O maser variability was dominated by early monitoring programs including mostly Miras and SR variables. In these stars with low mass loss rates ($<10^{-6}\,M_\odot$/yr) the maser shell is located at $\leq$50 AU (see also Bowers & Johnston 1994), within the zone of radial acceleration of the shell. There masers are beamed preferentially in a tangential direction (Chapman & Cohen 1985), leading to profiles peaked close to the stellar radial velocity. Irregular intensity fluctuations prevail, leading to a rather weak correlation with the optical lightcurve. The lifetime of individual maser spots appears to be $\leq$1 year, while the distribution of maser sites in the shell survives several years.

Bowers & Johnston (1994) found variations of the maser shell sizes comparing their maps to earlier interferometric data. We did not find such variations for RX Boo and U Her. They can be mimicked by varying sensitivities of the maps as the weaker emission spots tend to have larger radial distances and might not be detected close to minimum phase.

In OH/IR stars, which have higher mass loss rates ($>10^{-5}\,M_{\odot}/yr$), the maser shell is shifted to larger distances, at which the final expansion velocity is (almost) reached. Radial beaming leads to double-peaked profiles. In one case, OH 39.7+1.5, we discovered a "mode switch" from radial to tangential beaming during minimum phase (Engels et al. 1997). Regular intensity variations following the bolometric variations of the star lead to rather regular maser lightcurves. Often the masers are "extinguished" over a considerable fraction of the variation cycle, as the excitation temperatures seem to be high enough only during the phases close to maximum. The stability of the maser profiles over several variability cycles suggests, that the shell structure will be maintained over longer timescales than in Miras and SR variables.

M-Supergiants have even higher mass loss rates and luminosities, and larger H_2O maser shells than OH/IR stars (Richards & Yates 1998). Their variability properties are similar to those of OH/IR stars. SR variables, Miras, OH/IR stars, and M-Supergiants form a sequence of increasing maser shell sizes and decreasing strength of irregular variability.

Acknowledgments. We acknowledge the support by the Deutsche Forschungsgemeinschaft (Re 353/25, En 176/2) and by the Swedish Natural Science Research Council. The observations were partially obtained at the VLA, which is part of the NRAO operated by Associated Universities Inc., under cooperative agreement with the National Science Foundation of the USA.

References

Bowers P.F., Johnston K.J., 1994, ApJS 92, 189

Buhl D., Snyder L.E., Schwartz P.R., Barrett A.H., 1969, ApJ 158, L97

Chapman J.M., Cohen R.J., 1985, MNRAS 230, 415

Cohen R.J., 1989, Rep. Prog. Phys. 52, 881

Dickinson D.F., Bechis K.P., Barrett A.H., 1973, ApJ 180, 831

Engels D., Schmid-Burgk J., Walmsley C.M., 1986, A&A 167, 129

Engels D., Schmid-Burgk J., Walmsley C.M., 1988, A&A 191, 283

Engels D., Winnberg A., Brand J., Walmsley C.M., 1992, in Astrophysical Masers, eds. A. Clegg and G. Nedoluha, p. 403

Engels D., Winnberg A., Walmsley C.M., Brand J., 1997, A&A 322, 291

Johnston K.J., Spencer J.H., Bowers P.F., 1985, ApJ 290, 660

Knowles S.H., Mayer C.H., Cheung A.C., Rank D.M., Townes C.A., 1969, Science 163, 1055

Lane A.P., Johnston K.J., Bowers P.F., Spencer J.H., Diamond P.J., 1987, ApJ 323, 756

Marvel K., 1997, PASP 109, 1286

Richards A.M.S., Yates J.A., 1998, Irish Astr. J. 25, 7

Schwartz P.R., Harvey P.M., Barrett A.H., 1974, ApJ 187, 491

Yates J.A., Cohen R.J., 1994, MNRAS 270, 958

Asymptotic Giant Branch Stars
T. Le Bertre, A. Lèbre and C. Waelkens, eds.
I.A.U. Symposium 191, 1999

Long-term evolution of AGB wind envelopes: Insights from hydrodynamical models

M. Steffen and D. Schönberner

Astrophysikalisches Institut Potsdam, An der Sternwarte 16, 14482 Potsdam, Germany

R. Szczerba

N. Copernicus Astronomical Center, Rabianska 8, 87-100 Torun, Poland

Abstract. Up to now, *hydrodynamical* models of dust-driven AGB winds do *not* generally take into account the 'long-term' changes of the stellar parameters (on stellar evolution time scales of 10^3 to 10^5 yrs), although it is well known that the luminosity and (very likely) the *mass loss rate* undergo significant variations when so called 'thermal pulses' occur on the upper AGB. In this review we demonstrate that time-dependent radiation hydrodynamics calculations are needed to understand the formation, structure, and spectral energy distribution of detached *dust* shells detected by IRAS and ISO. Combined with appropriate models, these observations can reveal part of the previous mass loss history on the AGB and allow an empirical check of presently adopted mass loss laws.

Based on insights from hydrodynamical simulations, we discuss the two competing scenarios that have been put forward to explain the origin of the very thin molecular shells recently discovered around some carbon stars. We find that the signature of a short *mass loss 'eruption'* broadens quickly with time due to the related velocity gradient across the shell. Hence, this scenario is not considered a likely explanation of detached CO shells. On the other hand, the alternative mechanism, *interaction of winds*, is shown to be capable of producing very thin shells of greatly enhanced gas density in the dusty outflows from AGB stars by sweeping up matter at the interface between both type of winds.

1. Introduction

It is now generally accepted that the mechanism responsible for the high mass loss rates during the AGB evolution is based on the efficiency of radiation pressure on dust grains. However, it is largely unknown how the mass loss rate depends on the star's main sequence mass, its metallicity, its particular phase of evolution and pulsational properties. In order to determine the temporal behavior of the mass loss rate (exponentially increasing, interrupted, episodic ?) empirically, we have to study the radial structure of the circumstellar shells of AGB stars. In the following we briefly review different approaches to modeling circumstellar gas/dust shells and demonstrate that time-dependent model calculations including the effects of stellar evolution are needed to understand the formation, radial structure and spectral energy distribution of detached dust shells observed by IRAS and ISO. We present some example results obtained with a radiation hydrodynamics code recently developed as a diagnostic tool for the interpretation of IR observations of the dust emission around AGB and post-AGB stars. Comparison of such observations with appropriate models permits an empirical check of presently adopted mass loss laws, an important step towards revealing the detailed mass loss history on the AGB.

2. Different approaches to modeling circumstellar gas/dust shells

⋆ **Radiative transfer calculations:** The consistent solution of the radiative transfer equation is much more demanding. One has to specify the density structure of the circumstellar shell (usually assuming spherical symmetry, $\rho \sim r^{-\alpha}, \alpha \approx 2$) and the optical dust properties. Given the parameters of the central star (L_*, T_*) and the inner radius of the dust shell, r_1, the requirement of *dust radiative equilibrium* yields the radial temperature distribution of the dust, $T_d(r)$, the emergent spectral energy distribution, λF_λ, and the spatial surface brightness distribution projected onto the sky, $I_\lambda(p)$. This approach is widely used for the analysis of the SEDs of stars with dusty envelopes (e.g. Hashimoto 1994, 1995; Groenewegen 1995, 1997). Some authors take into account a dust size distribution (Justtanont & Tielens 1992; Szczerba et al. 1997), and there are even codes for treating non-spherical, disk-like configurations (Men'shchikov & Henning 1996). Lacking hydrodynamical models, some authors have tried to mimic the effect of mass loss variations by considering a sequence of density structures modified according to some adopted temporal mass loss behavior, assuming that the outflow velocity is basically independent of the mass loss rate, $v \neq f(\dot{M})$ (e.g. Bedijn 1987; Groenewegen & de Jong 1994; Suh & Jones 1997).

⋆⋆ **Steady state hydrodynamical models:** The above mentioned pure radiative transfer models ignore the fact that the problem of momentum transfer (hydrodynamics) is inherently coupled with the problem of radiative transfer: radiation pressure on dust determines the outflow velocity and hence the density structure; at the same time, the density structure determines – via radiative transfer effects – the spectrum of the photons and hence the effective radiation pressure. Over the last years, this circular problem has been addressed by Netzer & Elitzur (1993), Habing et al. (1994), Justtanont et al. (1994), Ivezić & Elitzur (1995), and Steffen et al. (1997a), who describe the gas/dust shell as a *hydrodynamical multi-fluid system* and allow for a relative motion of gas and dust. Steffen et al. (1997a) include the dynamical effects of the gas pressure which is usually ignored. With stellar parameters (M_*, L_*, T_*), mass loss rate ($\dot{M}$) and dust properties (grain size, a, mass density of the grain material, ρ_I, dust condensation temperature, T_c, absorption and scattering cross sections, C_λ^{abs}, C_λ^{sca} and the dust-to-gas mass ratio, δ) as input data, r_1, $v(r)$, $\rho(r)$, $v_d(r)$, $\rho_d(r)$, $T_d(r)$ as well as λF_λ and $I_\lambda(p)$ are found from the solution of the coupled hydrodynamics/radiative transfer problem.

For simplicity, all these model calculations are based on the assumption of a *spherically symmetric steady state* flow. It has been shown that, for a given sort of grains, a sequence of steady state models with increasing mass loss rate is mapped on a simple color-color relation, with optical depth of the dust shell as the single independent parameter (Ivezić & Elitzur 1995). Indeed the majority of mass losing AGB stars can be understood in terms of such steady state models characterized by a constant value of $\dot{M}$. However, even if one assumes a large variety of different grain compositions, it is impossible to explain the observed distribution of AGB objects with excess 60 and 100 μm emission in the IRAS two-color-diagram (cf. Fig. 2, top).

⋆⋆⋆ **Time-dependent radiation hydrodynamics models:** In order to properly take into account the evolutionary changes of the stellar parameters and a variable mass loss rate, it is necessary to solve the *time-dependent* coupled equations of hydrodynamics and radiative transfer. Given the stellar parameters ($M_*(t)$, $L_*(t)$, $T_*(t)$), and the mass loss rate ($\dot{M}(t)$) as time-dependent external boundary values, the solution of the equations of a dust-driven stellar wind describes the structure and dynamics of the circumstellar envelope as a function of time yielding: $r_1(t)$, $v(r,t)$, $\rho(r,t)$, $v_d(r,t)$, $\rho_d(r,t)$, $T_d(r,t)$ as well as $\lambda F_\lambda(t)$ and $I_\lambda(p,t)$. Model calculations of this type have first been performed by Vassiliadis & Wood (1992) and Szczerba & Marten (1993). More detailed results of time-dependent two-component simulations of the dusty winds of AGB stars in their final stages of evolution have recently been presented by Steffen et al. (1998). These models are further discussed in the next section.

⋆ ⋆ ⋆⋆ **Fully self-consistent radiation hydrodynamics models:** The models discussed above are far from being fully self-consistent: they cover only those parts of the circumstellar envelope that lie outside the sonic point and therefore cannot make predictions about the mass loss rate which is determined in the subsonic regions near the stellar surface. Clearly, the ultimate goal is to develop fully self-consistent physical models of circumstellar AGB shells, combining stellar pulsations, dust chemistry, hydrodynamics and stellar evolution. The most advanced models in this direction are those by Fleischer et al. (1992, & this volume; see also the review by Sedlmayr & Dominik 1995). Similar models have been computed by Höfner & Dorfi (1997, & this volume). Given the stellar parameters, the surface C/O ratio, and the pulsational properties of the star, these models give, among other things, the mass loss rate and the outflow velocity as a result. Up to now, these kinds of models are only available for carbon-rich chemistry. They are time-dependent in the sense that they include Mira-type pulsations (enforced by a piston), but they cannot yet account for the changes of the stellar parameters and surface composition on stellar evolution time scales. Clearly, a lot of work is still needed before truly self-consistent models of circumstellar AGB shells will be available.

3. Time-dependent AGB wind models coupled to stellar evolution calculations

We have applied time-dependent two-component hydrodynamics/radiative transfer calculations to model the dusty winds of AGB stars in their final stages of evolution. This approach allows us to take into account the evolutionary changes of the stellar parameters, including the mass loss rate, and to investigate the dynamical response of the AGB wind envelope and of the emergent spectral energy distribution to the temporal variations of L_*, T_*, and $\dot{M}$. The basic assumptions, numerical method and input data as well as first results have been

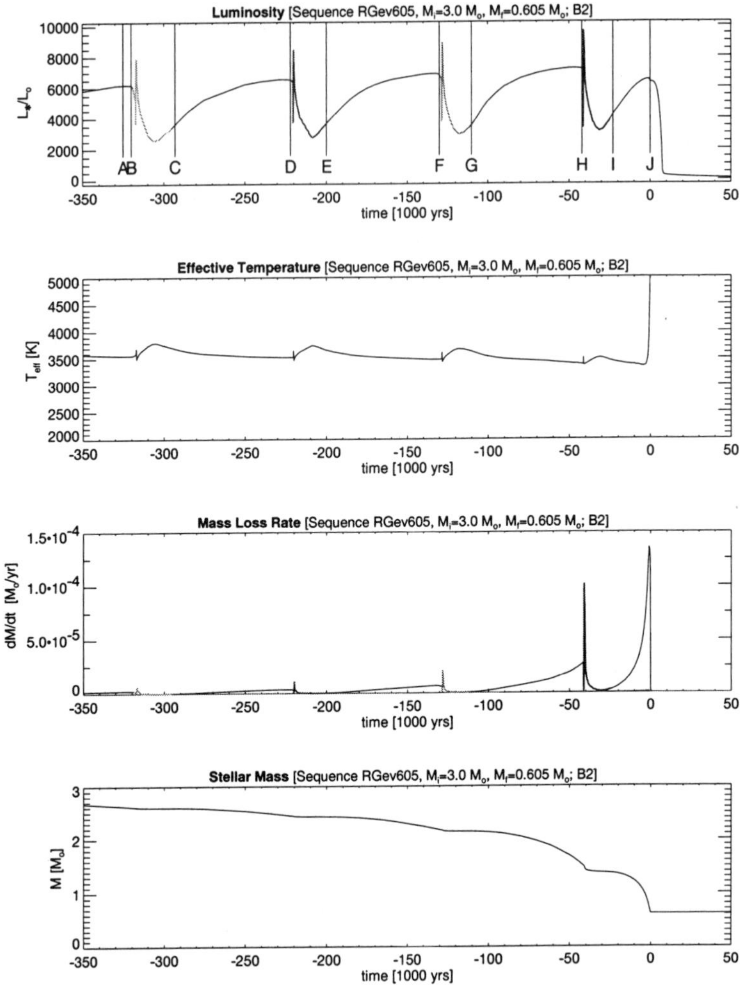

Figure 1. Luminosity, effective temperature, mass loss rate, and stellar mass versus time for the final four thermal-pulse cycles of a $3\mathrm{M}_\odot$ model sequence after Blöcker (1995). Time zero marks the beginning of the model's post-AGB evolution. For further details see text.

described by Steffen et al. 1998 (see also Steffen & Szczerba 1997; Steffen et al. 1997a,b; Schönberner et al. 1997, 1998).

The example results presented in the following are based on a stellar evolution sequence computed by Blöcker (1995) for a solar composition star of initially $3\ \mathrm{M}_\odot$. The temporal variation of the stellar parameters and the mass loss rate over a time interval of about 350 000 years, covering the final stages of evolution on the AGB, are shown in Fig. 1. The mass loss rate adopted for computing this sequence depends on the stellar parameters according to $\dot{M} = 5.85\,10^{-14}\,(L_*/L_\odot)^{4.2}\,(M_*/M_\odot)^{-3.1}\,(T_*\,[\mathrm{K}])^{-2}$. Note that the bulk of the mass loss occurs during the final four thermal pulse cycles. These stellar evolution data are used as an inner boundary condition for the hydrodynamical simulation of the long-term evolution of the circumstellar gas/dust shell. Mira-type stellar pulsations (with periods of a few 100 days) cannot be included explicitly (the time step of the simulations is typically 50 yrs) but only as an additional acceleration term representing the time-averaged pressure of shock waves.

What happens to the circumstellar envelope during a thermal pulse cycle? *Before the onset of a helium shell flash* (e.g. time 'D' in Fig. 1), luminosity and mass loss have been increasing slowly and steadily over several 10 000 years. Hence, the velocity and density structure of the wind closely resembles that of a steady state outflow: constant velocity over a wide radial range, and $\rho \sim r^{-2}$ (except for the outermost regions beyond $r \approx 5\,10^{17}$ cm that still exhibit the signature of the previous thermal pulse). At the same time, the emergent SED is typical for steady state conditions, characterized by a monotonically decreasing flux towards longer wavelengths. *At the subsequent mass loss rate minimum* (e.g. between times 'D' and 'E' in Fig. 1), the coupling between gas and dust has become insufficient to drive the outflow due to the very low matter densities in the inner regions. The slow (3 km/s) gas flow is supported only by the pressure of shock waves (see above). It is only in the outer regions ($r > 10^{17}$ cm) that we still see the signature of the previous phase of high mass loss rate: the gas velocity still reaches about 10 km/s, and the dust density increases distinctly, showing a local maximum near $r \approx 3 \cdot 10^{17}$ cm. This *detached dust shell* is the result of the sharp decrease of the mass loss rate by more than one order of magnitude (from $3\,10^{-6}$ $M_{\odot}$/yr at 'D' to $1\,10^{-7}$ $M_{\odot}$/yr about 12 000 years later). In the emergent spectrum, this excess concentration of cold dust is seen as a conspicuous excess emission at 60 and 100 μm. In this phase, the computed *radial intensity distribution* (e.g. λ 100 μm) shows a local maximum at distances of a few times 10^{17} cm from the central star, corresponding to a ring-like structure in the surface brightness. Indeed, such detached dust shells have been found by IRAS (Waters et al. 1994, Izumiura et al. 1997 [multiple shell]) and ISO (Izumiura et al. 1996 & this volume). Similar results have been obtained from models with oxygen-rich circumstellar shells.

As our model runs through a thermal pulse, the emergent spectral energy distribution changes with time. This temporal variation of the SED is translated into an extended loop in the IRAS two-color-diagram. Over the time interval of 350 000 yrs covered by Fig. 1, each of the four thermal pulses produces a distinct loop in the two-color-diagram (lower panel of Fig. 2). Comparison with the observed distribution of carbon stars in the IRAS two-color-diagram (top panel of Fig. 2) suggests that the objects located in region VIa are carbon stars just having suffered a thermal pulse which reduced the mass loss by at least an order of magnitude and lead to the detachment of the dust shell. Objects with steady state wind envelopes are expected to follow the simple color-color relation indicated by the dashed line (top panel of Fig. 2). Clearly, steady state models cannot explain the carbon stars observed in region VIa. For further details see Steffen et al. (1997b, 1998) and Steffen & Szczerba (1997).

Since the location, shape and population density of the loops depend sensitively on the adopted mass loss law, the comparison of observed and computed IRAS two-color diagrams is a powerful tool to put constraints on the mass loss history on the AGB, provided that the data sets used for comparison are statistically representative.

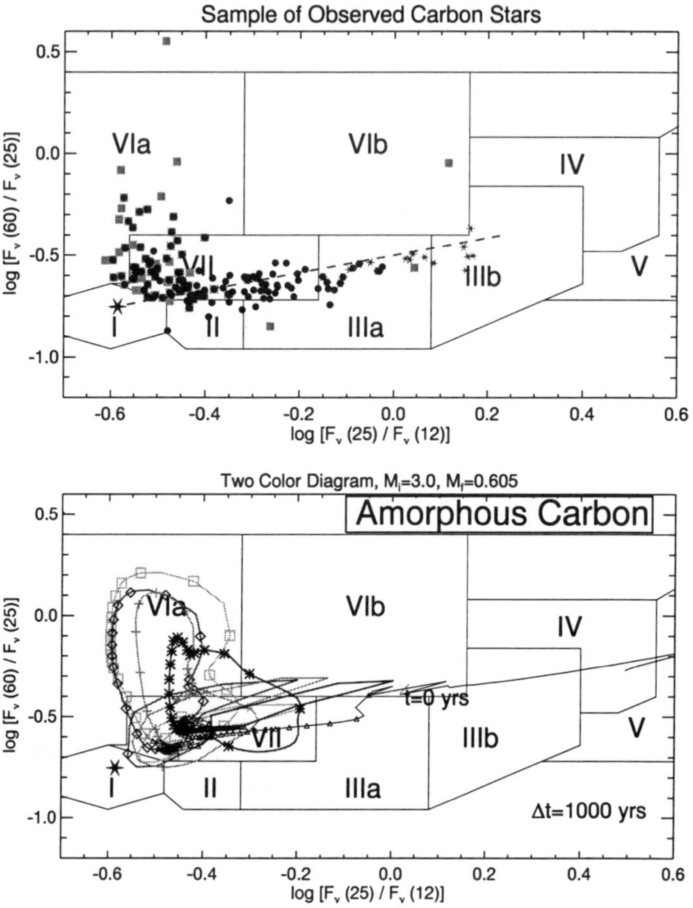

Figure 2. **Top**: Distribution of observed carbon stars in the IRAS [60]-[25]) vs. [25]-[12] two-color-diagram. The data for classical C-stars (circles), optical C-stars (squares) and extreme C-stars (asterisks) are courtesy of K. Volk (priv. comm.); only objects with fluxes of quality 3 at 12, 25, 60, and 100 μm are plotted. The big 'star' in region I is the Rayleigh Jeans point for $T = 3\,000$ K. The dashed line delineates the color-color relation for steady state models (cf. Steffen et al. 1997a). **Bottom**: Computed evolution of a carbon star in the same IRAS two-color-diagram, based on the stellar evolution sequence shown in Fig. 1. The hydrodynamical simulation covers 4 thermal pulse cycles during the final 350 000 years of AGB evolution, each of which produces a corresponding loop in the two-color-diagram, distinguished here by different symbols: squares (B-C), diamonds (D-E), plus-signs (F-G), and asterisks (H-I), plotted in equidistant time intervals of 1 000 years, while triangles are used otherwise. The sequence starts somewhere in region VII, reaches the position marked 't=0' at the end of the AGB and leaves the diagram to the right when entering the post-AGB phase.

4. On the origin of thin detached CO shells

Searching for evidence of episodic mass loss of evolved stars on the AGB, Olofsson et al. (1996) noted the existence of detached narrow shells (width-to-radius $\approx 0.3 \ldots 0.4$) in molecular line emission of CO around the carbon stars R Scl, U Ant, S Sct, and TT Cyg. Very recently, Olofsson et al. (1998) have succeeded in spatially resolving the CO shell around TT Cyg through interferometric observations and found an even 10 times smaller width-to-radius value of only ≈ 0.04. Absolute radii of the CO shells are estimated to lie between about 1 and 5 times 10^{17} cm. Since their discovery, the origin of the circumstellar CO shells has been a matter of debate. Two competing scenarios have been put forward to explain their existence: *Mass loss 'eruption'* versus *two-wind interaction*.

Even though no physical mechanism is presently known to be responsible for producing very short mass loss 'eruptions' (see however recent claims by Schröder et al. 1998 & this volume), it is nevertheless tempting to relate the existence of the observed very thin CO shells to this presumed phenomenon. According to Olofsson et al. (1998) the CO shell of TT Cyg has a width of 2 10^{16} cm and a radius of 5.1 10^{17} cm. Assuming a constant expansion velocity of 13 km/s (the measured Doppler velocity inside the CO shell), this implies a mass loss 'eruption' that happened about 12000 yrs ago and lasted for about 500 yrs, at a rate of $\dot{M} \approx 10^{-4}$ $M_\odot$/yr (estimated shell mass ≈ 0.024 $M_\odot$).

We have constructed a schematic mass loss history in accordance with the above mentioned data for TT Cyg ($\dot{M} = 5\,10^{-5}$ $M_\odot$/yr during the 'eruption'), and used it as a boundary condition for a time-dependent hydrodynamical simulation of the mass loss 'eruption' scenario. The results indicate that the outflow velocity decreases sharply from about $u_1 = 21$ km/s at the onset to $u_2 = 6$ km/s at the end of the 'eruption' due to the increasing optical depth of the ejected dust. The produced gas shell moves outward at an average speed of ≈ 15 km/s. However, the substantial velocity gradient across the shell leads to a considerable broadening of the density structure as it recedes from the star, and to a corresponding decline of its density contrast. Beyond $r = 10^{17}$ cm, the relative thickness of the shell remains approximately constant at a value of $\approx 0.6 \approx (u_1 - u_2)/(u_1 + u_2)$. This is more than one order of magnitude larger than the width observed in CO emission. We further point out that the amount of dust ejected by this 'eruption' is insufficient to account for the observed excess emission at 100 μm (cf. e.g. Kerschbaum & Hron 1996 for the observed SED of TT Cyg).

Interestingly enough, our simulations provide another mechanism creating thin shells of enhanced gas density: interaction of two winds of different velocity and density. While analyzing the simulations, we discovered that the wind velocity is an almost bi-modal function of the mass loss rate (cf. Fig. 3 for the behavior of mass loss rate and wind velocity during a typical thermal pulse). Briefly, once a critical mass loss rate is exceeded, the coupling between dust and gas becomes efficient and causes a sudden transition from a slower, dense, 'shock-driven' wind to a faster, less dense, 'dust-driven' wind, both type of winds carrying approximately the same mass loss rate ($r^2 \rho u \approx$ const.). This behavior leads to *two-wind interaction*: the faster wind in the inner parts of the shell runs into the slower wind in the outer parts, acting like a snow plow piling up matter into a thin shell at the interface between the different kind of winds. The

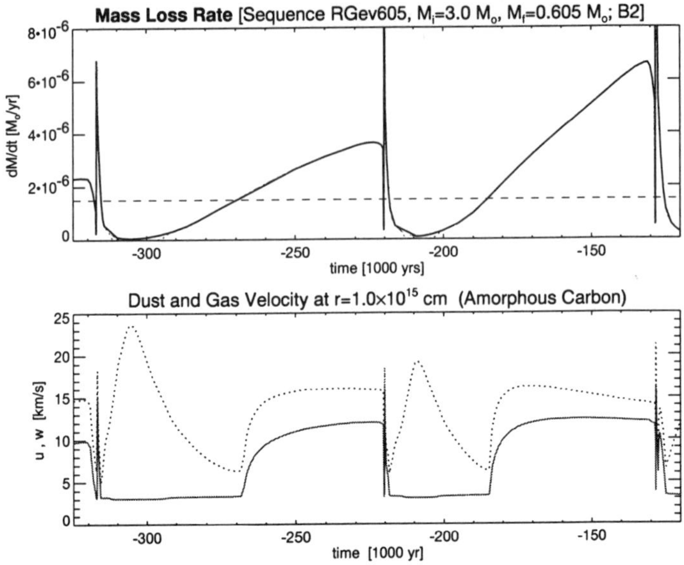

Figure 3. **Top**: Adopted mass loss rate over two thermal pulse cycles (Blöcker 1995). **Bottom**: Gas (solid) and dust (dotted) velocity in the inner wind shell ($r = 10^{15}$ cm) obtained from a two-component hydrodynamical simulation as function of time, showing a characteristic bi-modal behavior of the *gas velocity*. When $\dot{M} > 1.6\,10^{-6}$ $M_{\odot}$/yr (dashed line in upper panel) dust/gas coupling becomes efficient and the gas outflow velocity jumps from ≈ 3 to ≈ 12 km/s.

critical mass loss rate is $\dot{M} = 1.6\,10^{-6}$ $M_{\odot}$/yr in this example, but is expected to depend on the adopted dust properties and on the stellar parameters.

We have computed the resulting density, velocity, and temperature structure of the interaction region in a numerical experiment, using a one-component hydrodynamics code based on a high-resolution second-order Godunov-type advection scheme including a Riemann solver. This scheme adequately treats the propagation and interaction of non-linear waves in spherical geometry. Radiative cooling is treated in a simplistic way according to: $\log(Q_{\rm rad}/\rho) = 4.466 * (1 - \exp\{-0.7177\,T/1000\})$. Starting with a steady-state model with a mass loss rate of $\dot{M} = 3\,10^{-6}$ $M_{\odot}$/yr and a constant outflow velocity of $u_o = 5$ km/s, we abruptly increase the outflow velocity at the inner boundary by a factor of 3 and decrease the density of the wind by the same factor in order to keep the mass loss rate constant. We then follow the time evolution of the interaction region for about 15 000 years until it has propagated outward to several 10^{17} cm. The result is shown in Fig. 4. Indeed, a very narrow shell ($\Delta R_s/R_s \approx 0.03$) of strongly (factor of 30) enhanced gas density has developed. Note the small velocity gradient *within* the compressed shell. The gas behind the inner shock is heated to about 300 K, while the net heating behind the outer shock is negligible (a minimum temperature of 100 K was enforced everywhere). Probably, radiative cooling is still underestimated by our simple cooling function which we used here for illustration only.

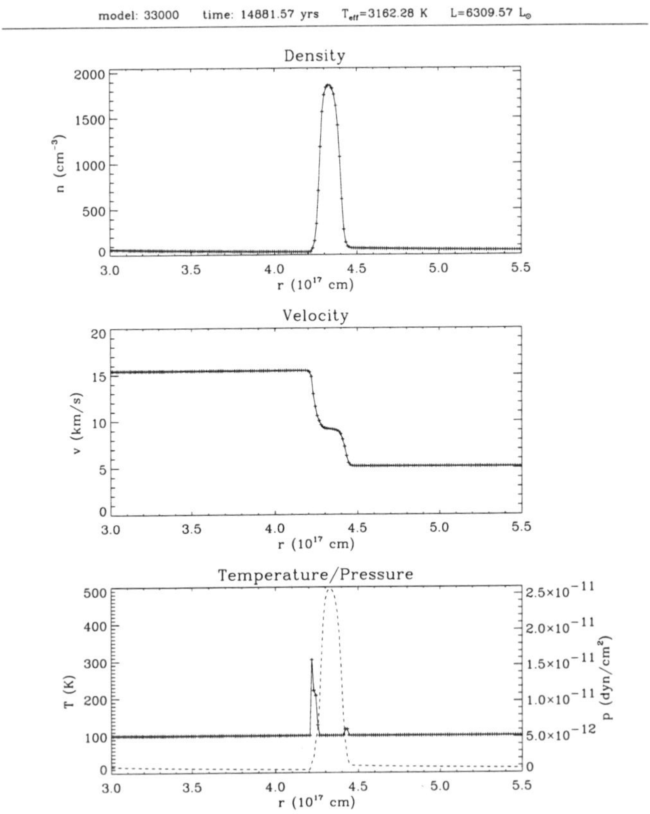

Figure 4. Particle number density (top), gas velocity (middle) and gas temperature/pressure (bottom, solid/dashed) resulting from a numerical simulation of idealized two-wind interaction. (+)–symbols indicate the numerical grid.

5. Conclusions

The time-dependent hydrodynamical models presented here provide a natural explanation for the existence of *detached dust shells*: mass loss decreases by more than one order of magnitude in the aftermath of a thermal pulse for a sufficiently long time to allow the formation of an extended, almost dust-free inner cavity. The origin of geometrically thin *detached gas shells*, however, is not yet sufficiently well understood. Hydrodynamical test calculations suggest that it may be difficult to produce very narrow shells by a very short mass loss 'eruption', because matter will not be ejected at a uniform velocity. On the other hand, our two-component hydrodynamical models suggest that two-wind interaction can occur in the circumstellar envelopes of AGB stars due to the almost bi-modal dependence of the outflow velocity on the mass loss rate. This behavior naturally leads to the formation of very thin shells of greatly enhanced gas density. Clearly, further research is necessary to clarify the details of this mechanism. Perhaps, the correct explanation of the remarkable CO shells turns out to be some combination of mass loss 'eruption' and two-wind interaction.

References

Bedijn P.J., 1987, A&A 186, 136
Blöcker T., 1995, A&A 297, 727
Fleischer A.J., Gauger A., Sedlmayr E., 1992, A&A 226, 321
Groenewegen M.A.T., 1995, A&A 293, 463
Groenewegen M.A.T., 1997, A&A 317, 503
Groenewegen M.A.T., de Jong T., 1994, A&A 282, 115
Habing H.J., Tignon J., Tielens A.G.G.M., 1994, A&A 286, 523
Hashimoto O., 1994, A&AS 107, 445
Hashimoto O., 1995, ApJ 442, 286
Höfner S., Dorfi E., 1997, A&A 319, 648
Ivezić Ž., Elitzur M., 1995, ApJ 445, 415
Izumiura H., Hashimoto O., Kawara K., et al., 1996, A&A 315, L221
Izumiura H., Waters L.B.F.M., de Jong T., et al., 1997, A&A 323, 449
Justtanont K., Tielens A.G.G.M., 1992, ApJ 389, 400
Justtanont K., Skinner C.J., Tielens A.G.G.M., 1994, ApJ 435, 852
Kerschbaum F., Hron J., 1996, A&A 308, 489
Men'shchikov A.B., Henning T., 1996, A&A 318, 879
Netzer N., Elitzur M., 1993, ApJ 410, 701
Olofsson H., Bergmann P., Eriksson K., Gustafsson B., 1996, A&A 311, 587
Olofsson H., Bergmann P., Lucas R., et al., 1998, A&A 330, L1
Schönberner D., Steffen M., Stahlberg J., Kifonidis K., Blöcker T., 1997, in *Advances in Stellar Evolution*, eds. R.T. Rood and A. Renzini, Cambridge University Press, p. 146
Schönberner D., Steffen M., Stahlberg J., Kifonidis K., Blöcker T., 1998, in *The Carbon Star Phenomenon*, Proc. IAU Symp. 177, ed. R.F. Wing, Kluwer Academic Publishers, in press
Schröder K.-P., Winters J.M., Arndt T.U., Sedlmayr E., 1998, A&A 335, L9
Sedlmayr E., Dominik C. 1995, Ap&SS Rev. 73, 211
Steffen M., Szczerba R., 1997, Ap&SS 251, 131
Steffen M., Szczerba R., Men'shchikov A., Schönberner D., 1997a, A&AS 126, 39
Steffen M., Szczerba R., Men'shchikov A., Schönberner D., 1997b, in *Advances in Stellar Evolution*, eds. R.T. Rood and A. Renzini, Cambridge University Press, p. 154
Steffen M., Szczerba R., Schönberner D., 1998, A&A 337, 149
Suh K.-W., Jones T.J., 1997, ApJ 479, 918
Szczerba R., Marten H., 1993, in *Mass Loss on the AGB and Beyond*, ed. H.E. Schwarz, ESO Conference and Workshop Proceedings No. 46, p. 90
Szczerba R., Omont A., Volk K., Cox P., Kwok S., 1997, A&A 317, 859
Vassiliadis E., Wood P.R., 1992, Proc. Astron. Soc. Aust. 10 (1), 30
Waters L.B.F.M., Loup C., Kester D.J.M., et al., 1994, A&A 281, L1

Asymptotic Giant Branch Stars
T. Le Bertre, A. Lèbre and C. Waelkens, eds.
I.A.U. Symposium 191, 1999

The final 10^5 years of stellar AGB evolution in the presence of a pulsating, dust-induced "superwind"

K.-P. Schröder, J.M. Winters and E. Sedlmayr

Technische Universität Berlin, Inst. f. Astronomie u. Astrophysik, Sekr. PN 8-1, Hardenbergstr. 36, D-10623 Berlin, Germany

Abstract. We have computed mass-loss histories and tip-AGB stellar evolution models in the presence of a dust-induced, carbon-rich "superwind", in the initial mass-range of 1.1 to about 2.5 solar masses and for nearly solar composition (X=0.28, Y=0.70, Z=0.02). Consistent, actual mass-loss rates are used in each time-step, based on pulsating and "dust-driven" stellar wind models for carbon-rich stars (Fleischer et al. 1992) which include a detailed treatment of dust-formation, radiative transfer and wind acceleration. Our tip-AGB mass-loss rates reach about $4 \cdot 10^{-5} M_{\odot} \mathrm{yr}^{-1}$ and become an influencial factor of stellar evolution.

Heavy outflows of 0.3 to 0.6 $M_{\odot}$ within only 2 to $3 \cdot 10^4$ yrs, exactly as required for PN-formation, occur with tip-AGB models of an initial stellar mass $M_i \gtrsim 1.3 M_{\odot}$. The mass-loss of our "superwind" varies strongly with effective temperature ($\dot{M} \propto T_{\mathrm{eff}}^{-8}$, see Arndt et al. 1997), reflecting the temperature-sensitive micro-physics and chemistry of dust-formation and radiative transfer on a macroscopic scale. Furthermore, a thermal pulse leads to a very short (100 to 200 yrs) interruption of the "superwind" of these models.

For $M_i \lesssim 1.1 M_{\odot}$, our evolution models fail to reach the (Eddington-like) critical luminosity L_c required by the radiatively driven wind models, while for the (initial) mass-range in-between, with the tip-AGB luminosity L_{tAGB} near L_c, thermal pulses drive bursts of "superwind", which could explain the outer shells found with some PN's. In particular, a burst with a duration of only 800 yrs and a mass-loss of about 0.03 $M_{\odot}$, occurs right after the last AGB thermal pulse of a model with $M_i \approx 1.1 M_{\odot}$. There is excellent agreement with the thin CO shells found by Olofsson et al. (e.g., 1990, 1998) around some Mira stars.

1. Introduction

Theoretical models of tip-AGB mass-loss are of considerable interest in several respect: from the stellar evolution point of view, for the understanding of the structure of circumstellar envelopes (CSE) and planetary nebulae (PN), and for the chemical evolution of the Galaxy and galaxies in general.

The term "superwind" was coined by Renzini (1981), referring to the heavy tip-AGB mass-loss ($\gtrsim 10^{-5} M_{\odot} \mathrm{yr}^{-1}$), which is required to form a PN of typically a few tenths of a solar mass within several 10^4yrs. It is supposed to develop rather gradually, by an accelerating increase of mass loss, from a long history of less massive AGB mass-loss. Such a picture is in good agreement with the findings of cool, dust- and CO-rich CSE's around PN's (Kwok 1981) and the mass-loss rates of about $10^{-4}\ M_\odot \mathrm{yr}^{-1}$ as modeled on LPV (long period variable) observations.

While the general picture of PN formation is now certainly understood, well observed details of that process still await an explanation by more detailed models of the tip-AGB evolution and superwind mass-loss. We may remind of the well-known outer shells seen around PN's in deep exposures. A probably related phenomenon seem to be the thin, detached CO shells found by Olofsson et al. (e.g., 1990, 1998) around a few carbon stars, having kinematic ages of 3 to 13 thousand years and masses of 0.4 to $5{\cdot}10^{-2}\ M_\odot$. These shells suggest very short (about a thousand years or less) episodes of strongly (by 2 orders of magnitude) increased mass-loss.

Because of a significant interaction between the stellar structure and a heavy mass-loss, both have to be computed hand in hand on the tip-AGB. Earlier approaches have been published by, e.g., Vassiliadis & Wood (1993) and Blöcker (1995), which both use a Bowen-type wind-model (Bowen 1988) with a period – mass-loss relationship, i.e., the mass-loss rate depends strongly on the surface gravity g. That leads to a gradual but strong enhancement of the tip-AGB mass-loss, in good agreement with observational evidence. Those simple wind-models fail, however, to treat the important problem of dust-formation, with its complex and highly temperature- and density-dependent physics and chemistry, in sufficient detail.

A much different approach to tip-AGB mass-loss has been achieved by a consistent treatment of a dust-induced wind generation – including the detailed description of hydrodynamics, thermodynamics, chemistry, radiative transfer, dust formation and growth (see Sedlmayr 1994 for a review). Based on such extensive computations, Fleischer et al. (1992) have introduced consistent dynamical wind models for pulsating, C-rich AGB stars.

2. The evolution code and our approach of the "superwind" problem

We use the most recent version of an evolution code as described by Pols et al. (e.g., 1995, 1998), which is based on the original evolution program of Eggleton (1971, 1972). The essential features, in which this very economic and robust code differs from other evolution programs, are: (1) the use of a self-adaptive, non-Lagrangian mesh (only 200 mesh points required), (2) the treatment of both convective and semi-convective mixing as a diffusion process with a diffusion rate adopted as a function of $(\nabla_{\mathrm{rad}} - \nabla_{\mathrm{ad}})$ (otherwise, standard mixing length theory is used) and (3) the simultaneous and implicit solution of both the stellar structure equations and the diffusion equations for the chemical composition. A trade-off for the fastness of the code is, however, that surface abundance changes through dredge-up's are, in the present version, not computable and thermal pulses are picked up only later on the AGB as with other codes.

We here use the same set of parameters as discussed and recommended by Schröder (1998), and which are the result of a variety of critical tests by empirical methods.

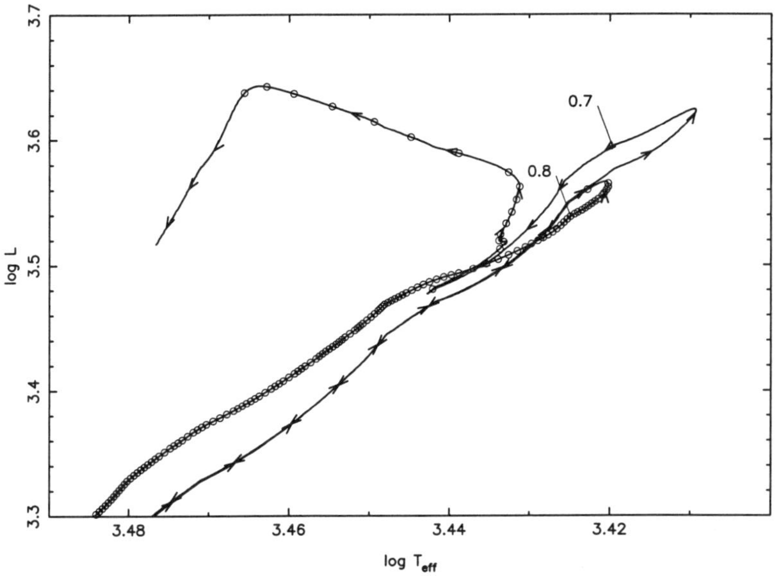

Figure 1. Tip-AGB evolution for $M_i = 1.10 M_\odot$ and solar composition. Only the last AGB thermal pulse is shown. Circles mark time-steps of 5000 yrs, arrows indicate direction of fast evolution and numbers mark the actual mass.

To incorporate realistic superwind mass-loss rates, we use analytic representations of $\dot{M}$ as obtained from a large number of detailed wind-model computations (after Fleischer et al. 1992). A minimum outward directed force is required to drive this type of wind, which translates into a critical *Eddington*-like luminosity.

Furthermore, there is a very strong dependence of $\dot{M}$ on effective temperature which has its origin in the temperature-critical micro-physics and chemistry of the dust formation process (i.e., the grain nucleation rate) and is a characteristic property which we expect from any truly detailed model of a dust-induced wind.

As it turns out, these models lead to a rapidly increasing tip-AGB mass loss, with a magnitude that qualifies for a superwind. From a grid of 48 detailed wind models, Arndt et al. (1997) derived an approximative mass-loss formula for the tip-AGB region of the HR diagram which we have adopted for our computations: $\log \dot{M} = 17.16 - 8.26 \cdot \log T_{\rm eff}/K + 1.53 \cdot \log L - 2.88 \cdot \log M$

Our first results have been published by Schröder et al. (1998). This work now incorporates several improvements:

(1) A very well defined *Eddington-like* luminosity L_c and its variation with effective temperature $T_{\rm eff}$. For that purpose, we computed a number of specific wind models in the critical parameter space and find, e.g., $L_c = 5\,500 L_\odot \cdot M_*/M_\odot$ at $T_{\rm eff} = 2\,600\,{\rm K}$, or $9\,500 L_\odot \cdot M_*/M_\odot$ at $T_{\rm eff} = 3\,000\,{\rm K}$, respectively.

(2) A realistic, gradual transition of the mass-loss rate at the onset of the "superwind". When L_{tAGB} approaches L_c, there is a ramp with a width of 0.04 in $\log L$, which matches the actual sensitivity of our wind models to small luminosity differences near L_c.

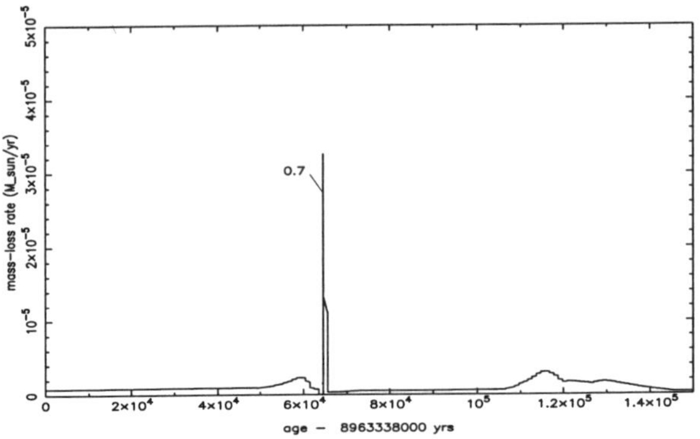

Figure 2. Tip-AGB mass-loss history for the evolution model ($M_i = 1.10 M_\odot$) shown in Fig. 1. Actual masses are marked by numbers. When such a star briefly reaches the critical luminosity on the tip-AGB with its last thermal pulse on the AGB, a short (about 800 yrs) burst of superwind occurs.

(3) A realistic, well constrained pre-AGB mass-loss: a critical figure is the minimum initial mass M_i of stars which, after their mass-losses on the giant branch (GB), can just reach their central He-burning – i.e. 0.95 $M_\odot$ with our parameterization. In particular, we use the "Reimers formula" (Reimers 1975) with $\eta_{\rm RML} = 2 \cdot 10^{-13}$.

(4) Consistently reduced time-steps, when larger mass-loss rates occur, to keep the mass lost in every time-step well below $10^{-3} M_\odot$, and to avoid any artificial reaction of the stellar model, even under extreme conditions.

3. Results: short bursts of heavy mass-loss and regular "superwinds"

When the tip-AGB luminosity $L_{\rm tAGB}$ reaches only just L_c, as with masses $\leq 1.3\,M_\odot$, then the stellar model is driven into and out of the "superwind" mass-loss by its last one or two thermal pulses on the AGB. The response of the then very thin ($\lesssim 0.2\,M_\odot$) stellar shell to thermal pulses, especially in luminosity, is very pronounced, and both gravity and $T_{\rm eff}$ are very low.

With the improvements mentioned above, we here can draw an accurate picture of the dramatic changes of mass-loss in this range of initial stellar mass: The lowest mass model ($M_i = 1.1\,M_\odot$), which reaches L_c only marginally, is plotted in Fig. 1 – its mass-loss history is plotted in Fig. 2. An isolated superwind burst is driven by the steep but very short (800 yrs) luminosity peak immediately after the last thermal pulse on the tip-AGB, during the re-ignition of the H-burning shell. The resulting superwind burst removes 0.025 $M_\odot$, at a rate of $3 \cdot 10^{-5} M_\odot {\rm yr}^{-1}$. That meets exactly the mass-loss requirements of the thin detached shells found in CO observations by Olofsson et al. (eg., 1990, 1998).

Models with slightly increased M_i already probe L_c more easily, resulting in several bursts of superwind with a typical duration of Δt of several thousand years, as with the mass-loss history of a star with $M_i \approx 1.2\,M_\odot$.

By contrast, from $M_i \approx 1.3\,M_\odot$ onwards, tip-AGB luminosities consistently reach beyond L_c and all our evolution models show a fairly similar, regular

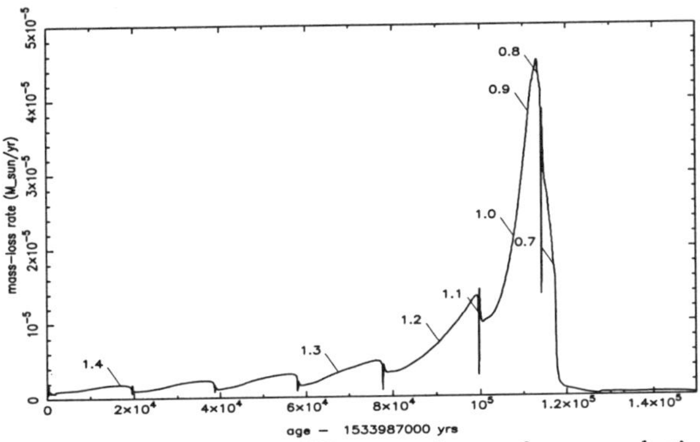

Figure 3. Tip-AGB mass-loss of an evolution model with $M_i = 2.25 M_\odot$, showing a typical "superwind" history.

superwind mass-loss record during their final 2 to $3 \cdot 10^4$ years (see Fig. 3). The superwind duration and the total mass loss in that phase varies surprisingly little with M_i: from about 0.3 $M_\odot$ with $M_i = 1.3\, M_\odot$ to 0.5 $M_\odot$ with $M_i = 2.25 M_\odot$.

During the brief drop of luminosity during each thermal pulse, combined with an increased $T_{\rm eff}$, for only 100 to 200 years, the mass-loss rate is reduced by 1 to 2 orders of magnitude. That strong contrast is the result of both the strong luminosity and temperature sensitivity of the dust-induced mass-loss. Again there is matching evidence from detached shell observations, which suggest mass-loss interruptions at similar time scales – see, e.g., Hashimoto et al. (1998).

The sensitive dependence of $\dot{M}$ on $T_{\rm eff}$ becomes noticeable in the mass-loss history. By contrast to other superwind models (e.g., Blöcker 1995), we find a *gradual* decline of $\dot{M}$ instead of an abrupt end of the superwind phase. It is caused by the rise of $T_{\rm eff}$ in the turn-off from the tip-AGB. This aspect should have consequences for the density profiles of CSE's and proto-PN's – i.e., density gradients of the inner, yet undisturbed, cool envelopes should be less steep.

4. Discussion

We consider this work as a first step to combine the so far separate worlds of stellar evolution models and detailed, self-consistent models of dust-driven winds to compute the final tip-AGB evolution and mass-loss history. In general, there is excellent agreement between our computations and a variety of observational facts, such as outer shell structure found in deep exposures of PN's in general, and detached circumstellar shells and envelopes in particular. To a significant extent, these results are, on a macroscopic scale, the direct consequence of the complex, temperature and density dependent micro-physics, dust-formation chemistry and radiative transfer – leading, in particular, to (1) a $T_{\rm eff}$ and mass-dependent (Eddington-like) critical luminosity L_c and (2) a mass-loss rate which is strongly $T_{\rm eff}$-sensitive.

Since the present version of our evolution code does not compute mixing associated with the dredge-up's, we cannot use the actual C:O surface ratio of our stellar models to define the mass-loss rate in a strictly self-consistent way. Fortunately, once a C-rich, dust-induced wind is established it is almost independent of the C:O ratio. Nevertheless, a critical point remains to be inspected: is a C:O ratio > 1 indeed reached *before* L reaches L_c? In fact, observational data support that assumption: for a tip-AGB stellar mass of 0.75 $M_\odot$ we obtain a minimum $\log L_c$ of about 3.6, that is $M_{bol} \approx -4.3$, while the bulk of carbon star M_{bol} values derived from HIPPARCOS parallaxes (Alksnis et al. 1998) already starts from about -3.7.

To conclude, we may say that despite the complexity of the problem, on both a macroscopic and a microscopic scale, and with the numerous different physical and chemical processes net-working together in a pulsating, C-rich and dust-induced superwind, we are able to compute quantitative models of the crucial final 10^5 years of stellar tip-AGB life, which are in excellent agreement with a variety of empirical facts and observational data.

Acknowledgments. This work has been supported by the DFG grants Se 420/12-1 and Se 420/8-1, and by the BMBF grant 05 3BT13A 6. We are grateful to P.P. Eggleton for providing his evolutionary code to us. We also thank D. Schönberner for very helpful suggestions.

References

Alksnis A., Balklavs A., Dzervitis U., Eglitis I., 1998, A&A 338, 209

Arndt T.U., Fleischer A.J., Sedlmayr E., 1997, A&A 327, 614

Blöcker T., 1995, A&A 297, 727

Bowen G.H., 1988, ApJ 329, 299

Eggleton P.P., 1971, MNRAS 151, 351

Eggleton P.P., 1972, MNRAS 156, 361

Fleischer A.J., Gauger A., Sedlmayr E., 1992, A&A 266, 321

Hashimoto O., Izumiura H., Kester D.J.M., Bontekoe Tj.R., 1998, A&A 329, 213

Kwok S., 1981, in *Physical Processes in Red Giants*, I., Jr., Iben and A. Renzini (eds.), D. Reidel Publ. Co., Dordrecht, p. 421

Olofsson H., Carlström U., Eriksson K., Gustafsson B., Willson L.A., 1990, A&A 230, L13

Olofsson H., Bergman P., Lucas R., Eriksson K., Gustafsson B., Bieging J.H., 1998, A&A 330, L1

Pols O.R., Schröder K.-P., Tout C.A., Hurley J.R., Eggleton P.P., 1998, MNRAS 298, 525

Reimers D., 1975, in *Problems in Stellar Atmospheres and Envelopes*, B. Baschek and W.H. Kegel (eds.), Springer, Berlin, p. 229

Renzini A., 1981, in *Physical Processes in Red Giants*, I., Jr., Iben and A. Renzini (eds.), D. Reidel Publ. Co., Dordrecht, p. 431

Schröder K.-P., 1998, A&A 334, 901

Schröder K.-P., Winters J.M., Arndt T.U., Sedlmayr E., 1998, A&A 335, L9

Sedlmayr E., 1994, in *Molecules in the Stellar Environment*, IAU Colloquium 146, U.G. Jørgensen (ed.), Springer, Berlin, p. 163

Vassiliadis E., Wood P.R., 1993, ApJ 413, 641

Asymptotic Giant Branch Stars
T. Le Bertre, A. Lèbre and C. Waelkens, eds.
I.A.U. Symposium 191, 1999

Link between Mass-loss and Variability Type for AGB Stars?

Željko Ivezić, Gillian R. Knapp

Princeton University, Department of Astrophysical Sciences, Princeton, NJ 08544-1001, USA

Abstract.
We find that AGB stars separate in the 25-12 vs. 12-K color-color diagram according to their chemistry (O, S vs. C) and variability type (Miras vs. SRb/Lb). While discrimination according to the chemical composition is not surprising, the separation of Miras from SRb/Lb variables is unexpected.

We show that "standard" steady-state radiatively driven models provide excellent fits to the color distribution of Miras of all chemical types. However, these models are incapable of explaining the dust emission from O-rich SRb/Lb stars. The models can be altered to fit the data by postulating different optical properties for silicate grains, or by assuming that the dust temperature at the inner envelope radius is significantly lower (300-400 K) than typical condensation temperatures (800-1 000 K), a possibility which is also supported by the detailed characteristics of LRS data. While such lower temperatures are required only for O- and S-rich SRb/Lb stars, they are also consistent with the colors of C-rich SRb/Lb stars.

The absence of hot dust for SRb/Lb stars can be interpreted as a recent (order of 100 yr) decrease in the mass-loss rate. The distribution of O-rich SRb/Lb stars in the 25-12 vs. K-12 color-color diagram shows that the mass-loss rate probably resumes again, on similar time scales. It cannot be ruled out that the mass-loss rate is changing periodically on such time scales, implying that the stars might oscillate between the Mira and SRb/Lb phases during their AGB evolution as proposed by Kerschbaum et al. (1996). Such a possibility appears to be supported by recent HST images of the Egg Nebula obtained by Sahai et al. (1997), the discovery of multiple CO winds reported by Knapp et al. (1998), and long-term visual light-curve changes detected for some stars by Mattei (1998).

1. Introduction

AGB stars are long-period variables (LPV) and show a variety of light curves. Based on visual light curves, the General Catalog of Variable Stars (GCVS) defines regular variables, or Miras, semiregular (SR) variables, and irregular variables (L). The distinctive features are the regularity of light curves, their amplitude and their period. Some types are further subdivided based on similar criteria (e.g. SRa, SRb, SRc,...).

AGB stars of different variability types cannot be distinguished by considering IRAS PSC data alone. Kerschbaum, Hron and collaborators (Kerschbaum & Hron 1996, and references therein; hereafter KH) have studied IR emission of various types of LPVs by combining IRAS broad-band fluxes with near-IR observations and spectra from the IRAS LRS database. They find that SRa variables appear to be a mixture of two distinct types: Miras and SRb variables. SRb variables have somewhat higher stellar temperatures and smaller optical depths than Miras, and can be further divided into "blue" and "red" subtypes, the former showing much less evidence for dust emission than the latter.

Particularly intriguing are differences between the LRS spectra of SRb variables and Miras with "10" μm silicate emission feature (LRS class 2n). The peak position of the "10" μm feature for SRb sources is shifted longwards, relative to the peak position for Miras, by 0.2-0.3 μm. In addition, the ratio of the strengths of silicate emission features at 18 μm and 10 μm, F_{18}/F_{10}, is larger for SRb variables than for Miras. These results are confirmed at high statistical significance by Marengo (1998), who determines the "10" μm feature peak position by fitting a polynomial to the 9-11 μm spectral region. He finds a difference in median peak positions of 0.2 μm, with probability larger than 0.9999 that the two distributions are different.

Motivated by these results, we further analyze correlations between IR emission and variability type for AGB stars, and their implications for the models of steady-state radiatively driven outflows. We base our analysis on IRAS data and a set of JHKLM photometric observations for about 600 LPVs, kindly made available to us by Franz Kerschbaum.

2. Distribution of AGB stars in K-12-25 Color-color Diagram

Following the results of KH, we consider only SRb, Lb, and Mira variables, and further divide them according to grain chemistry. Fig. 1 displays the distribution of sources in the K-12-25 diagram [1].

Stars in our sample clearly separate according to grain chemistry (O vs. C stars, S stars cannot be distinguished from O stars). Such a clear separation is not very surprising since the optical properties of silicate and carbon grains, which by and large determine the spectral shape of dust emission, are significantly different at these wavelengths. Stars also separate according to their GCVS variability type into two well defined regions, Miras and non-Miras (i.e. SRb and Lb, which exhibit similar behavior). This is in sharp contrast with the distribution of the same sample in the IRAS 12-25-60 diagram, which shows no such division.

A scheme which classifies stars into 4 groups according to their variability type (Miras, m, vs. non-Miras, n) and grain chemistry (O stars with silicate grains and C stars with carbonaceous grains) is shown in Fig. 1 by solid lines.

---

[1] We use colors defined as $\lambda_2 - \lambda_1 = \log(F_\nu(\lambda_2)/F_\nu(\lambda_1))$, where $F_\nu(\lambda_1)$ and $F_\nu(\lambda_2)$ are fluxes at wavelengths λ_1 and λ_2.

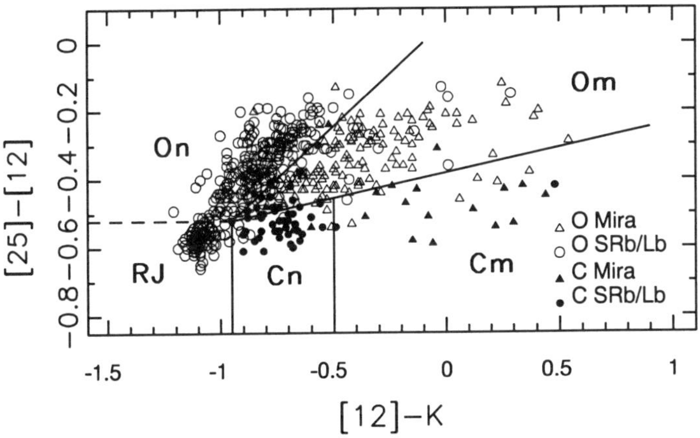

Figure 1. K-12-25 color-color diagram for mass losing AGB stars. Various symbols correspond to source variability type and grain chemistry, as marked. Solid lines display classification into 4 regions according to grain chemistry (O and S stars with silicate grains vs. C stars with carbonaceous grains), and variability type (Miras, m, vs. non-Miras, n). Dashed line separates On sources near the Rayleigh-Jeans point (RJ) which have no, or very little dust.

The boundaries, required to be straight lines, are determined by maximizing the efficiency and reliability of the resulting classification. The dashed line separates On sources around the Rayleigh-Jeans point and is determined by assuming a comparable scatter in both colors.

3. Interpretations of the Source Distribution in K-12-25 Color-color Diagram

3.1. Predictions of the "Standard" Models

We first attempt to model the source distribution by employing spectra expected for a steady-state radiatively driven outflow with either silicate or amorphous carbon grains (Ivezić & Elitzur 1995, and references therein). We define "standard" models by 2 500 K black body stellar spectral shape, and dust temperature at the inner envelope edge of 1 200 K for carbon grains and 700 K for silicate grains. These two temperatures mark the "standard" model tracks in Fig. 2. For each type of grain, the model track starts at the Rayleigh-Jeans point, with the positions along the track parametrized by dust optical depth.

The track for carbon grains passes through the distribution of stars with carbon dust, i.e. through the regions Cn and Cm. The scatter of points is marginaly consistent with the photometric errors, and non-Mira stars appear to be smaller optical depth counterparts of Miras. The model track for silicate dust passes through the distribution of Miras in Om region, although not through the region with the highest density of sources. The offset of about 0.5 mag. (0.2 on plotted log scale) is somewhat large, but still not sufficient to clearly rule

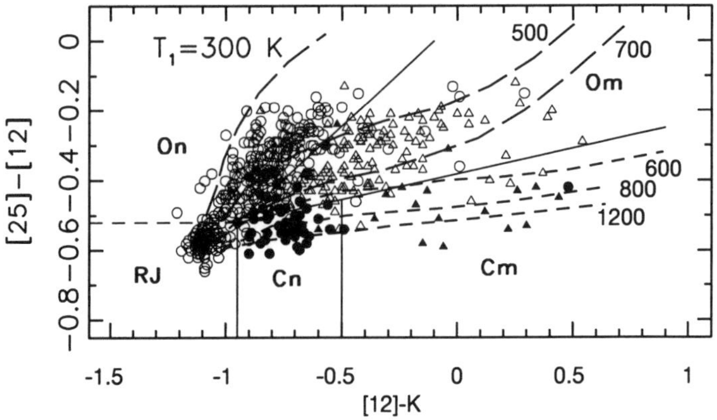

Figure 2. K-12-25 color-color diagram for mass losing AGB stars. Symbols correspond to source variability type and grain chemistry, as marked in Fig. 1. Long-dashed lines are model tracks for silicate grains with various dust temperatures at the inner envelope radius, T_1, as marked in the figure. Models for amorphous carbon grains are displayed by short-dashed lines. "Standard" models have T_1=700 K (silicate grains) and T_1=1 200 K (carbon grains).

out the model. However, the model prediction is in dire disagreement with the distribution of non-Miras with silicate dust (region On). For example, for 25-12 color of -0.3, the discrepancy in 12-K color is about 0.75 on log scale, or almost 2 magnitudes.

3.2. Variations of the "Standard" Models

There are several ways to augment the "standard" models without significantly changing the paradigm of steady-state radiatively driven outflow.

Stellar spectral shape can effect the model spectrum when dust optical depth is not large. For example, increasing the stellar temperature moves a model track to the left in the K-12-25 color-color diagram, as required by the data. However, we are unable to produce satisfactory model tracks by varying the stellar temperature within a plausible range ($<$ 4 000 K). Another possibility is that a black-body spectral shape is not an adequate description of the true stellar spectrum. This can be ruled out by considering the K-L-M color-color diagram: differences in median near-IR colors between Miras and non-Miras are about 3-4 times smaller than required to account for the observed differences in 12-K color. Such small differences in near-IR data also rule out an absorption feature in the K band, e.g. H_2O feature as seen in the data by Matsuura et al. (1998), as an explanation for redder 12-K colors observed for Miras. Furthermore, similar separation of stars into 4 regions can also be seen in the L-12-25 color-color diagram.

Models in agreement with data for non-Miras can be produced by altering the spectral shape of absorption efficiency for silicate grains. There are two required changes: the ratio of the strengths of silicate features at 18 μm and 10 μm has to be increased by a factor of about 2-3, and the peak position of

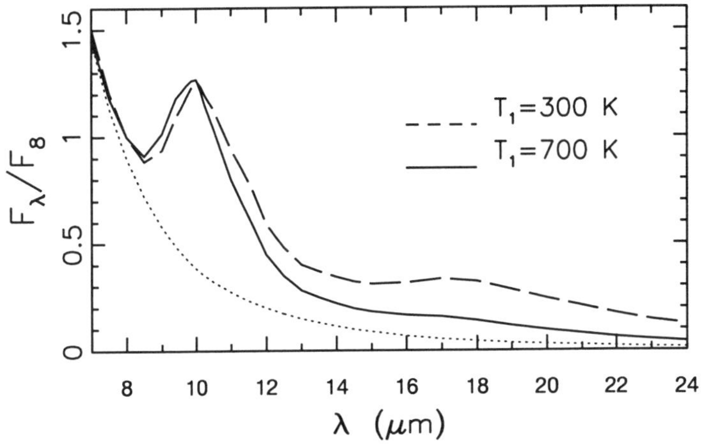

Figure 3. Model spectra in the LRS region obtained for silicate grains and 2 values of the dust temperature at the inner envelope radius, T_1 (solid and dashed lines). The dotted line shows the stellar contribution to the spectra. Note the horizontal shift between the features, although both are calculated with the same grain optical properties.

the "10" micron feature has to be shifted longwards by about 0.2-0.3 μm. Such absorption efficiency results in a model track which leaves the Rayleigh-Jeans point at a larger angle than before, and passes exactly through the observed source distribution.

Another way to make "standard" models agree with the data is to change T_1, the assumed dust temperature at the inner envelope edge. The model tracks for (unaltered) silicate grains with T_1 = 300, 500 and 700 K are shown in Fig. 2 by solid lines. They demonstrate that the distribution of stars with silicate dust in the K-12-25 color-color diagram can be reproduced by varying T_1 in the range 300–700 K, without any change in the adopted absorption efficiency for silicate grains. It is remarkable that these models are also capable of explaining the differences between Miras and non-Miras found in LRS data. Fig. 3 displays two model spectra obtained with visual optical depth of 0.6, and T_1 = 300 K (dashed line) and 700 K (solid line). The dotted line shows the stellar spectrum. Because the hot dust has been removed in the model with T_1 = 300 K, both the blue and red edges of the silicate 10 μm emission feature are shifted longwards by about 0.2-0.3 μm, causing the whole feature to appear shifted although the peak position is the same in both models. In addition, this model has a larger F_{18}/F_{10} flux ratio, as observed. This success and the application of Occam's razor hints that the absence of hot dust is the most likely explanation for differences in IR spectra between Mira and non-Mira stars with silicate dust.

The distribution of Miras and non-Miras with carbon dust is fairly well described by the "standard" model with T_1 = 1 200 K, and thus the assumption of significantly lower T_1 is not required by the data. However, such models are not ruled out by the data as shown by model tracks for T_1 = 600, 800 and 1 200 K displayed by dashed lines in Fig. 2.

4. Discussion

Absence of hot dust, as implied by low T_1, might be a natural consequence of a significant recent decrease in mass-loss rate. From the distribution of sources in the On region, we infer a time scale of about 100 years. What happens after that? The distribution of On sources ends quite sharply at its blue edge. If envelopes continued to freely expand, distribution would extend smoothly to bluer 12-K, and redder 25-12 colors. The absence of these sources argues that mass-loss resumes after about 100 years. This idea, that stars oscillate between the SRb and Mira phases, was proposed by Kerschbaum et al. (1996) based on similar number densities and scale heights.

This possibility seems to be supported by a number of new observations. Recent HST images of the Egg Nebula obtained by Sahai et al. (1997) show concentric shells whose spacing corresponds to a dynamical time scale of about 100 years. The discovery of multiple CO winds reported by Knapp et al. (1998) is in agreement with the hypothesis of variable mass-loss. Furthermore, they find that the velocity and mass-loss rate for each individual component follows the relation $v \propto \dot{M}^{1/3}$ found for stars with single CO winds (Young 1995), and theoretically expected for radiatively driven outflows (Ivezić, Knapp & Elitzur 1998). Finally, there is mounting direct evidence that long-term visual light curve changes are consistent with the transition between Mira and non-Mira variability types (Mattei 1998).

If proven correct, the mass-loss variations on time scales of about 100 years would significantly change our understanding of the stellar evolution on the AGB. What type of observations could rule out this hypothesis? The most obvious and direct test is mid-IR imaging. When T_1 decreases by a factor of two, the inner envelope radius increases by about a factor of four. Such a difference should be easily discernible for a few dozen candidate stars with largest angular sizes, which are already within the reach of the Keck telescopes (Monnier 1999).

Acknowledgments. We are grateful to Franz Kerschbaum for his JHKLM observations without which this work would not have been possible. Massimo Marengo and Mikako Matsuura have endebted us by communicating their results prior to publication. This work was supported by NSF grant AST96-18503.

References

Ivezić Ž., Elitzur M., 1995, ApJ 445, 415

Ivezić Ž., Knapp G.R., Elitzur M., 1998, astro-ph/9805003

Kerschbaum F., Hron J., 1996, A&A 308, 489

Knapp G.R., Young K., Lee E., Jorissen A., 1998, ApJS 117, 209

Marengo M., 1998, in preparation

Matsuura M., et al., 1998, in preparation

Mattei J.A., 1998, private communication

Monnier J., 1999, this volume

Sahai R., Trauger J.T., Watson A.M., et al., 1998, ApJ 493, 301

Young K., 1995, ApJ 445, 872

Asymptotic Giant Branch Stars
T. Le Bertre, A. Lèbre and C. Waelkens, eds.
I.A.U. Symposium 191, 1999

Extended Dust Shells Surrounding AGB Stars Revealed with ISO

H. Izumiura

Okayama Astrophysical Observatory, NAOJ, Kamogata, Asakuchi, Okayama 719-0232, JAPAN

O. Hashimoto

Gunma Astronomical Observatory 1-18-7 Ohotomo, Maebashi, Gunma 371-0847, JAPAN

Abstract. We obtained high resolution far-infrared images of the circumstellar dust shells of several AGB carbon stars using ISOPHOT on board the ISO. We used the C-100 and C-200 detector arrays in PHT32 oversampling mode at 90 μm and 160 μm except for U Ant, which was observed at 60 μm and 90 μm. AFGL 3068 is very compact in both bands. R Scl is marginally resolved at 90 μm. U Ant shows a double shell structure, a compact but well resolved shell surrounded by a very extended envelope. Y CVn exhibits a very extended, hollow dust shell. We analyse the structure of the dust shells using a simple dust shell model and discuss the mass-loss behaviors, particularly in conjunction with the thermal pulses on the AGB.

1. Introduction

Low- and intermediate-mass stars are known to undergo intense mass-loss when they are on the asymptotic giant branch (AGB). The mass-loss is considered to be one of the essential processes of the AGB evolution. It is, thus, important to understand when and how the AGB mass-loss occurs in individual stars. Since the AGB mass-loss is thought to be closely related to thermal pulses, it is crucial to trace the mass-loss history on time scales at least comparable to the pulse periods which are predicted to be about $10^4 - 10^5$ years. Observing extended dust shells in the far-infrared is the only way to study the mass-loss history on such time scales (Waters et al. 1994; Izumiura et al. 1996, 1997). In this paper we give a brief summary of our mapping observations of AGB carbon stars in the far-infrared and discuss the implications of the results.

2. Observations

We have observed stars listed in Table 1 in the far-infrared using ISOPHOT (Lemke et al. 1996) in a Guaranteed-Time program, HIZUMIUR_REDSTAR2 and an open-time program, HIZUMIUR_RYDRA of the Infrared Space Observatory (ISO, Kessler et al. 1996). The fifth column gives the time spent for each observation in seconds. In Fig. 1, we show the position of the program stars in the IRAS two-color diagram (cf. van der Veen & Habing 1988). We

Table 1. Log of observations of program carbon stars

Object	R.A.(J2000)	Decl.(J2000)	Filter	TDT	Map Size	Date
R Scl	1h 26m 58.0s	$-32°32'34.0''$	90	3462	$35' \times 10'$	21 Jul 1996
			160	3278	$36' \times 14'$	21 Jul 1996
R For	2h 29m 15.1s	$-26°05'52.9''$	90	3462	$35' \times 10'$	5 Feb 1998
			160	2998	$30' \times 14'$	5 Feb 1998
IRC +10216	9h 47m 57.4s	$13°16'43.6''$	90	4932	$50' \times 10'$	8-9 May 1996
			160	4818	$51' \times 14'$	9 May 1996
U Ant	10h 35m 12.9s	$-39°33'44.7''$	60	2463	$24' \times 10'$	26 Jul 1996
			90	2433	$24' \times 10'$	26 Jul 1996
Y CVn	12h 45m 07.8s	$45°26'24.0''$	90	3462	$35' \times 10'$	25 Apr 1996
			160	3278	$36' \times 14'$	25 Apr 1996
AFGL 3068	23h 19m 12.4s	$17°11'35.4''$	90	3462	$35' \times 10'$	22 Jun 1996
			160	3278	$36' \times 14'$	22 Jun 1996
RY Dra	12h 56m 25.7s	$65°59'38.9''$	90	3462	$35' \times 10'$	12 May 1997
			160	3698	$36' \times 14'$	12 May 1997

used the C-100 and C-200 cameras and the PHT32 astronomical observation template (AOT, Klaas et al. 1994), which provides oversampled maps with the focal plane chopper. We selected the C_60 and C_90 filters for the C-100 camera and C_160 filter for C-200 camera. C-100 and C-200 cameras consist of 3×3 array of $43''.5$ square pixel and 2×2 array of $89''.4$ square pixel, respectively. Oversampling factors in Y- and Z-directions (Klaas et al. 1994) for C-100 were 3 and 2/3, respectively, which resulted in spatial sampling frequencies of $15''$ and $23''$ with the array configuration. For C-200 the oversampling factors in Y- and Z-directions were 3 and 1, respectively, and the sampling frequencies were $30''$ and $92''$. C_60, C_90 and C_160 filters are centered at 60.8 μm, 95.1 μm, and 174 μm, respectively. In each map the object was placed at the center. The raw data were processed with standard procedures using PHOT Interactive Analysis (PIA, Gabriel et al. 1997) version 7.1 released in 1998. The obtained images were analysed using a simple dust shell model described in Izumiura et al. (1997). In the following we discuss the results on R Scl, U Ant, Y CVn, RY Dra, and AFGL 3068. We do not, however, describe here neither IRC +10216 nor R For because there are some calibration problems in the data.

3. AFGL 3068 & R Scl

In Fig. 2 we plot the brightness profile of R Scl at 90 μm as well as those of AFGL 3068, IRC −10529, and WX Psc scaled to R Scl. The last two stars are oxygen-rich AGB stars that were also observed in our GT program. The profiles of AFGL 3068, IRC −10529, and WX Psc are almost indistinguishable, indicating that AFGL 3068 is apparently as compact as the two O-rich AGB stars which have been known to possess a very thick and compact dust shell (Hashimoto & Izumiura 1997). Probably AFGL 3068 is much smaller than the ISOPHOT beam at 90 μm for which the FWHM is estimated to be about $50''$.

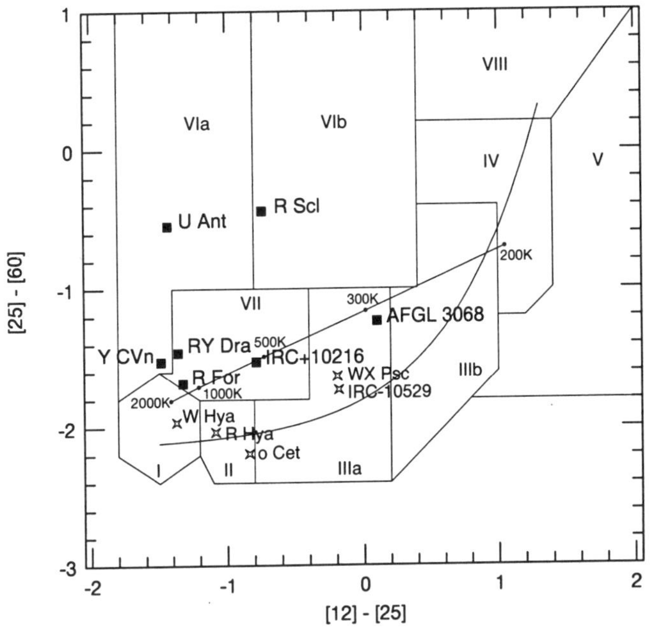

Figure 1. Location of observed carbon stars (filled squares) on the IRAS two-color diagram. Oxygen-rich AGB stars observed in our program are also shown (crosses)

No obvious extended emission is present either around this star. We have found, however, that R Scl is marginally resolved with the ISOPHOT beam at 90 μm. Besides, a hollow dust shell is postulated in order to reproduce the observed brightness profile on the basis of our dust shell model. The apparent inner radius is estimated to be about $11'' - 16''$, and the corresponding shell thickness, which is less well determined, to be about $37'' - 12''$. The hollow dust shell is consistent with the optical dust shell (Izumiura et al. 1998) as well as with the gas shell seen in mm-wave CO lines (Olofsson et al. 1996). The inner radius of the dust shell is also consistent with those of the CO and optical shells. The mass-loss rate in the dust shell has been estimated to be around 10^{-6} $M_\odot$ yr^{-1} for a reasonable set of stellar parameters. R Scl is, however, not associated with a very extended dust shell like the one seen in Y CVn. This is an important result since this star is one of the most likely candidates for a multiple shell source if every thermal pulse on the AGB gives rise to a significant modulation of mass-loss. We will further discuss this later.

4. U Ant

In Fig. 3, we show the 90 μm map of U Ant in an intensity scale which emphasizes faint extended emission. The image shows a central bright component of 100″ FWHM surrounded by a very extended, faint emission component with 10′ diameter. The central component is well resolved by the ISOPHOT beam

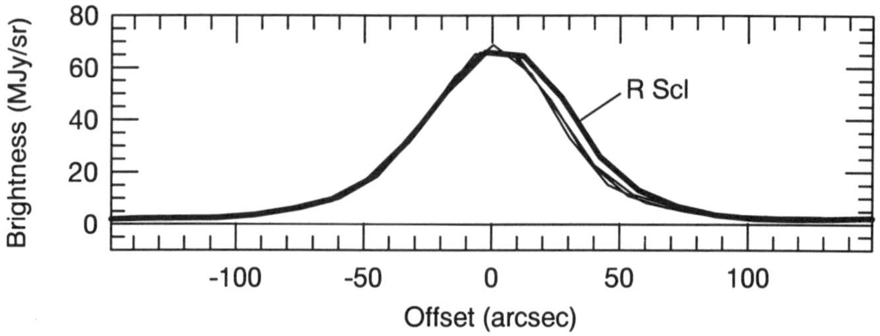

Figure 2. Brightness profile of R Scl at 90 μm (thick line), and those of AFGL 3068, IRC −10529, and WX Psc scaled to R Scl (thin lines)

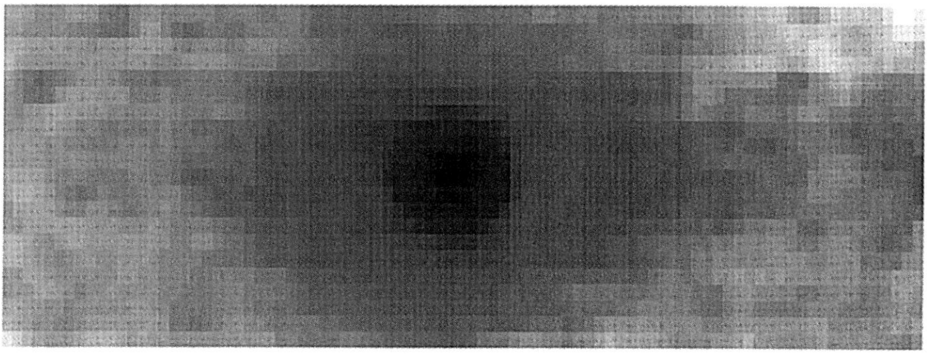

Figure 3. ISOPHOT image of U Ant at 90 μm, which measures $1155'' \times 483''$ with a pixel size of $15'' \times 23''$

(cf. Fig. 2). The two emission components are also observed in the 60 μm image. On the basis of our model analysis it turns out necessary to introduce at least two concentric, hollow dust shells in order to reproduce the observed brightness distribution. The inner radii of the two shells are well determined as $39''$ and $170''$. The outer radii are difficult to determine, but we find necessary to place a significant gap between the outer radius of the inner shell and the inner radius of the outer shell, i.e. the outer shell is detached from the inner shell. The inner radius of the inner shell agrees well with the mean radius of $43''$ of the detached, geometrically thin CO gas shell of U Ant (Olofsson et al. 1996). The CO gas shell is probably the direct counterpart of the inner dust shell. The presence of clear outer and inner boundaries of the CO gas shell due primarily to the actual variation of mass-loss strongly supports that the inner dust shell is hollow and that the outer dust shell is detached from the inner shell. This double shell structure has been proposed for this star by Izumiura et al. (1997). Using the distance of 256 pc (HIPPARCOS parallax of 3.90 milli-arcseconds, ESA 1997) we obtain mass-loss rates on the order of 10^{-6} $M_{\odot}$ yr^{-1} for both shells, which is again consistent with the mass-loss rate derived for the CO gas shell (cf. Olofsson et al. 1996). Our ISOPHOT observations have revealed that U Ant underwent at least two high mass-loss phases, each of which was followed by a period of a significantly lower mass-loss, in the last $(1–2)\times10^4$ years.

5. Y CVn and RY Dra

We found a very extended emission of 10′ diameter surrounding Y CVn at both 90 μm and 160 μm. The brightness distribution is well reproduced by a hollow dust shell surrounding the central star. Main results were already published in the A&A ISO special issue (Izumiura et al. 1996). With the PIA ver. 7.1 (Gabriel et al. 1997) the intensity calibration has been improved, which results in scaling up the brightness in the 90 μm and 160 μm maps by factors of 1.7 and 1.0, respectively. The model fitting gives a better power index of the dust opacity law of 1.1 instead of the previous value of 0.83. The improved calibration, however, does not change significantly the shape of the dust shell, hence, little influences our previous conclusions. Taking into account the shell asymmetry, we obtained the inner radius of the shell as $150'' - 200''$ or $(4.9–6.5)\times10^{17}$cm, and the mass-loss rate in the shell as $(5–9)\times10^{-6}$ $M_\odot$ yr^{-1} with the new calibration and the distance of 218 pc (HIPPARCOS parallax of 4.59 milli-arcseconds, ESA 1997).

RY Dra is the second brightest J-type carbon star next to Y CVn. We do not find, however, any obvious extended dust shell like the one seen in Y CVn. Since RY Dra is estimated to be 1.5–2 times more distant than Y CVn (Groenewegen et al. 1992; ESA 1997), it is possible that an extended dust shell is indeed present but is too close to the central star to be detected easily. Furthermore, cirrus confusion is significant for RY Dra, in particular at 160 μm, which indeed hampers detecting faint extended emission even if present. We have found that the 90 μm image is well fitted with our model by adopting an extended dust shell of inner radius around 68″. The mass-loss rate in the shell is found to be about 10^{-6} $M_\odot$ yr^{-1}. But the presence of the shell is not confirmed in the 160 μm image because of the cirrus contamination. Therefore the presence of an extended dust shell is inconclusive in the present analysis. It is necessary to remove the influence of the cirrus confusion before we draw any conclusions about the presence of an extended dust shell around RY Dra.

6. Some implications of the results

Reliable HIPPARCOS parallaxes are available for U Ant and Y CVn among our program stars, and give distances of 256 pc and 218 pc, respectively. The bolometric lumonosities of U Ant and Y CVn, then, become 5 890 $L_\odot$ and 5 360 $L_\odot$ (Groenewegen et al. 1992). The difference is smaller than the observational uncertainties. We therefore assume in the following discussion that the program stars have the same bolometric luminosity and are in similar evolutionary stages.

R Scl has hollow dust and gas shells which are similar to the inner shells of U Ant. Since the distances to R Scl and U Ant are also similar (Groenewegen et al. 1992), the shells of R Scl are smaller than those of U Ant in linear size, which probably means that the shells of R Scl are younger than those of U Ant. Then, one would expect that another dust shell, somewhat smaller than the outer shell of U Ant, could be seen around R Scl if each dust shell formation has a 1 to 1 correspondence with the thermal pulse in the carbon star phase. However, we find that it is not the case. This implies that the formation of a hollow dust shell could be episodic having nothing to do with the thermal pulse, or that the thermal pulse could occur not very periodically. The latter case requires a period fluctuation as long as 10^4 years. Otherwise the above assumptions are no longer valid and some differences in the evolutionary status or in the initial mass which cause a difference in the pulse period may be responsible for the difference in the mass-loss behavior of U Ant and R Scl.

In order to further elucidate the mass-loss behavior of individual stars on the AGB, much more observations with higher spatial resolutions as well as higher sensitivities in the far-infrared are required. Infrared Imaging Surveyor (IRIS), a Japanese infrared satellite which will be launched in 2003, would greatly enrich the study of such mass-loss behaviors in individual stars.

Acknowledgments. This study is based on observations with ISO, an ESA project with instruments funded by ESA Member States with the participation of ISAS and NASA. PIA is a joint development by the ESA Astrophysics Division and the ISOPHOT Consortium led by the Max Planck Institute for Astronomy (MPIA), Heidelberg. Contributing ISOPHOT Consortium institutes are DIAS, RAL, AIP, MPIK, and MPIA.

References

ESA, 1997, The HIPPARCOS Catalogue, ESA SP-1200

Gabriel C., Acosta-Pulido J., Heinrichsen I., Skaley D., Morris H., Tai W.-M., 1997, in Astronomical Data Analysis Software and Systems VI, eds. G. Hunt & H.E. Payne, ASP Conf. Ser. 125, p. 108

Groenewegen M.A.T., de Jong T., van der Bliek N.S., Slijkhuis S., Willems F.J., 1992, A&A 253, 150

Hashimoto O., Izumiura H., 1997, Ap&SS 251, 207

Izumiura H., Hashimoto O., Kawara K., Yamamura I., Waters L.B.F.M., 1996, A&A 315, L221

Izumiura H., Waters L.B.F.M., de Jong T., Loup C., Bontekoe Tj.R., Kester D.J.M., 1997, A&A 323, 449

Izumiura H., Waters L.B.F.M., de Jong T., Loup C., Hashimoto O., 1998, in The Carbon Star Phenomena, ed. R. Wing, IAU Symposium 177, in press

Kessler M.F., Steinz J.A., Anderegg M.E., et al., 1996, A&A 315, L27

Klaas U., Krüger H., Heinrichsen I., Heske A., Laureijs R. (eds), 1994, ISOPHOT Observer's Manual, version 3.1

Lemke D., Klaas U., Abolins J., et al., 1996, A&A 315, L64

Olofsson H., Bergman P., Eriksson K., Gustafsson B., 1996, A&A 311, 587

van der Veen W.E.C.J., Habing H.J., 1988, A&A 194, 125

Waters L.B.F.M., Loup C., Kester D.J.M., Bontekoe Tj.R., de Jong T., 1994, A&A 281, L1

Part 5. Non-Spherical Mass Loss, Binarity, Post-AGB Evolution

Gerd Weigelt commenting on high-angular resolution

Asymptotic Giant Branch Stars
T. Le Bertre, A. Lèbre and C. Waelkens, eds.
I.A.U. Symposium 191, 1999

AGB and post-AGB stars at high angular resolution

Bruno Lopez

Observatoire de la Côte d'Azur, Département Fresnel UMR 6528, BP 4229, F-06034 Nice Cedex 4, France

Abstract.
High angular resolution observations have provided some valuable information on the circumstellar environment of AGB and post-AGB stars.

We review the recent results obtained in this field and focus on observations showing evidence of non-sphericity at different spatial scales (envelope, jets, binarity, maser clumps, stellar "photosphere", hot spots, etc.). Such studies provide many informations/constraints not yet fully understood and seldom included in models.

Although the available observations are very nice and interesting, they sometimes show limited performances with regards to what is needed to analyse the onset of the winds. We try to show, here, what must be gained to improve our present knowledge.

1. Introduction

The mass loss of Asymptotic Giant Branch (AGB) stars plays an important role during this late stage of their evolution. One of the major questions raised in this field of research concerns the mechanisms producing the mass loss. There are, in my opinion, two obstacles which prevent to easily observe what causes the mass loss phenomenon (see Fig. 1):

- the mass loss is a combination of several mechanisms → their different contributions are difficult to disentangle;

- the signatures of the wind(s) begin at the stellar surface → ideally the photospheric layers should be angularly resolved.

To explain the mass loss phenomenon, a general picture including two mechanisms is presently admitted: the gas is, in a first step, levitated at a distance large enough to allow the formation of dust particles, then, the radiation pressure drives away the dust which drags the gas in its expanding outward motion. Despite the interest of this picture, the physical conditions that favour the mass loss activity and create the commonly observed asymmetries, are still not well understood.

In this review, we are pointing out a number of issues to address with the high angular resolution methods:

- When do asymmetries occur during the stellar evolution ?
- How are interpreted the non-spherical envelopes ?
- What may be the causes of departures from spherical symmetry ?
- What is observed during the superwind phase ?

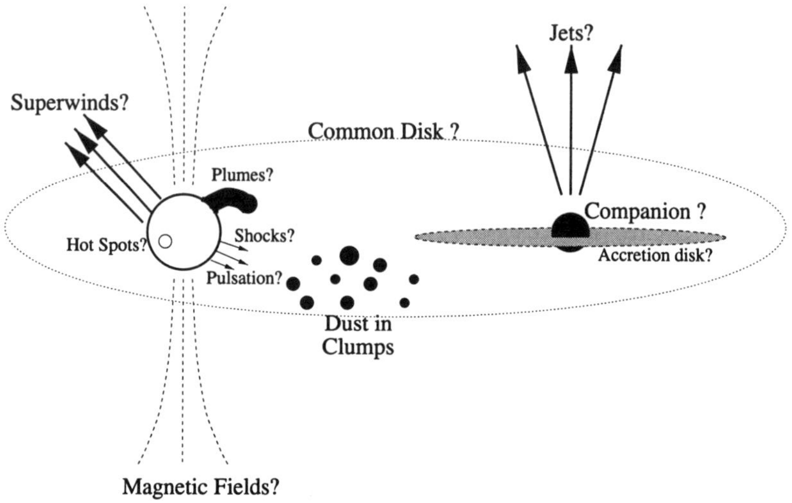

Figure 1. Sketch of the expected circumstellar environment of an AGB star. One stellar radius is between about 10 and 20 mas for the most nearby stars (located at a distance of a few hundred parsecs). Dust grain nucleation occurs at a distance of a few stellar radii from the star.

- What kind of clumpiness is observed ?
- What is observed at the stellar surface level ?

We will try, whenever possible, to link the observations and the theoretical models. However, this presentation will not be exhaustive. Observational results based on polarimetric measurements will not be covered here (see Magalhães & Nordsieck 1998), nor the observational results on massive cool supergiant stars, although some of the physical conditions in their circumstellar environment are probably similar to those of AGB stars.

2. When do asymmetries occur during the stellar evolution ?

Several observational studies show that an important fraction of AGB and/or post-AGB sources are asymmetric. For example, the ^{12}CO atlas of circumstellar envelopes of AGB and post-AGB stars (Neri et al. 1998) contains 46 objects. Interferometric visibility curves obtained with the IRAM interferometer are combined with short spacing observations from the IRAM 30 m telescope. Within the sensitivity of these observations, there is evidence for a significant asymmetry in about 30 percents of the detected envelopes. One can also refer to other recent papers such as the study of Sánchez Contreras et al. (1998)

in which the extended cold dust emission at 1.3 mm has been observed for a sample of 16 evolved stars, the mid-infrared observations of Busso et al. (1996), the mid-infrared survey of Meixner et al. (1998a, 1998b) showing two typical morphologies (toroïdal versus elliptical) for 66 PPNe.

Aside from these different observations which reveal that asymmetries occur for a non-negligible fraction of the sources, a deeper analysis is presently required to understand when (and why) are asymmetries occurring during the stellar evolution ? More precisely, does the asymmetric matter distribution rather occur at the tip of the AGB in the transition phase to a planetary nebula or does it occur during the AGB at the first mass loss stages ?

Some important clues for an envelope morphology possibly changing at the tip of the AGB are encountered in the envelope of CW Leo (IRC +10216). The envelope is detected out to a distance of 50 arcsec in the 1.3 mm continuum emission (Groenewegen et al. 1997) and about twice this distance in the visible wavelengths (de Laverny et al. 1998), corresponding to a dynamic time scale of about 6 000 years. The large outer shell of CW Leo is spherical in contrast with the inner core of asymmetric shape (Le Bertre et al. 1989; Hannif & Busher 1998; Weigelt et al. 1998a; Weigelt et al. 1999; Tuthill et al. 1998b; Tuthill et al. 1999). This may suggest that the geometry of the wind has changed during its history and may indicate that the asymmetries seen in most planetary nebulae and preplanetary nebulae are already established before the progenitor stars have left the AGB.

Asymmetries are also detected in other carbon stars such as CIT6 (Trammel & Goodrich 1998; Tuthill et al. 1998c) and IRAS 06088+1909 (Richichi et al. 1998). The source OH 231.8+4.2 which is the host of a Mira variable star has also a very well marked bipolar shape (Kastner et al. 1998; Kastner et al. 1999). In K band, its northern and southern lobes are resolved into jet like-fingers. As for CW Leo and CIT6, this source is probably at the tip of the AGB, in a brief transition stage of evolution: just before the central AGB star reaches a warmer photospheric effective temperature that characterizes the post-AGB stars.

One may wonder if similar asymmetrical features are encountered at the first mass loss stages, if they are common and what are the epoch of their onset ?

To our knowledge, asymmetries are not yet clearly detected before the star reaches the tip of the AGB (high mass loss rate). As a personnal experience, we performed a study using the University of California at Berkeley Interferometer on the Mira *o* Ceti. We had to consider several non-spherical geometries of the envelope in our radiative transfer model in order to fit the 10 μm visibility curves (Lopez et al. 1997b). But these geometries remain hypothetical because the observational contraints are not strong enough to determine a "unique" solution. In other words, during the first stages of the mass loss, when the mass loss rate is low or moderate, the dust shell is very thin and cannot be as easily observed as it is for the envelopes of PPNs or objects like the carbon-rich star CW Leo.

Answering to when does asymmetries occur, therefore requires to observe the dust shell geometry of Mira stars during the first stages of their mass loss episode. This goal will be actually achieved by using the next generation infrared or optical interferometers. These interferometers will allow a good Fourier plane coverage with possibly an imaging capability.

3. How are interpreted the non-spherical envelopes ?

It appears that the geometry of the envelopes cannot be easily defined, even when the large spatial structures of the envelopes are angularly well resolved. It is the case, for example, of the nearest post-AGB nebulae such as the Egg Nebula and the Red Rectangle. Their morphology has been interpreted in various ways by several authors, proposing:

- a bipolar dust shell with a thick equatorial disc of sharp edges (Dyck et al. 1987);
- a dust shell with an azimuthal density distribution varying in a smooth way above and below the equatorial plane (Collison & Fix 1991; Lopez et al. 1997a);
- an hybrid dust envelope composed by an initially spherically symmetric AGB wind combined with an axially symmetric superwind phase of enhanced mass loss in the equatorial plane, or in the perpendicular direction ? (Meixner et al. 1997);
- a spherical dust envelope with bipolar wind blown holes (Whitney & Hartmann 1993; Sahai et al. 1998a).

Among these various interpretations, a commonly accepted characteristic is the presence of a torus or a disk-like distribution of matter, consistent with the shape of many nebulae. The carbon-rich Egg Nebula, for example, has been observed using aperture synthesis method in several molecular lines (Nguyen-Q-Rieu & Bieging 1990). These observations are consistent with a toroidal geometry for the expanding envelope, that was originally inferred from optical scattering model (Yusef-Zadeh et al. 1984) and confirmed by more recent investigations (Skinner et al. 1997; Lopez & Perrin 1998a). HST high angular resolution near-infrared and visible images of the bipolar protoplanetary nebula CRL 2688 are reported in Sahai et al. (1998a). The emergence of the bipolarity of objects like the Egg Nebula is believed to be intimately connected to the formation and preexistence of a dusty cocoon concentrated toward the equatorial plane and with the interaction of a high velocity collimated outflow. The pre-cited data seem to provide evidence for both phenomena.

Some axisymmetric radiative transfer modelling were used to understand the geometry of the Red Rectangle nebula (Weigelt et al. 1998b; Weigelt et al. 1999; Lopez et al. 1997a). These models are based on high angular resolution observations, such as those obtained in near-infrared by Cruzalèbes et al. (1996) and in visible light by Osterbart et al. (1997). The Red Rectangle appears as an extended (up to 40 arcsec) reflecting nebula with a pronounced bipolar structure. The extension of the dark lane in the East-West direction allows to derive a lower limit for the outer radius of the dust disk of at least 65 AU and since there is no reddening of the fainter northern lobe, an upper limit of 200 AU can roughly be estimated (Weigelt et al. 1998b).

Furthermore, Waters et al. (1998) report ISO observations of the Red Rectangle revealing the presence of oxygen-rich materials similar to those found in dusty disks surrounding young stars. Grain processing and perhaps even planet formation may therefore also be occurring in the circumbinary disk of this evolved star. An evolutionary scenario is proposed in which the disk is the result of extensive mass transfer and/or mass loss while the star was still oxygen-rich. In the circumstellar environment of the Red Rectangle, Jura et al. (1997) have detected some big grains and have shown that the most plausible interpretation of the CO velocity profile is that it signals the presence of a long lived orbiting disk.

It can be concluded that the large scale structures observed in all the bipolar nebulae (the lobes) are rather well modelled and related to a concentration of material in an equatorial plane. However the present observations and the plausible models do not yet give precisely the exact inner geometry: disk, torus or cocoon ? Understanding the inner geometry remains a goal of importance since it should provide some relevant signatures on the origin(s) of asymmetries.

4. What may be the causes of departure from spherical symmetry ?

Asymmetries in the envelopes could have different origins which can be connected to:

- the rotation of the star,
- the magnetic fields,
- the presence of a companion.

The CO lines of V Hya suggest that it is rotating with a *vsini* between 10-20 km/s (Kahane et al. 1988). In addition, the radio map of its circumstellar envelope is clearly anisotropic and presents a bipolar component. According to Kahane et al. (1988), the anisotropy in the outflow from V Hya seems to result from the rotation of the star.

Magnetic field probably combined with stellar rotation may also produce asymmetries (Soker 1998). The SRa star W Hya shows strong circular polarization (Szymczak et al. 1998) in its OH emission spectrum. Masers with close spatial coincidence and opposite circular polarization lead to an estimate of the magnetic field of 0.6 mG at 80-130 AU from the star. The strength of the magnetic field at the stellar surface could reach about 10 G. Furthermore, in W Hya the blue-red shifted velocity gradient appears to be caused by a density enhancement towards the equatorial plane.

Aside, the most often invoked cause of asymmetry is the binarity. Direct evidence is found supporting a binary star formation mechanism for the Frosty Leo nebula (Roddier et al. 1995). According to Morris (1987), Morris (1981), Iben et al. (1993) and recently Mastrodemos (1998), a second star would attract the material ejected by the red giant and form an accretion disk. This has two consequences. Firstly, dust tends to accumulate in the orbital plane, obscuring both stars when the orbit is seen edge-on. Secondly, a fast polar wind is created on the second star and could blow out dust on each side of the disk. This second wind may be the cause of formation of the search-light beam observed in the Red Rectangle and the Egg Nebula (Sahai et al. 1998a). An indirect detection by spectrometric Doppler shift of a binary at the heart of the Red Rectangle has also been reported (Van Winckel et al. 1995; Walkens et al. 1996).

Jura & Turner (1998) report high angular resolution observations at millimeter and submillimeter wavelengths of the dust disk associated with the Red Rectangle. A dust clump with an estimated mass near that of Jupiter in the outer region of the disk is observed. The clump located at 4.3 arcsec from the primary (1600 AU) is larger than the Solar System and its nature is at present unclear. Sahai et al. (1998) report high resolution near-infrared images and polarimetric images of the inner region of the bipolar protoplanetary nebula CRL 2688. The polarization map yields the probable detection of a companion. Its measured projected separation is 750 AU. This may lend additional weight to binary based models.

Furthermore, the indirect evidence of an accretion disk around a companion has been realized for Mira (*o* Ceti). Warner (1972) has shown that the companion luminosity is flickering and suggests the presence of an accreting material. It is shown that considering the circumstellar density of the gas, the kinetic energy transformed in the accretion process could easily produce the companion luminosity. This binary has also been observed by Karovska et al. (1997) using the HST (see also Karovska 1999). The dust shell observed in the J band by means of adaptive optics possibly shows that the companion has some effects on the shape of the dust envelope, but this has to be confirmed (Lopez et al. 1998b).

Presently, among the proposed mechanisms (stellar rotation, magnetic field, binarity), the main cause of envelopes departures from spherical symmetry is not clearly defined. The binarity however seems to be the easiest hypothesis to check at the present time from an observational point of view.

5. What is observed during the superwind phase ?

Observations of the superwind phase show an enhanced mass loss rate associated with a high velocity wind. Atomic and molecular emission lines are produced by the superwind when it encounters the low velocity AGB wind. The geometry of the envelopes observed during this phase is asymmetrical. It is not known if the geometry of the superwind is intrinsically related to the properties of the star itself or if it mostly results from an interaction with the remnants of the AGB envelopes.

The well resolved morphology of IRAS 07134+1005 shows an elliptical outer shell surrounding two aligned peaks. They are interpreted as limb brightened peaks of an optically thin elliptical shell with an equatorial density enhancement (Meixner et al. 1997). Using a code describing the axial symmetry of the dust shell, the dust emission images and the spectral energy distribution of four other PPNe such as IRAS 07134+1005 are modelled. The dynamical age estimates indicate that these stars have left the AGB 300-1400 years ago, just after the superwind phase. In order to match with the models the 60 and 100 μm photometry, the AGB mass loss rate has to be about 20 times smaller than the superwind mass loss rate. Taking into account the distance uncertainties, the superwind mass loss rate ranges between 10^{-4} and 10^{-6} $M_{\odot}yr^{-1}$.

M1-92 is probably the PPN for which the structure and properties of the wind interaction, the dominant formation process of PN, are being studied in most details (Alcolea & Bujarrabal 1999). HST imaging of continuum and atomic lines (H_{α}, [OI], [SII] and [OIII]) are presented in Bujarrabal et al. (1998), see also Trammel & Goodrich (1996). Ground based imaging of the 2 μm continuum and H_2 ro-vibrational (S(1) v = 1–0 and v = 2–1) lines has also been performed. The optical line emissions come mainly from two chains of shocked knots placed along the symmetry axis of the nebula and inside those cavities for which relatively high excitation is deduced allowing the estimate of a shock velocity of about 200 km/s. The hollow shape of the CO lobes strongly suggests the effect of a wind interaction between the fast bipolar outflow presently ejected by the star and the slow dense envelope ejected during the past AGB-phase. The line emission images show several knots of emitting material that are not present in the continuum map. Such clumps are mostly located in the axis of the nebula and placed symmetrically at about 2.2 and 3.3 arcsec from the star. The clumpy image of this bipolar flow is similar to the sinuous chain of knots often mapped in jets associated to young stellar objects (Reipurth 1989). The big contrast between the emission from the bipolar jet and the emission from the bow leading shock may indicate that the jet density is smaller than that of the AGB shell by a factor larger than 10.

6. What kind of clumpiness is observed ?

Movies were presented during the Symposium by Diamond & Kemball (1999) and Richards et al. (1999) for two types of maser spots (SiO and H_2O, resp.). These very beautiful results allow the study of the velocity fields near the stars, and the analysis of the acceleration process on dust by radiation pressure.

Observations of TX Cam and U Her at 43 GHz (SiO masers) using the VLBA reveal well defined rings of spot emissions at 2-4 stellar radii (Diamond & Kemball 1999; Diamond et al. 1994). The masers appear to lie in the so called extended atmosphere, the region between the photosphere and the dust formation point. IR interferometry on such AGB stars (Danchi et al. 1994) indeed reveals that dust condenses at a few stellar radii (3-5) from the center of the star, while SiO masers are not located beyond 4 stellar radii. Presumably, this is the radius at which, Si atoms begin to condense into dust.

Filaments of water maser spots (about 1 km/s and 10-20 mas long) are observed in the dust shell of RT Vir (once the dust has already condensed). The distribution of the 22 GHz water maser spots (Bains 1995) shows a systematic red-shift of the masers to the East and a blue shift to the West. This suggests that the source may have an intrinsic bipolarity.

Because of instrumental limitations, while hundreds of maser spots and their motions are observed, only a few clumps of dust are clearly detected close to a star (i.e. at a distance were dust condensation has occurred during the recent mass loss history). About 4 clumps of dust are observed around the carbon star CW Leo (Haniff & Busher 1998; Tuthill et al. 1998b; Tuthill et al. 1999; Weigelt et al. 1998a; Weigelt et al. 1999). Considerable changes have occurred in the morphology of CW Leo in the past 8 years: whereas in 1989 both the K and L band emissions were apparently dominated by a single core, in 1997 there were four compact components.

These results are exciting and the still improving observational methods should allow in the future to study the dust formation region and its clumpiness. The clumps of CW Leo appear to move outward and offer the perspective to better understand the contribution of the radiation pressure to the mass loss and the physical conditions under which dust forms near the star.

7. What is observed at the stellar surface level ?

Observations of the stellar surface and its changes with time are crucial to understand the origin of mass loss. To the best of my knowledge there are only a few relevant direct observations of the stellar surfaces. This is because ideally they require the use of long baseline interferometry methods which are not yet able to perform image reconstructions at optical or infrared wavelengths. The presence of shock waves, hot spots, stellar distorsions or plumes is expected and need to be studied in deep details (see Fig. 1).

The photospheric surfaces of five long period variables (o Ceti, R Leo, W Hya, χ Cyg and R Cas) have been imaged in the optical/near-IR at the 4.2 m William Herschel Telescope (Tuthill et al. 1998a). All these sources exhibit departures from circular symmetry. Good fits of the data could be attained with bright spots models, but the detailed parameters - brightness and locations of the spots - are very poorly constrained. The use of a bright-spots model does not rule out the interpretation of a possible elliptical photosphere. Moreover, stellar morphologies change markedly from one epoch to another. Such changes are graphically evident in the Fourier data obtained at two epochs on o Ceti. Secular changes in the shape of o Ceti have been also interpreted by elliptical models (Karovska et al. 1991; Quierrenbach et al. 1992) giving a variable position angle and axial ratio. The observations of Tuthill et al. (1998a) have revealed that changes in the appearance of o Ceti are the result of the evolution of bright features rather than gross distorsions of the overall stellar atmosphere. The time scale for this evolution would seem to lie somewhere between 3 and 14 months.

Observations of the photospheric surface of Miras made with large aperture telescope are also presented by Hofmann et al. (1998) and by Karovska et al (1997). The image of o Ceti observed in the UV light by the HST is spectacular. It shows a plume like feature that points toward the companion position. To my knowledge, the direct imaging of such a plume feature is unique for a star other than the Sun. The physical reasons for the occurrence of this feature is not clear.

8. Conclusion

It is of crucial importance to study the geometry of the envelopes during the first episodes of the mass loss process, when the mass loss rate is low or moderate. This leads to a better understanding of when and why the asymmetries occur.

To my opinion, very nice observational results have been obtained up to now by means of high angular resolution methods. However, the best remains to be done. The interferometers such as the ISI, the VLTI and the Keck are expected to bring strong observational constraints on the stellar photospheres and on the circumstellar regions near the photosphere, where the mass loss starts.

I am impressed by the very nice movies obtained by radio interferometric method on maser clump motions. Because the objects we study are alive, it seems necessary not only to image the circumstellar environments with as many details as possible, but also to monitor all the physical processes encountered and to determine their typical lifetime scales.

Acknowledgments. Thanks to G. Niccolini for help in the reading of some papers during the preparation of this review. N. Epchtein and P. de Laverny were kind to read and comment this paper before publication.

References

Alcolea J., Bujarrabal V., 1999, this volume

Bains I., 1995, MSc thesis, University of Manchester

Danchi W.C., Bester M., Degiacomi C.G., Greenhill L.J., Townes C.H., 1991, ApJ 367, L27

Bujarrabal V., Alcolea J., Sahai R., Zamorano J., Zijlstra A.A., 1998, A&A 331, 361

Busso M., Origlia L., Marengo M., et al., 1996, A&A 311, 253

Collison A., Fix J.D., 1991, ApJ 368, 545

Cruzalèbes P., Lopez B., Bester M., Gendron E., Sams B., 1996, A&AS 116, 597

de Laverny P., Mauron N., Lopez B., 1998, poster contribution, this conference

Diamond P.J., Kemball A.J., Junor W., et al., 1994, ApJ 430, L61

Diamond P.J., Kemball A.J., 1999, this volume

Dyck H.M., Zuckerman B., Howell R.R., Beckwith S., 1987, PASP 99, 99

Groenewegen M.A.T., van der Veen W.E.C.J., Lefloch B., Omont A., 1997, A&A 322, L21

Haniff C.A., Buscher D.F., 1988, A&A 334, L5

Hofmann K.-H., Weigelt G., Balega Y.Y., Scholz M., 1998, AGM 14, 52

Iben I., Jr., Livio M., 1993, PASP 105 1373

Jura M., Turner J., Balm S.P., 1997, ApJ 474, 741

Jura M., Turner J., 1998, Nature 395, 144

Kahane C., Maizels C., Jura M., 1988, ApJ 328, L25

Karovska M., 1999, this volume

Karovska M., Hack W., Raymond J., Guinan E., 1997, ApJ 482, L175

Karovska M., Nisenson P., Papaliolios C., Boyle R.P., 1991, ApJ 374, L51

Kastner J.H., Weintraub D.A., Merrill K.M., Gatley I., 1998, AJ 116, 1412

Kastner J.H., Henn L., Weintraub D.A., Gatley I., 1999, this volume

Le Bertre T., Magain P., Remy M., 1989, The ESO Messenger 55, 25

Lopez B., Tessier E., Cruzalèbes P., Lefèvre J., Le Bertre T., 1997a, A&A 322, 868

Lopez B., Danchi W.C., Bester M., et al., 1997b, ApJ 488, 807

Lopez B., Perrin J.M., 1998a, A&A, in preparation

Lopez B., Cruzalèbes P., Aristidi E., et al., 1998b, A&A, in preparation

Magalhães A.M., Nordsieck K.H., 1998, in "The carbon star phenomenom", IAU Symposium 177, ed. R. Wing, in press

Mastrodemos N., 1998, Ph.D. thesis, University of California

Meixner M., Skinner C.J., Graham J.R., et al., 1997, ApJ 482, 897

Meixner M., Ueta T., Dayal A., et al., 1998a, submitted to ApJ
Meixner M., Ueta T., Dayal A., et al., 1998b, poster contribution, this conference
Morris M., 1987, PASP 99, 1115
Morris M., 1981, ApJ 249, 572
Neri R., Kahane C., Lucas R., Bujarrabal V., Loup C., 1998, A&AS 130, 1
Nguyen-Q-Rieu, Bieging J.H., 1990, ApJ 359, 131
Osterbart R., Langer N., Weigelt G., 1997, A&A 325, 609
Quierrenbach A., et al., 1992, A&A 259, L19
Reipurth B., 1989, ESO Workshop on "Low mass stars formation and pre-main sequence objects", ed. B. Reipurth, p. 247
Richards A.M.S., Cohen R.J., Bains I., Yates J.A., 1999, this volume
Richichi A., Stecklum B., Herbst T.M., Lagage P.-O., Thamm E., 1998, A&A 334, 585
Roddier F., Roddier C., Graves J.E., Northcott M.J., 1995, ApJ 443, 249
Sahai R., Hines D.C., Kastner J.H., et al., 1998, ApJ 492, L163
Sánchez Contreras C., Alcolea J., Bujarrabal V., Neri R., 1998, A&A 337, 233
Skinner C.J., Meixner M., Barlow M.J., et al., 1997, A&A 328, 290
Soker N., 1998, MNRAS 299, 1242
Szymczak M., Cohen R.J., Richards A.M.S., 1998, MNRAS 297, 1151
Trammel S.R., Goodrich R., 1996, ApJ 468, L107
Trammel S.R., Goodrich R., 1998, private comunication
Tuthill P.G., Haniff C.A., Baldwin J., 1998a, submitted to MNRAS
Tuthill P.G., et al., 1998b, in preparation
Tuthill P.G., et al., 1998c, in preparation
Tuthill P.G., Monnier J.D., Danchi W.C., 1999, this volume
Van Winckel H., Waelkens C., Waters L.B.F.M., 1995, A&A 293, L25
Waelkens C., Van Winckel H., Waters L.B.F.M., Bakker E.J., 1996, A&A 314, L17
Warner B., 1972, MNRAS 159, 95
Waters L.B.F.M., Waelkens C., Van Winckel H., et al., 1998, Nature 391, 868
Weigelt G., Balega Y.Y., Bloecker T., et al., 1998a, A&A 333, L51
Weigelt G., Balega Y.Y., Men'shchikov A.B., Osterbart R., 1998b, submitted to New Astronomy
Weigelt G., Blöcker T., Hofmann K.-H., et al., 1999, this volume
Whitney B.A., Hartmann L., 1993, ApJ 402, 605
Yusef-Zadeh F., Morris M., White R.L., 1984, ApJ 278, 186

Asymptotic Giant Branch Stars
T. Le Bertre, A. Lèbre and C. Waelkens, eds.
I.A.U. Symposium 191, 1999

Imaging the two wind post-AGB interaction in M 1-92

J. Alcolea and V. Bujarrabal

Observatorio Astronómico Nacional, Apdo. 1143, E-28800 Alcalá de Henares, Spain

Abstract. M 1-92, Minkowski's Footprint, is a well studied protoplanetary nebula (PPN). Here we present high resolution observations of this object in atomic and molecular lines, as well as in the continuum, at optical and radio wavelengths. From these observations we conclude that the present bipolar nebula is the old AGB shell, strongly modified by the interaction with some powerful post-AGB bipolar ejections, which started suddenly about 900 years ago. The nature of these post-AGB ejections as well as its comparison with the present bipolar fast flow are discussed. We conclude that a mechanism different from photon pressure must be invoked in order to explain the morphology and dynamics of M 1-92 and other similar objects.

1. Introduction

One of the crucial phases of the stellar evolution that are still not well known is the post-AGB evolution of late-type stars toward the Planetary Nebula (PN) stage. Today we know that the evolution of both the central star and its surrounding envelope is driven by the processes of copious mass loss occurring at the very end of the AGB phase. In a few thousand years the central star becomes a hot white dwarf, ionizing the AGB shell which then shines as a PN. This evolution is accompanied by drastic changes in the chemistry, shape and kinematics of the circumstellar envelopes. We note in particular the break-up of the spherical symmetry characteristic of AGB circumstellar shells (Neri et al. 1998), that is replaced in many post-AGB objects by a clear bipolar appearance.

For a better understanding of this process, we must study in detail the few objects that are suffering these deep transformations right now, i.e. the objects in the transition phase, usually known as protoplanetary nebulae (PPNe). Here we present a summary of our results on one of these few objects, M 1-92, Minkowski's Footprint (see also the contribution by Sánchez Contreras et al. in this volume). These results, which have been very fruitful in part because of the explicitness of the object, have allowed us to draw a global but also very detailed picture of the post-AGB processes in this source.

M 1-92 is a typical bipolar reflection nebula, 10″ long, with a clear symmetry axis oriented in the sky nearly in the NW-SE direction (Minkowski 1946, Herbig 1975, see Fig. 1c). The two lobes are separated by an equatorial disk of dust, partially obscuring the South lobe (the farthest one) and the central star. For a distance of 3 kpc, the total luminosity of M 1-92 is about $10^4\ L_\odot$ (Calvet &

Cohen 1978), and from the star light scattered by the dust grains, a temperature of 20 000 K has been derived for the central source (Herbig 1975; Cohen & Kuhi 1977). Some authors claim that this central source is in fact a binary system in view of its composite spectrum (Feibelman & Bruhweiler 1990), however no direct observation of the companion exists. M 1-92 presents many signs of being a relatively young PPNe, and therefore it constitutes a very good candidate for studying the post-AGB processes when they are occurring or are still very recent. We have conducted on this source a series of high-resolution observations at optical and radio wavelengths, in order to probe the different components of the circumstellar envelope.

2. Molecular line observations

The observations of the molecular envelope of M 1-92 were conducted using the IRAM Plateau de Bure interferometer, from 1993 to 1997. We observed the ^{12}CO and ^{13}CO J=1–0 lines with a spatial resolution of $\sim 2''$, and the ^{13}CO J=2–1 line with $1''$ resolution. See Bujarrabal et al. (1994, 1997 and 1998b) for more details. From these CO observations we have been able to derive a very detailed view of the structure and kinematics of the neutral envelope, which consists in three different components: an equatorial disk, dividing two elongated emptied shells (each one spatially coincident with each lobe of the nebula), ended in two high velocity knots expanding along the symmetry axis at a deprojected velocity of 70 km s^{-1} (Fig. 1a-b, see also Figures 1 and 2 in Bujarrabal et al. 1998b). The total size of the nebula is $4''\times11''$ ($1.5\times5\ 10^{17}$ cm deprojected), to be compared with the small width of the walls of the emptied shells, of only $0\farcs6$ ($2\ 10^{16}$ cm). This width is also comparable to the thickness of the equatorial disk and to the size of the high velocity knots, i.e. the nebula is relatively large but consists of several very thin sheet-like components.

From a model fitting of our maps obtained in the three CO lines, we have estimated the main physical parameters of the envelope, under the assumption that all three lines are thermalized and adopting a $^{13}CO/H_2$ relative abundance ratio of $2\ 10^{-5}$. We have obtained that the density gradually decreases from $3\ 10^5$ cm^{-3}, at the equatorial plane, to $5\ 10^4$ cm^{-3}, at both high-velocity tips. The kinetic temperature is very low, 15 K in the central disk and only 10 K in the outer parts of the nebula. We also derive a total mass of 0.9 $M_\odot$, this high value being quite comparable to the expected total mass lost during the AGB-phase. The kinematics of the whole molecular envelope is dominated by a strong and nearly constant velocity gradient along the nebular axis, of 7 km s^{-1} per arcsec (deprojected), from which we derive a dynamical age of 900 yr for the present CO kinematics. The mean gas expansion velocity in the lobes, about 40 km s^{-1}, is found to be in agreement with the mean dust expansion velocity, obtained from the Doppler shift measurements of scattered stellar line emission (Solf 1994). The equatorial disk is found to be in expansion at a moderate (typical AGB) velocity of $\sim$ 8 km s^{-1}; no sign of rotation around the symmetry axis is found.

In view of these results we conclude that this molecular envelope is the remnant of the AGB shell; a mass loss as copious as about 10^{-4} $M_\odot$ yr^{-1} at the end of the AGB is needed in order to account for the computed total mass.

The structure and kinematics of the old AGB envelope would have been strongly modified as a result of a strong wind interaction which occurred some 900 yr ago. The nearly constant velocity gradient suggests that, during most of that time, the acceleration has been negligible, i.e. the process responsible for the present dynamics probably lasted for less than or about 100 yr. (This Hubble-like velocity field is expected in the case of a sudden interaction.) Using this result, and thanks to the high quality of our interferometric maps, we have also been able to obtain a direct 3-D model of the nebula by simply assuming that the velocity field is proportional to the radial distance to the center of the nebula. For this simple velocity field, one can use the observed projected velocity shift to recover the position of the emitting gas along the line of sight (see Alcolea & Bujarrabal 1999). Following this procedure we can visualize the structure of the envelope, i.e. the shape of the equatorial disk and of the CO walls, and project it onto the plane of the sky (see Fig. 1b). This image is very useful for comparisons with the other images taken for the nebula.

The shock interaction between the old AGB wind and some very powerful post-AGB bipolar ejection occurred via a momentum driven bow shock, after which a strong cooling has taken place (a quasi-isothermal shock). This would explain the present low kinetic temperature and highly ordered structure of the nebula. Most remarkable is the amount of axial (along the symmetry axis) momentum carried by the molecular envelope. Assuming an AGB radial expansion velocity of 8 $\mathrm{km\,s^{-1}}$ (i.e., assuming that the equatorial disk is indeed the only part of the nebula which has not been modified by the post-AGB interaction), we estimate that the linear momentum gained by the envelope as a result of the shock interaction is 3 10^{39} $\mathrm{g\,cm\,s^{-1}}$. This is about 1 000 times the total mechanical momentum of the photons emitted by the central star during the time we have estimated the interaction lasted for. We must then conclude that photon pressure could not power the strong post-AGB ejections responsible for the present kinematics. This discrepancy still holds if we relax the interaction time up to the total post-AGB life of M 1-92 or if we consider multiple scattering (Bujarrabal et al. 1998b). Another mechanism, turned on right at the end of the AGB phase, must be invoked. This axial momentum excess is not exclusive of the neutral envelope of M 1-92, but on the contrary it is a relatively common feature of the envelopes around post-AGB sources (see also the contribution by Sánchez Contreras et al. in this volume). Although the nature of the mechanism responsible for the post-AGB dynamics is still unknown, the similarities between the post-AGB ejections and those from low mass young stellar objects, like for instance the momentum excess and high collimation (see next section), suggest that a similar process, ultimately powered by mass accretion, might be at work also at the beginning of the post-AGB era.

3. HST-WFPC2 observations

The present post-AGB ejections have been imaged through the observations of forbidden atomic lines using the WFPC2 of the HST. We obtained images of the SII and OI lines, two well known tracers of shocks, as well as of the OIII line to probe the excitation of the different regions. The continuum at 547 nm and the Hα line were also observed (Fig. 1c-d, Bujarrabal et al. 1998a). The results show

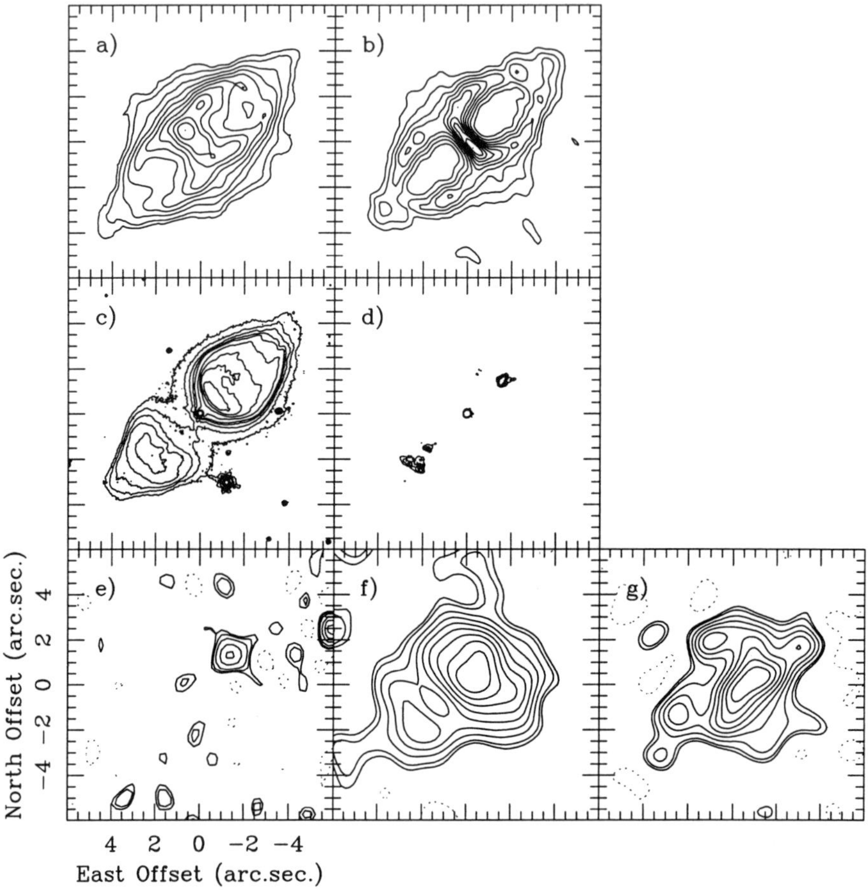

Figure 1. Different views of M 1-92. Dashed lines represent negative contours. Fig. 1a: the molecular shell of M 1-92 from the ^{13}CO J=2–1 total integrated emission. Fig. 2b: structure of the molecular shell (the CO walls plus the equatorial disk) projected onto the plane of the sky, obtained after a 3-D reconstruction of nebula from the ^{13}CO J=2–1 data. Fig. 1c: continuum image at 547 nm. Fig. 1d: image of the shocked gas in M 1-92 as seen in the SII forbidden line. Fig. 1e: continuum emission at 6 cm. Fig. 1f: continuum emission at 2.6 mm. Fig. 1g: continuum emission at 1.3 mm.

that the present bipolar post-AGB ejection is highly collimated. The emission in the SII and OI lines originates from a chain of compact knots along the axis of the nebula. Long slit observations of these lines (Solf 1994) show that this shocked atomic gas is expanding along the axis at mean velocities of about 90 km s^{-1}, about twice the expansion velocity of CO at similar distances from the star. When comparing these observations with the structure of the CO shell (Figs. 1b and 1d), it is clearly seen that the optical emission can not be directly due to shocks between the faster bipolar outflow and the molecular shell. The OI and SII knots are located just in the middle of the CO cavities, therefore this emission must be mostly due to shocks taking place within the jet itself. Only in the South lobe a structure perpendicular to the nebular axis is detected, just at the inner boundary of the CO shell. This could be the Mach disk, the point where the fast jet and molecular shell collide today. The results from the OIII observations suggest that the excitation is higher in the North lobe. No sign of the leading bow shock, produced in the strong interaction responsible for the present kinematics, has been found. This shock, if still present, must be in front of the accelerated molecular gas, but remains undetected till now.

The present post-AGB ejection is not affecting any more the kinematics of the molecular shell: apparently the two-wind interaction now is much less energetic than it was when it strongly accelerated the nebular gas 900 yr ago. For the ionized gas, we estimate a total axial momentum much lower, between 10^{37} and 10^{38} g cm s^{-1}. We must therefore conclude that present post-AGB ejection is substantially different (at least much less energetic) from what it was before, although we can not rule out that the mechanisms producing both post-AGB flows could be similar.

4. cm and mm continuum observations of M 1-92

6 cm (4.8 GHz) continuum observations of M 1-92 were also performed, using the VLA interferometer in A configuration in October 1995, the resulting synthesized HPBW being 0.″9. As it is shown in Fig. 1e the emission comes from two well separated spots of similar intensity located on the North-West side of the nebula. Both spots are unresolved by our beam. One of the spots is coincident with the high excitation knot seen in optical forbidden lines along the North side of the nebular axis. On the contrary, the other spot (just at the right border of the panel) originates from a position well outside M 1-92 and very probably is not associated to our source. Our map is compatible with the one previously obtained by Bowers & Knapp (1989) in which, due to the lower resolution, both spots appeared as an East-West elongated structure at the North lobe of the nebula. Most remarkable is the good correspondence of the first spot (the only one related to M 1-92) with the emission from shock excited atomic gas. The 6 cm continuum emission seems to be due to thermal free-free emission by electrons (no polarization has been detected). Since no dust neither gas opacities are expected at this low frequency, the fact that no 6 cm continuum emission is detected in the South lobe confirms that the present post-AGB bipolar ejections in M 1-92 are not symmetric, in agreement with the results obtained from the OIII/OI line ratio.

Simultaneously to our Plateau de Bure CO line observations, we have also mapped the continuum emission of M 1-92 at 2.6 and 1.3 mm (113 and 223 GHz respectively). As it is shown in Fig. 1f-g, the emission is resolved in the two bands, being elongated in the direction of the symmetry axis. The total detected emission is 16 mJy at 2.6 mm, and 74 mJy at 1.3 mm. From these values we derive a spectral index of 2.3, fully compatible with optically thin dust emission. We also note that at 2.6 mm, the location of the maximum is not coincident with the center of the nebula, it is slightly displaced to the North along the axis of symmetry. This can not be explained in terms of opacity, that must be very low also in this band. A possible explanation would be that at 2.6 mm there is still some contribution from the free-free emission detected at 6 cm. This would be possible only if this free-free emission is due to an ionized jet (for which a spectral index of about 0.6 is expected) and does not come from a standard HII-like region.

Acknowledgments. This work has been partially supported by the Spanish DGES project PB96-0104.

References

Alcolea J., Bujarrabal V., 1999, in preparation

Bowers P.F., Knapp G.R., 1989, ApJ 347, 325

Bujarrabal V., Alcolea J., Neri R., Grewing M., 1994, ApJ 436, L169

Bujarrabal V., Alcolea J., Neri R., Grewing M., 1997, A&A 320, 450

Bujarrabal V., Alcolea J., et al., 1998a, A&A 331, 361

Bujarrabal V., Alcolea J., Neri R., 1998b, ApJ 504, 915

Calvet N., Cohen M., 1978, MNRAS 182, 687

Cohen M., Kuhi L.V., 1977, ApJ 213, 79

Feibelman W.A., Bruhweiler F.C., 1990, ApJ 354, 262

Herbig G.H., 1975, ApJ 200, 1

Minkowski R., 1946, PASP 58, 305

Neri R., Kahane C., Lucas R., Bujarrabal V., Loup C., 1998, A&AS 130, 1

Solf J., 1994, A&A 282, 567

Asymptotic Giant Branch Stars
T. Le Bertre, A. Lèbre and C. Waelkens, eds.
I.A.U. Symposium 191, 1999

Infrared and Millimeter Views of the Helix: the Nearest, Massive, Neutral Remnant of a Circumstellar Envelope

P.J. Huggins

Physics Department, New York University, New York NY 10003, USA

P. Cox

IAS, Université de Paris XI, 91405 Orsay, France

T. Forveille

Observatoire de Grenoble, B.P. 53X, 38041 Grenoble Cedex, France

R. Bachiller

IGN OAN, Apt 1143, 28800 Alcalá de Henares, Spain

K. Young

SAO, 60 Garden Street, Cambridge, MA 02138, USA

Abstract.

We present new, infrared and millimeter views of the Helix nebula which illustrate the critical role of remnant, neutral AGB envelopes in the formation and evolution of planetary nebulae. Large scale $\sim 1\,000''$ mapping of the entire nebula in the CO ($J = 2-1$) line with the CSO reveals the global structure of the envelope. The CO emission forms the familiar ring structure seen in optical images of the Helix, and indicates a massive remnant with $\gtrsim 50\%$ the mass of the ionized nebula. High resolution CO mapping with the IRAM 30 m telescope shows that the whole envelope is fragmented into an intricate array of small clumps, closely related to the cometary globules seen in the central, ionized cavity. 5–17 μm spectroscopy of the Helix with ISOCAM reveals a remarkable near infrared spectrum, dominated by the pure ($v = 0-0$) rotational lines of H_2. The H_2 lines are excited to a temperature of $\sim$ 900 K, and likely arise in warm, outer layers of the small clumps seen in CO. Imaging of the H_2 emission with ISOCAM over the whole nebula provides a striking portrait of the fragmented neutral envelope. 3-dimensional views of the envelope are also presented, based on CO mapping and using 3-dimensional visualization techniques. Point symmetries dominate the toroidal structure, and suggest an origin for the Helix in equatorial mass-loss on the AGB, shaped by the action of bipolar outflows or jets.

1. Introduction

This paper is concerned with the final evolutionary stages of the circumstellar envelopes of AGB stars and their relation to the formation of planetary nebulae (PNe). It is well established from survey observations in CO that neutral remnants of the envelopes persist well into the ionized nebula phase in a significant number of PNe (Huggins et al. 1996). Here we focus on one example, the Helix nebula, NGC 7293. At a distance of $\sim$ 150 pc, the Helix is easily the nearest PN with a massive remnant envelope, and is therefore an important case for detailed study. We present new views of the Helix in molecular emission lines of CO and H_2, which highlight the structure of the envelope and its role in the formation of the nebula.

The general appearance of the Helix is familiar from optical images of the ionized nebula which can be found in many popular astronomy texts. For comparison with views at other wavelengths we include in Fig. 1 (left panel) an R-band image, where the dominant line contributions are from Hα and [N II]. The diameter of the main nebula is $\sim$ 1 pc, and because of its proximity, the angular size is extremely large $\sim 1\,000''$. The expansion time scale from the kinematics and the size is $\sim 8\,000$ yr. The central star is now well evolved from the AGB, with $T_* \sim 120\,000$ K and $L_* \sim 100\ L_\odot$.

If one compares the Helix nebula with a standard AGB envelope, it has obviously undergone dramatic changes in the several thousand years since the central star left the AGB. For example, the envelope of the AGB star IRC+10216 (which is at about the same distance from Earth as the Helix) is almost completely molecular, and the gas density and molecular emission decrease rapidly away from the central star (see Lucas & Guélin, this volume). The most extended emission is seen in CO, which can be detected at low levels out to $\sim 200''$ (Huggins et al. 1988). In contrast, in the Helix, the center is fully ionized, the bulk of the gas has moved much farther from the central star, and the density peaks in the periphery. Nevertheless, a substantial part of the circumstellar gas in the Helix still remains in neutral form.

2. Large Scale Structure in CO

The first new view of the neutral envelope of the Helix comes from a large scale mapping project (Young et al. 1999) carried out with the 10.4 m telescope of the Caltech Sub-millimeter Observatory (CSO) in the 230 GHz CO ($J = 2 - 1$) line. 3425 spectra have been used to map the entire face of the nebula, at a resolution of $31''$ (with $15''$ sampling). The map of the velocity integrated CO line intensity is shown in Fig. 2 (left panel).

The large scale structure of the envelope seen in CO shows the double ring or "helix" morphology familiar from optical images (see Fig. 1), and when they are superposed they show some striking similarities. It is evident that the ionized gas abuts the neutral gas, and that the ionized nebula has formed through photo-ionization of the surface of the neutral envelope by the radiation field of the central star. By the same line of reasoning, it is evident that the shape of the neutral envelope is a key to understanding the morphology of the ionized nebula.

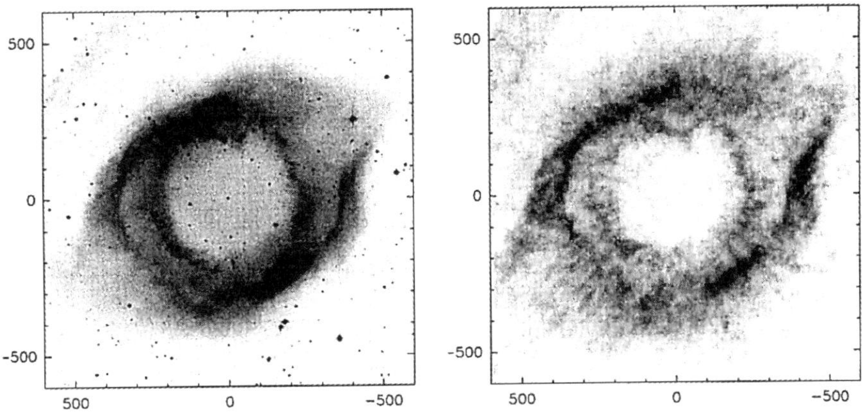

Figure 1. Images of the ionized and neutral components of the Helix nebula. *Left panel:* R-band image of the ionized gas. *Right panel:* ISOCAM LW2 (5–8.5 μm) image, dominated by the pure rotational lines of H_2. The offsets are in arcseconds from the center.

The complete CO map is also useful in refining earlier estimates of the mass of the neutral envelope (Huggins & Healy 1986). The flux can be used to determine the number of CO molecules in the envelope, and with an appropriate CO/H abundance ratio leads to an estimate of the mass. For a CO/H ratio in the molecular gas equal to the C/H ratio in the ionized gas, the mass of the envelope is $\gtrsim$ 10% of the ionized mass. A rough correction for the presence of carbon in other forms can be obtained from the recent detection of C I emission in the envelope (Young et al. 1997) which increases the mass estimate of the envelope to $\gtrsim$ 50% of the ionized mass. Thus the envelope is an important dynamical component of the nebula, as well as a massive reservoir of neutral gas, for much of the evolution of the nebula.

3. Small Scale Structure in CO

The second view of the Helix explores the smaller scale structure of the envelope in more detail. Using the IRAM 30 m telescope we have mapped a $200'' \times 300''$ region of the west limb in the CO ($J = 2 - 1$) line, at an angular resolution of $12''$ (with $5''$ pixels) and a velocity resolution of 0.2 km s^{-1}. The linewidth of the whole nebula due to expansion is $\sim$ 60 km s^{-1}.

One example of the CO velocity channel maps of this region is shown in Fig. 2 (right panel). It can immediately be seen that the gas is highly structured into small clumps, most of which are not resolved by the telescope beam. In addition, in other velocity channel maps of the same area offset by more than 1–2 km s^{-1}, the substructure is completely different, indicating very narrow widths to the components. Rough estimates of the masses of the clumps are of order $\sim 10^{-4}\,M_{\odot}$. Similar clumping is seen in a second field mapped to the North of the central star, so the whole envelope appears to be fragmented into clumps.

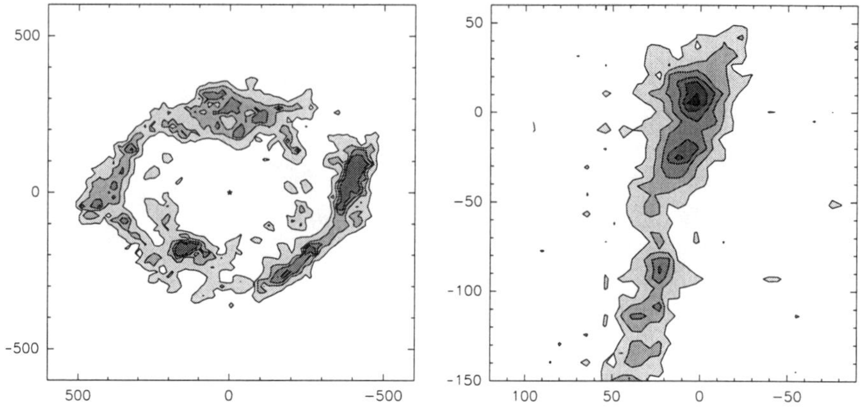

Figure 2. The neutral envelope of the Helix nebula in the CO (2–1) line. *Left panel:* Large scale structure in velocity integrated intensity from CSO. *Right panel:* Small scale structure: channel map 0.2 km s^{-1} wide from the IRAM 30 m telescope, centered on the west limb.

These structures are obviously close cousins of the well known cometary globules found in the ionized cavity of the Helix (for recent optical images, see Meaburn et al. 1998; O'Dell & Handron 1996). The fact that we previously found the cometary globules to have cores of molecular gas (Huggins et al. 1992) seals the relationship. Thus the cometary globules require no very special formation scenario, but are rather the innermost remnants of the fragmented neutral envelope.

4. Molecular Hydrogen

The third view of the Helix which complements those seen in CO has been made possible by the ISOCAM instrument on board ISO (Cox et al. 1998). Spectroscopy of the Helix with the CVF (5–16.6 μm) reveals that the near infrared spectrum is dominated by the pure rotational ($v = 0 - 0$) lines of molecular hydrogen. As one moves from the ionized cavity to the neutral envelope, the spectrum changes from one with weak ionized lines, to being dominated by relatively strong lines of H_2, from S(2) to S(7) (see Cox et al. 1998, Fig. 3). The ortho-para alternation can be seen in the spectra, and from plots of the line intensities vs. excitation levels, the temperature of the gas can be determined. In the small regions studied, the temperature is $\sim$ 900 K. The H_2 is much warmer than the CO, which is typically $\sim$ 20–50 K in evolved nebulae (Bachiller et al. 1997), and has a lower column density than the cooler gas, so it is probably present in thin, warm, outer layers around the cool clumps.

The H_2 line emission is strong enough to completely dominate the ISOCAM LW2 (5–8.5 μm) filter so it has also been possible to image the large scale H_2 emission over the whole nebula, with a resolution of 6″. The result is shown in Fig. 1 (right panel). The image confirms the basic picture seen in CO and the higher resolution shows finer details, including clumps of globules around the

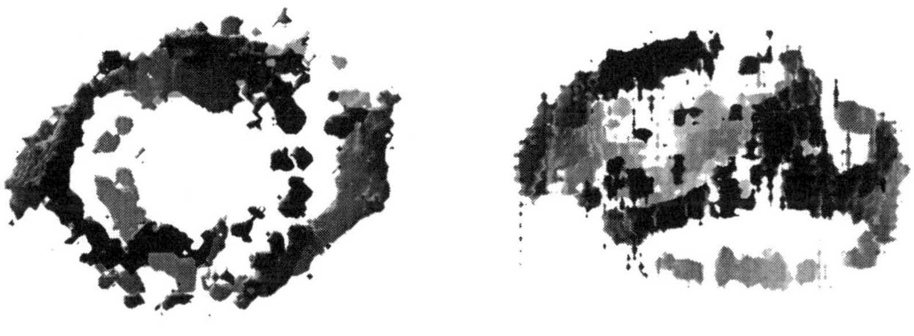

Figure 3. 3-D views of the Helix envelope. A surface contour of the CO data cube is illuminated diffusely from the rear. *Left panel:* front view. *Right panel:* view from below.

inner periphery and flocculent, radial rays extending to large distances. Thus it dramatically underscores the fragmented nature of the shell.

5. 3-D Structure

The final view of the Helix presented here is a 3-dimensional one. The global geometry is of particular interest for understanding the shaping mechanisms in PNe, and the morphology of the apparent helix has long been a target for model building (e.g., Carranza et al. 1968; Taylor 1977; Huggins & Healy 1986).

If we return to the large scale CO observations from the CSO, they contain information on the third dimension through the velocities of the CO components in the spectra. For a homologous radial expansion, expected to dominate in evolved PNe, the observed velocity difference of each component from the systemic velocity is proportional to the displacement from the plane of the sky along the line of sight. Thus the data cube can be used to obtain an approximate 3-dimensional picture of the envelope. In order to visualize the data, a surface contour has been constructed in the 3-dimensional data cube at 10% of the peak emission, and is illuminated from the rear with diffuse light to show the structure.

The left panel in Fig. 3 shows the front view of the surface. It resembles in outline the conventional map in Fig. 2, but with the illumination the 3-dimensional structure begins to become apparent. The darker sections are nearer the observer. An interesting aspect of this approach to visualization is that one can rotate the Helix to view the structure from different directions. Turned to the east or west, the general appearance is that of a distorted torus. The structure is more apparent on turning it through 90° to view it from below (Fig. 3, right panel). These views show the Helix torus includes a ring, with arcs blown out to each side. The structure also shows a remarkable point symmetry about the center: for each feature in the structure, there is a similar one on the opposite side of the central star.

The likely origin of this morphology is mass loss on the AGB towards a preferred equatorial plane, forming the basic toroidal structure, followed by the impact of off-axis, bipolar outflows or jets on the envelope. A striking example of the interaction of bipolar jets with a molecular torus has recently been reported in the young PN KjPn 8 by Forveille et al. (1998), and the Helix may well be a more evolved example of such a system.

6. Conclusions

The views of the Helix presented here illustrate the importance of remnant, neutral AGB envelopes in the evolution of PNe. The envelope of the Helix is not only a massive reservoir of neutral gas for the development of the ionized nebula, but is also a key element in both the large and small scale structural features of the nebula. The dominance of the neutral envelope in this and in many similar PNe, also means that the morphology of these nebulae is largely to be understood as the result of processes which shape the neutral envelopes.

Acknowledgments. It is a pleasure to thank the following collaborators on various parts of this work: F. Boulanger, D. Cesarsky, A.P. Jones, B. Lefloch, P.R. Rolfsema, and A.G.G.M. Tielens. This work was supported in part by NSF grant AST96-17941 (to P.J.H.). CSO is funded by the NSF under contract AST96-15025.

References

Bachiller R., Forveille T., Huggins P.J., Cox P., 1997, A&A 324, 1123

Carranza G., Courtès G., Louise R., 1968, in Planetary Nebulae, eds. D.E. Osterbrock & C.R. O'Dell, Reidel, Dordrecht, p. 249

Cox P., Boulanger F., Huggins P.J., Tielens A.G.G.M., Forveille T., Bachiller R., Cesarsky D., Jones A.P., Young K., Roelfsema P.R., Cernicharo J., 1998, ApJ 495, L23

Forveille T., Huggins P.J., Bachiller R., Cox P., 1998, ApJ 495, L111

Huggins P.J., Healy A.P., 1986, ApJ 305, L29

Huggins P.J., Bachiller R., Cox P., Forveille T., 1992, ApJ 401, L43

Huggins P.J., Bachiller R., Cox P., Forveille T., 1996, A&A 315, 284

Huggins, P.J., Olofsson H., Johansson L.E.B., 1988, ApJ 332, 1009

Meaburn J., Clayton C.A., Bryce M., Walsh J.R., Holloway A.J., Steffen W., 1998, MNRAS 294, 61

O'Dell C.R., Handron K.D., 1996, AJ 111, 1630

Taylor K., 1977, MNRAS 181, 475

Young K., Cox P., Huggins P.J., Forveille T., Bachiller R., 1997, ApJ 482, L101

Young K., Cox P., Huggins P.J., Forveille T., Bachiller R., 1999, in preparation

Asymptotic Giant Branch Stars
T. Le Bertre, A. Lèbre and C. Waelkens, eds.
I.A.U. Symposium 191, 1999

High-resolution, near-IR spectroscopy and imaging of the Egg and Rotten Egg nebulae (AFGL 2688 and OH 231.8+4.2)

Joel H. Kastner, LeeAnn Henn

Massachusetts Institute of Technology, NE80-6007, Cambridge, MA 02139 USA

David A. Weintraub

Vanderbilt University, Nashville, TN 37235 USA

Ian Gatley

Rochester Institute of Technology, Rochester, NY 14623 USA

Abstract.

Velocity-resolved 2.12 μm H_2 maps of the Egg Nebula (AFGL 2688), obtained with the NOAO Phoenix spectrometer, elaborate on previous observations of large velocity gradients in molecular emission both along and perpendicular to the polar axis of the system. The measured gradients along the polar axis support the notion that the polar H_2 emission regions are formed in shocks as fast, collimated winds collide with material ejected while the central star was still on the AGB. The kinematics of H_2 emission along the equatorial plane, meanwhile, are consistent with a model combining spherical expansion with rotation about the polar axis, although a model invoking multiple outflow axes cannot be ruled out.

We also obtained the first direct near-infrared images of QX Pup, a Mira variable embedded within the evolved bipolar nebula OH 231.8+4.2. The inferred absolute K magnitude of the central Mira appears "normal" given its $\sim$ 700 day period, which is remarkable in light of its position at the heart of such an unusual object.

1. Near-IR H_2 velocity mapping of the Egg Nebula (AFGL 2688)

The Egg Nebula (AFGL 2688) is among the best-studied examples of objects in transition from AGB star to planetary nebula. Recent HST images of the Egg reveal an extraordinary juxtaposition of circularly and axially symmetric structures (Sahai et al. 1998a, b). Perhaps most remarkable is the quadrupolar morphology of 2.122 μm H_2 emission as seen in an HST/NICMOS image (Fig. 1; Sahai et al. 1998a). The NICMOS image shows that the linear extents and bow-shock-like structures of the polar and equatorial plane H_2 emission regions are quite similar. The sharp, limb-brightened boundary of the H_2 emission is suggestive of a sudden event in the object's recent past.

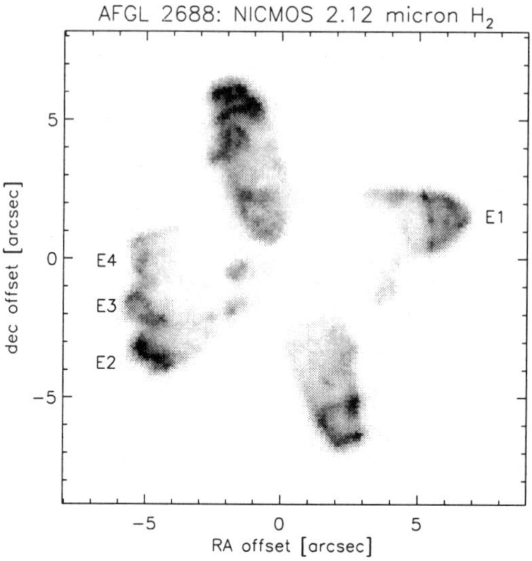

Figure 1. HST/NICMOS image of 2.12 μm H_2 emission from AFGL 2688 (Sahai et al. 1998a). Features identified E1–E4 by Sahai et al. are indicated.

The kinematics of molecular emission from AFGL 2688 have been the subject of numerous investigations. Interferometric observations of HCN by Bieging & Nguyen-Q-Rieu (1988, 1996) yielded the rather surprising result of velocity gradients both along *and perpendicular to* the polar axis of the system, a result confirmed by Cox et al. (1997) via a kinematic study of the 2.12 μm H_2 line. These gradients, which are of similar magnitude, were interpreted by Bieging & Nguyen-Q-Rieu (1988, 1996) as arising out of a combination of spherical expansion and rotation about the polar axis. Cox et al., however, intepret the H_2 kinematics in terms of a system of multiple jets.

1.1. Observations

To investigate the H_2 velocity gradients in detail, we obtained a series of high-resolution, long-slit spectra of AFGL 2688 in the 2.12 μm region with the NOAO[1] Phoenix near-infrared spectrometer (Hinkle et al. 1998). Data presented here were obtained with Phoenix on the 2.1 m telescope at Kitt Peak, AZ, in 1997 June. Phoenix illuminates a 256 $\times$ 1024 section of an Aladdin InSb detector array; the spectrograph slit was $\sim 30'' \times 1.2''$ oriented approximately east-west. From the widths of OH airglow lines present in the raw spectra, we estimate the velocity resolution as $\sim$ 4 km s^{-1}. This resolution is comparable to that of the mm-wave molecular line interferometry (e.g., Bieging & Nguyen-Q-Rieu 1996), and represents an order of magnitude improvement over previous imaging

[1]National Optical Astronomy Observatories is operated by Associated Universities for Research in Astronomy, Inc., for the National Science Foundation.

spectroscopy of H_2 emission from AFGL 2688 (Cox et al. 1997). The spatial resolution was $\sim 2''$. A spectral image centered near the 2.1218 μm $S(1)$, $v = 1-0$ transition of H_2 was obtained at each of 13 spatial positions as the slit was stepped from south to north across AFGL 2688. The step size, $1.0''$, provided coverage of the entire H_2 emitting region (Fig. 1) with spatial sampling approximating the slit height. Integration times were 1200 sec per position. Spectral images were reduced and wavelength calibrated as described in Weintraub et al. (1998). The reduced images were then stacked in declination to produce a (RA, dec, velocity) data cube of H_2 emission.

1.2. Results

In Fig. 2 we display selected velocity planes from the AFGL 2688 H_2 data cube. The four principal "lobes" of H_2 emission are apparent, with one pair oriented parallel to the polar axis (roughly N-S) and one perpendicular (roughly E-W). To first order, these pairs display velocity gradients of similar magnitude, with the N and E lobes blueshifted by up to ~ 30 km s^{-1} and the S and W lobes similarly redshifted. In detail, however, the N-S and E-W H_2 lobe pairs differ in their kinematic signatures. The polar H_2 lobes each display velocity gradients along the polar axis. This velocity gradient is clearly delineated in a plot of the velocity centroid of H_2 emission as a function of position along the polar axis (Fig. 3). Relatively low velocity emission is seen at the tips of the lobes, while the highest velocity emission in each lobe is found deep in its interior, nearest the position of the central star.

In contrast, the E lobe displays a gradient from north to south, *perpendicular* to the line joining it to the central star's position. Hence, the Phoenix data appear to resolve kinematically the spatially distinct equatorial plane emission components resolved by the NICMOS H_2 image of AFGL 2688 (Fig. 1). The W lobe, meanwhile, does not reveal resolvable velocity structure, consistent with its highly localized appearance in the NICMOS image. In Table 1, we list radial velocities (with respect to the systemic velocity of AFGL 2688, -30 km s^{-1}; e.g., Bieging & Nguyen-Q-Rieu 1996) measured for specific H_2 emission knots identified by Sahai et al. (1998a) in the E and W lobes.

Table 1. Radial velocities of H_2 equatorial emission knots

	west lobe	east lobe		
	E1	E2	E3	E4
Measured (km s^{-1})	$+7 \pm 2$	-13 ± 2	-9 ± 2	-3 ± 2
Model (km s^{-1})	$+7$	-12	-7	0

1.3. Discussion

The morphology of near-IR H_2 emission from AFGL 2688 strongly supports previous suggestions that the emission arises in shocks formed by the interaction of recently-developed, fast-moving winds with slower-moving material (Sahai et al. 1998a). This relatively slow-moving material likely was ejected during the AGB phase of the progenitor star. The deprojected expansion velocities at the tips of the N and S H_2 lobes (~ 20 km s^{-1}; Fig. 3) are typical of AGB

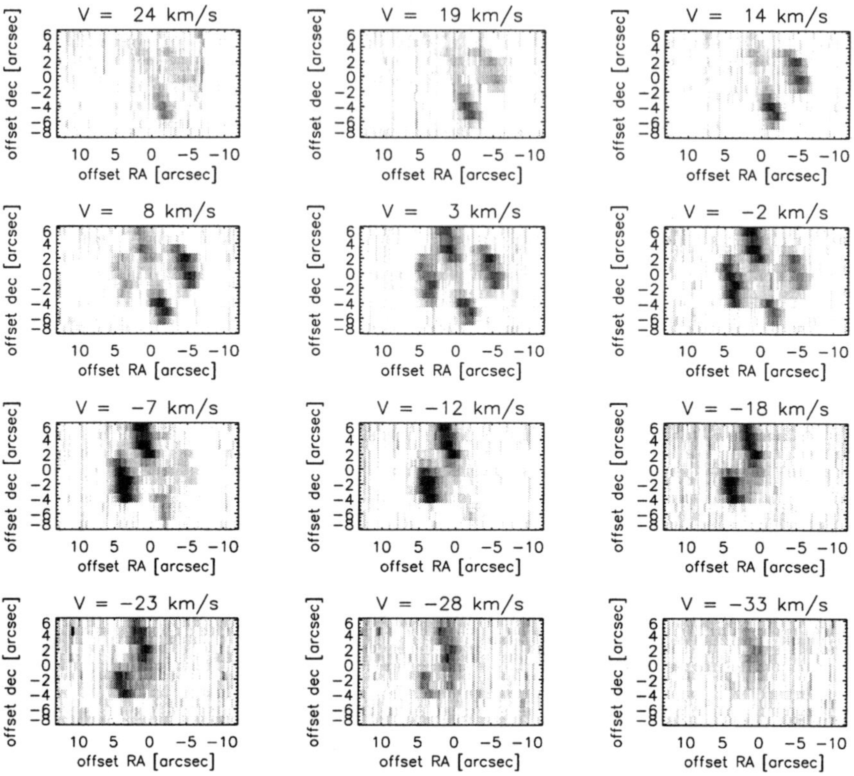

Figure 2. Velocity-integrated H_2 images of AFGL 2688, extracted from the Phoenix data cube. Each image is integrated over 5 km s^{-1}. Indicated radial velocities are relative to the systemic velocity of AFGL 2688.

star winds, bolstering this interpretation. Furthermore, the velocity gradients along the polar axis appear to trace a smooth, albeit rapid, transition from the "slow," spherically symmetric, AGB wind to a faster, collimated, post-AGB wind (Fig. 3). Perhaps remnant AGB ejecta that are closer to the central star are accelerated with increasing efficiency, as the star is "unveiled" and previously ejected dust is exposed to a progressively hotter photosphere (see, e.g., Gottlieb & Liller 1976). Such conditions would appear to be conducive to the formation of H_2 shocks during the post-AGB, pre-planetary nebula phase of bipolar nebulae, a scenario consistent with the conclusions of Weintraub et al. (1998). Assuming non-dissociative shocks, the shock fronts at the H_2 lobe tips must be traveling at < 50 km s^{-1} in the observer's frame. The projected displacement from the lobe tips to the central star (~ 6000 AU) then suggests the post-AGB "unveiling" of the central star began at least ~ 600 yr ago. Fig. 3 suggests that, subsequent to this event, the gas outflow velocity has increased by a factor of ~ 4.

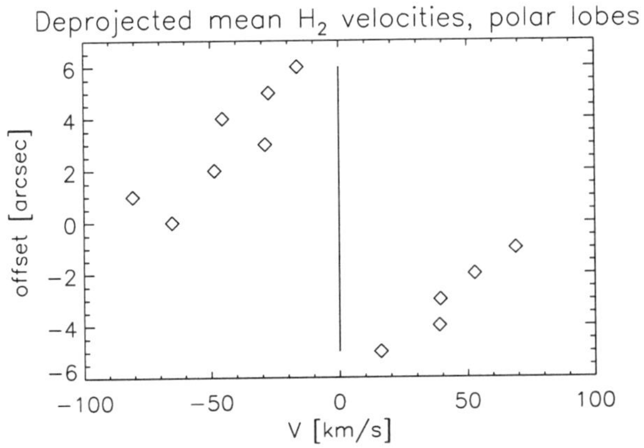

Figure 3. Deprojected radial velocity centroid, with respect to systemic velocity, of H_2 emission vs. position along the polar axis of AFGL 2688. Velocities have been deprojected assuming the polar axis is inclined by 15° out of the plane of the sky.

The different natures of the velocity gradients along and perpendicular to the N-S reflection lobes (§1.2) rule out the model proposed by Skinner et al. (1997), in which the polar axis is tilted by 45° with respect to these structures. We then are left to choose between two (potentially fatally flawed) alternative kinematic models to explain the velocity gradients projected along the equatorial plane:

(1) Multipolar outflows (e.g., Cox et al. 1997). The Phoenix and NICMOS data tightly constrain models in which the gradients parallel and perendicular to the polar axis result from a multipolar system of jets. Specifically, the fan-like eastern H_2 emission region would be directed well "above" the equatorial plane of the system, toward the observer, while the western H_2 emission region is more tightly confined and directed "below" the equatorial plane, away from the observer. Such a model geometry seems incompatible with the similar intensities of knots E1 and E3 (Fig. 1). Furthermore, in this model, the striking near-orthogonality of the four principle lobes of H_2 emission is merely a result of our fortuitous line of sight.

(2) Expanding, rotating equatorial disk (e.g., Bieging & Nguyen-Q-Rieu 1996). The Phoenix data are consistent with this model, if one postulates that all of the east-west H_2 emission is actually confined to a narrow region along the equatorial plane. The detailed H_2 morphology in the NICMOS image (Fig. 1) supports this hypothesis. In this case, one can adequately reproduce the velocities measured for knots E1 through E4 given a simple model consisting of rotation about the polar axis at ~ 7 km s^{-1} plus expansion along the equatorial plane at ~ 10 km s^{-1} (Table 1). The Achilles heel of this model is the implied angular momentum which, taken at face value, requires an unreasonably massive central star or an exceedingly close, massive binary companion. Bieging & Nguyen-Q-Rieu (1988, 1996) have speculated that the angular momentum instead might be supplied by very large magnetic fields, perhaps generated by a companion[2].

2. Diffraction-limited near-IR imaging of OH 231.8+4.2

The evolved bipolar nebula OH 231.8+4.2 is host to a Mira, QX Pup, at its core (Kastner et al. 1992). We used the Diffraction-Limited Near-infrared Imaging (DLIRIM) system at Kitt Peak to obtain the first direct images of QX Pup in the near-infrared (Kastner et al. 1998). In subarcsecond resolution DLIRIM images at K (2.2 μm) and L$'$ (3.8 μm), the star lies midway between the lobes of the OH 231.8+4.2 reflection nebula, confirming previous inferences based on polarimetric imaging and nebular colors (Kastner & Weintraub 1995). The inferred absolute K magnitude of the star (~ -10.2) is comparable to Mira variables of similar ($\sim$ 700 days) period in the Large Magellanic Cloud. In this and other respects, the central Mira appears remarkably "normal" given its position at the heart of such an unusual object.

Acknowledgments. The authors wish to acknowledge conference participants John Bieging, Michael Jura, Raghvendra Sahai, and Noam Soker for their stimulating questions and incisive comments.

References

Bieging J., Nguyen-Q-Rieu, 1988, ApJ 324, 516

Bieging J., Nguyen-Q-Rieu, 1996, AJ 112, 706

Cox P., et al., 1997, A&A 321, 907

Gottlieb E.W., Liller W., 1976, ApJ 207, L135

Hinkle K.H., Cuberly R., Gaughan N., Heynssens J., Joyce R., Ridgway S., Schmitt P., Simmons J.E., 1998, Proceedings of S.P.I.E. 3354, 810

Kastner J.H., Weintraub D.A., 1995, AJ 109, 1211

Kastner J.H., Weintraub D.A., Merrill K.M., Gatley I., 1998, AJ 116, 1412

Kastner J.H., Weintraub D.A., Zuckerman B., Becklin E.E., McLean I., Gatley I., 1992, ApJ 398, 552

Sahai R., Hines D., Kastner J.H., Weintraub D.A., Trauger J.T., Rieke M.J., Thompson R.I., Schneider G., 1998a, ApJ 492, L163

Sahai R., et al., 1998b, ApJ 493, 301

Skinner C.J., et al., 1997, A&A, 328, 290

Weintraub D.A., Huard T., Kastner J.H., Gatley I., 1998, ApJ, in press

---

[2] A widely separated companion recently has been detected (Sahai et al. 1998a), but this star is unlikely to be the required source of angular momentum or magnetic field activity.

Asymptotic Giant Branch Stars
T. Le Bertre, A. Lèbre and C. Waelkens, eds.
I.A.U. Symposium 191, 1999

AGB stars in binaries and their progeny

A. Jorissen
Institut d'Astronomie et d'Astrophysique, Université Libre de Bruxelles C.P. 226, Boulevard du Triomphe, B-1050 Bruxelles, Belgium

Abstract. An AGB star in a binary system is likely to pollute its companion with carbon- and s-process-rich matter. After the AGB star has faded into an unconspicuous white dwarf, the polluted companion enters the zoo of stars with chemical peculiarities. In this paper, the progeny of AGB stars in binary systems are identified among existing spectroscopic classes (Abell 35-like, binary post-AGB, WIRRing, dwarf Ba and C, subgiant CH, Ba, CH, S, yellow symbiotics) and their filiation is discussed from the properties of their eccentricity – period diagrams.

1. Introduction

Although binarity and AGB evolution are in principle disconnected concepts (a star ought not to be member of a binary system to evolve along the AGB!), a rich world flourishes at their contact. Its existence was first suggested by McClure et al. (1980; see also McClure 1984 and McClure & Woodsworth 1990) in relation with the discovery of the binary nature of barium stars. Overabundances of Ba, Sr and other heavy elements produced by the s-process are observed at the surface of these G and K giants. Their origin remained a mystery until McClure suggested that the envelopes of barium stars may have been polluted by matter accreted from a former AGB companion (being now a dim white dwarf).

That scenario raised at first several questions (Is the current companion of the barium star really a white dwarf? Where are the predicted dwarf barium stars? What is the mass transfer mode? Is wind accretion efficient enough? ...) that have now largely and satisfactorily been answered (see the recent reviews by Jorissen & Van Eck 1998, and Jorissen et al. 1998). The mass-transfer scenario appears to be of great potential to unify the zoo of red stars with chemical peculiarities, since many such families now fit within that scenario as reviewed in Sect. 2.

2. The mass-transfer scenario and the zoo of peculiar red stars

Fig. 1 depicts the full evolutionary sequence involving mass-transfer from an AGB star in a binary system (phase 5). As indicated in Sect. 1, it was originally introduced to account for the chemical peculiarities exhibited by barium stars (phase 11) which could not be understood in the framework of internal nucleosynthesis in a single star. As reviewed in this section, much progress has been made since, and each phase of this binary evolutionary sequence may now be

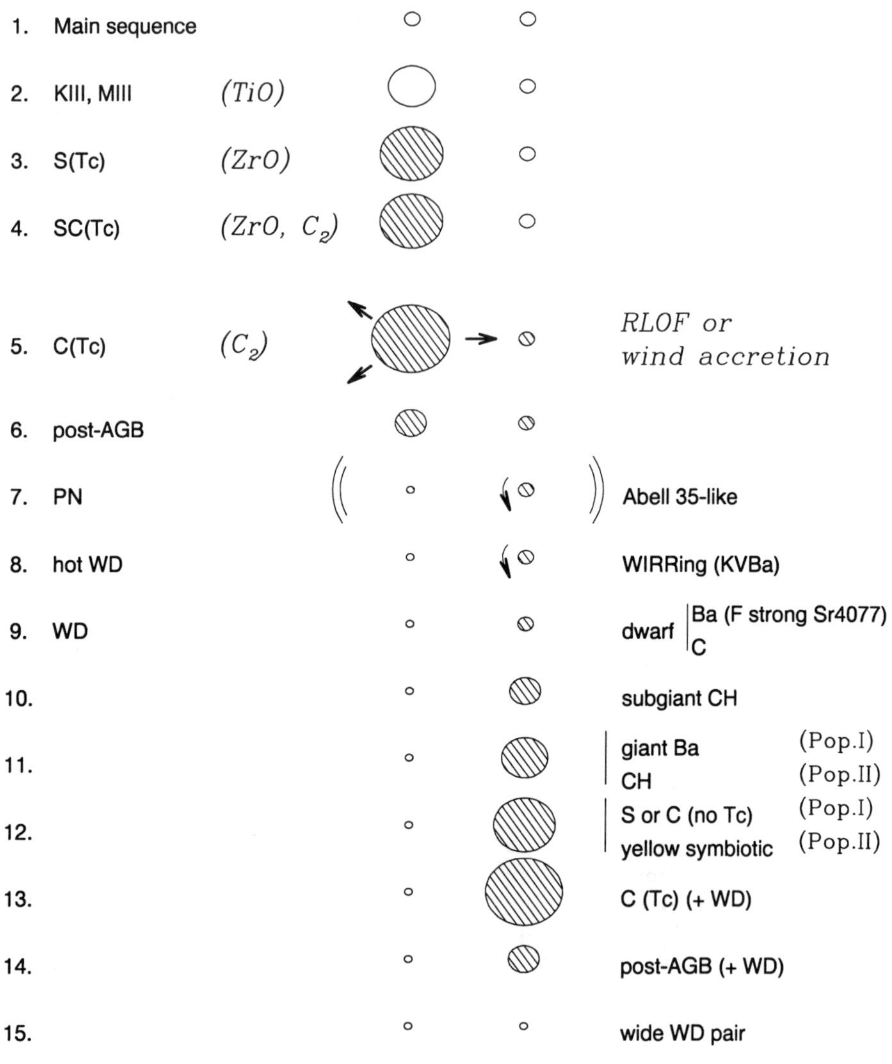

Figure 1. Binary evolution involving mass transfer from an AGB star (phase 5). Spectral families not requiring binarity (i.e. the normal evolutionary sequence for *single* stars) are listed in the left column, whereas the right column identifies classes of PRS *requiring binarity*. Stars with excess C and Ba at their surface are represented by hatched disks

associated with an existing class of peculiar red stars (PRS). This assignment requires both the typical chemical signature of pollution from an AGB star (i.e. excess C and Ba) to be found in the PRS, and the PRS family to consist exclusively of binary stars.

Phase 2: Normal giants in binary systems. Catalogues of binary systems involving giant stars are provided by Boffin et al. (1993) for field stars and by Mermilliod (1996) for open-cluster stars. The pre-mass-transfer status of these systems is generally not known for sure on an individual basis (since most are SB1 systems, leaving the main-sequence or WD nature of the companion uncertain). However, the comparison of their orbital elements with those of the barium stars reveals clear differences (further discussed in Sect. 3.) which suggest that most of the G-KIII systems are indeed pre-mass-transfer systems.

Phases 3-5: S(Tc)-SC(Tc)-C(Tc). Binarity is not required for these PRS exhibiting lines of Tc, an element with no stable isotopes (Little et al. 1987), as they represent the normal evolutionary sequence on the AGB (Iben & Renzini 1983).

Phase 5: Mass transfer from a C-rich AGB star. The pollution of the companion occurs at this phase, either through wind accretion (Boffin & Jorissen 1988; Han et al. 1995; Theuns et al. 1996; Karakas et al. 1998) or through Roche lobe overflow (RLOF) (Iben & Tutukov 1985; Webbink 1986). The efficiency of wind accretion seems to be sufficient to account for the longest periods observed among PRS (about 30 y, see Fig. 2 and Jorissen et al. 1998), and is moreover consistent with the non-circular nature of the orbits of those post-mass-transfer systems (Karakas et al. 1998). The occurrence of RLOF seems however unavoidable in the systems with circular orbits and periods shorter than a few hundred days. That question is discussed further in Sect. 3.

Phase 6: Binary post-AGB. The binary post-AGB stars identified by Van Winckel et al. (1995 and this volume) correspond most likely to this evolutionary phase. The mass functions of these SB1 systems do not always allow to distinguish between a main sequence and a WD companion. Some systems might thus correspond to phase 14 instead, although the systems with the largest mass functions (> 0.2 $M_{\odot}$) very likely host main sequence companions. Most (though not all) binary post-AGB stars are very metal-deficient (Van Winckel et al. 1995), and this correlation has been interpreted as the result of re-accretion of gas depleted in refractory elements from a stable circumbinary disk (Waters et al. 1992 and this volume). Because of this chemical fractionation, the photospheric chemical content of these stars does not bear any more the signature of the third dredge-up. Whether the same (or the complementary) chemical fractionation alters as well the composition of the matter accreted by the companion (i.e. the future Ba star) is currently unknown.

Phase 7: Binary nucleus of Planetary Nebulae (Abell 35-like). Among planetary nebulae with binary nuclei, there is a class consisting of the three objects Abell 35 (BD−22°3467 = LW Hya), LoTr 1 and LoTr 5 (HD 112313 = IN Com = 2RE J1255+255) whose optical spectra are dominated by late-type (G-K) stars, but whose UV spectra indicate the presence of extremely hot ($> 10^5$ K), hence young, WD companions (see e.g. Bond & Livio 1990). The late-type star is chromospherically active and rapidly rotating. This rapid rotation is likely to result from some interaction between the binary components. Since the orbital periods are still uncertain (Jasniewicz et al. 1994; Jeffries & Stevens 1996 and references therein; Gatti et al. 1997), the evolutionary history of these systems cannot unfortunately be identified with certainty: close binaries emerging from a common envelope phase (Bond et al. 1993) or wide binaries whose accreting star has been spun up by wind accretion (Jeffries & Stevens 1996)? In any case, the detection of moderate overabundances of Ba, Y and Sr in the G stars of Abell 35 and LoTr 5 (Thévenin & Jasniewicz 1997; Gatti et al. 1997 do not confirm, however, that conclusion which is based on very few broad lines) indicates that these stars have been polluted by mass transfer from their former AGB companion. The similarity between these systems and the WIRRing systems described next may be in favour of a wide binary system. Furthermore, HST observations of Abell 35 with the Planetary Camera revealed a measurable extension (0.08 arcsec, translating into a projected separation of about 18 AU at the minimum estimated distance of 160 pc) of the nucleus at 300 nm (Gatti et al. 1998), thus ruling out the close-binary hypothesis. Another argument favouring wide systems is provided by HD 128220, a related system consisting of a sdO and a rapidly rotating G star, which is the only one among those systems having a reliably determined period (872 d; Howarth & Heber 1990).

Phase 8: WIRRing systems (rapidly-rotating KVBa + hot WD). The class of WIRRing stars ('Wind-induced rapidly rotating') stars has been introduced by Jeffries & Stevens (1996) to describe a small group of rapidly rotating, magnetically-active K dwarfs with hot WD companions uncovered by the ROSAT Wide Field Camera or EUVE surveys (Kellet et al. 1995; Jeffries et al. 1996). From the absence of short-term radial velocity variations, it may be concluded that these systems have periods of a few months at least. Moreover, several arguments, based on proper motion, WD cooling time scale, and lack of photospheric Li, indicate that the rapid rotation of the K dwarf cannot be ascribed to youth. Jeffries & Stevens (1996) suggest that the K dwarfs in these wide systems were spun up by the accretion of the wind from their companion, when the latter was a mass-losing AGB star. The possibility of accreting a substantial amount of spin from the companion's wind has since been confirmed by *Smooth Particle Hydrodynamics* simulations (Theuns et al. 1996). A clear signature that mass transfer has been operative in 2RE J0357+283 is again provided by the detection of excess barium in that star (Jeffries & Smalley 1996).

The class of WIRRing stars may also perhaps include the barium star HD 165141, since it shares the properties of RS CVn and barium systems (Jorissen et al. 1996). The very long orbital period of the system ($P \sim 4\,800$ d; Fekel, private communication) forbids the barium star from having been spun up by tidal effects as is the case for RS CVn systems. This system moreover hosts the hottest WD ($T \sim 35\,000$ K) among barium systems (Fekel et al. 1993). The cooling time scale of the WD is about 10 My, shorter than the magnetic braking time scale of the giant star (about 100 My; see Theuns et al. 1996). The mass transfer thus occurred recently enough for the magnetic braking having not yet slowed down the giant star in a substantial way. HD 165141 is probably one of the very few barium systems where the mass transfer from the former AGB companion occurred when the accreting star was already a giant.

Phase 9: Dwarf Ba/C stars. About a dozen dwarf C stars are now known (Liebert et al. 1979, 1994; Dearborn et al. 1986; Green et al. 1991, 1992, 1994; Warren et al. 1993; Heber et al. 1993), and more may be expected from on-going dedicated surveys (MacConnell 1997). Many dC stars appear to have halo kinematics. Their binary nature has not been demonstrated in all cases.

Dwarf Ba stars (some with $C/O > 1$) with orbital elements similar to those of the giant barium stars were identified by North & Duquennoy (1991, 1992) and North et al. (1994) among the stars classified as FV strong Sr $\lambda 4077$ in the Michigan spectral survey. The family of CH subgiants identified by Bond (1974) comprises stars near and above the main sequence (Luck & Bond 1991). The high occurrence of binaries among subgiant CH stars has been demonstrated by McClure (1997).

Phase 12: S(no Tc) or C(no Tc) in Pop.I, yellow symbiotics in Pop.II. The identification of the Tc-poor S stars as descendants of the barium stars was first suggested by Iben & Renzini (1983) and Little et al. (1987), and later confirmed from the similarity of their orbital elements by Jorissen et al. (1998; see Fig. 2 and Sect. 3.). As S(no Tc) stars consist of mass-losing giants and WD companions in systems with orbital periods similar to those of the symbiotic systems, the question of the relationship between these two families must be raised. This question is actually twofold: (i) Do S(no Tc) stars exhibit some symbiotic activity? (ii) Is the barium syndrome observed among symbiotic systems? These questions have been reviewed by Jorissen (1997), and their answers may be summarized as follows: (i) Some S(no Tc) stars indeed exhibit (weak) symbiotic activity (see below); (ii) no S stars are known among 'red' symbiotics (involving M giants), but all 'yellow' symbiotics (involving G or K giants) studied thus far are halo objects exhibiting the barium syndrome (Schmid 1994; Smith et al. 1996, 1997; Pereira et al. 1998). As the RGB is shifted towards the blue in low-metallicity populations, yellow symbiotics may be regarded as the Pop.II analogs of the S(no Tc) stars (Smith et al. 1996). The absence of S stars among red symbiotics is more puzzling, but most likely is a population/metallicity effect: carbon, barium and S stars seem difficult to form in a high-metallicity environment (Jorissen & Boffin 1992; Jorissen & Van Eck 1998), and carbon symbiotics are indeed frequent in the Magellanic Clouds (Mikołajewska 1997). A detailed comparison of the kinematic properties of red symbiotics and S(no Tc) stars may shed light on that question.

A recent high-resolution survey of Hα emission among the 40 binary S(no-Tc) stars from Henize's (1960) sample uncovered only 2 stars with strong, broad (base width ~ 400 km s^{-1}) and double-peaked Hα emission lines (Van Eck et al., in preparation) resembling the Hα profiles of symbiotic stars (of type S3; Van Winckel et al. 1993) and of Abell 35 (Fig. 2 of Acker & Jasniewicz 1990). The distinctive parameter responsible for the symbiotic activity in these two S stars is currently unknown.

The existence of two kinds of S stars (Tc and no Tc), as predicted by the mass-transfer paradigm (phases 3-5 vs. 12 on Fig. 1), is now well established (Jorissen et al. 1993; Groenewegen 1993; Jorissen & Van Eck 1998). Whether the same dichotomy applies to C-N stars as well is not yet firmly demonstrated, although Barnbaum (1993) found 16 Tc-poor stars in a sample of 78 C-N stars with Ba excess. However, their binary nature remains to be established, not an easy task given the confusion introduced by the envelope pulsation on the radial-velocity curve (Udry et al. 1998; Hinkle et al., poster contribution, this conference). Nevertheless, the recognition of this dichotomy is important for a correct interpretation of the luminosity functions of AGB stars in external systems. As shown by Van Eck et al. (1998) from HIPPARCOS data, S(no Tc) stars are on average less luminous than S(Tc) stars. They thus add a low-luminosity tail on the luminosity function of the genuine AGB S(Tc) stars. If these S(no Tc) [and possibly C(no Tc) stars] are not properly removed from the luminosity functions observed in external systems, they may lead to erroneous constraints on the luminosity threshold for the occurrence of the third dredge-up in AGB stars. This danger is best illustrated by the luminosity function of carbon stars in the SMC obtained by Westerlund et al. (1995), where a formerly unrecognized low-luminosity tail extends down to $M_{\rm bol} \sim -2$. This low-luminosity tail precisely matches the luminosity range of S(no-Tc) stars in the solar neighbourhood (see Figure 7 of Van Eck et al. 1998), possibly suggesting that the SMC may contain an important fraction of binary C(no Tc) stars (see also Wood et al., this volume, for a similar conclusion)!

3. The $(e, \log P)$ diagrams of peculiar red stars

Fig. 2 compares the $(e, \log P)$ diagrams for the families of binaries described in Sect. 2. Orbital elements for *G-KIII* stars in open clusters are taken from Mermilliod (1996), for *post-AGB* systems (along with the sdO+G system HD 128220) from Van Winckel (this volume) and Howarth & Heber (1990), for *dwarf Ba and subgiant CH* stars from North et al. (in preparation) and McClure (1997), respectively, for *Ba* and *S* systems from Jorissen et al. (1998), and for *CH* systems from McClure & Woodsworth (1990). *Yellow symbiotic* systems displayed in Fig. 2 include TX CVn, BD$-21°$3873, LT Del (=He2-467) and AG Dra, with orbital elements from various sources as listed by Smith et al. (1996).

The comparison of the $(e, \log P)$ diagrams of these various families allows to check their filiation, and to get clues on the mass transfer process.

The $(e, \log P)$ diagram of G-K giants in open clusters differs from all the others in two clear ways: (i) many cluster giants are found with $P < 200$ d, whereas such short systems are quite rare among all PRS classes; (ii) the average eccentricity at any given period is much smaller for the PRS, a fact already pointed out by Webbink (1986) for barium stars, but which now turns out to hold true for all the PRS classes. These differences between cluster giants and PRS are a clear indication that mass transfer (and possibly tidal interaction) have taken place in the PRS systems. Moreover, the non-zero eccentricities of most of the PRS systems argue against Roche lobe overflow (RLOF) as the mass-transfer mode, since this mode requires the mass-loser star to fill its Roche lobe, a situation where tidal interactions will very rapidly circularize the orbit. Wind accretion is therefore preferred, as investigated by Boffin & Jorissen (1988), Han et al. (1995) and Karakas et al. (1998). However, wind accretion by definition requires the system to remain always detached, but as discussed e.g. by Jorissen et al. (1998) and Karakas et al. (1998), this requirement may be difficult to satisfy in systems with periods as short as 300 d. RLOF seems thus

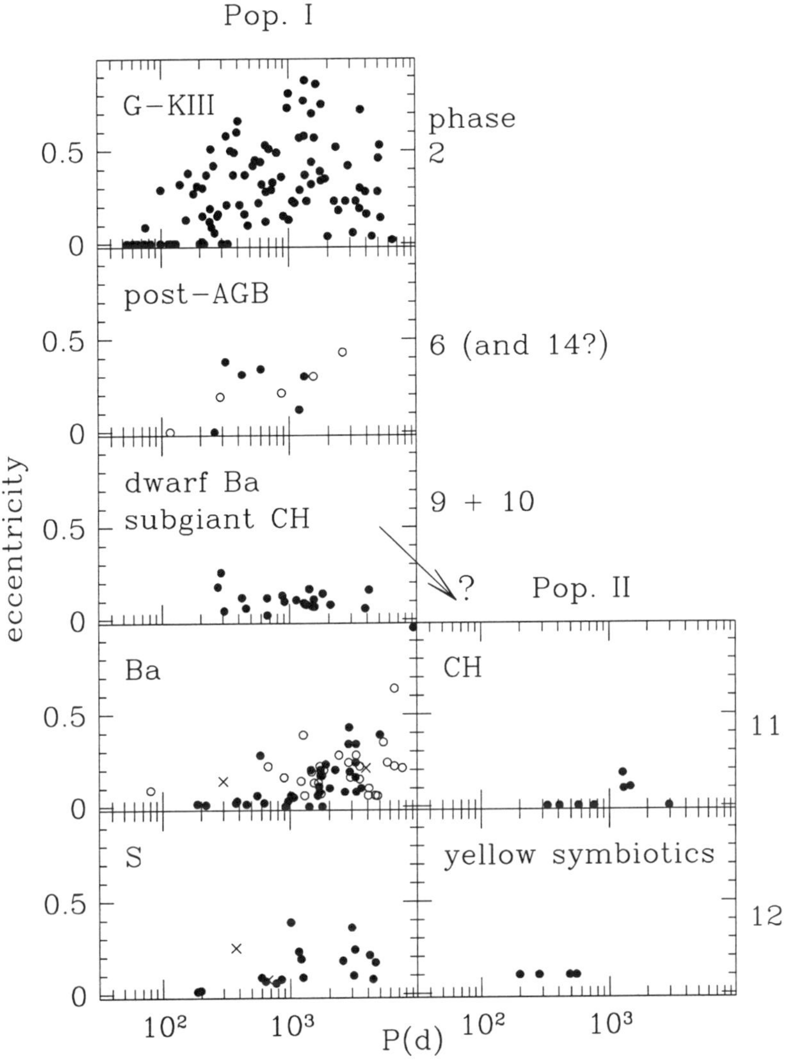

Figure 2. Comparison of the $(e, \log P)$ diagrams for the families described in Sect. 2., referred to by the phase label of Fig. 1 (written to the right of each panel). Left and right columns correspond to Pop.I and Pop.II systems, respectively. Post-AGB systems depleted in refractory elements are represented by black dots; systems not exhibiting the depletion pattern, or with no abundance data, are represented by open dots. The crosses in the Ba panel correspond to the two pairs of the triple system BD+38°118. Ba1,2,3 (i.e. mild Ba) and Ba4,5 (i.e. strong Ba) stars are represented by open and black dots, respectively. In the S panel, crosses correspond to two systems with unusually large mass functions. As the eccentricities of symbiotic systems are generally poorly determined, they have all been set to 0.1

unavoidable in these short-period systems with circular orbits (at least among Ba systems, but see below). That conclusion is troublesome, however, because RLOF from an AGB star with a deep convective envelope is generally believed to lead to a common envelope phase (at least when the mass-losing AGB star is the more massive component), and to a dramatic orbital shrinkage (e.g. Paczyński 1976; Meyer & Meyer-Hofmeister 1979; Iben 1995). The end product of such an evolution is a cataclysmic variable with an orbital period much shorter than the shortest among PRS systems. The very existence of PRS systems with periods too short to be accounted for by the wind accretion process indicates that these systems must have avoided the dramatic fate outlined above (see also Wood, this volume). Several ways out of this channel (which sometimes does operate, though, since cataclysmic variables *do* exist!) have been sketched by Jorissen & Van Eck (1998) and Jorissen et al. (1998; see also Han et al. 1995), but detailed simulations to confirm these ideas are badly needed.

The simple picture outlined above of two mass-transfer modes (RLOF for the short-period, circular orbits, and wind accretion for the long-period, eccentric orbits) that seems to emerge from the $(e, \log P)$ diagram of barium systems (after having excluded the triple system BD+38°118), runs into problems when confronted to the $(e, \log P)$ diagrams of post-AGB, dwarf Ba and subgiant CH systems. The latter systems have significantly non-zero eccentricities, even at short periods! The picture may in fact be complicated by the following two physical processes: (i) the interaction of a binary system with a circumbinary disk (a common feature in post-AGB systems; see e.g. Waters et al., this volume) has been shown to pump the eccentricity upwards (see e.g. Lubow & Artymowicz 1992); (ii) tidal circularization of the closest systems may occur when the Ba and S giants ascend the RGB, and account for the difference between the eccentricities of dwarf and giant Ba systems with the *shortest* periods.

A mismatch nevertheless remains for the widest systems, barium stars having larger eccentricities on average than CH subgiants. At this point, an observational selection effect against large eccentricities for CH subgiants with periods close to the time span of the monitoring cannot be ruled out as a possible origin of this mismatch (one long-period, high-eccentricity orbit is indeed pending in McClure's sample). An alternative possibility is that dwarf Ba and subgiant CH systems evolve into CH stars rather than into Ba stars. This second possibility is supported by the similarity of the eccentricity and mass function distributions of dwarf Ba stars, CH subgiants and CH stars (McClure & Woodsworth 1990; North et al. 1994, 1998; McClure 1997), and by the often large space velocities observed for CH subgiants (Bond 1974), which point towards subgiant CH stars belonging to a population intermediate between the halo CH systems and the old-disk barium systems.

Finally, a fully satisfactory match is observed between Ba and S systems, which are distributed identically in the $(e, \log P)$ diagram, thus confirming the hypothesis that they only differ in terms of effective temperature. The same conclusion is reached from the similarity of their mass-function distributions (Figs. 8 and 10 of Jorissen et al. 1998).

Acknowledgments. Pierre North and Hans Van Winckel are thanked for communicating orbital elements in advance of publication. A.J. is Research Associate, F.N.R.S. (Belgium).

References

Acker A., Jasniewicz J., 1990, A&A 238, 325

Barnbaum C., 1993, BAAS 182, #46.17

Boffin H.M.J., Jorissen A., 1988, A&A 205, 155

Boffin H.M.J., Cerf N., Paulus G., 1993, A&A 271, 125

Bond H.E., 1974, ApJ 194, 95

Bond H.E., Livio M., 1990, ApJ 355, 568

Bond H.E., Ciardullo R., Meakes M., 1993, in Planetary Nebulae, IAU Symp. 155, eds. R. Weinberger & A. Acker, Kluwer, Dordrecht, p. 397

Dearborn D.S.P., Liebert J., Aaronson M., Dahn C.C., Harrington R., Mould J., Greenstein J.L., 1986, ApJ 300, 314

Fekel F.C., Henry G.W., Busby M.R., Eitter J.J., 1993, AJ 106, 2370

Gatti A.A., Drew J.E., Lumsden S., et al., 1997, MNRAS 291, 773

Gatti A.A., Drew J.E., Lumsden S., et al., 1997, in Planetary Nebulae (IAU Symp. 180), eds. Habing H.J., Lamers H., Kluwer, Dordrecht, p. 105

Gatti A.A., Drew J.E., Oudmaijer R.D., et al., 1998, MNRAS 301, L33

Green P.J., Margon B., MacConnell D.J., 1991, ApJ 380, L31

Green P.J., Margon B., Anderson S.F., MacConnell D.J., 1992, ApJ 400, 659

Green P.J., Margon B., Anderson S.F., Cook K.H., 1994, ApJ 434, 319

Groenewegen M.A.T., 1993, A&A 271, 180

Han Z., Eggleton P.P., Podsiadlowski P., Tout C.A., 1995, MNRAS 277, 1443

Heber U., Bade N., Jordan S., Voges W., 1993, A&A 267, L31

Henize K.G., 1960, AJ 65, 491

Howarth I.D., Heber U., 1990, PASP 102, 912

Iben I., Jr., 1995, Phys. Rep. 250, 1

Iben I., Jr., Renzini A., 1983, ARA&A 21, 271

Iben I., Jr., Tutukov A.V., 1985, ApJS 58, 661

Jasniewicz G., Acker A., Mauron N., et al., 1994, A&A 286, 211

Jeffries R.D., Smalley B., 1996, A&A 315, L19

Jeffries R.D., Stevens I.R., 1996, MNRAS 279, 180

Jeffries R.D., Burleigh M.R., Robb R.M., 1996, A&A 305, L45

Jorissen A., 1997, in Physical Processes in Symbiotic Stars and Related Systems, ed. J. Mikołajewska, Copernicus Foundation for Polish Astronomy, Warsaw, p. 135

Jorissen A., Boffin H.M.J., 1992, in Binaries as tracers of stellar formation, eds. A. Duquennoy & M. Mayor, Cambridge Univ. Press, p. 110

Jorissen A., Frayer D.T., Johnson H.R., et al., 1993, A&A 271, 463

Jorissen A., Schmitt J.H.M.M., Carquillat J.M., et al., 1996, A&A 306, 467

Jorissen A., Van Eck S., 1998, in The Carbon Star Phenomenon, IAU Symp. 177, ed. R.F. Wing, Kluwer, in press

Jorissen A., Van Eck S., Mayor M., Udry S., 1998, A&A 332, 877

Karakas A.I., Tout C.A., Lattanzio J., 1998, MNRAS, in press
Kellet B.J., Bromage G.E., Brown A., et al., 1995, ApJ 438, 364
Liebert J., Dahn C.C., Gresham M., Strittmatter P.A., 1979, ApJ 233, 226
Liebert J., Schmidt G.D., Lesser M., et al., 1994, ApJ 421, 733
Little S.J., Little-Marenin I.R., Hagen Bauer W., 1987, AJ 94, 981
Lubow S.H., Artymowicz P., 1992, in Binaries as tracers of stellar formation, A. Duquennoy & M. Mayor (eds.), Cambridge Univ. Press, p. 145
Luck R.E., Bond H.E., 1991, ApJS 77, 515
MacConnell D.J., 1997, Baltic Astronomy 6, 105
McClure R.D., 1984, PASP 96, 117
McClure R.D., 1997, PASP 109, 536
McClure R.D., Fletcher J.M., Nemec J.M., 1980, ApJ 238, L35
McClure R.D., Woodsworth A.W., 1990, ApJ 352, 709
Mermilliod J.-C., 1996, in The Origins, Evolution, and Destinies of Binary Stars in Clusters, G.F. Milone & J.C. Mermilliod, ASP Conf. Ser. 109, p. 373
Meyer F., Meyer-Hofmeister E., 1979, A&A 78, 167
Mikołajewska J., 1997, in Physical Processes in Symbiotic Stars and Related Systems, ed. J. Mikołajewska, Copernicus Foundation for Polish Astronomy, Warsaw, p. 3
North P., Duquennoy A., 1991, A&A 244, 335
North P., Duquennoy A., 1992, in Binaries as tracers of stellar formation, eds. A. Duquennoy & M. Mayor, Cambridge Univ. Press, p. 202
North P., Berthet S., Lanz T., 1994, A&A 281, 775
North P., Jorissen A., Mayor M., 1998, in The Carbon Star Phenomenon, IAU Symp. 177, ed. R.F. Wing, Kluwer, in press
Paczyński B., 1976, in Structure and Evolution of Close Binary Systems, IAU Symp. 73, P.P. Eggleton, S. Mitton & J. Whelan (eds.), Reidel, p. 75
Pereira C.B., Smith V.V., Cunha K., 1998, AJ 116, 1977
Schmid H.-M., 1994, A&A 284, 156
Smith V.V., Cunha K., Jorissen A., Boffin H.M.J., 1996, A&A 315, 179
Smith V.V., Cunha K., Jorissen A., Boffin H.M.J., 1997, A&A 324, 97
Theuns T., Boffin H.M.J., Jorissen A., 1996, MNRAS 280, 1264
Thévenin F., Jasniewicz G., 1997, A&A 320, 913
Udry S., Jorissen A., Mayor M., Van Eck S., 1998, A&AS 131, 25
Van Eck S., Jorissen A., Udry S., Mayor M., Pernier B., 1998, A&A 329, 971
Van Winckel H., Duerbeck H.W., Schwarz H.E., 1993, A&AS 102, 401
Van Winckel H., Waelkens C., Waters L.B.F.M., 1995, A&A 293, L25
Warren S.J., Irwin M.J., Evans D.W., et al., 1993, MNRAS 261, 185
Waters L.B.F.M., Trams N.R., Waelkens C., 1992, A&A 262, L37
Webbink R.F., 1986, in Critical Observations vs. Physical Models for Close Binary Systems, eds. K.-C. Leung & D. Zhai, Gordon and Breach, p. 403
Westerlund B.E., Azzopardi M., Breysacher J., et al., 1995, A&A 303, 107

Asymptotic Giant Branch Stars
T. Le Bertre, A. Lèbre and C. Waelkens, eds.
I.A.U. Symposium 191, 1999

How Binary Stars affect Galactic Chemical Evolution

Christopher A. Tout

Institute of Astronomy, Madingley Road, Cambridge CB3 0HA, UK

Amanda I. Karakas

Dept of Mathematics, Monash University, Clayton, Vic. 3168, Australia

John C. Lattanzio

Dept of Mathematics, Monash University, Clayton, Vic. 3168, Australia

Jarrod R. Hurley

Institute of Astronomy, Madingley Road, Cambridge CB3 0HA, UK

Onno R. Pols

Instituto de Astrofísica de Canarias, Via Láctea, 38200 La Laguna, Tenerife, Spain

Abstract. At least 60% of stars appear to be binary and about half of these are close enough to interact. Because of the enormous expansion on the AGB, many of these interactions will involve an AGB star and a relatively compact companion, anything from a low-mass main-sequence star to a degenerate remnant. Mass loss plays the dominant role in determining the lifetime and the extent of nuclear processing of the AGB phase. Binary interaction will increase the mass loss from the AGB star and curtail its evolution, either through Roche-lobe overflow, common-envelope evolution or the driving of an enhanced stellar wind. These processes will tend to reduce the metals, particularly carbon, returned to the inter-stellar medium. On the other hand merged systems or companions that accrete a substantial amount of mass themselves evolve into AGB stars that can synthesize and return more carbon than the two individuals would have alone. By synthesizing large populations of stars, with nucleosynthesis and binary interaction, we estimate a reduction in carbon yield owing to binary star evolution of as much as 15%.

1. Introduction

The metallicity of the ISM is gradually increasing as stars form, process material through nuclear burning and then return it either in a stellar wind or supernova explosion. The most important contributors to the metallicity are the CNO catalytic elements, carbon, nitrogen and oxygen. These will determine how hydrogen burning proceeds for the next generation of stars. Hydrogen burning itself only mixes these elements, changing their ratios and isotopic fractions. It is not until helium ignites that new CNO elements are produced.

However, helium burning takes place deep inside the star, and is usually encased within a shell of hydrogen burning, so that it is not easy to bring the products to the surface whence they can be returned to the ISM. Three processes can overcome this. First, massive stars ($> 8\,M_\odot$), that end there lives in supernovae throw off a C/O region between helium- and carbon-burning shells. Because of the ease with which a fourth He^4 nucleus is added to C^{12} much of this material has been processed to O^{16} and beyond. Second, very massive stars ($> 25\,M_\odot$) may enter a Wolf-Rayet phase and develop a very strong stellar wind, particularly once helium has ignited. Observations indicate that this wind can strip off the hydrogen envelope and eat into a helium burning core. This core was once convective so its surface is carbon rich. Some calculations (Maeder & Meynet 1994) indicate that Wolf-Rayet stars are the major source of carbon enrichment of the ISM. The other process, that competes for effectiveness, is third dredge-up on the AGB. As the unstable helium-burning shell thermally pulses, the deep convective envelope reaches into a region rich in carbon which can be mixed to the surface. Models differ significantly in the predicted extent of this dredge-up (Frost & Lattanzio 1996) so the actual carbon enrichment remains uncertain. AGB stars also have relatively strong winds so this processed and dredged material is readily returned to the ISM. When almost all the envelope has blown off, the star cools to a white dwarf. Thus mass loss determines the AGB lifetime, its maximum core mass and the number of thermal pulses and dredge-up events.

A binary companion will curtail this evolution by interaction that increases mass loss. The only well quantified effect is Roche-lobe overflow. The initially more massive component of a binary system evolves and grows until it fills its last stable potential surface, most often as a red giant or AGB star, whereupon it begins to transfer mass to its companion. As it loses mass the envelope of a giant expands. If it is still the more massive component of the binary, and mass transfer is conservative, the orbit and Roche lobe shrink. Consequently the process of mass transfer leads, on a dynamical timescale, to the giant overfilling its Roche lobe yet more. The overflow rate rapidly rises and the companion, typically a lower-mass main-sequence star, cannot accrete the material. Its own Roche lobe is quickly filled and a common envelope engulfs the whole system. The two cores, the relatively dense companion and the core of the giant, are then assumed to spiral together by some, as yet undetermined, frictional mechanism. Some fraction of the orbital energy released is available to drive off the envelope. If all of it is ejected while the cores are still well separated we are left with a closer binary system comprising the unscathed companion and a white dwarf which may evolve to a cataclysmic variable. Alternatively, if some of the envelope still remains when the companion reaches the denser depths of the common envelope, they can merge leaving a single, rapidly rotating, giant. FK Comae may be such a merged system. Magnetic braking quickly spins down these merged giants and it is possible that the R stars, none of which appear to be binary (McClure, private communication), may be merged AGB stars. When the entire envelope is lost there can be no further AGB evolution or carbon enrichment. If part of the envelope remains AGB evolution is still curtailed, unless the companion, in

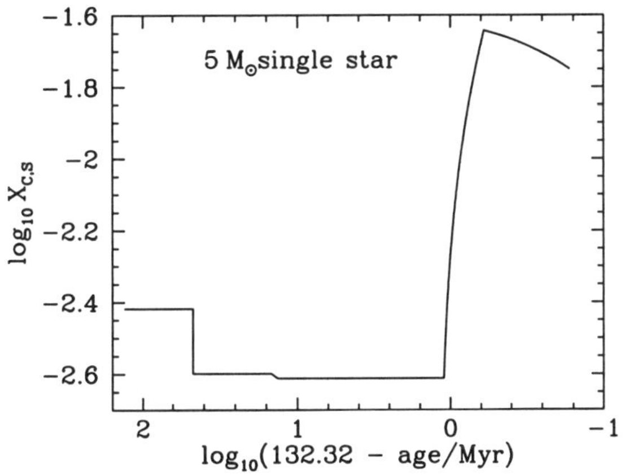

Figure 1. The mass fraction of carbon at the surface of a $5\,M_\odot$ star throughout its evolution. Age is measured from the zero-age main-sequence and the time axis is chosen to expand the AGB.

merging with the envelope, adds even more mass than was lost. In this case AGB evolution is prolonged and the merged star returns more carbon to the ISM than a single AGB star of the same initial mass.

Similar events follow mass transfer that begins on the first giant branch, though the denser envelope makes coalescence more likely. Any Roche-lobe overflow that begins before the giant branch proceeds in a stable manner and may eliminate AGB evolution altogether. However the companion that accretes the transferred mass can itself go through an AGB phase at a later stage. Another effect of a binary companion was proposed by Tout & Eggleton (1988) to account for inverted mass ratios in some RS CVn binaries and for the formation of wider Algol-like systems. They proposed that the companion, probably by forcing the giant to spin faster, can enhance its stellar wind by orders of magnitude. This appears to be necessary to account for the distribution of Barium stars (Karakas, Tout & Lattanzio 1998). Though some of the material is accreted by the companion, the AGB lifetime is generally shortened and less carbon is ejected. Indeed this affects much wider systems than Roche-lobe overflow.

With account taken of all these processes we can expect half of the 60% of stars that are binary systems to experience interaction during their lifetimes. To test the overall effect of the many convoluted paths we have developed a population-synthesis model with both binary interaction and nucleosynthesis.

2. Nucleosynthesis in single stars

Fig. 1 shows the evolution of the surface carbon abundance of a $5\,M_\odot$ star. The nucleosynthesis is based on the model developed by Groenewegen & de Jong (1993). For a population I star the initial mass fraction of carbon is $X_{\rm C,I} = 0.00382$. The surface abundance remains at this value until central hydrogen is exhausted and the star crosses the Hertzsprung gap to join the red giant branch, where a deep convective envelope reaches into regions previously

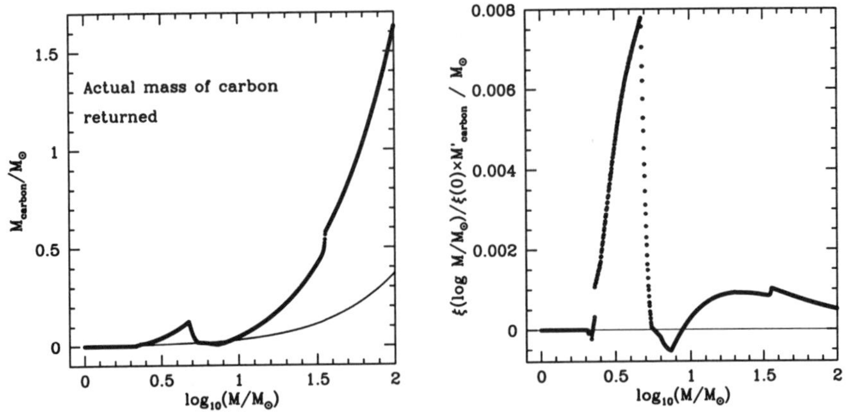

Figure 2. Carbon yields over 10^9 yr for stars with metallicity $Z = 0.02$ in the range $1 - 100\,M_\odot$. On the left merging solid circles show the mass of carbon ejected while the thin line represents the total mass ejected multiplied by $X_{\rm C,I}$. On right they are the difference between these two $M'_{\rm carbon}$, or the yield, weighted by the initial mass function.

affected by hydrogen burning. Here CNO equilibrium has converted much of the carbon to nitrogen and dredging this material lowers the surface abundance. Any wind from the star at this stage contains less carbon than the original ISM.

At the beginning of the AGB a second dredge-up event takes place for stars more massive than about $4.5\,M_\odot$. As the helium shell burns out to catch up with the hydrogen burning shell the latter is temporarily extinguished and the deep convective envelope reaches down into the intershell region, further reducing the surface carbon fraction. When the core mass reaches a critical value ($0.562\,M_\odot$ in our model) third dredge-up following thermal pulses begins and carbon enriched material is dredged to the surface. However, in stars initially over about $4\,M_\odot$, the temperature at the base of the convective envelope eventually reaches temperatures at which hydrogen burning can take place. This hot-bottom burning converts a substantial fraction of the dredged carbon to nitrogen and, as is apparent in the figure, some of the envelope carbon too. The total mass of carbon $M_{\rm carbon}$ ejected is the product of the mass-loss rate and the surface abundance $X_{\rm C,S}$ integrated over the lifetime of the star. In these exploratory models we have used a Reimers' (1975) mass-loss rate throughout the post-main-sequence evolution. Alternatives that concentrate mass loss towards the tip of the AGB would return more carbon.

Fig. 2 shows the integrated mass of carbon returned as a function of initial mass. Massive stars return the most carbon but they are fewer so we weight the ejected mass of carbon by the initial mass function $\xi(\log M)$ of Kroupa, Tout & Gilmore (1993) to give the relative contributions. We look at the yields over the first 10^9 yr since the stars formed because this is the typical time for ejected material to be recycled into new stars. Stars less massive than $2\,M_\odot$ do not evolve at all in this time. Between $2\,M_\odot$ and $2.2\,M_\odot$ they have not begun third dredge-up and so eject only material deficient in carbon owing to first dredge-

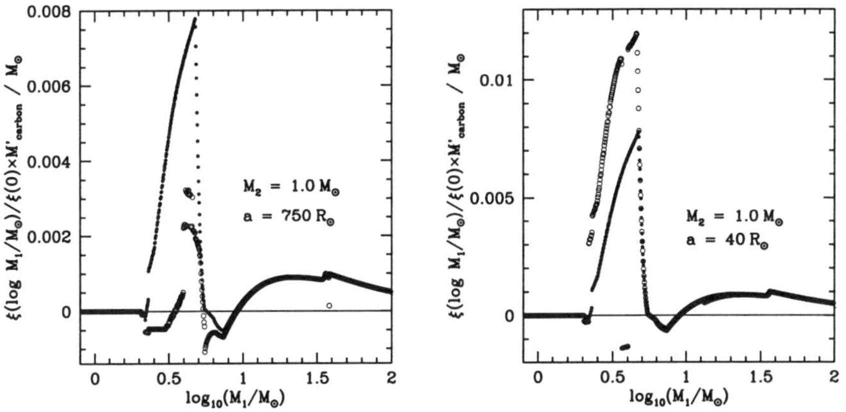

Figure 3. Open circles are the weighted yields from binary systems in which the primary mass M_1 varies as before. The companion has mass $M_2 = 1\,M_\odot$ and the separations are $a = 750\,R_\odot$ (left) and $40\,R_\odot$ (right). Solid circles are the yields from a single star of mass M_1.

up. From here up to $5\,M_\odot$ third dredge-up and mass loss on the AGB lead to a large peak in the carbon yield. The rise is due to the increasing amount of dredge-up while the decline is due to hot-bottom burning which eventually leads to the negative valley between 5 and $8\,M_\odot$. Above $8\,M_\odot$ supernovae begin to contribute. We take their carbon yields from Maeder (1992). Just above $30\,M_\odot$ there is an additional contribution from a Wolf-Rayet phase but we include somewhat less than predicted by Maeder & Meynet (1994). Their contribution would raise the yields by a factor of two or so above $25\,M_\odot$. The AGB peak contributes about twice as much as the massive stars. Integrating over the whole single-star population we find that $X_{\mathrm{C,I}} = 0.00382$ is raised to $X_{\mathrm{C}} = 0.0125$.

3. The effect of a binary companion

To illustrate the effects of binary evolution we consider two particular cases illustrated in Fig. 3. To each of the stars in Fig. 2 we give a companion of mass $M_2 = 1\,M_\odot$ at a separation of either $a = 750\,R_\odot$ or $40\,R_\odot$. In the former case all the stars fill their Roche lobes on either the first giant branch or the AGB and experience common-envelope evolution. Below $3\,M_\odot$ the entire envelope is lost on the giant branch after first dredge-up. The substantial positive yield of the single star becomes a negative one. For more massive primaries the cores merge while a fraction of the envelope still remains and so partial carbon enrichment is subsequently possible. The area under the AGB peak is reduced by a factor of six, substantially decreasing the average carbon yield for such a hypothetical population. At $a = 40\,R_\odot$ all the stars fill their Roche lobes before or early on the red giant branch while their envelopes are still relatively dense. Apart from primaries in the small range from 3.5 to $4\,M_\odot$ these stars go on to experience common-envelope evolution but the companions merge with the envelope before very much mass has been ejected. The AGB lifetime is increased and carbon enrichment enhanced so that the peak is doubled in area. The exceptions between 3.5 and $4\,M_\odot$ come about because above $3.5\,M_\odot$ the stars first fill their Roche lobes while crossing the Hertzsprung gap where mass transfer begins stably on a thermal timescale. Above $4\,M_\odot$ these stars go on to common-envelope evolution as the star becomes a fully fledged giant. Below $4\,M_\odot$ sufficient mass has been transferred by this point that the mass ratio has fallen below the critical value (about $M_1/M_2 = 0.7$) for the Roche lobe to expand faster than the giant envelope. Common-envelope evolution is avoided. However the, now quite massive, companion itself evolves up the AGB and fills its own Roche lobe. Common-envelope evolution ensues but the orbit is now so much wider and the envelope less dense that it is all ejected before carbon enrichment can begin. Above $8\,M_\odot$ most of the carbon comes from the final supernovae and, at these separations, the binary companion has little effect.

In practice binary systems have a wide range of masses and separations and the overall picture is rather convoluted. To model it we synthesize populations of two million systems in which the masses are each chosen randomly from the IMF and the separations from a distribution ranging from 5 to $5 \times 10^6\,R_\odot$ that peaks around $5\,000\,R_\odot$. The procedure used to model binary evolution is described by Tout et al. (1997) and Karakas et al. (1998). If we do not include Tout & Eggleton's (1988) binary enhanced mass loss the average mass fraction of carbon in all the material ejected by 10^9 yr is $X_C = 0.0119$ or a drop of 7% in the yield. If we do include the enhanced mass loss these figures become $X_C = 0.0112$ and 15%.

4. Conclusion

Neither of these differences is particularly large especially when compared with the likely errors in our single star yields. However they do represent a significant systematic effect. It will be necessary to include binary evolution whenever we believe we can predict carbon yields to within 20% or so.

Acknowledgments. We thank the UK PPARC, the British Council and the Australian Research Council for making our visits to Montpellier possible.

References

Frost C.A., Lattanzio J.C., 1996, ApJ 473, 383

Groenewegen M.A.T., de Jong T., 1993, A&A 267, 410

Karakas A.I., Tout C.A., Lattanzio J.C., 1998, MNRAS, submitted

Kroupa P., Tout C.A., Gilmore G., 1993, MNRAS 262, 545

Maeder A., 1992, A&A 264, 105

Maeder A., Meynet G., 1994, A&A 287, 803

Reimers D., 1975, in *Problems in Stellar Atmospheres and Envelopes*, eds. B. Baschek, W.H. Kegel & G. Traving, Springer, Berlin, p. 229

Tout C.A., Aarseth S.J., Pols O.R., Eggleton P.P., 1997, MNRAS 291, 732

Tout C.A., Eggleton P.P., 1988, MNRAS 231, 823

Asymptotic Giant Branch Stars
T. Le Bertre, A. Lèbre and C. Waelkens, eds.
I.A.U. Symposium 191, 1999

The Nature of RV Tauri Stars

T. Lloyd Evans

South African Astronomical Observatory, P. O. Box 9, Observatory 7935, South Africa

Abstract. The infrared colours of the RV Tauri stars fall in a well-defined and thinly-populated region of the IRAS [12] – [25], [25] – [60] diagram. This provides an alternative method of searching for such stars. Stars with these IRAS colours as well as the K and L excesses which are also typical of RV Tauri stars have spectral types from B to M. The characteristic variability of RV Tauri-type has been found for several stars with spectral types typical of known RV Tauri stars, while stars of different spectral types do not vary in this way. RV Tauri stars are the stars with these colours, probably indicative of a dusty circumstellar disk, which lie within the Cepheid instability strip.

1. Introduction

The RV Tauri stars are defined as radially-pulsating stars with alternating deep and shallow minima. An additional criterion, that they are of late spectral type (Kholopov et al. 1985), is required to avoid possible confusion with eclipsing variables of the β Lyrae type which could arise if the light curve was the only observational datum. The Type II Cepheids, or W Virginis stars, in globular clusters begin to show alternating minima at periods longer than about 25 days and a few are recognised as being of RV Tauri type. A similar situation holds for stars in the Large Magellanic Cloud (Pollard et al. 1996; Alcock et al. 1998). The galactic field population includes some RV Tauri stars with stronger metal lines and molecular features than those in globular clusters. Many of these strong-lined stars were found to have very large infrared excesses by Gehrz & Woolf (1970); later work showed that their colours in both the IRAS [12] – [25], [25] – [60] and the $J - K$, $K - L$ diagrams distinguish them from the great majority of the red variables which form the bulk of the stars in the IRAS catalogue (Lloyd Evans 1985; Raveendran 1989).

These infrared colours provide a means to discover new examples with selection effects which are quite different from those - primarily optical brightness and large amplitude - imposed by the traditional variable star surveys using blue-sensitive plates. The known RV Tauri stars in the IRAS Point Source Catalogue (Beichman et al. 1988) are largely confined to a rectangle bounded by [12] – [25] = 1.00 – 1.55, [25] – [60] = 0.20 – 1.00, which we call the RV Tauri box. Two RV Tauri stars with strong carbon features in their optical spectra (Preston et al. 1963; Lloyd Evans 1974), AC Her and RU Cen, have small near infrared excesses but much redder IRAS colours and suggest an extension to $[12]-[25]=2.4$, $[25]-[60]=1.2$.

2. Observations

A list of all the stars with good quality photometry in these two regions in the IRAS Point Source Catalogue was drawn up using a programme provided by Dr R. M. Catchpole. The positions of over 1000 sources were then examined on the Palomar, ESO and SRC Sky Survey prints and candidate stars were identified; many fields were empty or too crowded while others contained faint red stars which are plausible candidates. Infrared photometry was obtained for 370 stars and spectra were taken for many of them. Most of the stars in the RV Tauri box are M stars, while the extension contains many visually-bright young stars which are less easy to distinguish from the stars which are sought. Stars with typical excesses at K and L are found in all spectral classes from B to M; those of types B and A are Herbig Ae/Be stars and in many cases had already been identified by others. The M stars with infrared excesses are clearly not RV Tauri stars. We consider the 20 stars of types F, G and K as possible RV Tauri stars.

Repeated infrared observations over the period since initial identification (the project started in 1989) and the IRAS estimate of the probability of variability give an indication of the variability of a star, while more intensive observations in the $UBVRI$ passbands were made by the service observers on the 0.5m telescope over one or more observing seasons for some of the stars. Several stars are too faint or too crowded for aperture photometry with this telescope, however. Four new RV Tauri stars have been identified in this way, while a known variable, LR Sco (P = 104.4 days), which was recently identified as a G supergiant (Giridhar et al. 1990, 1992) is also considered to be one. Five stars do not match the photometric properties of RV Tauri stars in one respect or another and are of uncertain status. Many of the others are certainly variable but five stars are either non-variable or have such small amplitudes that they are very unlikely to be RV Tauri stars.

Most of the spectra are consistent with classification as members of group A (Preston et al. 1963), that is they have quite strong metallic lines with CH and CN bands which are of normal strength compared to the rest of the spectrum. Two of the four new RV Tauri variables are of the Preston B type which have CH and CN bands which are much too strong for the rest of the spectrum; one of them has C_2 bands as well. Two stars, one of which is IRAS 20056+1834 which has already been described (Menzies & Whitelock 1988), have emission in some of the resonance lines of the alkali earths. Both are variable; IRAS 20056+1834 has rather a small amplitude for an RV Tauri star but requires further study, while the other is faint but may have a large amplitude.

3. Discussion

The spectral types which have been determined for the IRAS stars observed here and those for a large sample of known RV Tauri stars (Lloyd Evans, in preparation) may be used to define the instability strip for the RV Tauri stars. Different types have been determined for various spectral features by reference to a set of standard star spectra, most of which are of supergiants of luminosity class Ib. The spectral type derived from the Balmer series of hydrogen seems most appropriate as a temperature indicator, as the other features are more affected by vagaries of chemical composition and the spectrographic resolution is not adequate to determine the type from the ratios in line pairs. The hydrogen lines are often affected by emission, however, so greatest weight is placed on higher members of the Balmer series. The steep decrement of the emission in these stars makes this a viable proposition. A simple average has been taken for the spectra of each star; this is perhaps more satisfactory for the known RV Tauri stars, which were usually observed more often and for which the already known periods enabled good coverage of the cycle to be obtained.

The known RV Tauri stars fall in the range F5 to G2, with very few earlier than F7. There are a very few stars between G5 and K2; their membership of the group may be questioned. Four of the non-variables among our new sample are earlier than F5, while amplitudes increase from F5 to G3 with a doubtful variable at F8. The late types (G3) for two stars, both confirmed RV Tauri stars, may result from relatively poor phase coverage. One of the five stars noted as possibly not of RV Tauri type from photometric considerations is F5, two are G5 and one, a non-variable, is K5. The spectral types of the new confirmed RV Tauri stars and the three inadequately studied variables are consistent with membership in the class.

The two stars with resonance emission lines are at the tip of the distribution of the F, G, K stars in the $J-K$, $K-L$ diagram. A similar situation is found for M and carbon stars with similar L excesses and has been interpreted in terms of a dusty disk which is seen edge-on in the stars with emission as the star is hidden by the disk and can only be seen by reflection from material out of the plane of the disk. T Tauri stars show similar K and L excesses which are attributable to the presence of a disk (eg. Strom et al. 1989). The material which lies above the plane of the disk in these two IRAS stars is the source of the resonance fluorescence emission (Lloyd Evans 1997, following Herbig 1969). It may be concluded that the RV Tauri variables are those members of a class of luminous F - K stars, probably with low masses as indicated by their kinematics (Joy 1952), which have dusty disks and which pulsate if they lie in the Type II Cepheid instability strip. There is evidence of continuing mass loss from some of these stars (Preston 1964; Pollard et al. 1997) but it seems unrealistic to relate the totality of the dust around them to maintenance by this current activity, as a disk is a stable configuration which will not dissipate as rapidly as an expanding shell.

Lloyd Evans (1985) found that the stars classified on the basis of their light curves as RVb, that is stars with a light variation of much longer period as well as the pulsational variation (Kholopov et al. 1985), were all among the redder stars in the near infrared $J-H$, $H-K$ and $J-K$, $K-L$ diagrams. Van Winckel (1999) has explained this correlation as the consequence of an assumed binary nature of the RV Tauri stars, which have circumbinary dust disks. There will be no long-period variation if the system is seen pole-on, whereas the star will be partly obscured when nearest to us in its orbit if the disk is seen from a low angle. The observation that the reddest objects show resonance emission as a

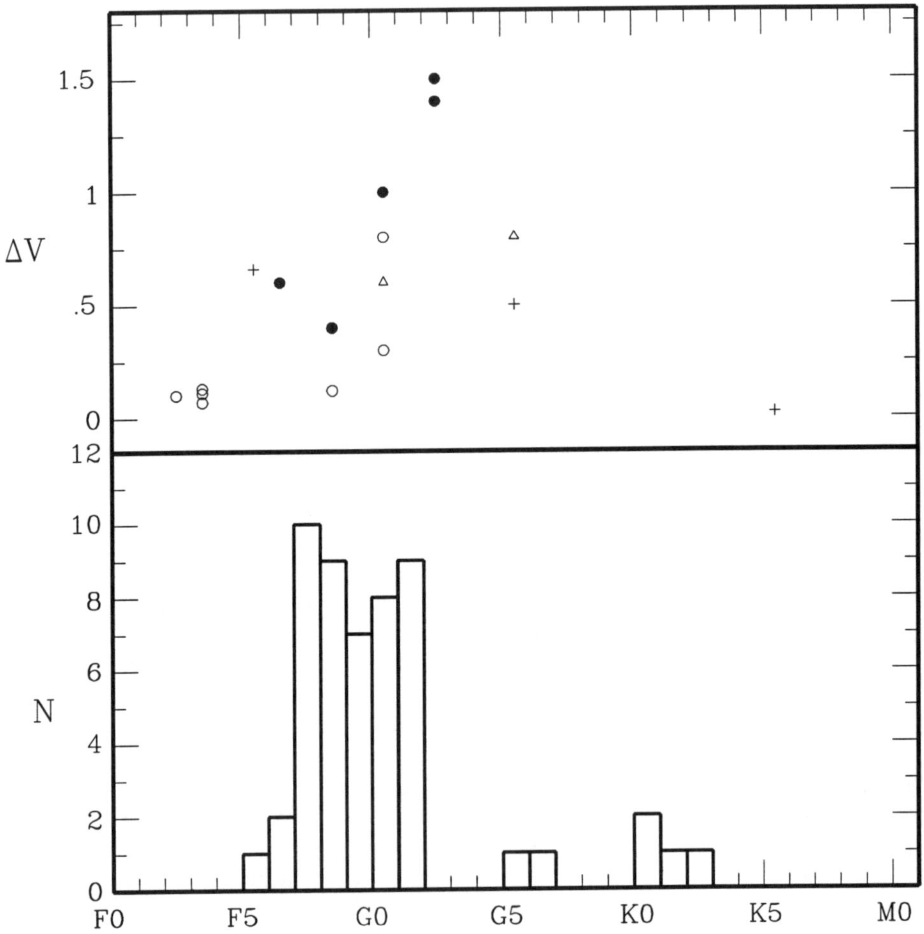

Figure 1. The RV Tauri instability strip. The lower panel shows the number distribution with spectral type of known RV Tauri and related stars, while the upper panel plots the estimated visual amplitude of the candidate RV Tauri stars as a function of spectral type. The symbols are: filled circles, confirmed RV Tauri stars; open circles, stars which are non-variable or of unknown period; open triangles, stars with resonance fluorescence emission; plus signs, stars whose photometric properties differ from those typical of RV Tauri stars.

result of the star being hidden from direct view confirms the correlation between infrared colour and viewing angle required by Van Winckel's hypothesis.

Acknowledgments. I am most grateful to Dr D. M. Kilkenny who arranged for the *UBVRI* observations to be made and for supplying the results in a convenient form, and to the 0.5m observers, especially Francois van Wyk, who did the observing. I am indebted to Mrs A. A. van der Wielen for assistance in the preparation of this report.

References

Alcock C., et al., 1998, AJ 115, 1921

Beichman C.A., Neugebauer G., Habing H.J., Clegg P.E., Chester T.J., 1988, Infrared Astronomical Satellite (IRAS) Catalog, NASA, Washington DC

Gehrz R.D., Woolf N.J., 1970, ApJ 161, L213

Giridhar S., Rao N.K., Lambert D.L., 1990, Observatory 110, 120

Giridhar S., Rao N.K., Lambert D.L., 1992, JA&A 13, 307

Herbig G.H., 1969, Contr. Lick Obs. No 302

Joy A.H., 1952, ApJ 115, 25

Kholopov P.N., Samus N.N., Frolov M.S., Goranskij V.P., Gorynya N.A., Kireeva N.N., Kukarkina N.P., Kurochkin N.E., Medvedeva G.I., Perova N.B., Shugarov S.Yu., 1985, General Catalogue of Variable Stars, Nauka Publishing House, Moscow

Lloyd Evans T., 1974, MNRAS 167, 17P

Lloyd Evans T., 1985, MNRAS 217, 493

Lloyd Evans T., 1997, Ap&SS 251, 239

Menzies J.W., Whitelock P.A., 1988, MNRAS 233, 697

Pollard K.R., et al., 1996, in Variable Stars and the Astrophysical Returns of Microlensing Surveys, 12th IAP Astrophysics Colloquium, R. Ferlet, J.-P. Maillard and B. Raban (eds.), Edition Frontières, Gif-sur-Yvette, p. 219

Pollard K.R., Cottrell P.L., Albrow M.D., Tobin W., 1997, MNRAS 286, 1

Preston G.W., 1964, ApJ 140, 173

Preston G.W., Krzeminski W., Smak J., Williams J.A., 1963, ApJ 137, 401

Raveendran A.V., 1989, MNRAS 238, 945

Strom K.M., Strom S.E., Edwards S., Cabrit S., Skrutskie M. F., 1989, AJ 97, 1451

Van Winckel H., 1999, this volume, p. 465

Xander Tielens inspecting the posters

Asymptotic Giant Branch Stars
T. Le Bertre, A. Lèbre and C. Waelkens, eds.
I.A.U. Symposium 191, 1999

RV Tauri stars and Type II Cepheids in the LMC

Karen R. Pollard

Department of Physics and Astronomy, University of Canterbury, Private Bag 4800, Christchurch, New Zealand

T. Lloyd Evans

South African Astronomical Observatory, PO Box 9, Observatory 7935, South Africa

Abstract. A review of the main results from the MACHO project for RV Tauri stars and Type II Cepheids in the LMC is presented. In light and colour curve behaviour, the RV Tauri stars appear to be a direct extension of the Type II Cepheids to longer periods. A single period–luminosity–colour relationship is seen to describe both the Type II Cepheids and RV Tauri stars in the LMC.

Follow-up low resolution spectroscopy of the LMC RV Tauri stars has been undertaken. The spectra of the longest period RV Tauri star in the MACHO sample show that this star has strong C_2 bands with a pronounced phase dependence, their strength being greater at minimum light. This and other spectral features are similar to those observed in proto-planetary nebulae. This LMC star therefore appears to combine the main spectral anomaly of the proto-planetary nebulae with RV Tauri-like photometric variability. This is an intriguing result which indicates a possible link between RV Tauri stars and proto-planetary nebulae, and strengthens the post-AGB interpretation of the RV Tauri stars.

1. Introduction

RV Tauri stars are luminous, long-period members of the Type II class of Cepheid variables. The shorter-period stars of the W Vir class are hydrogen- and helium-shell burning stars (Schwarzchild & Harm 1970) while the RV Tauri stars are believed to be in a more advanced stage of evolution at the end of the asymptotic giant branch (AGB) (Gingold 1974; Jura 1986). The AGB stars consist of a C–O degenerate core with a thin overlying helium layer inside a large tenuous convective hydrogen envelope. The star evolves off the AGB after severe mass loss from the envelope. An RV Tauri variable may be a post-AGB star with a necessarily low-mass but extended envelope. The surface composition of such a star would be expected to reflect that of its cool AGB predecessor, but with a far simpler spectrum due to its warmer temperature.

Recent high-resolution abundance studies of galactic RV Tauri stars (Gonzalez, Lambert & Giridhar 1997a, b) have shown, however, that many field RV Tauri stars have atmospheres that are preferentially depleted of elements of high condensation temperature. Intriguingly, further studies to examine these abundances (Gonzalez & Lambert 1997; Russell 1997) have found evidence that this preferential depletion does not appear to occur in globular cluster RV Tauri variables or high-velocity field stars, indicating that the low initial metallicity of the progenitor star may be an inhibiting factor in this depletion process.

1.1. RV Tauri stars in the Large Magellanic Cloud

The discovery of RV Tauri stars in the Large Magellanic Cloud (LMC) was announced recently (Pollard et al. 1997; Alcock et al. 1998). This discovery has finally allowed us to make a definitive assessment of the luminosities of these stars which will help us in our investigation of the other fundamental parameters. The derived absolute magnitude of about −4.5 for these variables at a fundamental period of ~50 days lends strong support to the proposal that these stars are luminous post-AGB stars evolving to the left in the HR diagram. Further results from this analysis of the MACHO database were that there was a progression in behaviour from the short period Type II Cepheids to the longer period RV Tauri stars: in terms of their light- and colour-curve properties the RV Tauri stars were found to be a direct extension of the Type II Cepheids to longer periods. A period-luminosity-colour relationship of the form

$$M_V = -0.61 - 2.95 \log P + 5.49 < (V - R)_0 > \tag{1}$$

gave an extremely tight relation for both the Type II Cepheids and RV Tauri stars in the LMC. Further details can be found in Pollard et al. (1997) and Alcock et al. (1998).

This paper presents the first spectroscopic results obtained for an RV Tauri star in the LMC. MACHO*04:55:43.2–67:51:10 has a formal period of 112.45 days and is the longest period variable from the Alcock et al. (1998) sample of 33 Type II Cepheid and RV Tauri stars in the LMC. The light curve is shown in Fig. 1. The solid curve plotted in this figure is the 5th order Fourier fit to the optical (V_M) photometry from the MACHO collaboration. The light and colour variations in this star are similar to the behaviour typical of galactic RV Tauri variables (Pollard et al. 1996). However, the low-resolution spectra of this object are most unusual and unlike any spectra that have been obtained previously of galactic RV Tauri stars (Pollard et al. 1997 and references therein).

2. Observations

Low-resolution optical spectra and infrared JHK photometry were obtained of the LMC RV Tauri variable star MACHO*04:55:43.2–67:51:10 in order to examine the overall spectral characteristics and infrared properties. Three spectra were obtained with the Unit Spectrograph and RPCS detector on the 1.9-m reflector at the South African Astronomical Observatory (SAAO) using a 7 Å resolution over the range 3500–6500 Å approximately. These spectrograms were obtained during 1996 December to 1997 February at widely-separated phases of 0.61, 0.87 and 1.16 (where a phase of 0.0 corresponds to the deep or 'primary' minimum). Infrared JHK observations of MACHO*04:55:43.2–67:51:10 were taken during 1996 December using the Mk III infrared photometer on the SAAO 1.9-m telescope, when the star was close to maximum light (phase 0.69).

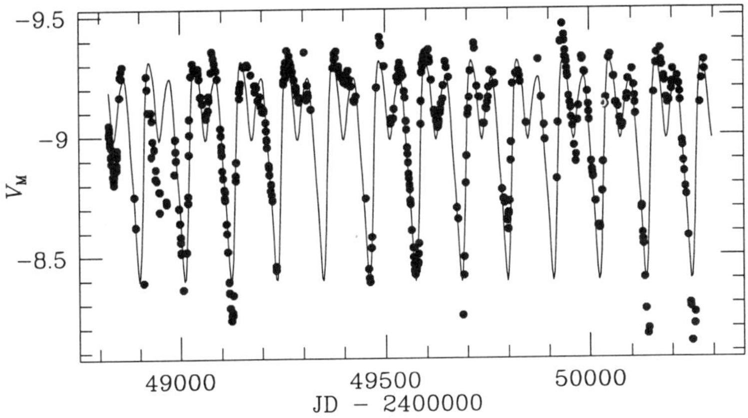

Figure 1. The raw V_{M} light curve for MACHO*04:55:43.2–67:51:10 plotted together with the 5th order Fourier fit to the photometric data. This star has a double or 'formal' period of 112.45 days.

3. Results

3.1. Spectral classification

The three spectra of the LMC RV Tauri star MACHO*04:55:43.2–67:51:10 were compared to a sequence of standard star spectra of supergiants of luminosity class Ib, using the Balmer lines of H, Ca II, CH and CN. The low resolution and S/N of these spectra does not permit the use of the weaker metallic lines. The features used do not yield a consistent result, a common situation with Type II Cepheids and RV Tauri stars whose spectra are affected by shock-wave phenomena as well as by the abundance anomalies referred to in the Introduction.

The weak Ca II lines in the December spectrum indicate an early F type, perhaps F3, and the weak CN bands indicate a type earlier than G0. The weak Balmer lines of H are not consistent with any type earlier than G2 while the G-band of CH suggests G0, however. The two subsequent spectra, especially that of 10 February, are a little later in spectral type (that is, apart from the carbon bands). The estimated $(V - R_{\mathrm{c}})_0$ colour of 0.27 on 10 December gives a spectral class of F9 Ib (Cousins 1981), while the $(B-V)_{sp,0}$ of 0.74 on the same date is consistent with this result, giving a spectral class of G0 Ib.

Comparison was made with the spectra of other carbon-rich variables. The three low-resolution spectra of the LMC RV Tauri star MACHO*04:55:43.2–67:51:10 are shown in Fig. 2, together with those of the R Coronae Borealis (RCB) star V517 Oph (Kilkenny et al. 1992) and the galactic RV Tauri star SX Cen. The principal absorption features are marked. Neither of the galactic stars are a particularly good match for MACHO*04:55:43.2–67:51:10.

No galactic RV Tauri star is known to have C_2 bands as strong as those in MACHO*04:55:43.2–67:51:10. Such strong C_2 bands normally imply very strong CN bands at 4215 and 3883 Å. Unusually, the presence of CN in the spectrum is only marginal. In addition, CH is represented in this star only by a weak G band. The Balmer lines of H are very weak in the two spectra with

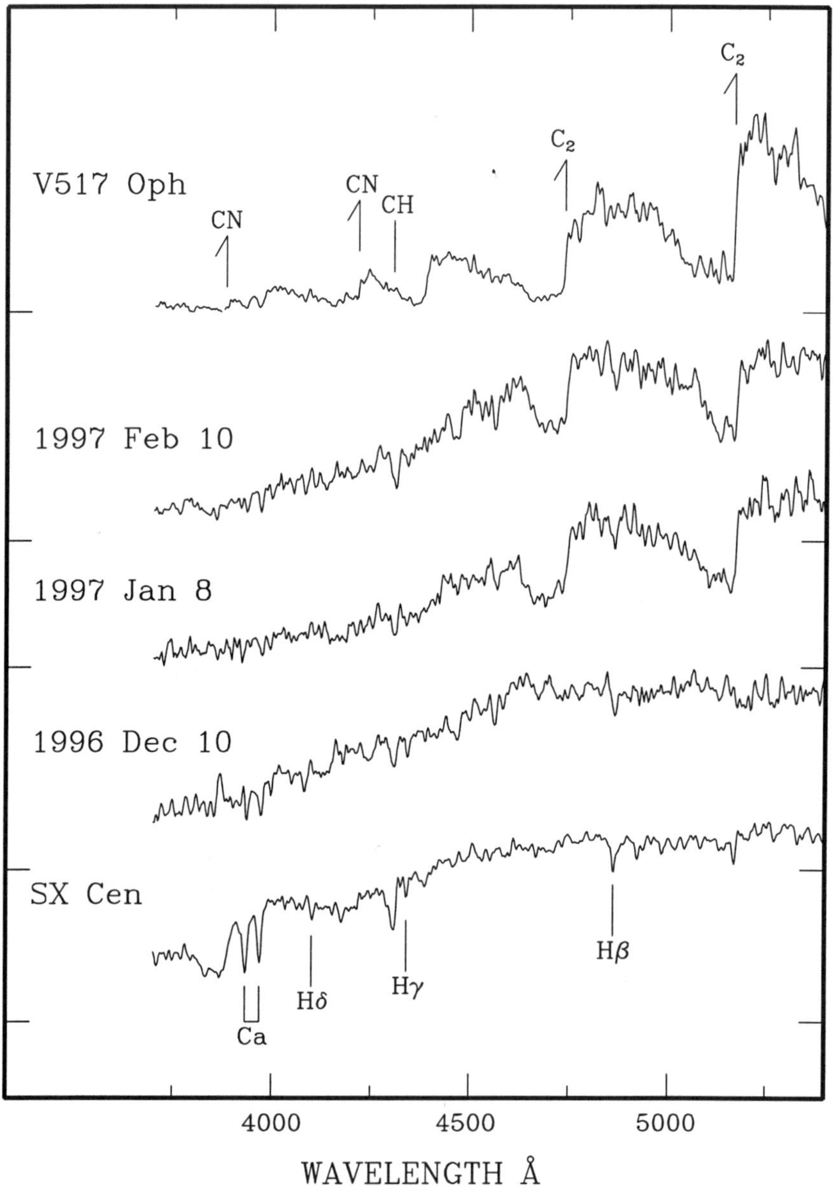

Figure 2. The three low-dispersion spectra of MACHO*04:55:43.2–67:51:10 (indicated with their dates of observation) are compared with SX Cen, a galactic type 'B' (that is, carbon-rich) RV Tauri star, and V517 Oph, a cool R Coronae Borealis star that displays relatively strong C_2 molecular bands. The ordinate is flux on an arbitrary scale.

strong C_2 bands, but are present and quite pronounced in the spectrum taken near maximum light. The spectra have a low S/N in the 4000 Å region and the presence of C_3 is uncertain – the strongest bandhead near 4050 Å may be present on the third observation.

The unusual features of the spectrum of MACHO*04:55:43.2–67:51:10 can be matched by stars of another type, the proto-planetary nebulae (Hrivnak & Kwok 1991; Hrivnak 1995). One of these stars has been found to be variable but the type of variability and the dependence of the spectrum on brightness is unknown. These stars were identified on the strength of large infrared excesses, especially at wavelengths beyond 10 microns, so that there must be a dust shell with a temperature of 150–300 K at some distance from the star. The JHKL photometry is not expected to show a large excess, being dominated by radiation from the photosphere.

3.2. Infrared photometry

The *JHK* data from SAAO indicates that MACHO*04:55:43.2–67:51:10 has little infrared excess at K and therefore no substantial warm or hot circumstellar dust. Observations at longer wavelengths would be most useful in detecting the possible presence of cool ($\sim$ 150–300 K) dust from mass-loss episodes in the past.

4. Discussion and future work

MACHO*04:55:43.2–67:51:10 is an RV Tauri variable with a double period of 112.45 d. Three low-resolution (210 Å mm^{-1} or 7 Å resolution) spectrograms obtained of this star show strong C_2 bands with a pronounced phase dependence, their strength being greatest at minimum light. The other spectral features observable with this resolution are all weak and it is impossible to assign a consistent spectral type.

The spectrum of MACHO*04:55:43.2–67:51:10 is similar to those observed in proto-planetary nebulae. It appears that this may be a unique star which combines the main spectral anomaly of the pre-planetary nebulae (PPN) as defined by Hrivnak & Kwok (1991) and Hrivnak (1995), with RV Tauri-like variability. One might have predicted the existence of such a star, given that, in some cases, the PPN must lie close to the Cepheid instability strip while the RV Tauri variables have been interpreted as post-AGB objects.

The strength of C_2 while CN and CH are relatively weak is hard to understand. Hrivnak (1995) suggested that C_2 and C_3 arose in the extended atmospheres of the pre-planetary stars. The exact relationship of the C_2 band-strength to phase in MACHO*04:55:43.2–67:51:10 is not clearly defined by our observations. However, the large intensity variations of the C_2 bands suggest strong coupling with the light variations, indicating they probably arise near the stellar photosphere.

The luminosity of MACHO*04:55:43.2–67:51:10 ($M_V = -3.87$, Alcock et al. 1998) is higher than those typical of the RV Tauri stars present in galactic globular clusters, but there is a rich intermediate-age population in the LMC, typified by the globular clusters, from which such a star might come. Carbon enrichment on the extended AGB of these clusters ultimately results in N-type carbon stars (Bessell, Wood & Lloyd Evans 1983), so the enhancement of carbon is readily understood. The occurrence of a luminous, carbon-rich RV Tauri-like star may be explained by assuming that it is a star of moderately low mass which has moved through the AGB and is now evolving to the blue in the pre-planetary nebula stage. This star may therefore represent a link between these two types of star and strengthens the post-AGB interpretation of RV Tauri stars.

Additional observations are required to clarify the status of this unusual star. One useful datum would be the carbon isotope ratio. A number of RV Tauri stars appear to have enhanced ^{13}C in the cases studied to date (Bujarrabal, Alcolea & Bachiller 1990; Shenton et al. 1992; Gonzalez, Lambert & Giridhar 1997a). The various degrees of enhancement of ^{13}C result from different degrees of processing by the CN-cycle and place important constraints on the past evolution of a star. The phase-dependence of the molecular band strengths needs to be delineated. Spectra of improved resolution and S/N are required to look for C_3 and enhanced s-process elements.

Acknowledgments. We are indebted to Doug Welch for drawing our attention to this star and to the MACHO collaboration for the use of their photometric data. Audrey van der Wielen kindly prepared Fig. 2. We would also like to thank Dave Kilkenny for providing comparison spectra of galactic RCB stars and for a number of very useful discussions.

References

Alcock C., et al., 1998, AJ 115, 1921

Bessell M. S., Wood P.R., Lloyd Evans T., 1983, MNRAS 202, 59

Bujarrabal V., Alcolea J., Bachiller R., 1990, A&A 234, 355

Cousins A.W.J., 1981, SAAO Circ. 6, 4

Gingold R.A., 1974, ApJ 193, 177

Gonzalez G., Lambert D.L., 1997, AJ 114, 341

Gonzalez G., Lambert D.L., Giridhar S., 1997a, ApJ 479, 427

Gonzalez G., Lambert D.L., Giridhar S., 1997b, ApJ 481, 452

Hrivnak B.J., 1995, ApJ 438, 341

Hrivnak B.J., Kwok S., 1991, ApJ 371, 631

Jura M., 1986, ApJ 309, 732

Kilkenny D., Lloyd Evans T., Bateson F.M., Jones A.F., Lawson W.A., 1992, The Observatory 112, 158

Pollard K.R., et al., 1997, in Variable Stars and the Astrophysical Returns of Microlensing Surveys, R. Ferlet, J.P.Maillard & B. Raban (eds.), Proc. 12th IAP Astrophysics Colloq., Editions Frontières, Gif-sur-Yvette, p. 219

Pollard K.R., Cottrell P.L., Kilmartin P.M., Gilmore A.C., 1996, MNRAS 279, 949

Pollard K.R., Cottrell P.L., Lawson W.A., Albrow M.D., Tobin W., 1997, MNRAS 286, 1

Russell S.C., 1997, A&A 326, 1069

Schwarzchild M., Harm R., 1970, ApJ 160, 341

Shenton M., et al., 1992, A&A 262, 138

Asymptotic Giant Branch Stars
T. Le Bertre, A. Lèbre and C. Waelkens, eds.
I.A.U. Symposium 191, 1999

Post-AGB Evolution

Hans Van Winckel
Instituut voor Sterrenkunde, K.U. Leuven, Celestijnenlaan 200B, 3001 Heverlee, Belgium

Abstract.
This review focuses on the evolution of single and binary optically bright post-AGB stars and the relation between the photospheric chemical abundances and the circumstellar material.

Surface abundance determinations of optically bright post-AGB stars offer a powerful tool, not only to clarify the evolutionary nature of the objects themselves, but also to constrain theoretical calculations for chemical evolution. However, a survey of recent results shows that the chemical patterns observed are very diverse and that several different classes can be distinguished. The most striking result is that only a minor fraction, namely the 21 μm objects, are conform to standard post third dredge-up theory.

In recent years it became clear that binarity can influence the late evolutionary stages considerably. A second part therefore discusses the varied and fundamental impact of the presence of a companion on the evolution of post-AGB systems.

1. Introduction

Post-AGB stars, also commonly named proto-planetary nebulae, are generally defined as objects having evolved off the AGB but not (yet) to high enough effective temperatures to ionise a Planetary Nebula (PN). As direct descendants of AGB stars, they are not only interesting objects by themselves, but in principle also ideal tracers to gain better insight into the complex physical and chemical changes that occurred during the previous (late) AGB phase. This is, however, not straightforward since theoretical (chemical) evolutionary AGB predictions are not only dependent on fundamental parameters like initial mass and metallicity, but also on less well constrained parameters like mass-loss history and geometry, details of the dredge-up phenomenon and thermal pulse nucleosynthesis, interpulse phase at which the post-AGB phase starts etc. (see e.g. the contributions of Lattanzio & Forestini, Blöcker, Mowlavi, Herwig et al. and Marigo in this volume). Moreover, since the post-AGB evolution is short (typically 10^3 to 10^4 years: e.g. Blöcker 1995), not many post-AGB stars are known and observational results on individual sources are rather difficult to interpret in a more general picture.

This review focuses on many recent observational results on optically bright post-AGB stars and tries to emphasise the most important findings in the general context of stellar evolution theory, together with some warning against observational bias introduced by selection procedures. Optically bright objects have the extra advantage that the broad spectral domain of their emission allows us not only to study the stellar F-K type photospheres in detail, but also the complex physical and chemical interplay between the central star and the circumstellar material. Recent results on optically bright post-AGB stars experiencing a late thermal pulse will not be covered in this paper but I refer to the contributions of Asplund, Cohen et al., Yudin & Tatarnikov and Werner et al. in this volume.

2. SED: detached dust-shells versus broad IR excesses in binaries

The IRAS Point Source Catalogue has been used extensively to search for new post-AGB stars (e.g. Pottash & Parthasarathy 1988; Hrivnak et al. 1989; Oudmaijer et al. 1992). The Spectral Energy Distributions (SEDs) obtained for the optically bright post-AGB candidate stars fall into two categories: the first one contains objects showing a double-peaked energy distribution determined by the UV-optical emission of the reddened stellar photosphere and the thermal re-radiation of the circumstellar, detached, and expanding dust-shell with a characteristic dust temperature of $T_{dust} \sim$ 100–200 K (for examples of SEDs e.g.: Hrivnak et al. 1989; Trams et al. 1991; Szczerba et al. 1997).

A significant fraction of post-AGB candidate stars also show the presence of a hot dust component with $T_{dust} \sim$ 500–1300 K making the IR excess very broad (for examples of SEDs e.g.: Trams et al. 1991; Waters et al. 1993). Subsequent radial velocity measurements revealed a tight correlation between the occurrence of *hot dust and binarity* (Waters et al. 1997 and references therein). Since the binary nature of the object can have fundamental influence not only on the SED but also on the evolution of the whole object, we will cover the binary post-AGB stars in a separate section (Sect. 4). The next section will cover the double peaked post-AGB objects for which no binary motion has yet been discovered.

3. (Single) Post-AGB evolution

3.1. 21 μm objects as post-carbon stars

The IR emission feature at 21 μm was first identified on IRAS LRS spectra by Kwok et al. (1989) and it was realised that the feature only occurs in spectra of C-rich post-AGB stars as evidenced by the presence of PAH bands in the mid-IR and often strong HCN fluxes (e.g. Kwok et al. in this volume). The 21 μm feature is sometimes accompanied by a strong 30 μm emission feature which is observed over a broader spectral type domain (Omont et al. 1995).

The carbon rich nature of the circumstellar environment of these objects was corroborated by the detection of C_2, C_3 and CN bands in the optical spectra (Hrivnak 1995). Detailed analysis of high-resolution spectra of these bands indicates that the ro-vibrational lines are formed in a relatively narrow shell in the circumstellar envelope (Bakker et al. 1996; 1997). Bakker & Lambert (1998) estimated on the basis of ultra high-resolution spectroscopy of these bands a $^{12}C/^{13}C$ ratio of 67 for HD 56126, demonstrating again an efficient 3rd dredge–up.

Chemical studies of the stellar photospheres confirmed the post-3rd dredge-up character of known 21 μm stars since they show not only a high C/O ratio but also large overabundances of s-process elements (Klochkova 1995; Začs et al. 1995; Van Winckel et al. 1996; Reddy et al. 1997; Decin et al. 1998). But also the reverse still holds: all known post-AGB stars showing strong s-process enhancements in their optical spectra display a 21 μm feature in their IR spectra (Decin et al. 1998). The F-G spectral types of these supergiants make them ideal objects to study a wide variety of species by their atomic photospheric lines, making them ideal sources to study the AGB nucleosynthesis and dredge-up phenomena. The sample of 21 μm objects studied in detail till now displays a surprisingly large variety of integrated neutron exposures, as parameterised by the heavy-to-light s-process index ([hs/ls] value, see also Smith in this volume) in between −0.6 and +0.5, independent on metallicity.

Most 21 μm objects are clearly resolved at mid-IR wavelengths and show axial symmetry with a $\dot{M}_{equator}$ / $\dot{M}_{pole}$ up to 20 (e.g. Meixner et al. 1997; Dayal et al. 1998). The resolved axial symmetry is certainly not limited to 21 μm objects and is in fact very common among double-peaked post-AGB stars. How this axial symmetry develops during the superwind phase is still a matter of debate (e.g. review by Meixner 1997) but is, however, fundamental since the future PN is shaped during this phase.

3.2. Dredge-up efficiency versus metallicity

The 21 μm stars are, however, only a fraction of the double-peaked post-AGB stars and the chemical composition of other sources are much more difficult to interpret (Luck et al. 1990; Van Winckel 1997; Začs et al. 1996). To illustrate the dredge-up efficiency Fig. 1 shows the overabundance of the s-process element Zirconium (Zr) as a function of the metallicity for a sample of homogeneously analysed objects. In other respects like IR emission, kinematics and metallicities these stars are similar, but there is a clear difference in photospheric composition. The absolute luminosity differences are of course very difficult to trace but may be crucial for understanding the different chemical evolutions.

This chemical diversity is emphasized further by the characteristic chemical patterns displayed by the hot B-type post-AGB stars. These optically bright post-AGB stars are not discovered by their IR properties but are identified among samples of high-galactic latitude B stars selected to study early-type stars in the halo of our Galaxy (e.g. Conlon et al. 1991; McCausland et al. 1992; Napiwotzki et al. 1994; Moehler & Heber 1998). Although the Fe abundance is very difficult to trace for the hottest atmospheres, the estimated metallicities determined from other species (Mg, Al, Si and S) range from −1 to −2. The most characteristic chemical signature is, however, the extreme low carbon abundance in all these stars. The low metallicity, position in the HR-diagram and kinematics point to a late post-AGB evolution of low-initial-mass objects. The post-AGB nature is further strengthened by the detection of remnants of the circumstellar dust around a few objects and the very similar chemical patterns displayed around C-poor halo PN (Conlon et al. 1993). Similar objects were recently discovered towards the Galactic Centre (Venn et al. 1998). It is remarkable that the special chemical patterns displayed by these stars are not found among the F-type post-AGB stars !

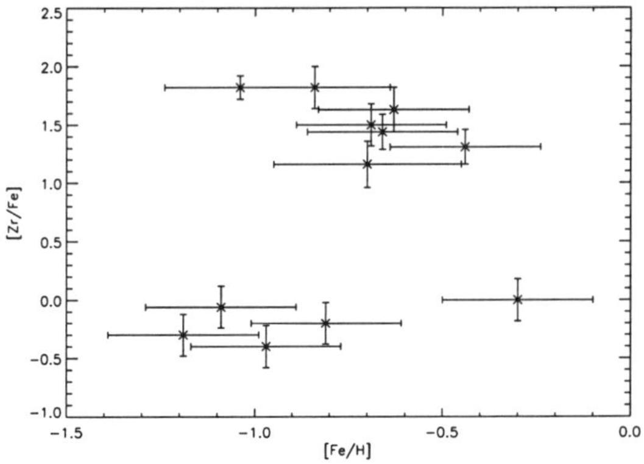

Figure 1. The [Zr/Fe] abundance ratio as a measure of the 3rd dredge-up efficiency versus metallicity for post-AGB stars showing a double-peaked energy distribution. The enriched objects are the 21 μm stars, while the lower ones are very similar objects but without chemical enrichment due to the 3rd dredge-up.

4. Binary post-AGB evolution

In this section we focus on the optically bright post-AGB stars showing a broad IR excess consisting not only of a cool dust emission but also of a significant hot dust component which subsequently turned out to be binaries (see Sect. 4.3). If one assumes that the broad IR excess is caused by hot circumstellar dust in radiative equilibrium with the radiation field of the central object, the dust must be very close to the central star. Since in most objects there is no evidence for a significant present day post-AGB mass-loss, the dust must be stored somewhere in the system presumably in a circumbinary dusty disc (Waters et al. 1993). Photometric monitoring campaigns showed that the fundamental period in the photometry for some objects is the binary period and the photometric behaviour can be understood as caused by variable circumstellar reddening in the case of HR 4049 and HD 213985 (Waelkens et al. 1991; Waelkens et al. 1995) or in terms of a changing scattering angle during orbital motion for HD 44179 (Waelkens et al. 1996). The extrema of the photometry occur at conjunction.

While the formation of this circumstellar dust geometry is not clear, binary interactions during phases of heavy mass-loss must be involved.

4.1. Case study: the Red Rectangle

The best studied circumbinary disc is certainly the one in the centre of the famous X-shaped reflection nebula, the Red-Rectangle (RR, Cohen et al. 1975). It has recently become clear that many of the remarkable phenomena observed in the RR are closely related to the special binary nature of the central object HD 44179 and the presence of its circumbinary disc (e.g. Waelkens et al. 1996). The disc is resolved in ground-based high-spatial-resolution imaging at optical and near-IR wavelengths (e.g. Roddier et al. 1995; Osterbart et al. 1997) as well as in HST snapshots (Bond et al. 1997). The central object is not seen directly at optical wavelengths but only in scattered light escaping from the poles of the thick disc. We give a short overview of the observational characteristics linked to this binarity: the atmosphere of HD 44179 is severely depleted in refractory elements (Waelkens et al. 1992) making the star one of the most iron deficient objects known ([Fe/H] = −3.3) in our Galaxy (see section 4.2); the extremely weak microwave CO emission with a small velocity width observed in the RR (Jura et al. 1995) together with the presence of very large grains inferred from cm continuum measurements (Jura et al. 1997) are distinct observational characteristics that can best be explained by long-term processing of dust in a stable circumstellar disc; the CO kinematics points to a Keplerian motion of this disc instead of a gas-outflow from the system. The stability of the disc was recently dramatically confirmed by the detection of *crystalline silicates* in this notorious carbon-rich nebula (Waters et al. 1998). The most likely scenario for the presence of oxygen-rich material is that the silicates in the long-lived disc were formed before HD 44179 became carbon-rich and expelled the C-rich nebula (Waters et al. 1998). Also the high crystallisation fraction of the silicates is a distinct signature of the strong processing of dust in stable disc-like geometries (Waters & Molster in this volume).

Although the interacting binary nature of the Red Rectangle is now well established, the carriers of the unique presumably molecular emission features and nebular emission, for which the nebula is also famous (Schmidt et al. 1980), are still largely unknown.

4.2. Depletion versus deficiency

It is now generally accepted that the extremely low photospheric iron abundances down to [Fe/H] = −4.8 of some post-AGB stars do not reflect initial conditions but are acquired during stellar evolution. During this process, dust formation selectively removes elements with a high dust-condensation temperature from the gas phase and subsequent gas-dust separation and gas accretion leave the photosphere depleted of these refractories. This depletion process mimics 3rd dredge-up nucleosynthetic photospheric abundance patterns by increasing the C/O ratio (since the CO molecule will be abundant in the gas-phase) and C/Fe, N/Fe and O/Fe abundance ratios, but can be identified by a high S abundance relative to adjacent α-process elements and a high Zn abundance relative to other Fe-peak elements (Venn & Lambert 1990; Bond 1991; Van Winckel et al. 1992). Since the s-process elements have high dust-condensation temperatures, they will also be depleted making the dredge-up history of depleted evolved atmospheres difficult to trace on the basis of photospheric spectra. IR circumstellar dust chemistry and molecular envelope chemistry are much more reliable tracers for these objects.

'Depleted' photospheric abundances therefore refer to the lack of refractory elements obtained by a chemical process during stellar evolution while 'deficient' abundances reflect sub-solar initial conditions or evolutionary decreased abundances from nucleosynthetic origin.

The binarity of *all* the extremely depleted objects (Van Winckel et al. 1995) and the observed presence of dust discs around some of them (e.g. Waelkens et al. 1991; Roddier et al. 1995), and the similar patterns observed in λ Bootis stars (Venn & Lambert 1990) led Waters et al. (1992) to suggest that the gas-dust separation needed to account for the depletion pattern is likely to take place in a (long-lived) disc. Binarity is in this scenario for post-AGB stars, a necessary condition for the depletion process to be effective.

4.3. Orbital elements of binary post-AGB stars

In Table 1 an overview is given of these post-AGB stars for which the binary period could be determined. The very wide spread in binary periods, in between 115 and 2600 days is apparent and since the broad IR excess is observed in all source (except BD +39.4926 where no IR excess is observed till now and HD 46703 where only cool dust is detected) it seems that binaries with a large spread in orbital periods are able to store at least part of the dust expelled during phases of heavy mass-loss in the system. The dust disc stability implies that the time elapsed since these objects have left the AGB cannot be derived from the inner radius of the dust shell !

For the shorter-period binaries, the actual orbit is too small to accommodate a normal AGB star and the actual orbital period as well as the often surprisingly high eccentricity set strong constraints on previous evolution. It is not clear how these systems avoided fatal spiral-in on the AGB. The carbon-rich nature of the circumstellar environment around several sources is the best observational evidence that there has indeed been internal chemical enrichment. These systems are by no means the only evolved binaries where the same problem occurs since also the shorter-period binaries from the Ba-star group have very similar periods and eccentricities. For more detailed comparisons between these post-AGB orbital elements and other classes of evolved binaries we refer to the review by Jorissen in this volume.

It is clear that these stars represent a peculiar stage in binary evolution rather than a genuine post-AGB evolutionary stage. This binarity has very fundamental consequences on the evolution of the system since it affects not only the SED evolution but also the chemical evolution, the mass loss and mass-loss history, the gas/dust ratio of the circumstellar environment and even the dust processing (see also Waters & Molster in this volume). Non-standard phenomena connected to the specific binary nature must therefore be explored to gain a better understanding of these systems.

4.4. RV Tauri stars and binarity

RV Tauri stars are relatively rare F,G or K variable supergiants characterised by alternating deep and shallow minima in their light curves (see also Lloyd Evans and Pollard & Lloyd Evans in this volume). With their high luminosity and often large IR-excesses due to thermal radiation of circumstellar dust, there is general agreement that RV Tauri stars are low-mass objects in an evolved, probably post-AGB, evolutionary stage (Jura 1986; Pollard & Lloyd Evans in this volume).

In a series of papers, Giridhar et al. (1994, 1998), Gonzalez et al. (1997a, 1997b) and Van Winckel et al. (1998b) have shown that the chemical composition of many field RV Tauri stars are also determined by depletion rather than dredge-up processes. If indeed binarity is a necessary condition for the depletion process to occur (see Sect. 4.2), the binary nature of these depleted RV Tauri stars should be explored. Recently AC Her, one of the prototypical RV Tauri stars, was indeed found to be a member of a wide binary system (Van Winckel et al. 1998) but binary motion is also observed in the RV Tauri stars IW Car (Pollard et al. 1996), EP Lyr (Gonzalez et al. 1997a), U Mon (Pollard & Cottrel 1995). Moreover ST Pup, a W Vir star showing also chemical deple-

Table 1. Binary post-AGB stars for which the period has been determined. All the objects (except BD+39 4926 and HD 46703) show the presence of hot dust which is interpreted as due to a dusty disc in the systems. Note the large spread in binary periods. The upper part consists of those objects where photospheric patterns are determined by depletion, the lower part where no chemical analysis was performed yet or no depletion patterns could be identified. Besides the orbital period, eccentricity and mass function, the chemical type of the circumstellar material is indicated. References are 1) Kodaira et al. 1970; 2) Pollard et al. 1996; 3) Waters et al. 1993; 4) Van Winckel et al. 1995; 5) Van Winckel et al. 1998a; 6) Hrivnak 1998; 7) Van Winckel et al. 1998b; 8) Van Winckel et al. 1999

Name	Period days	e	F(M) $M_\odot$	CS-chemistry
HR 4049[4]	429	0.31	0.143	C-rich
HD 44179[4]	318	0.38	0.049	C+O-rich
HD 52961[8]	1305	0.3	0.46	C-rich
BD+39.4926[1]	775			?
HD 213985[4]	259	0.0	0.97	C+O-rich
HD 46703[6]	600	0.34	0.28	?
AC Her[7]	1194	0.12	0.25	O-rich
SAO 173329[5]	115.9	0.0	0.026	?
89 Her[3]	288.4	0.19	0.00083	O-rich
HD 95767[5]	± 2050	± 0.3		?
HD 131356[8]	1534	0.3	0.72	?
U Mon[2]	2597	0.43	0.92	O-rich

tion patterns, is also a binary with a period of 410 days and the only B-type post-AGB star with a high S abundance relative to the other α-elements, is also a binary (Hambly et al. 1996).

Since large-amplitude pulsational radial-velocity variations and strong line-asymmetries and even line-splitting at some pulsational phases render direct detection of orbital motion cumbersome for RV Tauri stars, alternative tracers of the possible binary nature of these stars should be explored. One of the possibilities is a detailed analysis of the SED. Lloyd Evans (this volume) has shown that the dust-envelope of RV Tauri stars are characterised by a considerable near-IR excess. This disc-like appearance is very similar to the stars discussed in Sect. 4.3 which are known to be binaries.

Another possibility is a detailed analysis of the dust and molecular emission of the circumstellar material. The best studied example is AC Her for which there is strong observational evidence from CO rotational line flux and velocity measurements, sub-mm continuum measurements and from a full 2.3-40 μm SWS ISO spectrum that it is also surrounded by a long-lived dust-disc (Van Winckel et al. 1998b and references therein) very similar to the oxygen-rich dust-disc observed in the Red Rectangle (Waters & Molster in this volume). In

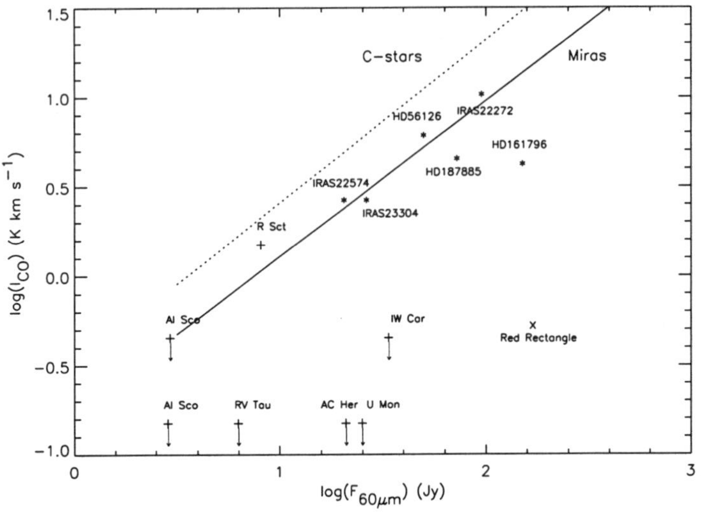

Figure 2. The CO (1-0) microwave integrated intensity versus the 60 μm flux both in log of post-AGB objects. The relation for C-stars (O-rich Miras) as determined by Nyman et al. (1992) is given by a dotted (full) line. The asterisks give the positions of double-peaked post-AGB stars, the +-signs indicate the RV Tauri stars. The arrows indicate upper limits. Data are from Bujarrabal et al. 1988; Alcolea et al. 1991; Loup et al. 1993; van der Veen et al. 1993; Jura et al. 1995

order to investigate the molecular emission of the circumstellar environment of RV Tauri stars we compiled in Fig. 2 microwave CO emission measurements of these stars and of the double-peaked post-AGB stars. It clearly shows that the circumstellar environment around RV Tauri stars is deficient in CO while other known and suspected single post-AGB stars like the 21 μm-objects or HD 161796 do follow the CO/60 μm relation known for AGB objects confirming that in the latter objects the IR excess is formed by the ejected AGB superwind. A similar and even stronger CO deficiency is observed in the Red Rectangle (Jura et al. 1995). The depletion is clearly reflected in the gas/dust ratio of the cool circumstellar dust envelope. This indicates again that the depletion process is intimately related to the special geometry of the circumstellar material.

Assuming that indeed the circumstellar material is trapped in the system, it is quite possible that the viewing angle onto the binary orbit determines the photometric class of the RV Tauri star: objects with a low orbital inclination would be the RVa objects with their constant mean magnitude, while those with a high inclination so that the disc is seen nearly edge-on would be the class RVb objects where the variability of the mean magnitude is due to variable circumstellar extinction during orbital motion.

The commonly observed photospheric chemical depletion pattern, the peculiar SEDs and the generally weak and narrow microwave CO emission of RV Tauri stars together with the difference in photospheric class of the RV Tauri objects can all be naturally explained by assuming that the circumstellar material is not freely expanding, but trapped in the binary system.

5. Conclusions

We can conclude that the 21 μm objects are the successors of the carbon stars. The wide spread in integrated neutron exposure they display together with the absence of 3rd dredge-up yields in the photospheres of very similar stars is still not well understood.

There is strong observational evidence that the final evolution of binary stars with a very wide period distribution (from 116 to 2600 days) is strongly affected by this binarity. The fundamental property seems to be that at least part of the circumstellar material is trapped in a circumbinary disc which has a strong impact on the chemical and dynamical evolution of the system. There is still good observational evidence that *depletion of refractory elements only occurs in binaries.* The high pole to orbital-plane density gradient due to the dusty disc in these binaries will certainly affect also future evolution during the PN phase. Waters & Molster (in this volume) shows that also the dust evolution and processing is very different from the processing in normal outflows.

Since binaries display a prolonged IR lifetime due to the presence of this stable circumbinary dusty disc, any sample of post-AGB objects defined on IR measurements is bound to be biased towards binaries. Also in the large sample of post-AGB stars with a very weak optical counterpart, binaries must be present. The best candidates are those with a very large pole-to-equator density gradient in the circumstellar envelope.

Acknowledgments. The many instructive discussions with Rens Waters and Christoffel Waelkens are warmly acknowledged. HVW is post-doctoral fellow of the Fund for Scientific Research, Flanders (Belgium)

References

Alcolea J., Bujarrabal V., 1991, A&A 245, 499
Bakker E.J., Waters L.B.F.M., Lamers H.J.G.L.M., et al., 1996, A&A 310, 893
Bakker E.J., Van Dishoeck E.F., Waters L.B.F.M., Schoenmaker T., 1997, A&A 323, 469
Bakker E.J., Lambert D.L., 1998, ApJ 502, 417
Blöcker T., 1995, A&A 299, 755
Bond H.E., 1991, IAU Symp. 145, p. 341
Bond H.E., Fullton L.K., Schaefer K. et al., 1997, IAU Symp. 180, p. 211
Bujarrabal V., Bachiller R., Alcolea J., Martín-Pintado, 1988, A&A 206, L17
Cohen M., Anderson C.M., Cowley A., et al., 1975, ApJ 196, 179
Conlon E.S., Dufton P.L., Keenan F.P., McCausland R.J.H., 1991, MNRAS 248, 820
Conlon E.S., Dufton P.L., Keenan F.P., et al., 1993, A&A 272, 243
Dayal A., Hoffmann W.F., Bieging J.H., et al., 1998, ApJ 492, 603
Decin L., Van Winckel H., Waelkens C., Bakker E.J., 1998, A&A 332, 928
Giridhar S., Rao N.K., Lambert D.L., 1994, ApJ 437, 476
Giridhar S., Lambert D.L., Gonzalez G., 1998, ApJ in press
Gonzalez G., Wallerstein, G., 1996, MNRAS 280, 515
Gonzalez G., Lambert D.L., Giridhar S., 1997a, ApJ 479, 427
Gonzalez G., Lambert D.L., Giridhar S., 1997b, ApJ 481, 481
Hambly N.C., Dufton P.L., Keenan F.P., Lumsden S.L., 1996, MNRAS 278, 811
Hrivnak B.J., 1995, ApJ 438, 341

Hrivnak B.J., 1998, in "*Carbon Stars*", IAU Symp. 177, in press
Hrivnak B.J., Kwok S., Volk K.M., 1989, ApJ 346, 265
Jura M., 1986, ApJ 309, 732
Jura M., Balm S.P., Kahane C., 1995, ApJ 453, 721
Jura M., Turner J., Balm S.P., 1997, ApJ 474, 741
Klochkova V.G., 1995, MNRAS 272, 710
Kodaira K., Greenstein J.L., Oke J.B., 1970, ApJ 159, 485
Kwok S., Volk K.M., Hrivnak B.J., 1989, ApJ 345, 51
Loup C., Forveille T., Omont A., Paul J.F., 1993, A&AS 99, 291
Luck R.E., Bond H.E., Lambert D.L., 1990, ApJ 357, 188
McCausland R.J.H., Conlon E.S., Dufton P.L., Keenan F.P., 1992, ApJ 394, 298
Meixner M., 1997, IAU Symp. 180, p. 325
Meixner M., Skinner C.J., Graham J.R., et al., 1997, ApJ 482, 897
Moehler S., Heber U., 1998, A&A 335, 985
Napiwotzki R., Heber U., Köppen J., 1994, A&A 292, 239
Nyman L.-Å., Booth R.S., Carlström U.U., et al., 1992, A&AS 93, 121
Omont B.E., Moseley S.H., Cox P., et al., 1995, ApJ 454, 819
Osterbart R., Langer N., Weigelt G., 1997, A&A 325, 609
Oudmaijer R.D., van der Veen W.E.C.J., Waters L.B.F.M., et al., 1992, A&AS 96, 625
Pollard K.R., Cottrel P.L., 1995, ASP Conf. Ser. 83, 409
Pollard K.R., Cottrell P.L., Kilmartin P.M., Gilmore A.C., 1996, MNRAS 279, 949
Pottasch S.R., Parthasarathy M., 1988, A&A 192, 182
Reddy B.E., Parthasarathy M., Gonzalez G., Bakker E.J., 1997, A&A 328, 331
Roddier F., Roddier C., Graves J.E., Northcott M.J., 1995, ApJ 443, 249
Schmidt G.D., Cohen H., Margon B., 1980, ApJ 239, L133
Skinner C.J., Meixner M.M., Hawkins, G.W., et al., 1994, ApJ 423, L135
Szczerba R., Omont A., Volk K., Cox P., Kwok S., 1997, A&A 317, 859
Trams N.R., Waters L.B.F.M., Waelkens C., et al., 1991, A&AS 87, 361
van der Veen W.E.C.J., Trams N.R., Waters L.B.F.M., 1993, A&A 269, 231
Van Winckel H., 1997, A&A 319, 561
Van Winckel H., Mathis J.S., Waelkens C., 1992, Nature 356, 500
Van Winckel H., Waelkens C., Waters L.B.F.M., 1995, A&A 293, L25
Van Winckel H., Waelkens C., Waters L.B.F.M., 1996, A&A 306, L37
Van Winckel H., Waelkens C., Waters L.B.F.M., 1998a, IAU Symp. 177, in press
Van Winckel H., Waelkens C., Waters L.B.F.M., et al., 1998b, A&A 336, L17
Van Winckel H., Waelkens C., Fernie J.D., Waters L.B.F.M., 1999, A&A, accepted
Venn K.A., Lambert D.L., 1990, ApJ 363, 234
Venn K.A., Smartt S.J., Lennon D.J., Dufton P.L., 1998, A&A 334, 987
Waelkens C., Lamers H.J.G.L.M., Waters L.B.F.M., et al., 1991, A&A 242, 433
Waelkens C., Van Winckel H., Trams N.R., Waters L.B.F.M., 1992, A&A 256, L15
Waelkens C., Waters L.B.F.M., Van Winckel H., Daems K., 1995, Ap&SS 224, 357
Waelkens C., Van Winckel H., Waters L.B.F.M., Bakker E.J., 1996, A&A 314, L17
Waters L.B.F.M., Trams N.R., Waelkens C., 1992, A&A 262, L37
Waters L.B.F.M., Waelkens C., Mayor M., Trams N.R., 1993, A&A 269, 242
Waters L.B.F.M., Waelkens C., Van Winckel H., 1997, IAU Symp. 180, p. 211
Waters L.B.F.M., Waelkens C., Van Winckel H., et al., 1998, Nature 391, 868
Začs L., Klochkova V.G., Panchuk V.E., 1995, MNRAS 275, 764
Začs L., Klochkova V.G., Panchuk V.E., Spēlmanis R., 1996, MNRAS 282, 1171

Asymptotic Giant Branch Stars
T. Le Bertre, A. Lèbre and C. Waelkens, eds.
I.A.U. Symposium 191, 1999

Evolutionary connection between C-rich AGB stars and C-rich central stars of PNe

M. Parthasarathy

Indian Institute of Astrophysics, Bangalore - 560034, India

Abstract. On the basis of C/O ratios the evolutionary connection between C-rich AGB/post-AGB stars and C-rich central stars of PNe is discussed. The proportion of carbon-rich post-AGB stars is comparable to that of carbon-rich PNe. The post-AGB stars with 21μm emission are found to be over abundant in carbon and s-process elements indicating that they have gone through the third dredge-up and carbon star phase on the AGB. As these stars evolve to higher temperature they will form carbon-rich PNe with carbon-rich central stars. Analysis of C/O ratios of PNe suggests that 68 percent of PNe have C/O > 1 indicating that the progenitors have gone through the third dredge-up and carbon-rich phase on the AGB. The Type I PNe can be divided into two groups. Group I Type I PNe are those with C/O < 1 indicating that their massive progenitors have gone through the third dredge-up and hot bottom burning on the AGB. Group II Type I PNe are those with C/O > 1. Their progenitors have gone through third dredge-up only and may not have experienced hot bottom burning. Carbon stars with silicate dust and or carbon-rich post-AGB stars (such as post-AGB stars with 21μm emission) may be the progenitors of PNe with [WC] central stars. The evolutionary sequence seems to be: carbon-rich AGB → carbon-rich post-AGB → [WC late type] → [WC early type] → [WC-PG 1159] → PG 1159.

1. Introduction

Red giant stars are believed to evolve along the asymptotic giant branch (AGB) through the M → MS → S → C sequence. Along this sequence the abundance of C^{12} increases by mixing of Helium-burning products in the stellar surface. Thus the evolution along the AGB and third dredge-up transform an oxygen-rich AGB star into a carbon-rich AGB star with overabundances of s-process elements and C^{12}. The question of burning and mixing along the AGB is intimately linked to the carbon star mystery. Observations of the LMC and SMC prove that most of the carbon and s-process element enriched AGB stars are found at rather low luminosities indicating that they also have rather low masses (1 to 3 $M_{\odot}$). On the other hand, AGB models predict the 3rd dredge up i.e the mixing of the interior carbon to the surface during the recurrent He-shell flashes on the AGB, mostly for much higher luminosities (i.e. larger (core) masses). The reason for the absence of luminous carbon stars is most likely due to the occurrence of hot bottom burning and mass loss on the AGB.

The hypothesis that carbon-rich AGB stars with high mass loss rates evolve into planetary nebulae (Zuckerman et al. 1976) is well supported by the result that approximately half of the mass loss from AGB stars in the solar neighbourhood results from carbon stars (Jura & Kleinmann 1989) and that half of all planetary nebulae in the solar neighbourhood are carbon-rich (Zuckerman & Aller 1986).

Planetary nebulae (PNe) reflect the evolutionary histories of the AGB stars that preceded them and carry the by-products of nuclear burning - enriched levels of helium, nitrogen and carbon. PNe allow to determine how the production of chemical elements depends on initial mass or metallicity and how the products are dredged-up during the late stages of stellar evolution.

Carbon abundances of post-AGB stars and PNe are of particular interest in this respect. The link between AGB stars and PNe is most directly seen from the presence of the remnants of the AGB molecular envelopes around several PNe and post-AGB stars.

From an analysis of the IRAS data several post-AGB supergiant like stars/ proto-planetary nebulae (PPNe) have been detected. The cooler (K, G, F) and the hotter (A, B) post-AGB stars seem to form an evolutionary sequence in the transition region from the tip of the AGB into early stages of PNe (Parthasarathy & Pottasch 1986; Parthasarathy 1993a, 1993b, 1994; Parthasarathy et al. 1993, 1995).

2. Carbon-rich post-AGB stars

The chemical composition analysis of post-AGB stars with 21 μm emission shows that they are all carbon-rich and over-abundant in s-process elements. The CO and HCN data (Omont et al. 1993) and infrared spectra indicate that the dust shells around these stars are also carbon-rich. Recently Webster (1995) suggested that the 21 μm emission observed in these carbon-rich post-AGB stars is due to fullerences C_{60} Hm (m = 1 to 60). Emission at 3.3 and 3.4 μm, commonly attributed to polycyclic aromatic hydrocarbons (PAHs), is seen in several carbon-rich PPNe.

Recently, using mid-infrared spectra of several carbon-rich post-AGB stars with 21 μm emission Justtanont et al. (1996) found in addition to the standard UIR features (6.2, 7.7, 8.6, 11.3 μm attributed to PAHs) new features at 7.9, 8.2, 10.6, 11.5 and 12.2 μm, which they attribute to PAH molecule chrysene. The UIR features were also observed in several young PNe and in [WC] PNe. The 21 μm feature is found only in carbon-rich post-AGB stars. There are 12 carbon-rich post-AGB stars with 21 μm emission feature (Kwok 1993; Decin et al. 1998). All of them show overabundance of carbon and s-process elements indicating that they have gone through the third dredge-up and carbon star phase on the AGB (Parthasarathy et al. 1992; Hrivnak 1995; Decin et al. 1998; Reddy et al. 1997; Začs et al. 1996; Van Winckel 1997). As these stars evolve to higher temperatures they form carbon-rich PNe with carbon-rich central stars.

3. Carbon-rich PNe

Zuckerman & Aller (1986) from an analysis of C/O ratios of large sample of PNe find that 62 percent of PNe have C/O > 1. This is in agreement with the relative numbers of carbon and oxygen-rich AGB stars with large mass loss rates and is consistent with the idea that the progenitors of PNe are this class of AGB stars. Among PNe with [WC] type central stars about 80% percent of the nebulae are carbon-rich. Among 68 PNe studied by Zuckerman & Aller (1986) 41 have bipolar or elliptical symmetry. Of these 59 percent are carbon-rich. Therefore, bipolar or elliptical symmetry is equally probable regardless of C/O ratio. Rola & Stasinska (1994) reexamined the problem of the discrepancy between carbon abundances in PNe derived from the CII 4267 Å and CIII] 1909 Å line intensities. They find that about 50 percent of PNe are carbon-rich (C/O > 1). Rola and Stasinka (1994) found that 29 percent of Type I are carbon-rich. Slightly lower percentage of the carbon-rich PNe in the sample of Rola & Stasinka (1994) may be due to the fact that they use C/O > 1.2 to consider a PN as definitely carbon-rich.

Recently Kholtygin (1998) reconsidered the problem of the difference between the abundances of carbon and oxygen in galactic planetary nebulae derived from the intensities of the recombination and collisionally excited lines. Kholtygin (1998) redetermined the C/O ratios of a large number of PNe. The C/O ratios of PNe determined by Kholtygin are more reliable than the previous estimates. From this data we find that 68 percent of PNe are carbon-rich. Jura & Kleinman (1989) find that about one half of all the mass losing AGB stars are carbon-rich. They conclude that the carbon star phase occurs in about one half of all main-sequence stars of 1 to 1.5 $M_\odot$.

4. Type-I and bipolar PNe

Their kinematic properties and galactic distribution indicate that the masses of the progenitor stars were higher than those of other PNe. Most of the galactic He-N rich (Type I) PNe are bipolar and most bipolar PNe are He-N rich. PNe with He/ H > 0.125 or N/O > 0.5 were defined as Type I (Peimbert 1978). Recently Kingsburgh & Barlow (1994) defined Type I PNe as those with N/O ratio (> 0.8) higher than their initial (C+N)/O ratio implying a complete conversion of the initial C into N and also of the C produced by the third dredge-up.

We find that about 50 percent of Type I PNe have C/O> 1 indicating that they have experienced third dredge-up. On the basis of C/O ratios of Type I and bipolar PNe we find that the Type I PNe can be classified into two groups. Group I Type I PNe are those with C/O < 1 and Group II Type I are those with C/O> 1. These two groups may be related to the hot bottom burning and third dredge-up of progenitor stars during their evolution on the AGB. Several of the carbon-rich PNe, C-rich Type I PNe, bipolar PNe and PNe with [WC] central stars have CO molecular envelopes (Huggins et al. 1996) and molecular hydrogen emission (Kastner et al. 1996) indicating a range in masses of the progenitors for the carbon-rich PNe.

Jura (1991) examined the published data for carbon stars losing more mass than $10^{-6} M_\odot$ yr^{-1} in the solar neighborhood. The carbon stars with higher terminal velocities have significantly higher fluxes of the J = 1-0 emission line of HCN relative to their infrared continuum emission at 60 μm. Jura (1991) argued that this result is probably a reflection of a higher nitrogen abundance by a factor of 10 in these outflows. Such a high nitrogen abundance is consistent with the hypothesis that the high terminal velocity carbon stars have more massive main-sequence progenitors. It is likely that these high terminal velocity carbon stars may, themselves, be the progenitors of Type I PNe.

Recently Clegg et al. (1997) determined $^{12}C/^{13}C$ isotope ratio in the ionized gas in 2 PNe, NGC 3918 and SMC N2. Clegg et al. (1997) derived $^{12}C/^{13}$C ratios to be 15 and 7 respectively. NGC 3918 and SMC N2 are carbon-rich (C/O 1.6 and 3.6 respectively) and have probably descended from C-rich AGB stars. The $^{12}C/^{13}C$ ratio in these nebulae is lower than the typical values in carbon stars. This may be due to hot bottom burning in the progenitors.

Leisy & Dennefeld (1996) determined carbon abundances of several Type I PNe in LMC and SMC. From the study of carbon abundances in LMC and SMC it is found that the third dredge-up has been occurring in all PNe. The third dredge-up efficiency is higher with lower metallicities which is in agreement with the higher number of carbon stars found in metal-deficient galaxies. Leisy & Dennefeld (1996) study of carbon abundances in LMC and SMC PNe shows that the third dredge-up carbon is sometimes transformed into nitrogen by hot-bottom burning, but in a few objects only and not even in all the Type I PNe. This in agreement with our classification of Type I PNe into two groups.

Type I PNe with C/O< 1 and with over-abundance of nitrogen have gone through the third dredge-up and hot-bottom burning on the AGB. They are relatively more massive than Type I PNe with C/O>1 which have probably gone through third dredge-up and a carbon-rich phase on the AGB but not experienced hot-bottom burning. The question whether the progenitors of the Type I PNe have gone through the carbon star stage is governed by the competition between dredge-up and hot-bottom burning, i.e. the efficiencies of both processes which, in turn, depend on convection and mass loss (see the review by Blöcker in this volume).

5. PNe with [WC] central stars

About 7 percent of the known central stars of planetary nebulae (CSPN) show Wolf-Rayet (WR) spectrum falling all in the [WC] class (Acker et al. 1992, 1996). The [WC] central stars are helium and carbon rich. More than 80 percent of the nebulae with [WC] central stars are carbon-rich (C/O > 1). The distribution of the spectral types of [WC] stars shows that most of them are either early type [WC2-4] and late type [WC9-11]. The reason for the paucity of [WC5-7] types is not clear. The temperatures of the [WC] central stars range from 20 000K for the [WC11] stars to as much as 150 000K for the [WC2-3]. The over-abundance in carbon of the PNe with [WC] central stars shows that the progenitors have gone through the third dredge-up and carbon-rich phase on the AGB. Some of the PNe with [WC] central stars are Type I and or bipolar and have CO molecular envelopes (Huggins et al. 1996) and molecular hydrogen emission (Kastner et al. 1996) indicating a range in the masses of progenitors. It is quite likely that post-

AGB stars of different initial masses can pass through a [WC] stage with similar atmospheric parameters.

The PNe with [WC] central stars in LMC and SMC are also found to be extremely carbon-rich (Pena et al. 1997). The high electron densities of these PNe indicate that they are formed only recently from the termination of the AGB phase of progenitor carbon-rich stars.

The formation of [WC] central star with extreme under-abundance of hydrogen and over-abundance of C and He (C/He $\simeq$ 1) may be the result of thermal pulse either at the very end of the carbon-rich AGB phase or during the young PNe stage, or when the central star is already on the white dwarf cooling track (very late thermal pulse or born-again AGB scenario, Iben 1984).

Tylenda, Acker & Stenholm (1993) determined the [WC] subclasses for a large number of CSPN. They found several "weak emission line central stars" (WELS) which have the CIV 5805Å line systematically weaker and narrower than early [WC] CSPN and also the CIII 5696Å line very weak or absent. Parthasarathy, Acker & Stenholm (1998) found that the characteristics of the spectra of weak emission line stars are very similar to that of [WC]-PG1159 objects. The broad HeII/CIV absorption complex some times with NLTE emission cores is the spectroscopic hallmark in the optical to identify PG1159 stars. On the basis of the spectra Parthasarathy, Acker & Stenholm (1998) discovered 30 new [WC]-PG1159 stars. These stars mark the transition from [WC] central stars of planetary nebulae to the PG1159 (pre)white dwarfs. LM1-64 is most likely a [WC]-PG1159 CSPN in the LMC.

The presence of nebulae and dust shells around these stars and also the significant percentage of [WC]-PG1159 stars among the [WC] central stars indicate that very late thermal pulse or born-again AGB scenario is not the principal mechanism for the post-AGB evolution of [WC] stars towards the non-DA white dwarf stage.

Recent ISO observations have revealed that a significant fraction of [WC] PNe have cool oxygen-rich dust (Waters et al. 1998). In the wavelength region shortward of 15 μm the spectra of PNe with [WC] central stars show PAH features indicating carbon-rich dust, whereas longward of 18 μm solid state features from crystalline silicates indicating oxygen-rich dust are seen. The transition to [WC] type (from oxygen-rich state to carbon-rich state) may have taken place during the evolution from the tip of the AGB and before the central star attained a temperature of 20 000K (within a time span of the order of 1 000 years).

In the infrared the spectral features of PNe with [WC] central stars are found to be similar to that observed in the spectra of carbon-rich post-AGB stars and post-AGB stars with 21μm emission (UIR and PAH features). Carbon stars with silicate dust and or carbon-rich post-AGB stars (such as post-AGB stars with 21μm emission) may be the progenitors of PNe with [WC] central stars. The evolutionary sequence seems to be carbon-rich AGB $\rightarrow$ carbon-rich post-AGB $\rightarrow$ [WC late] $\rightarrow$ [WC early] $\rightarrow$ [WC]-PG1159 $\rightarrow$ PG1159.

References

Acker A., et al., 1992, Strasbourg-ESO Catalogue of Galactic Planetary Nebulae, ESO, Garching
Acker A., et al., 1996, A&A 305, 944
Clegg R.E.S., et al., 1997, MNRAS 284, 348
Decin L., et al., 1998, A&A 332, 928
Hrivnak B.J., 1995, ApJ 438, 349
Huggins P.J., 1996, A&A 315 284
Jura M., 1991, ApJ 372, 208
Jura M., Kleinmann S.G., 1989, ApJ 341, 359
Justtanont K., et al., 1996, A&A 309, 612
Kastner J.H., 1996, ApJ 462, 777
Kholtygin A.F., 1998, A&A 329, 691
Kingsburgh R.L., Barlow M.J., 1994, MNRAS 271, 257
Kwok S., 1993, ARA&A 31, 63
Leisy P., Dennefeld M., 1996, A&A Suppl. 116, 95
Omont A., et al., 1993, A&A 267, 515
Parthasarathy M., 1993a, in "Luminous High Latitude Stars", ed. D.D. Sasselov, ASP Conf. Ser. 45, 173
Parthasarathy M., 1993b, ApJ 414, L109
Parthasarathy M., et al., 1993, A&A 267, L19
Parthasarathy M., 1994, in "The MK process at 50 years", ed. C.J. Corbally et al., ASP Conf. Ser. 60, 261
Parthasarathy M., et al., 1995, A&A 300, L25
Parthasarathy M., Acker A., Stenholm B., 1998, A&A 329, L9
Parthasarathy M., Garcia-Lario P., Pottasch S.R., 1992, A&A 264, 159
Parthasarathy M., Pottasch S.R., 1986, A&A 154, L16
Peimbert M., 1978, IAU Symposium 76, ed. Y. Terzian, Reidel, p. 215
Pena M., Ruiz M.T., Torres-Peimbert S., 1997, A&A 324, 674
Reddy B.E., Parthasarathy M., Gonzalez G., Bakker E.J., 1997, A&A 328, 331
Rola C., Stasinska G., 1994, A&A 282, 199
Tylenda R., Acker A., Stenholm B., 1993, A&A Suppl. 102, 595
Van Winckel H., 1997, A&A 319, 561
Waters L.B.F.M., et al., 1998, A&A 331, L61
Webster A., 1995, MNRAS 277, 1555
Začs L., et al., 1995, MNRAS 275, 764
Zuckerman B., et al., 1976, ApJ 205, L15
Zuckerman B., Aller L.M., 1986, ApJ 301, 772

Asymptotic Giant Branch Stars
T. Le Bertre, A. Lèbre and C. Waelkens, eds.
I.A.U. Symposium 191, 1999

Sakurai's object – stellar evolution in real time

Martin Asplund

Nordita, Blegdamsvej 17, DK-2100 København Ø, Denmark

Abstract. The born-again giant Sakurai's object is currently experiencing its *second* stage as an AGB-star. Furthermore, Sakurai's object has shown an unprecedented rapid stellar evolution since discovery in 1996, both in terms of a continued cooling of the photosphere and spectacular changes in chemical composition on a time-scale of a mere few months. The surface cooling and abundance alterations of H, Li and the *s*-elements are the direct consequences of the expansion, mixing and nucleosynthesis which has ensued as a result of a final He-shell flash occurring while the star was descending the white dwarf cooling track. Sakurai's object shows striking similarities with the R CrB stars in both chemical composition and visual variability.

1. Introduction

The incongruity of the time-scales of stellar and human life-times has throughout much of human evolution resulted in the impression that stars are immutable. Today we know that stars are born, live and die, though the time-scale for this evolution is typically million or billions of years. Our understanding of stellar evolution thus predominantly comes from observing many stars in different evolutionary stages and relying on theoretical modelling. Witnessing stellar evolution in real time is therefore extraordinarily rare. Supernovae and novae provide of course notable exceptions, though such cataclysmic events only represent the ultimate end as normal stars. "Born-again" giants, however, offer a unique opportunity to study stellar evolution on a time-scale more appropriate to the life-time of astronomers, which gives important direct clues to how stellar evolution and nucleosynthesis proceeds during late stages of stellar evolution.

During the descent of the white dwarf cooling track a final He-shell flash (thermal pulse) may occur, which balloons the star back to supergiant dimensions due to the released energy from the once-again initiated nuclear burning. The star thus makes a second appearance as an AGB-star: a born-again giant. Though rarely observed, such a scenario is not unlikely for intermediate- and low-mass stars: theoretical estimates suggest that about 10-20% of such stars experience a phase as born-again giants. The life-time as a rejuvenated giant is, however, short – typically only 100-1000 years – before the star re-traces its previous evolution towards the planetary nebulae phase and eventually the white dwarf domain, this time with no turning back.

Only a few candidates for having been born-again giants are known. Best known is perhaps FG Sge which has evolved from being an O star a century ago to now being a late-type giant. V605 Aql and a few central stars of PNe may also have experienced a similar final He-shell flash, and the R CrB stars may be descendants of such events (Asplund et al. 1998a; Lambert et al. 1998). The remarkably rapid evolution of Sakurai's object, however, outpaces all other stars, and provides a unique opportunity to study stellar evolution during the very act, using all modern means of observations. We have monitored the evolution of Sakurai's object since discovery in 1996 using high-resolution optical spectra (Asplund et al. 1998b), which has allowed us to trace the rapidly changing appearance of the star both in terms of surface temperature and chemical composition.

2. Observations and analysis

Several studies of the chemical composition of Sakurai's object have been presented during the last years (Asplund et al. 1997b; Kipper & Klochkova 1997; Shetrone & Keane 1998). However, since the various analyses have relied on very different types of model atmospheres and have utilized different spectral features, it has been difficult to directly compare the results. Here we present a homogeneous re-analysis of all available published spectra of the star from 1996 with appropriate model atmospheres (Asplund et al. 1997a) and whenever possible the same spectral lines to avoid systematic differences.

The spectra have been obtained with various telescopes at varying resolution and S/N. The May and October 1996 spectra were acquired with McDonald Observatory 2.1 and 2.7 m telescopes at a resolving power of 30-60 000 (Asplund et al. 1997b). Sakurai's object was also covered with the CTIO 4 m telescope in April 1996 (Shetrone & Keane 1998) and with the SAO 6 m telescope in July 1996 (Kipper & Klochkova 1997) though with slightly inferior resolution. The S/N of the individual spectra varies between 75 and 200. We defer the analysis of the 1997 and 1998 spectra to a later study, mainly due to the added complications arising from the terribly rich atomic and molecular line spectrum of later dates.

Especially constructed line-blanketed, H-deficient model atmospheres similar to those of Asplund et al. (1997a) formed the basis of the analysis. The stellar parameters were estimated using a range of ionization balance (mainly N I/N II, Si I/Si II, Cr I/Cr II, and Fe I/Fe II), excitation balance (O I, Fe I and Fe II) and line strengths of sensitive spectral features (C_2, C II and He I). Further information was provided by the hydrogen Balmer line profiles and the $(B-V)_0$ colours (Duerbeck et al. 1997). The microturbulence parameter was estimated from C I, Ti II, Fe I and Fe II lines of various strengths. The spectroscopically derived $T_{\rm eff}$ agrees very well with the photometric estimates (Duerbeck et al. 1997) using the same model atmospheres as those adopted in the present study, as evident from Fig. 1. Typical uncertainties are $\Delta T_{\rm eff} = \pm 300$ K, $\Delta \log g = \pm 0.3$ dex, $\Delta\xi_{\rm turb} = \pm 1.0\,{\rm km\,s^{-1}}$. From the C II and He I lines in the April and May spectra, C/He=10% (by number) was estimated. It should be noted, however, that the derived abundance from the C I lines result in a discrepancy of 0.6-0.7 dex: the same magnitude of the infamous "carbon problem" as those hampering all analyses of R CrB stars and related objects (Asplund et al. 1997a; Asplund et al. 1997b; Asplund et al. 1998a; Asplund et al. 1998b; Kipper & Klochkova 1997;

Table 1. The chemical compositions of Sakurai's object in 1996 (normalized to $\log \Sigma\mu_i\epsilon_i = 12.15$). The errors quoted in the table are the formal statistical errors; no error is given in case the abundance is derived from a single line. Uncertain values are marked with ":"

Element	Sun	Sakurai's object in 1996				$\Delta\log\epsilon_i$
		April 20-25	May 5-9	July 3	October 7	
H	12.00	10.0	9.7	9.6	9.0	⇓ 1.0
He	10.93	11.4	11.4	11.4	11.4	0.0
Li	3.31	3.6	3.6	4.0	4.2	⇑ 0.6
C	8.52	9.7 ± 0.2	9.7 ± 0.2	9.7 ± 0.3	9.8 ± 0.3	0.1
N	7.92	9.0 ± 0.3	8.9 ± 0.4	9.2 ± 0.3	8.9 ± 0.2	0.3
O	8.83	9.2 ± 0.2	9.5 ± 0.3	9.3 ± 0.1	9.4 ± 0.2	0.3
Ne	8.08	9.4 ± 0.2	9.4 ± 0.3	9.5 ± 0.3		0.1
Na	6.33	6.6 ± 0.1	6.7 ± 0.1	6.6 ± 0.2	6.8 ± 0.1	0.2
Mg	7.58	6.5 ± 0.4	6.6 ± 0.4	$6.3 : \pm 0.4$	6.5 ± 0.3	0.3
Al	6.47	6.5 ± 0.2	6.6 ± 0.2	6.6 ± 0.3	6.3	0.3
Si	7.55	7.3 ± 0.0	7.1 ± 0.2	7.1 ± 0.0	7.5 ± 0.2	0.4
P	5.45	6.2	6.2 ± 0.4	6.3		0.1
S	7.33	6.8 ± 0.1	6.6 ± 0.1	6.7 ± 0.1	6.9 ± 0.1	0.3
K	5.12		4.9 ± 0.0	5.0 ± 0.1	5.0 ± 0.0	0.1
Ca	6.36	5.2 ± 0.1	5.6 ± 0.3	5.6 ± 0.4	5.5 ± 0.4	0.4
Sc	3.17	3.1 ± 0.1	3.1 ± 0.1	3.3 ± 0.2	3.9 ± 0.2	⇑ 0.8
Ti	5.02	4.0 ± 0.1	4.1 ± 0.2	4.4 ± 0.2	4.6 ± 0.2	⇑ 0.6
Cr	5.67	4.5 ± 0.1	4.5 ± 0.2	4.8 ± 0.2	5.1 ± 0.2	⇑ 0.6
Fe	7.50	6.4 ± 0.2	6.4 ± 0.2	6.6 ± 0.2	6.6 ± 0.3	0.2
Ni	6.25	6.1 ± 0.3	6.1 ± 0.4	6.0 ± 0.2	6.2 ± 0.2	0.2
Cu	4.21		5.0 ± 0.3	5.1 ± 0.0	5.0 ± 0.1	0.1
Zn	4.60	4.9 ± 0.2	4.8 ± 0.2	5.1	5.4	⇑ 0.6
Rb	2.60		< 3.7		4.6	⇑ > 0.9
Sr	2.97	$4.7 : \pm 0.1$	$4.9 : \pm 0.2$		$5.4 : \pm 0.0$	⇑ 0.7
Y	2.24	3.2 ± 0.3	3.3 ± 0.3	3.7 ± 0.2	4.2 ± 0.2	⇑ 1.0
Zr	2.60	3.1 ± 0.2	3.0 ± 0.3	3.3 ± 0.2	3.5 ± 0.3	⇑ 0.5
Ba	2.13	1.5 ± 0.1	1.5 ± 0.2	1.8 ± 0.1	1.9 ± 0.4	0.4
La	1.17		< 1.6	1.3	1.5	0.2

Lambert et al. 1998). The reason for this anomaly is still unknown but it may reflect inadequacies in the modelling of the stellar photospheres (Lambert et al. 1998). Fortunately, the derived abundance *ratios* such as [X/Fe] are relatively immune to the carbon problem (Lambert et al. 1998) and therefore important conclusions regarding the evolution of Sakurai's object can still be drawn, in particular in light of the very extreme abundance ratios presented by Sakurai's object.

3. Elemental abundances

The derived LTE abundances in Sakurai's object throughout 1996 are presented in Table 1. Two important conclusions are immediately obvious from an inspection of the table. First, the abundances of most elements (He, C, N, O, Ne, Na, Mg, Al, Si, P, S, K, Ca, Fe, Ni, Cu and La) have remained the same within the uncertainties (about 0.3 dex), which is reassuring and strengthens our confidence in the estimated stellar parameters. Needless to say, the stellar parameters were obtained before considering the abundances of individual elements, and no a posteriori modifications have been allowed. Second, several elements show abundance alterations which significantly exceed the statistical scatter and the estimated errors, and thus likely reflect real variations. The elements having undergone such alterations are H (decrease by 1.0 dex), Li (+0.6 dex), Sc (+0.8 dex), Zn (+0.7 dex), Rb (> +0.9 dex) and Y (+1.0 dex), and probably also Ti (+0.6 dex), Cr (+0.6 dex), Sr (+0.7 dex) and Zr (+0.5 dex). There are also some indications that the heavy *s*-element Ba has increased to a smaller extent but this conclusion is less secure. As will be shown below, all but the variations for Ti and Cr can be naturally explained by the mixing and nucleosynthesis which has ensued due to the final He-shell flash.

Considering the drastic abundance alterations for some elements, one has first to investigate whether they indeed reflect real variations. It is important to realize that all changes will not disappear with any alternatively adopted stellar parameters. The required modifications of $T_{\rm eff}$ ($\approx$ 1 000 K) to annul most of the alterations would be in severe conflict with the various $T_{\rm eff}$-$\log g$ criteria. Furthermore, most of the elements showing an abundance trend are not more sensitive to the adopted stellar parameters than other elements. Hence different parameters would introduce other changes less easily explainable in the context of a final He-shell flash event. Also the fact that all elemental abundances in question show a definite trend with time supports the conclusion that the variations are real. We have also investigated the effects of hyperfine and isotope splitting, since many of the analysed lines are by necessity relatively strong and are able to dismiss those as causing the changes. Finally, we believe our assumption of LTE in the analysis is unlikely to severely affect our interpretation of the reality of the variations, since in most cases we are utilizing many different lines which should be little influenced by departures from LTE. Also it seems less likely that the relatively modest changes in the adopted stellar parameters can cause *differential* NLTE effects of $\sim$ 1.0 dex.

4. Changes in chemical composition and $T_{\rm eff}$

The abundance alterations of certain elements in 1996 can be naturally explained by the mixing and nucleosynthesis which was initiated in connection with the final He-shell flash. Most striking is the diminishing H-content, which reflects the ingestion and burning of the H-rich envelope following the development of a convection zone due to the He-shell flash. H-burning is now in fact likely the dominating nuclear energy source; thus we are witnessing the nuclear fuel of the star being consumed. The generated energy from H-burning not only provides the luminosity but has also caused the significant expansion of the star

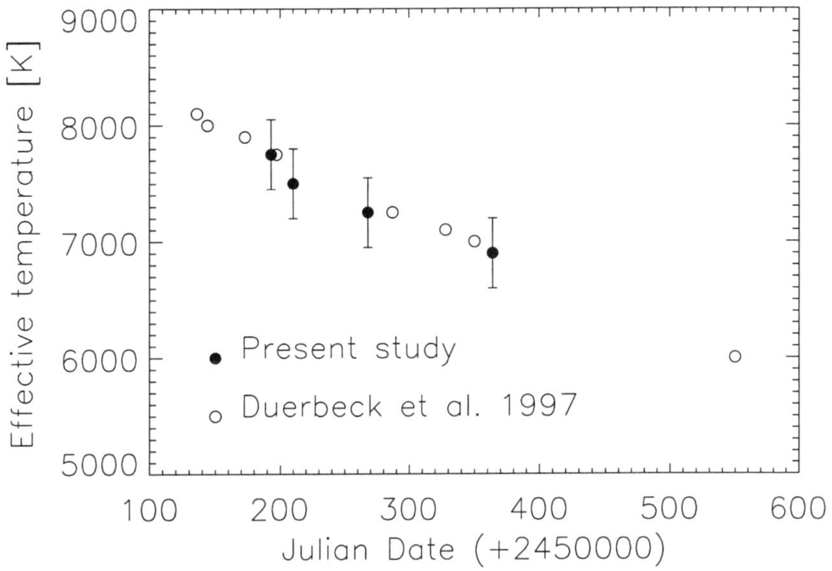

Figure 1. The evolution of the effective temperature of Sakurai's object as judged from spectroscopy (present study) and *UBVRiz* photometry (Duerbeck et al. 1997)

to giant dimensions. This expansion has resulted in a continued cooling of the stellar surface, as evident both from spectroscopy and photometry (Fig. 1); the cooling has continued throughout 1997 and 1998. The time-scale for the observed variations in $T_{\rm eff}$ and chemical composition ($\sim$ months) is significantly faster than predictions from the only published numerical calculation of such an event (Iben & MacDonald 1995) but it can likely be accounted for with a slightly larger stellar mass (say $M_* \approx 0.8\,{\rm M_\odot}$ instead of $M_* = 0.6\,{\rm M_\odot}$).

Further support for a second, current stage of CNO-cycling comes from the low observed $^{12}C/^{13}C$ (≈ 3); following He-burning, which is required to explain the over-abundances of C, N, O and Ne, the ratio should be very high. If the proton supply is exhausted before completion, CNO-cycling will produce vast amounts of ^{13}C, which can be brought down to He-burning temperatures and there release free neutrons. Sakurai's object shows evidence of very prominent *s*-processing, in particular of the light *s*-elements (including Sc, Ni, Cu and Zn). The abundance distribution of the *s*-elements is best explained by a *single* neutron exposure, which suggests that the *s*-processing occurred during the final He-shell flash. However, the alternative interpretation that the *s*-elements stem from thermal pulses during the previous AGB-phase and now are being mixed to the surface can not be convincingly ruled out either. It should be noted that the seemingly simultaneous productions of ^{13}C and *s*-elements pose a serious challenge to theoretical modelling if correct, in particular how the ^{13}C is brought down to the He-burning environment, since the initiated H-burning should cause the convection zone due to He-burning to split into two.

Li is most easily explained by production through the Cameron-Fowler mechanism from ingested ^{3}He in the H-rich envelope. Since H-burning necessarily destroys Li, the observed Li can not have been inherited from a previous AGB-phase but must be currently synthesized. The increases in Ti and Cr abundances, however, presently lack a nucleosynthetic explanation.

Sakurai's object has thus evolved significantly both in terms of surface temperature and in chemical composition within only six months. To our knowledge, Sakurai's object represents the fastest case of stellar evolution ever observed, when excluding complete stellar disruptions such as supernovae and novae.

5. Similarities with the R CrB stars

Sakurai's object shows striking similarities with the R CrB stars in terms of chemical composition. The R CrB stars are distinguished by their prominent H-deficiency and C-enrichment. Sakurai's object not only shares these properties, but also closely resemble the R CrB stars in the observed [X/Fe] ratios, though it has more pronounced over-abundances of the light *s*-elements and have a low ^{12}C/^{13}C ratio. Furthermore, Sakurai's object has recently been reported to experience the typical irregular visual variability of the R CrB stars (H. Duerbeck and E. Guinan, private communication). Thus, Sakurai's object has evolved into being an R CrB star, which is the first time ever this has been observed. Sakurai's object therefore lends strong support to the suggestion that final He-shell flashes in pre-white dwarfs can produce these stars. It is, however, likely premature to conclude that *all* R CrB stars have been formed similarly.

Acknowledgments. The author greatly appreciates the rewarding collaborations with B. Gustafsson, K. Jonsell, T. Kipper, D.L. Lambert, N.K. Rao, M. Shetrone and B. Smalley on the topic of Sakurai's object and related stars. Photometric data from H. Duerbeck and E. Guinan is acknowledged.

References

Asplund M., Gustafsson B., Kiselman D., Eriksson K., 1997a, A&A 318, 521

Asplund M., Gustafsson B., Lambert D.L., Rao N.K., 1997b, A&A 321, L17

Asplund M., Gustafsson B., Rao N.K., Lambert D.L., 1998a, A&A 332, 651

Asplund M., Lambert D.L., Kipper T., Shetrone M.D., 1998b, A&A, submitted

Duerbeck H.W., Benetti S., Gautschy A., et al., 1997, AJ 114, 1657

Iben I., Jr., MacDonald J., 1995, in White dwarfs, D. Koester and K. Werner (eds.), Springer, Berlin, p. 48

Kipper T., Klochkova V.G., 1997, A&A 324, L65

Lambert D.L., Rao N.K., Gustafsson B., Asplund M., 1998, A&A, submitted

Shetrone M.D., Keane M., 1998, in Faint blue stars, A.G. Davis-Philip (ed.), L. Davis Press, Schenectady, in press

Asymptotic Giant Branch Stars
T. Le Bertre, A. Lèbre and C. Waelkens, eds.
I.A.U. Symposium 191, 1999

Dust formation events in the envelopes of the peculiar post-AGB stars FG Sge and V4334 Sgr (Sakurai's object)

B. F. Yudin and A. M. Tatarnikov
Sternberg Astronomical Institute, Universitetskii pr. 13, 119899, Moscow, Russia

Abstract. Distinctive features of dust condensation events in the circumstellar shells of the born-again AGB stars FG Sge and V4334 Sgr are described together with updated optical, IR and bolometric light curves.

1. FG Sge

In August 1992 FG Sge entered a qualitatively new phase of its evolution. Suddenly, in ~ 50 days, its visual brightness dropped by $\sim 5^m$; during the summer of 1998 already a 5th deep decline has been observed. In Fig. 1 we display the recent light curves, for which the optical and IR data were taken from Arkhipova et al. (1996, references therein and private communication), and from Tatarnikov et al. (1998a, and references therein), respectively. All these deep declines in the visible were caused by the formation of dust grains around the star; in a sense, they all were unpredictable. The star itself did not change appreciably neither before the first deep decline nor since; it has the appearance of a late-F-type supergiant with rather strong Swan bands in its spectrum; it continues pulsating on a time scale of ~ 115 days.

The dust condensation events in FG Sge are very similar to those observed in R CrB type stars. An important advantage with FG Sge is that the recent declines started after a period when its circumstellar shell was dust free, so that the effects related to the dust formation events can be studied more clearly than in other cases.

The first condensation event in FG Sge, which does not differ markedly from the subsequent ones, is very similar to those observed in classical novae (Fig. 2). The nucleation and grain growth occur on a short timescale (about half a pulsation period for FG Sge) and result in a steep extinction of visible starlight.

The temperature of the newborn dust grains is $\sim 1\,000$ K, at least at the moment when their radiation begins to stand out against the background of the star or the old dust shell. Thus, the grains are located quite far from the star, at a distance corresponding to more than 200 days assuming an expansion velocity of 100 km s^{-1} for FG Sge. Possible hotter grains do not cause an observable near-IR excess.

The newborn IR excess is so large that it requires the dust clouds to cover an appreciable part of the star, not less than 50% for FG Sge. No significant reddening, as measured by the $B-V$ color, was observed in the visible during

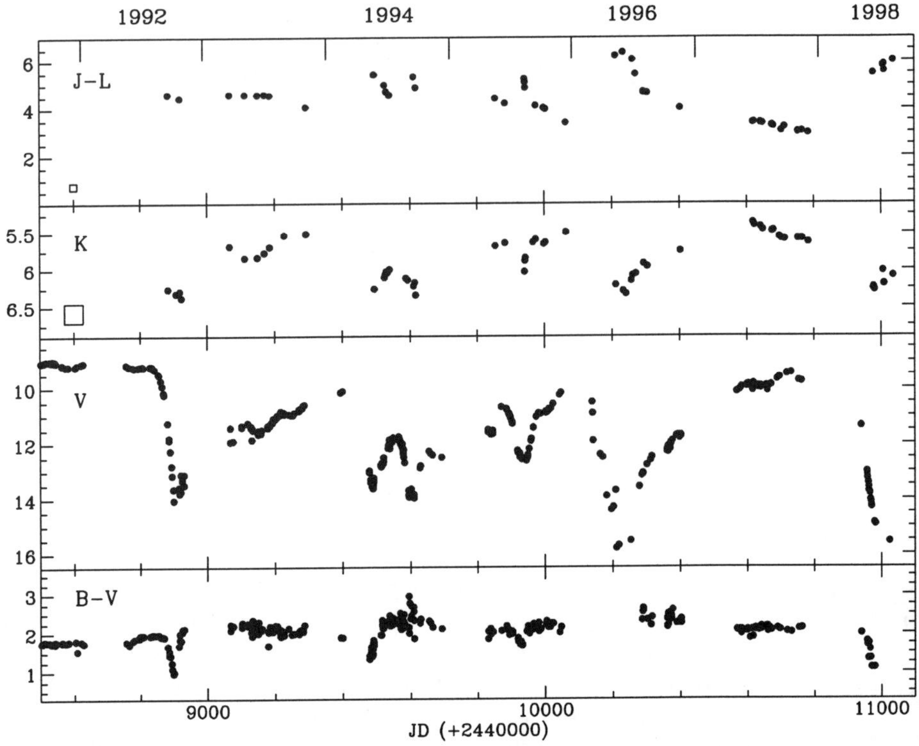

Figure 1. Optical and IR light curves of FG Sge in 1992-1998.

an overall cycle, i.e. between the start of the brightness decline and the recovery. This means that the optical depth of the dust layer from its formation to dispersal is governed by grains with a size of ~ 0.1 μm or larger and which by mass contain the bulk of the dust circumstellar matter. Comparing the dimming of the star in the V and J spectral bands, we come to the same conclusion.

It should be noted that even in such a case fine graphite grains could produce a 2200 Å absorption band. Moreover, if the important blueing of FG Sge in the visible, appearing on the descent to the deep minimum, is attributed to reflection of the direct star light by graphite grains, the most suitable grain size for creating such an effect is ~ 0.05 μm.

Paying attention to the similarity between the dust condensation episodes observed in classical novae and FG Sge, we recall that also in the former objects the grain nucleation takes place quite far from the star. Thus, the photometric behavior of FG Sge is fully consistent with a far-from-the-star model for dust formation.

Correcting for the blueing of the starlight in the visible, the spectral energy distribution of FG Sge in the range 0.36-10 μm can be fitted with a model with a spherically symmetric dust layer composed of 0.05, 0.1 and 0.25 μm graphite grains. Such a set of grains of different sizes is necessary and sufficient for a

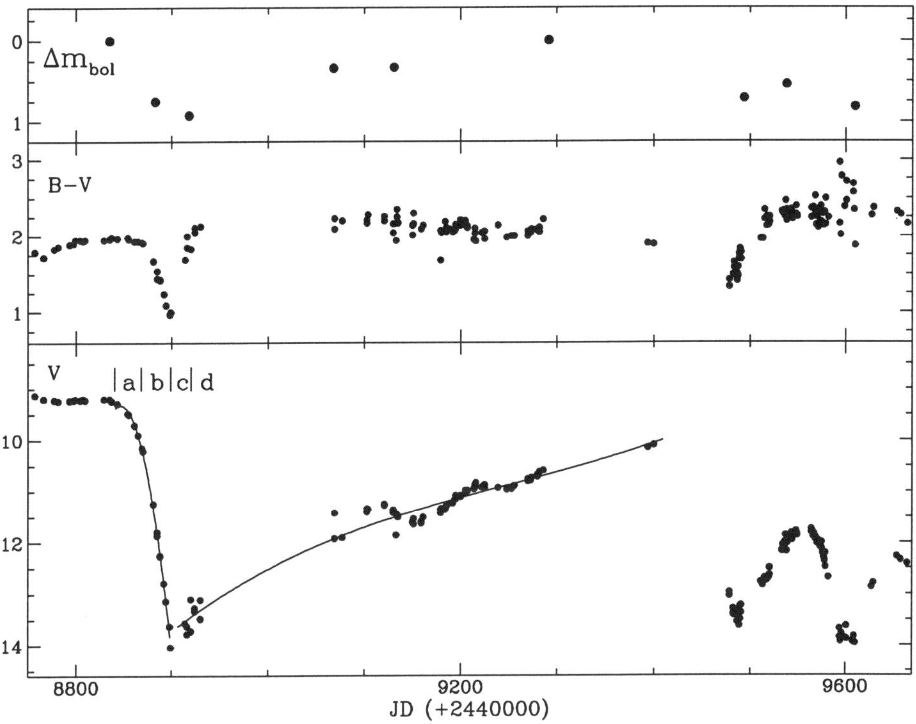

Figure 2. The first deep decline in the visual brightness of FG Sge.
a - initial decline (ΔV≈ 1 for 25 days), no change of $B-V$ color.
b - final decline (ΔV≈ 3.5 for 30 days), decrease of $B-V$ color.
c - short minimum (ΔV≈ 0 for 20 days), recovery of $B-V$ color.
d - recovery (ΔV≈ −3.5 for 500 days), no change of $B-V$ color.

good fit (Tatarnikov et al. 1998a). The spherically symmetric model reproduces also the case when the dust clouds cover the entire star.

During the deep decline the bolometric luminosity of FG Sge dropped by a factor ~ 2.5 relative to its mean value prior to this event. This factor could be compatible with the value of ~ 1.8 observed during the strongest pulsation episode in the 1980s, by accounting for holes in the dust cloud coverage then. However we cannot conclude that the deep decline was not accompanied by a decrease of the star's bolometric luminosity. It is also important to compare the IR excess before, during and after deep decline. For example, during the first deep decline the IR excess was smaller by a factor ~ 2.1 than after brightness recovery in 1993, although the optical depth of the dust shell had decreased noticeably by this time. The observed correlation between apparent bolometric magnitude and L-band magnitude suggests a connection between deep decline and decrease of the stellar bolometric luminosity.

We note that it is possible to fit the observed spectral energy distribution also with a nonspherical model in a way which is not worse than in the spherically symmetric case, by changing the grain size distribution. Thus, it is not possible to find out whether the extinction by dust clouds is homogeneous or clumpy from energy distribution fitting only. However, after even partial brightness recovery in the visible, the apparent bolometric magnitude returns to its mean value observed before the first deep decline (Fig. 2). Thus, it might be that by this time the dust clouds spread over the entire sphere. Before the second deep decline, which occurred during the spring of 1994, the optical depth of the dust shell was ~ 1.3, and it was caused mainly by large grains, with a size of ~ 0.1 μm or larger. Such a dust shell is a very efficient scatterer. However, the obscuration by such a cloud of the direct starlight would have reduced the visual brightness of FG Sge by $\leq 1.^m5$, while the observed decrease during the second deep decline was by $\geq 3^m$. Hence, it cannot have been caused by a single dust cloud in the line of sight towards the star.

Usually, on the descent to deep minimum one observes important blueing of the optical starlight (Fig. 2); the decrease of the $B - V$ color index can reach $\sim 0.^m7$. If the blueing is due to scattering by dust, such a large value means that the scattered component of the light accounts for almost the entire starlight in the visible and it should be produced by Rayleigh scatterers. As in scattered light the star's spectrum is not altered, the emission lines and emission Swan bands, which appear on the descent to the deep minimum, should originate in the shell. To avoid this conclusion we should reject dust scattering as the cause of the blueing effect.

We could fit the spectral energy distribution of FG Sge, observed at the moment when the object was bluest, only with a model in which the optically thin dust clouds (with visual optical depth ranging from 0.2 to 0.3) cover about a half of sphere around the star. These clouds consist of ~ 0.05 μm graphite grains and provide the visible brightness of FG Sge. The observed IR excess is created by the optically thick dust clouds composed of ~ 0.1 μm graphite grains which do not cause the reddening. On the line of sight to the star their optical depth is ~ 5.2 (Tatarnikov & Yudin 1998b). If silicate grains are used as the scatterers the range for the possible values of model parameters becomes wider.

2. Sakurai's object

In February 1996 Sakurai discovered a nova-like object in Sagittarius with an outburst amplitude of $\sim 10^m$. Subsequent photometric and spectroscopic observations have shown that it is surrounded by a faint old planetary nebula and that it looks like an early-F supergiant with noticeably faint Balmer lines. Thus, it turned out that Sakurai had discovered a new "born-again AGB star". Apparently, the star underwent the final helium flash in late 1994 (Fig. 3).

The evolution of Sakurai's object after its discovery is characterized by a gradual decrease of the stellar surface temperature and increase of the stellar apparent bolometric magnitude (Arkhipova et al. 1998 and references therein; Whitelock & Feast 1998). The bolometric flux was derived by the integration of the spectral energy distribution from U to M bands of Johnson photometric system, and corrected for radiation out of this spectral range. This correction does not exceed 20%; since the dust shell formation it came mainly from radiation at wavelengths longer than 5 μm. From early 1996 to today the apparent

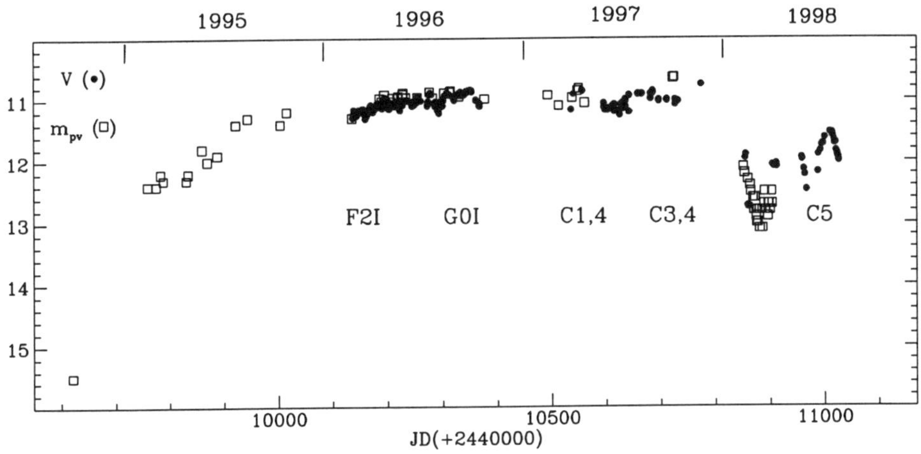

Figure 3. Visual light curve of Sakurai's object.

bolometric flux increased by more than a factor 4. During this time the stellar surface cooled down from $\sim 8\,000$ K to $\sim 4\,000$ K.

During the autumn of 1996 Sakurai's object looked still as a yellow supergiant with a dust-free circumstellar envelope. However, during the winter of 1997 the Swan bands and an IR excess appeared (Fig. 3, Fig. 4). Thus, during the spring of 1997 it became a carbon star surrounded by an optically thick dust shell (Arkhipova et al. 1998). However, in the visible there were not yet any indications for the dust shell formation. The visual brightness was essentially constant (Fig. 2) and the increase of $B - V$ color, observed at that time, could be attributed entirely to the cooling of the star. Hence, without IR observations we could not suspect the existence of the dust shell and could have missed the remarkable increase of the bolometric magnitude of Sakurai's object.

Without noticeable changes in the visual brightness in reply to the dust condensation, we could suggest that in 1997 there was still a hole in the blanket of dust clouds in the line of sight towards the star, although, judging the value of the IR excess, there should only be few such holes. With such a dust shell structure we could connect the sudden decline in visual brightness of Sakurai's object discovered in February 1998 (Liller et al. 1998) with the appearance of a dust cloud in the line of sight. Accordingly, after this event the apparent bolometric flux can not exceed the one of the star itself, and we can safely conclude that its bolometric luminosity has increased by more than a factor 4 during the last 3 years.

There are two main differences in the dust condensation events observed on Sakurai's object and on FG Sge. The first one is that in the former the transition from a dust-free shell to a one crowded with cool dust grains was much smoother than in the latter, though more IR photometric observations of FG Sge in 1992 but before the first deep decline would be useful to verify this conclusion. The second one is that the profound fading of Sakurai's object in the visible was not accompanied by the noticeable changes in bolometric flux that were observed in the case of FG Sge (Fig. 2, Fig. 4). These usually are

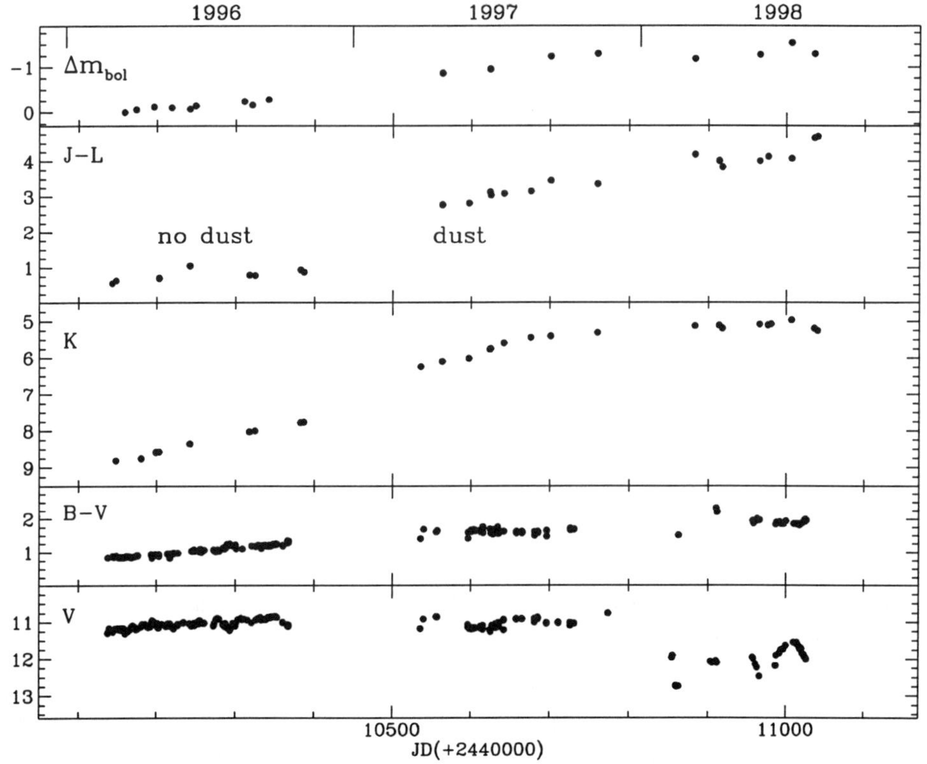

Figure 4. Optical, IR and bolometric light curves of Sakurai's object in 1996-1998.

considered as an argument in favor of spherically symmetric dust condensation around the star. The presence or absence of blueing during the descent to the minimum could clarify this point but it occurred in the dead season for Sakurai's object observations.

Acknowledgments. We thank V. P. Arkhipova and P. A. Whitelock for providing optical and IR data on FG Sge and Sakurai's object in advance of publication.

References

Arkhipova V.P., Zaitseva G.V., Ikonnikova N.P., et al., 1996, Pis'ma Astron. Zh. 22, 836

Arkhipova V.P., Esipov V.F., Noskova R.I., et al., 1998, Pis'ma Astron. Zh. 24, 297

Liller W., del Mar V., Janson M., et al., 1998, IAU Circ. No. 6825

Tatarnikov A.M., Shenavrin V.I., Yudin B.F., 1998a, AZh 75, 428

Tatarnikov A.M., Yudin B.F., 1998b, Pis'ma Astron. Zh. 24, 359

Whitelock P.A., Feast M.W., 1998, private communication

Asymptotic Giant Branch Stars
T. Le Bertre, A. Lèbre and C. Waelkens, eds.
I.A.U. Symposium 191, 1999

Born-again AGB stars: Starting point of the H-deficient post-AGB evolutionary sequence?

K. Werner, S. Dreizler, T. Rauch

Institut für Astronomie und Astrophysik, Univ. Tübingen, Germany

L. Koesterke

Institut für Physik, Universität Potsdam, Germany

U. Heber

Sternwarte Bamberg, Universität Erlangen-Nürnberg, Germany

Abstract. We discuss the origin of hot hydrogen-deficient post-AGB stars. Evidence is growing that the "born-again AGB star scenario" as a consequence of a late He shell flash cannot explain the diverse observations. Analyses of the stars and their associated planetary nebulae suggest that the H-deficiency is not formed during the second post-AGB phase but already while the star is still on the AGB. There are even hints that the H-deficiency can occur before the first departure from the AGB.

1. Introduction

Hot post-AGB stars ($T_{\rm eff}$>20 000 K) can be grouped into either H-deficient or H-"normal" objects. The evolution of the latter group is in principle understood, so we concentrate here on the problems with H-deficient objects. Spectroscopic analyses during the recent years have firmly established an evolutionary sequence formed by the Wolf-Rayet central stars of planetary nebulae (CSPN) and the PG 1159 stars. This is suggested by their location in the g–$T_{\rm eff}$-plane (Fig. 1) as well as the similarity of their chemical abundance patterns. The Wolf-Rayet CSPN, dominated by He, C, and O emission lines, are subdivided into late and early spectral types, [WCL] and [WCE]. The link between the [WCE] and PG 1159 stars is marked by a few [WC]–PG 1159 transition objects which show a mixed emission/absorption line spectrum. The PG 1159 stars display essentially absorption line spectra. While [WC] stars are known for a long time, the PG 1159 stars have been established as a spectral class more recently (Wesemael et al. 1985). We know 31 PG 1159 stars and every other is a CSPN. Note that the so-called "OVI" classification of CSPN is no longer used because objects having very different spectra were subsumed under it. All former "OVI" stars are now [WCE] stars of different sub-classes according to the classification scheme of Méndez & Niemela (1982). An exception is the well known CSPN NGC 246, also a former "OVI" object, which is now classified as PG 1159.

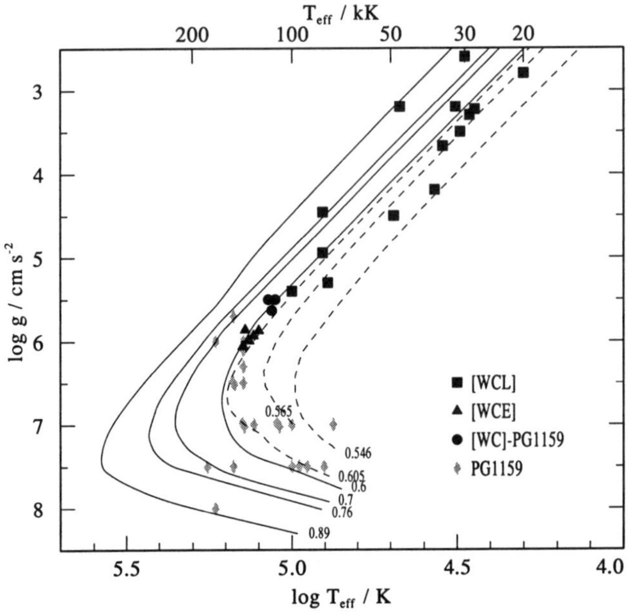

Figure 1. Hydrogen-deficient post-AGB stars in the g–$T_{\rm eff}$–plane. We identify Wolf-Rayet CSPN (Hamann 1997), PG 1159 stars as well as [WC]–PG 1159 transition objects. Evolutionary tracks are from Schönberner (1983, dashed lines) and Wood & Faulkner (1986) (labels: mass in $M_{\odot}$). They are only shown for a rough comparison, because they predict normal, H-rich surface abundances. $T_{\rm eff}$ for the [WC] stars is defined related to the stellar radius at $\tau_{\rm Ross}$=20.

For detailed spectroscopic results on the [WC] stars we refer to a review by Hamann (1997). An overview about results on PG 1159 stars is found in Werner et al. (1996), and latest results based on HST spectroscopy were published by Dreizler & Heber (1998).

2. Origin of H-deficient post-AGB stars

The typical chemical composition of PG 1159 atmospheres is displayed by the prototype PG 1159-035 itself. We found He:C:O=33:50:17 (mass fractions) while H is not detectable. A comparison of this abundance pattern with the interior structure of post-AGB stellar models suggests that the PG 1159 stars must have lost their entire H-rich envelope ($\approx 10^{-4}\,M_{\odot}$) and also most of their He-rich intershell layer ($\approx 10^{-2}\,M_{\odot}$) because the high C and O abundances are found only at its bottom. With the first analyses of PG 1159 stars (Werner et al. 1991), we suggested that their H-deficiency is a consequence of a late He shell flash. Originally, Iben (1984) introduced this scenario in order to explain the discovery of H-deficient zones in the inner parts of the PNe Abell 30 and Abell 78 (whose nuclei are today classified as [WC]–PG 1159 transition objects). In this scenario He shell burning re-ignites when the post-AGB star is already on the WD cooling track. As a consequence, the star moves back to the AGB ("born-again" AGB star). The star suffers another superwind phase and during its second descent from the AGB the star looses its remaining H envelope by a fast wind.

There are several problems with this scenario, which suggest that the H-deficiency is not only formed during the second post-AGB evolution, but already during the second AGB phase. In spite of the rapid evolutionary rate, the existence of cool (i.e. young) [WC] stars requires the complete ejection of the H-rich envelope on the AGB. This is also corroborated by the fact that the ejection of H-deficient material in the PNe Abell 30 and Abell 78 occurs with low expansion velocities ($< 100\,\mathrm{km\,s^{-1}}$, Borkowski et al. 1995), which means that already the AGB precursor of the CSPN must have been H-deficient.

The born-again scenario is faced with another problem, which is related to the expansion ages of the nebulae of [WC] stars. Their PNe should be systematically older than those of the CSPN with normal surface chemistry, however, this is not the case (Górny & Stasińska 1995). This finding suggests that the H-deficiency occurs possibly at the first departure from the AGB, i.e. without invoking the born-again scenario at all. What then is the origin of the H-deficiency? One possibility is close binary evolution during which the stellar envelope could be stripped off by interaction with the companion star. However, for [WC] stars binarity is extremely difficult to detect directly from radial velocity or photometry. For PG 1159 stars the binary frequency does not appear to be unusually high (Napiwotzki 1998). Bipolarity of PNe has frequently been invoked as a tracer of binarity. Here we face a puzzling result: while the frequency of bipolar PNe around [WC] stars appears to be "normal" (Górny & Stasińska 1995), the fraction of PG 1159 stars with a bipolar PN appears to be high (3/4, Napiwotzki et al. 1996). The latter result, however, is limited by small number statistics and inhomogeneous observational material.

A further problem is the high photospheric O abundance found in many PG 1159 and [WC] stars, which approaches almost 20% by mass. As already mentioned above, this means that not only the H envelope must be removed from the star, but also the 100 times more massive He envelope. This conflicts even more severely with the existence of the young [WCL] stars: removal of the He-rich intershell during the post-AGB evolution can be enforced in theoretical calculations by high mass-loss rates, but envelope removal is not completed until the star is very hot ($T_{\rm eff}$>100 000 K), because the main effect of mass-loss is to increase the evolutionary rate (Schönberner & Blöcker 1992). However, new evolution calculations show that this problem is probably much less severe. These calculations, which include a hydrodynamically based overshoot prescription (Herwig et al. 1997), provide a very high O abundance throughout the He-rich intershell, and not only at its bottom as suggested by earlier computations. Though the problem of envelope removal seems less serious now, it is still unsolved. Nevertheless, we have to consider the exceptional case of H 1504+65, a PG 1159 star which is completely devoid of H *and* He, i.e. its atmosphere is composed only of C and O, by equal parts (Werner 1991; Kruk & Werner 1998). Obviously this naked C/O stellar core must have lost also its massive He-rich intershell.

Further evidence that the H-deficiency is formed on the AGB or at least very early on the post-AGB track comes from the analysis of ISO spectroscopy of two [WCL] CSPN (Waters et al. 1998). C-rich dust (PAHs) in the inner part of the PN and crystalline (O-rich) silicates in the outer part suggest that the chemical composition of both stars changed to a C-rich mixture very recently, i.e. less than 1 000 years ago. The H-rich nature of the C-rich dust suggests that still some H was left on the star, which is in accordance with the identification of H in one of these CSPN (Leuenhagen & Hamann 1998).

The detection of residual H in some [WC] and PG 1159 stars (Dreizler et al. 1997) as well as the detection of N in a few cases suggest that mixing and incomplete burning of the H envelope take place in these stars. According to evolutionary calculations this can only occur after a very late He shell flash (i.e. on the WD cooling track, where H shell burning is extinct) and not on or soon after the AGB phase (where H shell burning is active; Iben & MacDonald 1995). This is in clear contradiction with the PN observations of H-deficient gas and dust around the stars (see above).

To conclude, the true origin for H-deficiency in post-AGB stars is not known. Further progress can be expected from evolutionary calculations which must include both overshoot as described above (Herwig et al. 1997) and time-dependent mixing as presented in an exploratory study by Iben & MacDonald (1995).

3. Another H-deficient post-AGB sequence

A few hot H-deficient post-AGB stars cannot be classified as PG 1159 stars, because their absorption line spectra are exhibiting almost exclusively He lines and, if at all, only weak C or other metal lines. This small group of stars comprises four objects, two of which have an associated PN. The stars are classified as "O(He)" according to the scheme of Méndez et al. (1986) and the prototype is the CSPN K1-27. Recent analyses by Rauch et al. (1994, 1996, 1998) confirm that only trace amounts of metals are present in the He-dominated atmospheres but, interestingly, two of them have a considerable fraction of H left (10% and 30%, by number). They are very hot ($T_{\rm eff}$=100 000–140 000 K) and have relatively low surface gravities ($\log g$ =5.5–6.5), hence, their location in the g–$T_{\rm eff}$–plane is among the PG 1159 and [WCE] stars. Rauch et al. suggest that the O(He) stars are the successors of He-rich sdO stars, which are on post-AGB evolutionary tracks (Husfeld et al. 1989). The variety of abundance patterns among PG 1159 stars and [WC] stars is rather diverse and the existence of O(He) stars makes an explanation for the origin of H-deficient stars even more difficult. We can at present only speculate that, among other details, the phase in the thermal pulse cycle at the moment of AGB departure plays an important role.

4. Mass-loss from H-deficient post-AGB stars

From the above discussions it is clear that the knowledge about mass-loss from [WC] and PG 1159 stars is essential to our understanding of the H-deficient post-AGB stars. Mass-loss rates from [WC] stars can be determined by the analyses of the optical emission line spectra. Good spectral resolution is required in order to eliminate the contribution from nebular emission lines to the photospheric features. As a result, high mass-loss rates are found, ranging between 10^{-5}–$10^{-6.5}$M$_\odot$/yr, with rates decreasing along the post-AGB evolution. The mass-loss rate of two [WC]–PG 1159 transition objects amounts to $10^{-7.5} M_\odot/yr$ (Werner & Koesterke 1992; Hamann 1997). For the PG 1159 stars one expects even smaller rates. As a consequence, the optical spectrum is formed in the hydrostatic layers of the photosphere and only the strong UV resonance lines of C IV and O VI can be used to study the properties of the weak winds. We have analyzed IUE, HST and ORFEUS spectra of four low gravity PG 1159 stars, i.e. immediate successors of [WCE] stars, and found $\dot{M}=10^{-7}$–$10^{-8} M_\odot/yr$ (Koesterke et al. 1998; Koesterke & Werner 1998). Comparison with theoretical studies (Pauldrach et al. 1988) suggests that the winds are radiation driven. For the [WC] stars the highly stratified ionization structure of metals in our photospheric models lends further support to the hypothesis that the strong [WC] winds are radiation driven, too. This might explain why the supposed progenitors of the O(He) stars, the He-rich and metal poor sdO-stars, do have "normal" mass-loss rates (only observable in the UV spectra) in contrast to the C- and O-rich [WC] stars which are found in the same place of the g–$T_{\rm eff}$–plane.

5. Occurrence of super-hot winds at the end of pre-WD evolution

The bifurcation of post-AGB evolution into H-deficient and H-normal channels seems to continue into the WD evolution, dividing the cooling sequence into non-DA and DA white dwarfs. It is natural to assume that the hottest He-rich white dwarfs (spectral type DO) are descendants of the PG 1159 stars, however, some H-deficient post-AGB stars will become DA white dwarfs, namely those who have retained at least some H in their envelope. The transition of post-AGB stars into the hot WD phase is encompassed by a spectacular phenomenon, which has been discovered only recently. We have found that a large fraction (50%) of the hot DO white dwarfs shows signatures of a super-hot wind (Werner et al. 1995; Dreizler et al. 1995) in the optical spectrum. We have discovered high ionization absorption lines of the CNO elements (e.g. C VI, N VII, O VIII, and even Ne X). The high excitation potentials involved require temperatures approaching almost one million K and the triangular shaped line profiles suggest their formation in a rapidly accelerating wind from the WD, reaching a terminal speed of the order of 10 000 km s^{-1}. Preliminary model calculations suggest mass-loss rates between 10^{-7} and $10^{-10} M_\odot/yr$. However, at present these radiative equilibrium models are crude at best, as they simply assume a very high $T_{\rm eff}$ in order to produce the high ionization stages. Since the post-AGB age is of the order of 10^5-10^6 yr, consequences of the stellar wind on evolutionary times scales and chemical surface composition are possible, if the mass-loss rates are in fact as high as estimated right now.

Finally we emphasize that this wind phenomenon is not restricted to DO white dwarfs. We have also detected one DAO white dwarf with similar high ionization features (Dreizler et al. 1995). This is the only case discovered among DA/DAO white dwarfs so far, and the puzzle arises why there is such a strong preference among DOs to show this phenomenon.

Acknowledgments. HST and ORFEUS data analysis is supported by the Deutsches Zentrum für Luft- und Raumfahrt (DLR) under grants 50 OR 9705 and 50 QV 97054 (in Tübingen) and 50 OR 96029 and 50 QV 97026 (in Bamberg).

References

Borkowski K.J., Harrington J.P., Tsvetanov Z.I., 1995, ApJ 449, L143
Dreizler S., Heber U., 1998, A&A 334, 618
Dreizler S., Heber U., Napiwotzki R., Hagen H.-J., 1995, A&A 303, L53
Dreizler S., Werner K., Heber U., 1997, IAU Symp. 180, Kluwer, p. 103
Górny S.K., Stasińska G., 1995, A&A 303, 893
Hamann W.-R., 1997, IAU Symp. 180, Kluwer, p. 91
Herwig F., Blöcker T., Schönberner D., El Eid M., 1997, A&A 324, L81
Husfeld D., Butler K., Heber U., Drilling J.S., 1989, A&A 222, 150
Iben I., Jr., 1984, ApJ 277, 333
Iben I., Jr., MacDonald J., 1995, in White Dwarfs, eds. D. Koester and K. Werner, Springer, Lecture Notes in Physics 443, p. 48
Koesterke L., Werner K., 1998, ApJ 500, L55
Koesterke L., Dreizler S., Rauch T., 1998, A&A 320, 1041
Kruk J.W., Werner K., 1998, ApJ 502, 858
Leuenhagen U., Hamann W.-R., 1998, A&A 330, 265
Méndez R.H., Niemela V.S., 1982, IAU Symp. 99, p. 457
Méndez R.H., Miguel C.H., Heber U., Kudritzki R.P., 1986, IAU Coll. 87, Reidel, Dordrecht, p. 323
Napiwotzki R., 1998, in The Third Conference on Faint Blue Stars, eds. A.G.D. Philip, J. Liebert and R.A. Saffer, in press
Napiwotzki R., Haas, S., Schönberner, D., 1996, in Hydrogen-Deficient Stars, eds. U. Heber and C.S. Jeffery, ASP Conference Series 96, p. 213
Pauldrach A., Puls J., Kudritzki R.P., Méndez R.H., Heap S.R., 1988, A&A 207, 123
Rauch T., Köppen J., Werner K., 1994, A&A 286, 543
Rauch T., Köppen J., Werner K., 1996, A&A 310, 613
Rauch T., Dreizler S., Wolff B., 1998, A&A 338, 651
Schönberner D., 1983, ApJ 272, 708
Schönberner D., Blöcker T., 1992, in The Atmospheres of Early-Type Stars, eds. C.S. Jeffery and U. Heber, Springer, Lecture Notes in Physics 401, p. 305
Waters L.B.F.M., Beintema D.A., Zijlstra A.A., et al., 1998, A&A 331, L61
Werner K., 1991, A&A 251, 147
Werner K., Koesterke L., 1992, in The Atmospheres of Early-Type Stars, eds. C.S. Jeffery and U. Heber, Springer, Lecture Notes in Physics 401, p. 288
Werner K., Heber U., Hunger K., 1991, A&A 244, 437
Werner K., Dreizler S., Heber U., Rauch T., Wisotzki L., Hagen H.-J., 1995, A&A 293, L75
Werner K., Dreizler S., Heber U., Rauch T., 1996, in Hydrogen-Deficient Stars, eds. U.Heber and C.S.Jeffery, ASP Conference Series 96, p. 267
Wesemael F., Green R.F., Liebert J., 1985, ApJS 58, 379
Wood P.R., Faulkner D.J., 1986, ApJ 307, 659

Part 6. AGB Stars as a Population of Various Galaxies

Andrejs Alksnis evocating the memory of Prof. Jurij Frantsman (1939–1998)

Asymptotic Giant Branch Stars
T. Le Bertre, A. Lèbre and C. Waelkens, eds.
I.A.U. Symposium 191, 1999

AGB stars and galactic dynamics

H. Dejonghe and K. Van Caelenberg

Sterrenkundig Observatorium, Ghent University, Krijgslaan 281, B-9000 Gent, Belgium

Abstract. AGB stars, seen as a stellar population, can be used to probe the dynamical state of galaxies. The relevant data are mostly positions and line-of-sight velocities, sometimes together with information on chemical composition and/or age. As of now, dynamical models have been made for OH/IR stars and Planetary Nebulae. Other candidates are C stars, S stars, and Miras. We review the methods used and the results obtained so far, for the Milky Way and for (relatively nearby) extragalactic stellar systems.

1. Introduction

It happens sometimes that data acquired for one purpose can be very useful for other purposes as well. A case in point are the AGB stars, which are studied in this volume mainly for their physical properties. If the data provide information on systemic velocities and if the set is sufficiently large and reasonably free of biases, it can be useful for dynamical analysis of our Galaxy or extragalactic systems. For obvious reasons, this contribution is kept as non-technical as possible.

Let us firstly recall the concept of stellar population: any class of stars that can be defined on the basis of physical characteristics of its members. These characteristics are such that they indicate a common origin, in place, in time, or both, given some a priori paradigm on the evolution of the stellar system to which they belong. The kinematics is not part of this definition; here it is seen rather as a property, which can be used to define subpopulations once a dynamical model is made, or to characterize the dynamical evolution.

The above definition is easily applied to, say, young stars, where the common origin is fairly obvious, and which do not yet have much of a history. However, by the time a star reaches the AGB phase, its very existence and evolution has been determined by the star formation rate, the initial mass function and the binary frequency, which all three depend on density, pressure, temperature and chemical properties of the interstellar medium. Moreover, our understanding of an AGB star depends on the theory of stellar evolution, of binary evolution and of the AGB phase itself in order to properly interpret the ages and chemical composition. Hence, there probably does not exist something like 'conservation of stellar population' in time, and AGB stars as a class are presumably the result of a mix of stellar populations. In short, it would be surprising to see that different classes of *old* AGB stars show different kinematics. The main thesis of this contribution indeed is that, at least for our Galaxy, these differences do not seem to exist.

This does not mean, of course, that there is little to be known from studying the dynamics of AGB stars. For instance, since the global gravitational potential is blind for all problems mentioned before, an AGB star is in principle as good a probe into the potential as any star. In fact, Planetary Nebulae (PNe) provide a very direct means of assessing the amount of dark matter in the halos of extragalactic systems. Because of space limitations, we will not discuss this very important topic any further.

2. AGB star roundup

In this section we will briefly review the properties of the most important classes of AGB stars, with special attention to the features that are relevant for a dynamical analysis. In general, their luminosity is a big advantage, because dynamical models need kinematical information over large volumes. Hence also infrared and radio astronomy occupy a privileged place. We will focus on AGB stars in those stellar systems that appear sufficiently relaxed so that a dynamical analysis seems possible.

2.1. OH/IR stars

These strong infrared emitters (several thousand $L_\odot$) shine right through the dusty Galactic Plane, a property which is helpful for a good spatial coverage. The characteristic double peak maser at 1612 MHz, caused by an expanding shell, is easily recognizable, and provides a simple way of measuring the systemic line-of-sight velocity. In addition, the velocity of the expanding circumstellar shell may indicate an age, in the (statistical) sense that larger expansion velocities are associated with younger stars. A metallicity effect may also be present however. The expansion velocity can thus be used to define OH/IR classes which may be identified with populations.

OH/IR stars have been detected in 2 ways. One method uses selection criteria on infrared colors (most often based on IRAS data). This method has been applied by Eder et al. (1988), Sivagnanam et al. (1989), te Lintel Hekkert et al. (1991a), with many more references, and David et al. (1993). The selected candidates are subsequently radio-checked for the 1612 MHz emission. The first dynamical models for these stars have been constructed by te Lintel Hekkert et al. (1991b). This early work established that dynamical models could reproduce the data, using a potential which did not differ in any substantial way from what was known at that time. There is also evidence for 2 distinct populations, based on differences in outflow velocity. See also Dejonghe (1993).

The alternative method simply rounds them up via a blind radio survey at 1612 MHz, which is particularly effective around the galactic plane and the Galactic Center (see te Lintel Hekkert et al. 1989, for a compilation of data before 1983). An early attempt to survey the Galactic Center was initiated by Habing et al. 1983), and has been substantially improved in a more limited region by Lindqvist *et al.* (1992a). Dynamical analysis can be found in Lindqvist *et al.* (1992b). The authors find evidence for a central mass concentration of the order of $10^7 M_\odot$ at 4 pc. Subsequent analysis by Sevenster et al. (1995) compares this sample with the te Lintel Hekkert et al. (1991b) data. The authors confirm the existence of 2 populations of OH/IR stars. The Lindqvist sample is a rather pure set of the younger population, while the te Lintel Hekkert et al. data consists of both populations. Further survey work includes seven fields in the galactic plane, surveyed by Blommaert et al. (1994). Sevenster et al. (1997a, b) have surveyed and published the strip $|b| \leq 3°$, $-45° \leq l \leq 10°$. Other data in other regions may be forthcoming.

A few caveats must be mentioned. (1) OH/IR stars are variable, and any survey is thus in principle a snapshot of an (unknown) fraction of the total population, which is obviously dependent on detector sensitivity. Monitoring of the Galactic Center region has been done by van Langevelde et al. (1993), and been used to obtain additional detections by Sjouwerman et al. (1998). (2) In some cases there is clearly maser emission at the right wavelength, but without the characteristic double peak. (3) OH/IR stars can be detected only in our Galaxy. Distances are in general unknown, and it is not clear how completeness varies as function of distance.

2.2. Planetary Nebulae

Their abundant radiation in emission, most notably in the [O III] line at 500.7 nm, good for about 15% of the radiative power, makes them real lighthouses. Detection of individual stars is possible, even at large distances (up to 20-30 Mpc), and analysis of the spectrum yields the determination of a radial velocity and abundances. The latter quantity can be used to identify different populations. For our purposes, we only consider the Planetary Nebulae (PNe) for which radial velocities are known.

Galactic PNe PNe are also strong infrared emitters, and occupy a fairly well defined place in infrared colour-colour diagrams (Pottash et al. 1988; Ratag et al. 1990). The catalog by Acker et al. (1992) is a critical compilation of all known measurements in the (vast) literature at that time. A complete compilation of radial velocities of galactic PNe is performed by Durand et al. (1998), with references therein. There are very important but unquantifiable selection biases present in these catalogs, since they originate in large part from literature searches. Add to this problem the galactic extinction in the plane (most of it is optical work), and it will be clear that these catalogs are unsuitable to yield star counts (see also later). Nevertheless, if the COBE 2μm map is substituted for the star counts, useful dynamical models can be made (Durand *et al.* 1996). These authors could not obtain a fit with two-integral models (see later), thus demonstrating the need for a global dependence on the third integral.

Extragalactic PNe The search for extragalactic PNe has now only begun in earnest. Most often candidates are found by blinking two images, one of them taken in a narrow band filter around [O III], the other in a sufficiently wide band that does not contain [O III]. Subsequent multifiber spectroscopy yields the radial velocities. This method has been fully exploited by Ciardullo, Jacoby, Hui, Ford and coworkers (see e.g. Hui & Ford 1993). Variants of this technique include selection on the basis of colour-colour diagrams (Theuns & Warren 1997). A quite different approach uses a Fabry-Perot instrument (e.g. Tremblay et al.

1995). A new and very promising technique searches with a dedicated instrument using the Counter Dispersed Imaging technique (Douglas et al. 1997). Very briefly, PNe will appear as shifted point sources through a (slitless) spectrograph. After rotation of the instrument by 180°, the PNe can be identified, and the line-of-sight systemic velocity follows after calibration with a foreground star.

2.3. C stars, S stars and Miras

We will consider these 3 classes together, not because there is an astrophysical reason to do so, but because the line-of-sight systemic velocity is obtained for all three of them in the same way: a spectrum is cross-correlated with a template. The spectrum evidently yields chemical information, which may be used to identify stellar populations.

C stars have been searched for in the Galaxy through objective prism surveys (Nassau & Velghe 1964; Aaronson et al. 1989, 1990), colour-colour or colour-magnitude diagrams (Green et al. 1994; Totten & Irwin 1998). S stars have been surveyed by means of Hα or near-IR surveys (Henize 1960; MacConnell 1982; Van Eck et al. 1998). The most extensive catalogue is that of Stephenson (1984). As far as the Miras are concerned, detection obviously is based on their variability, which makes them the longest period known AGB stars. The General Catalog of Variable Stars is the standard source for Galactic Miras. IRAS follow up has resulted in many Miras being found in the Galaxy as well (e.g. Whitelock 1994).

As all this work is optical, it is of course vulnerable to extinction effects. Template mismatch is always a worry with cross-correlation techniques, and the errors on the line-of-sight velocities are generally larger than those that can be obtained with emission line work.

3. Dynamical models

The basic question of stellar dynamics is: (1) how does one distribute stars on orbits, in order to reproduce the observations? (2) How does one interpret success or failure to do so?

The conceptually simplest way to go about this problem is to perform N-body simulations: stars are put on orbits, and these are integrated. It is a very general method, and the gravitational effects of gas, dark matter, etc... are easily incorporated (in principle). However, in order to start, one needs the initial conditions of the stars, and the gravitational potential, if stars are not the only players. All samples, except the Miras, have in common that the distances to individual stars are poorly, if at all, known. In most cases, proper motions are also absent. Hence it is clear that a dynamical analysis based on orbit reconstruction is a risky business, especially if the sample size is not large (a few hundreds). This problem can be alleviated by assuming a priori information on the nature of the orbits. This is hard to do in this case: e.g., it is a bit of a stretch to assume that an intermediate or old population moves on circular orbits.

The concept of an orbit therefore must be simplified in order to be useful. Since every sample is a snapshot in time, one intuitively sees that time-independent characteristics of the orbits, if these exist, are probably most useful (see also Dejonghe 1992, for a different approach). Such quantities, which remain constant along an orbit, are called integrals of the motion. For example, if the gravitational potential $V(\vec{r})$ is stationary, the energy $E = \frac{1}{2}\vec{v}^2 + V(\vec{r})$ is such a quantity. If the potential also has axial symmetry along the z-axis, it follows from general theory that the angular momentum component $L_z = (\vec{r} \times \vec{v})_z$ is also an integral. Numerical experiences show that most orbits in stationary potentials conserve also a third integral I_3 fairly well, be it possibly over more limited periods in time. A general theorem does not exist.

The basic question now becomes: (1) how do we obtain the distribution of the stars over these integrals, and what is the distribution function $F(E, L_z, I_3)$? (2) How does one interpret success or failure to do so?

The first question is more technical, especially its first part, and will not be discussed here. We refer the reader to te Lintel Hekkert et al. (1991b), or Dejonghe (1992), Dejonghe (1993), Sevenster et al. (1995) or Durand et al. (1996), and note that the method used in these papers does not depend on distance information to the individual stars.

There are 2 obvious aspects to the second question. (1) The quality of the fit depends on the adopted potential: a dynamical model allows a discussion on the total mass of the stellar system, and possibly its shape. (2) The form of the distribution function provides information on the dynamical state of the system: are there clumps in integral space that indicate different dynamical histories? Is the function $F(E, L_z)$ (i.e. two-integral) rather than $F(E, L_z, I_3)$? The former is believed to represent a population that is more mixed than the latter, which is an indication on age.

When the gravitational potential varies in time, the concept of integral of the motion loses its meaning, and one must try to assume special time dependences for the distribution function and work from there. A so-called mode analysis is such a method, but a discussion is beyond the scope of this contribution.

4. A comparison of a few galactic AGB samples

We will now check our hypothesis that the different AGB classes considered in Sect. 2 are unlikely to be very different dynamically, though differences probably exist between subclasses of one particular class (for OH/IR stars, see Habing 1993, te Lintel Hekkert *et al.* 1991b and Sevenster et al. 1995). This is not in contradiction with our conjecture here, because fine-tuning within a class may very well depend on age and metallicity.

Therefore, we compare the kinematics of 4 samples that we had available in electronic form: 885 mostly IRAS based OH/IR stars from the te Lintel Hekkert et al. (1991b) sample 'OH/IR(tL)', 507 OH/IR stars from the Sevenster et al. (1997a, b) survey 'OH/IR(S)', 867 PNe from the Strasbourg collaboration (Acker et al. 1992) 'PNe', and 478 carbon stars from the Aaronson et al. (1998, 1990) survey 'C' which covers 12 fields in the strips $|b| \leq 3°$, $-117° \leq l \leq -47°$ and $55° \leq l \leq 138°$.

Clearly not every sample is everywhere complete. We will use as our reference the COBE $2\,\mu$m data, which are generally believed to provide a good and reasonably extinction-free representation of the intermediate and old age

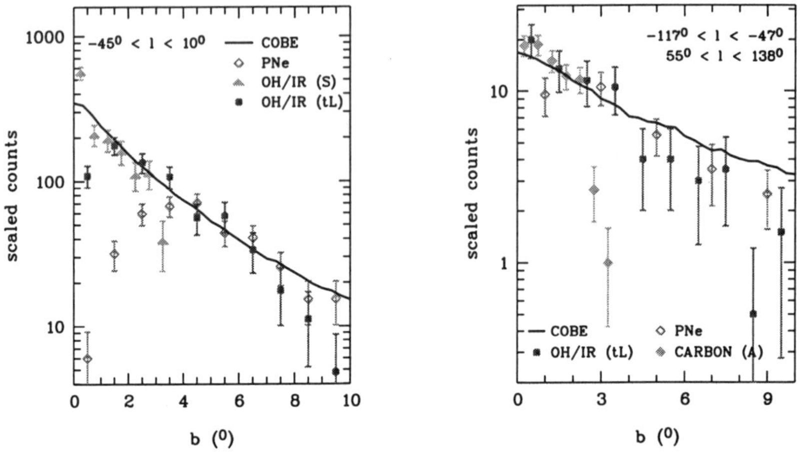

Figure 1. Star counts, summed over the indicated interval in galactic longitude, and scaled to the COBE counts.

population in the Galaxy. We will ignore asymmetries in galactic longitude l in the COBE data that are known to be present and generally believed to be due to the bar, because the relatively poor number statistics of the AGB samples do not allow detection of such an effect. Hence we will use a $2\,\mu$m map that is symmetrised in l and symmetrised in galactic latitude b, and apply the same symmetrisation to the 4 samples.

In Fig. 1 we compare the star counts, summed in longitude as indicated, as a function of $|b|$. The samples were scaled to coincide with each other and with the COBE counts as much as possible. We judge a sample complete if it follows the COBE counts. The results can be summarized in the following table.

	OH/IR(tL)	OH/IR(S)	PNe	C
$-45° \leq l \leq 10°$	$2° \leq \|b\| \leq 6°$	$\|b\| \leq 3°$	$\|b\| \geq 4°$	—
$-117° \leq l \leq -47°$ $55° \leq l \leq 138°$	$\|b\| \geq 2°$	—	$\|b\| \geq 3°$	$\|b\| \leq 2°$

Table 1. Regions in the sky where the four samples follow the COBE $2\,\mu$m data. In all, about 1 000 stars qualify.

Having established where a comparison of the 4 samples makes sense, we now turn to the kinematics. As is clear from Fig. 2, one would be hard pressed to argue that any difference between samples might exist.

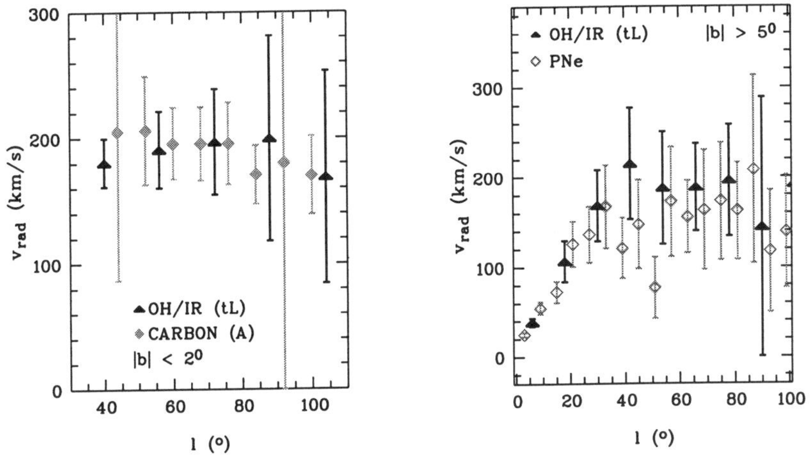

Figure 2. A comparison of line-of-sight velocities, summed over the indicated interval in galactic longitude, for different samples.

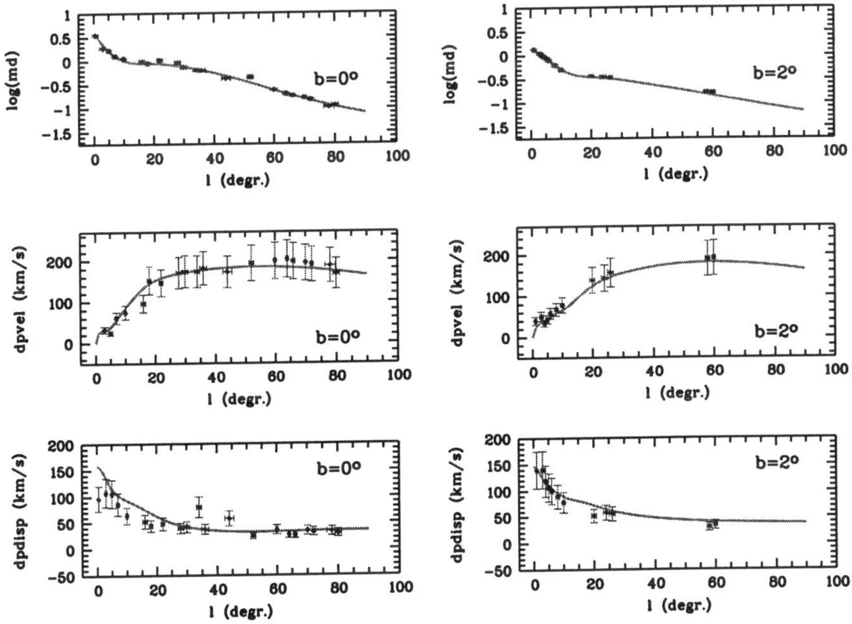

Figure 3. The COBE $2\,\mu$m density (upper panels), the projected mean velocity (middle panels) and the projected velocity dispersions (lower panels) of the model (full curves), superimposed on the data.

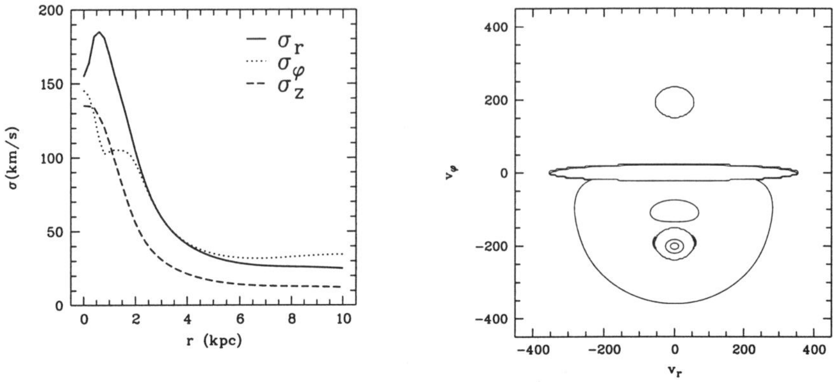

Figure 4. The intrinsic velocity dispersion as a function of galactocentric radius (left panel), and the velocity distribution in the Galactic Plane at the solar radius (right panel)

5. A dynamical model

Since we want to make genuine three-integral dynamical models, we assume a Galaxy potential that admits a third integral for all orbits: a so-called Stäckel potential. Batsleer & Dejonghe (1994) have constructed a large number of acceptable and simple potentials, of which we use the one with (in their notation) $a_D/c_D = 75$, $a_H/c_H = 1.01$ and $k = 0.07$. The distribution function is a sum of elementary and analytical functions, with coefficients determined by a Quadratic Programming technique. The method follows essentially the procedures used by Sevenster et al. (1995), except that the 3-integral components are of Abel type (Dejonghe & Laurent 1991). Fig. 3 compares data and model.

The construction of a dynamical model can be seen as a sophisticated deprojection. For example, the model has a mass density, that is the distribution of the mass in three dimensional space, while the original data only cover a range in l and b, with no distance information. By projecting the model back onto the sky, we get a density that is compatible with the observed one. More importantly however, the deprojected model is not a simple deprojection, but is supported by a distribution function: it can be realized by putting stars on orbits.

The left panel of Fig. 4 shows the velocity dispersion of our model in the r, φ and z directions (cylindrical coordinates) in the Galactic Plane. We see that the dispersion σ_z is smaller than σ_r, as could be expected. The curves for σ_φ and σ_r on the other hand, seem to indicate a dispersion in v_φ that is higher than that in the radial direction. This is of course not a very plausible result. However, the second plot of Fig. 4 represents the underlying distribution function in the (v_r, v_φ) plane, at $r = 8$ kpc, and makes clear that the velocities indeed do behave the way they are supposed to do. The velocity distribution clearly is more complex than can be inferred from the intrinsic velocity dispersions. In order to get the proper picture of the dynamical structure of the Galaxy, one needs to look at the underlying distribution function, since the dynamical structure is too complex to be 'summarized' in only a few numbers.

6. Conclusions

We argue that it is unlikely that large differences in the kinematics of different classes of old AGB stars exist in the Galaxy, and demonstrate that this is indeed true for OH/IR stars, PNe and C stars. We review the detection methods for these stars, with special attention to datasets with dynamical relevance. We produce, for the first time, a global dynamical model that incorporates different classes of AGB stars. We discuss the intrinsic velocity dispersions, and show that they may give misleading information on the dynamical structure, hence demonstrating the need to also consider the distribution function. Finally, it can safely be stated that upcoming large infrared surveys (such as DENIS and 2MASS) will turn up many more AGB stars, yielding large samples with well-understood biases.

Acknowledgments. The data used for this contribution were kindly made available in electronic form by A. Acker, S. Durand, M. Sevenster, P. Schechter and P. te Lintel Hekkert. I thank M. Feast, H. Habing and A. Zijlstra for useful comments.

References

Aaronson M., Blanco V.M., Cook K.H., Schechter P.L., 1989, ApJS 70, 637

Aaronson M., Blanco V.M., Cook K.H., Olszewski E.W., Schechter P.L., 1990, ApJS 73, 841

Acker A., Ochsenbein F., Stenholm B., Tylenda R., Marcout J., Schon C., 1992, Strasbourg-ESO catalogue of galactic planetary nebulae, ESO

Batsleer P., Dejonghe H., 1994, A&A 287, 43

Blommaert J.A.D.L., van Langevelde H.J., Michiels W.F.P., 1994, A&A 287, 479

David P., Le Squeren A.M., Sivagnanam P., 1993, A&A 277, 453

Dejonghe H., 1992, in Planetary Nebulae, R. Weinberger & A. Acker (eds.), Reidel, Dordrecht, p. 541

Dejonghe H., 1993, in Galactic Bulges, H. Dejonghe & H. Habing (eds.), Reidel, Dordrecht, p. 73

Dejonghe H., Laurent D., 1991, MNRAS 252, 606

Douglas N.G., Taylor K., Freeman K.C., Axelrod T.S., 1997, in Planetary Nebulae, H. Habing & H. Lamers (eds.), Reidel, Dordrecht, p. 493

Durand S., Dejonghe H., Acker A., 1996, A&A 310, 97
Durand S., Acker A., Zijlstra A., 1998, A&AS 132, 13
Eder J., Lewis B.M., Terzian Y., 1988, ApJS 66, 183
Green P.J., Margon B., Anderson S.F., Cook K.H., 1994, ApJ 434, 319
Habing H.J., Olnon F.M., Winnberg A., Matthews H.E., Baud B., 1983, A&A 128, 230
Henize K.G., 1960, AJ 65, 491
Hui X., Ford H.C., 1993, in Planetary Nebulae, IAU Symp. 155, R. Weinberger & A. Acker (eds.), Kluwer, p. 533
Lindqvist M., Winnberg A., Habing H.J., Matthews H.E., 1992a, A&AS 92, 43
Lindqvist M., Habing H.J., Winnberg A., 1992b, A&A 259, 118
MacConnell D.J., 1982, A&AS 48, 355
Nassau J.J., Velghe A.G., 1964, AJ 139, 190
Pottasch S.R., Bignell C., Olling R., Zijlstra A.A., 1988, A&A 205, 248
Ratag M.A., Pottasch S.R., Zijlstra A.A., Menzies J., 1990, A&A 233, 181
Schneider S.E., Terzian Y., Purgathofer A., Perinotto M., 1983, ApJS 52, 399
Sevenster M.N., Chapman J.M., Habing H.J., Killeen N.E.B., Lindqvist M., 1997a, A&AS 122, 79
Sevenster M.N., Chapman J.M., Habing H.J., Killeen N.E.B., Lindqvist M., 1997b, A&AS 124, 509
Sevenster M.N., Dejonghe H., Habing H.J., 1995, A&A 299, 689
Sivagnanam P., Braz M.A., Le Squeren A.M., Tran Minh F., 1989, A&A 211, 341.
Sjouwerman L.O., van Langevelde H.J., Winnberg A., Habing H.J., 1998, A&AS 128 35
Stephenson C.B., 1984, Publ. Warner & Swasey Observ. 3, 1
te Lintel Hekkert P., Caswell J.L., Habing H.J., Norris R.P., Haynes R.F., 1991a, A&AS 90, 327
te Lintel Hekkert P., Dejonghe H., Habing H.J., 1991b, Proc. Astron. Soc. Austr. 9, 20
te Lintel Hekkert P., Versteege-Hensel H.A., Habing H.J., Wiertz M., 1989, A&AS 78, 399
Theuns T., Warren S.J., 1997, MNRAS 284, L11
Totten E.J., Irwin M.J., 1998, MNRAS 294, 1
Tremblay B., Merritt D., Williams T.B., 1995, ApJ 443, L5
Van Eck S., et al., 1998, poster contribution, IAU Symposium 191
van Langevelde H.J., Janssens A.M., Goss W.M., Habing H.J., Winnberg A., 1993, A&AS 101, 109
Whitelock P.A., 1994, MNRAS 267, 711
Zijlstra A.A., Acker A., Walsh J.R., 1997, A&AS 125, 289

Asymptotic Giant Branch Stars
T. Le Bertre, A. Lèbre and C. Waelkens, eds.
I.A.U. Symposium 191, 1999

Mass-losing AGB stars in Galactic Bulge ISOGAL fields

J.A.D.L. Blommaert

ISO Data Centre, Astrophysics Division, Space Science Dept. of ESA, Villafranca, PO Box 50727, E-38080 Madrid, Spain

S. Ganesh*, A. Omont, M. Schultheis

Institut d'Astrophysique de Paris, CNRS, France
**: on leave from Physical Research Laboratory, Ahmedabad, India*

D. Ojha

TIFR, Bombay, India

C. Alard, G. Simon

Observatoire de Paris, DASGAL, France

and The ISOGAL Collaboration[1,2]

Abstract. First results on AGB stars with the ISOGAL mid-infrared survey are presented. We show that M6 and later spectral type giants are detected at 7 and 15 μm up to the distance of the Galactic Centre. Analysis of the observed magnitudes and colours indicates that most of the detected AGB stars in the Galactic Bulge are losing mass. ISOCAM CVF spectra on a field in the Bulge confirm the mass loss and also show a large diversity in the shape of the silicate feature.

1. Introduction, the ISOGAL survey

ISOGAL is a survey at 7 and 15 μm of $\approx$ 22 degree^2 of the inner Galaxy. It is performed with ISOCAM (Cesarsky et al. 1996), the mid-infrared imaging device of the Infrared Space Observatory (ISO[3]) satellite. In combination with IJK_s DENIS (Epchtein et al. 1997) data this survey allows studies of cold stellar

[1] ISOGAL Co-Investigators include: C. Alard, J. Blommaert, C. Cesarsky, N. Epchtein, M. Felli, P. Fouqué, R. Genzel, G. Gilmore, F. Guglielmo, H. Habing, A. Omont (PI), M. Pérault, S. Price, A. Robin and G. Simon

[2] The ISOGAL data reduction team includes: C. Alard, T. August, J. Blommaert, E. Copet, S. Ganesh, D. Ojha, M. Pérault, M. Schultheis, G. Simon and M. Unavane

[3] ISO, an ESA project with instruments funded by ESA Member States (especially the PI countries: France, Germany, the Netherlands and the United Kingdom) with the participation of ISAS and NASA

populations and galactic structure in regions which are highly obscured in the visible. In comparison with IRAS, the sensitivity and spatial resolution are almost 100 times better. This allows us to study AGB stars in regions which were too confused for IRAS like the galactic plane and the inner Galactic Bulge. Apart from AGB studies, the survey wants to serve several other scientific goals like galactic structure, a characterisation of the small scale interstellar extinction, dusty young stars, diffuse mid-infrared emission. For a further description of ISOGAL see Omont et al. (1998).

ISOGAL will give us a possibility to study large numbers of AGB stars at different positions in the Galaxy and enable us to study dependences of local environment on their evolution (for instance different metallicities). In this contribution we focus on a study of the Bulge and demonstrate the possibilities offered with this new mid-infrared survey.

Systematic cross-identifications with near-infrared I,J,K_s DENIS sources is performed for all ISOGAL detected sources. It is intended to make a five-wavelength ISOGAL catalogue available by mid-1999 (through CDS and on `http://www-isogal.iap.fr/`).

2. Observations and data reduction

ISOGAL observations were performed throughout the whole lifetime of the ISO satellite (November 1995 - April 1998). Unfortunately a large fraction of the fields were not observed simultaneously at 7 and 15 μm. The observed regions of $\approx$ 22 degree^2 were distributed along the inner galactic disk (mostly within $|l| < 30^o, |b| < 1^o$). The central region ($|l| < 1.5^o, |b| < 0.5^o$) was almost entirely observed, except very close to the centre itself as that would saturate the detector. For comparison and to extend our study, some fields at larger longitude and at higher latitude (in the Bulge) were observed. A major difficulty for the measurements in the galactic plane was to avoid bright ($\approx$ 6 Jy at 12μm) sources which would saturate the detector. For this reason, the observations were performed in small rasters ($\approx$ 0.1 deg^2). The pixel field of view selected was 6$''$ (resulting in a total field of view of $3' \times 3'$ per raster position). The raster step size was 1.5$'$. Generally broadband filters were used: LW3 (12-18μm) and LW2 (5-8.5 μm) and only short integration times (40×0.28 seconds). In special cases, like near the Galactic Centre or on active star formation regions, narrowband filters at the same central wavelengths were used to allow for higher fluxes. The overall sensitivity is better than 10 mJy (corresponding to 10 mag. for LW2 and 8.5 mag. for LW3). Additionaly, 18 ($3' \times 3'$) fields were selected for ISOCAM CVF (5-17 μm) measurements. One example of this spectroscopic mode will be discussed in section 4. For a further description of the observations see Omont et al. 1998.

The data reduction consists of several "standard" steps like the removal of cosmic ray impacts, dark current subtraction and flat fielding. Special care is taken for the time-dependent behaviour of the detector, but the corrections for this are still under investigation. In the often crowded fields we use sophisticated source extraction methods. The photometry is done by psf-fitting. For the ISO results we are presenting here, we have a photometric accuracy better than 0.2 mag. (rms).

3. AGB stars near the Galactic Centre

A [15] vs [7]–[15] diagram of sources selected in the direction of the Bulge is shown in Fig. 1. A clear linear sequence of increasing [7]–[15] colour for brighter 15 μm fluxes can be seen. The interpretation is that we see here a sequence of increasing mass-losing AGB stars. This hypothesis is investigated by an observation of one particular field (called C32) which is at 1 degree north of the Galactic Centre and which we discuss here further.

The DENIS K_s vs J–K_s diagram of the C32 field shows a well-defined red-giant Bulge sequence shifted by a fairly uniform extinction of $A_v = 6$ mag. Most of the ISOGAL sources have $(K_s)_o < 8.2$ mag. (Fig. 2) and are thus above the RGB limit (Tiede et al. 1995). The $(K_s)_o$ range of stars also indicates that the spectral type is M5 or M6 and later (Frogel & Whitford 1987). This is further confirmed in Fig. 2 when comparing with the DENIS data of Miras and Semi-Regulars in the Bulge PG3 field (centered on (l, b) = (0, −8), Schultheis 1998).

Fig. 3 shows again the [15] vs [7]–[15] diagram but now only for the C32 field. The 5 sources with relatively blue colours ([7]–[15] < 0.5) but high 15 μm flux densities ([15] < 6.5 mag.) were identified through SIMBAD and are most likely foreground stars. In the same figure two theoretical sequences of M-giants with increasing mass-loss rates are overplotted (sequences provided by Martin Groenewegen). The lower line shows the model prediction of a M5 giant (Fluks et al. 1994) with $L_* = 2\,000\ L_\odot$ and a total mass loss rate ranging from 10^{-9} to $10^{-7}\ M_\odot$/yr. The upper line shows the range for the same mass loss rates but now for a M8 giant of $L_* = 2\,900\ L_\odot$. It clearly follows the observed sequence of points well. In both cases silicate dust (David & Papoular 1990) with a gas to dust ratio of 200 was used.

The sequence seen in the [15] vs [7]–[15] diagram can thus well be explained in terms of an increasing mass loss rate. The relationship between the $(I–J)_o$ colour and the [7]–[15] (Fig. 4) suggests a relationship between later spectral type and higher mass loss rates. It is interesting to point out that at the mass-loss rates mentioned, there is no sign of an increase of the near-infrared colours. The mass-loss is only noticeable in our mid-infrared data.

Mass loss rates for late M giants have been studied and discussed in papers as early as Reimers (1975) and Gehrz & Woolf (1971) but it is for the first time that we can observe the onset of the mass loss on the AGB at a distance of 8 kpc. Together with the DENIS data we will be able to get a complete census of the mass loss on the AGB which should give us a better understanding of the importance of the mass loss on the evolution.

It is also interesting to put the number of mass losing stars in the perspective of the variable stars like Miras or Semi-Regulars (SRV), known to undergo mass loss. On the basis of an extrapolation of the number of Miras in the Bulge (Blommaert 1992) we expect to find not more than 20 SRVs + Miras in the C32 field. About 180 ISOGAL sources are known, of which most are in the sequence of mass loss, so much more than the number of Long Period Variables. Interestingly, results of the very sensitive MACHO project (Alves et al. 1998) in the direction of the Magellanic Clouds indicate that 90% of their discovered red variables (considered to be AGB) have an R-amplitude of less than 0.5 magnitude and would have been undetected in previous LPV surveys.

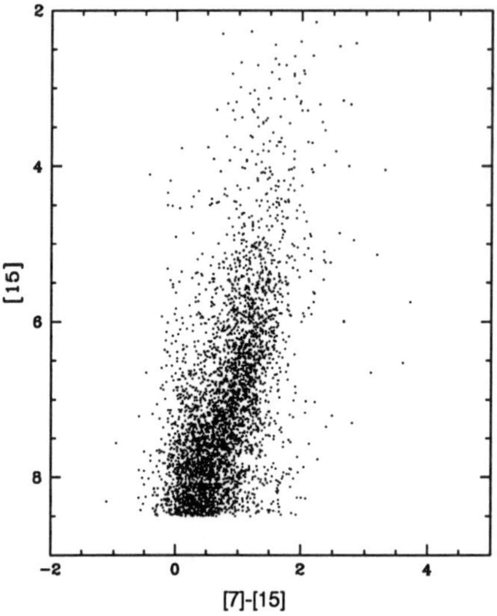

Figure 1. [15] vs [7]–[15] diagram for all ISOGAL fields in the Bulge

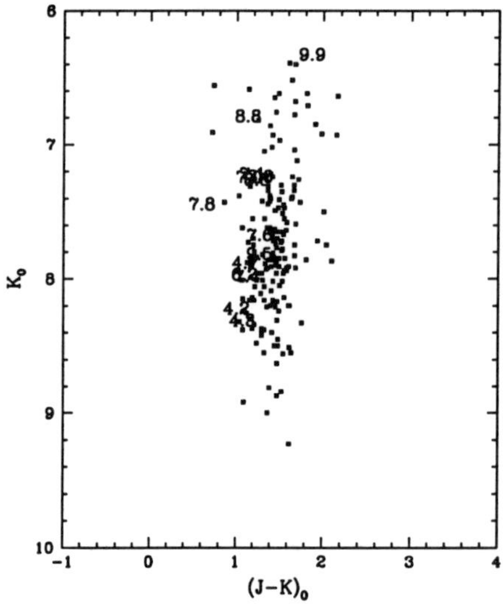

Figure 2. $(K_s)_o$ vs $(J–K_s)_o$ diagram of the ISOGAL sources in C32. Overplotted are the spectral types of LPV stars of the PG3 field (shifted by -0.5 in $(J–K_s)_o$).

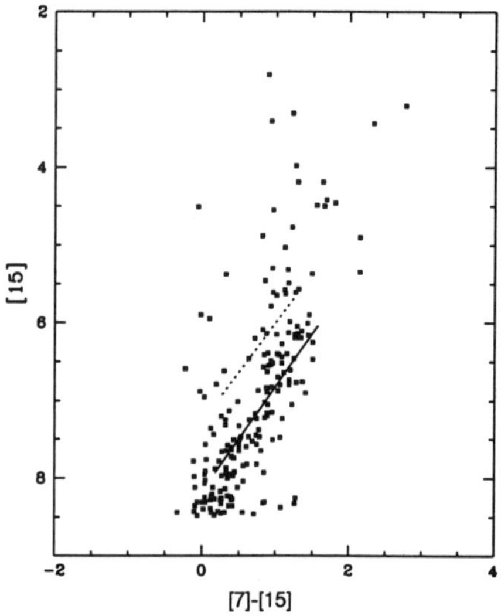

Figure 3. [15] vs [7]–[15] diagram of the C32 field with the theoretical sequences of M giants with increasing mass loss rate.

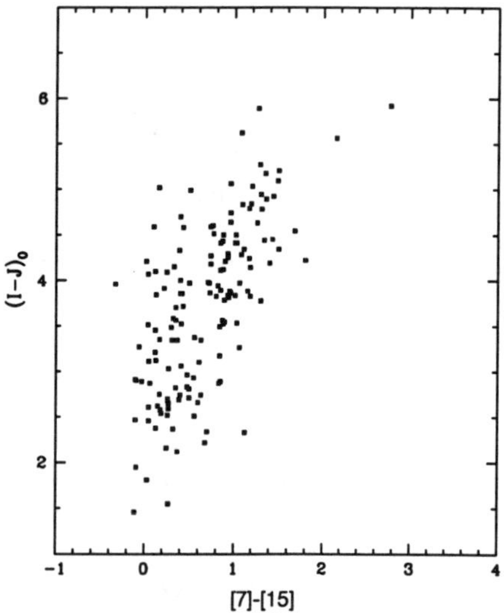

Figure 4. $(I–J)_o$ vs [7]–[15] diagram of the C32 field.

4. CVF measurements

To further substantiate our claim that we see the mass loss of the stars observed in the Bulge we now discuss the $5 - 15\mu m$ spectra taken in an ISOCAM CVF measurement which was performed on a $3' \times 3'$ subfield of C32. Around 30 objects were detected of which half had a significant flux up to 16 μm. All these latter sources are also in the catalogue of our ISOCAM filter measurements. In Figs. 5–7 examples are shown of stars in different parts of the [15] vs [7]–[15] diagram. The first object comes from the bottom of the sequence. The dashed line is a M5 giant model (Fluks et al. 1994) shifted to fit the spectrum at 6 μm. There is no mid-IR excess. One also clearly notices the SiO band absorption common in late type giants, but not included in the model. The second example shows a source from the middle of the sequence. Here the model of the M5 giant is reasonable well followed up to $\approx 9\mu m$ where a silicate feature can be seen, clearly indicative of mass loss. The last example is of a source above the sequence. One further remarkable fact is that we see a very clear diversity of silicate features in our sample.

Acknowledgments. We thank Martin Groenewegen for providing us the model sequences. For the ISOCAM data reduction we used "CIA", a joint development by the ESA Astrophysics Division and the ISOCAM Consortium. The ISOCAM Consortium is led by the ISOCAM PI, C. Cesarsky, Direction des Sciences de la Matière, C.E.A., France.

References

Alves D., Alcock C., Cook K., et al., 1998, in "Pulsating Stars - Recent Developments in Theory and Observation", XXIII IAU Gen. Assembly, Takenti and Sasselov (eds.), astro-ph/9804003

Blommaert J.A.D.L., 1992, thesis, University of Leiden

Cesarsky C., Abergel A., Agnèse P., et al., 1996, A&A 315, L32

David P., Papoular R., 1990, A&A 237, 425

Epchtein N., et al., 1997, The ESO Messenger 87, 27

Fluks M.A., Plez B., Thé P.S., et al., 1994, A&ASS 105, 311

Frogel J.A., Whitford A.E., 1987, ApJ 320, 199

Gehrz R.D., Woolf N.J., 1971, ApJ 165, 285

Omont A., the ISOGAL Collaboration, 1998, in "Astrophysics with Infrared Surveys: A Prelude to SIRTF", Pasadena, ASP Conference Series, in press

Reimers D., 1975, in "Problems in Stellar Atmospheres and Envelopes", B. Bascheck, W.H. Kegel and G. Traving (eds.), Springer, Berlin, p. 229

Schultheis M., 1998, thesis, University of Vienna

Tiede G.P., Frogel J.A., Terndrup D.M., 1995, AJ 110, 2788

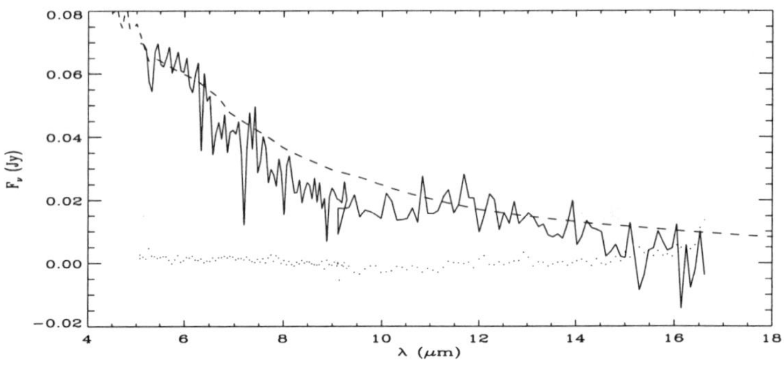

Figure 5. CVF spectrum of a source at the bottom of the sequence seen in the [15] vs [7]–[15] diagram (Fig. 3)

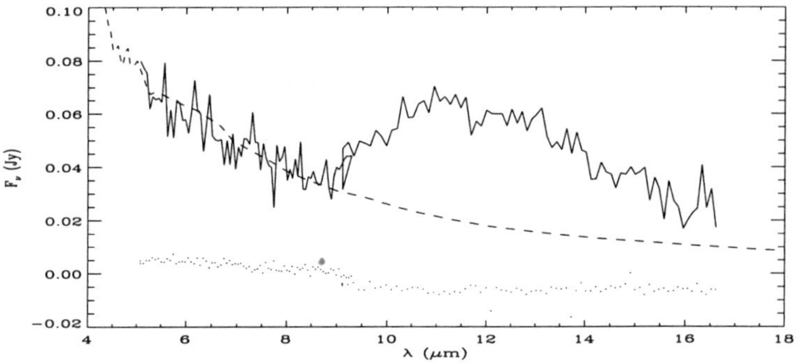

Figure 6. CVF spectrum of a source in the middle of the sequence

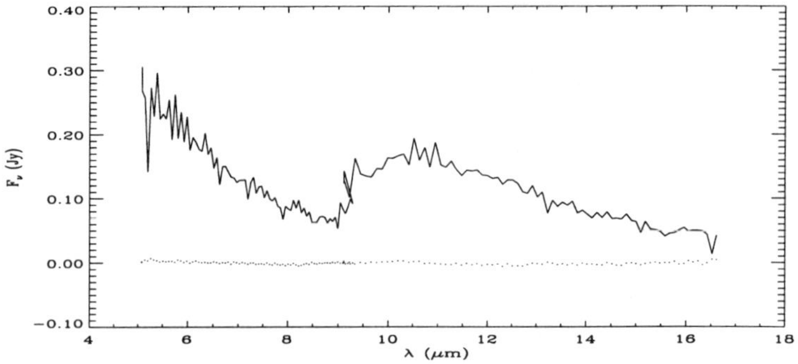

Figure 7. CVF spectrum of a source above the sequence

Asymptotic Giant Branch Stars
T. Le Bertre, A. Lèbre and C. Waelkens, eds.
I.A.U. Symposium 191, 1999

The Galactic Disk Distribution of Dust Emission Features in Planetary Nebulae

S. Casassus and P.F. Roche

Astrophysics, Oxford University, Keble Road, Oxford OX1 3RH, UK

Abstract. The properties of the carbon and oxygen chemical balance in planetary nebulae (PNe) are analysed through mid infrared spectroscopy of warm dust emission features in a sample of 72 objects. The adoption of a statistical distance scale shows that the galactic disk distribution of warm dust types in PNe is rather homogeneous with height over the galactic plane, and that there is a trend for the proportion of PNe with O rich dust signatures to decrease with galactocentric radius. Models of the galactic distribution of PNe compositions require a minimum progenitor mass of $1.2\,M_{\odot}$, although the observational constraints suffer from the smallness of the sample. This initial investigation is however an incentive to pursue the use of warm dust emission in PNe to study their progenitor population in various galactic environments.

1. Introduction

In the context of single star evolution, planetary nebulae (PNe) compositions reflect the processing undergone by the progenitor stars in their evolution up the Asymptotic Giant Branch. The dredge up of carbon synthesised in the He shell flashes alters the chemical balance in the envelope. But if the temperature at the bottom of the convective envelope is high enough, CNO processing enhances ^{13}C and ^{14}N at the expenses of ^{12}C and ^{16}O, and prevents dredged up carbon from reaching the surface. This process, often referred to as 'hot bottom burning' (HBB), produces radical changes in the predicted surface compositions. Thus a sequence is expected in nebular C/O ratio which initially increases with main sequence mass, but then decreases again as HBB occurs. Another consequence of HBB is a strong enhancement of the nitrogen surface abundance; HBB is the only mechanism for single star evolution that can explain the highest N/O ratios in PNe.

The C/O chemical balance in PNe can be determined through gas phase abundances, but also through the warm dust emission signatures at $10\,\mu m$. The dust signatures fall into 4 groups (see Roche 1987), and correspond to silicate grains, silicon carbide (SiC), the unidentified infrared bands (UIR), or result in either weak or unclassified continuum emission. The link with the carbon to oxygen ratio in the nebulae has been established by comparing with the gas phase abundances (e.g. Barlow 1983, Roche 1989). Silicate dust bands indicate an O rich environment, SiC a C rich environment, and UIR bands correlate with a strong overabundance of C relative to O.

We used the dust signatures in a sample of 72 PNe as a systematic indicator of the C/O chemical balance, and obtained their galactic disk distribution (Sect. 2.). The selection criteria were that the nebulae be compact, less than 10" in diameter, and that their IRAS 12μm flux be higher than 0.5 Jy. The distribution of dust emission types is rather homogeneous with height above the galactic plane, and there seems to be a trend for a decreasing fraction of silicate emission PNe towards the outer Galaxy. We also confirmed, through crude galactic disk and synthetic AGB models, that the properties of the distribution can be understood in terms of single star evolution (Sect. 3.).

Although the warm dust emission types are likely to be affected by phenomena other than AGB evolution (e.g. interacting binaries), these results support the statistical interpretation of warm dust emission types in samples of PNe to investigate the progenitor populations.

2. Warm dust emission features in a sample of compact PNe

Figure 1a is a comparison between PNe types (Peimbert 1978) and dust signatures, for the PNe in our sample with known nitrogen abundance (with references in Casassus & Roche 1998). PNe of type II are sub-divided according to the classification described in Faúndez-Abans & Maciel (1987). The nitrogen rich PNe are the type I PNe, with log(N/O)>-0.3, and the type IIa PNe which have log(N/O)>-0.6. PNe with no nitrogen enrichment are the type IIb PNe, with Vlsr deviating from galactic disk circular motion by less than 60 km s^{-1} , or else the type III PNe. The sequence in Peimbert types is indicative of progenitor mass, the highest being associated with type I. One would therefore expect a stratification in height above the galactic plane as a function of Peimbert type, as in Maciel & Dutra (1992). Such a stratification is indeed present in our sample, as is illustrated in Fig. 1b, based on the distance scale by Zhang (1995). Inside the solar circle, the root mean square height over the plane, z_{rms}, is 0.14 kpc for type I, 0.37 for type IIa, 0.48 for type IIb, and 0.67 for type III. This stratification is also an indication that our sample is big enough for a statistical study. Out of 50 PNe in our sample with known N/O ratios, 30$\pm$5% are type I, which is typical of PNe catalogues (e.g. Maciel & Dutra 1992), and we seem to sample homogeneously the disk PNe population.

Thus, UIR emission is related to higher progenitor masses, silicate emission is found mainly for low mass progenitors, but also for the most massive ones. SiC emission is related to intermediate nitrogen abundances, whereas 'weak' continuum PNe are equally represented.

Fig. 2 shows the spatial distribution of dust emission types, with distances on the scale by Zhang (1995). We list the properties of the dust type distributions in Table 1, where N stands for number, and the total number of sources was Ntot=53, 33 for R<Ro, and 20 for R>Ro. Although there is a concentration of UIR emission PNe at low latitudes, the vertical distribution of PNe is rather homogeneous when compared to the Peimbert types, with z_{rms}=0.45$\pm$.1 kpc for O rich PNe, and z_{rms}=0.35$\pm$.04 kpc for C rich PNe. There seems to be a trend towards a decreasing relative proportion of silicate dust type nebulae from 27% in the inner Galaxy, to 15% outside the solar circle, but as the standard error is $\sim$10% this decrease corresponds to only one sigma.

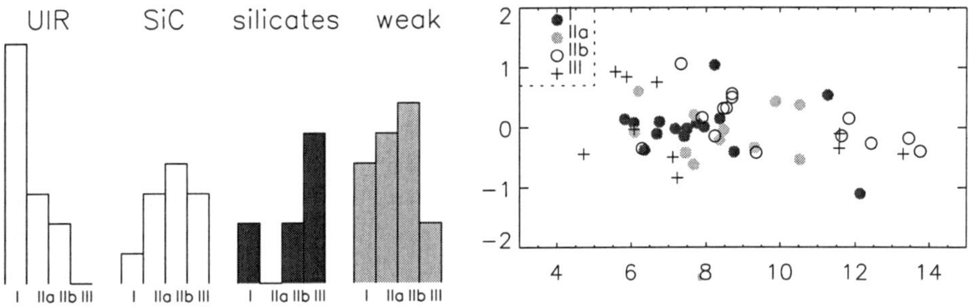

Figure 1. The relationships between the dust signatures and progenitor mass. Left hand plot: the number of PNe of each dust emission types as a function of Peimbert types. Right hand plot: the vertical distribution of PNe types, with R (galactocentric radius) in x, and z (height over the plane) in y.

Table 1. The properties of the distribution of planetary nebulae dust types.

type	$b_{\rm rms}$ [degrees]	$z_{\rm rms}$ [100pc]	N/Ntot	R<Ro	R>Ro
Silicates	**8.0**	**4.5**	**0.23**	0.27	0.15
SiC	**7.9**	**3.8**	**0.30**	0.21	0.45
UIR	**4.2**	**3.0**	**0.47**	0.52	0.40

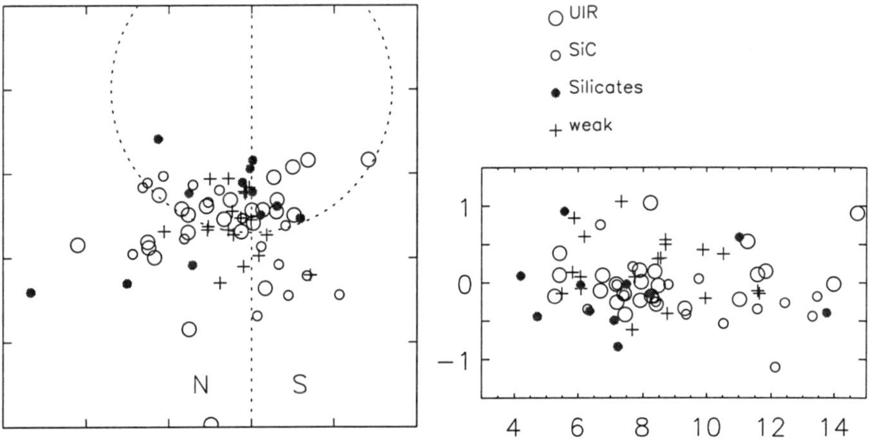

Figure 2. The galactic disk distribution of dust emission types in PNe. Left: face on view of the disk (ticks every 5 kpc). Right: vertical distribution, in kpc.

It was mentioned that PNe with no warm dust are equally distributed among Peimbert types. They also have high 6 cm fluxes, which together with the selection criteria of compact angular size, explains why they are on average closer. PNe with weak continuum probably correspond to later evolutionary stages. Their gas phase C/O ratios reflect the proportions for the whole sample, with an average at 1.1 and a dispersion of 1.2 FWHM, again supporting that these are more evolved PNe (as opposed to PNe with C/O$\sim$1).

Although all known IR bright and compact PNe were included in this study, further observations are required to improve the statistics.

3. Synthetic distribution

We constructed a synthetic AGB model that should be considered as an observational requirement only. The model will be fully described elsewhere (Casassus & Roche 1998), but it follows the prescriptions compiled by Groenewegen & de Jong (1993). Our aim was to check whether AGB evolution can explain the trends in the chemical composition of PNe. But we noted that existing AGB models cannot account for the continuous range of nitrogen abundances observed in LMC PNe (Leisy & Dennefeld 1996), and the fact that 60% of the type I PNe in our sample have C rich dust (confirmed by the gas phase abundances). We resolved this discrepancy by implementing HBB through a simple CNO tri-cycle at temperature T_B, with a typical exposure time t_{HBB} and cumulative exposures every τ_{conv}=0.5 yrs ($\sim$ convective turnover timescale). We fixed $t_{HBB}=10^{-5}$yrs so as to obtain $^{12}C/^{13}C$ ratios at CN equilibrium values for the most massive stars. We account for N enrichment in C rich PNe by a linear dependence of T_B with core mass, $dT_B/dM_c = 2\times10^9\,K\,M_\odot^{-1}$. An increase of T_B on the AGB strong enough to produce C and N rich PNe was obtained by Renzini & Voli (1981), although their mass loss rate was underestimated. The model was also required to reproduce the luminosity function of C stars in the LMC (Cohen et al. 1981) and the initial-final mass relation (Weidemann & Koester 1983; Weidemann 1987), which fixed the minimum core mass for dredge-up $M_c^{min}=0.58\,M_\odot$, the dredge-up parameter λ=0.6, and mass loss parameter η=5 (Reimers law). The resulting PNe abundances, averaged over the last 25 000 yrs of AGB evolution, are shown in Fig. 3.

We also needed a crude galactic disk model, including as much as possible observed distributions rather than model results. The star formation rate as a function of galactocentric radius was extrapolated from the UCHII region survey in Bronfman et al. (1998). We used the results presented in Köppen & Cuisinier (1997) for the age of each star and the treatment of radial diffusion. For the vertical structure of the disk, we used the approximations in Talbot & Arnett (1975) and the age-velocity relation by Wielen (1977). As in Carraro et al. (1998), we took a radial metallicity gradient of -0.07 dex/kpc constant in time. Z(R=Ro,t) was required to match the age metallicity relation by Meusinger et al. (1991). The model is therefore a very simple one, but it should be kept in mind that we are modelling a PNe distribution based on a statistical distance scale. This distribution cannot be regarded as absolute, only the relative distributions of the different PNe types contain valuable information.

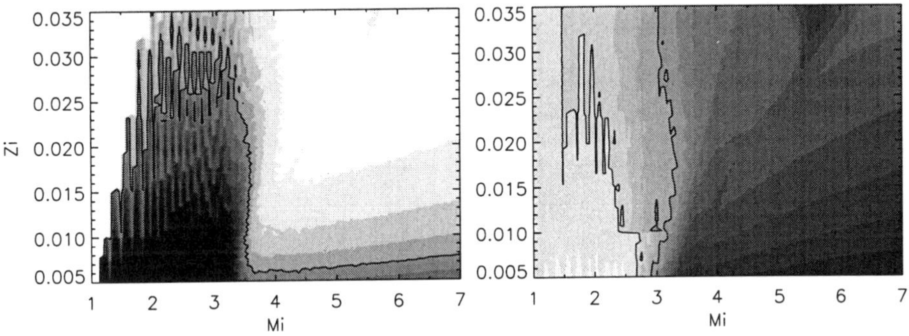

Figure 3. PNe abundances as a function of initial mass in x, initial metallicity in y. The left hand plot shows the C/O ratio in grey scale, with darker contours at .2,.4,.6,... and the solid line is at C/O=1. The right hand plot shows log(N/O), with darker grey scale contours at -.9,-.8,...,1 and the solid lines represent the contours for type IIa and type I PNe.

In order to link the modelled and observed PNe galactic distributions, we averaged the fraction of flux emitted in the IRAS 12μm band, and applied our flux selection criterion to the synthetic population. But many selection effects come from the PNe catalogues themselves, and we adopted a minimum galactocentric radius of 0.5 Ro (as can be inferred from Fig. 2, see also Durand et al. 1998). Since PNe with no warm dust correspond to later evolutionary stages, the link between the computed C/O ratios and the dust features is very tight. Silicates represent C/O<1, and SiC and UIR bands C/O>1.

The properties of the synthetic distribution that we would like to compare to the observations are shown in Fig. 4. PNe compositions were computed by averaging the stellar mass loss over the last 2 000 years of evolution on the AGB to compare with the dust signatures, and over the last 25 000 years to compare with the gas phase abundances. The minimum mass for the progenitors of compact and infrared bright PNe, M^{min}, was kept as a free parameter. Fig. 4a and 4b are the results from t_{PN}=2 000 yrs; Fig. 4a shows the fraction of O rich PNe and Fig. 4b is the ratio of the root mean square height over the plane for O rich and C rich PNe. Fig. 4c shows the proportion of N rich PNe, as obtained with t_{PN}=25 000 yrs. Values of M^{min} between 1.15 $M_\odot$ and 1.35 $M_\odot$ can account for all the observed properties of PNe within their standard errors. With M^{min}=1.2 $M_\odot$, the fraction of type I PNe with C rich dust signatures is 70%, whereas the fraction with gas phase C/O>1 is 40%.

4. Conclusions

We undertook a systematic study of the C/O chemical balance at the end of the AGB through warm dust emission in a sample of 72 compact and infrared bright PNe. The properties of the galactic disk distribution of the dust signatures in PNe are consistent with a simple AGB model, and require a minimum progenitor mass of about 1.2 $M_\odot$ and an ejection age of 2 000 years. These results are

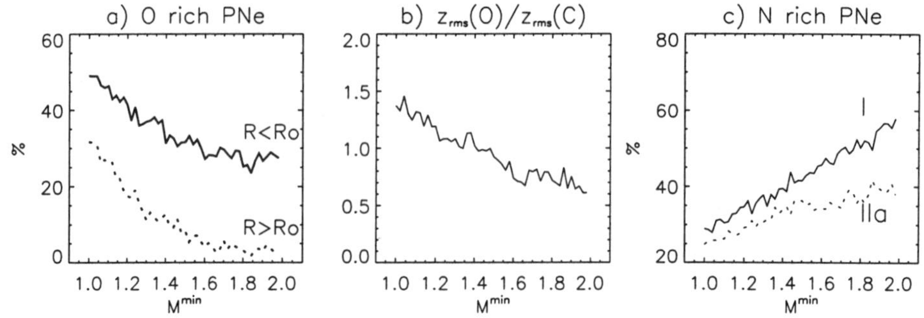

Figure 4. Properties of the synthetic PNe population as a function of minimum main sequence progenitor mass, M^{min}.

an incentive to pursue the use of warm dust emission in PNe to study their progenitor population in various galactic environments.

References

Barlow B.M., 1983, in "Planetary Nebulae", IAU Symposium 103, ed. D.R. Flower, p. 105

Bronfman L., Casassus S., May J., Nyman L.-Å., 1998, in preparation

Carraro G., Ng Y.K., Portinari L., 1998, MNRAS 296, 1045

Casassus S., Roche P.F., 1998, in preparation

Cohen J.G., Frogel J.A, Persson S.E., Elias J.H., 1981, ApJ 249, 481

Durand S., Acker A., Zijlstra A., 1998, A&A 132, 13

Faúndez-Abans M., Maciel W.J., 1987, A&A 183, 324

Groenewegen M.A.T., de Jong T., 1993, A&A 267, 410

Köppen J., Cuisinier F., 1997, A&A 319, 98

Leisy P., Dennefeld M., 1996, A&AS 116, 95

Maciel W.J., Dutra C.M., 1992, A&A 262, 271

Meusinger H., Reimann H.-G., Stecklum B., 1991, A&A 245, 57

Peimbert M., 1978, in "Planetary Nebulae", IAU Symposium 76, Reidel, p. 215

Roche P.F., 1987, in "Planetary and Proto-Planetary Nebulae: From IRAS to ISO", A. Priete Martinez (ed.), p. 45

Roche P.F., 1989, in "Planetary Nebulae", IAU Symposium 131, ed. S. Torres Peimbert, p. 117

Renzini A., Voli M., 1981, A&A 94, 175

Talbot R.J., Arnett W.D., 1975, ApJ 197, 551

Weidemann V., 1987, A&A 188, 74

Weidemann V., Koester D., 1983, A&A 121, 77

Wielen R., 1977, A&A 60, 263

Zhang C.Y., 1995, ApJS 98, 659

Asymptotic Giant Branch Stars
T. Le Bertre, A. Lèbre and C. Waelkens, eds.
I.A.U. Symposium 191, 1999

Long-period Variable Stars Near the Galactic Centre

I.S. Glass

S. A. Astr. Obs., PO Box 9, Observatory, 7935 South Africa

S. Matsumoto

Inst. of Astr., University of Tokyo, Mitaka, Tokyo 181, Japan

B.S. Carter

Carter Observatory, PO Box 2909, Wellington, New Zealand

K. Sekiguchi

Subaru Telescope, NAOJ, 650 N A'ohoku Pl, Hilo, HI 96720, USA

Abstract. A second report on our 4-year survey for long-period variables (LPVs) in the central 24×24 arcmin^2 of the Galactic Bulge is presented. All four seasons of data have now been reduced and ~ 350 variables have been found. Preliminary periods have been obtained for most of them. Fifty-five out of 103 known OH/IR stars in the field have been recovered in the K-band.

1. Introduction

The variable stars of the inner Bulge are of great interest both as population and distance indicators. The easiest of them to identify are the long-period AGB variables, comprising Miras and OH/IR stars. Because of the high extinction in their direction, fields near the Galactic Centre are best observed in the near-infrared, between H (1.6 μm) and L (3.5 μm). At longer wavelengths the photospheric component is swamped by emission from circumstellar and interstellar dust. So far, detailed knowledge of stellar properties near the Centre has largely been confined to the GC Cluster, the nearby "Quintuplet" and the cluster of hot stars near G0.121+0.017.

Surveys for LPVs have been made in the central cluster in the K-band and in the low-obscuration Sgr I and NGC 6522 windows using visible and near-IR photographic techniques. The most recent survey for OH sources near the Galactic Centre has been made by Sjouwerman et al. (1998), who also lists previous work. Some infrared identifications of OH sources and period determinations have been made by Wood, Habing & McGregor (1998). Periods of several OH sources were also found by Blommaert (1998). In addition, Jones et al. (1994) have given the infrared-derived periods of a few Bulge OH/IR stars.

The present programme seeks to uncover the variable star content of the Inner Bulge using infrared techniques. Use is made of a very large-area array of limited sensitivity to study the most luminous part of the late-type stellar population. A first report, containing the findings of the first two seasons' work, was given by Glass et al. (1996).

2. Observations

The observations were made over four southern winters, viz 1994-1997 inclusive. An area of 24 $\times$ 24 arcmin^2, centered near the Galactic Centre, was divided into 25 fields of 300 $\times$ 300 arcsec^2 so that there was an overlap of 15 arcsec between adjacent fields. The PANIC (PtSi Astronomical Near-Infrared Camera), based on a Mitsubishi 1040 $\times$ 1040 PtSi array with pixels of 17 $\times$ 17 μm^2, was used. A complete description of this camera has been given by Glass, Sekiguchi & Nakada (1995).

3. Data-reduction

Standard data-processing involved principally the removal of telescope and sky radiation, amplifier emission and dark current. An empty "sky" frame was constructed by combining groups of five (occasionally three) exposures. First, the five exposures were averaged. The difference between each individual and the average exposure was found and an additive correction was made so that each frame had the same modal pixel value. The median average of the five frames was then found and treated as an empty frame.

The extremely long exposures (usually 10 minutes) necessary because of the low quantum efficiency of the array means that the background level may vary considerably between the first and last of a five-frame series. For this reason, the quality of the "sky" frame was not always perfect and may have led to increased photometric scatter for faint stars when the most crowded fields were reduced.

The coordinates and intensities of the stars were extracted from each exposure using DoPHOT (Schechter, Mateo & Saha 1993). The output of each exposure was called a *sum* file. Further processing depended on the "STAR" suite of programmes written by Dr L.A. Balona (SAAO), whom we thank. Each of the 25 fields was treated separately.

For each field a particular *sum* file was selected as a fiducial one for cross-identification purposes. A new set of files - the *cat* files - on the coordinate system of the fiducial exposure was then created, corresponding to each of the *sum* files, omitting images classified by DoPHOT as non-stellar. This process allowed for translation and rotation of the coordinates and depended on the recognition of at least five bright stars in each frame.

In order to form some idea of the accuracy of the positional measurements, the x and y coordinates of the brightest star ($K \sim 5.8$) and the 200th brightest star ($K \sim 10.8$) in field 1 were extracted from each *cat* file and their standard deviations were calculated. Taking the scale as 3.5 pixels arcsec^{-1} in the x direction and 1.75 pixels arcsec^{-1} in the y direction, the s.d.'s amounted to (s.d.x, s.d.y) = (0.02, 0.02) arcsec for the bright star and (0.09, 0.07) for the faint one.

Thus measurement error does not contribute significantly to the overall error in the position determinations.

The next step was to make a table of photometry for each star, still at this point on the natural systems of the individual exposures. Around 10 to 15 bright stars from the field were compared through all the available *cat* files and used to place the individual exposures on a uniform photometric scale. Because the brightest stars are often variable, the residuals were examined and individual stars were rejected before iterating. After a few iterations, a suitable set of non-variable stars is found, with standard deviations typically around 0.04 mag.

4. Astrometry

Astrometric positions of stars in each of the 25 fields were obtained by reference to stars included within the USNO-A1.0 astrometric catalogue (Monet et al. 1996). The coverage of this catalog is sufficiently dense that many tens of stars are found in each of our fields. Its typical astrometric error is believed to be about 0.25 arcsec. For cross-identification, overlays of USNO-A stars were prepared on the same scale as charts printed out from a DoPHOT *cat* file. Because the USNO stars are often very faint in the K-band, recourse was sometimes necessary to the J-band for guidance in locating the infrared counterparts. The DoPHOT coordinates of the selected stars were then entered into an astrometry programme with RA and Dec (2000) positions from USNO-A. The residuals from the output were examined to eliminate misidentifications (occasionally due to chance correspondences of different visible and infrared objects). Proper motions, crowding and low signal-to noise may also have contributed to the residuals. The final standard errors in the positions were usually estimated to be around 0.5 arcsec, based on the residuals of the local standards from the USNO-A catalogue.

5. Extraction of variable stars

For each field, the light curves of the brightest 200 stars were plotted out and examined by eye. This tedious method was unfortunately necessary because of accidental mis-identifications produced when DoPHOT refused to classify an image as stellar and the cross-identification programme chose a neighbouring faint image instead. This problem occurred once in a few hundred measurements.

Other causes of spurious variability included crowding, such as in the Galactic Centre cluster and the nearby hot star cluster, and proximity to the edge of the array. Crowding was checked for by creating overlays of the variable star positions which matched the scale of the K-image print-outs. Whenever an apparent variable star appeared to be within 40 pixels of the edge in x, or 20 pixels in y, this fact was flagged.

The standard deviations of individual non-variable stars in two of the fields, one crowded and the other less so, were examined as a function of K magnitude (see Fig. 1), after removal of the accidental points referred to. Bright stars have standard deviations around 0.04 mag. but fainter ones have standard deviations that reach 0.1 at $K \sim 10.4$ and 0.2 mag. at ~ 11.4. These figures give an

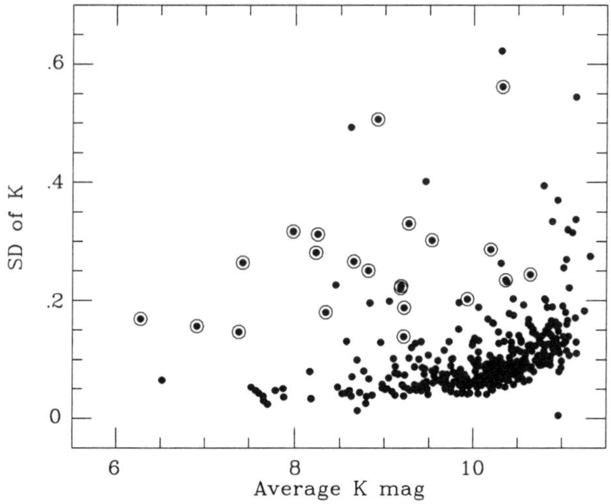

Figure 1. Variability statistics for 400 of the brightest stars in the 5 × 5 $arcmin^2$ sub-field no. 13, which contains the "Quintuplet" or AFGL 2004 cluster. Stars believed to be genuine variables are circled. Other high standard deviations are attributable to crowding, proximity to the edges of the array and, at the fainter end, to imperfections in sky-subtraction.

indication of the level at which sinusoidally variable stars of full amplitudes 0.3 and 0.6 mag., of unknown period and with few data points, might be detected.

Periods were determined from power spectra of the data for each star. About 250 of the 350 variables found so far appear to be regular LPVs. However, the present method of analysis may not be optimal, and it is intended to use alternative methods to refine the results.

6. Period Distribution

Fig. 2 shows the period distribution of the LPVs found to date in the overall survey field as well as a similar histogram for the Sgr I Baade's Window, where the extinction is < 0.2 mag. at K. The proportion of longer-period LPVs at first glance appears to be higher in the Centre. However, systematic effects must be guarded against. Firstly, the stars examined so far for variability represent the brightest 200 of each field at *K at a particular time*, and the limiting magnitude will vary from field to field according to the overall numbers of stars. Secondly, if $A_K \sim 3.0$, as it is over much of the survey region, one would expect 200-day Mira variables to appear at about $K = 10.7$ and 600-day variables at about $K = 9.0$, based on their properties in the Sgr I field (Glass et al. 1995). Thus some incompleteness will very probably be present in the census of the shorter-period LPVs. Further analysis will attempt to address this problem.

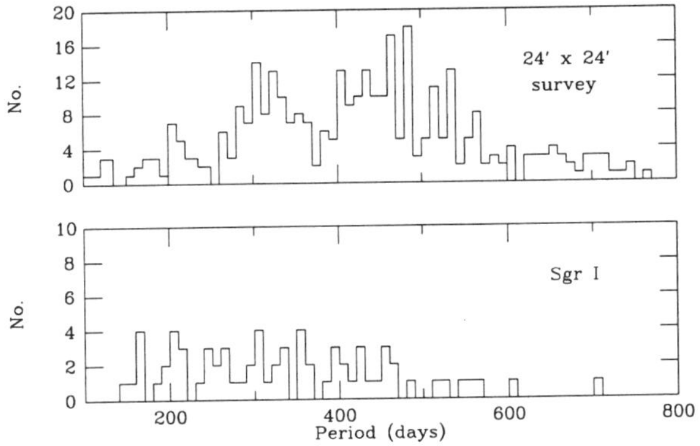

Figure 2. The period distribution of long-period variables *top* in this work and *bottom* in the Sgr I field (Glass et al. 1995).

7. OH/IR Stars

The positions of known OH/IR stars from Sjouwerman et al. (1998) and variables from Wood et al. (1998) were searched specifically if they had not already been located during routine processing. Because the OH/IR star amplitudes are usually very large, it was often possible to find them even if they were faint. Of the ~ 103 OH/IR stars known in the survey area from radio studies, 55 have so far been recovered in our searches. The remainder may be self-obscured by thick dust shells, may be obscured by interstellar matter (which is known to exceed $A_K = 6$ mag. in places (Glass, Catchpole & Whitelock 1987) or may be too crowded for regular DoPHOT analysis, particularly in and around the GC Cluster.

Fig. 3 shows the sources detected in our field as well as known OH/IR sources and variables located by Wood et al. (1998). A small number of the latter were below the faintness limit of this survey. Reference to a K-map of the area (e.g., Glass, Catchpole & Whitelock 1987) shows that detectability is, as expected, reduced in the areas hidden by dense molecular clouds.

References

Blommaert J.A.D.L., van der Veen W.E.C.J., van Langevelde H.J., Habing H.J., Sjouwerman L.O., 1998, A&A 329, 991

Glass I.S., Catchpole R.M., Whitelock P.A., 1987, MNRAS 227, 373

Glass I.S., Matsumoto S., Ono T., Sekiguchi K., 1996, in *The Galactic Center*, Fourth ESO/CTIO Workshop, R. Gredel (ed.), PASP Conf. Series 102, p. 312

Glass I.S., Sekiguchi K., Nakada Y., 1995, in *New Developments in Array Technology and Applications*, A.G. Davis Philip et al. (eds.), IAU Symposium 167, Kluwer, Dordrecht, p. 109

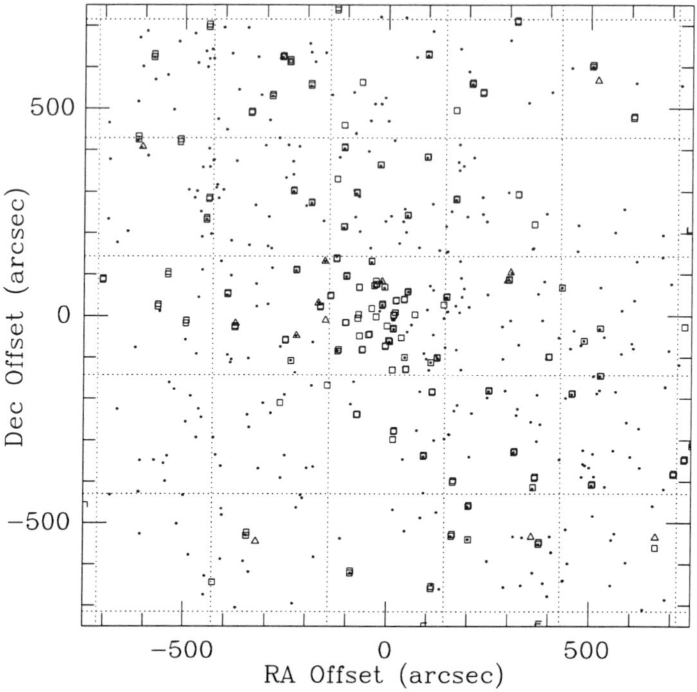

Figure 3. Variables and OH/IR sources. Dots indicate the newly-found variables. Squares are OH/IR stars from Sjouwerman et al. (1998) and triangles are variables found by Wood et al. (1998). The dotted lines indicate the boundaries of the 25 sub-fields.

Glass I.S., Whitelock P.A., Catchpole R.M., Feast M.W., 1995, MNRAS 273, 383

Jones T.J., McGregor P.J., Gehrz R.D., Lawrence G.F., 1994, AJ 107, 1111

Monet D., et al., 1996, *USNO-A1.0: A Catalog of Astrometric Standards*, US Naval Observatory, Washington DC (CD-ROM set)

Schechter P.L., Mateo M., Saha A., 1993, PASP 105, 1342

Sjouwerman L.O., van Langevelde H.J., Winnberg A., Habing H.J., 1998, A&AS 128, 35

Wood P.R., Habing H.J., McGregor P.J., 1998, A&A 336, 925

Asymptotic Giant Branch Stars
T. Le Bertre, A. Lèbre and C. Waelkens, eds.
I.A.U. Symposium 191, 1999

Iron Abundances in AGB Stars and M Supergiant Stars at the Galactic Center

S. V. Ramírez[1], K. Sellgren, D. M. Terndrup

The Ohio State University, Columbus OH 43210, USA

J. S. Carr

Naval Research Laboratory, Washington, DC 20375, USA

S. Balachandran

University of Maryland, College Park, MD 20742, USA

R. D. Blum

CTIO, La Serena, Chile

Abstract. We are measuring Fe abundances of cool, luminous stars within 30 pc of the center of the Milky Way. Our sample contains both AGB stars and M supergiants. Low-resolution ($\lambda/\Delta\lambda = 500$) H and K band spectra are used to estimate temperatures and gravities. Stellar Fe abundances are determined from high-resolution ($\lambda/\Delta\lambda = 40\,000$) K band spectra obtained on the IRTF using CSHELL. We find that Fe abundances of stars in the Galactic Center are consistent with the solar Fe abundance.

1. Introduction

We have measured iron abundances of cool luminous stars located within 30 pc of Galactic Center (GC), using K−band (2.2 μm) high-resolution spectra. Infrared techniques are required because of the large optical extinction towards the GC ($A_V \sim 30$ mag.). Our sample includes both AGB stars and M supergiant stars.

Chemical abundances in the disks of spiral galaxies are observed to reach their highest values at the center (see Pagel & Edmunds 1981 and Shields 1990). H II regions in the Milky Way also show a radial metallicity gradient (Peimbert et al. 1978; Hawley 1978; Shaver et al. 1983; Pipher et al. 1984; Simpson & Rubin 1990; Simpson et al. 1995; Afflerbach et al. 1997). Chemical evolution models strive to explain these gradients by considering the relative star formation and gas infall/outflow rates, the star formation history, and the metal abundance of the gas compared to the stars (see Audouze & Tinsley 1976 and Rana 1991). The gas metal abundance of the GC varies between 1 and 5 times solar depending

[1]US Gemini Fellow.

on the element (Aitken et al. 1974, 1976; Willner et al. 1979; Lacy et al. 1980; Lester et al. 1981; Lester et al. 1987; Rubin et al. 1988; Shields & Ferland 1994; Simpson et al. 1995; Afflerbach et al. 1997), but Fe has not been measured in the gas phase. Our measurements of [Fe/H] are the first stellar abundance determinations in the GC.

2. Observations

We have observed cool luminous stars in the GC using the CSHELL spectrograph at the NASA Infrared Telescope Facility, a 3–m telescope located at Mauna Kea Observatory in Hawaii. The $K-$band spectra taken with this instrument have a spectral resolution of $\lambda/\Delta\lambda = 40\,000$ or 7 km s^{-1}. We have chosen iron lines which are free of molecular contamination, especially CN in M supergiants and water in M giant and AGB stars.

We have observed all cool stars that have low water absorption and that are brighter than $K = 9$, which is the limit set by the sensitivity of CSHELL. Our sample consists of seven M supergiants and three AGB stars. Nine of these stars, including the three AGB stars, are located in the central cluster within 2.5 pc of the GC. The remaining M supergiant belongs to the Quintuplet cluster, located 30 pc away from the central cluster.

3. Abundance analysis

For our abundance analysis we are using model atmospheres for late type giants and supergiants from Kurucz (1993) and Plez (1992), with the spectral synthesis program MOOG (Sneden 1973). We have found a good agreement between abundances derived from both sets of models in the range of stellar effective temperature (T_{eff}) and surface gravity (log g) where they overlap. Our abundance analysis requires as input parameters: T_{eff}, log g, the microturbulent velocity of the star (ξ), and the oscillator strengths of the lines we are studying. We have verified our analysis technique for the GC stars by observing and analyzing cool luminous stars of known abundances (Smith & Lambert 1985, 1986, 1990; Luck & Bond 1989; McWilliam 1990).

Effective temperature: For all our GC stars, we use the CO and H_2O absorption observed in the H and K bands at low resolution ($\lambda/\Delta\lambda = 500$) to get T_{eff}. The CO strength increases with decreasing T_{eff} and decreasing log g. The H_2O strength has the same T_{eff} dependence but opposite surface gravity dependence. These two features together provide 2-dimensional spectral classification (Kleinmann & Hall 1986). Once we establish the luminosity class of each star, we use the CO strength to get T_{eff}. In addition, we can use lines of CO present in $H-$band (1.6 μm) high resolution spectra to get T_{eff}. Lines with different lower excitation potentials give the same carbon abundance when T_{eff} is correct. We have observed these CO lines only in IRS 7, which is the brightest star in the GC. Both methods give the same value of T_{eff} for IRS 7.

Surface gravity: We derive the bolometric magnitude (M_{bol}) of the GC stars using the observed K magnitude, extinction, distance, and T_{eff}. We locate the

GC stars in the HR diagram, using T_{eff} and M_{bol}. We estimate their masses by comparing their location in the HR diagram with evolutionary tracks for stars of different masses (Schaller et al. 1992; Schaerer et al. 1993a, b; Charbonnel et al. 1996). We compute the surface gravity using the estimated mass, luminosity and T_{eff}.

Microturbulent velocity: For all our GC stars, we estimate ξ from log g (McWilliam et al. 1995). In addition, we can use lines of CO present in the $H-$band to get ξ spectroscopically. Lines of different equivalent widths give the same carbon abundance when ξ is correct. We have observed these CO lines only in IRS 7. These two techniques agree for IRS 7.

Oscillator strengths: We have derived empirical oscillator strengths for our abundance analysis. We use the spectrum of α Boo (Wallace & Hinkle 1996), the abundance measurements of Peterson et al. (1993) for α Boo, and the synthetic spectrum generated by MOOG to derive gf-values.

We have studied how errors in T_{eff} and log g would affect our Fe abundance determination. The Fe abundance is quite insensitive to errors in log g, since a typical error of $\pm$ 0.2 in log g corresponds to an error in [Fe/H] of $\pm$ 0.05 dex. The Fe abundance is more sensitive to errors in T_{eff}, since a typical error of $\pm$ 200 K represents an error in [Fe/H] of $\pm$ 0.1 dex.

4. Results

In the Table we present results for our sample of 10 stars within 30 pc of the GC. Note that IRS 7 is the brightest star in the GC, and VR 5-7 is the star located at the Quintuplet cluster.

A statistical analysis gives us an unweighted mean of [Fe/H] = +0.03 and a mean weighted by the errors of [Fe/H] = +0.13 . The estimate of the dispersion (σ) is 0.20 dex. This is not significantly different from the error in [Fe/H] for each star, given in the Table.

Star	# lines	[Fe/H]
IRS 7	4	$+0.08 \pm 0.18$
IRS 11	3	$+0.15 \pm 0.20$
IRS 19	3	$+0.30 \pm 0.09$
IRS 22	4	$+0.14 \pm 0.23$
RCS 72	2	-0.16 ± 0.34
RCS 114	2	-0.35 ± 0.27
RCS 124	2	$+0.19 \pm 0.09$
RCS 129	2	$+0.05 \pm 0.23$
RCS 140	2	$+0.09 \pm 0.43$
VR 5-7	2	-0.18 ± 0.13

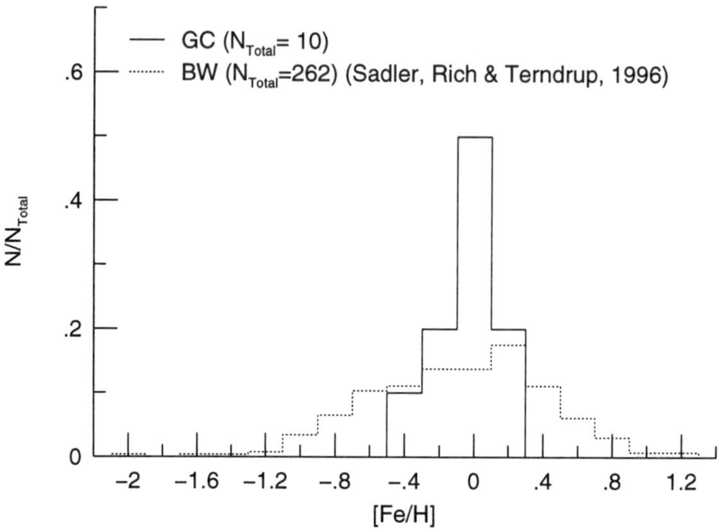

Figure 1. Fractional Fe abundance distribution of the GC compared with that of Baade's Window ($l, b = 1°, -4°$) in the Galactic Bulge.

5. Discussion

In Fig. 1 we compare our Fe abundance distribution in the GC with the Fe abundance distribution of Baade's window in the Bulge (Sadler et al. 1996). Sadler et al. (1996) resolve the abundance spread of the Bulge distribution. We have not detected a GC [Fe/H] distribution that is wider than the errors, in spite of having similar errors per star as Sadler et al. (1996). Thus, the dispersion of the GC distribution is significantly narrower than that of the Bulge [Fe/H] distribution.

In Fig. 2 we compare our GC Fe abundance distribution with the Fe abundance distribution of supergiants in the solar neighborhood (Luck & Bond 1989). We see that the GC abundance distribution is consistent with the abundance distribution of supergiants in the solar neighborhood.

We find that the average GC stellar Fe abundance is near solar. This result differs from the gas abundances in the GC, which vary between 1 and 5 times solar, depending on the element. In the future, we will measure stellar abundances of α–elements (Ca, Si, Ti) to address the issue of selective enrichment in the GC.

6. Conclusions

- Galactic Center Fe abundances are consistent with the solar value.

- Galactic Center Fe abundance distribution is consistent with the Fe abundance distribution of supergiants in the solar neighborhood.

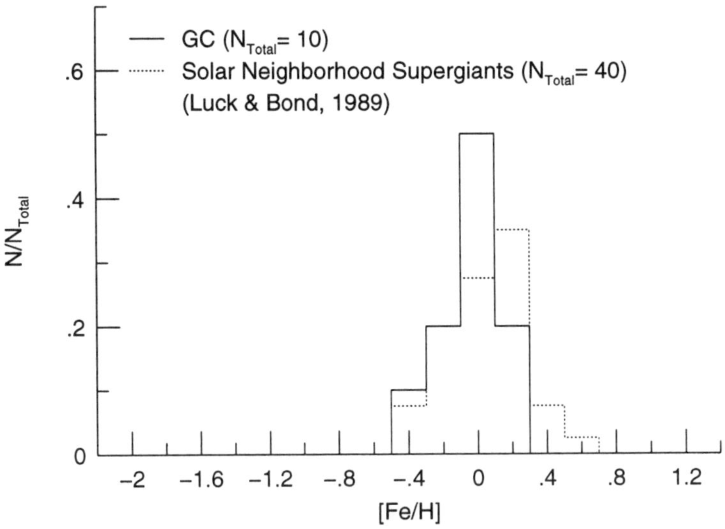

Figure 2. Fractional Fe abundance distribution of the GC compared with that of supergiants in the solar neighborhood.

- VR 5-7 (located 30 pc away from the center of the Galaxy) is not different in [Fe/H] from stars located in the central cluster.

Acknowledgments. Support for this work was provided by the National Science Foundation through the grant # GF-1003-97 from the Association of Universities for Research in Astronomy, Inc., under NSF cooperative agreement AST-8947990 and from Fundación Andes under project C-12984, and from NSF grant AST-9619230.

References

Afflerbach A., Churchwell E., Werner M.W., 1997, ApJ 478, 190

Aitken D.K., Jones B., Penman J.M., 1974, MNRAS 169, 35P

Aitken D.K., Griffiths J., Jones B., Penman J.M., 1976, MNRAS 174, 41P

Audouze J., Tinsley B.M., 1976, ARA&A 14, 43

Charbonnel C., Meynet G., Maeder A., Schaerer D., 1996, A&AS 115, 339

Hawley S.A., 1978, ApJ 224, 417

Kleinmann S.G., Hall D.N.B., 1986, ApJS 62, 501

Kurucz R.L., 1993, Kurucz CD-ROM No. 13 "ATLAS9 Stellar Atmosphere Programs and the 2 $km\,s^{-1}$ Grid", Smithsonian Astrophysical Observatory, Cambridge

Lacy H.J., Townes C.H., Geballe T.R., Hollenbach D.J., 1980, ApJ 241, 132

Lester D.F., Bregman J.D., Witteborn F.C., Rank D.M., Dinerstein H.L., 1981, ApJ 248, 524

Lester D.F., Dinerstein H.L., Werner M.W., Watson D.M., Genzel R., Storey J.W.V., 1987, ApJ 320, 573

Luck R.E., Bond H.E., 1989, ApJS 71, 559

McWilliam A., 1990, ApJS 74, 1075

McWilliam A., Preston G.W., Sneden C., Searle L., 1995, AJ 109, 2757

Pagel B.E.J., Edmunds M.G., 1981, ARA&A 19, 77

Peimbert M., Torres-Peimbert S., Rayo J.F., 1978, ApJ 220, 516

Peterson R.C., Dalle Ore C.M., Kurucz R.L., 1993, ApJ 404, 333

Pipher J.L., Helfer H.L., Herter T., Briotta D.A., Houck J.R., Willner S.P., Jones, B., 1984, ApJ 285, 174

Plez B., 1992, A&AS 94, 527

Rana N.C., 1991, ARA&A 29, 129

Rubin R.H., Simpson J.P., Erickson E.F., Haas M.R., 1988, ApJ 327, 377

Sadler E.M., Rich R.M., Terndrup D.M., 1996, AJ 112, 171

Schaerer D., Meynet G., Maeder A., 1993a, A&AS 98, 523

Schaerer D., Charbonnel C., Meynet, G., Maeder A., Schaller G., 1993b, A&AS 102, 339

Schaller G., Schaerer D., Meynet G., Maeder A., 1992, A&AS 96, 269

Shaver P.A., McGee R.X., Newton L.M., Danks A.C., Pottasch S.R., 1983, MNRAS 204, 53

Shields G.A., 1990, ARA&A 28 525

Shields J.C., Ferland G.J., 1994, ApJ 430, 236

Simpson J.P., Rubin R.H., 1990, ApJ 354, 165

Simpson J.P., Colgan S.W.J., Rubin R.H., Erickson E.F., Haas M.R., 1995, ApJ 444, 721

Smith V.V., Lambert D.L., 1985, ApJ 294, 326

Smith V.V., Lambert D.L., 1986, ApJ 311, 843

Smith V.V., Lambert D.L., 1990, ApJS 72, 387

Sneden C., 1973, Ph.D. thesis, University of Texas

Wallace L., Hinkle K., 1996, ApJS 107, 312

Willner S.P., Russell R.W., Puetter R.C., Soifer B.T., Harvey P.M., 1979, ApJ 229, L65

Asymptotic Giant Branch Stars
T. Le Bertre, A. Lèbre and C. Waelkens, eds.
I.A.U. Symposium 191, 1999

Carbon stars in populations of different metallicity

M.A.T. Groenewegen

Max-Planck-Institut für Astrophysik, Karl-Schwarzschild-Straße 1, D-85740 Garching, Germany

Abstract. Our current knowledge of carbon stars in the Local Group and beyond, is discussed. Although many carbon stars and late M-stars have been identified in external galaxies, a coherent understanding in terms of the chemical evolution- and star-formation-rate history of a galaxy is still largely lacking. Issues that need to be addressed are: 1) for some of the larger galaxies only a small fraction in area has been surveyed so far, 2) surveys have been conducted using different techniques, and may be incomplete in bolometric magnitude, 3) only for some galaxies is there information about the late M-star population, 4) not all galaxies in the Local Group have been surveyed, 5) only for a sub-set of stars are bolometric magnitudes available.

From the existing observations one can derive the following: the formation of carbon stars is both a function of metallicity and star formation. In galaxies with a similar star-formation-rate history, there will be relatively more carbon stars formed in the system with the lower metallicity. On the other hand, the scarcity of AGB-type carbon stars in some systems with the lowest metallicity indicates that these galaxies have had a low, if any, star-formation-rate history over the last few Gyrs.

1. Introduction

Carbon stars are tracers of the intermediate age population in galaxies. Either they are currently undergoing third dredge-up (the cool and luminous N-type carbon stars), or have been enriched with carbon-rich material in a binary system when the present-day white dwarf was on the AGB (the carbon dwarfs, CH-stars. The R-stars may be the result of coalescing binaries [McClure 1997]).

Since their spectral signature is very different from the one of oxygen-rich stars, it is relatively easy to identify carbon stars even at large distances. In Sect. 2 the different techniques to identify carbon stars are discussed, and in Sect. 3 the various surveys for carbon stars in external galaxies are summarised. The results are discussed in Sect. 4.

2. Methods

Basically three methods have been used to identify carbon stars in external galaxies: (optical) spectroscopy of red stars, a combination of a grating and a prism (GRISM), and narrow-band filters. The first method was used in the "early days" when follow-up spectroscopy was done on red stars identified in colour-magnitude diagrams. This led to the discovery of the first known carbon stars

Table 1: Carbon stars in external galaxies

Name	D (kpc)	M_V (mag)	[Fe/H]	N_C	Area (kpc^2)	N_M	Refs.
M31	770	−21.2	0.0	243	12	5254 (?)	1,2,3,4
Galaxy		−20.6	0.0	81	1.00		5
M33	850	−19.0	−0.6	15	0.20	60 (3+), 15 (5+)	2
LMC	50	−18.1	−0.6	1045	4.8	5200(2+),1300(5+),475(6+)	6
NGC 6822	540	−16.4	−0.8	36	0.18	16 (3+), 1 (5+)	2
NGC 205	770	−16.4	−0.85	7	0.30	17 (2+)	8
SMC	63	−16.2	−1.0	789	5.4	1250(2+),180(5+),60(6+)	6
				1707	12.2		7
IC 1613	765	−14.9	−1.2	15	0.52	6 (3+), 0 (5+)	2
WLM	940	−14.0	−1.5	14	0.28	6 (3+), 0 (5+)	2
Sagittarius dw	24	−14.0	−0.85	26			9,10,11
Fornax	131	−13.0	−1.4	104	0.44	25 (2+), 4 (5+)	12
Leo I	270	−12.0	−2.0	16	0.45		13
And I	810	−11.7	−1.45	0			16
And II	700	−11.7	−1.3	8		1 (?)	15, 16
Sculptor dw	78	−10.7	−1.8	8	0.63	40 (2+), 20 (3+)	13
Leo II	230	−10.2	−2.0	6	0.59		14
And III	760	−10.2	−2.0	0			16
Sextans	86	−10.0	−2.1	(0)			17
Phoenix	420	−9.9	−2.0	2	0.44		18
Carina	87	−9.2	−1.5	9	0.23		13
Ursa Minor	69	−8.9	−2.5	1	0.58		13
Draco	76	−8.6	−2.3	3	0.71		13
NGC 55	2150	−20.0	−0.6	14	3.2	7 (5+)	19
NGC 300	2150	−18.7	−0.4	16	3.2	23 (5+)	20
NGC 2403	3180	−20.3	0.0	4	1.6	7 (?)	21

References: 1=Richer & Crabtree (1985), 2=Cook et al. (1986), 3=Richer et al. (1990), 4=Brewer et al. (1995), 5=Groenewegen et al. (1992), 6=Blanco & McCarthy (1983), 7=Rebeirot et al. (1993), 8=Richer et al. (1984), 9=Ibata et al. (1995),10=Ng & Schultheis (1997), 11=Whitelock (1998), 12=Azzopardi et al. 1998, 13=Azzopardi et al. 1986, 14=Azzopardi et al. 1985, 15=Aaronson et al. (1985), 16=Cook, Olszewski & Suntzeff (unpublished, quoted in Armandroff 1994), 17=Irwin et al. (1990), 18=Da Costa (1994), 19=Pritchet et al. (1987), 20=Richer et al. (1985), 21=Hudon et al. (1989)

in e.g. the Carina dwarf (Cannon et al. 1980), or Fornax (Aaronson & Mould 1980).

Both the GRISM and the narrow-band filter technique allow for a more systematic search for carbon stars over a large area. The former method aims at identifying the prominent C_2 band heads at 4737 and 5165 Å. An illustration of this can be found in Westerlund (1979). The latter method uses standard V and I filters in combination with two narrow-band filters near 7800 and 8100 Å, which are centered on a CN-band in carbon stars, and on a TiO band in oxygen-rich stars, respectively. In a [78-81] versus $[V - I]$ colour-colour plot, carbon stars and late-type oxygen-rich stars clearly separate. For an illustration see Cook & Aaronson (1989), or Kerschbaum et al. (1998).

3. Surveys

In this section the surveys for carbon stars in external galaxies are described. References to the carbon star surveys are listed in Table 1. Labels in the text ("[1]") refer to these references.

The Magellanic Clouds: I reviewed the subject of AGB stars in the Magellanic Clouds at the "ISO's view on stellar evolution" meeting (Groenewegen 1998) and refer to that publication for more details. In brief, based on [6] and [7] the total estimated number of carbon stars in the LMC is about 11 000 and about 3 100 in the SMC.

Sagittarius dwarf galaxy: [9] identified 4 carbon stars which are members of the Sagittarius dwarf. Infrared photometry for them are listed in Whitelock et al. (1996). This data was transformed from the SAAO to the CTIO system, and bolometric corrections from Frogel et al. (1980) were used to derive

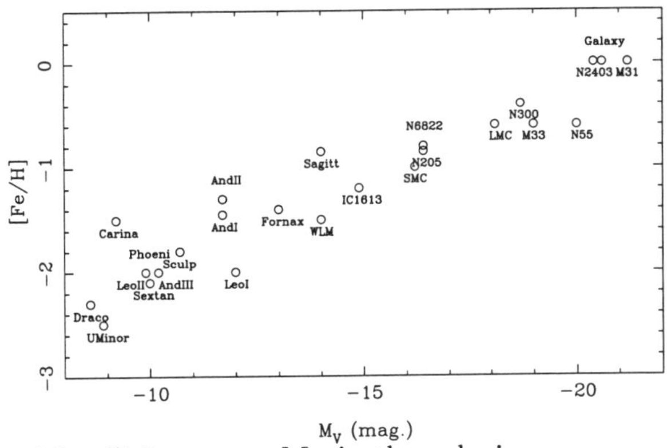

Figure 1. Metallicity versus M_V in the galaxies.

m_{bol}. [10] suggested another carbon star. Since then, [11] reported on recent work confirming membership for a total of 26 carbon stars, including 8 large- and 7 small-amplitude variables. This marks the discovery of the first carbon-Mira in any of the galactic dwarf spheroidals.

Fornax dwarf galaxy: Fornax has the largest number of C-stars of the galactic dwarf spheroidals and the study of its red population dates back to Demers & Kunkel (1979), and Aaronson & Mould (1980). The latest survey ([12]) should be area complete. Published accurate apparent bolometric magnitudes are taken from Frogel et al. (1982) and Aaronson & Mould (1980) for 15 different stars, based on IR photometry. For three stars in common the average value was taken. [12] mention that they have obtained *JHK* photometry for 94% of these stars and find a minimum, mean and maximum absolute bolometric magnitude of −1.2, −3.7 and −5.6, respectively. Frogel et al. (1982), in a smaller field, find 25 C, 6 M2+ and 0 M5+ stars. Scaling this to the total number of carbon stars I estimate that there should be in total about 25 M2+ and possibly 4 M5+ stars in Fornax.

Leo I dwarf galaxy: The survey by [13] is area complete. It should be noted that Regulus is very close to this galaxy and therefore some carbon stars may have escaped detection. Lee (1993) lists the apparent bolometric magnitudes for 15 objects.

Sculptor dwarf galaxy: The survey by [13] is area complete. Apparent bolometric magnitudes for 2 stars are listed in Frogel et al. (1982). In a smaller area Frogel et al. (1982) found 2 C-, 10 M2+ and 5 M3+ stars. Scaling this to the total number of carbon stars results in the number for the M-stars in Table 1.

Leo II dwarf galaxy: The survey by [14] is area complete. Apparent bolometric magnitudes are listed in Aaronson & Mould (1985).

Sextans dwarf galaxy: In the discovery paper of this galaxy by [17] it is mentioned that there is no indication for an AGB in the colour-magnitude diagram. On the other hand, none of the methods mentioned in Sect. 2 was used to actively look for carbon stars.

Phoenix dwarf galaxy: Van de Rydt et al. (1991) provide a detailed colour-magnitude diagram that shows a few candidate AGB stars and covers

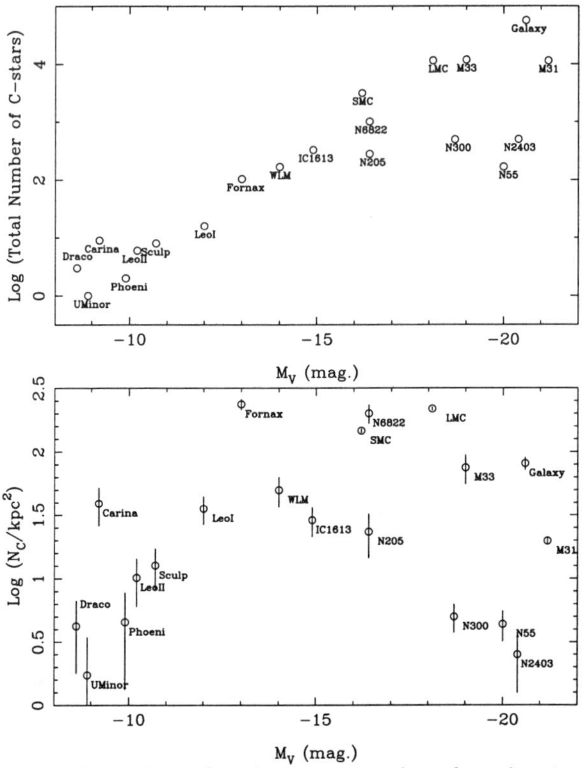

Figure 2. Total number of carbon stars and surface density of carbon stars versus M_V. In the latter diagram no correction for projection effects has been made.

the whole of the galaxy. [18] took spectra of 5 of these candidates and confirmed two as being carbon stars. I have calculated $m_{\rm bol}$ using the photometry in van de Rydt et al., their adopted reddening and bolometric corrections (BCs) following Bessell & Wood (1984, hereafter BW84) .

Carina dwarf galaxy: The survey by [13] is area complete. Apparent bolometric magnitudes are available for 6 stars from Mould et al. (1982).

Ursa Minor dwarf galaxy: The survey by [13] is area complete. The apparent bolometric magnitude is listed in Aaronson & Mould (1985).

Draco dwarf galaxy: The survey by [13] is area complete. Apparent bolometric magnitudes for the 3 stars are listed in Aaronson et al. (1982).

M31: This galaxy was surveyed for carbon stars by [1, 2, 3] and most extensively by [4]. The latter find a total of 243 C- and 5254 late-M stars (using their colour and magnitude criterion). The V, I data on the individual stars is available on CD-ROM, and was used together with their adopted reddening (depending on the field) and BW84 to obtain $m_{\rm bol}$. The Brewer et al. survey covers about 2.1% of the galaxy (this, and similar numbers below is derived from the ratio between survey area and area of the galaxy, as given in the NED database). In a follow-up paper, Brewer et al. (1996) took spectra of some of

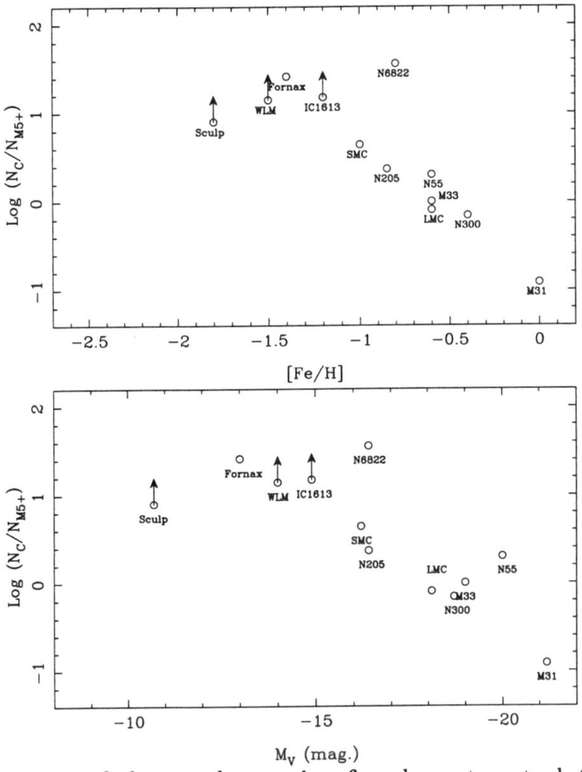

Figure 3. Log of the number ratio of carbon stars to late M-stars versus metallicity and M_V.

the stars previously identified and discovered an S-star, and C-stars enhanced in ^{13}C (J-type) and in Lithium.

M33: The area surveyed by [2] covered about 0.13% of the galaxy.

NGC 205: This galaxy is a companion of M31. The area surveyed by [8] covered about 2.5% of the galaxy. Apparent bolometric magnitudes were derived by me from the listed *V*, *I* photometry, their adopted reddening, and BCs from BW84.

And I, II, III: And I, II, III are dwarf spheroidal companions to M31. Searches for carbon stars are reported by [15] for And II, and by [16] for all three systems. Aaronson et al. (1985b) found 1 certain, 1 possible C-star and a M-giant from spectroscopy of red stars. [16] used the narrow-band filter technique to report on no C-stars in And I and III, and 8 candidate C-stars and 3 candidate S-stars in And II, including the one by [15]. The area that was covered in the observations is not quoted by Armandroff.

NGC 6822: The survey by [2] covered about 3.5% of the galaxy. This galaxy also contains an S-star (Aaronson et al. 1985a)

IC 1613: The survey by [2] covered about 4.5% of the galaxy.

WLM: The survey by [2] covered about 8.3% of the galaxy.

NGC 55: This galaxy is in the Sculptor group. [19] surveyed a field covering about 8.3% of the galaxy. The bolometric magnitude was derived by me from

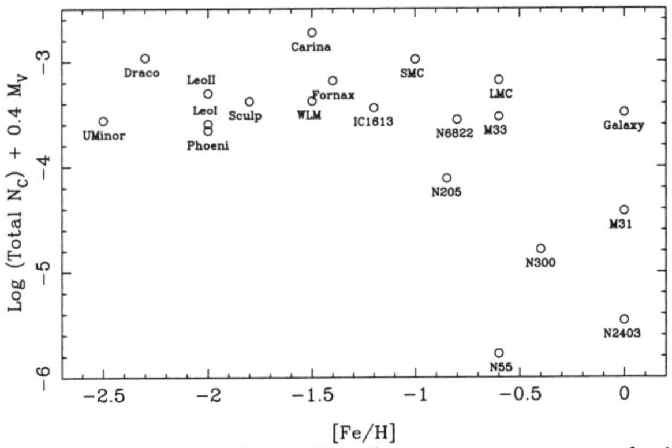

Figure 4. Log of the number of carbon stars over total visual luminosity versus metallicity

the listed V, I photometry, their adopted reddening and the BCs from BW84, except for five stars with a lower limit on $V - I$.

NGC 300: This galaxy is in the Sculptor group. [20] surveyed a field covering about 4% of the galaxy. The bolometric magnitude was derived by me from the listed V, I photometry, their adopted reddening and the BCs from BW84, except for three stars with a lower limit on $V - I$.

NGC 2403: This galaxy is in the M81 group, and the most distant galaxy where carbon stars have been identified so far. [21] tentatively identified 4 C- and 7 M-stars, and give the apparent bolometric magnitude based on R, I photometry. Their survey covered about 0.8% of the galaxy.

Table 1 summarizes the number of known carbon stars in external galaxies. The last three entries are galaxies outside the Local Group. Listed are the adopted distance, absolute visual magnitude, metallicity, number of known carbon stars, the area over which they were found, and the number of late-type M-stars, when known (the symbol '3+' meaning stars of spectral type M3 and later, etc.). The entry for the number of carbon stars in our galaxy is the local surface density of TP-AGB stars by Groenewegen et al. (1992).

4. Discussion

Fig. 1 shows the well-known correlation between metallicity and absolute visual magnitude. The interpretation being that the more luminous galaxies are also the more massive ones that have had more gas mass available to transform into stars and enrich the interstellar medium.

Fig. 2 shows the number of carbon stars in a galaxy versus M_V represented in two ways. First, the total number of C-stars in a galaxy was estimated by multiplying the known number by the ratio of total surface area of a galaxy to the survey area. As for some galaxies the survey area is less than a few percent, this correction factor can be quite large (and uncertain). To circumvent this, the bottom panel shows the surface density of carbon stars in the particular survey. The drawback of this approach is that it does not take into account the spatial variation of the density of carbon stars within a galaxy. In neither

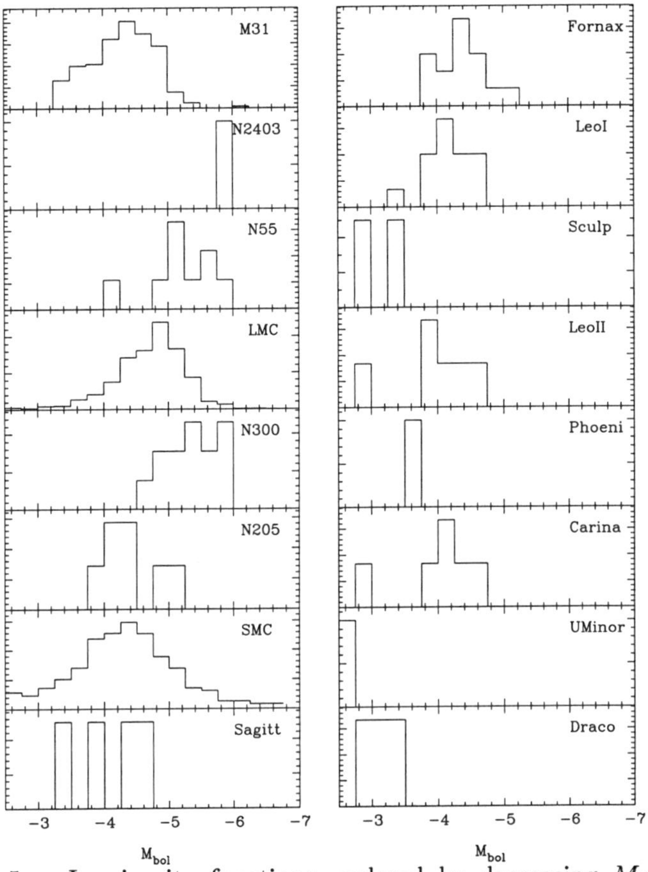

Figure 5. Luminosity functions, ordered by decreasing M_V. The total number of stars plotted per galaxy is listed in column 6 of Table 2. The lowest luminosity bin in the SMC and LMC are cumulative.

approach did I correct for the fact that we do not see these galaxies face-on. Some interesting things can be noticed. There is a clear relation between the (estimated) total number of C-stars and M_V, and there seems to be a maximum surface density of about 200 kpc^{-2} averaged over a galaxy. In both plots NGC 55, 300 and 2403 are clear outliers. These are the most distant galaxies surveyed, and one might suppose that the surveys have been incomplete. For NGC 55 the explanation probably lies as well in the fact that we see this galaxy almost disk-on, and so both the total number as well as the surface density have been underestimated. Reddening within the galactic disk of the galaxy can also play a role. For NGC 2403 the small number of carbon stars lies in the fact that the survey has been incomplete. All 4 C stars have luminosities that are much higher than the average in galaxies for which we know the luminosity function (LF) in more detail. The same is true for NGC 300. A last note of caution is that I did not try to make a distinction between carbon stars on the TP-AGB and lower-luminosity carbon stars. For example, the data for our Galaxy represents

Table 2. Carbon stars: the luminosity function

Name	$M_{\rm bol}^{\rm max}$ (mag.)	$M_{\rm bol}^{\rm min}$ (mag.)	$M_{\rm bol}^{\rm mean}$ (mag.)	spread (mag.)	$N_{\rm all}$	$N_{{\rm M_{bol}}>-3.5}$
M31	-6.14	-3.34	-4.31	0.50	243	16
NGC 2403	-5.91	-5.75	-5.83	0.08	4	0
NGC 55	-5.79	-4.18	-5.21	0.52	9	0
LMC	-6.30	-1.25	-4.70	0.55	923	25
NGC 300	-5.91	-4.66	-5.37	0.41	13	0
NGC 205	-5.01	-3.91	-4.38	0.44	7	0
SMC	-8.17	-1.62	-4.32	0.81	1626	227
Sagittarius	-4.61	-3.29	-4.08	0.60	4	1
Fornax	-5.22	-3.81	-4.35	0.37	15	0
Leo I	-4.68	-3.38	-4.17	0.32	15	1
Sculptor	-3.33	-2.95	-3.14	0.27	2	2
Leo II	-4.51	-2.91	-3.91	0.55	6	1
Phoenix	-3.71	-3.55	-3.63	0.11	2	0
Carina	-4.55	-2.84	-4.00	0.61	6	1
Ursa Minor	-2.74	-2.74	-2.74	0.00	1	1
Draco	-3.45	-2.75	-3.10	0.35	3	3

TP-AGB stars only, while on the other hand the SMC is known to contain a large fraction of low-luminosity carbon stars (see later).

Fig. 3 shows the ratio of C- to late M-stars, and it confirms the well known trend with metallicity (Cook et al. 1986; Pritchet et al. 1987). The interpretation is that a star with a lower metallicity needs fewer thermal pulses to turn from an oxygen-rich star into a carbon star.

Fig. 4 shows the ratio of the total estimated number of carbon stars over the visual luminosity of the galaxy (e.g. Aaronson et al. 1983). Most of the galaxies scatter between a value of –3 and –4, with a few outliers which are the same as noticed in Fig. 2.

Fig. 5 shows the C-stars bolometric LF for the galaxies for which it could be constructed (Sect. 2); for the Magellanic Clouds see Groenewegen (1998) for details. Table 2 lists the maximum, minimum and mean magnitude, as well as the spread, calculated from the rms deviation from the mean. Also listed are the number of stars that went into the calculation and are plotted in Fig. 5, and the number of C-stars that are fainter than $M_{\rm bol} = -3.5$. The latter is an approximate borderline (which is in fact dependent on metallicity) between TP-AGB and low-luminosity C-stars. The data show that in well populated LF, the mean $M_{\rm bol}$ is between –4 and –5. It also shows that the mean in NGC 300 and NGC 2403 is much higher. Unless one would invoke a large uncertainty in the distance or a burst of recent star formation, the most natural explanation lies in the incompleteness of the surveys in these distant galaxies. Finally, the data shows that in the fainter galaxies the mean magnitude increases and that a fair number of C-stars are of the low-luminosity type. This is more clearly seen in Fig. 6 where the mean magnitude is plotted. This indicates a low star formation over the last few Gyrs.

In principle, the overall carbon star LF contains information about the star-formation-rate history from, say, 10 Gyr ago (the low-luminosity C-stars in

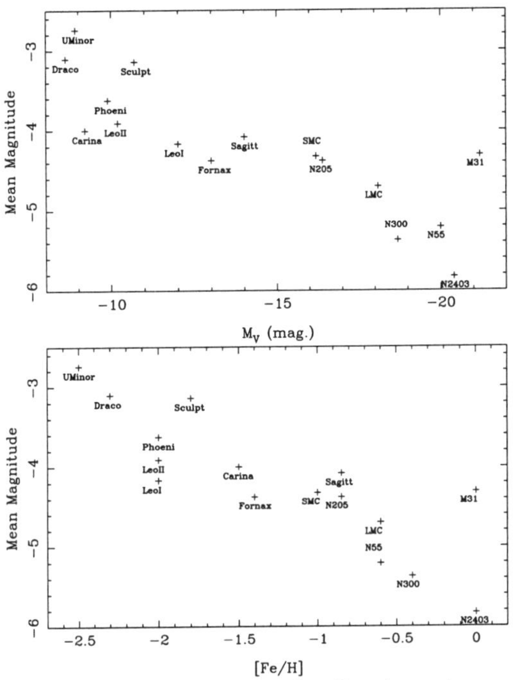

Figure 6. Mean bolometric magnitude of carbon stars versus M_V and metallicity.

binaries) to a few-hundred Myr ago (the high luminosity tail of the LF). It is a challenge to theoretical models to use these constraints together with other data to derive the chemical evolution- and star-formation rate history of these galaxies.

References

Aaronson M., Mould J., 1980, ApJ 240, 804

Aaronson M., Mould J., 1985, ApJ 290, 191

Aaronson M., Liebert J., Stocke J., 1982, ApJ 254, 507

Aaronson M., Mould J., Cook K.H., 1985a, ApJ 291, L41

Aaronson M., Olszewski E.W., Hodge P.W., 1983, ApJ 267, 271

Aaronson M., Gordon G., Mould J., Olszewski E.W., Suntzeff N., 1985b, ApJ 296, L7

Aaronson M., Da Costa G.A, Hartigan P., Mould J.R., Norris J., Stockman H.S., 1984, ApJ 277, L9

Armandroff T.E., 1994, in "Dwarf galaxies", eds. G. Meylan and P. Prugniel, ESO, Garching, p. 211

Azzopardi M., Lequeux J., Westerlund B.E., 1985, A&A 144, 388

Azzopardi M., Lequeux J., Westerlund B.E., 1986, A&A 161, 232

Azzopardi M., Breysacher J., Muratorio G., Westerlund B.E., 1998, in "Stellar populations in the local group", IAU Symposium 192, in press

Bessell M.S., Wood P.R., 1984, PASP 96, 247 (BW84)

Blanco V.M., McCarthy M.F., 1983, AJ 88, 1442

Brewer J., Richer H.B., Crabtree D.R., 1995, AJ 109, 2480

Brewer J., Richer H.B., Crabtree D.R., 1996, AJ 112, 491

Cannon R.D., Niss B., Nørgaard-Nielsen H.U., 1980, MNRAS 196, 1P

Cook K.H., Aaronson M., 1989, AJ 97, 923

Cook K.H., Aaronson M., Norris J., 1986, ApJ 305, 634

Da Costa G.S., 1994, in "Dwarf galaxies", eds. G. Meylan and P. Prugniel, ESO, Garching, p. 221

Demers S., Kunkel W.E., 1979, PASP 91, 761

Frogel J.A., Persson S.E. Cohen J.G., 1980, ApJ 239, 495

Frogel J.A., Blanco V.M., McCarthy M.F., Cohen J.G., 1982, ApJ 252, 133

Groenewegen M.A.T., 1998, in "ISO's view on stellar evolution", eds. L. Waters, C. Waelkens, K. A. van der Hucht and P.A. Zaal, Kluwer, Ap&SS 255, p. 379

Groenewegen M.A.T., de Jong T., van der Bliek N.S., Slijkhuis S., Willems F.J., 1992, A&A 253, 150

Hudon J.D., Richer H.B., Pritchet C.J., Crabtree D.R., Christian C.A., Jones J., 1989, AJ 98, 1265

Ibata R.A., Gilmore G., Irwin M.J., 1995, MNRAS 277, 1354

Irwin M.J., Bunclark P.S., Bridgeland M.T., McMahon R.G., 1990, MNRAS 244, 16P

Kerschbaum F., Nowotny W., Schultheis M., Hron J., 1998, in "Stellar populations in the local group", IAU Symposium 192, in press

Lee M.G., Freedman W., Mateo M., Thompson I., Roth M., Ruiz M.-T., 1993, AJ 106, 1420

McClure R.D., 1997, PASP 109, 256

Mould J., Cannon R.D., Aaronson M., Frogel J., 1982, ApJ 254, 500

Ng Y.K., Schultheis M., 1997, A&AS 123, 115

Pritchet C.J., Richer H.B., Schade D., Crabtree D.R., Yee H.K.C., 1987, ApJ 323, 79

Rebeirot E., Azzopardi M., Westerlund B.E., 1993, A&AS 97, 603

Richer H.B., Crabtree D.R., 1985, ApJ 298, L13

Richer H.B., Crabtree D.R., Pritchet C.J., 1984, ApJ 287, 138

Richer H.B., Crabtree D.R., Pritchet C.J., 1990, ApJ 355, 448

Richer H.B., Pritchet C.J., Crabtree D.R., 1985, ApJ 298, 240

van de Rydt F., Demers S., Kunkel W.E., 1991, AJ 102, 130

Westerlund B.E., 1979, ESO Messenger 19, 7

Whitelock P.A., 1998, in "Stellar populations in the local group", IAU Symposium 192, in press

Whitelock P.A., Irwin M., Catchpole R.M., 1996, New Astr. 1, 57

Asymptotic Giant Branch Stars
T. Le Bertre, A. Lèbre and C. Waelkens, eds.
I.A.U. Symposium 191, 1999

Are there carbon stars in the Bulge ?

Yuen Keong Ng

Padova Observatory, Vicolo dell'Osservatorio 5, I-35122 Padua, Italy

Abstract. The Bulge carbon stars have been a mystery since their discovery by Azzopardi et al. (1991), because they are about $2^m_\cdot5$ too faint to be regarded as genuine AGB stars, if located inside the metal-rich Bulge ($m - M = 14^m_\cdot5$). Part of the mystery can be solved if these carbon stars are related to the Sagittarius dwarf galaxy (SDG; $m - M \simeq 17^m_\cdot0$). They are in that case not old and metal-rich, but young, ~0.1 Gyr, with SMC-like metallicity (Ng 1998).

The $\sigma_{RV} = 113 \pm 14$ km s^{-1} (Tyson & Rich 1991) radial velocity dispersion of the stars appears to be consistent with Bulge membership. On the other hand, a similar velocity dispersion could be the result from an induced star formation event when the SDG crosses the galactic midplane. It is suggested that the carbon stars are tracers of such an event and that they therefore are located at distances related to the SDG. However, the majority of the carbon stars are not member of the SDG, nor are they similar to the C-stars which are member of the SDG.

The radial velocities can be used to determine a possible membership to the SDG. However, they do not give information about the distance of the stars. In particular, if the stars are located at a distance comparable to the SDG. This implies that only the period-luminosity relation (Groenewegen & Whitelock 1996) can be used to distinguish unambiguously if the carbon stars are located at Bulge-like or SDG-like distances. Thus far only carbon stars with reliable periods have been identified at a SDG related distance (Ng & Schultheis 1997; Whitelock 1998).

1. Introducing the mystery

After a long search 34 carbon stars were found by Azzopardi et al. (1991 and references cited therein – ALRW91) in the direction of the Galactic Centre. If located in the metal-rich Bulge they are about $2^m_\cdot5$ fainter than the carbon stars from the Magellanic Clouds or the dwarf spheroidal galaxies (Lequeux 1990; Tyson & Rich 1991 – TR91; Westerlund et al. 1991).

The Bulge is older than the maximum age of $\sim$4 Gyr for a carbon star to be regarded as a genuine TP-AGB star (see Ng 1997 & 1998 and references cited therein). The ALRW91 C-stars are therefore thought to be located along the RGB as a result of binary evolution. Fig. 1 indicates that for the latter case the stars do not reach the RGB tip. Moreover, the origin of this truncation has not been explained.

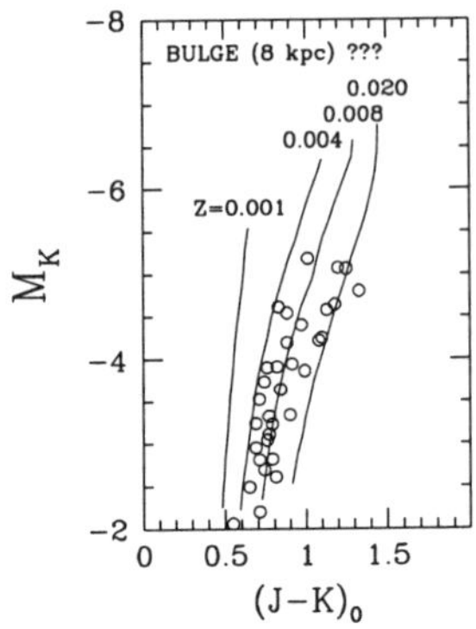

Figure 1. The $(M_K, J\text{–}K)_0$ CMD of the RGB for 10 Gyr isochrones for the metallicities Z = 0.001, 0.004, 0.008, and 0.020 (Bertelli et al. 1994). The absolute magnitudes of the ALRW91 C-stars are determined under the assumption that they are located in the Bulge at a distance of 8 kpc. Note that the ALRW91 C-stars do not reach the RGB tip and no explanation has been provided thus far for this behaviour.

The questions raised and answered partly to provide a solution to this standing mystery are: are these stars really located in the Bulge ? Is there an alternative ? If so: where are they ? and what are the implications ?

2. Relation to the Sagittarius dwarf galaxy ?

The serendipitous identification of the Sagittarius dwarf galaxy (SDG) was made by Ibata et al. (1994, 1995). Accurate distance determinations from RR Lyrae stars belonging to this galaxy range from 22.0 – 27.3 kpc (Ng 1997 and references cited therein). The $\sim 2\overset{m}{.}5$ difference of the distance modulus between the dwarf galaxy $(m-M = 17\overset{m}{.}0)$ and the Galactic Centre at 8 kpc $(m-M = 14\overset{m}{.}5)$ lead Ng & Schultheis (1997) to suggest that the ALRW91 C-stars could actually be located at the distance of the dwarf galaxy. Note that this does not necessarily imply a SDG membership. Its presence was unknown at the time when the C-stars were identified and a different location could solve the standing question about the origin of the 'Bulge' carbon stars.

3. Not so metal-rich after all !

Ng (1997 & 1998) analysed this possibility by assuming a relation of the ALRW91 C-stars with the SDG. Once more it is stressed that *related* $\neq$ *membership*, except for a small fraction of the C-stars. Ng demonstrated that the photometric sequence of the ALRW91 C-stars is in that case not exceptional, but comparable with the sequence found for the SMC. In addition, the ALRW91 carbon stars are not metal-rich as previously thought. The metallicity was initially estimated to $Z \simeq 0.008$ (Ng 1997), but a revision has been made to $Z \simeq 0.004$ (Ng 1998). Ng further argued that the stars have an age of ~ 0.1 Gyr.

4. Elusive radial velocities

The radial velocities from TR91 for the carbon stars appear to be consistent with a Bulge membership. However, an alternative explanation for the large velocity dispersion is available: induced star formation from the material residing in either the SDG and/or the impact spot in the Galactic disc when the SDG

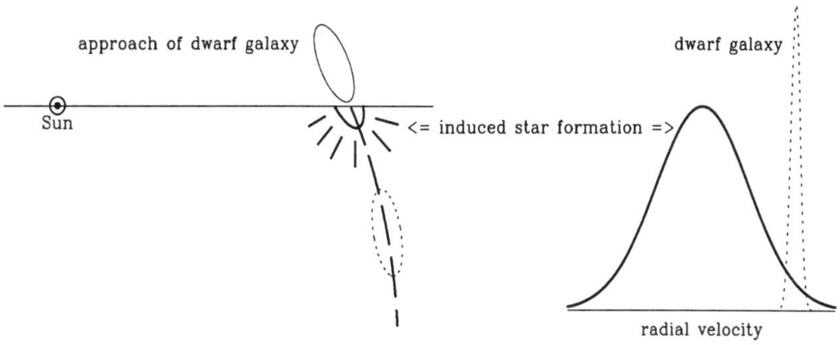

Figure 2. Schematic view of the broad, Bulge like, radial velocity distribution for the material from an induced star formation event, caused by the crossing from a dwarf galaxy through the Galactic mid-plane. Note that the majority of the new material will not be moving along the dwarf galaxy orbit. The dotted ellipse and the dotted line indicate respectively the current position and radial velocity (not on scale) for the dwarf galaxy

respectively approaches and crosses the galactic plane (Ng 1997, 1998). Fig. 2 displays a schematic view of the process, where the SDG pushes material out of the Galactic disc[1]. Only a small fraction of the newly formed stars is dragged along the orbit, while another fraction is moving away from the SDG. Note that another fraction of these young stars is moving towards us. This implies that the distribution of radial velocities is considerably broader than expected from a SDG membership alone, i.e. $\sigma_{RV} >> 11.4 \pm 0.7$ km s^{-1}. The TR91 velocity dispersion of $\sigma_{RV} = 113 \pm 14$ km s^{-1} for the ALRW91 C-stars may have been interpreted mistakingly as an indication for a Galactic Bulge membership[2]. Alternatively the large velocity dispersion can also be a signature from an induced star formation event.

5. Gateway: the galactic disc and the Sagittarius dwarf galaxy

The impact position at the galactic mid-plane is obtained from the unweighted least-squares fit through the distances determined from the SDG RR Lyrae stars as a function of the galactic latitude (Ng 1998) and yields 22.8 kpc. Adopting

[1]Numerical calculations of the structure of the H I disc by Ibata & Razoumov (1998) indicate that just after impact (their Fig. 2b) material is scattered in all directions above and below the Galactic plane. This should result in a broad radial velocity of the material scattered after the crossing of the Galactic plane. Their Fig. 4 further indicates that part of the gaseous material with log T in the range 2 – 4 should have been converted into stars; some of them possibly comparable to the ALRW91 C-stars.

[2]The uncertainties in the individual velocity values may be as large as 90 km s^{-1}. However, if the uncertainties follow a random pattern the mean velocity and its dispersion will not change drastically if the radial velocity values are improved.

a Solar galacto-centric distance of $R_0 = 8.0$ kpc implies that the crossing of the SDG occurred at 14.8 kpc distance from the Galactic centre.

To determine the disc metallicity at impact position one has to take the radial dependence of [Fe/H] into account. Carraro et al. (1998) determined from open clusters the radial metallicity gradient in different age ranges. The disc metallicity at the crossing position is [Fe/H] $= -0.61 \pm 0.20$ ($Z \simeq 0.0050$) or [Fe/H] $= -0.73 \pm 0.10$ ($Z \simeq 0.0035$) with respectively the present day and the average radial metallicity gradient.

Irrespective of the present day or average metallicity gradient it is estimated that the disc metallicity at the crossing position is $Z \simeq 0.0045 \pm 0.0010$. Within the uncertainties this is in agreement with $Z \simeq 0.004$.

Considering that an induced star formation event can explain the large velocity dispersion for the ALRW C-stars and taking into account the similarity between the metallicity from a photometric estimate and the disc impact position, it is suggested that the C-stars do not originate from the SDG itself, but are likely formed from material originating from our own Galactic disc.

The average $(J-K)_{0,SAAO}$ colour of the carbon stars which are members of the dwarf galaxy is $1^m\!.40$ (Whitelock et al. 1996), while the average $(J-K)_{0,ESO}$ colour of the ALRW91 C-stars is $0^m\!.85$. This immediately shows the marked difference between two groups of carbon stars found in the same direction at a comparable distance. The redder SDG C-stars are likely older and metal-richer than the ALRW91 C-stars, because the redder SDG C-stars should be associated with a SDG population comparable or younger than ~4 Gyr. The youngest population identified for the SDG has an age of about 4 Gyr and a metallicity $Z \simeq 0.008$ (Mighell et al. 1998).

6. Why are they still around Sagittarius dwarf galaxy?

The average radial velocity of -44 km s^{-1} (TR91) provides a first indication of the distance traveled by the ALRW91 C-stars. Considering a smaller velocity in the past the distance traveled in 0.1 Gyr is ~2.2–3.1 kpc. The traveled distance is within the 10%–15% uncertainty of the 22.8 kpc distance to the impact position.

The velocity dispersion of the C-stars provides further an indication about the average separation. Taking into account that the dispersion is smaller, due to a combination of turbulence and a lower velocity in the past, the separation is estimated to ~5.6–8.0 kpc. For comparison: the distance between 2 of the 4 globular clusters associated with SDG is about 10 kpc, i.e. Terzan 8 at 21.1 kpc and Arp 2 at 31.0 kpc (Da Costa & Armandroff 1995), while the distance between the two carbon stars with well determined periods is about 5 kpc, i.e. 21.9 kpc for a carbon semiregular variable (Ng & Schultheis 1997), and 26.7 kpc for a carbon Mira (Whitelock 1998).

Both the average distance traveled by the ALRW91 C-stars and the separation between them are such that most of the stars can still be associated with the Sagittarius dwarf galaxy.

7. Still mysterious ???

The general accepted models of the SDG orbit indicate (see Figure 11 in Ibata et al. 1997 and references cited therein), together with the proper motion reported by Irwin et al. (1996), that the SDG is still moving towards the galactic mid-plane. On the other hand, the ALRW91 C-stars appear to form the factual evidence that the SDG has already crossed the galactic mid-plane. In addition, the orbit at low galactic latitudes (Figure 4 in Ng 1997) is not consistent with the position obtained from RR Lyrae stars, which might partly be due to uncertainties in the extinction. A study of the Galactic globular cluster Palomar 5 (Scholz et al. 1998) further indicates that the cluster is moving in the opposite direction with respect to the Ibata et al. orbital motion of the SDG.

Instead of looking for alternative explanations for the apparent contradiction, one might consider a re-analysis of the orbit combined with an independent verification of the proper motion. A revised orbit and proper motion might show that the globular cluster Palomar 5 is tidally stripped from the SDG (Lin 1996) and also associate the Smith high velocity Cloud with the SDG (Bland-Hawthorn et al. 1998).

The models thus far consider only the gravitational interaction between the Galaxy and the SDG (Ibata & Lewis 1998; Nair & Miralda-Escudé 1998), but do not take into account an encounter, induced, star formation event. Improved models should therefore be employed to trace the stars from such an event. In addition, a detailed proper motion study and a renewed determination of the radial velocities of the ALRW91 C-stars are still required to explain and to understand better the presence of the carbon stars.

8. Conclusion

The 'Bulge' (ALRW91) carbon stars are located at a distance comparable to the Sagittarius dwarf galaxy. However, the majority of these stars ($Z \simeq 0.004$, $t \simeq 0.1$ Gyr) are not member of the Sagittarius dwarf galaxy nor are they comparable with the carbon stars found in the Sagittarius dwarf galaxy ($Z \simeq 0.008$, $t \simeq 4$ Gyr; see for details Ng 1998).

The radial velocities cannot be used reliably as an argument in favour of Bulge membership, due to a possible confusion with the velocity dispersion of the stars resulting from a crossing induced star formation event which scatters new stars in all directions[3].

If there are carbon stars in the Bulge then the carbon star period-luminosity relation (Groenewegen & Whitelock 1996) should be employed to determine their distances. Up to date no carbon semiregular or Mira variable has been reported with a period-luminosity which unambiguously associates it with the Galactic Bulge.

---

[3]Bonus: the microlensing events between the stars from the Bulge/bar and the induced semi-sphere of young stars, due to the crossing of the SDG through the galactic midplane, could well prove to contribute significantly to the surprisingly high event rate found towards the Galactic Centre (Paczyński 1996 and references cited therein).

Acknowledgments. Ng acknowledges a travel grant received from the IAU. The research is supported by grant ERBFMRX-CT96-0086 from the TMR network *'Formation and evolution of galaxies'* of the European Community.

References

Azzopardi M., Lequeux J., Rebeirot E., Westerlund B.E., 1991, A&AS 88, 265 (ALRW91)

Bertelli G., Bressan A., Chiosi C., Fagotto F., Nasi E., 1994, A&AS 106, 275

Bland-Hawthorn J., Veilleux S., Cecil G.N., et al., 1998, MNRAS 299, 611

Carraro G., Ng Y.K., Portinari L., 1998, MNRAS 296, 1045

Da Costa G.S., Armandroff T.E., 1995, AJ 109, 2533

Groenewegen M.A.T., Whitelock P.A., 1996, MNRAS 281, 1347

Ibata R.A., Lewis G.F., 1998, ApJ 500, 575

Ibata R.A., Razoumov A.O., 1998, A&A 336, 130

Ibata R.A., Gilmore G., Irwin M.J., 1994, Nature 370, 194

Ibata R.A., Gilmore G., Irwin M.J., 1995, MNRAS 277, 781

Ibata R.A., Wyse R.F.G., Gilmore G., et al., 1997, AJ 113, 635

Irwin M., Ibata R., Gilmore G., et al., 1996, in *'Formation of the galactic halo ... inside and out'*, ASP Conf. Series 92, H. Morrison and A. Sarajedini (eds.), p. 84

Lequeux J., 1990, in *'From Miras to planetary nebulae: which path for stellar evolution?'*, M.-O. Mennessier and A. Omont (eds.), Editions Frontières, p. 273

Lin D.N.C., 1996, in *'Gravitational dynamics'*, O. Lahav, E. Terlevich and R.J. Terlevich (eds.), Cambridge University Press, p. 15

Mighell K., Armandroff T., Sarajedini A., et al., 1998, in *'Galaxy interaction at low and high redshift'*, IAU Symposium 186, ed. D.B. Sanders, Kluwer, in press

Nair V., Miralda-Escudé J., 1998, ApJ *submitted* (`astro-ph/9805278`)

Ng Y.K., 1997, A&A 328, 211

Ng Y.K., 1998, A&A 338, 435

Ng Y.K., Schultheis M., 1997, A&AS 123, 115

Paczyński B., 1996, ARA&A 34, 419

Scholz R.-D., Irwin M., Odenkirchen M., Meusinger H., 1998, A&A 333, 531

Tyson N., Rich R.M., 1991, ApJ 367, 547 (TR91)

Westerlund B.E., Lequeux J., Azzopardi M., Rebeirot E., 1991, A&A 244, 367

Whitelock P.A., 1998, in *'Pulsating Stars - Recent Developments in Theory and Observation'*, M. Takeuti and D. Sasselov (eds.), Universal Academic Press, Tokyo, in press

Whitelock P.A., Irwin M., Catchpole R.M., 1996, New Astronomy 1, 57

Asymptotic Giant Branch Stars
T. Le Bertre, A. Lèbre and C. Waelkens, eds.
I.A.U. Symposium 191, 1999

Mass loss and AGB evolution in extra-galactic stellar populations

Albert A. Zijlstra

Department of Physics, UMIST, P.O. Box 88, Manchester M60 1QD, UK

Abstract. The effect of metallicity on the AGB mass loss is reviewed. Observations have mainly been limited to the Magellanic Clouds but are observationally feasible throughout the Local Group. Expansion velocities are predicted to depend strongly on Z but the peak mass-loss rates appear not to. The Mira PL relation shows no evidence for a Z dependence, giving a powerful potential distance indicator. I derive a distance modulus to the LMC of $m - M = 18.63 \pm 0.09$ as well as a bright calibration for the Horizontal Branch in globular clusters. Finally, the predicted initial-final mass relation at low Z is shown to give rise to higher mass remnants. This would result in an increased supernova rate in young, low-Z populations, such as found in the early Universe.

1. Introduction

AGB mass loss is generally believed to be a two-step process, involving pulsation-driven atmospheric shock waves, followed by radiation pressure on the dust driving the outflow. The dust production rate, and the eventual dust-to-gas ratio of the ejected material, will depend on the metallicity of the star. The details of the mass loss may therefore be different for different stellar populations.

Existing theoretical formalisms for the mass loss are generally derived from the Reimers' relation: $\dot{M}_R = -4\,10^{-13} LR/M$. This equation is adjusted to fit existing observations as well as constraints such as the initial–final mass relations. Examples are Blöcker's law ($\dot{M} = -4.8\,10^{-9}\,(LR/M)\,(L^{2.7}/M)$), Vassiliadis & Wood's (1993) formalism and the relation from Judge & Stencel ($\dot{M} = (2.3+/-1.3)\,10^{-14}\,(R^2/M)^{1.43+/-.23}$). Because observations of AGB mass loss have been largely limited to galactic stars, these formalisms have limited predictive power for different stellar populations. So far, only Bowen & Willson (1992) have attempted to predict mass-loss rates at low Z from theoretical pulsation models.

The study of extra-galactic sources is thus of interest to extend the parameter space of mass loss. At the same time, since the majority of stars with life-times less than the Hubble time will have passed through the AGB, a better understanding of mass loss especially at low Z is also important for our knowledge on stellar remnants and on the recycling of stellar gas into the ISM. The lack of observational data is mainly due to the difficulty to detect dust at any large distance, and the paucity of 'dusty' AGB stars due to the short life-time of the superwind phase.

2. Extra-galactic observations

2.1. Capabilities

Existing and planned instrumentations have the capability to detect AGB stars to distances far outside the Local Group. Observations of mass loss of individual AGB stars is observationally feasible at least for the satellites of the Galaxy. Of these, studies of the Magellanic Clouds have already been initiated (Reid 1991; Wood et al. 1992; Zijlstra et al. 1996; van Loon et al. 1997,1998; Groenewegen & Blommaert 1998).

Table 1 lists the observational limits of some present and future instrumentation. The last column gives the maximum distance to which a luminous AGB star may be detected given these limits. The K-magnitudes refer to an unobscured, high-mass AGB star with K= −9. The N-band is scaled from an obscured luminous AGB star in the LMC. The NGST value refers to low-resolution spectroscopy with $R = 200$: the value is very preliminary and only based on possible instrumentation/telescope design.

Table 1. Obervational capabilities.

Band	Instrument/telescope	point-source sensitivity (5-σ in 1 hour)	distance limit (K= −9)
K	SOFI/NTT	20.9 mag	10 Mpc
K	ISAAC/VLT	22.0 mag	16 Mpc
K	NGST	26.0 mag	100 Mpc [a]
N	TIMMI2/3.6m	2.5 mJy	1 Mpc [b]
N	VISIR/VLT	0.25 mJy	3 Mpc
Q	VISIR/VLT	10: mJy	0.5 Mpc

[a] Low-resolution spectroscopy
[b] Scaled from F(12 μm)=1.0 Jy at the LMC

2.2. Individual stars

The AGB population in the Magellanic Clouds is the only extra-galactic population for which extensive observations are available. ISO observations give the first reliable information on AGB mass loss and are presented elswhere in this volume (van Loon; Blommaert et al.).

AGB stars in more distant galaxies can be detected using infrared arrays, however this is so far limited to non-mass-losing stars. Davidge et al. (1998) present JHK images of the Sculptor group galaxies, and detect indications for luminous AGB stars below K=16, near the detection limit. Alonso et al. (1998)

present VIJHK data for the dwarf galaxy NGC 3109. The AGB is clearly detected for ages down to 1Gyr, but the survey is not deep enough to detect older AGB stars in the infrared. The new generation of wide-field infrared arrays is expected to give rapid progresses in the next few years.

Evolved AGB stars can also be identified through their variability, using I-band observations. Tolstoy et al. (1996) observed suspected Cepheid variables in Gr-8 (d = 2.4Mpc). One star in their sample (V-1) both has a possible long period of 110 days and very red colours: it is presently the best candidate for a Mira-like variable outside the Local Group.

In the galactic dwarf spheroidals, the presence of carbon stars is indicative of intermediate-age populations (younger than globular clusters). Especially Fornax shows evidence for star formation less than 2 Gyr ago (Beauchamp et al. 1995) and may well contain mass-losing stars. Whitelock et al. (1996) study carbon stars in the nearby Sagittarius dwarf: one star of their sample has since been shown to be a carbon Mira. A review of extra-galactic carbon stars is given by Groenewegen (these proceedings).

2.3. Integrated properties

Although individual mass-losing stars can only be detected at limited distances, their integrated flux over a galaxy may be appreciable. Knapp et al. (1992) studied elliptical galaxies in Virgo: they showed that on average 40 per cent of their 12-μm emission may be due to circumstellar dust. The expected photospheric contribution is $S_{12} = 0.08\, S_{2.2}$ (with S in Jy), whereas the observations for Virgo ellipticals give $S_{12} = 0.138 +/- 0.014\, S_{12}$. This is in contrast to globular clusters where the 12-μm emission is consistent with purely photospheric emission (Knapp et al. 1995; Penny et al. 1997).

Recently, Bressan et al. (1998) have used the Padova isochrones to predict integrated IR colours of stellar populations for various mid-infrared bands. Their results rely heavily on assumed relations, e.g. between Z and $\dot{M}$, but can easily be updated if required.

3. Expansion velocities at low metallicity

The expansion velocity v_{exp} of the gas is a result of the radiation pressure on the dust, and the 'drag' or coupling between the dust and the gas. This will depend on several parameters: $\dot{M}$ (or the optical depth through the envelope), the stellar luminosity L and the dust-to-gas ratio. Habing et al. (1994) calculated outflow velocities as function of these parameters and showed that v_{exp} reaches a maximum for intermediate values of $\dot{M}$ but continuously increases with L: $v_{exp} \propto L^{0.35}$. However, a very large effect on the dust-to-gas ratio was found: the expansion velocities drop to only a few km/s if the dust-to-gas ratio drops by a factor of 10 compared to AGB stars of solar metallicity, although the effect differs in magnitude between carbon and oxygen-rich stars.

Few observations of expansion velocities outside the galactic disk are available. Wood et al. (1992) determined v_{exp} for five OH/IR stars in the LMC. They found values of only 60 % of similar galactic OH/IR stars. This conclusion however depends on the way v_{exp} is measured: Wood et al. used the two strongest peaks in the OH spectrum which may not be appropriate for noisy spectra. If instead the outermost emission peaks are used, Zijlstra et al. (1996) find that the four confirmed AGB stars have $v_{exp} = 12$ km/s versus the galactic comparison stars $v_{exp} = 14$ km/s, a difference of only 15%. The fifth star of Wood et al. is a luminous supergiant: SiO emission showed that in this case the OH was not centred on the stellar velocity and therefore v_{exp} was greatly underestimated (van Loon et al. 1996).

Groenewegen et al. (1997) detected CO for one of the two known mass-losing carbon-Miras in the galactic halo. The star has $v_{exp} = 3.2 \pm 0.2$ km/s. In comparison, galactic disk stars average around 10 km/s (Groenewegen et al. 1998). A metallicity of [Fe/H]= −0.7 is estimated for the halo carbon star. The very low v_{exp} is qualitatively consistent with the Habing et al. predictions, but further data on halo stars would be required for any conclusions.

The present data suggests that v_{exp} may be lower at subsolar Z, but quantative information for a range of mass-loss rates and luminosities is lacking. The difference between the LMC and the Galaxy may be less than predicted by Habing et al. (1994), but this could easily be due to compensatory effects, such as higher luminosity and/or different pulsation properties in the LMC stars.

4. Mass-loss rates at low metallicity

The mass-loss formalisms mentioned in the Introduction do not include a direct Z-dependence. As the temperature on the early AGB is Z-dependent, this may indirectly introduce lower mass-loss rates at lower Z. (For the mass-losing part of the AGB we do not have data on how the temperature scale depends on Z.) On the other hand, the formalisms were not designed for different metallicities and cannot be used for predictions. If the mass loss is driven by radiation pressure on dust, a possibly quite large effect could be expected.

There are a number of observations of mass loss at low Z. Up to 50 mass-losing AGB stars are now known in the LMC, with several having sufficiently high mass-loss rates to have significant extinction even at K. Zijlstra et al. (1996) conclude that for these stars, there is no evidence that $\dot{M}$ in the LMC is systematically lower than in the Galaxy. Also the SMC stars in their sample show no such effect. Groenewegen & Blommaert (1998) find evidence for further highly obscured AGB stars in the SMC which also exhibit high mass-loss rates similar to those of galactic OH/IR stars.

Further data comes from Mira variables in globular clusters and in the galactic halo. Groenewegen et al. (1998) derive dust mass-loss rates of a few $10^{-9}\,M_{\odot}\,yr^{-1}$ for two distant halo Miras with low Z. Mira variables in 47 Tuc and Omega Cen have been detected at 12 μm and show similar mass-loss rates (Gillet et al. 1988; Origlia et al. 1995, 1997). The corresponding gas mass-loss rates are of order $10^{-6}\,M_{\odot}\,yr^{-1}$ (Origlia et al. derive much lower values but these are based on a Reimers' law. Their dust detection implies a much higher rate.) Frogel & Elias (1988) find from a more complete sample an *average* $\dot{M} > 10^{-7}\,M_{\odot}\,yr^{-1}$ for long-period variables in globular clusters with [Fe/H,> 0.1. Thus, variables in globular clusters show mass-loss rates within normal ranges. It should be noted that a correction may have to be made if the expansion velocities are very low.

The data, within the limited accuracy, shows no evidence for low mass-loss rates at low Z. This may be related to the fact that an AGB star will continue to evolve to higher luminosities untill the mass-loss rate exceeds the nuclear-burning rate. If mass loss is less efficient at low Z, the star will reach higher luminosity which will again increase the mass-loss rate (Habing et al. 1994). Thus, stars in different populations are likely to reach similar mass-loss rates, but may do so at different phases of their evolution (Bowen & Willson 1991).

5. Mira Period–Luminosity relations

An important result obtained from extra-galactic AGB stars is the existence of a well-defined period–luminosity relation for Miras in the LMC. The relation was discovered by Glass & Evans (1981). Feast et al. (1989) showed that the relation is especially narrow (1-$\sigma = 0.13$ mag.) when using the K-band. Not all Miras follow the *PL* relation. LMC Miras with periods longer than 400 days tend to be overluminous (e.g. Zijlstra et al. 1996). In contrast, OH/IR stars near the Galactic Centre appear underluminous (Blommaert et al. 1998; Wood et al. 1998), although there are uncertainties in the bolometric corrections. A few nearby Miras are also underluminous (van Leeuwen et al. 1997).

Bedding & Zijlstra (1998) show that nearby Miras and semi-regulars (SR) with accurate HIPPARCOS parallaxes define three sequences: one sequence agrees with the LMC Mira *PL* relation, a broader sequence contains SR variables which is shifted by about a factor of two towards shorter periods (e.g. Wood & Sebo 1996), and a few stars fall well below the relation. (A number of semiregulars show two periods, one of which agrees with the Mira relation and the other with the less narrow SR sequence.) The few stars below the relation have luminosities similar to the semiregulars (although having much longer periods) and may be evolutionary related to this group. The Miras show slightly higher luminosities and appear therefore more evolved. The luminosity difference between SR's and Miras is consistent with an evolution along a Whitelock (1986) track. From the HIPPARCOS data, life-times of the order of $5\,10^5$ yr can be estimated for the Mira phase and similar or larger for the SR's.

Wood (1990) predicts that the *PL* relation should depend on Z, with galactic Miras being less luminous at K by 0.25 mag compared to LMC Miras of the same period (or, galactic Miras have longer periods). This is however not confirmed by available observations (e.g., van Leeuwen et al. 1997, Feast 1996, Whitelock et al. 1994). Alvarez et al. (1998) find some evidence for an offset between the HIPPARCOS Miras and the LMC *PL* relation, but the more accurate data of Bedding & Zijlstra (1998) does not confirm this.

Fig. 1 compares the LMC Miras with HIPPARCOS Miras and with globular cluster Miras. The HIPPARCOS Miras (open circles) all have parallax uncertainties less than 15%, and the resulting Lutz–Kelker bias is less than 0.05 mag (Koen 1992). A direct fit between the LMC and HIPPARCOS Miras (applying a bias correction according to Koen 1992) gives an LMC distance modulus of 18.63 ± 0.09, not inconsistent with other determinations. A shift of 0.25 mag would put the LMC at an unlikely distance of 18.9 and seems excluded.

Globular cluster Miras are shown in Fig. 1 as triangles. All occur in clusters with $[Fe/H] > 0.1$ (Whitelock 1986). To scale the data, two Hori-

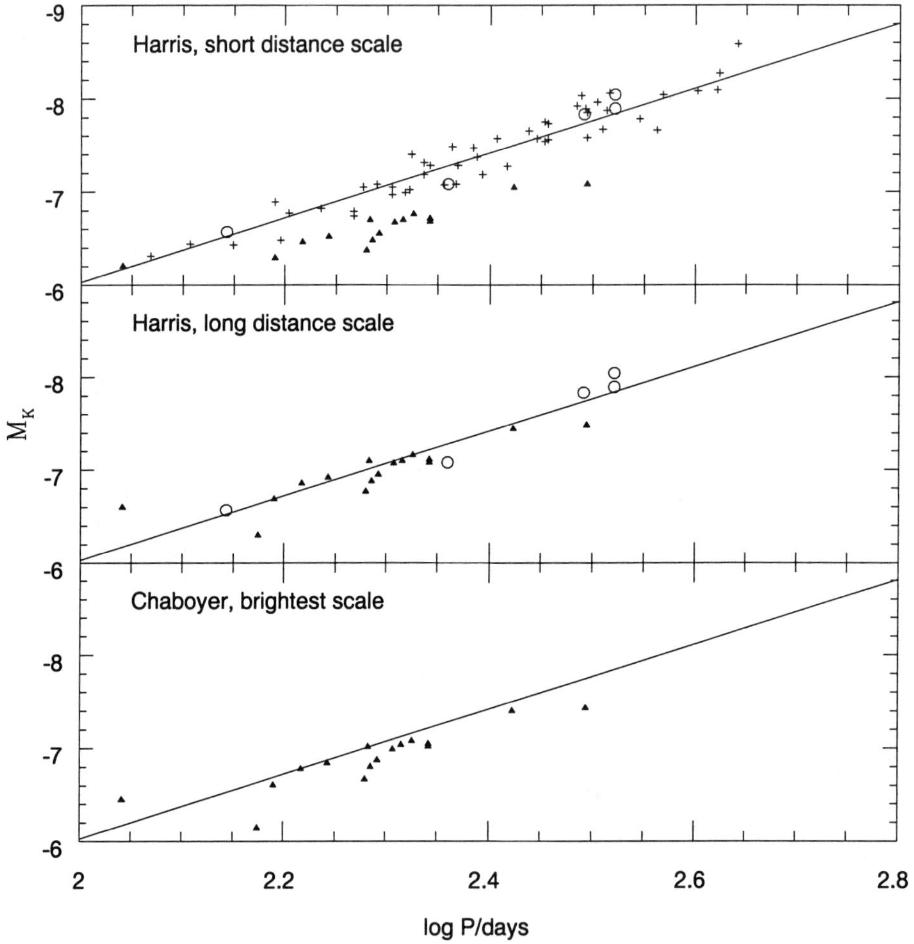

Figure 1. The Mira PL relations. Plus signs indicate LMC Miras (Feast et al. 1989), circles galactic HIPPARCOS Miras (Bedding & Zijlstra 1998) and triangles Miras in globular clusters (Whitelock 1986). LMC Miras are scaled to a distance modulus of 18.57. Three different distance scales for globular clusters are shown, based on different calibrations for the horizontal branch magnitude.

zontal Branch distance scales of Harris (1996) are used, a short scale (derived from RR Lyrae statistical parallaxes) and a long one (field subdwarfs). The average scale corresponds to $M_V(HB) = 0.15\,[Fe/H] + 0.85$. The bottom panel uses the Chaboyer (1998) relation $M_V(HB) = 0.23\,([Fe/H] + 1.6) + 0.46$ which again is 0.1 mag. brighter than an average including the statistical parallax. It is clear that only the brightest scale gives good agreement with the Mira *PL* relation. A direct fit to the LMC data using the Harris scale gives $M_V(HB) = 0.15\,[Fe/H] + 0.60 \pm 0.03$. (This would yield globular cluster ages of 11 +/− 2 Gyr.)

There is thus little evidence for a dependence of the *PL* relation on Z, when using K-band magnitudes. If the relation is indeed invariant, it would make Mira variables a highly useful distance indicator.

6. Mass loss and the PL relation

The position of a Mira in the *PL* diagram is determined by the core-mass–luminosity relation and by the pulsation equation:

$$\log P = 1.5 \log R - 0.5 \log M + \log Q \tag{1}$$

where P is the period in days, R and M are the radius and mass in solar units and Q is the pulsation constant. Using the AGB relation of Wood (1990):

$$M_{\rm bol} = 15.7 \log T_{\rm eff} + 1.884 \log z - 2.65 \log M - 59.1 - 15.7\Delta \tag{2}$$

(where the disturbance term Δ is usually ignored), one finds (Feast 1996):

$$M_{\rm bol} = -2.036 \log P + 0.73 \log z - 2.049 \log M + 2.881 + 2.036 \log Q \tag{3}$$

Here $z = Z/Z_\odot$. The existence of a narrow *PL* relation implies P is not an independent parameter: $M_{\rm bol} = f(z, M)$. Note that z is not *a priori* known for an LMC star but may vary between stars of different ages. M is not constant, due to mass loss. In addition, $M_{\rm bol}$ is determined by the core mass *independently from M*, but will change by as much as 1 mag. over the phase of the thermal-pulse cycle. These separate effects should lead to a large broadening of the *PL* relation (e.g. Vassiliadis & Wood 1993). From the narrow *PL* relation we find:

$$dP/dM = 0; \quad d\log P/dM_{\rm bol} = -0.33; \quad dP/dz = 0, \tag{4}$$

valid during the mass loss phase. (The second relation is the observed *PL* relation.) To keep a star on the *PL* relation during the Mira phase, the stellar radius must adjust to compensate for the changing luminosity and mass. The AGB relation above, valid for the earlier AGB, may thus not fully describe Miras.

The observed radius is wavelength dependent, due to molecular bands in the extended atmosphere. This re-introduces a metallicity effect, seen as a colour term dependent on z (Feast 1996). The *PL* relation is best defined at K, where the atmospheric bands do not contribute and the observed radius is closest to the photospheric radius.

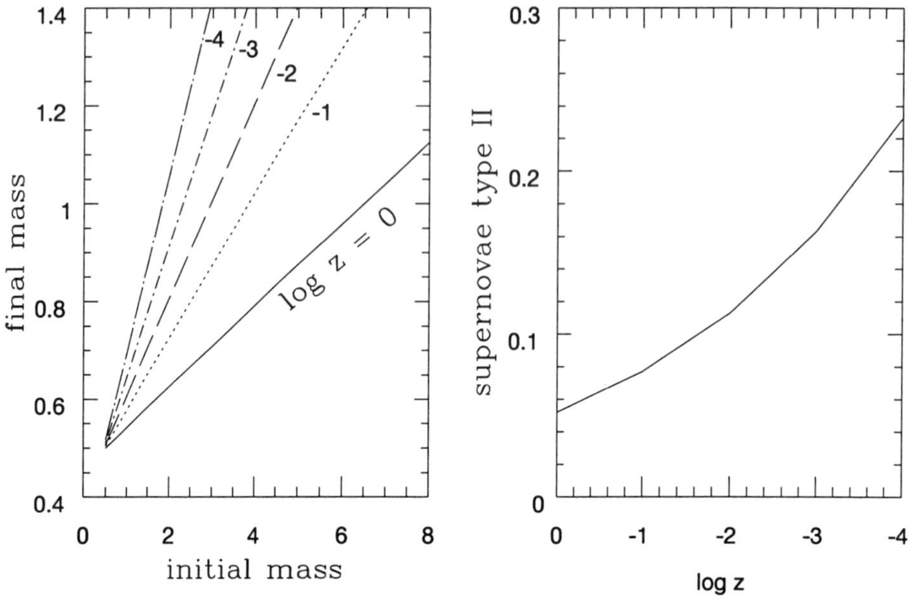

Figure 2. Left: initial-final mass relations for intermediate-mass stars at low z. Right: The inferred relative number of type-II (post-main-sequence) supernovae as function of $\log z$, integrated over the first Gyr.

A possible explanation for the relations above is a feedback between the stellar pulsations and mass loss. Such a mechanism could break down either at high luminosities or at high mass-loss rates. The latter case could correspond to the underluminous stars at the Galactic Centre.

7. Initial–Final Mass relations at low Z

The AGB mass-loss formalisms include an R-dependence. Using the $R \propto Z^{0.088}$ relation of Iben (1984) for the AGB predicts that a low-z AGB star will have a slightly lower $\dot{M}$ at the same luminosity. For $z = 0.1$, the Reimers mass loss would be 20 % lower. To compensate, a low-z star would require a higher luminosity (and therefore a higher core mass) to reach the same $\dot{M}$. Assuming the evolution is terminated when $\dot{M}$ exceeds the nuclear burning rate, a low-z star is expected to reach a higher final core mass and thus yield a more massive white dwarf than a star with the same initial mass but higher z.

For $z < 0.1$, the effect of radiation pressure of dust becomes very small and the mass loss is probably purely driven by the stellar pulsation. For such a dustless wind, Bowen & Willson (1991) find that $M_{\rm bol}$ should be 0.3 mag. brighter to reach the same mass loss as a dusty wind. The total increase in brightness as function of Z to reach the same mass-loss rates is $-\Delta M_{\rm bol} = 0.30 - 0.33 \log z$, for $z = Z/Z_\odot < 0.1$.

The core mass–luminosity relation $L = 59\,250(M_c - 0.495)$ (e.g. Boothroyd & Sackman 1988) can be used to quantify the effect on the final mass of the resulting white dwarf. The required increase is very small for low core mass, but becomes significant for core masses corresponding to the heaviest white dwarfs. We use an approximate present initial–final mass relation of $M_f = 0.5 + M_i/12$, and a Salpeter IMF $dN/dM = m^{-2.35}$. Fig. 2 shows the resulting M_i–M_f relation as function of z. At $z < 0.01$, the heaviest remnants reach the Chandresekhar mass giving rise to a class of AGB supernovae. At the extreme z, the contribution of these AGB-SN exceeds the type II SN by a factor of a few, over the first Gyr of the stellar population. If confirmed, such an effect would have dramatic consequences for the evolution of the earliest stellar populations in the high-redshift Universe.

References

Alonso V.M., Minniti D., Zijlstra A.A., Tolstoy E., 1998, A&A, submitted

Alvarez R., Mennessier M.-O., Barthès D., Luri X., Mattei J.A., 1997, A&A 327, 656

Beauchamp D., Hardy E., Suntzeff N.B., Zinn R., 1995 AJ 109, 1628

Bedding T.R., Zijlstra A.A., 1998, ApJ 506, L47

Blommaert J.A.D.L., van der Veen W.E.C.J., van Langevelde H.J., Habing H.J., Sjouwerman L.O., 1998, A&A 329, 991

Boothroyd A., Sackman I.-J., 1988, ApJ 328, 641

Bowen G.H., Willson L.A., 1991, ApJ 375, L53

Bressan A., Granato G.L., Silva L., 1998, A&A 332, 15

Chaboyer B., 1998, http://xxx.lanl.gov/abs/astro-ph/9808202

Davidge T.J., 1998, ApJ 497, 650

Feast M.W., 1996, MNRAS 278, 11

Feast M.W., Glass I.S., Whitelock P.A., Catchpole R.M., 1989, MNRAS 241, 375

Frogel J.A., Elias J.H., 1988, ApJ 324, 822

Gillet F.G., de Jong T., Neugebauer G., Rice W.L., Emerson J.P., 1988, AJ 96, 116

Glass I.S., Evans T.L., 1981, Nature 291, 303

Groenewegen M.A.T., Whitelock P.A., Smith C.H., Kerschbaum F., 1998, MNRAS 293, 18

Groenewegen M.A.T., Blommaert J.A.D.L., 1998, A&A 332, 25

Groenewegen M.A.T., Oudmaijer R.D., Ludwig H.-G.,1997, MNRAS 292, 686

Habing H.J., Tignon T., Tielens A.G.G.M., 1994, A&A 286, 523

Harris W.E., 1996, AJ 112, 1487 (http://www.physics.mcmaster.ca/Globular.html)

Iben I., Jr., 1984, ApJ 277, 333

Koen C., 1992, MNRAS 256, 65

Knapp G.R., Gunn J.E., Wynn-Williams C.G., 1992, ApJ 399, 76

Knapp G.R., Gunn J.E., Connolly A.J., 1995, ApJ 448, 195

Origlia L., Ferraro F.R., Fusi Pecci F., 1995, MNRAS 277, 125

Origlia L., Scaltriti F., Anderlucci E., Ferraro F.R., Fusi Pecci F., 1997, MNRAS 292, 753

Reid I.N., 1991, ApJ 382, 143

Tolstoy E., Daha A., Hoessel J.G., Danielson G.E., 1996, AJ 109, 579

van Leeuwen F., Feast M.W., Whitelock P.A., Yudin B., 1997, MNRAS 287, 955

van Loon J.Th., Zijlstra A.A., Whitelock P.A., te Lintel Hekkert P., Chapman J.M., Loup C., Groenewegen M.A.T., Waters L.B.F.M., Trams N.R., 1998, A&A 329, 169

van Loon J.Th., Zijlstra A.A., Whitelock P.A., Waters L.B.F.M., Loup C., Trams N.R., 1997, A&A 325, 585

van Loon J.Th., Zijlstra A.A., Bujarrabal V., Nyman L.-Å., 1996, A&A 390, L29

Vassiliadis E., Wood P.R., 1993, ApJ 413, 641

Whitelock P.A., 1986, MNRAS 219, 525

Whitelock P.A., Menzies J.W., Feast M.W., Marang F., Carter B.S., Roberts G., Catchpole R.M., Chapman J.W., 1994, MNRAS 267, 711

Whitelock P.A., Irwin M., Catchpole R.M., 1996, New Astronomy 1, 57

Wood P.R., 1990, in From Miras to Planetary Nebulae: Which path for stellar evolution?, M.-O. Mennessier & A. Omont (eds.), Editions Frontières, p. 32

Wood P.R., Habing H.J., MacGregor P.J., 1998, A&A 336, 925

Wood P.R., Whiteoak J.B., Hughes M.B., Bessell M.S., Gardner F.F., Hyland A.R., 1992, ApJ 397, 552

Wood P.R., Sebo K.M., 1996, MNRAS 282, 958

Zijlstra A.A., Loup C., Waters L.B.F.M., Whitelock P.A., van Loon J.Th., Guglielmo F., 1996, MNRAS 279, 32

Asymptotic Giant Branch Stars
T. Le Bertre, A. Lèbre and C. Waelkens, eds.
I.A.U. Symposium 191, 1999

ISOCAM and DENIS Survey of 0.5 square degrees in the Bar of the LMC. Detection of the whole TP-AGB Star Population

C. Loup, E. Josselin

Institut d'Astrophysique de Paris, CNRS, 98bis bd Arago, F-75014 Paris, France

M.-R. Cioni, H.J. Habing

Sterrewacht Leiden, Postbus 9513, 2300 RA Leiden, the Netherlands

J.A.D.L. Blommaert, N.R. Trams

ISO Data Centre, Astrophysics Div. of ESA, Villafranca, Spain

M.A.T. Groenewegen

Max-Planck-Institut für Astrophysik, Garching, Germany

C. Alard, P. Fouqué

Observatoire de Paris, France

F. Kerschbaum

Institut für Astronomie, A-1180 Wien, Austria

L.B.F.M. Waters, J.Th. van Loon

University of Amsterdam, the Netherlands

A.A. Zijlstra

UMIST, Manchester, UK

and the DENIS consortium

See aknowledgments

Abstract. We surveyed 0.5 square degrees in the Bar of the LMC with ISOCAM at 4.5 and 12 μm, and with DENIS in the I, J, and K_S bands. Our goal was to build a complete sample of Thermally-Pulsing AGB stars. Here we present the first analysis of 0.14 square degrees. In total we find about 300 TP-AGB stars. Among these TP-AGB stars, 9% are obscured AGB stars (high mass-loss rates); 9 of them were detected by IRAS, and only 1 was previously identified. Their luminosities range from 2 500 to 14 000 $L_{\odot}$, with a distribution very similar to the one of optical TP-AGB stars (i.e. those with low mass-loss rates). Such a luminosity distribution, as well as the percentage of obscured stars among TP-AGB stars, is in very good agreement with the evolutionary models of Vassiliadis & Wood (1993) if most of the TP-AGB stars that we find have initial masses smaller than 1.5 to 2 $M_{\odot}$.

1. Introduction

The so-called "obscured" AGB stars are AGB stars in the Thermal-Pulse phase (TP-AGB), losing mass at such rates that the dust formed in their circumstellar envelopes becomes optically thick to the stellar radiation. The maximum of their emission is in the infrared, typically between 2 and 20 μm. The first search for obscured AGB stars in the Large Magellanic Cloud (LMC) has been done by Frogel & Richer (1983), with a moderate success due to their magnitude limit of 11 in the K band, and to the small area surveyed. Later, searches based on the IRAS data have been much more successful, leading to the discovery of more than 50 obscured AGB stars in the whole LMC (Reid et al. 1990; Reid 1991; Wood et al. 1992; Zijlstra et al. 1996; van Loon et al. 1997, 1998). However, due to the sensitivity limit of IRAS (even using the Faint Star Catalogue and the pointed observations), only the brightest or/and reddest sources could be detected by IRAS (Loup et al. 1997). TP-AGB stars with intermediate mass-loss rates, as well as those with high mass-loss rates and low luminosities, were still to be discovered.

We surveyed 0.5 square degrees in the LMC in 5 IR bands, using the DENIS and ISOCAM instruments in order to build a complete sample of TP-AGB stars, with the whole range of mass-loss rates and luminosities, and with as less observational biases as possible. Here we present the first analysis of 0.14 square degrees.

2. Observations and data reduction

Seven fields of 16′x16′, located in the Bar of the LMC, have been surveyed with the infrared camera ISOCAM on board of the Infrared Space Observatory, in the LW1, 4.5μm, and LW10, 12μm, bands, with a pixel size of 6″. Here we present the analysis for 2 fields centered at (J2000) 05:25:22 –69:46:30 (the LMC Optical Center) and 05:29:56 –69:57:50. Data reduction has been performed with the CIA package. PSF reconstruction and point source extraction are done with a software especially developed by Alard to take into account undersampling. The final result is not expressed as a flux density following the IRAS/ISO conventions, but as a magnitude, such as all the infrared magnitudes of Vega (without infrared excess) are 0. We chose this convention in order to have a system consistent with ground-based infrared photometry. Limiting magnitudes are 11 (7 mJy) and 10.5–10 (2–3.5 mJy) in the LW1 and LW10 bands, respectively. The typical uncertainty is 0.2 mag.

The same fields have been surveyed with the three channels instrument of the Deep Near Infrared Southern Sky Survey, DENIS, mounted on the ESO 1m telescope at La Silla (Epchtein et al. 1997). Observations in the I, J, and

K_S bands are simultaneous. Data reduction has been done through the regular pipeline of the survey (Paris and Leiden Data Analysis Centers). Limiting magnitudes are 18, 16, and 14, in the I, J, and K_S bands, respectively. In summary, we thus present the analysis of 0.14 square degrees in the Bar of the LMC, observed in 5 bands from 0.8 to 12 μm.

3. Characterization and cross-identifications

AGB stars have been well studied in our Galaxy up to about 5 kpc, in particular with the IRAS data, and with classical JHKL'MN photometry. Depending on their mass-loss rate, their emission peaks typically between 1 and 20 μm. DENIS and ISO provide the jump in sensitivity required to study AGB stars in the Magellanic Clouds with a similar quality. We chose the LW1 and LW10 bands because they complement the IJK_S bands of DENIS at longer wavelengths, and because LW1 can be compared with the M band, and LW10 is very similar to the IRAS $12\mu m$ band. The combination of both DENIS and ISO data allows to detect all the LMC AGB stars in the Thermal-Pulse phase (TP-AGB) whatever their mass-loss rate.

To characterize AGB stars, and cross-identify DENIS and ISO observations, we use three colour-colour diagrams (CCD) : (I–J, J–K_S), (J–K_S, K_S–LW10), and (K_S–LW10, LW1–LW10), as well as colour-magnitude diagrams. The location of the DENIS-ISO candidates in these diagrams is compared with the location of : (i) galactic sources observed in the J, K, M, and 12 μm bands (from the database of Kerschbaum et al.), (ii) models performed by Groenewegen et al. (1998 and private communication) for C stars, and M stars with silicates, with a large range of dust opacities, and fitting satisfactorily galactic sources. Within some limits, the location in the 3 CCDs also allows to estimate their chemical type in most cases. With the sensitivity of the ISO minisurvey, except a few F-K foreground giants, all point sources detected in LW1 or LW10, and isolated from extended structures, are TP-AGB stars or M supergiants.

The great majority of the sources that we detect are AGB and RGB stars, with no or faint mass-loss, only detected with DENIS (which provides enough information in this case). The 82 sources detected by both DENIS and ISO are either bright stars with low mass-loss rates, or TP-AGB stars with intermediate or high mass-loss rates. In the following, we define obscured AGB stars such as : $\tau(0.5\mu m) \geq 1.0$, corresponding to $(K_S\text{–LW10})_0 \geq 2.3$ and ≥ 3.4 for C-rich and O-rich sources, respectively. In total we detect 2160 sources. About 300 can be statistically selected as TP-AGB stars from their location in the DENIS CCD (Loup 1998; Cioni et al. 1998). Among these TP-AGB stars, 28 are obscured, with $(K_S\text{–LW10})_0$ up to 7.3. Finally, among these 28 obscured AGB stars, only 9 were detected in the IRAS pointed observations (Schwering & Israel 1990), and only 1 was previously identified by Wood et al. (1992). Among the 9 IRAS sources, one finds some of the reddest objects, but not all of them.

4. The luminosity distribution of obscured AGB stars

Traditional methods to estimate bolometric luminosities of AGB stars from IR photometry first convert magnitudes into flux densities from the effective or isophotal zero flux densities, and then integrate over by fitting a blackbody (in general), extrapolating the fit at short and long wavelengths. This method uses two contradictory assumptions : in the first step, the spectral distribution is the same as Vega, and in the second step, it is similar to a blackbody (but much colder than Vega). In this work, we only use integrated in-band fluxes, which at least remove the first assumption in all cases. From the models for C-rich AGB stars and M stars with silicates, and for blackbodies, we compute the percentage of flux actually measured in the 5 bands compared to the total flux emitted by the source, as a function of the various IR colours. From the measured in-band fluxes, colours, and estimated chemical types of our observed sources, we can then estimate their bolometric luminosities by comparing with the models, assuming of course that their spectral distribution is close to the models. Blackbodies systematically underestimate bolometric luminosities. We also compute the uncertainty on each bolometric luminosity estimate from measurement uncertainties, the spread from the various IR colours, and the intrinsic spread of models on the relation between the percentage of measured fluxes and IR colours (typically 2%).

The result is presented in Fig. 1 for all the sources detected by ISO and DENIS, and DENIS only, as a function of $(J-K_S)_0$. Note that obscured AGB stars typically have $(J-K_S)_0 \geq 2.3$. Though we present statictics for only 2160 stars, one distinguishes the Tip of the Red Giant Branch at $M_{bol} \simeq -3.4$ (i.e. 1800 $L_\odot$, see Cioni et al. 1998), using a distance of 49 kpc. The bolometric luminosities of obscured AGB stars in the observed fields range between −3.8 and −5.6, i.e. between 2500 and 14000 $L_\odot$. Their luminosity distribution is very similar to the one of optical carbon stars (Blanco et al. 1980; Costa & Frogel 1996) and TP-AGB stars with low mass-loss rates (Loup 1998). The only difference which may be noticed is that the lower boundary is smaller for optical stars than for obscured stars.

Though we will need to get the statistics on the 7 fields, our first analysis of 2 fields seems to be in very good agreement with the evolutionary models of Vassiliadis & Wood (1993). For stars of initial mass smaller than 1.5-2.0 $M_\odot$, they predict that the average luminosity is almost constant along the TP-AGB phase, with a luminosity ranging between 2000 and 8000 $L_\odot$, in excellent agreement with our observations. They also predict that TP-AGB stars undergo important mass-loss during about 15% of the total TP-AGB phase, to be compared to about 9% in our observations.

Acknowledgments. DENIS consortium : C. Alard, E. Bertin, J. Borsenberger, B. de Batz, M.-R. Cioni, E. Copet, M. Dennefeld, E. Deul, N. Epchtein (PI), T. Forveille, P. Fouqué, F. Garzon, H.J. Habing, J. Hron, S. Kimeswenger, F. Lacombe, T. Le Bertre, C. Loup, G. Mamon, A. Omont, G. Paturel, P. Persi, A. Robin, D. Rouan, G. Simon, D. Tiphène, I. Vauglin, S. Wagner. This paper makes use of observations obtained at the European Southern Observatory (La Silla) in the framework of the DENIS project. DENIS has been financially supported by the EC and by national funding sources from France, the Netherlands, Italy, Austria, Germany and Brazil. This work makes use of observations with ISO, an ESA project with instruments funded by ESA Member States (especially the PI countries: France, Germany, the Netherlands, and the United Kingdom) with the participation of ISAS and NASA. The ISOCAM data presented in this

Figure 1. Bolometric luminosities of AGB and RGB stars in 0.14 square degrees in the Bar of the LMC. See Sect. 4. Sources only detected with DENIS are drawn with small dots, without errorbar for clarity (the typical uncertainty is 0.1 mag). Squares indicate C-rich sources, circles O-rich sources, both detected with ISO and DENIS; full squares or circles indicate that the chemical type is known (spectral type or OH maser); open symbols mean that the chemical type is an estimate from CCDs (see Sect. 3). Arrows indicate lower limits on (J–K_S) when J is not detected.

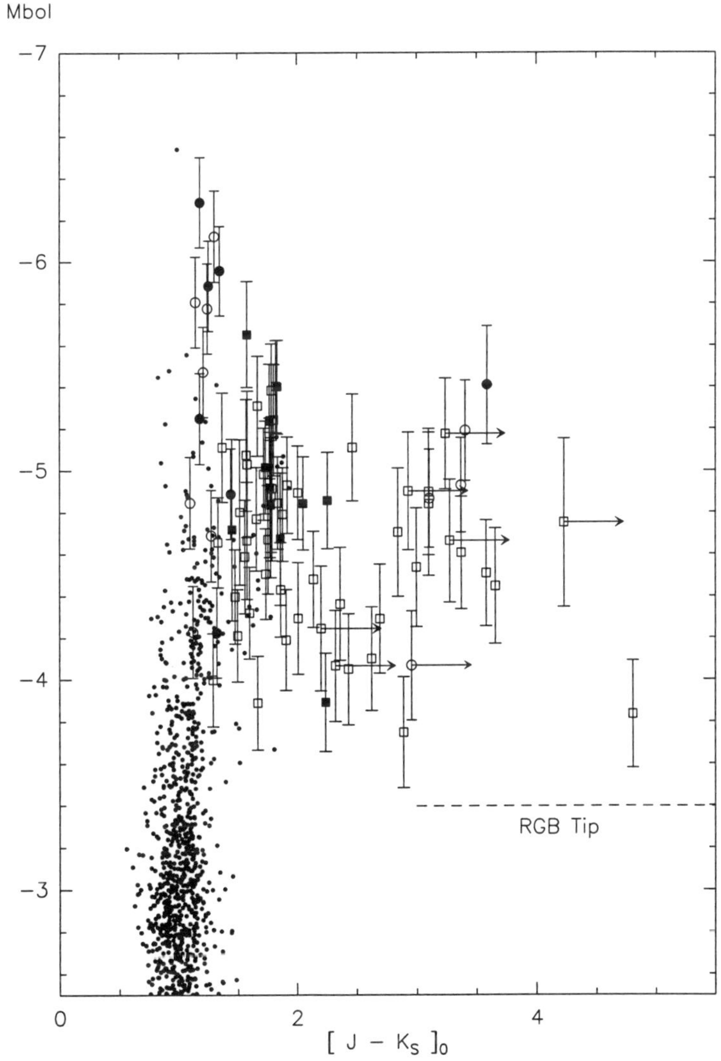

paper were analysed using "CIA", a joint development by the ESA Astrophysics Division and the ISOCAM Consortium. The ISOCAM Consortium is led by the ISOCAM PI, C. Cesarsky, Direction des Sciences de la Matière, CEA, France. F. Kerschbaum was supported by APART from the Austrian Academy of Sciences.

References

Bertelli G., Bressan A.G., Chiosi C., 1985, A&A 150, 33

Blanco V.M., McCarthy M.F., 1990, AJ 100, 674

Blanco V.M., McCarthy M.F., Blanco B.M., 1980, ApJ 242, 938

Cioni M.-R., Habing H.J., Loup C., Epchtein N., 1998, in "*New Views on the Magellanic Clouds*", IAU Symposium 190, in press

Costa E., Frogel J.A., 1996, AJ 112, 2607

Epchtein N., et al., 1997, The ESO Messenger 87, 27

Frogel J.A., Richer H.B., 1983, ApJ 275, 84

Frogel J.A., Blanco V.M., 1990, ApJ 365, 168.

Groenewegen M.A.T., Whitelock P.A., Smith C.H., Kerschbaum F., 1998, MNRAS 293, 18

Hughes S.M.G., 1989, AJ 97, 1634

Loup C., Zijlstra A.A., Waters L.B.F.M., Groenewegen M.A.T., 1997, A&AS 125, 419

Loup C., Duc P.-A., Fouqué P., Bertin E., Epchtein N., 1998, in "*The Impact of Near-Infrared Sky Surveys on Galactic and Extragalactic Astronomy*", ed. N. Epchtein, Kluwer, p.115

Loup C., 1998, in "*New Views on the Magellanic Clouds*", IAU Symposium 190, in press

Reid I.N., Tinney C., Mould J., 1990, ApJ 348, 98

Reid I.N., 1991, ApJ 382, 143

Richer H.B., 1981, ApJ 243, 744

Schwering P.B.W., Israel F.P., 1990, "*Atlas and catalogue of infrared sources in the Magellanic Clouds*", Kluwer

van Loon J.Th., Zijlstra A.A., Whitelock P.A., Waters L.B.F.M., Loup C., Trams N.R., 1997, A&A325, 585

van Loon J.Th., Zijlstra A.A., Whitelock P.A., te Lintel Hekkert P., Chapman J.M., Loup C., Groenewegen M.A.T., Waters L.B.F.M., Trams N.R., 1998, A&A 329, 185

Vassiliadis E., Wood P.R., 1993, ApJ 413, 641

Wood P.R., Whiteoack J.B., Hughes S.M.G., Bessell M.S., Gardner F., Hyland A.R., 1992, ApJ 397, 552

Zijlstra A.A., Loup C., Waters L.B.F.M., Whitelock P.A., Guglielmo F., 1996, MNRAS 279, 32

Asymptotic Giant Branch Stars
T. Le Bertre, A. Lèbre and C. Waelkens, eds.
I.A.U. Symposium 191, 1999

Obscured Asymptotic Giant Branch stars in the Magellanic Clouds

Jacco Th. van Loon

Astronomical Institute, University of Amsterdam, Kruislaan 403, NL-1098 SJ Amsterdam, The Netherlands

Abstract. We report on some recent advances in the study and understanding of heavily obscured AGB stars in the Magellanic Clouds.

1. Introduction

Asymptotic Giant Branch (AGB) stars develop strong mass loss at rates up to $\dot{M} \sim 10^{-5}$ $M_\odot$ yr^{-1} or more. The optical light of AGB stars with the highest $\dot{M}$ is almost entirely absorbed by their dusty circumstellar envelopes (CSEs), but re-emitted at longer wavelengths. They become very bright infrared (IR) objects, outshining any other star in a galaxy except for a few red supergiants (RSGs). During a brief period in their lives, they loose 50 to 80% of their initial mass. This makes them important contributors to the chemical enrichment of the interstellar medium. Obscured AGB stars are the main sources of dust particles in the Universe. Dust is a necessary condition for the formation of planets that may support life. Obscured AGB stars contain masers that emit at radio wavelengths, where a high velocity resolution is obtained (~ 1 km s^{-1}). Maser stars can thus be used to trace the gravitational potential in a galaxy.

Some questions remain to be answered: (1) How does an AGB star drive its mass loss, and what is the rôle of stellar pulsation in this matter? (2) How does an AGB star form a Planetary Nebula (PN), for which an enhancement of the already high $\dot{M}$ is required? (3) What is the most massive progenitor of an AGB star? (4) Which AGB stars become carbon stars, thereby completely changing the chemistry of molecule and dust formation in the CSE? (5) How do complex molecules and dust particles form in the CSE of an AGB star?

2. The search for obscured AGB stars in the Magellanic Clouds

The Large and Small Magellanic Clouds (LMC & SMC) have two major advantages over the Milky Way when it comes to studying AGB stars: their distances are well determined ($\sim$ 50 and 60 kpc) and interstellar extinction is relatively small both in front of the MCs and within. Furthermore, the average metallicities of the intermediate age populations in the LMC and SMC are factors of respectively 2 and 5 lower than in the Milky Way. Thus, the aforementioned astrophysical problems may be investigated on their metallicity dependence.

Previous studies of AGB stars in the MCs were based on optically visible stars, with only very few obscured objects identified. Nothing was known about

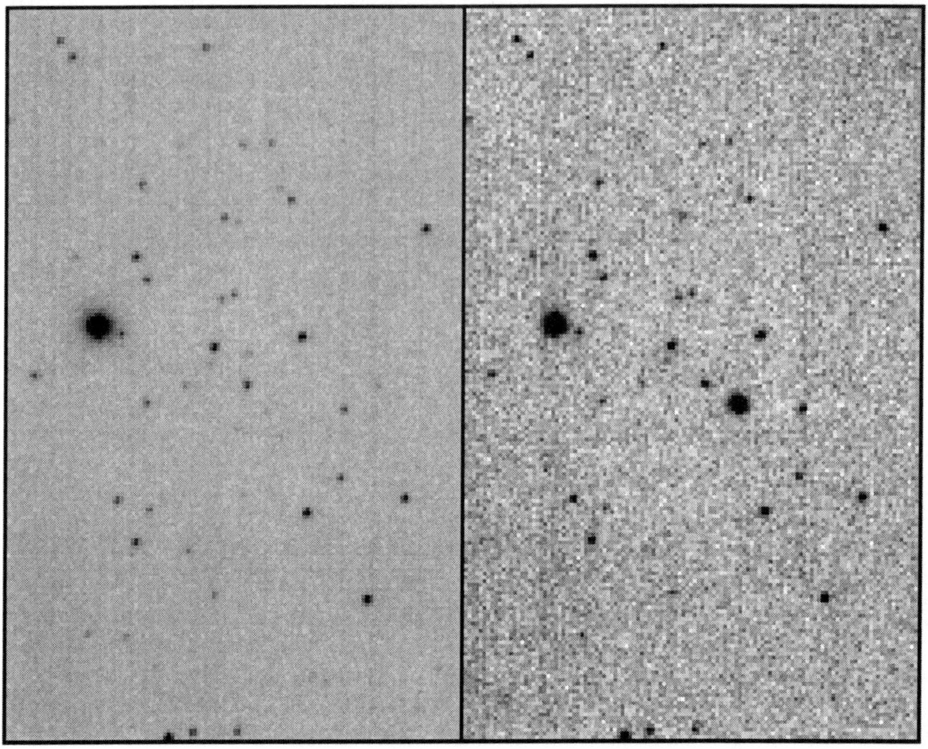

Figure 1. Identifying a magellanic AGB star as counterpart of an IRAS source. Invisible in the J-band image (at left), it rivals the brightest star in the K-band image (at right). The images were obtained using the IRAC2b near-IR camera at the ESO/MPI 2.2m telescope at ESO La Silla, Chile. South is up, and East is to the right. The frames shown here measure one arcminute along the short (E-W) axis.

how AGB stars continue to evolve after they become obscured. As an example, a long standing problem has been the observation that no carbon stars in the MCs were found at luminosities higher than 2×10^4 $L_\odot$ (Iben 1981; Hughes & Wood 1990; Costa & Frogel 1996). The burning of carbon into nitrogen and oxygen at the bottom of the convective mantle, Hot Bottom Burning (HBB), was proposed as a mechanism to prevent the formation of more luminous carbon stars. Alternative explanations include luminous carbon stars being dust-enshrouded. There was a clear need for a systematic (IR) search for obscured AGB stars in the MCs, and the determination of their luminosities and chemical types.

The first attempts to find obscured AGB stars in the MCs had resulted in a sample of evolved stars, comprising mainly RSGs and only a few AGB stars (Whitelock et al. 1989; Reid 1991; Wood et al. 1992). We successfully searched for counterparts of IRAS point sources in the direction of the LMC, mainly using near-IR imagery at La Silla (Loup et al. 1997; Zijlstra et al. 1996; van Loon et al. 1997, 1998a). We identified most of the counterparts as obscured AGB stars, all members of the LMC. As an example, we show J and K-band images of IRAS 06025−6712 obtained with the IRAC2b camera on the 2.2m ESO/MPI telescope at ESO La Silla, Chile (Fig. 1). Invisible in the J-band image down to 18th magnitude, it is very bright in the K-band image at $K = 11.52$ mag., and probably a carbon star. It is located in a small, anonymous open cluster. We obtained deep optical images to determine the age of the cluster, and hence the main-sequence mass of the progenitor of the obscured AGB star. We are doing the same for the luminous OH/IR star IRAS 05298−6957 in the double cluster HS327, that also contains a carbon star (van Loon et al. 1998a).

We increased the number of known obscured AGB stars in the LMC from only a few previously known objects to a sample of ~ 50 (van Loon et al. 1997, 1998a). In the SMC, ~ 25 such stars are now known (see also Groenewegen & Blommaert 1998). The K-band distribution function of IRAS-detected AGB stars is truncated at the faint end by detection limits in the near-IR ($K \sim 19$ mag.). Indeed, for some IRAS sources we failed to identify a red stellar counterpart in the near-IR. Furthermore, there are ~ 3 times as many IRAS sources left unexplored, and we predicted that ISO would detect more obscured AGB stars too faint for IRAS (see also the contribution of Loup et al. in these proceedings). A large population of obscured AGB stars may explain the apparent lack of AGB stars as compared to the number of PNe in the LMC (Frogel et al. 1990; Reid et al. 1990; Wood et al. 1992). In the process, we also found post-AGB star candidates in the LMC, and highly reddened background galaxies.

3. Chemical compositions

We established the chemical composition for many obscured AGB stars in the MCs, using IR colours and IR spectroscopy from the ground (ESO, SAAO, CTIO) and space (IRAS, ISO). In Fig. 2 we show an example of L-band spectra obtained with the IRS at the 4m Blanco telescope at Cerro Tololo, Chile, of the LMC carbon star IRAS 04496−6958 and M10 star IRAS 04509−6922 (van Loon et al. 1998a). HCN and C_2H_2 molecules cause strong absorption between 3.0 and 3.1 μm in carbon stars, whereas oxygen-rich (M) stars show a featureless continuum. With the VLT at ESO Paranal, Chile, it will be possible to determine in this way the chemical type of all obscured AGB stars in the MCs, and the brightest in other nearby dwarf galaxies. We also used ISO to obtain PHOT-S spectra between 2.5 and 12 μm of 15 objects, and N-band spectra with ISOCAM and a Circular Variable Filter of 12 objects.

Many (but not all) obscured AGB stars in the LMC are carbon-rich. Among these, we discovered the most luminous carbon star in the MCs, IRAS 04496−6958 having a luminosity of 4×10^4 $L_\odot$ (van Loon et al. 1998a). Thus we showed that carbon stars do exist up to the highest AGB luminosities. This inspired stellar evolution calculators to propose an explanation (Frost et al. 1998; Marigo et al. 1998): due to mass loss the stellar mantle is reduced, thereby switching off HBB that otherwise prevents massive stars from becoming carbon stars. We conclude that the population of optically visible AGB stars shows rather unevolved intermediate-age stars: they still have to pass through the phase where they will shed most of their mass and become obscured. Evolution of the chemi-

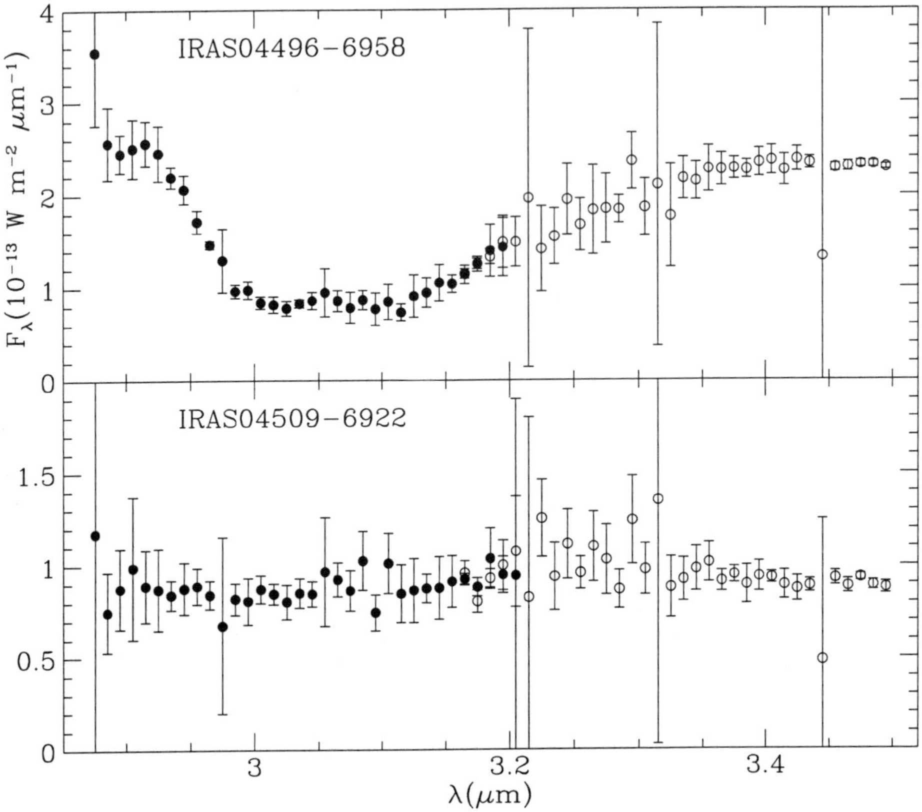

Figure 2. L-band spectra can be used to distinguish between carbon stars (IRAS 04496−6958 at top) and oxygen-rich M stars (IRAS 04509−6922 at bottom) because carbon stars show strong absorption between 3.0 and 3.1 μm due to HCN and C_2H_2 molecules. These spectra were obtained with the IRS at the 4m Blanco telescope at CTIO, Chile.

cal composition of the stellar photospheres and the circumstellar dust continues during the obscuration phase.

4. Metallicity dependence of mass-loss

AGB mass-loss is often assumed to decrease with lower metallicity, but this still awaits strong observational support. We noticed that very obscured AGB stars exist in both the LMC and the SMC, despite their low metallicities. Assuming a dust-to-gas ratio proportional to the metallicity, and accounting for the luminosity dependence of the dust formation radius, the near-IR colours suggest that AGB stars in the Milky Way, LMC, and SMC reach similar maximum $\dot{M}$. However, the optical depths of magellanic CSEs may also be large if the expansion velocity $v_{\rm exp}$ of the dust decreases with lower metallicity.

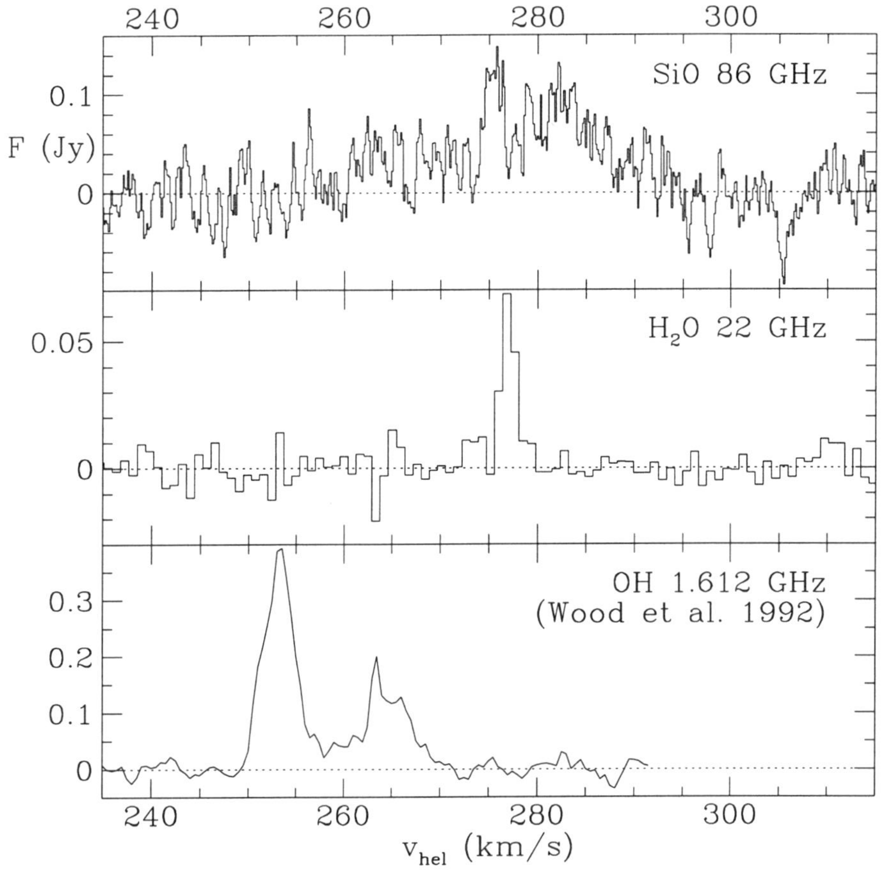

Figure 3. Maser emission can be detected from extra-galactic circumstellar envelopes. The LMC red supergiant IRAS 04553−6825 is the first star to have been detected in OH (Wood et al. 1986), SiO (van Loon et al. 1996), and H_2O (van Loon et al. 1998b).

We can measure v_{exp} of the gas component from circumstellar maser emission profiles. Wood et al. (1986, 1992) detected the first magellanic OH masers, that suggest v_{exp} to be proportional to metallicity. We have searched for more OH masers (van Loon et al. 1998a), and also for SiO and H_2O masers in the LMC. We used the 15m SEST at La Silla, Chile, to detect the first extra-galactic SiO maser (van Loon et al. 1996), and we predicted and subsequently confirmed the first extra-galactic circumstellar H_2O maser using the 64m telescope at Parkes, Australia (van Loon 1998; van Loon et al. 1998b). We show the maser spectra of this LMC supergiant (IRAS 04553−6825) to illustrate that it has become feasible to study masers in extra-galactic CSEs (Fig. 3). We argue that v_{exp} of magellanic CSEs may have been generally underestimated. More maser searches need to be performed in the MCs before definitive conclusions can be drawn on the metallicity dependence of v_{exp} and $\dot{M}$.

5. Mass-loss evolution of AGB stars

We show that AGB mass-loss does not evolve gradually, but undergoes a phase transition. In the early phase, an AGB star has a short pulsation period (P < 400 days) and small bolometric amplitude ($\Delta M_{\rm bol}$). Containing little dust, the CSE is optically thin and expands slowly ($v_{\rm exp} \sim$ few km s^{-1}), possibly driven by radiation pressure on molecules. In the late phase, the AGB star has a long pulsation period (P> 600 days) and large $\Delta M_{\rm bol}$. The dusty CSE is optically thick and expands with $v_{\rm exp} \sim 10$ to 20 km s^{-1}, efficiently driven by radiation pressure on dust. The mass-loss characteristics change suddenly when the star evolves from the early into the late phase, possibly due to the sensitivity of dust formation on the conditions in the CSE (e.g. Arndt et al. 1997).

Energy is stored and released during a stellar pulsation cycle. Part of the energy is thought to be used in initiating stellar mass loss. We argue from first principles that this concept naturally results in two observable facts: (1) a minimum $\Delta M_{\rm bol}$ is needed to create a dusty CSE and high $\dot{M}$, and (2) supergiants need smaller $\Delta M_{\rm bol}$ than AGB stars to support the same $\dot{M}$.

Acknowledgments. We thank the organisers for a pleasant and interesting conference, and for giving me the opportunity to present my work. Milhões de beijinhos para o anjinho Joana.

References

Arndt T.U., et al., 1997, A&A 327, 614
Costa E., Frogel J.A., 1996, AJ 112, 2607
Frogel J.A., et al., 1990, ApJ 352, 96
Frost C.A., et al., 1998, A&A 332, L17
Groenewegen M.A.T., Blommaert J.A.D.L., 1998, A&A 332, 25
Hughes S.M.G., Wood P.R., 1990, AJ 99, 784
Iben I., Jr., 1981, ApJ 246, 278
Loup C., et al., 1997, A&AS 125, 419
Marigo P., et al., 1998, A&A 331, 564
Reid I.N., 1991, ApJ 382, 143
Reid I.N., et al., 1990, ApJ 348, 98
Van Loon J.Th., 1998, in "ISO's View on Stellar Evolution", Ap&SS 255, p. 405
Van Loon J.Th., et al., 1996, A&A 306, L29
Van Loon J.Th., et al., 1997, A&A 325, 585
Van Loon J.Th., et al., 1998a, A&A 329, 169
Van Loon J.Th., et al., 1998b, A&A 337, 141
Whitelock P.A., et al., 1989, MNRAS 238, 769
Wood P.R., et al., 1986, ApJ 306, L81
Wood P.R., et al., 1992, ApJ 397, 552
Zijlstra A.A., et al., 1996, MNRAS 279, 32

Asymptotic Giant Branch Stars
T. Le Bertre, A. Lèbre and C. Waelkens, eds.
I.A.U. Symposium 191, 1999

Systematic study of AGB stars in the intermediate-age globular clusters in the Magellanic Clouds[1]

T. Tanabé, S. Nishida, and Y. Nakada

Institute of Astronomy, School of Science, The University of Tokyo, Mitaka, Tokyo, 181-8588, Japan

T. Onaka

Department of Astronomy, School of Science, The University of Tokyo, Bunkyo-ku, Tokyo, 113-0033, Japan

I. S. Glass

South African Astronomical Observatory, P.O.Box 9, Observatory 7935, South Africa

M. Sauvage

CEA/DSM/DAPNIA/Service d'Astrophysique, C. E. N. Saclay F-91191 Gif-sur-Yvette Cedex, France

Abstract.
We performed systematic infrared observations of the intermediate-age Magellanic Clouds clusters NGC 419, NGC 1783 and NGC 1978. Mid-infrared stars discovered in NGC 419 and NGC 1978 are very red and must be undergoing intense mass loss (comparable to superwinds). They are probably carbon stars but do not seem to show any FIR excesses. Three optically visible carbon stars as well as (at least) 2 near-infrared carbon stars observed with ISOPHOT show 60 μm excesses which may indicate mass loss in the past. It seems that the MIR stars are fainter than the AGB tip luminosity and that their $M_{\rm bol}$s are close to those of the transition luminosity from M type to C stars. Therefore, these MIR stars may not be in the final stage of the AGB phase. This may suggest that AGB stars lose mass heavily at some other time, possibly during the transition from M type to C stars.

1. Introduction

Globular clusters (GCs) are thought to be an ideal test field for the theory of stellar evolution. However, GCs in our Galaxy are all old and their number is

[1]Based on observations with ISO, an ESA project with instruments funded by ESA Member states (especially the PI countries: France, Germany, the Netherlands and the United Kingdom) and with participation of ISAS and NASA.

limited. In contrast, GCs associated with the Magellanic Clouds (MCs) span a wide range of age, which enables us to study the evolution of stars with various masses. Furthermore, GCs are numerous in the MCs and this richness makes them particularly suitable for studying the short-lived late stages of stellar evolution.

Toward the end of the AGB phase, AGB stars are thought to undergo heavy mass loss and to be surrounded by thick circumstellar dust shells. It is, therefore, expected that AGB stars become very red, perhaps not even visible in the near-infrared (NIR). Thus, observations in the mid-infrared (MIR) are likely to be important for the study of the late stages of stellar evolution.

We have performed systematic observations of AGB stars in GCs in the MCs with ISO (Kessler et al. 1996) in the MIR and with a ground-based telescope in the NIR. We here report the results of NIR and MIR observations of intermediate-age clusters, NGC 419, NGC 1783 and NGC 1978, and discuss the evolution of low- to intermediate-mass stars. The ages of the clusters are estimated to be 1–2 Gyr which roughly corresponds to a turn-off-mass of 1.5 to 2 $M_\odot$.

2. Observations

The ISO observations consist of CAM broad band (LW1, LW2, LW10) images and CAM CVF spectra. For 7 infrared bright AGB stars belonging to 3 clusters, we also performed PHOT 25 and 60 μm photometry. CAM data were reduced with the *ISOCAM Interactive Analysis (CIA)* [2] version 2.0 and the photometry was done with *IRAF/DAOPHOT*. PHOT data were reduced with *ISOPHOT Interactive Analysis (PIA)* [3] version 7.1.

NIR observations have been carried out at the South African Astronomical Observatory since 1993. The main results come from J, H and K broad band imaging with the PtSi Astronomical Near-Infrared Camera, (PANIC) (Glass et al. 1994; Glass et al. 1995; Tanabé et al. 1996), attached to the Cassegrain focus of the 0.75 m telescope. We also observed several red AGB stars with the IR photometer at 1.9 m telescope.

3. Results and discussion

In the course of our NIR observations, we found 3 infrared stars (hereafter NIR stars) in these 3 clusters (Tanabé et al. 1997). One of the stars found in NGC 1978 was first discovered by Frogel et al. (1990). Repeated observations show that they are long period variables with $P \sim$ 400–500 days (Nishida et al.). Furthermore, NIR spectroscopic observations show they are all carbon stars.

---

[2]The ISOCAM data presented in this paper was analyzed using "CIA", a joint development by the ESA Astrophysics Division and the ISOCAM Consortium led by the ISOCAM PI, C. Cesarsky, Direction des Sciences de la Matière, C.E.A., France.

[3]PIA is a joint development by the ESA Astrophysics Division and the ISOPHOT Consortium led by the Max Planck Institute for Astronomy (MPIA), Heidelberg. Contributing ISOPHOT Consortium institutes are DIAS, RAL, AIP, MPIK, and MPIA.

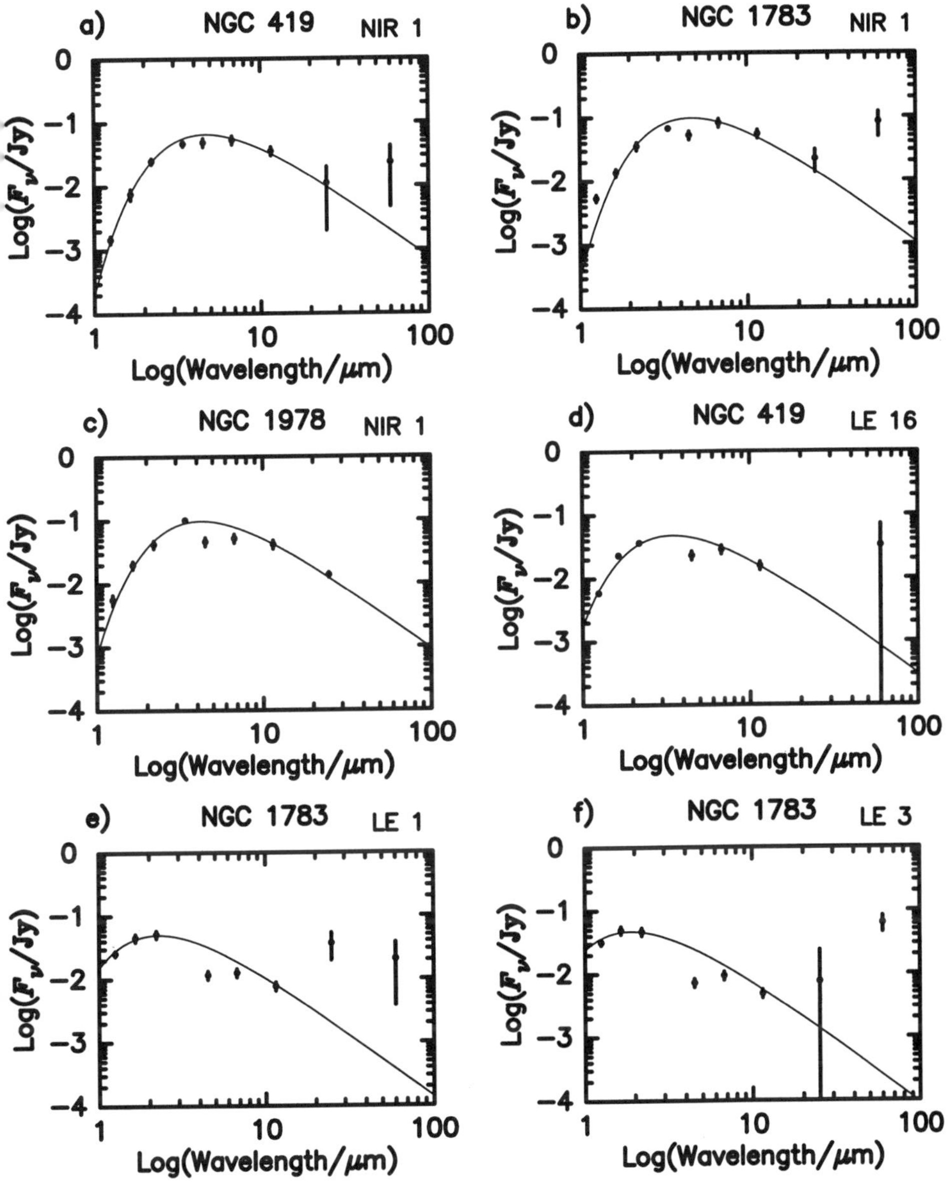

Figure 1. Spectral energy distributions of a) NGC 419 NIR 1, b) NGC 1783 NIR 1, c) NGC 1978 NIR 1, d) NGC 419 LE 16, e) NGC 1783 LE 1, and f) NGC 1783 LE 3. J, H, and K data of NIR stars are the mean values of the lightcurves. Those of the other stars are from Frogel et al. (1990). Each solid curve is a blackbody fit to the photometric data.

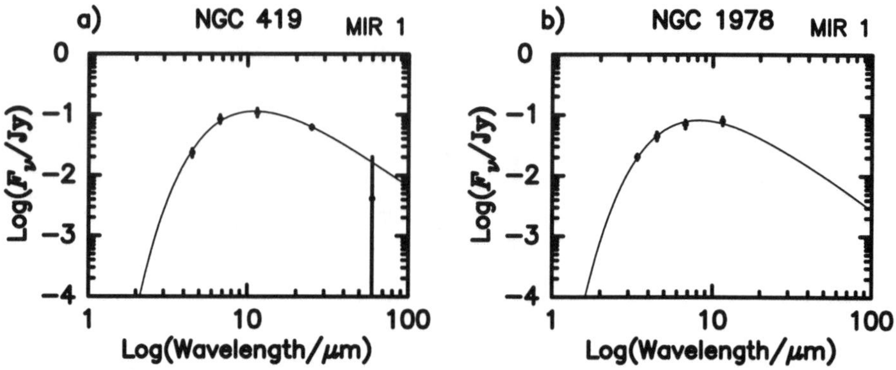

Figure 2. Spectral energy distribution of a) NGC 419 MIR 1 and b) NGC 1978 MIR 1. Each solid curve is a blackbody fit to the photometric data.

ISO PHOT observations of NGC 419 NIR 1 and NGC 1783 NIR 1 revealed 60 μm excesses. We observed 3 other optically identified infrared-bright carbon stars with PHOT and they also show 60 μm excesses. Fig. 1 shows the spectral energy distributions of these stars. We think that this far-infrared excess is indicative of mass loss in the past. This may be evidence that stars in the AGB phase repeat the thermal pulses which are associated with increased mass loss.

ISOCAM observations revealed much redder stars (hereafter MIR stars) in NGC 419 and NGC 1978 (Tanabé et al. 1998). NGC 1978 MIR 1 can be seen in the deep K-band image but NGC 419 MIR 1 has not yet been identified. Fig. 2 shows their spectral energy distributions. From their extreme redness, these stars must be experiencing intense mass loss, possibly caused by superwinds. CAM CVF spectra of these stars show broad emission features around 11.5 μm which are similar to the one observed in carbon stars. CVF spectra also show broad absorptions around 7–8 μm which are probably due to HCN and/or C_2H_2 molecules. Thus, these stars are thought to be carbon stars. However, contrary to the optically-visible and NIR carbon stars mentioned above, NGC 419 MIR 1 does not show any far-infrared excess.

We plot color-magnitude diagrams of AGB stars in these 3 clusters in Fig. 3. Bolometric magnitudes, $M_{\rm bol}$, of NIR and MIR stars in the 3 clusters were estimated by simply fitting blackbody curves to the flux data. We have not included the contributions of far-infrared excesses, if any, in the bolometric magnitudes. $M_{\rm bol}$ of optically-visible stars are taken from Frogel et al. (1990). Here, we have adopted $m - M = 18.5$ for the LMC and 18.9 for the SMC. We can clearly see the sequence from oxygen-rich through carbon stars to NIR stars and these stars constitute direct evidence that AGB stars go through a superwind phase (Renzini 1981). However, it seems that the MIR stars are fainter than the AGB tip luminosity and that their $M_{\rm bol}$s are close to those of the transition luminosity from M type to C stars. It has been established that the luminosity of a star increases toward the end of the AGB phase (see the core-mass luminosity relation of Paczyński 1970). Therefore, these MIR stars may not be in the final

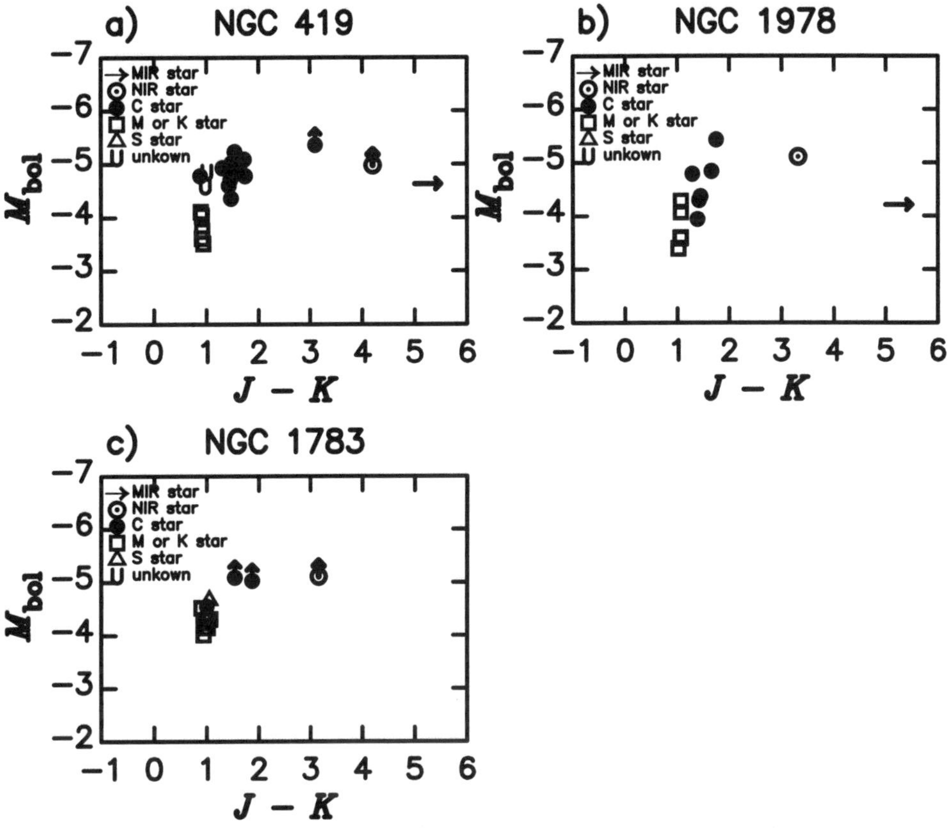

Figure 3. Color magnitude diagram of AGB stars in a) NGC 419, b) NGC 1978, and c) NGC 1783

stage of the AGB phase. This may suggest that AGB stars lose mass heavily at some other time, possibly during the transition from M type to C stars.

However, the conclusion above should await further observations, especially in the NIR. ISOCAM observations together with NIR photometry indicate that many AGB stars, both M type and C stars, have a rather deep absorption around 4.5 μm (see Fig. 1) possibly due to CO molecules and around 7–8 μm, the absorption already mentioned above. Thus, simple blackbody fitting may lead to a lower apparent luminosity. Also, it is possible that these MIR stars are variable and that we observed them at the minimum phase.

4. Summary

1. MIR stars discovered in the intermediate-age clusters NGC 419 and NGC 1978 are very red and must be undergoing intense mass loss (comparable to

superwinds). They are probably carbon stars but do not seem to show any FIR excesses.

2. Three optically visible carbon stars as well as (at least) 2 NIR carbon stars observed with ISO PHOT show 60 μm excesses which may indicate mass loss in the past. Their absolute luminosities are higher than those of MIR stars.

3. The bolometric magnitudes of MIR stars seem to be close to that of the transition luminosity from M type to C stars. This suggests that they are not in the final stage of the AGB.

Acknowledgments. T. T., Y. N. and T. O. are funded in part by a Grant-in-Aid from the Ministry of Education, Science, Sports and Culture in Japan.

References

Cesarsky C.J., Abergel A., Agnèse P., et al., 1996, A&A 315, L32

Frogel J.A., Mould J., Blanco V.M., 1990, ApJ 352, 96

Glass I.S., Carter D.B., Sekiguchi K., Nakada Y., 1994, in Infrared Astronomy with Arrays: The Next Generation, I.S. McLean (ed.), Kluwer Academic Publishers, Dordrecht, p. 285

Glass I.S., Sekiguchi K., Nakada Y., 1995, in New Developments in Array Technology and Applications, A.G. Davies Phillip, K.A. Janes and A.R. Upgren (eds.), Kluwer Academic Publishers, Dordrecht, p. 109

Kessler M.F., Steinz J.A., Anderegg M.E., et al., 1996, A&A 315, L27

Nishida S., Tanabé T., Nakada Y., et al., in preparation

Paczyński B., 1970, Acta Astron. 20, 47

Renzini A., 1981, in Physical Processes in Red Giants, I. Iben, Jr., and A. Renzini (eds.), Reidel, Dordrecht, p. 431

Tanabé T., Nishida S., Nakada Y., Matsumoto S., Onaka T., Sekiguchi K., Ono T., Glass I.S., Carter D.B., 1996, SPIE 2744, 110

Tanabé T., Nishida S., Matsumoto S., Onaka T., Nakada Y., Soyano T., Ono T., Sekiguchi K., Glass I.S., 1997, Nature 385, 509

Tanabé T., Nishida S., Nakada Y., Onaka T., Glass I.S., Sauvage M., 1998, Ap&SS 255, 407

Asymptotic Giant Branch Stars
T. Le Bertre, A. Lèbre and C. Waelkens, eds.
I.A.U. Symposium 191, 1999

A critical look at the role of AGB stars in stellar population synthesis

A. Lançon

Observatoire astronomique, Université L. Pasteur & CNRS, UMR 7550, 11 rue de l'Université, 67 000 Strasbourg, France

Abstract. Asymptotic giant branch stars are essential contributors to the near and mid-IR emission of intermediate age (10^8-10^9 yr old) stellar populations. Detecting this light will set constraints on the star formation history in galaxies and, conversely, the search for AGB signatures in well studied populations will help us reduce some of the still large uncertainties in AGB models. This paper reviews how AGB stars are currently included in population synthesis models and which spectral features can be used to identify their emission in galaxy light; targets for observational tests are suggested, and some observational and theoretical difficulties are discussed.

1. Introduction

Asymptotic Giant Branch stars (AGB stars) are rare members of stellar populations. However, they are among the most luminous cool stars and can therefore be detected sometimes even individually in external galaxies; they are also the coolest of the luminous stars and hence produce spectral signatures that we may search for in integrated near/mid-IR spectra. Finally, the carbon we are made of and the pre-solar meteorites found on Earth remind us of the role played by AGB stars in the evolution of galaxy matter.

Population synthesis techniques predict the evolution of stellar distributions in the HR diagram and the corresponding integrated emission properties. Nowadays they also keep track of the chemical evolution.

In this review, we will focus on the spectrophotometric aspect of population synthesis, at solar and LMC metallicities. We will discuss how we can predict the contribution of AGB stars to the integrated spectrum and how we can recognise these contributions in observed spectra. Much of this work has become feasible only recently thanks to advances in AGB theory and to more complete samples of AGB observations. We will then suggest extragalactic targets and discuss what the detection – or not – of AGB signatures can tell us about the galaxies or alternatively about AGB evolution. The relevant observations are becoming feasible now and are mostly work of the future.

2. The contribution of AGB stars to the bolometric luminosity

Predicting the contribution f_{bol} of AGB stars to the bolometric luminosity as a function of time for an evolving stellar population (SP) "only" requires a set of stellar evolution tracks. Such tracks used to be constructed semi-empirically (e.g. Charlot & Bruzual 1991) and are now provided by so-called synthetic AGB evolution models (Groenewegen & de Jong 1993; Marigo, this volume). The effects of thermal pulses are taken into account in these calculations, but the product used as input in population synthesis calculations is a set of smooth tracks that needs to be well understood as a representation of either the inter-pulse properties or some mean properties, e.g. the energy weighted averages over helium shell flash cycles. The precise definition determines further steps in the synthesis code. In the first case for instance, corrections should be applied to account for the post-flash luminosity dip.[1]

The critical parameter in setting f_{bol} is the mass loss $\dot{M}$, which determines AGB lifetimes and the range of initial masses for which the AGB phase occurs. Other parameters are implicit in $\dot{M}$: e.g., metallicity influences grain formation and the efficiency of radiation pressure, the mixing parameters and gas opacities influence the extension of the stellar envelopes and the Mira-type pulsation properties, that contribute to starting the winds. Available tracks usually differ in more than one of these parameters, which makes it difficult to interpret differences in lifetimes and energy distributions; at this stage, the scatter among the lifetimes found in the literature for the thermally pulsing AGB (TP-AGB) is about a factor of 2 and must be treated as an uncertainty.

The predictions reported in this paper are based on the population synthesis code of Fioc & Rocca-Volmerange (hereafter FRV; Z=0.02: 1997, other Z: 1998). A Salpeter IMF with a lower mass limit at $0.1\,M_\odot$ is adopted in all calculations. The TP-AGB evolution is included in the code following the prescriptions of Groenewegen et al. (1993, 1995); the resulting tracks follow those of Vassiliadis & Wood (1993) closely in the HR diagram. The FRV lifetimes are always close to the shortest ones found in the literature, therefore we do not expect to overestimate the effect of the AGB. A more complete study will be reported elsewhere.

After an instantaneous burst of star formation (IB), we find that the contribution f_{bol} due solely to the TP-AGB stars reaches maximal values ($\gtrsim 10\,\%$) between **0.3 and ~1.5 Gyr**, corresponding to AGB progenitor masses between 1.7 and $3.5\,M_\odot$. At 1.5 Gyr the complete AGB (E-AGB+TP-AGB) contributes about a quarter of the light; values of up to 40 % have been reported (Charlot & Bruzual 1993). At later stages, the light of the TP-AGB stars is progressively lost in the increasing contribution of E-AGB and first giant branch stars.

AGB lifetimes can in principle be tested with integrated V-K colours since the AGB contribution is essentially emitted at near-IR wavelengths. The TP-AGB by itself is responsible for more than 40 % (and up to 60 %) of the K band

[1] Synthetic evolution was first introduced as a means of simplifying population synthesis calculations; we note however that synthetic models become more and more complex, and that a return to the direct use of the output of full evolutionary calculations may become attractive again.

light **0.1 to 1 Gyr** after an IB. At these ages, the red supergiants responsible for very red colours earlier on have died. We refer to Charlot (1996) for an interesting model allowing comparison of colour and AGB contribution predictions.

Girardi & Bertelli (1998) show how changes in $\dot{M}$ affect V-K; in particular, the increased $\dot{M}$ and hence shorter lifetimes resulting from envelope burning in massive AGB stars may reduce V-K by up to 1 magnitude at the earliest AGB-dominated times. Some ambiguities and practical difficulties faced when performing tests based on colours are discussed in Sect. 3.2. and 4.

3. The effect of AGB stars on the near and mid-IR spectrum

3.1. Constructing a suitable stellar library

In order to predict the distribution of AGB light along the IR spectrum and to determine how AGB stars can be recognised among other cool contributors (red supergiants and first giant branch stars), stellar spectroscopic libraries are needed. Commonly used near-IR libraries include the empirical ones of Kleinmann & Hall (1986), Terndrup et al. (1991), Lançon & Rocca-Volmerange (1992), the theoretical one of Bessell et al. (1991) and the semi-empirical compilation of Lejeune et al. (1998). None of these considers the pulsating long period variables (LPVs) explicitely, despite the fact that pulsation is known to modify the spectral type – temperature – colour relations and to increase molecular absorption features (Johnson & Mendez 1970; Bessell et al. 1989), and that basically all the coolest AGB stars are LPVs. As a consequence, near-IR stellar energy distributions (SEDs) of galaxies have been interpreted in terms of a supergiant and giant dichotomy (mainly on the basis of CO absorption at 2.3 μm and of the Ca triplet at 850 nm), but no segregation among cool giant populations has been possible.

Lançon & Wood (1997, hereafter LW97) have obtained 0.5-2.5 μm spectra for ~100 cool stars, including instantaneous spectra of a sample of O-rich and C-rich LPVs observed at various phases. The observations confirm that only LPVs are able to produce the deepest H_2O bands around 1.4 and 1.9 μm (Mouhcine & Lançon 1998). Other characteristic near-IR molecular bands are those of VO (1.05 μm) and TiO (1.25 μm, in the latest spectral types), or of CN (1.1, 1.4 μm) and C_2 (1.77 μm) in C-stars.

Temperatures were assigned to individual Mira-type spectra of LW97 in two ways. On one hand, R. Alvarez compared the energy distribution (near-IR H_2O bands excluded) with the static giant models of Plez (model version of Alvarez & Plez 1998). On the other, colour-temperature relations based on angular diameter measurements were used (Feast et al. 1996). Below 3000 K, the first temperatures are systematically higher than the second, the difference increasing up to ~400 K around 2000 K. The LPV atmosphere models of Bessell et al. 1996 also indicate low temperatures (in the rather rare cases where the fit is satisfactory). The two latter methods are not independent: angular diameter interpretations require model atmospheres. In view of the difficult reconciliation between large observed radii and fundamental mode pulsation (Wood et al., this volume), there may be systematic errors related to the use of current AGB

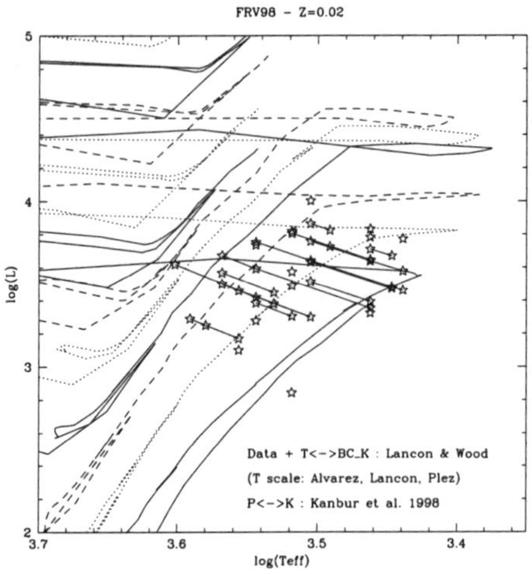

Figure 1. Plotted over the evolution tracks of FRV (1998) are the locations of observed spectra from LW97. T_{eff} is estimated according to the warmer scale described in the text; luminosities are derived from a P-L_K relation and the K band bolometric corrections are measured on the instantaneous spectra themselves (K band variability is neglected). Lines connect observations of single stars at various phases.

atmosphere models and the second temperature estimates may be too cool. Therefore, the results presented here are based on an intermediate scale.

The variations of individual stars are illustrated in Fig. 1. Averaging the spectra star by star is not a practical way of obtaining a library for population synthesis: it requires complete phase coverage for each star as well as an estimate of its fundamental parameters (initial mass, metallicity, evolutionary status). The latter are so uncertain that the resulting mean spectra are difficult to order. First attempts with the LW97 library have shown that the expected global spectral evolution from higher to lower temperature is lost in object to object variations. It is more convenient to group the spectra in instantaneous temperature bins. Although the dispersion in the spectral properties inside each bin remains to be explained (Mouhcine & Lançon, in preparation), this method provides a smoothly evolving sequence.

For population synthesis purposes, the LW97 spectra were combined with the library of Lejeune et al. (1998), as implemented by FRV (1998).

3.2. Results

As mentioned above, the AGB spectra that we are able to distinguish from those of older giants are those of *pulsating* stars.

Fig. 2 shows the predicted behaviour of colours and molecular indices at solar metallicity, with the assumption that the TP-AGB and LPV stages coincide and that LPVs remain visible and O-rich along most of the TP-AGB. The

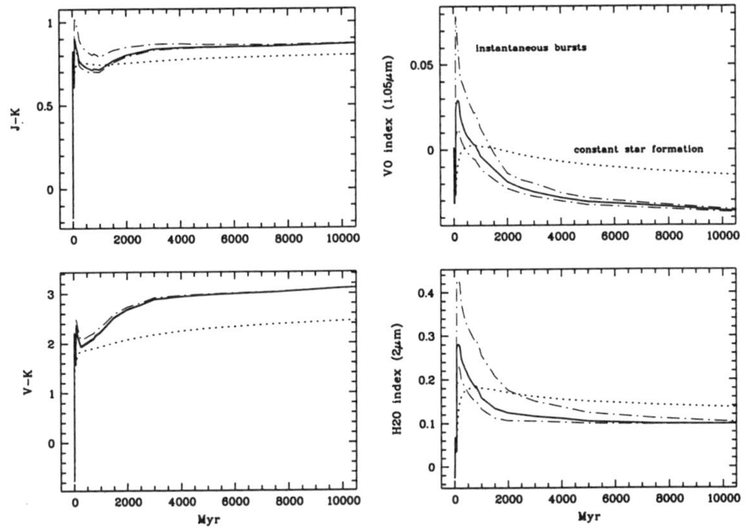

Figure 2. Evolution of selected properties with time. The dot-dashed lines illustrate the burst evolution for the two extreme spectral temperature scales discussed in the text. Molecular indices are in magnitudes.

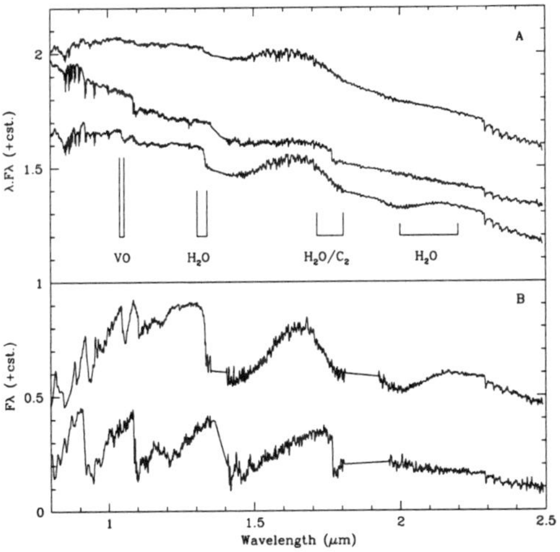

Figure 3. A: selected synthetic integrated spectra. From top to bottom: a static giant dominated 4 Gyr old population, and two 200 Myr old postbursts, when most TP-AGB stars are respectively C-rich and O-rich. Filter centers for useful molecular indices are indicated. B: typical cool O-rich and C-rich LPV spectra.

blue post-starburst colours are due to the inevitable contamination by luminous intermediate mass blue main sequence stars, coeval with the AGB population.

Several parameters are degenerate in studies based on colours alone: extinction, metallicity and age all redden the integrated spectra, continuous star formation keeps them blue over long times. As an example, the ratio of an IB spectrum at 500 Myr to the spectrum of a constantly star forming population at 2 Gyr is essentially flat between 0.5 and 2.5 μm. Spectral features must be observed to distinguish the two. As we know from previous models and from observations of star clusters, the Balmer and Paschen series of numerous A type stars characterise the younger of the two populations below 1 μm. When LPV spectra are included in the models, H_2O bands appear in the near-IR part of the ratio as broad absorption features at the same relative level as the Hα line (see also Fig. 3 A).

The right hand side of Fig. 2 demonstrates that the molecular bands of H_2O and of VO (1.05 μm) unambiguously characterise "postburst populations" 0.1 to 1 Gyr after an IB, with a peak at 150 Myr. The maximal numerical values of the molecular indices however depend strongly on the adopted T_{eff} scale. In the case of a constant star formation rate, only the warmer T_{eff} scale (which leads to the use of spectra with deeper features at a given point of an isochrone) leads to a significant increase of the molecular bands, when compared to the predictions of FRV (1998) without the specific LPV library.

These results remain qualitatively valid in LMC metallicity models.

3.3. The effects of various model assumptions

3.3.1 – Does the TP-AGB coincide with the instability strip for LPV pulsation? Period-luminosity (P-L) diagrams of observed Miras suggest that the instability strip is narrower. If we transport the blue edge of the strip into the population synthesis code using the prescription of Vassiliadis & Wood (1993), we find that the effect on the colours is negligible (LPV spectra are replaced by static giant spectra with very similar energy distributions) but that the depth of the H_2O features is reduced during the 2 Gyr following an IB (by ~ 0.03 mag for the index in Fig. 2). VO is not affected: it is produced by the cooler Miras.

In fact, recent data supports a more complex picture (Alvarez et al. 1997; Barthès 1998; Wood et al., this volume; Kiss & Szatmáry, this volume), that also shows through in theory (Fox & Wood 1982; Xiong et al. 1998; Tuchman, this volume). Pulsation may be intermittent, depending on phase in the thermal pulse cycle; but globally, AGB stars start varying as small amplitude high overtone pulsators and successively (with overlaps in time) become unstable to lower mode, higher amplitude and longer period pulsation. The Miras of the empirical P-L relations are only the largest optical amplitude pulsators. Since we have not excluded low amplitude LPVs from our spectral temperature bins, these bins represent a mixture of pulsation states that may reasonably approximate reality in models based on coinciding TP-AGB and LPV stages.

At the red edge of the LPV instability strip, stars become invisible even in the near-IR due to ejected material, and instead emit longwards of 3 μm. Bressan et al. (1998) studied the consequences on integrated broad band colours: during the AGB-dominated post-starburst stages, the optical/near-IR spectrum becomes bluer due to the missing late AGB contributions, while the K-λ colours with $\lambda > 5\,\mu$m increase dramatically due to OH/IR type infrared sources. At LMC metallicities, where the red edge is best defined, we find that V-K changes by a few percent and the molecular features are hardly affected at all. The effect could be stronger at higher metallicities.

3.3.2 – Some AGB stars will end their lives as C stars, with completely modified spectra. An exploratory study based on the relative C-rich TP-AGB lifetimes of Groenewegen & de Jong (1993), at the metallicity of the LMC, shows that this totally removes any oxide features apart from CO for most of the duration of the AGB-dominated post-starburst. Fortunately, the C_2 bandhead at $1.77\,\mu$m then replaces the H_2O features, and a narrow band photometric index can be defined that measures the time dependent AGB contribution, whether the AGB stars be O-rich or C-rich (Fig. 3; Lançon et al. 1998).

4. Observational perspectives: where, how and why?

4.1. Observational strategies

The presence of pulsating AGB stars modifies the near-IR spectral signatures and the mid-IR colours during several Gyr after an episode of enhanced star formation and more permanently if star formation continues. However, at the time of this conference the uncertainties on fundamental stellar inputs such as the Mira temperature scale (necessary to connect suitably averaged stellar spectra to evolutionary tracks), the mass-dependent TP-AGB and the LPV lifetimes, the O-star to C-star transition, limit the number of *robust* predictions directly applicable to extragalactic studies to a few: we expect unambiguous AGB signatures 0.1 to 1 Gyr after a starburst, and we know we may search them using combined near and mid-IR colours (Bressan et al. 1998) or near-IR molecular features (Lançon et al. 1998). Other diagnostic features will certainly appear as soon as the astronomical community will have digested the ISO data. It is important that the most robust predictions be tested observationally before more difficult data interpretations (mixed populations with a complex star formation history) be undertaken.

Significant populations with appropriate ages exist in star clusters of Local Group galaxies, in the bodies of some of these galaxies, in E+A galaxies and in post-starburst galaxies.

Star clusters, in particular the LMC age sequence, may look like ideal targets but each of them contains only a handful of luminous AGB stars. The integrated colours are significantly affected by the stochastic fluctuations in the AGB numbers (Ferraro et al. 1995; Santos & Frogel 1997; Marigo et al. 1996). Assuming Poisson statistics for these small numbers, we can write the relative, time and wavelength dependent luminosity fluctuations due to TP-AGB stars analytically as:

$$\frac{\sigma_L}{L} = \frac{1}{\sqrt{N}} . \frac{\sqrt{\alpha}\,(L_a - L_r)}{\alpha\, L_a + (1-\alpha)\, L_r}$$

Here, N is the total number of stars in the population, α is the (small) proportion of TP-AGB stars, $L_a(\lambda, t)$ and $L_r(\lambda, t)$ are respectively the average emissions per TP-AGB or non–TP-AGB star and are both obtained from population synthesis calculations. With the above mentioned IMF, more than 10^6 stars are required in order to keep these fluctuations below 10 % at K a few 100 Myr after an IB. They will then be 10 times smaller at 0.7 μm or in terms of bolometric luminosities, and become negligible at K within a Gyr.

Nevertheless, now that the AGB numbers in individual Local Group star clusters are known and their effect on the colours has been demonstrated it remains interesting to test whether their near-IR spectroscopic features show through. This requires either several slit positions per cluster, or wide field or imaging spectroscopy.

Larger AGB numbers are present in the bodies of selected nearby dwarf or E+A galaxies, but they are superimposed on older populations and their spectral signatures will be severely diluted. If we adopt the maximal value reached by a molecular absorption index in a constant star formation scenario as the threshold for the unambiguous detection of AGB contributions (Fig. 2), we find that postburst populations can be identified as long as they represent more than ~10 % of the total stellar mass in the observed area (Lançon et al. 1998). This result is relatively independent of the SF history of the supposedly old underlying population.

Counts should be used preferentially in the studies of Local Group galaxies (e.g. Gallart 1998), especially since integrated near-IR spectra are still difficult to obtain for extended objects that do not have the high surface brightnesses of currently star forming regions. One of the main observational difficulties is crowding (Martinez-Delgado & Aparicio 1997). It may also be possible to study AGB contributions with extended "pixel fluctuation methods" (as used in microlens searches, e.g. AGAPE, Ansari et al. 1997)[2], when pixels integrate the light of less than 10^6 stars.

It follows from the above arguments that the ideal targets for first tests should be *both massive and localised, 0.1-1 Gyr old postburst populations.* A few such objects have recently been identified: new clusters with estimated stellar masses of up to 10^7-$10^8\,M_\odot$ form in interacting galaxies and can apparently survive as bound objects for more than the 0.1 Gyr required for AGB stars to appear. Clusters with appropriate estimated ages are listed for NGC 7252, NGC 1275 and NGC 3921 (Miller et al. 1997 and references therein)[3]. Neither dilution nor statistical fluctuations are expected to represent serious hurdles there. Many of the more extended (but still bound) post-starburst areas predicted by the dynamical arguments of Kroupa (1998) would also satisfy the detection criteria.

Before closing this section, we wish to mention observational difficulties. The IR signatures of interest are located near the strong telluric absorption bands of H_2O and CO_2, calling for particularly frequent observations of a calibration star during ground-based observations. High signal-to-noise ratios per (low resolution) spectral element are required to detect molecular absorption bands. On the other hand, dust emission associated with ongoing star formation may hinder mid-IR searches for OH/IR emission in places. Spatial resolution must always be sought to limit dilution.

---

[2] An idea first mentioned to me by A.L. Melchior

[3] A proposal has been submitted to perform a near-IR search for the predicted AGB signatures

4.2. Perspectives

What can we learn about galaxies?

Once the first tests on populations with known ages will have confirmed the predictions regarding the IR signatures of AGB-rich populations, these observations will be useful in the study of the star formation history of galaxies, the propagation of star formation, the survival of starburst clusters and the potential precursors of globular clusters or dwarf galaxies. Recent star formation is often associated with extinction, in which case optical (e.g. Balmer line) studies present incomplete pictures. Post-starburst knots may be found that coexist with current starbursts in dusty, infrared lumionous objects like NGC 253, IC 342.

In more evolved objects, late stellar evolution paradoxally helps us understand the UV emission: UV-excesses, previously interpreted as signatures of very recent star formation, are now explained by relatively old AGB-manqués or Hot Horizontal Branch objects, closely related to the stars discussed in this conference (Bressan et al. 1994; Brown et al. 1995; FRV 1998).

Finally, as suggested by Bressan et al. (1998), AGB stars may be used to break the well-known age-metallicity degeneracy of galaxy colours. In brief, the red giant population of galaxies increases with time and with metallicity, leading to similarly red integrated colours in both cases; but while the mid-IR emission of OH/IR stars in galaxies is predicted to increase with metal abundance, it decreases with age. A similar argument holds for the near-IR molecular TP-AGB signatures, although the predicted effect is small.

What can we learn about the AGB?

The tests discussed in Sect. 4.1 will provide constraints on AGB lifetimes (i.e. on stellar evolution models), on temperature scales (i.e. on pulsating atmosphere models) and on the O- to C-star transition (i.e. on dredge-up). All these parameters badly need to be better determined.

The unique access to distant populations provided by integrated spectroscopic observations will then allow us to explore AGB evolution in more extreme environments. For instance, it has been claimed that stars with ages of ~ 0.1 Gyr dominate in the emission of the companion of quasar PG 1700+518 at redshift 0.29 (Canalizo & Stockton 1997). Are AGB stars numerous there? Are they carbon-rich or oxygen-rich? While more sensitive infrared instruments are being constructed for high spatial resolution telescopes, we have time to exploit Local Universe data and to improve upon existing AGB model predictions.

Acknowledgments. It is a pleasure to acknowledge the active collaboration of R. Alvarez, M. Fioc, M.A.T. Groenewegen, M. Mouhcine, B. Rocca-Volmerange, M. Scholz, D.R. Silva and P.R. Wood in making this review possible.

References

Alvarez R., Mennessier M.-O., Barthès D., et al., 1997, A&A 327, 656

Alvarez R., Plez B., 1998, A&A 330, 1109

Ansari R., Aurière M., Baillon P., et al., 1997, A&A 324, 843

Barthès D., 1998, A&A 333, 647

Bessell M.S., Brett J.M, Scholz M., Wood P.R., 1989, A&A 213, 209

Bessell M.S., Brett J.M., Scholz M., Wood P.R., 1991, A&AS 89, 335

Bessell M.S., Scholz M., Wood P.R., 1996, A&A 307, 481

Bressan A., Chiosi C., Fagotto F., 1994, ApJS 94, 63

Bressan A., Granato G.L., Silva L., 1997, A&A 332, 135

Brown T., Ferguson H., Davidsen A., 1995, ApJ 454, L15

Bruzual G.A., 1983, ApJ 273, 105

Canalizo G., Stockton A., 1997, ApJ 480, L5

Charlot S., Bruzual G.A., 1991, ApJ 367, 26

Charlot S., 1996, PASP Conf. Ser. 98, 275

Feast M.W., 1996, MNRAS 278, 11

Ferraro F.R., Fusi Pecci F., Testa V., et al., 1995, MNRAS 272, 391

Fioc M., Rocca-Volmerange B., 1997, A&A 326, 950 (FRV 1997)

Fioc M., Rocca-Volmerange B., 1998, in preparation (FRV 1998; cf. also Fioc M., 1997, PhD thesis, Univ. Paris XI)

Fox M.W., Wood P.R., 1982, ApJ 259, 198

Gallart C., 1998, ApJ 495, L43

Girardi L., Bertelli G., 1998, MNRAS 300, 533

Groenewegen M.A.T., de Jong T., 1993, A&A 267, 410

Groenewegen M.A.T., van den Hoek L.B., de Jong T., 1995, A&A 293, 463

Johnson H.J., Méndez M.E., 1970, AJ 75, 785

Kanbur S.M., Hendry M.A., Clarke D., 1998, in *A Half Century of Stellar Pulsation: A Tribute to A.N. Cox*, eds. P.A. Bradley & J.A. Guzik, Los Alamos

Kleinmann S.G., Hall D.N.B., 1989, ApJS 62, 501

Kroupa P., 1998, MNRAS 300, 200

Lançon A., Rocca-Volmerange B., 1992, A&AS 96, 593

Lançon A., Wood P.R., 1997, IAU Symp. 189 on *Fundamental Stellar Properties: the Interaction between Observations and Theory*, Poster proceedings, ed. T.R. Bedding, Publ.: School of Physics, Univ. of Sydney (`http://astro.u-strasbg.fr/~lancon`)

Lançon A., Mouhcine M., Fioc M., 1998, in preparation

Lejeune T., Cuisinier F., Buser R., 1998, A&AS 130, 65

Marigo P., Girardi L., Chiosi C., 1996, A&A 316, 1

Martinez-Delgado D., Aparicio A., 1997, ApJ 480, 107

Miller B.W., Whitmore B.C., Schweizer F., Fall S.M., 1997, AJ 114, 2381

Mouhcine M., Lançon A., 1998, in *Unsolved Problems in Stellar Evolution*, Poster proceedings, ed. M. Livio, Space Telescope Sc. Inst.

Santos J.F.C., Jr., Frogel J.A., 1997, ApJ 479, 764

Terndrup D.M., Frogel J.A., Whitford A.E., 1991, ApJ 378, 742

Vassiliadis E., Wood P.R., 1993, ApJ 413, 641

Xiong D.R., Cheng Q.L., Deng L., 1998, ApJ 500, 449

Part 7. Summary

The meteoritic inclusions in the AGB community, FANTASTIC !,
(Sachiko Amari)

Asymptotic Giant Branch Stars
T. Le Bertre, A. Lèbre and C. Waelkens, eds.
I.A.U. Symposium 191, 1999

AGB Theory — A Retrospective

Icko Iben, Jr.

University of Illinois, Astronomy Department, 1002 W. Green St., Urbana, IL 61801, USA

1. Introduction

My assignment was to listen to all of the talks at the symposium and to discover and summarize results that have appeared since the last conference on AGB stars. Since the entire meeting was devoted to new results, discovery was a simple task. Because of the large volume of new results, an adequate summary is impossible. I therefore do not attempt to provide a comprehensive survey, but confine my remarks to a small subset of the discoveries and advances which have enriched and modified our understanding of AGB stars over the past decade.

2. Interrelationships

One of the most delightful aspects of a conference such as this, at which experts from many different specialities interact, is to see unexpected interrelationships exposed. One which particularly intrigued me was the formation of ^{26}Al at the base of the convective envelope of relatively massive AGB models, as described by Lattanzio & Forestini, and the use which Glassgold made of the positrons from the decay of ^{26}Al in the circumstellar matter at 10^{17} cm from the star to convert molecular hydrogen and the molecule CO into the radical HCO^+. The abundance of ^{26}Al expected in matter ejected in the wind of the real analog of the AGB model appears to be sufficient to account for the abundance of HCO^+ inferred from spectroscopy and a simple spherically symmetric circumstellar shell formed by the wind.

And this is just the beginning! According to Zinner & Amari, it appears that, in meteoritic inclusions in which the ^{12}C/^{13}C ratio is similar to that found in AGB model envelopes, the ^{26}Mg/^{27}Al ratio is about the same as the ^{26}Al/^{27}Al ratio in the model envelopes.

These coincidences, if true, show remarkable correspondences between model predictions and observations. They imply further that isotopic abundances in some chunks of matter formed in stellar interiors and injected into a stellar wind manage to survive the arduous trip through space to the protosolar nebula and to survive further processing in this nebula. FANTASTIC!

Of course, the connection between nucleosynthesis in AGB stars and abundances in meteoritic inclusions extends far beyond this example. The beautiful work by Gallino and his coworkers on s-process nucleosynthesis has been matched by equally beautiful detective work by Amari, Anders, Lewis, Zinner, and others on inclusion compositions.

3. Luminosity core-mass relationships

In several talks, mention has been made of the "failure" of the luminosity core-mass relationship. This is misleading. First of all, since, at any given core mass, thermally pulsing AGB (TPAGB) stars vary quasi periodically in luminosity over the pulse cycle (by as much as a magnitude in low mass models), there is, strictly speaking, no such thing as a time-independent luminosity core-mass relationship.

On the basis of models by himself and by Uus (1970) in which thermal pulses were suppressed, Paczyński (1970) constructed a luminosity core-mass relationship, $L \sim 60\,000\,(M_{\rm core}/M_\odot - 0.5)\ L_\odot$, which is actually a very good approximation to the maximum luminosity of low mass thermally pulsing models near the end of the quiescent hydrogen shell burning phase, just prior to the helium shell flash which marks the onset of a thermal pulse. The Paczyński-Uus (PU) relationship has played an important role in the development of our understanding of real TPAGB stars by providing a useful approximation to one portion of the real time-varying relationship for low mass models. It also describes well the luminosities of the central star of a planetary nebula during the plateau phase. The relationship has on occasion been misused, when it has been forgotten that it describes only a particular point in the thermal pulse cycle.

As can be seen from several of Uus' models and those of subsequent researchers, the PU relationship underestimates the luminosity maximum when the base of the convective envelope extends into the hydrogen-burning region. This is particularly true of relatively massive TPAGB models, the more so the larger the mixing-length to scale-height parameter l/H. This is because, for l/H larger than some critical value, a considerable fraction of the luminosity of the model during the hydrogen-burning phase is produced within the convective envelope; the larger l/H, the larger is the luminosity of the model relative to that given by the PU relationship. However, this does *not* mean that there is not a luminosity core-mass relationship analogous to the PU relationship. When sufficiently many pulses are followed (typically ~ 30 pulses are enough), the pulse amplitude reaches a locally asymptotic value, and, for every given choice of l/H, one can construct a perfectly valid luminosity core-mass relationship in the spirit of the PU relationship (see, e.g., Iben 1977). Further, the more massive the core, the larger is the fraction of the thermal pulse cycle that this relationship faithfully describes. Thus, use of the phrase "failure of *the* luminosity core-mass relationship" should be avoided.

4. Nucleosynthesis, dredge-up, and carbon stars

Nice discussions of the third dredge-up process and its consequences for carbon-star formation were given by Lattanzio & Forestini and Mowlavi. It was, however, somewhat amusing to learn that "dredge-up cannot occur in models when the Schwarzschild criterion for convection is employed" while remembering that this criterion was in fact employed in the models used to discover the third dredge-up process in the first place (Iben 1974, 1975, 1976)!

Blöcker and Lattanzio & Forestini described element and isotope nucleosynthesis in the convective envelopes of AGB models (e.g., ^{7}Li, ^{13}C, ^{17}O, ^{18}O, ^{26}Al) and debated the relative effectiveness of hot bottom burning and dilution in the convective envelope in accounting for the apparent paucity of very luminous carbon stars. It was pointed out that luminous ("dust-enshrouded") carbon stars, although rare, do exist (Kastner et al. 1993), and that this can be explained by the fact that the temperature at the base of the convective envelope of a mass-losing model at first increases, but then decreases as the mass of the hydrogen-rich envelope decreases below some critical value.

Not highlighted at the symposium was the very important discovery by Straniero et al. (1995) that, during the quiescent hydrogen-burning phase, ^{13}C existing in the radiative helium zone between the CO or ONe core and the hydrogen-burning shell actually experiences the ^{13}C(α,n)^{16}O reaction and serves as a neutron source for s-process nucleosynthesis prior to the formation of the convective shell during a helium shell flash. Blöcker described the equally important discovery (Herwig et al. 1997) that overshoot at the base of the convective envelope during the third dredge-up mixes hydrogen and carbon together in such a way that a pocket of ^{13}C is formed by hydrogen burning after dredge-up, and this provides the first step in the now standard way of producing s-process isotopes in low mass, TPAGB stars of population I (Gallino et al. 1998).

Herwig et al. pointed out that, if overshoot occurs at the base of the convective envelope, it should also occur at the base of the convective shell during the helium shell flash. This allows carbon and oxygen to diffuse into the convective shell, decreasing the abundance by mass of ^{4}He and raising the abundances of both ^{12}C and ^{16}O above the abundances by mass ($X_4 \sim 0.75^+$, $X_{12} \sim 0.20^+$, $X_{16} \sim$ a few $\times$ 0.01) found when overshoot is neglected. This might help in understanding the abundances ($X_{12} \sim 0.5$, $X_4 \sim 0.35$, $X_{16} \sim 0.17$) typical of PG1159 stars (Werner et al., this volume), if they are made according to the born-again scenario (Iben et al. 1983; Iben & MacDonald 1995). It is reasonable to call this additional mixing process at the base of the convective shell the "fourth" dredge-up process.

5. Pulsation mode

The debate about the dominant radial pulsation mode of Mira variables has gone on for at least three decades and, judging from the presentations by Wood et al., Tuchman, and Feast, it promises to continue. During the seventies, Lee Ann Willson consistently and forcefully defended the fundamental mode and Peter Wood consistently favored the first overtone. At some point Peter experienced a conversion to the fundamental, and he presented observational evidence here to support this conversion. Tuchman described results of theoretical hydrodynamical calculations which suggest that, for model parameters chosen to fit those of Mira variables, the full-amplitude mode surviving after many pulsation periods tends to be the fundamental. However, the first overtone still has its champions, and Michael Feast made the case for the first overtone, using stellar angular diameters measured by interferometry and/or lunar occultations coupled with distances determined by HIPPARCOS to estimate stellar radii.

A possible resolution of the discrepancy in its present incarnation may lie in a fuller understanding of what "radius" the different types of observation are measuring, as compared to the theoretical radius R used in the determination of Q in the equation $P = Q\,R^{3/2}/M^{1/2}$. A large Q implies pulsation in the fundamental mode; a small Q implies pulsation in the first overtone. The theoretical pulsation equation assumes that R is the radius at optical depth 2/3, and this corresponds to the value of R obtained from $\sigma T_e^4 = L/4\pi\,R^2$, where T_e is estimated photometrically. On the other hand, the angular diameter measurements may estimate something larger than R at optical depth 2/3, for the same reason that, when viewing a thin spherical shell which is optically thin in the radial direction, one does not see a disk, but rather a ring which has a larger diameter than the inner diameter of the ring (optical images of the Ring Nebula are classical examples of this effect). A smaller R implies a larger Q, which moves in the direction of favoring pulsation in the fundamental mode.

6. Winds — Pulsations, shocks, dust, radiation pressure, etc.

The basic ingredients of the currently most popular theory of mass outflow from AGB stars were established by the early eighties. Acoustical pulsation calculations by Wood (1976, 1979), Hill & Willson (1979), and Willson & Hill (1979) demonstrated that, although shocks formed during pulsations can by themselves drive matter from the surface, they can not do so at a sufficiently high rate to account for the observations. Some other mechanism such as radiation pressure on grains formed in the atmosphere must assist in the mass-loss process. That dust grains form in atmospheres of cool giants was known from the observations (Woolf & Ney 1969; Gehrz & Woolf 1971), and the theory of grain nucleation and growth was evolving (e.g., Draine 1981). Bowen (1988) put all of the ingredients together to produce the first self-consistent theoretical paradigm for AGB winds: shocks due to large amplitude acoustical pulsations create a dynamic, inflated atmosphere in which grains can form and grow, radiation pressure drives the grains outward, and the grains drag the gas along.

At this meeting, Fleischer et al. and Höfner described results of extensive hydro-dynamical calculations which have developed the paradigm to a high degree of sophistication. For example, in the case of carbon-rich atmospheres, the interplay of shocks, dust formation, dust evaporation, radiative transfer and absorption, and grain acceleration can lead to multimode, asymmetric light curves, even when the "piston" which replaces the acoustical pulsations in the real AGB analog is driven with a perfectly symmetric sinusoidal velocity (Winters et al. 1994).

Crucial to the further development of the paradigm is a theory of grain nucleation and growth consistent with results of laboratory experiments. Thanks to the importance of carbon-rich molecules in the fuel industry, an extensive experimental foundation for understanding nucleation and grain growth is available for situations in which C/O > 1 (Henning, this volume). Unfortunately, the classical nucleation theory which has been used extensively in the context of AGB atmospheres apparently fails completely to explain the results of the laboratory experiments. An adequate laboratory foundation for a theory of grain nucleation and growth when O/C > 1 is not yet available, but the experience with carbon-rich environments suggests that classical theory may fail also when O/C > 1. The assessment by Henning led to spirited interchanges between members of the audience and speakers Cherchneff, Jeong, and Kozasa who presented theoretical models for nucleation and grain growth.

Lodders & Fegley gave a very nice review of the results of thermochemical equilibrium calculations under average conditions found in the detailed model atmospheres given by the hydrodynamical calculations, emphasizing that, for compounds which are not expected to be affected by processing in the ISM, the solar nebula, and in meteorites, the relative abundances found in some meteoritic inclusions are consistent with those given by the equilibrium calculations. This is encouraging support for the utility of the equilibrium calculations, but it does not mean that time-dependent calculations can be avoided in developing atmospheric models and understanding the process of mass loss and the observed spectral variations of real TPAGB stars. Winters (Winters, Le Bertre & Keady, this volume) showed quite convincing evidence for the temporal evolution of the grain-size distribution function in stars such as IRC +10216, and this variation must play an important role in the details of mass loss.

There may be mechanisms other than the pulsation-shock-grain-radiation pressure paradigm at work in driving mass loss from TPAGB stars. Bujarrabal pointed out that, for hundreds of variable AGB stars for which reliable estimates have been made, inferred mass-loss rates lie typically in the range of a few $\times 10^{-8}$ $M_\odot$ yr^{-1} to a few $\times 10^{-5}$ $M_\odot$ yr^{-1}, whereas indirect evidence provided by the distribution and kinematics of circumstellar matter in post-AGB, proto-planetary stars suggest a final mass-outflow rate of the order of $\sim 5\times10^{-4}$ $M_\odot$ yr^{-1}, perhaps 10-20 times larger than the mass-outflow rate of IRC +10216, everyone's favorite Mira wind. One solution to the puzzle may lie in the structure which develops near the hydrogen-helium discontinuity in the interior following a helium shell flash when the mass of the electron-degenerate core becomes large enough and the mass of the hydrogen-rich envelope declines below a critical value. In a poster paper which describes results of his theoretical quasi-static model calculations, Alan Sweigart shows that, under appropriate conditions of core mass and envelope mass, the gas pressure relative to the radiation pressure ($P_{\rm gas}/P_{\rm rad}$) at a point near the hydrogen-helium discontinuity reaches a minimum which decreases in value with each successive pulse. He infers that the minimum will go to zero (which numerical difficulties prevent him from demonstrating explicitly) and that matter beyond the minimum is effectively detached from the interior and flies off as the superwind proposed by Renzini (1979).

7. Mass transfer and chemical peculiarities in binary systems with TPAGB donors

The less evolved component in a close binary can acquire chemical peculiarities by accreting matter from a TPAGB star companion which has developed its own chemical peculiarities in consequence of nucleosynthesis during successive helium shell flashes followed by dredge-up. In turn, the nebular material ejected by the TPAGB star can be shaped by interaction with the companion star. If the TPAGB star is single or in a wide binary, the ejected material can also be shaped by a planetary system, a cometary system, or a combination of these systems. Here, I will review only the mass-transfer phenomenon.

Ba stars and their population II counterparts, the CH stars, occur in binary systems in which the secondary is typically a low mass white dwarf (McClure, Fletcher & Nemec 1980; McClure 1983; McClure & Woodsworth 1990). The fact that the typical orbital separation of binaries with Ba star components is ~ 2 AU (McClure 1983), coupled with the fact that a TPAGB star with a hot white dwarf core of mass $\sim 0.6 \pm 0.05\ M_{\odot}$ has a mean radius of the order of $\sim 1.06 \pm 0.37$ AU, shows that mass transfer from the TPAGB progenitor of the white dwarf to its companion does not, in general, result in a dramatic decrease in orbital separation, as might be expected if the donor fills its Roche lobe with a deep convective envelope and/or if the mass ratio q of donor to accretor exceeds ~ 0.67 during the mass-transfer episode. A possible inference is that, in some instances, the TPAGB star does not fill its Roche lobe during most of the mass-transfer episode, with mass transfer being primarily by accretion from the wind of the TPAGB star (Boffin & Jorissen 1988; Jorissen & Boffin 1992; Jorissen et al. 1998). In other instances, even when the TPAGB star fills its Roche lobe, the self-sustained wind from the donor may result in much more matter being lost from the system than is transferred to the accretor, with the tendency toward orbital shrinkage by common envelope (CE) action being countered by a tendency toward expansion occasioned by system mass loss; in the absence of any accretion, the product of the orbital separation and the total mass of the system would remain constant. For still other systems, because of extensive mass loss from the TPAGB star prior to Roche-lobe filling, $q < 0.67$ when the donor eventually fills its Roche lobe, permitting mass transfer to continue stably. In this case, the mass-transfer rate may be accelerated during the brief periods of radial expansion that follow helium shell flashes.

S stars (giants with C/O ratios intermediate between M stars and C stars) which do not show lines of Tc form another class of objects, all of which appear to be in binaries (Brown et al. 1990; Jorissen et al. 1993; Johnson, Ake & Ameen 1993). At this meeting, Jorissen reviewed the most recent findings regarding the binary characteristics of S stars, including the fact that, in the Galaxy, S stars showing lines of unstable Tc form a class whose members are, in general, brighter than members of the class of S stars which do not show Tc lines (Van Eck et al. 1998). Thus, S stars with no Tc are first giant branch stars with a white dwarf companion and S stars with Tc are TPAGB stars which have dredged up Tc from their own interiors. The S stars with Tc may be single stars, members of binaries with less evolved companions, or members of binaries with white dwarf companions.

The distributions of binary Ba, CH, and S stars in the eccentricity-period (e-$P_{\rm orb}$) diagram, when compared with the distribution in this same plane of binaries with spectroscopically normal G and K giant components, contain important information about the nature of mass transfer when the donor is a TPAGB star (Jorissen et al. 1998). For periods $P_{\rm orb} > 300$ days, binaries with normal G and K giants are distributed roughly uniformly in e over the range $0.05 < e < 0.9$, and, for periods $P_{\rm orb} < 300$ days, roughly half have $e \sim 0.0$. This shows that binary stars are born with a wide distribution of eccentricities and that orbits of binaries containing giants which do not support a strong wind are circularised by tidal forces when the radius of the giant is comparable to its Roche-lobe radius. Binaries containing Ba, CH, or S giants and having periods as large as $\sim 5\,000$ days have $e < 0.5$, suggesting that a frictional interaction between component stars and matter in the wind from a TPAGB precursor may have succeeded in reducing the eccentricity of the primordial binary (remember that the lifetime of a TPAGB star is ~ 100 times shorter than the lifetime of a first red giant branch star and that a typical TPAGB star in a 5 000 day binary is far from filling its Roche lobe).

Binaries containing Ba stars and with $P_{\rm orb} < 600$ days have $e < 0.2$. Roughly half of the binaries containing Ba stars and most binaries containing strong Ba stars and with $P_{\rm orb} = 200$-$2\,000$ days have orbits consistent with $e \sim 0.0$. Most binaries containing mild Ba stars and with $P_{\rm orb} > 600$ days and most binaries containing strong Ba stars and with $P_{\rm orb} > 2\,000$ days have $0.1 < e < 0.4$. For S stars and CH stars, the statistics are not as good, but there is a conspicuous break in the e distribution at $P_{\rm orb} \sim 1\,000$ days; binaries with $P_{\rm orb} < 1\,000$ days have distinctly smaller e than those with $P_{\rm orb} > 1\,000$ days. Even for $P_{\rm orb} > 1\,000$ days, however, binaries containing CH stars have $e < 0.2$ and binaries containing S stars have $e < 0.4$. These trends reinforce the suspicion that a frictional interaction between stellar components and the intrinsic TPAGB wind contributes to a reduction in the eccentricity of the primordial orbit.

Jorissen et al. (1998) argue that, in the Galaxy, strong Ba stars belong to a more metal-weak population than do mild Ba stars. This inverse correlation may be related, in part, to the fact that the frequency of carbon stars decreases in passing from the SMC to the LMC and then to the Galaxy (Blanco, McCarthy & Blanco 1980); the abundances of oxygen and of iron increase in passing through the stellar aggregates in the stated order, but the abundance of carbon (and possibly also of s-process isotopes) in matter dredged up following a thermal pulse is (may be) more or less independent of the metallicity. The relative ease with which metal-weak TPAGB stars produce carbon stars has an important consequence for comparisons between theoretical models of the third dredge-up and the distribution of carbon stars in low metallicity systems. As Jorissen emphasized, the carbon star distribution in the SMC (Westerlund et al. 1995) extends approximately two magnitudes fainter (to ~ -2.3 mag) than the distribution found in earlier, less deep surveys (Blanco et al. 1980; Westerlund et al. 1991). The earlier surveys gave a low luminosity cutoff which is more or less consistent with the luminosity of low mass TPAGB stars at minimum during a thermal pulse cycle at the onset of the TPAGB phase. The significance of the newer survey is that (1) most of the SMC carbon stars with luminosity in the range $M_{\rm bol} \sim$ -3.5 mag to -2.3 mag are first red giant branch stars with white dwarf companions and that (2) the carbon star distribution does not clearly define the low luminosity cutoff for TPAGB stars of low metallicity.

Among binaries consisting of a white dwarf and a chemically peculiar giant, there are many which are close enough that the TPAGB precursor of the white dwarf must have filled its Roche lobe, but which are wider than would be expected if unstable mass transfer led to the formation of a CE. HD 77247 with $P_{\rm orb} \sim 80.5$ days and $e = 0.09$ is a conspicuous example. A similar phenomenon may account for the orbital characteristics of the double white dwarf

L870-2, which has a relative long period of $\sim$ 1.5 days (Saffer, Liebert & Olszewski 1987, 1988) even though the precursor binary has passed through two phases of mass transfer during which the donor overfilled its Roche lobe. The mass donor in the first mass-transfer event was probably a TPAGB star, and the energy required to expel the envelope matter from the binary system is greater than can be explained by the increase in the orbital binding energy (Iben & Webbink 1989; Iben & Livio 1993). The inference is that, when the Roche-lobe filling star is a TPAGB star with a strong intrinsic wind, $\dot{M}_{\rm wind} >> \dot{M}_{\rm transferred}$, and the tendency for expansion associated with mass loss competes successfully with the frictional interaction between stars and wind matter to reduce the degree of orbital shrinkage which would occur in the absence of the intrinsic wind. Further support for this interpretation comes from the fact that close binary central stars in planetary nebulae originate typically in initial binaries in which the primary fills its Roche lobe in an early case C event, before the TPAGB, dusty-wind phase (Iben & Tutukov 1993).

Using data obtained from infrared and MACHO observations, Wood et al. presented the distribution of LMC long period variables in the magnitude-period plane. Stars in his sample fall into five rather clearly separated sequences, three of which correspond to radial pulsation in the fundamental mode and in the first two overtone modes. A fourth (poorly populated) sequence consists of contact binaries on the first giant branch for low mass stars. In addition, some objects with contact binary-like light curves exist in the equivalent evolutionary phase for intermediate mass stars, i.e., at the end of the core helium-burning loops. The fifth, longest period sequence, is interpreted by Wood et al. as an orbital period sequence, with median period ranging from $P_{\rm orb} \sim$ 100 days at the smallest luminosities to $P_{\rm orb} \sim$ 1 000 days at the largest luminosities. If one doubles the periods as plotted by Wood for members of the fourth sequence, these members define a low-luminosity extension of the fifth sequence. Fully 25% of all variables in the sample are in the fifth sequence. At any luminosity, the scatter in the logarithm of the orbital period about the median is only $\sim \pm 0.15$; stars in the fifth sequence appear to be binaries with a period such that the TPAGB star component only slightly overfills or underfills its Roche lobe (by, say, $\sim \pm 25\%$). Once again, the inference is that, when the donor is a TPAGB star with a strong intrinsic wind, and whether or not the donor fills its Roche lobe, the conventional picture of mass transfer through a Roche lobe is superseded by wind mass loss from the system and accretion by the secondary at a rate small compared with the wind mass-loss rate.

The most puzzling aspect of Wood's fifth (binary) sequence is the small dispersion in orbital period at a given luminosity. One might expect initial main-sequence binaries to be distributed rather evenly in $\log A$ (and therefore in $\log P_{\rm orb}$). Yet orbital separations of the evolved binaries in the fifth sequence are very sharply concentrated at a value apparently determined by the radius of the TPAGB component. What could funnel binaries into this narrow range? And how could there be so many of them?

8. AGB stars with Oxygen-Neon cores

Not discussed at this meeting except in this summary is work that has been done to estimate (1) the initial mass marking the transition between stars which become AGB stars with CO cores and those which become AGB stars with ONe cores and (2) the initial mass marking the transition between stars which ultimately become white dwarfs and those which become neutron stars after first being TPAGB stars.

There are only two sets of studies which address the first of these transitions by actually calculating the conversion by carbon burning of a partially electron-degenerate CO core into a fully electron-degenerate ONe core. The first set (Nomoto 1984, 1987) examines the evolution of the core and the helium layer above it, treating the hydrogen-rich envelope as a boundary condition. The second set explores the evolution of complete models for a population I composition and masses 9 $M_\odot$ (García-Berro, Ritossa & Iben 1997, paper III), 10 $M_\odot$ (Ritossa, García-Berro & Iben 1996, paper II), 10.5 $M_\odot$ (Iben, Ritossa & García-Berro 1997, paper IV), and 11 $M_\odot$ (Ritossa, García-Berro & Iben 1999, paper V). The paucity of relevant models is probably due to the fact that, when carbon ignites off center, as it does in most of the models, the burning front narrows significantly as it evolves toward the center, and resolving the front to the extent necessary for model convergence requires that the time step be reduced for each successive model; Zeno's paradox is in practice averted because the thickness of the front remains finite as the distance of the front from the center goes to zero. In any case, when $Z = 0.02$ and $Y = 0.28$, the transition mass for models which become TPAGB stars with ONe cores rather than CO cores is slightly smaller than 9 $M_\odot$.

The condition for the second transition (between white dwarf and neutron star products) is that, when the mass of the ONe core reaches $\sim 1.37\ M_\odot$, electron captures on ^{24}Mg and ^{16}O lead to core collapse to neutron star dimensions (Miyaji et al. 1980; Miyaji & Nomoto 1987). Paper V finds that, when $Z = 0.02$ and $Y = 0.28$, the second transition mass is slightly smaller than 11 $M_\odot$.

Table 1 summarizes several characteristics of the models described in papers II-V at the end of the carbon-burning phase. In this table, M_* is the model mass, $M_{\rm ONe}$ is the mass of the ONe core, defined as the point at which the abundances of ^{12}C and ^{20}Ne are equal, $\Delta M_{\rm CO}$ is the mass of the CO layer above the ONe core, and $\Delta M_{\rm Ne\ free}$ is the mass of a "neon-free" zone, defined as the region over which all neon isotopes are less abundant by number than 10^{-4}. All masses are in solar units. The neon-free zone resides between the CO layer and the hydrogen-rich envelope. Its significance is that, if a white dwarf of this composition is in a cataclysmic variable, there will be a sequence of nova outbursts such that Ne will be underabundant in the nova ejecta.

Also shown in Table 1 are the mass $M_{\rm WD}$ of the white dwarf expected if the real analog loses mass at the average rate of $10^{-4}\ M_\odot\ {\rm yr}^{-1}$ during the TPAGB phase. The real analog of the 11 $M_\odot$ model will become a neutron star before it loses its hydrogen-rich envelope.

The final entry in Table 1, $(X_{\rm C})_{\rm max}$, is the maximum abundance by mass of ^{12}C left in the ONe core at the end of carbon burning. Its significance is that, if a white dwarf of this composition is one of a merging pair of white dwarfs or if it can accrete matter from a companion with a hydrogen-rich or helium-rich envelope until its mass reaches the Chandrasekhar limit, heat released when carbon is ignited in the ONe core could trigger the ignition of oxygen, with a resulting disruption of the star rather than a collapse into a neutron star.

Table 1. Properties of models at the end of the carbon-burning phase

M_*	$M_{\rm ONe}$	$\Delta M_{\rm CO}$	$\Delta M_{\rm Ne\ free}$	$M_{\rm WD}$	$(X_{\rm C})_{\rm max}$
9	1.066	0.050	0.014	1.16	0.048
10	1.190	0.015	0.009	1.25	0.012
10.5	1.263	0.0065	0.0036	1.31	0.006
11	1.36764	0.00082	0.00065	NS	0.000018

Acknowledgments. It is a pleasure to thank Thibaut Le Bertre for providing financial assistance and to thank him and the other members of the scientific organizing committee for arranging a very stimulating conference. Thanks also to Peter Wood and Alain Jorissen for useful comments on the binary star section of this contribution.

References

Blanco V.M., McCarthy M.F., Blanco B.M., 1980, ApJ 242, 938

Boffin H.M.J., Jorissen A., 1988, A&A 205, 155

Bowen G.W., 1988, ApJ 329, 299

Brown J.A., Smith V.V., Lambert D.L., Dutchover E., Jr., Hinkle K.H., Johnson H.R., 1990, AJ 99, 1930

Draine B.T., 1981, in "Physical Processes in Red Giants", ed. I. Iben, Jr., and A. Renzini, Reidel, Dordrecht, p. 317

Gallino R., Arlandini C., Busso M., Lugaro M., Travaglio C., Straniero O., Chieffi A., Limongi M., 1998, ApJ 497, 388

García-Berro E., Ritossa C., Iben I., Jr., 1997, ApJ 485, 765 (paper III)

Gehrz R.D., Woolf N.J., 1971, ApJ 165, 285

Herwig F., Blöcker T., Schönberner D., El Eid M., 1997, A&A 324, 81

Hill S.J., Willson L.A., 1979, ApJ 229, 1029

Iben I., Jr., 1974, BAAS 6, 316

——1975, ApJ 196, 525

——1976, ApJ 208, 165

——1977, ApJ 217, 788

Iben I., Jr., Kaler J.B., Truran J.W., Renzini A., 1983, ApJ 264, 605

Iben I., Jr., Livio M., 1993, PASP 105, 1373

Iben I., Jr., MacDonald J., 1995, in "White Dwarfs", ed. D. Koester and K. Werner, Springer, Berlin, p. 48

Iben I., Jr., Ritossa C., García-Berro E., 1997, ApJ 489, 772 (paper IV)

Iben I., Jr., Tutukov A.V., 1993, ApJ 418, 343
Iben I., Jr., Webbink R.F., 1989, in "White Dwarfs", ed. G. Wegner, Springer, Berlin, p. 477
Johnson H.R., Ake T.B., Ameen M.M., 1993, ApJ 402, 667
Jorissen A., Boffin H.M.J., 1992, in "Binaries as Tracers of Stellar Formation", ed. A. Duquennoy and M. Mayor, Cambridge University Press, Cambridge, p. 110
Jorissen A., Frayer D.T., Johnson H.R., Mayor M., Smith V.V., 1993, A&A 271, 463
Jorissen A., Van Eck S., Mayor M., Udry S., 1998, A&A 332, 877
Kastner J.H., Forveille T., Zuckerman B., Omont A., 1993, A&A 275, 163
McClure R.D., 1983, ApJ 268, 264
McClure R.D., Fletcher J.M., Nemec J.M., 1980, ApJ 238, L35
McClure R.D., Woodsworth A.W., 1990, ApJ 352, 709
Miyaji S., Nomoto K., 1987, ApJ 318, 307
Miyaji S., Nomoto K., Yokoi K., Sugimoto D., 1980, PASJ, 32, 303
Nomoto K., 1984, ApJ 277, 791
——1987, ApJ 322, 206
Paczyński B., 1970, Acta Astron. 20, 47
Renzini A., 1979, in "Stars and Star Systems", ed. B.E. Westerlund, Reidel, Dordrecht, p. 155
Ritossa C., García-Berro E., Iben I., Jr., 1996, ApJ 460, 489 (paper II)
——1999, ApJ, in press (paper V)
Saffer R.A., Liebert J.W., Olszewski E., 1987, BAAS 19, 1041
——1988, ApJ 334, 947
Straniero O., Gallino R., Busso M., Chieffi A., Raiteri C.M., Salaris M., Limongi M., 1995, in "Nuclei in the Cosmos III", ed. M. Busso, R. Gallino and C.M. Raiteri, AIP, New York, p. 407
Uus U., 1970, Nauch. Inform. Acad. Sci. 17, 3
Van Eck S., Jorissen A., Udry S., Mayor M., Pernier B., 1998, A&A 329, 971
Westerlund B.E., Azzopardi M., Breysacher J., Rebeirot E., 1991, A&AS 91, 425 (erratum 1991, A&AS 92, 683)
——1995, A&A 303, 107
Willson L.A., Hill S.J., 1979, ApJ 228, 854
Winters J.M., Fleischer A.J., Gauger A., Sedlmayr E., 1994, A&A 290, 623
Wood P.R., 1974, ApJ 190, 609
——1979, ApJ 227, 220
Woolf N.J., Ney E.P., 1969, ApJ 151, L181

Thibaut Le Bertre, only now realizing the forthcoming work with the Proceedings

Asymptotic Giant Branch Stars
T. Le Bertre, A. Lèbre and C. Waelkens, eds.
I.A.U. Symposium 191, 1999

Recent and Future Studies of Circumstellar Matter – A Snapshot

M. Jura
Department of Physics and Astronomy, University of California, Los Angeles CA 90095-1562, USA

Abstract. At this meeting, powerful new images and spectroscopy of AGB stars were presented. Theoretical models have advanced, and we are learning more from studies of the pre-solar grains isolated from meteorites.

We suggest that several mass loss mechanisms may be operating in AGB stars: current images imply both spherical winds and highly flattened outflows. There are good arguments that in some cases, a companion may be critical in driving the final outflow. In the near future, a number of extremely powerful new instruments and techniques will be available that will lead to a much deeper understanding of these systems.

1. Introduction

One of our basic goals is to understand how stars die. As intermediate mass stars evolve, they eject matter into space. We want to know the composition and mass of this ejecta, and its fate as it is returned into the interstellar medium.

At the 1989 conference here in Montpellier on AGB stars (see Mennessier & Omont 1989), there had been considerable progress in understanding the circumstellar matter around AGB stars. The results from the IRAS satellite had been released, and we therefore had a much more comprehensive understanding of the dust around these stars. Furthermore, the IRAM 30m telescope had been in operation for a few years, and molecular line studies of these stars had been very successful. Some of the basic results that were well known in 1989 were: (1) AGB stars are luminous ($\sim 10^4$ $L_\odot$) highly evolved mass-losing red giants, descended from intermediate mass main sequence stars. These stars are about to become Planetary Nebulae and eventually will become white dwarfs. (2) The winds are composed mainly of dust and molecules. (3) Mass loss rates seem to be known to within a factor of 2 or 3 for many stars. Typical values range from 10^{-7} $M_\odot$ yr^{-1} to 10^{-5} $M_\odot$ yr^{-1} with a few stars having even higher mass loss rates. (4) Most AGB stars have relatively low mass loss rates, but most mass is ejected by the relatively few stars with the high mass loss rates. (5) Outflow velocities can be measured to perhaps 10% with typical values between 10 and 25 km s^{-1}.

Even though considerable progress had been made at that time, there were (and still remain) a number of unsolved problems. Although we have reasonable estimates of the mass loss rates from individual stars by modeling the infrared and radio data, we still do not have very good understanding of how dM/dt, varies as a function of the star's luminosity, mass, composition, pulsational properties, rotation, magnetic field, or binarity. All of these factors can produce some effect on the total mass loss rate. Furthermore, we do not have a very good idea on how dM/dt from a particular star varies as a function of direction (either as clumps or in bipolar flows) or as a function of time – producing multiple shells. We cannot start from first physical principles and compute the total amount of mass lost by the star on the AGB – a fundamental parameter for both stellar evolution and the evolution of the Milky Way.

Even though there are basic theoretical uncertainties, observations have provided considerable insight into the outflows, and we now have a much deeper understanding of the circumstellar matter around mass-losing AGB stars. Below, in the two following sections ("Spectroscopic Observations" and "Imaging Observations") I outline some of the advances that struck me during the conference. This discussion is not comprehensive, but it may at least highlight some of the progress that is being made. In the following two sections, I list some inferences that struck me as relevant to the topic. First, in "Multiple Mechanisms", I address the issue of whether we ought to consider more than one mass-loss mechanism. In the next section, "Some speculative questions", I list some topics which might be worth addressing. Finally, in the last section, I list some future instrumental capabilities.

2. Spectroscopic observations

One of our goals is to measure the amounts of the different molecules and particles that are ejected by an AGB star. For example, for a carbon star, we still do not have a very precise accounting of all the carbon – how much is lost in CO and other small molecules, how much is lost in grains of different sizes and shapes, and how much is lost in intermediate size clusters such as PAH's.

At this conference, results from new instruments – notably the spectrometers on ISO – were presented. The detection of CO_2 in the outflows from oxygen-rich stars (Justtanont et al. 1998) is striking. Infrared observations are also necessary to measure the amount of C_2H_2 – a major constituent in the outflows from carbon stars, and progress is now being reported for measuring the amount of this molecule in outflows (see Yamamura et al. 1998). While atomic lines have been detected in emission with the ISO data (Barlow 1998; Fong et al. 1998), a related advance is the ability to detect submillimeter wavelength CI emission by using large ground-based telescopes such as the CSO (see the poster by Crosas et al. 1998)

Another major advance during the past few years has been the study of the solid state features in the spectra of the circumstellar dust (Waters & Molster 1998). Particularly striking is the detection of crystalline silicates in the dust around AC Her, a result that seems to be best interpreted by assuming a long-lived orbiting disk of material (Van Winckel et al. 1998).

3. Imaging observations

Enormous progress has been made in mapping circumstellar matter around stars. Some of the highlights presented at this conference include:

1. The "movie" of the SiO masers around TX Cam achieved with the VLBA (Diamond & Kemball 1998) shows mostly outflow but also dramatic infall of matter during certain pulsational phases.

2. The near-IR interferometric maps of the Red Rectangle (Osterbart et al. 1997, 1998; Mékarnia et al. 1998) beautifully display both the dusty disk and the X shaped pattern of what may be a bipolar outflow from the disk. Other optical and near IR interferometry shows intense regions of activity on some stars (Tuthill, Monnier & Danchi 1998).

3. Radio interferometry at the Plateau de Bure (Lucas & Guélin 1998) shows molecular structure – including striking rings of molecules around IRC +10216 as expected in the photochemical models (Glassgold 1996).

4. Optical images of the reflection of the diffuse galactic light by the dust around IRC +10216 show multiple shells and extended spherical symmetry (de Laverny, Mauron & Lopez 1998). The extended emission line scattering around evolved stars with detached shells is also noteworthy (Delgado et al. 1998).

5. The HST optical image of the Egg Nebula (Sahai et al. 1998) shows both more-or-less spherical shells and a very flat disk. What the image shows, but was not clear from previous ground-based studies, is the appearance of two very distinct circumstellar regions. This suggests that two mechanisms may be at work in the mass loss from this object.

6. Ground-based images taken at mid-IR wavelengths often show axially symmetry in Proto Planetary Nebulae (Meixner et al. 1997, 1998).

4. Multiple mass-loss mechanisms?

If only for simplicity, there is a preference to think that each star loses mass because of a single mechanism. However, reality is not always this easy. Models demonstrating the coupling between radiation pressure on dust and pulsations are making great progress (Höfner 1998; Fleischer, Winters & Sedlmayr 1998). These models predict essentially spherically symmetric mass loss and can explain most of the observations of the extended circumstellar envelope around IRC +10216. In contrast to what is expected from these models, it is clear that some outflows are very axi-symmetric but non-spherical. In the case of the Egg Nebula, the image obtained with the HST points to a very marked contrast between the spherical and flattened circumstellar dust distributions, suggesting two different kinds of outflows.

Clearly, there must be at least some instances where a binary companion plays a key role in the evolution of the mass-losing star. Binaries such as the Red Rectangle (Van Winckel, Waelkens & Waters 1995; Waelkens et al. 1996) and AC Her (Van Winckel et al. 1998) have periods near 1-3 years, and in such evolved systems, the companion must have played an important role in driving the mass loss (see, for example, Taam 1996). While there have been extensive discussions of the role of binaries in outflows (Soker 1998; Jorissen 1998), we still do not have a full understanding of when and how this occurs.

In following the mass loss process, we should not only consider the energetics but also follow the angular momentum history as well. The frequency of axi-symmetric outflows suggests that there are often instances when angular momentum transfer plays a key role in the mass loss.

5. Some speculative questions

While we study the circumstellar envelopes around AGB stars, there are certain questions that come to mind. By posing and then answering these questions, we may learn more about these systems. I would include the following:

1. When and where do the gas and dust separate from each other in some circumstellar regions? Abundance analysis of several post-AGB stars shows a remarkable pattern: volatile elements (carbon, nitrogen, oxygen, zinc) have approximately solar abundances while refractory elements (such as iron) are depleted by as much as a factor of 10^3 (Waters, Trams & Waelkens 1992). The interpretation of these data is that the star has accreted gas but not dust. If so, then the gas and dust have separated.

2. Can planets form around post-AGB stars? Planets have been found around the pulsar PSR 1257+12 (Wolsczan & Frail 1992). There appear to be long-lived disks around some post-AGB stars as well, such as the Red Rectangle and AC Her (Jura, Balm & Kahane 1995; Van Winckel et al. 1998). The mass of dust in these disks is 10^{30} g – sufficient for planet formation. In the Red Rectangle, there appears to be a large structure in the dust cloud (Jura & Turner 1998). Can planets form in these objects?

3. Can we account for the powerful X-ray sources associated with the accretion from the outflows from AGB red giants? Accretion of matter onto a compact source such as a white dwarf, neutron star or black hole results in the copious emission of X-rays. There are large numbers of mass-losing red giants in the Milky Way. However, there are very few such red giants such as GX 1+4 (Chakrabarty, van Kerkwijk & Larkin 1998) associated with X-ray sources more powerful than 10 $L_\odot$. Since so many stars are binaries, there should be large numbers of binary red giants – some of which should have X-ray companions (see, for example, Hunsch et al. 1998). While some symbiotic stars do display X-ray emission, the total X-ray luminosity often is less than 1 $L_\odot$ (Jorissen et al. 1996; Muerset, Wolff & Jordan 1997). Also, it seems that a few of the "luminous supersoft X-ray Sources" include this class of object – white dwarfs powered by accretion of the wind from an AGB companion (Kahabka & van den Heuvel 1997). Further study of this class of wide binary where accretion from a red giant wind may occur seems appropriate.

4. Where are the large (diameters up to 10 μm) inclusions in meteorites formed? As discussed by Anders & Zinner (1993) and Ott (1993), there is now compelling evidence that some inclusions in meteorites are grains that were formed in the outflows from AGB stars and then survived both their passage through the interstellar medium and the pre-solar nebula. Some of these particles are nearly as large as 10 μm in diameter and the "average" size is well over 1 μm in diameter (Bernatowicz et al. 1996). Typically, the grains ejected by mass-losing carbon stars such as IRC +10216 have diameters near 0.1 μm (Jura 1994). Although there might be a natural selection effect to find larger particles because they are more durable, it is surprising that such large particles are created in standard winds. Could it be that such particles are formed in long-lived disks around post-AGB binaries? (Jura, Turner & Balm 1997)?

5. Are there non-standard mass-losing AGB systems? One of the surprising results from the IRAS data was the discovery that a few carbon-rich stars, such as BM Gem, possess oxygen-rich dust (Little-Marenin 1986; Willems & de Jong 1986). Are there other unexpected objects?

6. Can we identify the carriers of the diffuse interstellar bands in the AGB outflows? The carriers of the Diffuse Interstellar Bands is an unsolved mystery (Herbig 1995). The nebulosity in the Red Rectangle has a remarkable and nearly unique set of emission bands longward of 5800 Å (Schmidt, Cohen & Margon 1980). Similar emission has been found to date only in one other star, the R CrB star V854 Cen at minimum light (Kameswara-Rao & Lambert 1993). It seems possible that the carrier of these bands is related to at least some of the Diffuse Interstellar Bands (Sarre, Miles & Scarrott 1995). Can we learn more from extending this kind of work? A related topic is the question of the origin of the Extended Red Emission which has been found throughout the Milky Way (Gordon, Witt & Friedmann 1998; Szomoru & Guhathakurta 1998), and which may be related to the large continuum bump seen longward of 5800 Å in the spectrum of the Red Rectangle.

6. The future?

Progress in instrumentation almost inevitably will lead to advances in our astrophysical understanding. Consider some of the following which will improve our understanding of AGB stars and their winds:

1. New telescopes such as SIRTF, SOFIA, the sub-mm array and the millimeter array and various infrared and optical interferometers will provide extremely powerful tools for studying these stars.

2. New instruments such as SCUBA and the mid-IR imagers (such as MIRLIN and OSCIR) greatly enhance the capability of existing telescopes.

3. The 2MASS and DENIS surveys, as well as further exploitation of the *HIPPARCOS* data, and then the cross correlation of the surveys with

each other, should allow a much more comprehensive view of the AGB phenomena.

4. Progress in computers will allow for much more sophisticated modeling.

5. Studies of the meteoritic inclusions are progressing rapidly, and they provide a unique set of insights into the physics of these stars.

In summary, we have made such progress during the past ten years, and I am very optimistic that the next ten years will be a very exciting time to study these stars.

Acknowledgments. This work was partly supported by NASA. I thank the organizers of the conference for designing such an interesting and timely program.

References

Anders E., Zinner E., 1993, Meteoritics 28, 490

Barlow M.J., 1998, IAU Symposium No. 191, this volume

Bernatowicz T.J., Cowsik R., Gibbons P.C., Lodders K., Fegley B., Amari S., Lewis R.S., 1996, ApJ 472, 760

Chakrabarty D., Van Kerkwijk M.H., Larkin J.E., 1998, ApJ 497, L39

Crosas M., Young K., Ivezić Ž., Knapp G.R., 1998, poster contribution, IAU Symposium No. 191

de Laverny P., Mauron N., Lopez B., 1998, poster contribution, IAU Symposium No. 191

Delgado D.G., Olofsson H., Schwarz H., Eriksson K., Gustafsson B., 1998, poster contribution, IAU Symposium No. 191

Diamond P.J., Kemball A.J., 1998, IAU Symposium No. 191, this volume

Fleischer A.J., Winters J.M., Sedlmayr E., 1998, IAU Symposium No. 191, this volume

Fong D., Meixner M., Sutton E., et al., 1998, poster contribution, IAU Symposium No. 191

Gordon K.D., Witt A.N., Friedmann B.C., 1998, ApJ 498, 522

Glassgold A.E., 1996, ARA&A 34, 241

Herbig G.H., 1995, ARA&A 33, 19

Höfner S., 1998, IAU Symposium No. 191, this volume

Hunsch M., Schmitt J.H.M.M., Schröder K.-P., Zickgraf F.-J., 1998, A&A 330, 225

Jorissen A., 1998, IAU Symposium No. 191, this volume

Jorissen A., Schmitt J.H.M.M., Carquillat J.M., Ginestet N., Bickert K.F., 1996, A&A 306, 467

Jura M., 1994, ApJ 434, 713

Jura M., Balm S.P., Kahane C.,1995, ApJ 453, 721

Jura M., Turner J.L., 1998, Nature 395, 144

Jura M., Turner J.L., Balm S.P.,1997, ApJ 474, 741

Justtanont K., Feuchtgruber H., de Jong T., Cami J., Waters L.B.F.M., Yamamura I., Onaka T., 1998, A&A 330, L17

Kahabka P., van den Heuvel E.P.J., 1997, ARA&A 35, 69

Kameswara-Rao N., Lambert D.L., 1993, MNRAS 263, 27P

Little-Marenin I.R., 1986, ApJ 307, L15

Lucas R., Guélin M., 1998, IAU Symposium No. 191, this volume

Meixner M., Skinner C.J., Graham J.R., Keto E., Jernigan J.G., Arens J.F., 1997, ApJ 482, 897

Meixner M., Ueta T., Dayal A., Hora J., Fazio G., Hrivnak B., Skinner C.J., Hoffmann W.F., Deutsch L.K., 1998, ApJS, in press

Mékarnia D., Rouan D., Tessier E., Dougados C., Lefèvre J., 1998, A&A 336, 648

Mennessier M.O., Omont A. (eds.), 1989, From Miras to Planetary Nebulae: Which Path for Stellar Evolution?, Editions Frontières

Muerset U., Wolff B., Jordan S., 1997, A&A 319, 201

Osterbart R., Langer N., Weigelt G., 1997, A&A 325, 669

Osterbart R., Balega Y.Y., Langer N., Menshchikov A.B., Weigelt G., 1998, poster contribution, IAU Symposium No. 191

Ott U., 1993, Nature 364, 25

Sahai R., et al., 1998, ApJ 493, 301

Sarre P., Miles J.R., Scarrott S.M., 1995, Science 269, 674

Schmidt G.D., Cohen M., Margon B., 1980, ApJ 239, L133

Soker N., 1998, ApJ 496, 827

Szomoru A., Guhathakurta P., 1998, ApJ 494, L93

Taam R.E., 1996, in Compact Stars in Binaries, ed. J. van Padadijs, E.P.J. van den Heuvel and E. Kuulkers, IAU Symposium No. 165, Kluwer, Dordrecht, p. 3

Tuthill P.G., Monnier J.D., Danchi W.C., 1998, IAU Symposium No. 191, this volume

Van Winckel H., Waelkens C., Waters L.B.F.M., 1995, A&A 293, L25

Van Winckel H., Waelkens C., Waters L.B.F.M., Molster F.J., Udry S., Bakker E.J., 1998, A&A 336, L17

Waelkens C., Van Winckel H., Waters L.B.F.M., Bakker E.J., 1996, A&A 314, L17

Waters L.B.F.M., Molster F.J., 1998, IAU Symposium No. 191, this volume

Waters L.B.F.M., Trams N.R., Waelkens C., 1992, A&A 262, L37

Wolsczan A., Frail D.A., 1992, Nature 355, 145

Willems F.J., de Jong T., 1986, ApJ 309, L39

Yamamura I., de Jong T., Waters L.B.F.M., Cami J., Justtanont K., 1998, IAU Symposium No. 191, this volume

Franz Kerschbaum and Nathan Netzer, poster-sessioning

POSTER CONTRIBUTIONS

The abstracts of the poster contributions presented at I.A.U. Symposium n° 191 have been published within a separate book distributed to the participants during the Conference. An electronic version of these abstracts can be consulted at:

http://www.dstu.univ-montp2.fr/GRAAL/agb98/BOOK/prog/program.html

These abstracts are also archived in the Astrophysics Data System (ADS) of the N.A.S.A.

1. Basic Facts, Structure, Evolution, Nucleosynthesis

Abia C., de Laverny P., Pavlenko Y.
New and accurate Li abundance determinations in super Li-rich AGB stars

Amari S., Nittler L.R., Zinner E., Lewis R.S.
Isotopic analyses of presolar SiC grains of type Y

Cioni M.R., Habing H.J., Loup C., Epchtein N.
DENIS: Color-Magnitude diagrams and Luminosity Function towards the Magellanic Clouds

Decin L., Cohen M., Eriksson K., Gustafsson B., Huygen E., Morris P., Plez B., Sauval J., Vandenbussche B., Waelkens C.
Modelling of ISO-SWS spectra of red giants

Domínguez I., Straniero O., Isern J.
AGB stars as Astroparticle Laboratories

Driebe T., Blöcker T., Herwig F., Schönberner D.
Diffusive overshooting in hot bottom burning AGB models

Dufour E., Kahane C., Greaves J.C.
Modelling $^{12}C/^{13}C$ ratios

Gallino R., Busso M., Wasserburg G.J., Straniero O.
Short-lived Isotopes from a Close-by AGB Star Triggering the Protosolar Nebula

Girardi L.
Constraints to the AGB evolution provided by AGB and white dwarf stars in clusters, and integrated cluster colours

Herwig F., Schönberner D., Blöcker T.
Violation of the Core Mass – Luminosity Relation for AGB models which experience the third dredge-up

Josselin E., Blommaert J.A.D.L., Groenewegen M.A.T., Omont A.
Identification and Characterization of Red and Infrared Supergiants

Klochkova V.G.
Chemical Composition of atmospheres of stellar IRAS-sources

Lugaro M., Gallino R., Arlandini C., Busso M., Straniero O.
Sr Isotopic Composition in AGB SiC Grains

Mazzitelli I., D'Antona F., Ventura P.
Lithium formation in massive AGB stars: new models

Ohnaka K., Tsuji T.
Elemental Abundance Analysis in Carbon Stars

Rauscher T., Thielemann F.-K.
Astrophysical reaction rates for nucleosynthesis in AGB stars

Ryde N., Eriksson E., Gustafsson B., Olofsson H., Plez B.
Modelling of molecular bands of oxygen-rich AGB stars

2. Pulsation, Mass Loss, Cool Envelopes

Andronov I.L.
Multiple periodicities in long-period variables

Balklavs A., Dzervitis U., Eglitis I.
Analysis of absolute magnitudes of carbon stars from HIPPARCOS parallaxes

Barthès D., Luri X.
Luminosity calibrations and kinematics of Galactic oxygen-rich LPVs: derivation of P-L-C relations and modelling

Broc A., Lafon J.-P.J.
Non-equilibrium radiative hypersonic flows

Cami J., de Jong T., Tielens A.G.G.M., Justtanont K., Waters L.B.F.M., Yamamura I.
Gas-phase CO_2 around O-rich AGB stars

Carpenter K.G.
The Structure of the Outer Atmosphere and Wind of λ Vel

Chinarova L.L.
Time series analysis of the long-term variability of the giant components in symbiotic variables

Duari D., Cherchneff I., Willacy K.
Carbon-bearing species in the inner wind of oxygen-rich Miras

Gray M.D.
The Natural Saturated Beam Angles of Circumstellar SiO Masers

Helling Ch., Woitke P., Winters J.M., Sedlmayr E.
Influence of molecular opacities on the generation of cool stellar winds

Herpin F., Baudry A., Alcolea J., Cernicharo J.
On the high velocity SiO maser emission from late-type stars

Jiménez-Esteban F., García-Lario P., Manchado A., Engels D.
Identification and monitoring program of OH/IR stars

de Jong T., Yamamura I., Cami J., Onaka T., Waters L.B.F.M.
Discovery of the SO_2 *7.3* μm *band in Oxygen-rich Miras*

Jorissen A., Udry S.
Shock waves in Mira stars : the pseudo SB2 orbit of the CS star RZ Peg

Justtanont K., de Jong T., Tielens A.G.G.M., Cami J., Waters L.B.F.M., Yamamura I.
SWS Observations of Supergiants

Kudashkina L.S., Andronov I.L.
Atlas of the mean light curves of the Mira-type stars

Lebzelter T., Hinkle K.H., Hron J.
Atmospheric Kinematics of Long period Variables

Lewis B.M.
Mass-Loss and Gravitational Contraction

Linsky J.L., Harper G.M., Bennett P.D., Brown A., Valenti J.
A Critical Evaluation of Mass Loss Rates and Wind Properties of Evolved Late-type Stars

Marsakova V.I., Andronov I.L.
Correlations between the characteristics of the individual cycles of the Mira–type stars

Matsuura M., Yamamura I., Murakami H., Freund M.M.
The Observation of H_2O *in M-giants in the Near– and Mid– Infrared Region with the Infrared Telescope in Space (IRTS)*

Mattei J.A., Foster G.
Trend Analysis of Oxygen–Rich Long–Period Variables

McNamara D.H.
Period–Luminosity relations of Populations II Cepheids

Netzer N.
Radiative Transfer and Dynamics of Circumstellar Envelopes of Variable Red Giants

Perrin G., Coudé du Foresto V., Ridgway S.T., Mariotti J.-M., Ruilier C., Mennesson B., Traub W.A.
Direct Evidence of the Pulsation of the Photosphere of AGB Stars

Schultheis M., Aringer B., Höfner S., Jørgensen U.
TiO bands in Miras and Semi–regular Variables

van Belle G., The PTI Collaboration, Thompson R.R.
Evolved Star Sizes, Temperatures as Directly Measured with Interferometry

Wallerstein G., Barnbaum C.
U Equ: A Unique Variable with Strong TiO Emission

Windsteig W., Aringer B., Lebzelter T., Höfner S.
4 μm High Resolution Spectra of AGB stars

Wing R.F., Alvarez R., Plez B., Yuan Y.
Behavior of Mira Variables in Bandstrength-Color Diagrams. Comparison of Dynamic Models with Calibrated Spectral Scans

3. Formation, Composition, and Processing of Dust

Andersen A.C., Jørgensen U.G., Henning Th., Mutschke H.
Laboratory study of dust from AGB stars

Cau P., Cherchneff I.
The formation of aromatics and metal carbides in the inner wind of carbon stars

Chamaillard K., Lafon J.-P.J.
Effects of rough surfaces on optical properties of grains

Dirks U., Sedlmayr E.
The influence of temperature fluctuations on dust formation in cool stellar atmospheres

Ferguson J.W., Alexander D.R., Johnson H.R., Allard F., Hauschildt P.H.
The Formation of PAHs in Carbon Star Atmospheres

Jura M.
Scientific opportunities with SIRTF

Koike C., Suto H., Tuchiyama A., Shibai H., Tanabe T.
The spectra of pyroxenes in mid and far infrared

Le Bertre T., Winters J.M., Sedlmayr E.
The mass loss rates of AGB stars from near–infrared photometry

Liberatore S., Lafon J.-P.J.
Modelling of dynamics in circumstellar dust shell

Lobel A., Doyle J.G., Bagnulo S.
Modelling the Spectral Energy Distribution of Carbon and Oxygen–rich stars from stellar model spectra

Lodders K., Fegley B. Jr.
The influence of metallicity on dust condensation temperatures in circumstellar envelopes of giant stars

Marengo M., Busso M., Silvestro G., Fazio G.
AGB circumstellar envelope dust mineralogy from ground based mid-IR imaging and photometry

Molster F., Waters L.B.F.M., Jäger C., Henning Th.
The crystalline silicates of AFGL 4106

Patzer A.B.C., Chang Ch., John M., Sedlmayr E.
On the formation of inorganic clusters in oxygen–rich circumstellar envelopes of AGB stars

Pompeia L., Lorenz-Martins S.
Dust envelope models and spectroscopic study of oxygen-rich stars: corundum and silicate grains

Ryde N., Eriksson K., Gustafsson B., Olofsson H., Wolstencroft R., Nyman L.–Å.
Mid and far-infrared spectra of the third brightest carbon star IRAS 15194-5115

Simis Y.
Two fluid hydrodynamics of dust driven AGB winds

Speck A.K., Hofmeister A.M., Barlow M.J.
Resolution of the SiC problem: astronomical and meteoritic evidence reconciled

Suto H., Koike C., Shibai H., Tuchiyama A., Mizutani K.
The optical constants of crystalline silicate particles in mid- and far-infrared

Voshchinnikov N.V.
Diagnostics of dust properties from polarimetric observations of late-type stars

Woitke P., Sedlmayr E.
Thermal Bifurcations in Circumstellar Envelopes of C-Stars

4. Circumstellar Envelopes

Chapman J.M., Rudnitskij G.M.
Shock Waves and Radio Continuum in Miras

Colomer F., Reid M.J., Menten K.M., Bujarrabal V.
The spatial and frequency structure of circumstellar water masers

Crosas M., Young K., Ivezic Z., Knapp G.R.
New Atomic Carbon Detections in post-AGB stars

Delgado D.G., Olofsson H., Schwarz H., Eriksson K., Gustafsson B.
Images in scattered light of detached circumstellar shells

Etoka S., Szymczak M., Le Squeren A.-M.
Variability of OH masers in Semi-regulars

Gérard E.
The kinematics and dynamics of OH masers in circumstellar envelopes of AGB stars

Hashimoto O., Izumiura H.
Circumstellar dust envelopes of oxygen-rich AGB stars with intermittent mass loss

Irrgang P., Balega Y.Y., Gauger A., Osterbart R., Schniggenberg G., Weigelt G.
Speckle masking imaging and radiative transfer modeling of the oxygen-rich dust shells of AFGL 2290 and CIT 3 (WX Psc)

Käufl H.U., Stecklum B., Richichi A., Richter S.
AGB-Star Diameters Measured in the mid-Infrared Using Lunar Occultations

Kerschbaum F., Olofsson H., Larsen F., Bergman P.
mm/submm-spectroscopy and -interferometry imaging of envelopes of O-rich Semiregular and Irregular Variables

Larsen F., Olofsson H.
Circumstellar envelopes of carbon stars

de Laverny P., Mauron N., Lopez B.
Observations and models of AGB dust envelopes illuminated by ambient galactic light

Masheder M.R.W., Richards A.M.S., van Langevelde H.J.
VLBI Observation of OH stars

Nyman L.-Å., Olofsson H., Schwarz H., Sahai R.
The bipolar outflow from the M-giant HR 3126

Olofsson H., Bergman P., Lucas R., Eriksson K., Gustafsson B., Bieging J.H.
Episodic mass loss of the carbon star TT Cyg

Sivagnanam P., David P.
Anomalies of OH maser profiles in circumstellar envelopes: Overshoot of OH 1667 MHz masers

Szymczak M., Cohen R.J., Richards A.M.S.
Polarization structure of OH masers in circumstellar shells

van Langevelde H.J., Diamond P.J., Schilizzi R.T., Baudry A., Habing H.J.
VLBI measurements of the parallax and proper motion of U Herculis

Yates J., Sylvestor R.
Modelling the MIR and FIR radiation transport of CO and H_2O in the circumstellar envelopes of late-type stars

5. Non-spherical Mass Loss, Binarity, Post-AGB Evolution

Buscher D.F., Haniff C.A., Oudmaijer R.D.
Asymmetric structures in the circumstellar envelopes of planetary nebula progenitor stars

Dayal A., Sahai R.
Optical/Infrared Imaging and 3-Dimensional Spatial and Kinematical Modeling of Proto-Planetary and Planetary Nebulae

Fong D., Meixner M., Sutton E., Bujarrabal V., Kelly D., Haas M., Barlow M.,
Glassgold A., Nguyen-Q-Rieu, Skinner C., Tielens A.G.G.M.
Observations of Atomic Gas in Post-AGB Envelopes

Frankowski A., Tylenda R.
The AGB phase in binaries

Gledhill T., Chrysostomou A., Yates J., Efstathiou A., Ménard F., Chikami K.
Imaging Polarimetry of Circumstellar Dust Envelopes: The Detached Shell around IRAS 19114+0002

Harrington J.P., Borkowski K.J., Blondin J.M.
HST Polarization Observations of the Young Post-AGB Object He 3-1475 and the Origin of Bipolar Jets

Hinkle K., Fekel F., Joyce R.
Orbits for AGB binary systems

Hrivnak B.J., et al.
A Study of Variability in Post-AGB Stars

Hua C.T., Dopita M.A.
Detection of new emission structures around planetary nebulae

Hurley J., Pols O., Tout C.
AGB stars in Binaries

Karakas A.I., Tout C.A., Lattanzio J.C.
The eccentricities of the barium stars

Meixner M., Ueta T., Dayal A., Bobrowsky M., Hrivnak B.J., Skinner C.J., Hora J., Deutsch L.K., Hoffman W.F., Fazio G.
The Morphology of Proto Planetary Nebulae Dustshells

Mikolajewska J., Ivison R.I., Omont A.
Mass loss from red giants in symbiotic binaries

North P.
Fundamental parameters of Ba dwarfs and CH subgiants

Ortiz R., Castilho B.V.
Atmospheric parameters and abundance analysis of post-AGB candidates

Osterbart R., Balega Y.Y., Langer N., Men'shchikov A.B., Weigelt G.
75 mas Speckle Imaging and Radiative Transfer Modeling of the Red Rectangle

Pooley D.J., Albrow M.D., Pollard K.R., Cottrell P.L.
Spectroscopic monitoring of southern post-AGB stars

Rudnitskij G.M.
Effects of a Close Low-mass Companion on the Spectrum and Light Curve of a Mira-type Star

Sahai R.
Towards a New Understanding of the Formation of Aspherical Planetary Nebulae from AGB stars: The Impact of HST

Soker N.
Axisymmetrical final glory of dying stars

Su K.Y.L., Kwok S., Volk K., Hrivnak B.J.
WFPC2/HST Imaging of the Bipolar Proto-Planetary Nebulae

Sweigart A.V.
Radiation-Pressure Ejection of Planetary Nebulae in Asymptotic-Giant-Branch Stars

Szczerba R., Jeske K., Shematovich V., Zalfresso-Jundzillo M., Volk K.
Hydrodynamical modelling of the post-AGB phase of stellar evolution: dust and molecular gas emission

Van Eck S., Jorissen A., Mayor M., Udry S., Burnet M.
Extrinsic and intrinsic S stars in the Henize sample

Villaver E., Manchado A., García-Segura G., Guerrero M.
Hydrodynamical simulations of Multiple Shell Planetary Nebulae

Volk K., Kwok S., Hrivnak B.
ISO Observations of Proto-Planetary Nebulae and Carbon Stars

Začs L., Nissen P.E., Schuster W.J.
HD 196944: a carbon and s-process rich, very metal-poor star

6. AGB Stars as a Population of Various Galaxies

Blommaert J., Groenewegen M., Trams N., Cioni M.R., Habing H., Okumura K., van Loon J., Loup C., Waters R., Zijlstra A.A.
Evolution and mass loss of AGB stars in the Magellanic Clouds

Dopita M.A., Vassiliadis E., Wood P.R., Meatheringham S.J., Harrington J.P., Bohlin R.C., Ford H.C., Stecher T.P., Maran S.P.
HST observations of Magellanic Cloud Planetary Nebulae

Exter K., Barlow M., Clegg R., Walton N., Parker Q.
Chemistry of the Galactic Bulge

Jean C., Terzan A., Guibert J.
Characteristics and IRAS Identifications of 4 000 Long Period Variables detected in a field towards the Galactic Bulge

Jehin E., Magain P., Neuforge C., Noels A., Parmentier G., Thoul A.A.
The impact of AGB stars on the chemical evolution of globular clusters

Kholtygin A.F.
New CNO abundances: test for the Galaxy chemical evolution

Matsumoto S., Nakada Y., Glass I.S.
PANIC Survey of the Galactic Bulge - The Duration Time of High Mass Loss Phase

Ojha D.K., Omont A., Simon G., ISOGAL team
Mass-losing AGB stars in the ISOGAL survey

Pols O., Hurley J., Tout C.
An efficient evolution algorithm for population synthesis of AGB stars

QUOTATIONS FROM AGB98

• *Actually, it is a C-star model which we used for O-rich stars.*
(B. Plez, answering to a question about the selection of models that so nicely agree with the observations.)

• *We are still far from deriving stellar parameters from such modelling, but we are certainly getting closer to it.*
(B. Plez, this volume, p. 81)

• *We have to burn the haystack to get the needle.*
(K. Lodders, on what happens with stardust in the lab.)

• *Reaction rates are measured at room temperature, and then extrapolated at 30 K!*
(A. Glassgold)

• *Each observation amounted to one hour of ISO time; some yielded interesting results, others, not.*
(H. Izumiura)

• *The imagination of all of us is very large, and the constraints from observations are very scarce.*
(B. Lopez)

• *All R stars appear to be single. Thus, since it is not normal that all objects in a population are single, they must all have been binaries in the past.*
(C. Tout)

• *I didn't get the observing time and thus had to be a bit more creative, and so... I looked at the data again !*
(Y.K. Ng)

• *I made a mistake, you have twenty more minutes.*
(A. Omont, session chairman, to reviewer A. Zijlstra)

• *Non-sphericity doesnot mean chaos.*
(M. Jura)

Christoffel Waelkens publicizing the ABSTRACT Booklet

Janet Mattei and Bob Wing enjoying the buffet

Alfred Glassgold, Ariane Lançon, Ulli Käufl, Mustapha Mouhcine, etc.

Author Index

Object Index

Subject Index